Handbook of
Zinc Oxide and
Related Materials

Volume Two

Devices and
Nano-Engineering

Handbook of Zinc Oxide and Related Materials: Volume One, Materials

Handbook of Zinc Oxide and Related Materials: Volume Two, Devices and Nano-Engineering

Electronic Materials and Devices Series

Series Editors

Jean-Pierre Leburton and Yongbing Xu

The MOCVD Challenge: A Survey of GaInAsP-InP and GaInAsP-GaAs for Photonic and Electronic Device Spplications, Second Edition, *Manijeh Razeghi*

Handbook of Zinc Oxide and Related Materials: Volume One, Materials, *Edited by Zhe Chuan Feng*

Handbook of Zinc Oxide and Related Materials: Volume Two, Devices and Nano-Engineering, *Edited by Zhe Chuan Feng*

Handbook of
Zinc Oxide and
Related Materials

Volume Two

Devices and
Nano-Engineering

Edited by

Zhe Chuan Feng

CRC Press
Taylor & Francis Group
Boca Raton London New York

CRC Press is an imprint of the
Taylor & Francis Group, an **informa** business

A TAYLOR & FRANCIS BOOK

First published 2013 by CRC Press

Published 2019 by CRC Press
Taylor & Francis Group
6000 Broken Sound Parkway NW, Suite 300
Boca Raton, FL 33487-2742

First issued in paperback 2020

Version Date: 20120605

ISBN 13: 978-0-367-57668-4 (pbk)
ISBN 13: 978-1-4398-5574-4 (hbk)

Library of Congress Cataloging-in-Publication Data

Handbook of zinc oxide and related materials / editor, Zhe Chuan Feng.
    v. cm. -- (Electronic materials and devices series)
    Includes bibliographical references and index.
    Contents: v. 1. Materials -- v. 2. Devices and nano-engineering.
    ISBN 978-1-4398-5570-6 (v. 1 : alk. paper) -- ISBN 978-1-4398-5574-4 (v. 2. : alk. paper)
    1. Zinc oxide--Handbooks, manuals, etc. 2. Microelectronics--Materials--Handbooks, manuals, etc. 3. Electronic apparatus and appliances--Materials--Handbooks, manuals, etc. I. Feng, Zhe Chuan.

TK7871.15.Z56H36 2013
621.3815'2--dc23   2012015917

Contents

Preface...ix
Editor..xi
Contributors...xiii

Part I ZnO-Based Nanostructures

1 **Metalorganic Vapor Deposition and Characterization of ZnO-Based Nanostructures and Thin Films**...3
   Dong-Sing Wuu, Chia-Cheng Wu, and Ray-Hua Horng

2 **Solution-Grown *n*-Type ZnO Nanostructures: Synthesis, Microstructure, and Doping**...59
   Rodrigo Noriega, Saahil Mehra, and Alberto Salleo

3 **Heteroepitaxy of ZnO on SiC as a Route toward Nanoscale p–n Junction**..............83
   Volodymyr Khranovskyy and Rositza Yakimova

4 **ZnO-Based Nanostructures**...133
   José Ramón Durán Retamal, Cheng-Ying Chen, Kun-Yu Lai, and Jr-Hau He

Part II ZnO-Based Optoelectronic Devices

5 **Light-Emitting Diodes Based on *p*-GaN/*i*-ZnO/*n*-ZnO Heterojunctions**...............177
   Yichun Liu and Yanhong Tong

6 **Light-Emitting Diodes Based on *n*-ZnO/*n*-Si(GaAs) Isotype Heterojunctions**.....219
   S.T. Tan, J.L. Zhao, and X.W. Sun

7 **ZnO Thin Films, Quantum Dots, and Light-Emitting Diodes Grown by Atomic Layer Deposition**...237
   Miin-Jang Chen, Jer-Ren Yang, and Makoto Shiojiri

8 **Hybrid Light-Emitting Diodes Based on ZnO Nanowires**.....................................279
   Jinzhang Liu

9 **ZnBeMgO Alloys and UV Optoelectronic Applications**...309
   Hsin-Ying Lee, Li-Ren Lou, and Ching-Ting Lee

10 **Ultraviolet ZnO Random Laser Diodes**...339
   Siu Fung Yu and Hui Ying Yang

Part III   ZnO-Based Electronic Devices and Application

11 Metal-Semiconductor Field-Effect Transistors and Integrated Circuits Based on ZnO and Related Oxides.................................369
Heiko Frenzel, Michael Lorenz, Friedrich-L. Schein, Alexander Lajn, Fabian J. Klüpfel, Tobias Diez, Holger von Wenckstern, and Marius Grundmann

12 Growth of ZnO for Neutron Detectors 435
Eric A. Burgett, Elisa N. Hurwitz, Nolan E. Hertel, Christopher J. Summers, Jeff Nause, Na Lu, and Ian T. Ferguson

13 Amorphous In-Ga-Zn-O Thin Film Transistors: Fabrication and Properties 485
Toshio Kamiya and Hideo Hosono

Index...537

Preface

Zinc oxide (ZnO) is an "old" semiconductor that has attracted the attention of researchers for a long time because of its applications in science and industry such as piezoelectric transducers, optical waveguides, acousto-optic media, conductive gas sensors, transparent conductive electrodes, and so on.

ZnO, which crystallizes in the wurtzite structure, is a direct band-gap semiconductor with a room temperature band gap of 3.37 eV, an exciton binding energy of 60 meV, and other useful properties. ZnO can be grown at relatively low temperatures below 500°C. The band gap of ZnO can be tuned by forming alloys of ZnMgO, ZnCdO, etc. Magnetic semiconductors can be obtained from ZnMnO, ZnCrO, and so on, which have wonderful applications in spintronics and other fields.

Therefore, ZnO and related materials as well as quantum/nanostructures have now received increasing attention and have been recognized as promising candidates for efficient UV/blue light–emitting diodes (LEDs), sensors, photodetectors, and laser diodes (LDs). A strong research trend has formed. A large number of publications and books have now appeared, and conferences have been held. More new researchers, contributors, and especially new graduate students have devoted themselves to these fields.

In recent years, research and development on wide gap semiconductors, GaN-SiC-ZnO and related materials, and quantum/nanostructures have been very active. GaN-based LEDs are forming new industries worldwide. It is expected that LEDs may replace traditional lightbulbs and tubes to achieve a new lighting echo. SiC is recognized as the power electronic material for the twenty-first century. ZnO is rapidly emerging as a third class of promising wide gap semiconductors. ZnO and related materials—together with two other classes of wide gap semiconductors, GaN and SiC—are currently revolutionizing an increasing number of applications and bring apparent benefits to vast areas of development, such as lighting, communications, biotechnology, imaging, energy conversion, photovoltaic, and medicine, with energy-efficient/saving and environment-friendly devices.

I have recently published four review books on SiC and III-Nitrides. The current two volumes on ZnO and related materials, devices and nano-engineering, provide up-to-date, comprehensive reviews of various technological fields on ZnO.

The research and application on these materials and devices are developing very fast. Data, even if published recently, need to be updated constantly. This two-volume set covers the state of the art in the field. These books are oriented more toward engineering and materials science rather than pure science.

Handbook of Zinc Oxide and Related Materials: Volume One, Materials and *Handbook of Zinc Oxide and Related Materials: Volume Two, Devices and Nano-Engineering* are intended for a wide range of readers and covers each of the basic and critical aspects of ZnO science and technology. Each chapter, written by experts in the field, reviews the important topics and achievements in recent years, especially after 2005, discusses the progress made by different groups, and suggests further works needed. This volume provides useful information about the device and nanoscale process; the fabrication of LEDs, LDs, photodetectors, and nanodevices; and the characterization, application, and development of ZnO-based semiconductor devices and nano-engineering.

Handbook of Zinc Oxide and Related Materials: Volume Two, Devices and Nano-Engineering consists of 13 well-written chapters, and is divided into 3 parts: Part I—ZnO-Based

Nanostructures, Part II—ZnO-Based Optoelectronic Devices, and Part III—ZnO-Based Electronic Devices and Application. It presents the key properties of ZnO-based devices and nano-engineering, describes important technologies, and demonstrates the remaining challenging issues in nanomaterial preparation and device fabrication for R&D in the twenty-first century. It can serve well material growers and evaluators, device design and processing engineers, as well as potential users of ZnO-based technologies, including newcomers, postgraduate students, engineers, and scientists in the ZnO and related fields.

Zhe Chuan Feng
National Taiwan University
and
Feng Research Laboratories

Editor

Professor Zhe Chuan Feng received his BS (1962–1968) and MS (1978–1981) from Peking University, Department of Physics, Beijing, People's Republic of China. He engaged in semiconductor growth process, device fabrication and testing, semiconductor laser and waveguide optics, and teaching activities in China until 1982. He then moved to the United States and received his PhD in condensed matter physics from the University of Pittsburgh in 1987. He also worked at Emory University (1988–1992), the National University of Singapore (1992–1994), EMCORE Corporation (1995–1997), the Institute of Materials Research and Engineering (1998–2001), Axcel Photonics (2001–2002), and Georgia Tech (1995, 2002–2003), with much success. Since August 2003, Dr. Feng has been a professor at the Graduate Institute of Photonics and Optoelectronics and in the Department of Electrical Engineering, National Taiwan University. His current research interests include materials research and MOCVD growth of LED, III-Nitrides, and SiC, ZnO, and other semiconductors/oxides.

Dr. Feng has edited/coedited nine specialized review books on compound semiconductors and microstructures, porous Si, SiC, and III-Nitrides, ZnO devices, and nano-engineering (including the current two-volume ZnO books) and has published approximately 500 scientific papers with more than 190 selected by the Science Citation Index (SCI) and cited nearly 2300 times. He has been a symposium organizer and invited speaker at various international conferences and universities, has served as a reviewer for several international journals, and has been a guest editor of *Thin Solid Films* and *Surface and Coatings Technology*. He has also been a visiting/guest professor at South China Normal University, Huazhong University of Science and Technology, Nankai University, and Tianjin Normal University. He is currently a member of the International Organizing Committee of Asian Conferences on Chemical Vapor Deposition and serves on the board of directors for the Taiwan Association for Coating and Thin Film Technology (TACT). website: http://www.ee.ntu.edu.tw/profile?id=57.

Contributors

Eric A. Burgett
School of Engineering
Idaho State University
Pocatello, Idaho

Cheng-Ying Chen
Institute of Photonics and Optoelectronics
National Taiwan University
Taipei, Taiwan, Republic of China

Miin-Jang Chen
Department of Materials Science and
    Engineering
National Taiwan University
Taipei, Taiwan, Republic of China

Tobias Diez
Institute of Experimental Physics II
Universität Leipzig
Leipzig, Germany

Ian T. Ferguson
Department of Electrical and Computer
    Engineering
University of North Carolina at Charlotte
Charlotte, North Carolina

Heiko Frenzel
Institute of Experimental Physics II
Universität Leipzig
Leipzig, Germany

Marius Grundmann
Institute of Experimental Physics II
Universität Leipzig
Leipzig, Germany

Jr-Hau He
Institute of Photonics and Optoelectronics
National Taiwan University
Taipei, Taiwan, Republic of China

Nolan E. Hertel
George W. Woodruff School of Mechanical
    Engineering
Georgia Institute of Technology
Atlanta, Georgia

Ray-Hua Horng
Graduate Institute of Precision
    Engineering
National Chung Hsing University
Taichung, Taiwan, Republic of China

Hideo Hosono
Materials and Structures Laboratory
Tokyo Institute of Technology
Yokohama, Japan

Elisa N. Hurwitz
Department of Electrical and Computer
    Engineering
University of North Carolina at Charlotte
Charlotte, North Carolina

Toshio Kamiya
Materials and Structures Laboratory
Tokyo Institute of Technology
Yokohama, Japan

Volodymyr Khranovskyy
Department of Physics, Chemistry, and
    Biology
Linköping University
Linköping, Sweden

Fabian J. Klüpfel
Institute of Experimental Physics II
Universität Leipzig
Leipzig, Germany

Kun-Yu Lai
Institute of Photonics and Optoelectronics
National Taiwan University
Taipei, Taiwan, Republic of China

Alexander Lajn
Institute of Experimental Physics II
Universität Leipzig
Leipzig, Germany

Ching-Ting Lee
Department of Electrical Engineering
Institute of Microelectronics
National Cheng Kung University
Tainan City, Taiwan, Republic of China

Hsin-Ying Lee
Department of Photonics
National Cheng Kung University
Tainan City, Taiwan, Republic of China

Jinzhang Liu
School of Physics, Chemistry, and
    Mechanical Engineering
Queensland University of Technology
Brisbane, Queensland, Australia

Yichun Liu
Center for Advanced Optoelectronic
    Functional Materials Research
Northeast Normal University
Changchun, Jilin, People's Republic of China

Michael Lorenz
Institute of Experimental Physics II
Universität Leipzig
Leipzig, Germany

Li-Ren Lou
Department of Electrical Engineering
Institute of Microelectronics
National Cheng Kung University
Tainan City, Taiwan, Republic of China

Na Lu
Department of Engineering Technology
University of North Carolina at Charlotte
Charlotte, North Carolina

Saahil Mehra
Department of Materials Science and
    Engineering
Stanford University
Stanford, California

Jeff Nause
Cermet, Inc.
Atlanta, Georgia

Rodrigo Noriega
Department of Applied Physics
Stanford University
Stanford, California

José Ramón Durán Retamal
Institute of Photonics and Optoelectronics
National Taiwan University
Taipei, Taiwan, Republic of China

Alberto Salleo
Department of Materials Science and
    Engineering
Stanford University
Stanford, California

Friedrich-L. Schein
Institute of Experimental Physics II
Universität Leipzig
Leipzig, Germany

Makoto Shiojiri
Professor Emeritus
Kyoto Institute of Technology
Kyoto, Japan

Christopher J. Summers
School of Materials Science and
    Engineering
Georgia Institute of Technology
Atlanta, Georgia

X.W. Sun
School of Electrical and Electronic
    Engineering
Nanyang Technological University
Singapore, Singapore

S.T. Tan
School of Electrical and Electronic
    Engineering
Nanyang Technological University
Singapore, Singapore

Yanhong Tong
Center for Advanced Optoelectronic
    Functional Materials Research
Northeast Normal University
Changchun, Jilin, People's Republic of China

Holger von Wenckstern
Institute of Experimental Physics II
Universität Leipzig
Leipzig, Germany

Chia-Cheng Wu
Department of Materials Engineering
National Chung Hsing University
Taichung, Taiwan, Republic of China

Dong-Sing Wuu
Department of Materials Science and
    Engineering
National Chung Hsing University
Taichung, Taiwan, Republic of China

Rositza Yakimova
Department of Physics, Chemistry, and
    Biology
Linköping University
Linköping, Sweden

Hui Ying Yang
Pillar of Engineering Product Development
Singapore University of Technology and
    Design
Singapore, Singapore

Jer-Ren Yang
Department of Materials Science and
    Engineering
National Taiwan University
Taipei, Taiwan, Republic of China

Siu Fung Yu
Department of Applied Physics
The Hong Kong Polytechnic University
Kowloon, Hong Kong

J.L. Zhao
Department of Applied Physics
Tianjin University
Tianjin, People's Republic of China

Part I

ZnO-Based Nanostructures

1

Metalorganic Vapor Deposition and Characterization of ZnO-Based Nanostructures and Thin Films

Dong-Sing Wuu*, Chia-Cheng Wu, and Ray-Hua Horng

CONTENTS

1.1 Introduction..3
1.2 MOCVD Experimental Setup 4
1.3 MOCVD Growth of ZnO Nanostructures on Sapphire Substrates........................5
    1.3.1 Growth and Characterization of ZnO Nanowall Networks 5
    1.3.2 Growth and Characterization of ZnO Nanorod Arrays 8
    1.3.3 Growth and Characterization of ZnO Nanotube Arrays.............................17
    1.3.4 Growth Mechanism of Various ZnO Nanostructures.................................23
1.4 ZnO Thin Films and Related Ternary Compounds..34
    1.4.1 Repeated Growing and Annealing toward ZnO Films 34
    1.4.2 Growth and Characterization of $Al_xZn_{1-x}O$ Films................................43
    1.4.3 Growth and Characterization of $Mg_xZn_{1-x}O$ Films.............................45
1.5 Novel Applications 49
    1.5.1 ZnO Sacrificial Layers 50
    1.5.2 Al-Doped ZnO Transparent Conducting Layers 51
1.6 Conclusions and Outlook 52
Acknowledgments 53
References..53

1.1 Introduction

There has been a great deal of interest in zinc oxide (ZnO) in recent years. Much attention on this semiconductor material is fueled and fanned by its extensive applications in optoelectronic devices owing to its good piezoelectricity, direct wide band gap (~3.3 eV at 300 K), a large exciton binding energy (~60 meV), high carrier mobility, and low-temperature process [1]. Some application aspects of ZnO overlap with GaN, which is another wide gap semiconductor and has been widely used in ultraviolet, blue, green, and white light-emitting devices. Nevertheless, ZnO has some properties superior to GaN, which contains the availability of high-quality ZnO single bulk crystals and thermal energy sufficiently higher than that at room temperature (~27 meV). Moreover, many crystal-growth technologies have been exploited for ZnO bulk crystal and thin film formations, leading to potential low cost for ZnO-based devices.

---

* To whom correspondence should be addressed. Email: dsw@dragon.nchu.edu.tw

As mentioned already, ZnO bulk crystals have been grown by a number of methods. In general, growth of bulk ZnO crystals is mainly implemented by three methods: hydrothermal [2], vapor phase [3,4], and melt growth [5]. The low super-saturation of the solution during hydrothermal reaction favors crystal growth. As a result, its application to the large-scale single crystals is quite suitable. A method for producing very-high-quality bulk ZnO wafers is based on vapor transport. Additionally, another method for producing bulk ZnO is that of melt growth, which is adopted by Cermet, Inc. [5]. The Cermet, Inc. melt method is based on a pressurized induction melting apparatus.

For epitaxial ZnO and related films, it is critical to reduce the strains and dislocation density in it. Thereby, closely lattice-matched substrates are beneficial to growth. Sapphire substrates are mostly used for ZnO heteroepitaxy, primarily on the (0001) orientation (or c plane) and also on the (11–20) (or a plane). Additionally, ZnO and related oxides have also been grown on Si [6], SiC [7], GaAs [8,9], CaF_2 [10], and $ScAlMO_4$ [11]. Single-crystal ZnO films mounted on sapphire with good surface flatness have succeeded, which is crucial for device fabrication. Many kinds of techniques have been used for the deposition of ZnO layers on sapphire such as magnetron sputtering [12,13], atomic layer deposition [14], pulsed laser deposition (PLD) [15], molecular beam epitaxy (MBE) [16], metalorganic chemical vapor deposition MOCVD [17–19], and solution-phase process [20]. MOCVD, as a mature commercial technique for producing high-quality device material at high deposition rates, has been commonly used in many III–V compound semiconductor growths. However, the research on the MOCVD growth of ZnO is still in its very early stage [21]. The biggest obstacle in MOCVD process for growing ZnO crystal films lies in the fact that the high pre-reaction rate of Zn and O precursors prevents the improvement of ZnO film quality. Instead, nanocrystalline or polycrystalline ZnO is prone to immerse on the sapphire substrate, such as nanotubes, nanowires, nanorods, nanobelts, and other materials. Owing to the unique physical properties of quantum confinement effects, ZnO nanostructures have been demonstrated to be useful for many device applications such as surface acoustic wave devices, hydrogen-storage devices, transparent electrodes, transparent thin-film transistors, solar cells, sensors, and so on [22–25]. On the other hand, the ZnO films are actively exploited by various modified growth mode and the alternative methods, expecting more applications in other domains as well.

1.2 MOCVD Experimental Setup

In this chapter, the growth of ZnO-based materials were all carried out by the MOCVD system reassembled from the Emcore D-180 system as shown in Figure 1.1. It retains the original control system, vacuum system, and chamber. But the gas flow geometry has been modified as this system could precisely control over the inner, middle, and outer gas flow rate. The ZnO-based nanostructures and thin films were directly deposited on the c-sapphire substrate. DEZn ($Zn(C_2H_5)_2$ [99.9999% purity]), TMAl (TMAl [99.9999% purity]), CpMg (($(C_5H_5)_2$Mg [99.99% purity]), and oxygen gas (O_2 [99.999% purity]) were used as Zn, Al, Mg, and O precursors, respectively. Ar (99.9999% purity) was supplied into the reactor as the carrier gas. The DEZn, TMAl, and CpMg bubblers were kept at 17°C, −15°C, and 30°C, respectively. In order to reduce the gas phase reaction to a minimum level, which can cause the production of particles that thus degrades the quality of ZnO structures, DEZn and oxygen gas were introduced separately and mixed together immediately just before

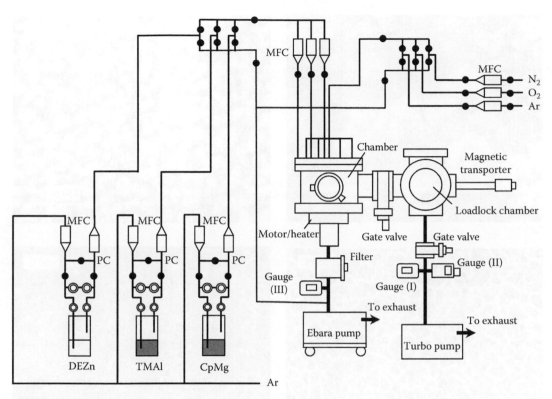

FIGURE 1.1
Schematic diagram of the refitted MOCVD system for the growth of ZnO-based nanostructures and thin films.

entering the chamber. The ZnO-based materials with different structures were deposited on sapphire substrates.

1.3 MOCVD Growth of ZnO Nanostructures on Sapphire Substrates

Since GaN-based materials and related devices have been successfully implemented depending on the great breakthroughs in MOCVD technique, people naturally expect similar achievement in ZnO-based materials. However, at present there are still no commercial MOCVD systems for the synthesis of ZnO, so researchers have to develop by themselves such systems for the growth of high-quality ZnO-based oxides. According to recent reports, the two-dimensional (2D) growth has not realized yet and the quality of ZnO structures is far lower than that for devices. Therefore, it should come as no surprise that there is a lot of research focused on this area. In this section, we mainly discuss the growth and characterization methods of several ZnO nanostructures, and the corresponding growth mechanisms are further clarified for better understanding.

1.3.1 Growth and Characterization of ZnO Nanowall Networks

ZnO nanowall networks with a honeycomb-like pattern on GaN/sapphire substrates were deposited by MOCVD without using any metal catalysts. In the experiment, the

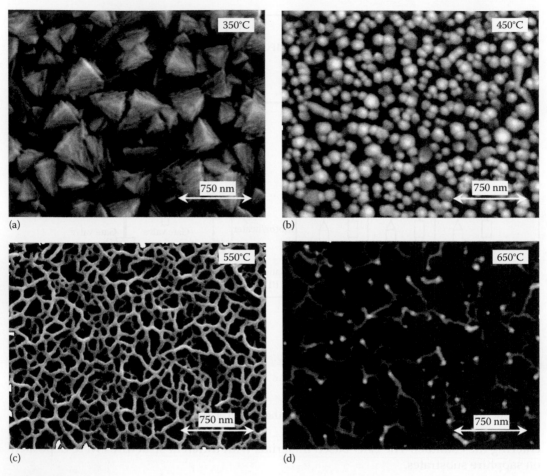

FIGURE 1.2
Top-view images of ZnO structures grown at various substrate temperatures: (a) 350°C, (b) 450°C, (c) 550°C, and (d) 650°C. (Reproduced from Wu, C.C. et al., *Jpn. J. Appl. Phys.*, 47, 746, 2008. With permission from Japan Society of Applied Physics.)

chamber pressure was set at 20 Torr with the total gas flow rate of DEZn, O_2, and Ar fixed at 2800 sccm. The growth time was 120 min. The substrate temperature rose from 350°C to 650°C. And the O_2/DEZn (VI/II) ratio was changed by varying the DEZn flow rate from 10 to 80 sccm, while the O_2 flow rate was kept at 200 sccm [26]. Results show that the growth temperature and VI/II ratio played a predominant role on the surface morphology and optical properties of ZnO nanowall networks.

Figure 1.2 presents the top-view scanning electron microscope (SEM) images of ZnO grown at different temperatures from 350°C to 650°C with the DEZn flow rate of 20 sccm. The following result shows that there are significant differences of the surface morphology of ZnO structures with increasing process temperature. As shown in Figure 1.2a, striped grains form on the GaN/sapphire substrate at low temperature of 350°C. Figure 1.2b exhibits the vertically well-aligned ZnO nanorod arrays (ZNRs) grown at 450°C. With increasing growth temperature, the morphology of ZnO appears like a nanowall-network-like structure and its density decreases, as shown in Figure 1.2c and d. This result keeps good agreement with those obtained by other researchers who synthesized

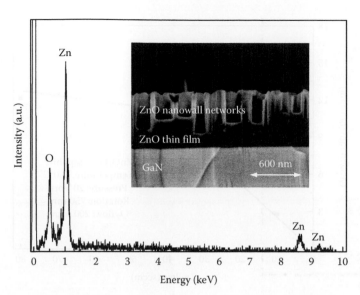

FIGURE 1.3
FE-SEM EDS result of ZnO thin film area marked with the arrow. The inset is the ZnO nanowall networks cross-sectional image. (Reproduced from Wu, C.C. et al., *Jpn. J. Appl. Phys.*, 47, 746, 2008. With permission from Japan Society of Applied Physics.)

ZnO nanowall networks using a vapor–liquid–solid (VLS) reaction process with Au catalysts at higher growth temperatures [27–30]. Therefore, it can be concluded that a higher temperature may be a key factor in the formation of ZnO nanowall networks. By comparison, the temperature for generating nanowall networks in our study is much lower than that employed in previous reports (about 875°C–950°C). ZnO nanowall networks could be directly formed at 550°C by MOCVD without any metal catalysts. So, we consider that metal catalysts may not be essential to the synthesis process of the nanowall networks.

The ZnO nanowall-network structures directly grown at 550°C on the GaN/sapphire substrate were also evaluated by FE-SEM. Figure 1.3 shows the energy-dispersive x-ray spectroscopy (EDS) result with the cross-sectional image of the ZnO nanowall networks as the inset. The EDS measurement result of the thin film area located beneath the nanowall networks clearly indicates that the film just contains two elements, Zn and O, which also confirms the fact that the ZnO nanowall networks follow the self-catalytic growth mode. In addition, from the inset we notice that a continuous thin film is formed between the GaN buffer layer and the nanowall networks.

The influences of DEZn flow rate on growth rate, surface morphology, and the optical properties of ZnO nanostructures on the GaN/sapphire have also been studied. The growth rates of ZnO at different DEZn flow rates ranging from 10 to 70 sccm and at an O_2 flow rate of 200 sccm are shown in Figure 1.4. The plot of growth rate versus DEZn flow rate in Figure 1.4 can be divided into three regions, that is, oxygen-rich (O-rich), stoichiometric, and zinc-rich (Zn-rich) regions. The growth rate increases quickly with increasing DEZn flow rate in the O-rich stage. There exists a turning point at the stoichiometry of Zn/O ratio in the curve. After the turning point, the increase in growth rate is slow and is almost fixed in the Zn-rich region. In summary, the growth rate increases with increasing DEZn flow rate up to 40 sccm and then decreases and saturates with further increase in DEZn flow rate from 40 to around 70 sccm. The maximum growth rate is about 12 nm/min.

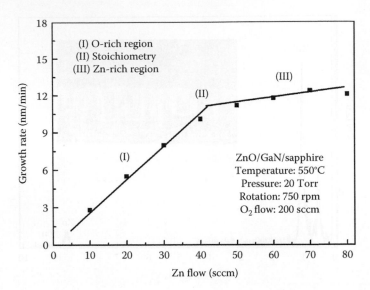

FIGURE 1.4
Relationship of the growth rate of ZnO with DEZn flow rate changing from 10 to 80 sccm. (Reproduced from Wu, C.C. et al., *Jpn. J. Appl. Phys.*, 47, 746, 2008. With permission from Japan Society of Applied Physics.)

This behavior is considered due to the limitation of DEZn flow rate in the O-rich region and the oxygen flow rate in the Zn-rich region.

Figure 1.5 displays the surface and structural morphologies of ZnO at various DEZn flow rates of 10, 40, and 70 sccm. As shown in Figure 1.5a, ZnO shows a nanorod-like structure at a low DEZn flow rate of 10 sccm. Arrays of vertically well-aligned ZnO nanorods are formed on the GaN/sapphire substrate (Figure 1.5b). When the DEZn flow rate increases to 40 sccm, the ZnO nanowall-network structure emerges on the substrate. In fact, we notice that the structure begins to form with the flow rate of DEZn as low as 20 sccm (result not shown here). By comparing Figure 1.5c and e, it is found that the ZnO nanowall networks grown at a DEZn flow rate of 70 sccm are larger in both wall width and network size. As illustrated in Figure 1.6, the average heights of the nanowall networks (including the ZnO thin films) are 600, 1000, and 1100, while the average widths are 34, 40, and 72 nm at DEZn flow rates of 20, 40, and 70 sccm, respectively. In other words, this could supply an effective way to control the size of nanowall networks by changing the DEZn flow rate.

The room-temperature PL spectra of ZnO nanostructures at different DEZn flow rates were performed to investigate the optical properties. Also included in Figure 1.7 are the emitted intensities as a function of wavelength in the range of 350–600 nm. The results showed that a strong and narrow ultraviolet peak was located at 380 nm (3.27 eV) both in the O-rich region and stoichiometry point except the ZnO-rich region. In all the cases, the highest intensity and the narrowest full width at half maximum (FWHM) were at the stoichiometry of ZnO. It is worth pointing out that no obvious green band emission, which is connected to the radial recombination of oxygen vacancy, and other structural defects in the surface and subsurface lattices [29–31] were noticed.

1.3.2 Growth and Characterization of ZnO Nanorod Arrays

We prepared the single-crystalline ZNRs by MOCVD on ZnO buffer/sapphire substrate without using any metal catalyst. Results indicate that the density of vertically aligned

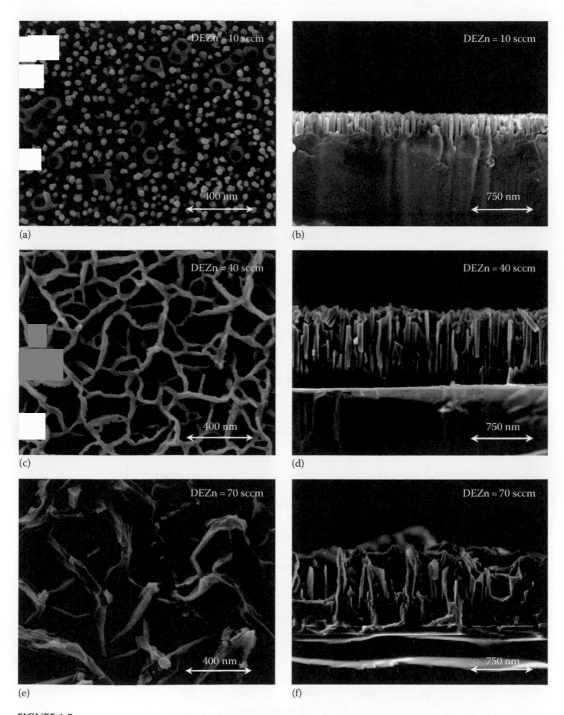

FIGURE 1.5
Top-view and cross-sectional images of ZnO nanostructures observed by FE-SEM with the DEZn flow rate of (a, b) 10, (c, d) 40, and (e, f) 70 sccm, respectively. (Reproduced from Wu, C.C. et al., *Jpn. J. Appl. Phys.*, 47, 746, 2008. With permission from Japan Society of Applied Physics.)

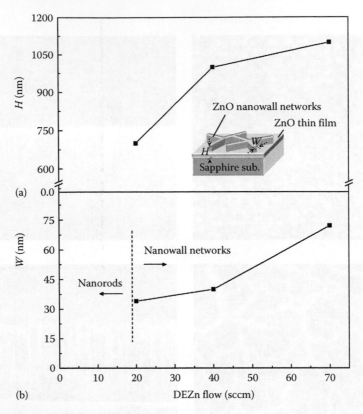

(a)

(b)

FIGURE 1.6
DEZn flow rates dependence of height (a) and width (b) for ZnO nanowall networks. (Reproduced from Wu, C.C. et al., *Jpn. J. Appl. Phys.*, 47, 746, 2008. With permission from Japan Society of Applied Physics.)

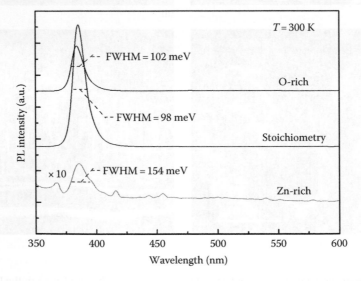

FIGURE 1.7
Room temperature PL spectra for ZnO structures grown under O-rich, stoichiometric, and Zn-rich conditions. (Reproduced from Wu, C.C. et al., *Jpn. J. Appl. Phys.*, 47, 746, 2008. With permission from Japan Society of Applied Physics.)

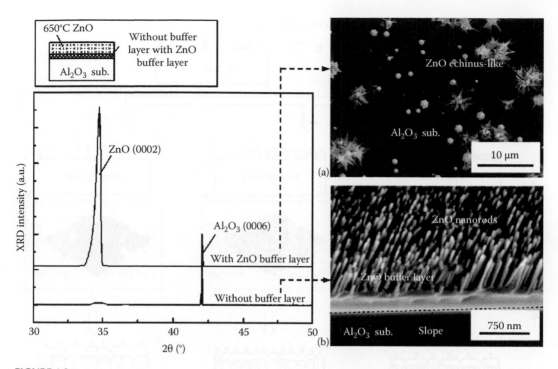

FIGURE 1.8
SEM images of ZnO nanostructures on the (a) sapphire and (b) ZnO buffer layer. (c) HRXRD patterns of ZnO nanostructures grown at 650°C with and without ZnO buffer layers.

ZNRs is determined by the morphology and thickness of the buffer layer [32]. Thus, the ZnO buffer layer can serve as the nucleation template to control the growth direction and the density of the ZNRs. Besides, by controlling the DEZn flow rate, we can manipulate the size, crystal, and optical quality of ZNRs.

Figure 1.8a and b shows the SEM images of ZnO nanostructures grown at 650°C on the sapphire substrate without and with ZnO buffer layer. The ZnO grown directly on sapphire substrate have an echinus-like morphology and random orientation. In contrast, the ZnO structures grown on ZnO buffer layer exhibit vertical evenly aligned ZNRs 40 nm in diameter and 700 nm in height, respectively. Obviously, the ZnO buffer layer provides nucleation sites for the formation of ZNRs [33]. As soon as a continuous buffer layer forms, the ZNRs start to grow on it. Testing of the high-resolution x-ray diffraction (HRXRD) indicates that the ZnO (0002) peak controlled by the hexagonal wurtzite structure is predominant in ZNRs, showing that the ZNRs are highly *c*-axis oriented and vertically well aligned. The intensity of the XRD (0002) peak is weak largely because the ZnO structures do not cover the whole surface of sapphire substrate and its growth direction is random.

In order to further realize the effect of buffer layer, ZnO buffer layers with different thickness are used in ZNRs growth at 650°C, as shown in Figure 1.9. If the ZnO is directly grown on the sapphire, it has low orientation. Figure 1.9a through c displays the surface morphologies of 10, 50, and 100 nm ZnO buffer, respectively. The SEM images of the *c*-axis oriented ZNRs grown on the ZnO buffer layers with thickness of 10–100 nm are shown in Figure 1.9d through f. The density of ZNRs on the 50-nm ZnO buffer layer is highest in all three samples.

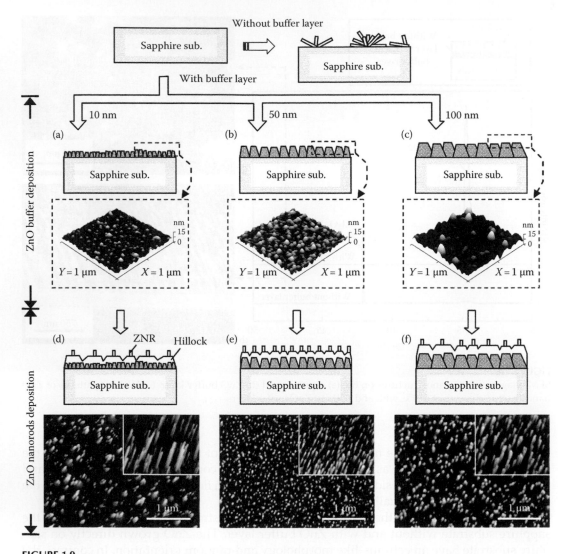

FIGURE 1.9
Sketches of ZnO buffer layer and ZNRs grown on it. The top three sketches are ZnO buffer layer deposited on sapphire with different thickness of (a) 10-, (b) 50-, and (c) 100-nm. The inset AFM images show the morphologies of ZnO surface. The bottom three sketches, (d), (e), and (f) are ZNRs growth on buffer layer show and the inset SEM pictures show the microstructures of them, respectively.

The density change of ZNRs along with the ZnO buffer layer thickness is shown in Figure 1.10. The density of ZNRs increases with increasing buffer layer thickness from 10 to 50 nm, and then decreases with further increase of buffer layer thickness up to 100 nm. In addition, the surface roughness of ZnO buffer layer measured by atomic force microscopy (AFM) appears to vary in a similar manner, as illustrated in Figures 1.9a through c and Figure 1.10. The root mean square (RMS) roughness of ZNRs is 1.35, 5.62, and 2.72 nm respectively, when the buffer layer thickness is 10, 50, and 100 nm. All the buffer layers show three-dimensional (3D) island morphology while the 50 nm buffer layer has the roughest surface. On the basis of these results, we deduce that the larger surface roughness may lead to higher nucleation density. In brief, the goal of controlling the density of

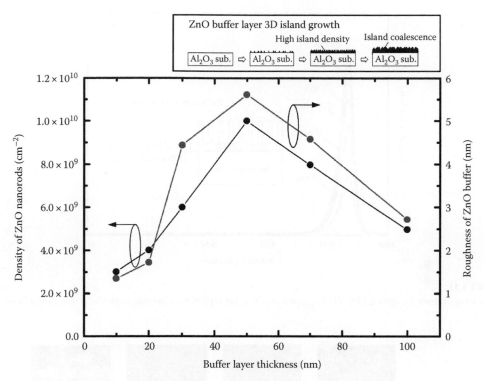

FIGURE 1.10
Relationship of ZNRs density and surface roughness of buffer layer versus ZnO buffer layer thickness.

ZNRs probably can be achieved by manipulating the densities of ZnO nuclei with different buffer thicknesses.

Room temperature PL spectroscopy was used to examine the optical properties of ZNRs, as shown in Figure 1.11. Generally, there is a strong ultraviolet near-band-edge (NBE) and a neglectable deep level emission (DLE) in the visible region under all kinds of buffer thicknesses, which indicates that all ZNRs have high optical quality. With increasing buffer layer thickness from 10 to 100 nm, the NBE peak of ZNRs increases first, and then decreases. The highest NBE peak is observed with the buffer layer thickness of 50 nm, which can be ascribed to the higher ZNRs density.

Figure 1.12a through d shows the cross-sectional SEM images of ZNRs grown on the ZnO buffer/sapphire under various DEZn flow rates of 10, 20, 30, and 40 sccm. A small quantity of ZNRs begins to deposit on the template at 10 sccm DEZn flow rate. With the increase of DEZn flow rates from 20 to 40 sccm, ZNRs constantly grow on the ZnO buffer/sapphire substrate as shown in Figure 1.12b through d. As indicated in Figure 1.12e, the average lengths of ZNRs increase whereas their average diameters decrease with the increasing DEZn flow rate from 10 to 40 sccm. The result indicated that under high DEZn flow rate, most of the zinc vapor can reach the substrate to perform the ZNRs growth rapidly, leading to long and slim nanostructures, while zinc atoms would have more oxygen atoms to react with to form short and wide ZNRs due to faster lateral growth under the low flow rate.

The typical θ–2θ XRD scan for the ZNRs grown at different flow rate is shown in Figure 1.13a. Strong (0002) and (0004) diffraction peaks can be seen, indicating that the ZNRs are of high purity and single-crystalline with hexagonal wurtzite structure. At the same time, there is an (0006) Al_2O_3 peak related to the substrate lying at about 41°.

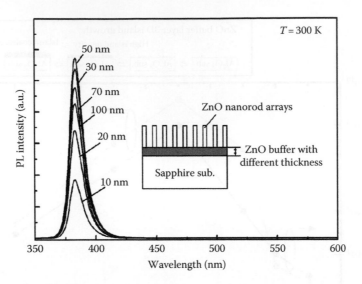

FIGURE 1.11
Room temperature PL spectra for ZNRs grown at 650°C on sapphire substrate with thickness range from 10 to 100 nm.

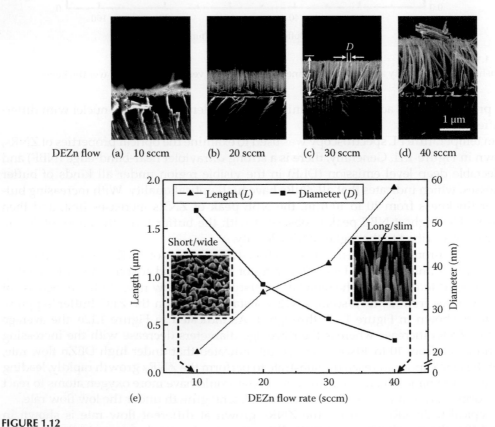

FIGURE 1.12
Cross-sectional micrographs of ZNRs deposited on ZNO buffer/sapphire substrate under the DEZn flow rates of (a) 10, (b) 20, (c) 30, and (d) 40 sccm. (e) DEZn flow rate dependence on the height and diameter of ZNRs. The inset figures are the tilt SEM images of ZNRs corresponding to flow rate of 10 and 40 sccm.

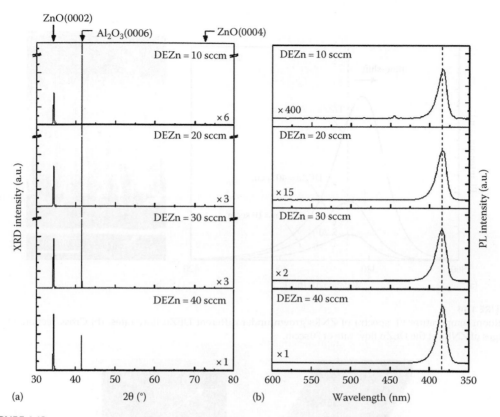

FIGURE 1.13
(a) XRD θ–2θ pattern and (b) room temperature PL of ZNRs grown at various DEZn flow rates ranging from 10 to 40 sccm.

These indicate that the ZNRs grown on sapphire substrate are along $(0001)_{ZnO}$ paralleling to $(0001)_{sapphire}$. We also find that the peak of XRD intensity increases and the FWHM decreases with increasing DEZn flow rate, which demonstrates that a better crystal quality can be obtained under high DEZn flow rate. The room temperature PL measurements were performed on these samples. As shown in Figure 1.13b, a strong single UV emission peaks at ~385 nm with the FWHM values in the range of 80–100 meV. No obvious peaks caused by defect-related emissions are observed within the range of 500–550 nm after the experiment data have been multiplied by 1, 2, 15, and 400, respectively. However, the optical characteristics of ZNRs are more commonly evaluated by the NBE-DLE intensity ratio value [34]. Results calculated from Figure 1.13b shows that the NBE-to-DLE intensity ratio of ZNRs increases with increasing DEZn flow rate, which implies that the optical quality of nanorods improves with the increase of DEZn flow rate.

In Figure 1.14a, it is found that the PL peak position systematically blue-shifts to a shorter wavelength with increasing DEZn flow rate, which could be due to the 1D quantum-confinement effect of ZNRs. Figure 1.14b shows the cross-sectional SEM image of the ZNRs grown at 70 sccm DEZn flow rate. The average diameter of nanorods is less than 20 nm and their average length is 2.7 μm.

The transmission electron microscopy (TEM) analyses of the ZNRs grown at the DEZn flow rate of 40 sccm have been performed for further investigation of the structural features as shown in Figure 1.15. It is clearly shown that the ZNR has a straight shape and

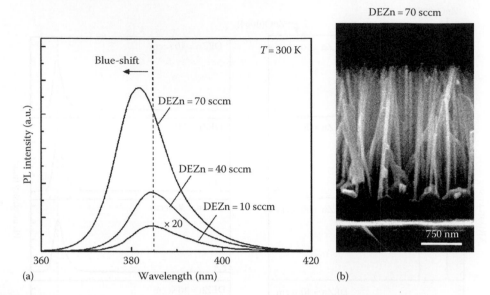

FIGURE 1.14
(a) Room temperature PL spectra of ZNRs grown under different DEZn flow rates. (b) Cross-sectional SEM images of ZNRs at the DEZn flow rate of 70 sccm.

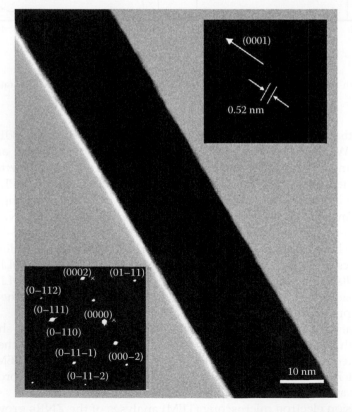

FIGURE 1.15
Cross-sectional TEM micrograph of ZNR grown at DEZn flow rate of 40 sccm. The two inset figures show the HRTEM image and SAED pattern of ZNRs. The spots marked by × are those due to forbidden reflections.

a uniform diameter. And the selected area electron diffraction (SAED) pattern, as illustrated in the inset, indicates that the ZNRs have high crystallinity in nature and preferred (0001) growth orientation, which keeps good agreement with the XRD analysis. The SAED pattern corresponds to the crystal plane of hexagonal structure, and the high-resolution transmission electron microscopy (HRTEM) image of the ZNRs presents a lattice spacing of about 0.52 nm, which corresponds to the *d*-spacing of (0001) plane in hexagonal ZnO crystal lattice.

1.3.3 Growth and Characterization of ZnO Nanotube Arrays

Vertically aligned ZnO nanotube arrays have been fabricated without any metal catalyst by MOCVD. The process follows a three-step route as illustrated in Figure 1.6a. First, the ZnO buffer layer was deposited on the sapphire substrate at 450°C, and then the sample was heated up to 650°C for the growth of the nanorod. Afterward, the sample was cooled to 450°C and the ZnO nanotube arrays were deposited on the ZNRs [35]. A timing chart of the corresponding growth process by MOCVD is given in Figure 1.16b.

SEM images in Figure 1.16 present the morphologies of ZnO at different growth steps. It is found that the ZnO nanonod arrays turn into nanotube arrays when the growth temperature is reduced from 650°C to 450°C. In order to further investigate the effect of ZnO nanonods on the formation of the nanotube arrays, the SEM image of ZnO structure grown directly on the ZnO buffer layer without the growth step of ZNRs is shown in Figure 1.17. It has been observed that ZnO nanorods act as nucleation sites. In Figure 1.18a, we can clearly see that the distribution of ZnO nanotube arrays with a hexagonal-like pattern is quite uniform over the entire substrate. Figure 1.18b is the cross-sectional SEM image of ZnO nanotube arrays. The thickness of the entire ZnO structure is around 2.5 μm. And the diameter of the structure starts to increase within a thickness of ~1.3 μm (dotted in Figure 1.18c), which indicates that the ZnO nanorods transfer to ZnO nanotube arrays. Moreover, the ZnO nanostructure is vertically well aligned on the buffer layer. The structure has also been characterized by EDS, as shown in Figure 1.18d. The SEM-EDS results of ZnO buffer and nanotube arrays indicate that the buffer and tubes are both made up of Zn and O without any metal catalysts or additives. It is suggested that the growth of ZnO nanotube arrays may follow the self-catalyzed mechanism. Additionally, the growth temperature of ZnO nanotubes is much lower than that used in a VLS mode with a metal catalyst. It is said that high growth temperature up to about 900°C employed in the VLS method [36,37] will seriously limit the device application and increase the thermal strain in the as-grown ZnO nanotube arrays.

A comparative study on the surface morphologies of ZnO nanostructures deposited on the ZNRs/sapphire substrates at different growth temperature in the third step of the growth process was carried out. Figure 1.19a through f shows the tilt and cross-sectional SEM images of the ZnO nanostructures deposited at 300°C, 450°C, and 550°C, respectively. It is noted that the surface morphologies of ZnO nanostructures grown under various temperatures vary dramatically. In Figure 1.19a and d, the ZnO nanostructure grown at relatively low temperature has lower thickness and tends to perform 2D growth [38]. However, in Figure 1.19c and f the ZnO nanostructure appears rod-like structure at 550°C. In MOCVD process, the growth temperature is considered to affect the structure most than any other growth parameters. Thereby, the growth temperature is the key factor which has significant influence on the structure of ZnO nanotube arrays.

The growth orientation and crystal alignment of ZnO nanotube arrays have been investigated by various XRD techniques such as θ–2θ, rocking curve and in-plane ordering

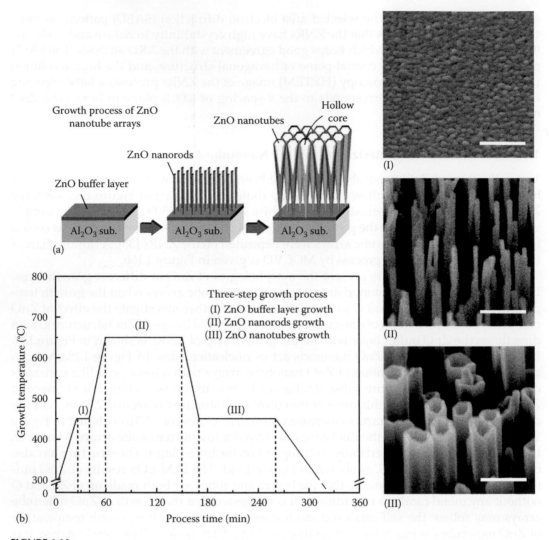

FIGURE 1.16

(a) Schematic diagram of the growth process of ZnO nanotube arrays. (b) A timing chart of MOCVD three-step growth process for ZnO nanotube arrays on sapphire substrate. The FE-SEM images to the right show the three-step growth process including (I) buffer layer, (II) nanorod arrays, and (III) nanotube arrays with all the same scale bar of 250nm. (Reproduced with permission from Wu, C.C., Wuu, D.S., Lin, P.R., Chen, T.N., and Horng, R.H., Three-step growth of well-aligned ZnO nanotube arrays by self-catalyzed metalorganic chemical vapor deposition method, *Cryst. Growth Des.*, 9, 4555–4561. Copyright 2009 American Chemical Society.)

mode. Figure 1.20a and b presents the θ–2θ XRD spectra of ZnO nanorod and nanotube arrays, respectively. Besides an Al_2O_3 (0006) peak related to the substrate, both results exhibit only a sharp diffraction peak which refers to the wurtzite structure of ZnO (0002) plane at about 34.4°. It means that the ZnO nanorod and nanotube arrays are highly (0002) oriented and include only a single phase which is completely perpendicular to the substrate. Therefore, the nanotube arrays grow along the same [0001] crystal orientation as that of the ZNRs below, which has been demonstrated by the SEM images shown in Figure 1.18. As seen in the inset of Figure 1.20a, the FWHM of the rocking curve on the (0002) shows ~1425 as. The narrow FWHM implies that the growth orientation of ZNRs

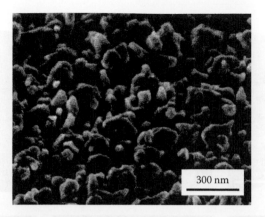

FIGURE 1.17
SEM image of ZnO structure grown directly on ZnO buffer/sapphire substrate without the pregrowth of ZNRs. (Reproduced with permission from Wu, C.C., Wuu, D.S., Lin, P.R., Chen, T.N., and Horng, R.H., Three-step growth of well-aligned ZnO nanotube arrays by self-catalyzed metalorganic chemical vapor deposition method, *Cryst. Growth Des.*, 9, 4555–4561. Copyright 2009 American Chemical Society.)

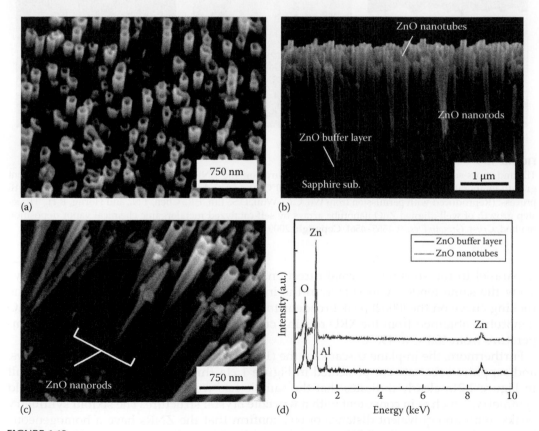

FIGURE 1.18
(a) 30° tilt, (b) cross-sectional, and (c) tilt SEM images of ZnO nanotube arrays. (d) EDS results of the ZnO buffer and nanotube arrays as indicated in Figure 1.18b. (Reproduced with permission from Wu, C.C., Wuu, D.S., Lin, P.R., Chen, T.N., and Horng, R.H., Three-step growth of well-aligned ZnO nanotube arrays by self-catalyzed metalorganic chemical vapor deposition method, *Cryst. Growth Des.*, 9, 4555–4561. Copyright 2009 American Chemical Society.)

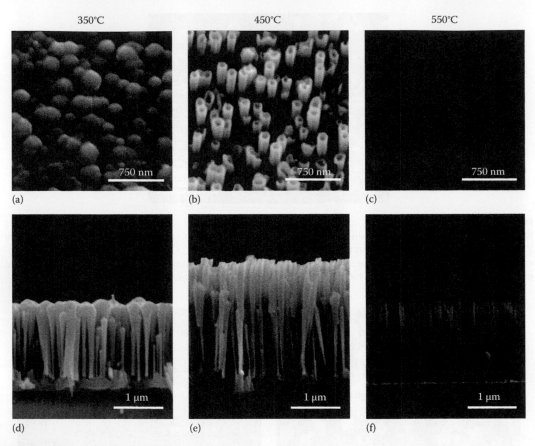

FIGURE 1.19
Tilt and cross-sectional SEM images of ZnO structures on the ZNRs/sapphire substrates under different growth temperatures: (a) and (d) 350°C, (b) and (e) 450°C, and (c) and (f) 550°C in the third step of the growth process. (Reproduced with permission from Wu, C.C., Wuu, D.S., Lin, P.R., Chen, T.N., and Horng, R.H., Three-step growth of well-aligned ZnO nanotube arrays by self-catalyzed metalorganic chemical vapor deposition method, *Cryst. Growth Des.*, 9, 4555–4561. Copyright 2009 American Chemical Society.)

is parallel to the substrate normal direction. The XRD result of ZnO nanotube arrays show the same tendency as shown in the inset of Figure 1.20b and the FWHM of the rocking curve on the (0002) peak broadens to ~1625 as. The lower crystal quality of ZnO nanotubes obtained from the XRD rocking curve is due to the reduction of growth temperature from 650°C to 450°C.

Furthermore, the in-plane ψ-scans for the (10–12) plane of ZNRs, ZnO nanotube arrays, and sapphire substrate are presented in Figure 1.21a through c, respectively. The result in Figure 1.21a clearly indicates that the sample has well-defined peaks with a sixfold symmetry, which is in consistent with a wurtzite crystal structure. The sixfold symmetry peaks, with an equivalent distance of 60°, confirm that the ZNRs have a homogeneous in-plane alignment on the (0001)-oriented sapphire. The same scan for the (10–12) plane-oriented sapphire substrate is also shown in Figure 1.21c. This measurement result corresponds to a 30° in-plane rotation of the ZnO basal plane with respect to that of sapphire crystals, which is the same as that of GaN on sapphire [39]. In Figure 1.21b, ZnO nanotube array (10–12) peaks agree with that of ZNRs, according to which the growth relationship

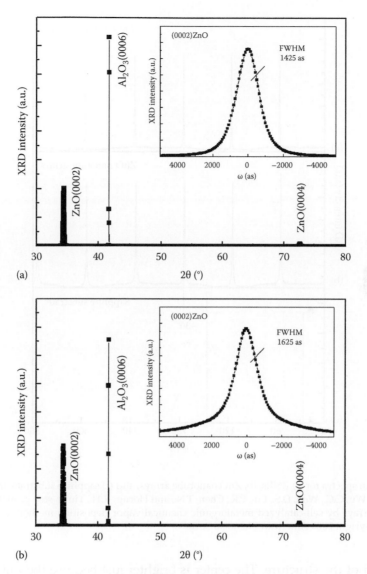

FIGURE 1.20
XRD θ–2θ scans for (a) ZnO nanorod and (b) ZnO nanotube arrays. Inset figures in (a) and (b) show the XRD rocking curve profiles on the (0002) plane reflection peak. (Reproduced with permission from Wu, C.C., Wuu, D.S., Lin, P.R., Chen, T.N., and Horng, R.H., Three-step growth of well-aligned ZnO nanotube arrays by self-catalyzed metalorganic chemical vapor deposition method, *Cryst. Growth Des.*, 9, 4555–4561. Copyright 2009 American Chemical Society.)

of the ZnO nanotube arrays on the ZNRs/buffer/sapphire substrate can be determined as ZnO nanotube arrays (0001), ZNRs (0001), Al₂O₃ (0001) and ZnO nanotube arrays [10–10], ZNRs [10–10], and Al₂O₃ [11–20].

The ZnO nanotube arrays on the ZNRs were also characterized by use of TEM. Figure 1.22a shows the general morphology of an individual nanotubes, and the average diameter of the *in situ* nanotube is in the range from 50 to 150 nm with a few micrometers in length. It can be found that the shape of the rod and tube are similar, making it difficult to identify the interface. However, there is a chiaroscuro difference between the edge and the center

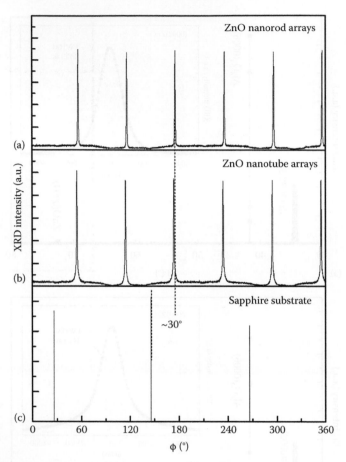

FIGURE 1.21

XRD ψ(10–12) scan spectra for (a) ZNRs, (b) ZnO nanotube arrays, and (c) sapphire substrate. (Reproduced with permission from Wu, C.C., Wuu, D.S., Lin, P.R., Chen, T.N., and Horng, R.H., Three-step growth of well-aligned ZnO nanotube arrays by self-catalyzed metalorganic chemical vapor deposition method, *Cryst. Growth Des.*, 9, 4555–4561. Copyright 2009 American Chemical Society.)

at the top end of the structure. The center is brighter just because the ZnO nanotube is hollow, as also reported by Shi and co-workers [40]. Figure 1.22b and c exhibits the SAED pattern and HRTEM graph which are acquired from the circled area, as shown in Figure 1.22a. The results indicate that the ZnO nanotube is in the single-crystal wurtzite structure and contains no defects such as dislocations and stacking fault. The lattice constant of the ZnO nanotube is about 0.52 nm corresponding to the (0001) plane of hexagonal ZnO, which signifies that the growth direction of the nanotube is along c-axial. As stated earlier, all the results confirm that the nanotube grows preferentially along the [0001] direction and the lattice planes of the hexagonal phase stacked in that direction are closed-packed [40,41], which are consistent with the XRD results.

In Figure 1.23, the PL spectrum of ZNRs shows a strong ultraviolet NBE at around 384 nm with a FWHM value of 110 meV, which is generally attributed to the excitonic recombinations [42]. Meanwhile, the green band DLE around 500–600 nm is hardly observed in the sample, indicating a good optical quality of the ZNRs. A similar result was found in the ZnO nanotube arrays grown on ZNRs. Since no noticeable DLE peak is observed from

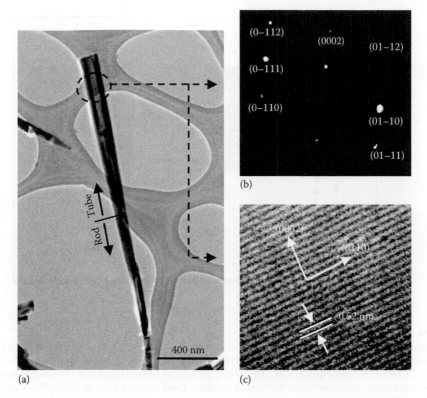

FIGURE 1.22
(a) TEM micrograph of the ZnO nanotube synthesized on the ZnO nanorod/sapphire substrate by MOCVD. (b) SAED pattern and (c) HRTEM micrograph of ZnO nanotube obtained from the area marked by dashed line in (a). (Reproduced with permission from Wu, C.C., Wuu, D.S., Lin, P.R., Chen, T.N., and Horng, R.H., Three-step growth of well-aligned ZnO nanotube arrays by self-catalyzed metalorganic chemical vapor deposition method, *Cryst. Growth Des.*, 9, 4555–4561. Copyright 2009 American Chemical Society.)

the nanotube structure, the intensity of DLE peak is magnified 1000 times as shown in the inset of Figure 1.23. In addition, the intensity ratios of the NBE to DLE are 820:1 and 2050:1 for ZnO nanorod and nanotube arrays, respectively, which suggest that the ZnO nanotube arrays have better optical property than that of ZNRs. This is due to the larger surface area of the tubular structures, which can be beneficial to the optical property.

1.3.4 Growth Mechanism of Various ZnO Nanostructures

In order to have a deep insight into the growth mechanism of ZnO nanowall networks, we illustrate it by investigating the influences of fabrication conditions (chamber pressure, O_2/DEZn ratio, growth time) on the properties of ZnO nanowall networks. For better comparison, the base parameters were maintained constant, such that growth temperature was 500°C, oxygen flow rate was 600 sccm, Ar gas flow rate through the DEZn vessel was 20 sccm, rotation of disk was 1000 rpm, and total growth time was 60 min.

It is shown that the growth rate of ZnO decreases gradually with increasing chamber pressure. Since the thickness of the boundary layer increases with the increase of chamber pressure, it becomes more difficult for the atoms diffusing onto the growing surface, and therefore leads to the reduction of growth rate. Figure 1.24a through f displays the SEM

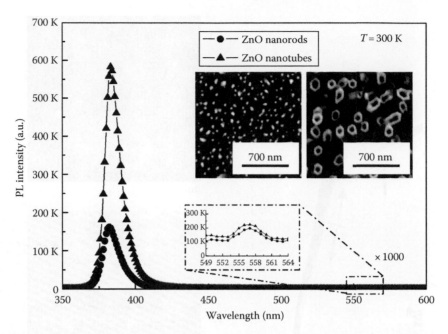

FIGURE 1.23

Room temperature PL spectra of ZNRs and ZnO nanotube arrays. Inset figures present the SEM images of ZnO nanorod and ZnO nanotube structure, respectively. (Reproduced with permission from Wu, C.C., Wuu, D.S., Lin, P.R., Chen, T.N., and Horng, R.H., Three-step growth of well-aligned ZnO nanotube arrays by self-catalyzed metalorganic chemical vapor deposition method, *Cryst. Growth Des.*, 9, 4555–4561. Copyright 2009 American Chemical Society.)

images of ZnO grown at various chamber pressures from 10 to 60 Torr. In Figure 1.24a and b, it is found that the morphologies of ZnO exhibit columnar grain and worm-like structure. With increasing chamber pressure, the ZnO structures turn into pyramid-like grain and the grain size reduces obviously as shown in Figure 1.24c through e. But in Figure 1.24f, ZnO does not uniformly distribute on the sapphire substrate simply because the gas molecules may prereact before absorbing onto the substrate. The generation of particles during gas phase reaction process would result in a high surface roughness of the sample. For this reason, the chamber pressure was eventually reduced to achieve stable flow and suppress the prereaction.

The ZnO samples grown at various chamber pressures were further characterized by double crystal XRD technique. As shown in Figure 1.25, the peak of XRD intensity disappears at 60 Torr. When the pressure is 50 Torr, the intensities of (002) and (101) diffraction peaks both emerge. At 40 Torr, the ZnO peak intensities of (002) and (101) are approximately the same. As the chamber pressure decreases from 40 to 10 Torr, the (101) diffraction peak of ZnO intensity vanishes while the intensity of (002) peak gradually increases. The rocking curve profile of the (002) peak of ZnO nanostructures is shown in the inset of Figure 1.25. With increasing the chamber pressure, the intensity of (002) ZnO diffraction peak is shrinking while the FWHM of that is broadening. At 10 Torr, the peak intensity of (002) ZnO is the strongest with a narrow FWHM of 175 as. From these results, it turns out that the small FWHM can be obtained with the ZnO grown at lower chamber pressure and the boundary layer starts to become thick at the chamber pressure above 10 Torr, causing less adsorption of reactants onto the growth surface, thus reducing the growth rate and crystal quality.

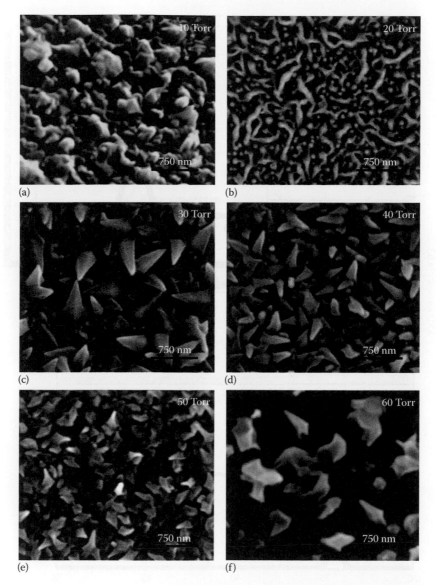

FIGURE 1.24
Top-view SEM images of ZnO grown on sapphire substrate under different chamber pressures: (a) 10, (b) 20, (c) 30, (d) 40, (e) 50, and (f) 60 Torr. (Reproduced with kind permission from Springer Science+Business Media: *Nanoscale Res. Lett.*, Effects of growth conditions on structural properties of ZnO nanostructures on sapphire substrate by metal-organic chemical vapor deposition, 4, 2009, 377, Wu, C.C., Wuu, D.S., Li, P.R., Chen, T.N., and Horng, R.H.)

The optical properties of ZnO were examined by PL measurements at room temperature. As shown in Figure 1.26, for all of the ZnO samples, an UV emission peaking at near 3.2 eV is dominantly observed and the FWHM values of the UV peak varies from 151 to 123 meV with chamber pressure changing from 60 to 10 Torr. Besides, a very broad and strong blue peak has emerged in the spectrum of ZnO grown above 50 Torr. It is said that the UV and blue peaks mainly result from the NBE and DLE, respectively. The formation of DLE is

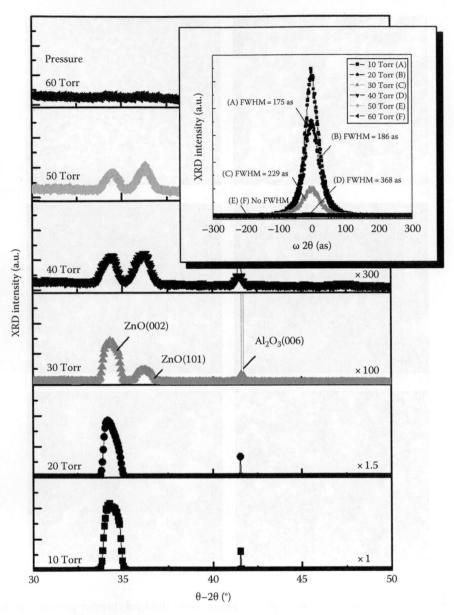

FIGURE 1.25
Double crystal XRD θ–2θ pattern of ZnO grown at the pressure range of 10–60 Torr. The inset figure shows the XRD rocking curve profile of (002) ZnO peak as a function of chamber pressure. (Reproduced with kind permission from Springer Science+Business Media: *Nanoscale Res. Lett.*, Effects of growth conditions on structural properties of ZnO nanostructures on sapphire substrate by metal-organic chemical vapor deposition, 4, 2009, 377, Wu, C.C., Wuu, D.S., Li, P.R., Chen, T.N., and Horng, R.H.)

associated with native defects such as zinc interstitial, oxygen vacancy, and zinc vacancy [43–45]. This can be explained by the suppression desorption of Zn from the surface at the higher reactor pressure, leading to Zn-rich conditions on the surface [46]. Therefore, this reinforces that the optical characteristics of ZnO can be improved by lowering the chamber pressure.

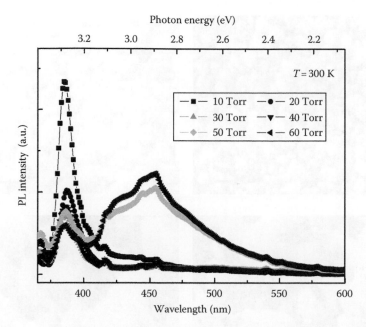

FIGURE 1.26
Room temperature PL spectra for ZnO grown at various chamber pressures. (Reproduced with kind permission from Springer Science+Business Media: *Nanoscale Res. Lett.*, Effects of growth conditions on structural properties of ZnO nanostructures on sapphire substrate by metal-organic chemical vapor deposition, 4, 2009, 377, Wu, C.C., Wuu, D.S., Li, P.R., Chen, T.N., and Horng, R.H.)

Furthermore, we have also studied the effects of O_2/DEZn (VI/II) ratio on the surface morphology and crystal quality of ZnO on the sapphire substrates. In this experiment, the variation of VI/II ratio was realized by changing the DEZn flow rates from 10 to 40 sccm while the O_2 flow rate was fixed at 600 sccm. The chamber pressure was kept at 10 sccm during the growth. Figure 1.27 shows the SEM images of ZnO grown at different DEZn flow rates. The ZnO samples are of columnar grain microstructure and the average grain size increases with increasing DEZn flow rate up to 20 sccm, as shown in Figure 1.27a and b. However, in Figure 1.27c and d, ZnO structures are nanowall-network structures with increasing DEZn flow rate above 30 sccm. The same change trend is found in the size of nanowall network structures, when DEZn flow rates increase from 30 to 40 sccm.

The crystal quality, growth rate, and surface roughness of ZnO grown at different DEZn flow rates were also investigated by XRD and AFM, respectively. In Figure 1.28a, it is found that the FWHM of (002) ZnO rocking curves increases from 126 to 175 as when the DEZn flow rate ranges from 10 to 20 sccm. However, FWHM values decrease slightly with the DEZn flow rate increasing from 20 to 40 sccm, which is attributed to the transformation of ZnO structure from columnar grain-like structure to nanowall networks. The improved crystal quality may result from the high *c*-axis-oriented nanowall-network structure. Such a result indicates that the O_2/DEZn ratio can have great effect on the structure of ZnO while it has no obvious effect on the crystal quality of ZnO. As shown in Figure 1.28b and c, with DEZn flow rate varying from 10 to 40 sccm, the growth rate increases from 9.1 to 22.3 nm/min and the surface roughness from 12.4 to 54.6 nm as well, which reveals that ZnO prefers to grow in a 3D mode under high DEZn flow rate resulting in increased surface roughness and the generation of the nanowall network structures [47]. In summary,

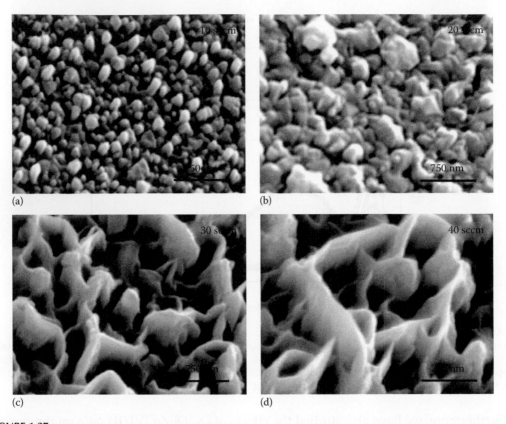

FIGURE 1.27
Tilt SEM images of ZnO grown on sapphire substrate at different DEZn flow rates of (a) 10, (b) 20, (c) 30, and (d) 40 sccm. (Reproduced with kind permission from Springer Science+Business Media: *Nanoscale Res. Lett.*, Effects of growth conditions on structural properties of ZnO nanostructures on sapphire substrate by metal-organic chemical vapor deposition, 4, 2009, 377, Wu, C.C., Wuu, D.S., Li, P.R., Chen, T.N., and Horng, R.H.)

the formation of columnar structures follows the 1D growth mode under higher VI/II ratio while the 3D growth mode becomes dominant under lower VI/II ratio.

Optical properties of ZnO grown at different DEZn flow rates were characterized by comparing the room temperature PL spectra. As seen in Figure 1.29, the PL intensity of NBE increases significantly with decreasing the DEZn flow rate and the FWHM values are 94, 112, 123, 110, and 115 meV while the DEZn flow rates are 10, 15, 20, 30, and 40 sccm, respectively. This suggests that the better optical quality can be obtained under the present condition by reducing the DEZn flow rate to 10 sccm. Moreover, the peak wavelength has a blue-shift tendency with decreasing DEZn flow rate. According to recent report, the shift of band gap energy associates with the change of strain [48,49]. The band gap energy decreased from 3.23 to 3.20 eV as DEZn flow rate increasing from 10 to 20 sccm, indicating the relaxation of strain, which may be essential in the formation of ZnO nanowall networks.

Recently, VLS has been widely used to explain the formation of ZnO nanostructures [28,50,51]. In this growth mode, the temperature for deposition was very high ($\sim$925°C) [50] and the typical metal catalyst like Au, Cu, Ni, and Sn was necessary. However, we successfully synthesized ZnO nanowall networks at a relatively low temperature of around 500°C without using any catalyst. For better understanding the growth mechanism, ZnO

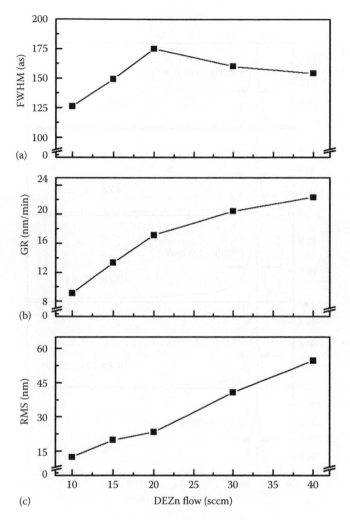

FIGURE 1.28
(a) FWHM of XRD rocking curve profile at the (002) reflection, (b) growth rate, and (c) surface roughness of ZnO grown as a function of DEZn flow rate. (Reproduced with kind permission from Springer Science + Business Media: *Nanoscale Res. Lett.*, Effects of growth conditions on structural properties of ZnO nanostructures on sapphire substrate by metal-organic chemical vapor deposition, 4, 2009, 377, Wu, C.C., Wuu, D.S., Li, P.R., Chen, T.N., and Horng, R.H.)

nanowall networks were fabricated at various time intervals of 25, 30, and 35 min. Figure 1.30 presents a set of SEM images for illustrating the formation of ZnO nanowall networks. The continuous grains appear at the initial stage, as shown in Figure 1.30a. With the passage of time, the surface becomes rough and the grains start to connect with each other by bridges (Figure 1.30b). Finally, the vertically well-aligned ZnO nanowall networks are formed on the sapphire substrate (Figure 1.30c).

It is well known that the growth of ZnO does not obey a 2D layer-by-layer mode due to the large lattice mismatch between ZnO and sapphire substrate. At high DEZn flow rate, the 3D growth mode causes the increase of the surface roughness. Then, the ZnO nanowall networks preferentially form at the tips of the ZnO surface so as to reduce the local surface energy [52,53]. Figure 1.31 presents the EDS analysis of ZnO nanowall networks grown at

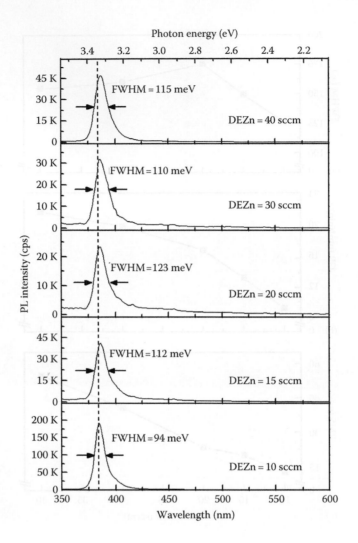

FIGURE 1.29
Room temperature PL spectra of ZnO grown at different DEZn flow rates ranging from 10 to 40 sccm. (Reproduced with kind permission from Springer Science+Business Media: *Nanoscale Res. Lett.*, Effects of growth conditions on structural properties of ZnO nanostructures on sapphire substrate by metal-organic chemical vapor deposition, 4, 2009, 377, Wu, C.C., Wuu, D.S., Li, P.R., Chen, T.N., and Horng, R.H.)

different growth times of 30 and 60 min and the inset figures show the morphologies of ZnO structure grown at 30 and 60 min with the FE-SEM measured regions marking by the "red cross." It is clearly shown that, at the growth time of 30 or 60 min, the nanowall networks are only composed of Zn and O, suggesting that the growth of ZnO nanowall networks follows the self-catalyzed growth mechanism rather than the common VLS mechanism.

It is found that ZnO buffer layer plays an important role in the forming process of ZNRs. With an intention to better understand the growth behavior of ZNRs on buffer layer, the morphologies of ZNRs grown on 50 nm ZnO buffer layer under different growth times of 0, 10, 20, and 40 min were studied by FE-SEM, the results of which are shown in Figure 1.32a through d. Figure 1.32a presents the surface appearance of 50 nm ZnO buffer layer. This buffer layer can provide nucleation sites for the formation of

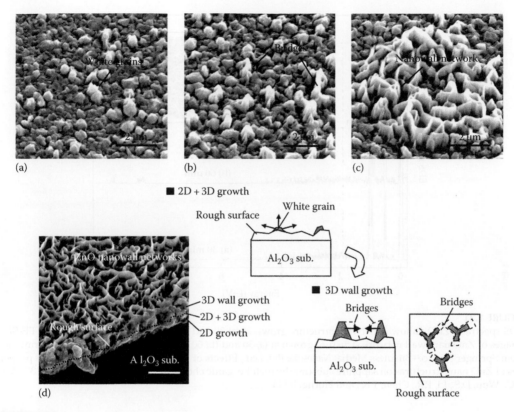

FIGURE 1.30
SEM images of ZnO nanowall networks grown at different time intervals of (a) 25, (b) 30, and (c) 35 min. (d) Schematic diagram of the growth mechanism of ZnO nanowall networks on sapphire substrate without using any metal catalysts. The inset shows the SEM image of nanowall networks grown at 60 min with the scale bar of 2.5 μm. (Reproduced with kind permission from Springer Science+Business Media: *Nanoscale Res. Lett.*, Effects of growth conditions on structural properties of ZnO nanostructures on sapphire substrate by metalorganic chemical vapor deposition, 4, 2009, 377, Wu, C.C., Wuu, D.S., Li, P.R., Chen, T.N., and Horng, R.H.)

nanorods with lower lattice mismatch. After the growth of buffer layer, ZnO is directly deposited on it. At the early stage of the growth process, ZnO on the buffer layer occurs in a hillock-like ZnO structure. Furthermore, rod-like structures that nucleate and grow at the apices of these hillocks are observed in Figure 1.32b. As growth time increases to 20 min, longitudinal growth of the rod-like structures seems to predominate as shown in Figure 1.32c. Meanwhile, we can still find some salient structures on the hillocks which are ready to form rod-like structures as shown in the inset of Figure 1.32c. When growth time increases to 40 min, we can only observe the growth of nanorods as shown in Figure 1.32d. Finally, well-aligned ZNRs are successfully synthesized on the buffer/sapphire substrate. A diagram illustrating the proposed growth mechanism of ZNRs on the ZnO buffer layer is shown in Figure 1.32e. The growth process of ZNRs can be simply divided into two stages including nucleation and growth. The predeposition of the ZnO buffer layer accelerates the nucleation of ZnO and formation of hillocks on the surface of the buffer layer. With increasing growth time, the ZNRs start to generate on the hillock structures. The tip of hillocks could act as an activation site for the formation of ZNRs due to the difference in surface-free energy.

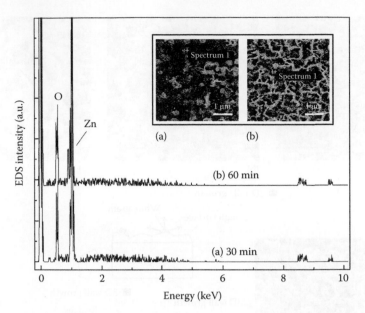

FIGURE 1.31

EDS spectra of ZnO nanowall network structure grown at (a) 30 and (b) 60 min. The insets show the FE-SEM images of ZnO nanowall network structure grown at (a) 30 and (b) 60 min. (Reproduced with kind permission from Springer Science+Business Media: *Nanoscale Res. Lett.*, Effects of growth conditions on structural properties of ZnO nanostructures on sapphire substrate by metal-organic chemical vapor deposition, 4, 2009, 377, Wu, C.C., Wuu, D.S., Li, P.R., Chen, T.N., and Horng, R.H.)

XPS (x-ray photoelectron spectroscopy) and TEM-EDS analyses were conducted for demonstrating the growth behavior of ZNRs. All the results discussed next are based on the ZNRs grown on ZnO buffer/sapphire substrate at 650°C under DEZn flow rate of 40 sccm. In XPS analysis, we can only observe the peaks corresponding to the elements of Zn, O, and C in the broad scan survey spectrum, as described in Figure 1.33a. It is reported that the presence of carbon-related peak is due to the foreign carbon contamination from air or other contaminations [54]. Moreover, in order to examine the chemical states of zinc in the ZNRs, XPS was used to study the elemental depth profiles. In Figure 1.33b, higher Zn concentration can be observed in the ZNRs, which suggests a possible Zn-terminated surface appearing on the top of ZNRs. Some recent studies have indicated that the Zn-terminated surfaces are catalytically active while the O-terminated surfaces are chemically inert [55]. As a result, ZNRs would keep growing and forming the Zn-terminated surfaces due to self-catalytic property. Besides, Figure 1.33c presents the TEM-EDS measurement result for the top surface of ZNR. As expected, it is found that the ZNRs contain only Zn and O, which is concordant with the XPS analysis. The Cu signal observed in the spectrum is originated from TEM copper mesh, contaminated during sample preparation. Both the XPS and EDS analyses reveal that no metal catalysts or additives exist during the growth of ZNRs. Therefore, ZNRs growth is considered to follow the self-catalyzed growth mechanism instead of the metal-catalyzed VLS mechanism.

Consequently, the transformation of nanorod arrays into nanotube arrays was especially studied further. Figure 1.34a exhibits the FE-SEM images of the nanotube arrays synthesized at 450°C on nanorod arrays under the growth time of 10 min. In the inset of Figure 1.34a, we can notice that the growth of ZnO nanotube initially starts from the edges of the structure. The growth sites 1, 2, and 3 on each growth surface of the ZnO nanorod are

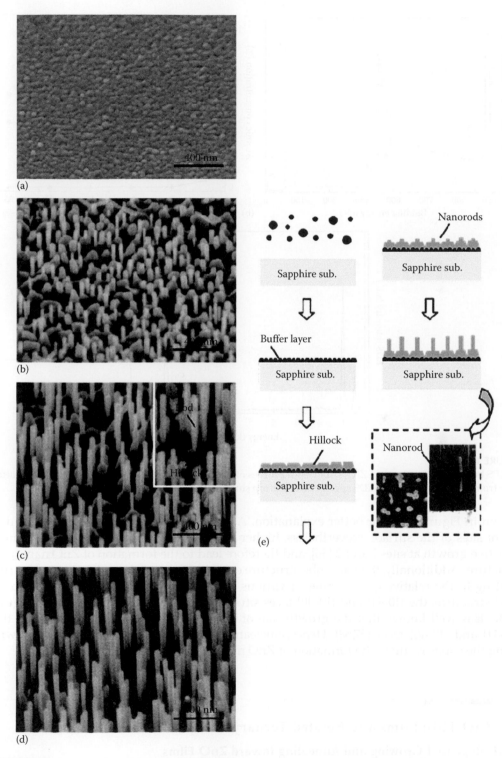

FIGURE 1.32
SEM images of ZNRs on 50 nm ZnO buffer/sapphire substrate under various growth times: (a) 0, (b) 10, (c) 20, and (d) 40 min. (e) Schematic illustration of growth mechanism of ZNRs on ZnO buffer layer.

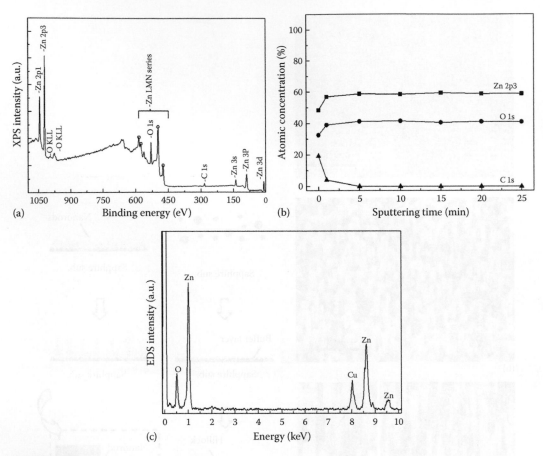

FIGURE 1.33
(a) XPS broad scan spectrum of ZNRs. (b) Depth profiles of zinc and oxygen in ZNRs as a function of sputter time from 0 to 25 min. (c) TEM-EDS spectrum for the top surface of ZNR.

shown in Figure 1.34b for better explanation. At first, atoms can adsorb on all three different sites of the surface. Nevertheless, higher binding energies at the edges will cause selective growth at sites 1 and 2 [56], and therefore lead to the formation of ZnO nanotube structure. Additionally, the nanotube structure exhibits different growth behaviors partly relating to the relative growth rates of various crystal facets. In the wurtzite hexagonal ZnO structure, the {10–11} and {10–10} faces situate at the lateral side as shown in Figure 1.34c. It is well known that the growth rate of the {0001} faces is faster than that of the {10–11} and {10–10} faces [57,58]. Hence, nucleating on the edges and preferring growth along the c-axis result in the formation of ZnO nanotube structures.

1.4 ZnO Thin Films and Related Ternary Compounds

1.4.1 Repeated Growing and Annealing toward ZnO Films

It is well-known that heteroepitaxial layers grown on large lattice-mismatched substrates usually form nanostructures. Due to the large lattice-mismatched sapphire substrate, it

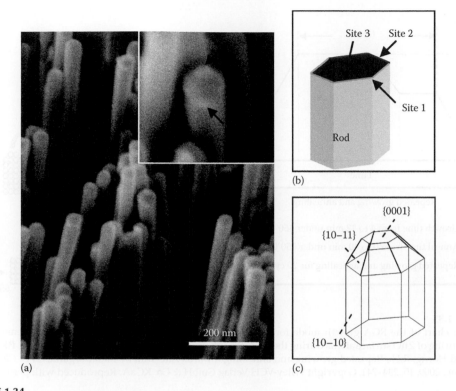

FIGURE 1.34
(a) FE-SEM image of ZnO nanotube arrays grown on ZNRs. The inset figure shows a magnified SEM image exhibiting the growth of ZnO nanotube directly on the ZnO nanorod. Schematic illustration of (b) nucleation sites on the surface of the nanorod and (c) various faces of the ZnO crystal. (Reproduced with permission from Wu, C.C., Wuu, D.S., Lin, P.R., Chen, T.N., and Horng, R.H., Three-step growth of well-aligned ZnO nanotube arrays by self-catalyzed metalorganic chemical vapor deposition method, *Cryst. Growth Des.*, 9, 4555–4561. Copyright 2009 American Chemical Society.)

is too hard to completely perform the 2D growth for flat ZnO epilayers. And the results from our previous studies have indicated that high-quality ZnO epitaxial structure can be obtained at high growth temperature while it is inclined to form nanostructure [59]. Therefore, in order to obtain a smooth surface, an annealing step was introduced in the growth process. We proposed the repeated growing and annealing (RGA) growth mode, in which the film growth and thermal annealing temperatures were repeatedly switched between 450°C and 650°C.

The corresponding time chart of the RGA process is presented in Figure 1.35a. The growth and annealing temperatures of ZnO were repeatedly switched between 450°C and 650°C for 20 cycles. The growth time was varied from 4 to 12 min while the annealing time was varied from 0 to 40 min. Figure 1.35b exhibits the schematic illustration of the RGA growth mode. When the growth and annealing temperatures of ZnO were repeatedly transferred between 450°C and 650°C, the surface atoms would restructure and move to a more stable state, resulting in a more flat surface during the annealing process.

The microstructures of the ZnO samples grown at various time intervals and after 20 times of RGA cycles were measured by AFM as shown in Figure 1.36. From the results, it is seen that the morphologies of ZnO structures are remarkably different with various

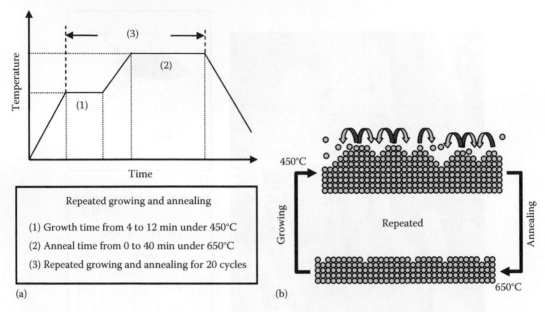

FIGURE 1.35
(a) Time chart of the RGA growth mode for ZnO film deposited on sapphire substrate. (b) Scheme for the restructuring of ZnO surface atoms during the RGA growth process. (From Wu, C.C., Wuu, D.S., Lin, P.R., Chen, T.N., and Horng, R.H., Repeated growing and annealing towards ZnO film by metal-organic CVD. *Chem. Vap. Deposition.* 2009. 15. 234–241. Copyright Wiley-VCH Verlag GmbH & Co. KGaA. Reproduced with permission.)

growth time. At the growth time of 4 min, the sparse island is obtained. With the growth time of 8 min, the 3D dense distribution island forms. And with the increasing growth time to 12 min, the ZnO islands start to coalesce to become bigger islands resulting in a larger height difference [60]. After annealing at 650°C in Ar atmosphere for 20 min, there has been improvement in surface morphology of all the ZnO samples. This contributes to the rearrangement of ZnO structure by supplying sufficient thermal energy. Particularly, 2D and 3D AFM surface images of the ZnO structure with the growth time of 8 min was flatter than those grown at 4 and 12 min. Therefore, we chose 8 min as the growth time of ZnO in the RGA growth mode for obtaining a smooth surface.

We also investigated the role of annealing time during the RGA growth mode. Figure 1.37a and b shows the time chart of ZnO annealed at 650°C for different time intervals from 0 to 40 min and the dependence of ZnO surface roughness at 650°C on the annealing time, respectively. With the increase of annealing time, the RMS surface roughness of ZnO continuously decreases, suggesting that the surface morphology gradually transits to 2D flat surfaces during the more complete coalescence process. It is found that the surface roughness is basically unchanged with annealing time over 20 min. Low surface roughness reflects a more complete coalescence and recrystallization process achieved by sufficient thermal energy. Thereby, we propose the RGA growth mode that includes 20 cycles of growth temperature at 450°C for 8 min and annealing temperature at 650°C for 20 min.

The effect of various annealing ambient environments on the morphology of ZnO film was studied as well. Figure 1.38a through f presents the SEM and AFM graphs of ZnO structures after 20 cycles RGA growth mode in Ar, N$_2$, and O$_2$ annealing ambient, showing that the surface morphologies of the samples are quite different. The RMS surface roughness and the thickness of ZnO films treated in O$_2$ ambient is the largest among the

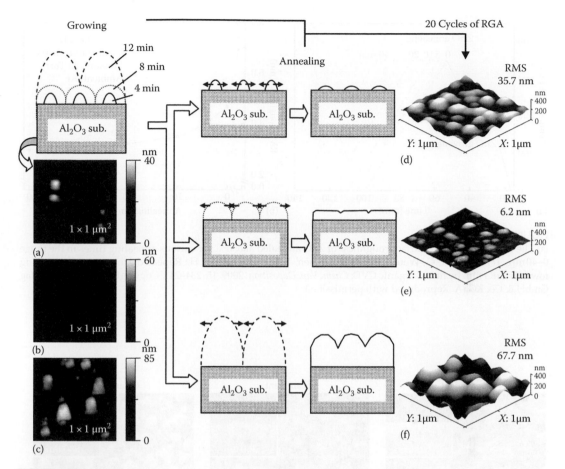

FIGURE 1.36

(a–c) 2D AFM surface morphologies of ZnO grown at 450°C for 4, 8, and 12 min. (d–f) 3D AFM surface morphologies of ZnO structure after 20 cycles RGA growth mode corresponding to different growth times of 4, 8, and 12 min. The insets are the illustration of the island growth and annealing behavior of ZnO/sapphire under various growth times. (From Wu, C.C., Wuu, D.S., Lin, P.R., Chen, T.N., and Horng, R.H.: Repeated growing and annealing towards ZnO film by metal-organic CVD. *Chem. Vap. Deposition.* 2009. 15. 234–241. Copyright Wiley-VCH Verlag GmbH & Co. KGaA. Reproduced with permission.)

samples annealing in three different ambient gas environments. This is probably due to the reaction of Zn atoms escaping from the ZnO film with O_2 atoms and the new generation of ZnO structure on the surface [61]. The increased thickness of ZnO film under O_2 ambient can be attributed to the reproduction of ZnO structure, as shown in the inset of Figure 1.38c. However, Dai et al. have proved that the RMS can be reduced from 20 to 2.5 nm by adjusting the growth parameters and structure design [62]. Under Ar annealing ambient, the RMS is about 6 nm, indicating that the disorder growth of ZnO can be effectively restrained.

Figure 1.39a through d shows the room temperature PL spectra of the as-grown ZnO at 450°C and the ZnO film grown by 20 cycles RGA growth mode with various annealing ambient of Ar, N_2, and O_2. For comparison, the inset of Figure 1.39a presents the PL spectrum of the bulk ZnO which has a narrow FWHM of 100 meV and a negligible DLE. The PL intensity of Figure 1.39a through d has been multiplied by 3, 2, 2, and 1, respectively.

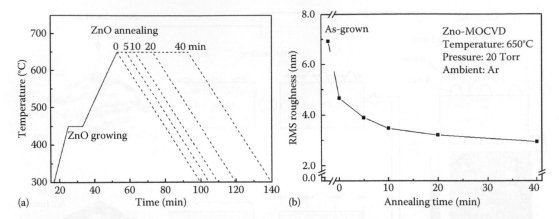

FIGURE 1.37
(a) The time chart and (b) surface roughness of ZnO annealed at 650°C for various annealing times in the range 0–40 min. (From Wu, C.C., Wuu, D.S., Lin, P.R., Chen, T.N., and Horng, R.H.: Repeated growing and annealing towards ZnO film by metal-organic CVD. *Chem. Vap. Deposition*. 2009. 15. 234–241. Copyright Wiley-VCH Verlag GmbH & Co. KGaA. Reproduced with permission.)

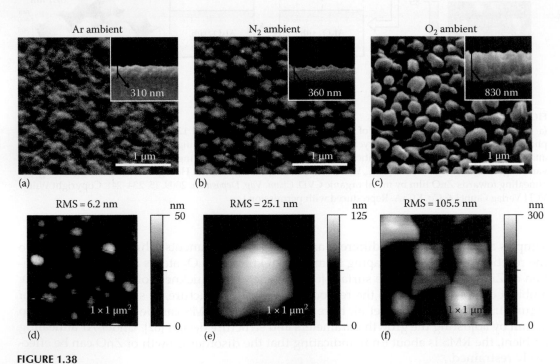

FIGURE 1.38
(a–c) Top-view SEM and (d–f) AFM images of ZnO film after 20 cycles of RGA growth mode under Ar, N_2, and O_2 annealing atmosphere. The insets of figures (a–c) exhibit the corresponding cross-sectional SEM images. (From Wu, C.C., Wuu, D.S., Lin, P.R., Chen, T.N., and Horng, R.H.: Repeated growing and annealing towards ZnO film by metal-organic CVD. *Chem. Vap. Deposition*. 2009. 15. 234–241. Copyright Wiley-VCH Verlag GmbH & Co. KGaA. Reproduced with permission.)

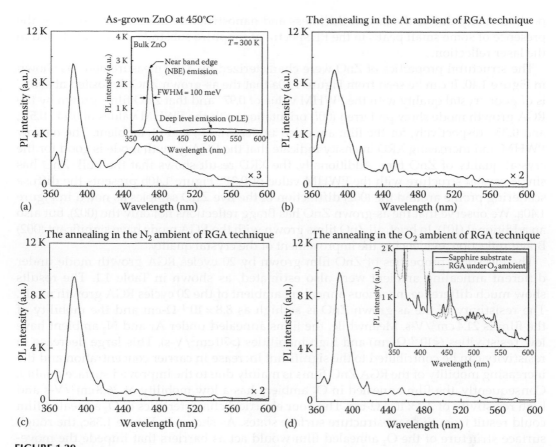

FIGURE 1.39

Room temperature PL spectra of (a) as-grown ZnO at 450°C, and after 20 cycles of RGA growth mode annealing in different atmospheres including (b) Ar, (c) N_2, and (d) O_2. Inset figures in (a) and (d) show the PL spectra of bulk ZnO substrate and sapphire substrate. (From Wu, C.C., Wuu, D.S., Lin, P.R., Chen, T.N., and Horng, R.H.: Repeated growing and annealing towards ZnO film by metal-organic CVD. *Chem. Vap. Deposition.* 2009. 15. 234–241. Copyright Wiley-VCH Verlag GmbH & Co. KGaA. Reproduced with permission.)

By comparing with the as-grown ZnO, it is concluded that the RGA growth mode is essential for determining the luminescent properties of the ZnO film. It is clear that the RGA growth mode can increase the intensity of NBE emission at about 380 nm and decrease the intensity of DLE between 440 and 600 nm under all kinds of annealing ambient. The DLE is ascribed to the impurities and structural defects in the crystal, which could result in shortening carrier/exciton lifetime and decreasing emission efficiency in the UV light devices. In Figure 1.39d, the NBE emission is enhanced and the DLE is quenched under O_2 ambient. This is because oxygen can diffuse into the lattice to fill the oxygen vacancies or react with zinc interstitials during high temperature and the O_2 annealing process. However, the intensity of DLE also decreases when samples are annealed in Ar and N_2 ambient, as shown in Figure 1.39b and c. Therefore, we consider the zinc interstitial to be the major defect in the ZnO because high annealing temperatures can reduce the defect of the zinc interstitial rather than that of oxygen vacancy [63]. Zhang et al. also reported that the luminescence in the range of 428–478 nm (2.9–2.6 eV) pertained to the zinc defects [64]. Among all the samples, ZnO structure annealed under O_2 ambient has the top PL peak intensity due to the largest thickness and nanostructure morphology. Besides, the presence of some small peaks in the PL spectra of sapphire substrate can be attributed to the laser reflection.

The structural properties of ZnO were characterized by XRD measurement, as shown in Figure 1.40. It can be seen from Figure 1.40a that the as-grown ZnO deposited at 450°C is of poor crystal quality with the FWHM value of 0.97° and that ZnO films grown by the RGA growth mode show preferred (00*l*) orientation with the FWHM values of 0.44°, 0.57°, and 0.33°, respectively, for the film annealed under Ar, N_2, and O_2 ambient. The narrow FWHM and increasing XRD intensity indicate that the RGA growth mode is good for the crystal quality of ZnO film. Additionally, the XRD result shows that the bulk ZnO has single crystal structure with the FWHM value of 0.31°. Figure 1.40b presents the diffuse scattering profile around the (002) diffraction of the five ZnO samples as noted in Figure 1.40a. We observe that the as-grown ZnO has Bragg reflections not only the (002), but also an additional (101). In brief, all ZnO films grown with the RGA mode have significant (002) Bragg reflection, evidencing the improvement of the crystal quality.

The electrical properties of ZnO film grown by 20 cycles RGA growth mode under different annealing ambient were also estimated, as shown in Table 1.1. The results show much difference in various annealing ambient of the 20 cycles RGA growth mode. The resistivity of the as-grown ZnO is as high as 8.8×10^{-1} Ω-cm and the mobility of the film is 21.4 cm²/V-s. Meanwhile, the films annealed under Ar and N_2 ambient have low resistivities (<10^{-2} Ω-cm) and high mobilities (>70 cm²/V-s). This large decrease in resistivity can be contributed to the significant increase in carrier concentration and the increasing mobility of the RGA ZnO films is mainly due to the improved surface quality. Consequently, the film annealed in O_2 ambient has a low mobility of 26.8 cm²/V-s and high resistivity of 2.2×10^{-1} Ω-cm. The poor electrical characteristics of O_2 annealed film could result from the nanostructure surface states. As shown in Figure 1.38c, the rough surface structure of the O_2 annealed film would act as barriers that impede the movement of electrons and thus reduce the electron mobility. The reduction of the electron mobility would partly result from the increase of the resistivity. This suggests that electrical characteristics of ZnO film can be improved after the RGA growth process. After comparing and analyzing the results of annealing ambient during the RGA growth mode, it is indicated that the ZnO film under N_2 condition has better hexagonal structure, low resistivity of 3.43×10^{-3} Ω-cm and high mobility of 85.2 cm²/V-s. In addition, all the ZnO structures exhibit n-type conductivity. As we know, undoped ZnO is n-type conductivity due to the existence of intrinsic defects such as zinc interstitial and oxygen vacancy, many reports have claimed that zinc interstitial and oxygen vacancy act as donor defects in ZnO structure [65,66]. This makes it difficult to achieve stable and reproducible p-type ZnO.

In order to understand the surfaces of zinc and oxygen in the ZnO samples, XPS measurements were performed. Figure 1.41 shows high-resolution XPS spectra of the O_{1s} core level for all the samples. The O_{1s} spectrum could be consistently deconvoluted with two Gaussian functions, peaking at 529 and 531 eV [46,49]. The lower binding energy component is ascribed to O^{2-} in the wurtzite structure of ZnO while the higher binding energy is associated with O^{2-} ions, which are in the oxygen deficient regions of the ZnO matrix. In Figure 1.41, the relative intensity of the higher binding energy component in ZnO film annealed under O_2 ambient is obviously enhanced as compared to that of the ZnO samples annealed under Ar and N_2 ambient. This suggests that the oxygen vacancy concentration increases significantly after annealing the ZnO in O_2 ambient but little oxygen defects exist on the surface of ZnO film.

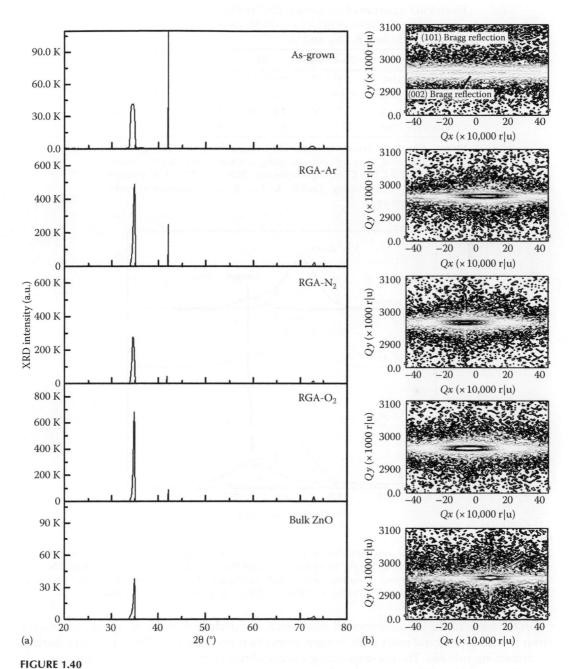

FIGURE 1.40
(a) XRD θ–2θ scan and (b) x-ray reciprocal space maps around the (002) diffraction peak of as-grown ZnO at 450°C, after 20 cycles of RGA growth mode with different annealing atmosphere including Ar, N₂, and O₂ and the bulk ZnO substrate. (From Wu, C.C., Wuu, D.S., Lin, P.R., Chen, T.N., and Horng, R.H.: Repeated growing and annealing towards ZnO film by metal-organic CVD. *Chem. Vap. Deposition.* 2009. 15. 234–241. Copyright Wiley-VCH Verlag GmbH & Co. KGaA. Reproduced with permission.)

TABLE 1.1

Electrical Properties of As-Grown ZnO at 450°C and ZnO Film
Grown with 20 Cycles of RGA Growth Mode under Various
Ambient including Ar, N_2, and O_2

	As-Grown	Ar	N_2	O_2
Resistivity (Ω-cm)	8.81×10^{-1}	9.92×10^{-3}	3.43×10^{-3}	2.20×10^{-1}
Mobility (cm²/V-s)	21.4	74.6	85.2	26.8
Conduction (cm⁻³)	3.34×10^{17}	4.41×10^{18}	1.16×10^{19}	6.921×10^{17}
Conductivity type	N	N	N	N

Source: Wu, C.C., Wuu, D.S., Lin, P.R., Chen, T.N., and Horng, R.H.:
Repeated growing and annealing towards ZnO film by metalorganic CVD. *Chem. Vap. Deposition.* 2009. 15. 234–241. Copyright
Wiley-VCH Verlag GmbH & Co. KGaA. Reproduced with
permission.

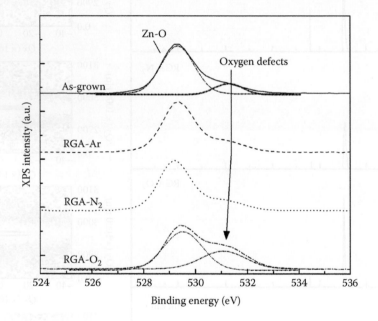

FIGURE 1.41

XPS spectra of O_{1s} core-level for as-grown ZnO at 450°C and ZnO films after 20 cycles of RGA growth mode
under various annealing ambient including Ar, N_2 and O_2. (From Wu, C.C., Wuu, D.S., Lin, P.R., Chen, T.N., and
Horng, R.H.: Repeated growing and annealing towards ZnO film by metal-organic CVD. *Chem. Vap. Deposition.*
2009. 15. 234–241. Copyright Wiley-VCH Verlag GmbH & Co. KGaA. Reproduced with permission.)

On the basis of the aforementioned discussion, we deduce that the escaping zinc atoms
from the surface would react with oxygen atoms to reproduce new ZnO structure during
O_2 annealing process. The corresponding chemical reaction formulas are as follows:

$$Zn_i + \frac{1}{2}O_2 \rightarrow ZnO \tag{1.1}$$

$$Zn_{Zn} + V_o + \frac{1}{2}O_2 \rightarrow ZnO \tag{1.2}$$

Since no decrease of oxygen vacancy is found in O_2-annealed ZnO film, we infer that the regrowth of new ZnO structure follows the reaction (1.1) instead of reaction (1.2). Namely, the zinc interstitials should be the major defects in the as-grown ZnO while excess oxygen vacancies only derive from the regrowth of ZnO structure. For the ZnO annealed in Ar or N_2 ambient, no regrowth ZnO structure is found on the film surface.

1.4.2 Growth and Characterization of $Al_xZn_{1-x}O$ Films

Al-doped ZnO (AZO) thin films are emerging as an alternative potential candidate for ITO (Sn-doped In_2O_3) films recently not only because of their comparable optical and electrical properties (high optical transmittance in the visible range, infrared reflectance and low DC resistivity) to ITO films, but also because of their higher thermal and chemical stability than ITO [67]. So, AZO film also can be utilized as the surface texturing, uniform current spreading, high transparency and thick window layer for achieving a highly efficient light-emitting diode (LED) [68]. In this study, we discuss the doping influences on the morphology, structural, optical, and electrical properties of AZO.

Figure 1.42 presents the variation of roughness at different Al doping concentrations in ZnO. It is found that the surface roughness decreases as the Al doping concentration increases. Once Al was induced into ZnO surface, it would serve as impurities during the growth process. The surface would become smoother due to the enhanced nucleation sites, which indicates that increasing Al doping concentration might degrade the light extraction efficiency of LEDs.

Figure 1.43 shows XRD θ–2θ pattern of the ZnO with various TMAl flow rate. It is observed that the intensities of ZnO (002) and (004) peaks are decreased with increasing TMAl flow rate. This results from the formation of stresses created by the different ion radii between zinc (Zn^{2+}) and the dopant (Al^{3+}) in the AZO. Meanwhile, with increasing TMAl flow rate from 0 to 4 sccm, the diffraction angles of (002) peaks are decreased from 34.55° to 34.26°, which indicates an increase in the c-axis-oriented lattice parameter and also in the interplanar spacing d as it is equal to $c/2$ for the (002) plane in the hexagonal wurtzite structure. The c-lattice constant calculated from the XRD results was

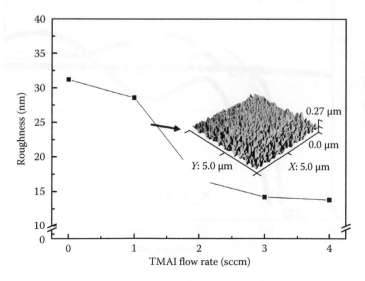

FIGURE 1.42
RMS roughness of the ZnO films grown at various TMAl flow rates.

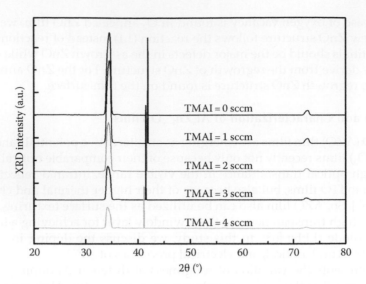

FIGURE 1.43
Double crystal XRD θ–2θ scan of the ZnO grown at different TMAl flow rates.

0.51922–0.52348 nm as TMAl flow rate increases from 0 to 4 sccm. This increase in lattice parameter of the AZO films could have been caused by incorporation of Al^{3+} ions in the interstitial positions [69–71].

The optical transmittance of the ZnO grown on double sides polished sapphire substrate with various TMAl flow rate were compared with that of the blank sapphire substrate, as presented in Figure 1.44. It is found that with the increasing of the Al doping concentration,

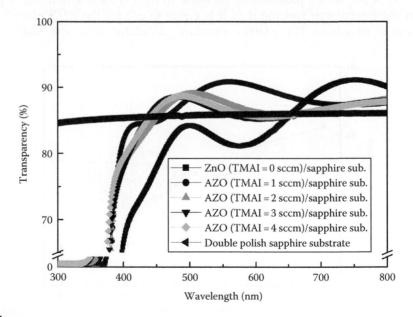

FIGURE 1.44
Optical transmission spectra of the ZnO of various TMAl flow rate deposited on double sides polished sapphire substrate.

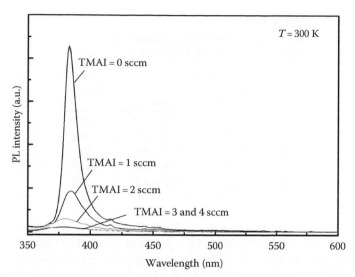

FIGURE 1.45

PL spectra at room temperature for ZnO samples grown at various TMAl flow rates.

the transparency of the ZnO also slightly raises at blue and green regions. When the Al doping concentration is more than 2 sccm, the transparency remains almost unchanged. Overall, the transparency is above 85% in the wavelength ranging from 400 to 700 nm with the increase of Al-doped concentration.

Figure 1.45 is the PL spectra taken at room temperature for Al-doped ZnO samples deposited under different Al doping concentrations. With increasing concentration of the Al dopants, the peak intensity of NBE emission decreases while no peak intensity of DLE are observed. It is also found that the peak position for the NBE emission of Al-doped ZnO shifts to shorter wavelengths with increasing doped concentration, which is well known as the Burstein–Moss effect [72,73].

The results summarized in Table 1.2 are the resistivity, carrier concentration, and mobility of AZO with different Al-doping concentrations. The resistivity monotonically decreases with increasing the Al concentrations, which is attributed to the donor electrons from Al dopants and associated oxygen vacancies. The AZO shows a minimum resistivity value of 2.05×10^{-3} Ω-cm at the TMAl flow rate of 4 sccm. Correspondingly, the carrier concentration increases with the increase of Al-doped concentration. The increased disorder of the crystal lattice induces phonon scattering and ionized impurity scattering, which leads to the decrease of mobility [69,70].

1.4.3 Growth and Characterization of $Mg_xZn_{1-x}O$ Films

In order to realize high-performance ZnO-based optoelectronic devices, two important requirements are necessary: one is p-type doping of ZnO, and the other is modulation of the band gap (E_g) [74,75]. While p-type doping of ZnO is studied intensively, the latter has been demonstrated by the development research on $Mg_xZn_{1-x}O$ allowing modulation of band gap in a wide range, from 3.34 to 7.8 eV [76]. In this section, we investigate the structural, optical, and electrical properties of the wurtzite-structure $Mg_xZn_{1-x}O$ films with five different compositions (*x* from 0 to 0.14) synthesized by MOCVD.

TABLE 1.2

Resistivity, Carrier Concentration, and
Mobility of ZnO with Different Al-Doping
Concentrations

	Resistivity (Ω-cm)	Concentration (cm^{-3})	Mobility (cm^{-3}/V-s)
0 sccm	6.5×10^{-2}	5.54×10^{18}	25.8
1 sccm	7.29×10^{-3}	3.08×10^{19}	29.4
2 sccm	4.19×10^{-3}	5.38×10^{19}	27.8
3 sccm	2.47×10^{-3}	9.46×10^{19}	26.8
4 sccm	2.05×10^{-3}	1.29×10^{20}	23.8

Figure 1.46 presents the XRD patterns of five samples with various Mg contents. The measurements show that the (002) diffraction peaks at about 34.4° are all detectable, indicating that the $Mg_xZn_{1-x}O$ films of different Mg contents are single wurtzite crystalline with the preferential orientation of (002). As the Mg content is up to 6%, the (101) peak of $Mg_xZn_{1-x}O$ feebly emerges. The intensity of (002) diffraction peak reduces gradually with increasing x (Mg), while the (101) peak intensity improves at the same time. Some recent studies suggest that for x (Mg) below 35% the alloy is wurtzite structure, and it becomes cubic structure when x (Mg) is higher than 35% [77]. Our results also claim that there is an intermediate polycrystalline phase before completely transforming to MgO cubic structure. Moreover, we notice that with increasing Mg content, the $Mg_xZn_{1-x}O$ (002) peak shifts slightly toward the higher diffraction angle due to the replacement of Zn^{2+} in the ZnO lattice partly by Mg^{2+} with smaller radius.

To better understand the effect of Mg content in ZnO, the transmittance spectra of $Mg_xZn_{1-x}O$ films grown at different x (Mg) were measured. It needs to be a detailed specification on the ZnO and $Mg_xZn_{1-x}O$ films grown on the double-sides polished sapphire substrate for more accurate result. Figure 1.47a shows the transmittance of each sample which was directly compared with that of double-sides polished sapphire substrate. The transmittance of ZnO films is over 90% in the visible wavelength range (400–800 nm). Similarly, $Mg_xZn_{1-x}O$ films at various Mg contents also present high transmittance over 90% in the most visible light range. The cut-off edge for transmittance spectra gradually shifts toward a shorter wavelength with increasing Mg content.

The relationship between transmittance and E_g can be expressed as [78,79]

$$\alpha = \left(\frac{1}{t}\right) \times \left[\ln\left(T\%\right)\right] \tag{1.3}$$

$$\left(\alpha h\nu\right)^2 = h\nu - E_g \tag{1.4}$$

where
  α stands for the absorption coefficient
  t is the film thickness
  $T(\%)$ is the transmittance
  $h\nu$ is the photon energy
  E_g is the band gap energy

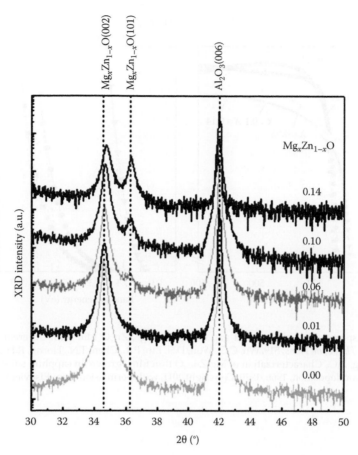

FIGURE 1.46

X-ray diffraction patterns of $Mg_xZn_{1-x}O$ (x from 0 to 0.14) grown on sapphire substrate. (Reproduced from *Thin Solid Films*, 519, Wu, C.C., Wuu, D.S., Lin, P.R., Chen, T.N., Horng, R.H., Ou, S.L., Tu, Y.L., Wei, C.C., and Feng, Z.C., Characterization of $Mg_xZn_{1-x}O$ thin films grown on sapphire substrates by metalorganic chemical vapor deposition, 1966–1970. Copyright 2011, with permission from Elsevier.)

According to formula (1.4), E_g values with various Mg content are calculated by extrapolating the linear region in a plot of $(\alpha h\nu)^2$ versus $h\nu$, as indicated in Figure 1.47b. The E_g values of $Mg_xZn_{1-x}O$ films reach 3.34, 3.40, 3.51, 3.58, and 3.64 eV with the x (Mg) = 0, 0.01, 0.06, 0.10, and 0.14, respectively. This result demonstrates that the band gap energy of $Mg_xZn_{1-x}O$ can be tailored by modifying the Mg content.

Combined PL and Raman scattering spectra were collected for these MgZnO films under excitation lines at 325 nm. In Figure 1.48a, we can clearly observe the PL emission bands with peak energy of 3.31, 3.38, and 3.53 eV for these samples with small Mg compositions of 0, 0.01, and 0.06, respectively, which presents a blue-shift as x (Mg) increases. However, for the highest Mg content of 0.14, the MgZnO-related PL band is beyond the instrument measurement range and the sharp Raman lines appear in the right-hand side of Figure 1.48a. From the enlarged view shown in Figure 1.48b, it can be determined that the band gap of the film for $x = 0.1$ is 3.65 eV.

In addition, the resistivity, carrier concentration and mobility of $Mg_xZn_{1-x}O$ thin films have also been studied as shown in Table 1.3. The experiment data indicate that the

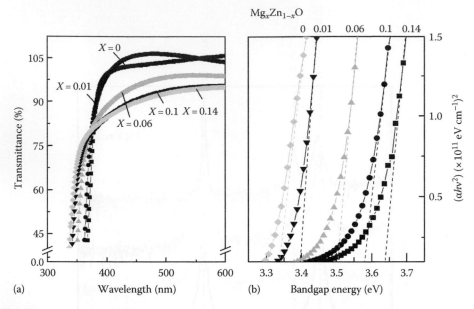

FIGURE 1.47
(a) Transmittance spectra and (b) the band gap energy of $Mg_{1-x}Zn_xO$ film with different x (Mg) contents. (Reproduced from *Thin Solid Films*, 519, Wu, C.C., Wuu, D.S., Lin, P.R., Chen, T.N., Horng, R.H., Ou, S.L., Tu, Y.L., Wei, C.C., and Feng, Z.C., Characterization of $Mg_xZn_{1-x}O$ thin films grown on sapphire substrates by metalorganic chemical vapor deposition, 1966–1970. Copyright 2011, with permission from Elsevier.)

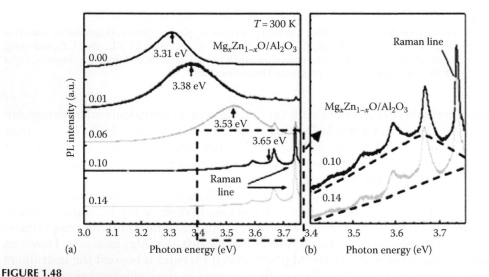

FIGURE 1.48
(a) UV-PL spectra for $Mg_xZn_{1-x}O$ films grown under 450°C with different x (Mg) contents. (b) The magnified UV-PL spectra of $Mg_{0.1}Zn_{0.9}O$ and $Mg_{0.14}Zn_{0.86}O$. (Reproduced from *Thin Solid Films*, 519, Wu, C.C., Wuu, D.S., Lin, P.R., Chen, T.N., Horng, R.H., Ou, S.L., Tu, Y.L., Wei, C.C., and Feng, Z.C., Characterization of $Mg_xZn_{1-x}O$ thin films grown on sapphire substrates by metalorganic chemical vapor deposition, 1966–1970. Copyright 2011, with permission from Elsevier.)

TABLE 1.3

Electronic Properties of $Mg_xZn_{1-x}O$ Thin Films with Different Mg Doping Concentrations

	$Mg_xZn_{1-x}O$				
	0.00	0.01	0.06	0.10	0.14
Resistivity (Ω-cm)	1.32×10^{-2}	2.47×10^{-2}	1.75×10^{-1}	1.82×10^{1}	1.47×10^{3}
Mobility (cm^2/V-s)	13.21	9.33	3.55	1.11	0.82
Concentration (cm^{-3})	3.54×10^{19}	2.39×10^{19}	5.25×10^{18}	2.37×10^{17}	4.22×10^{15}
Type conduction	N	N	Ambiguous	P	P

Source: Reproduced from *Thin Solid Films*, 519, Wu, C.C., Wuu, D.S., Lin, P.R., Chen, T.N., Horng, R.H., Ou, S.L., Tu, Y.L., Wei, C.C., and Feng, Z.C., Characterization of $Mg_xZn_{1-x}O$ thin films grown on sapphire substrates by metalorganic chemical vapor deposition, 1966–1970. Copyright 2011, with permission from Elsevier.

resistivity of $Mg_xZn_{1-x}O$ thin films increases from 1.32×10^{-2} to 1.47×10^{3} Ω-cm with increasing of Mg content ranging from 0 to 0.14. With the increase of Mg content, the carrier concentration gradually decreases, and the lowest p-type carrier concentration value of 4.22×10^{15} cm^{-3} is obtained when the Mg content rise to 0.14. Meanwhile the mobility of $Mg_xZn_{1-x}O$ films steadily reduces with increasing Mg content. All these results suggest that the films grown with lower Mg content have better electrical properties.

We find that the carrier concentration of $Mg_xZn_{1-x}O$ thin films decreases obviously as $x > 0.06$, which could be due to the formation of polycrystalline phase, resulting in more trap centers between the grain boundaries. This can be confirmed through the XRD patterns shown in Figure 1.46. Kim et al. have also reported a similar result where the MgZnO film was grown on the c-sapphire substrate with a thin inserting layer of MgO by MOCVD [80]. It is worth mentioning that near x (Mg) = 10%, the n-type $Mg_xZn_{1-x}O$ thin film would convert to p-type semiconductor. It is well known that Mg is typically used as an acceptor dopant for achieving the p-type GaN. However, when hydrogen is present in the Zn precursor, it could easily be incorporated in the ZnO film during its growth. Since H is found to act as a shallow donor and it readily binds with Mg in ZnO material, forming the Mg–H electrical neutral complexes [81]. Therefore, effectively compensating the H effects will be a major concern in preparing p-type conduction in ZnO.

1.5 Novel Applications

Most of the present applications for ZnO as a transparent conducting film for displays and photovoltaic panels have been generally reviewed [82–84]. Several studies describe the LEDs [85,86] and the inexpensive transistors [87,88] using this class of binary or ternary compounds. The role of 1D quantum effects on the properties of ZnO-based materials and their applications in devices such as gas sensing detectors [89], spintronics [90], smart textiles [91], etc., have also been reviewed. This section will primarily discuss applications in GaN-based epilayers and GaN-LEDs of these materials, which have already been written in other reviews or books.

1.5.1 ZnO Sacrificial Layers

It is well known that wet chemical etching is commonly used for uniform stripping or patterning the dielectric and metallization layers through a mask. Wet etching processes are usually anisotropic and highly selective. As a result, the ZnO sacrificial layer can be effectively removed by proper acid solution during the separation of GaN from sapphire.

The uniform GaN epilayer was successfully transferred from sapphire substrate to the copper template by electroplating and wet etching techniques. As shown in Figure 1.49a, the cross-shaped holes were used to increase the contact area between the etchant and the ZnO sacrificial layer, and substantially shorten the etching time. Figure 1.49b through d presents the SEM images for better understanding the patterned electroplating process. The ZnO sacrificial layer and GaN epilayer were first deposited on the c-oriented sapphire substrate by MOCVD. Then, the Au/Cr films were thermally evaporated onto the GaN as a seed layer. The surface was protected by the thick photoresist (PR) defined via photolithography while only the exposure area would be filled with thick copper during the electroplating process (Figure 1.49b). And then, the remaining

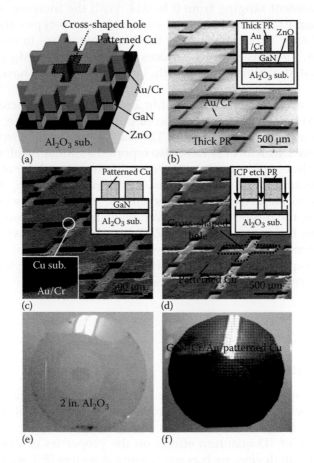

FIGURE 1.49

(a) Schematic diagram of the cross-shaped holes which end up at the surface of the ZnO sacrificial layer. SEM images of the electroplating process for obtaining GaN/patterned Cu structure: (b) PR pattern before electroplating, (c) electroplating the patterned copper, and (d) ICP etching the GaN epitaxial layer after removing PR. Images show the lift-off result of (e) sapphire substrate and (f) GaN/Cr/Au/patterned Cu.

PR could be removed by dipping into a solvent (Figure 1.49c). So, the patterned copper was electroplated on the top GaN surface as a metallic substrate. Next, the Au/Cr/GaN layer was etched through until the ZnO surface by inductively coupled plasma (ICP) system (Figure 1.49d). The sample was treated by etching in diluted HCl solution and finally thoroughly rinsed in DI water. The etching images that represented the sapphire substrate and GaN/Cr/Au/patterned Cu are exhibited in Figure 1.49e and f, respectively. The results indicate that the as grown 2-in. GaN epilayer on a patterned copper substrate can be easily obtained and the original sapphire substrate can be reused for epitaxial growth.

1.5.2 Al-Doped ZnO Transparent Conducting Layers

Other than good optoelectronic properties, AZO is favored more for the abundance of the materials, simple production procedure, low cost, highly thermal and chemical stability, nontoxicity, etc., making it an attractive replacement of ITO.

We fabricated the GaN-based LEDs with AZO as the transparent conductive layer (TCL). For the purpose of achieving a better performance, the AZO film of thickness 1200 nm was directly deposited on the top p-GaN layer at a TMAl flow rate of 1 sccm after the wafer cleaning. Then, the AZO was etched by diluted HCl solution, and the n-GaN layer was exposed by partially etching with inductively coupled Cl_2/Ar plasma. Next, the Ti/Au (12/1500 nm) bi-metal layers were deposited on the n-GaN and p-GaN layer by electron-beam evaporation as the n- and p-electrode, respectively. Finally, the SiO_2 passivation layers were deposited by plasma-enhanced chemical vapor deposition (PECVD) to cover the whole surfaces of LED chips.

The characteristics of GaN-based LEDs with different TCL of AZO and ITO layer were investigated comparatively. We noticed that their forward voltages (V_f) of the two GaN-based LED are almost the same (not shown). Figure 1.50 exhibits the output power-current

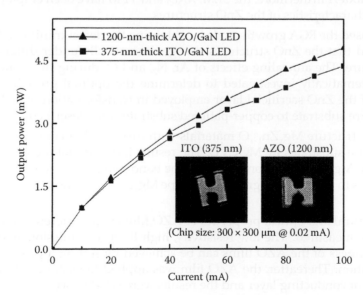

FIGURE 1.50
Output power-current characteristics of GaN LEDs with 1200-nm-thick AZO and 375-nm-thick ITO as the transparent conducting layer, respectively. The inset figures show the OM images of the two different GaN-based LEDs that worked at a low current of 0.02 mA.

curves of GaN-based LEDs with ITO and AZO layers deposited as TCL. At a bias of 20 mA, the optical output power was 1.6 and 1.7 mW with ITO and AZO, respectively. Additionally, it was clearly observed that the LED with 1200-nm-thick AZO can reach a uniform current spreading at a low current (0.02 mA), as shown in the inset. This significant enhancement results mainly from the uniform current distribution and thick window layer.

1.6 Conclusions and Outlook

In this chapter, we have introduced the work on growth and characterization of ZnO-based materials, especially on the results of ZnO nanostructures and ZnO films. The main achievements of this work are summarized in the following:

1. ZnO nanowall networks deposited on the GaN/sapphire substrates without using any metal catalysts were achieved by MOCVD. The surface morphologies and optical properties were investigated for better understanding the growth mechanism of ZnO nanowall networks.

2. Well-aligned and single-crystalline ZNRs were successfully synthesized. It has been demonstrated that the morphology and thickness of the ZnO buffer layer play a key role in the formation of ZNRs. And the quality of ZNRs also can be manipulated by controlling the DEZn flow rate.

3. Vertically well-aligned ZnO nanotube arrays were obtained using a three-step growth process by MOCVD. The results indicate that the growth of ZnO nanotube arrays depends greatly on the growth temperature and follows a self-catalyzed growth mode. Furthermore, the SEM, XRD, and TEM have been employed to evaluate the characteristics of the ZnO structures.

4. We proposed the RGA growth mode for obtaining the defect-minimized ZnO film. It is found that the ZnO structures change dramatically under different growth temperature. The annealing effects of Ar, N_2, and O_2 during the growth process were systematically investigated to determine the optimal growth parameters. Details of the ZnO sacrificial layer employed in transformation of GaN epilayers on sapphire substrate to copper-pattered substrate were described.

5. Wurtzite-structure $Mg_xZn_{1-x}O$ materials with different Mg contents were grown on sapphire substrates by MOCVD. The results indicate that increasing Mg content in the $Mg_xZn_{1-x}O$ not only extends the band gap (E_g) of the material but also contributes to the epitaxial growth of p-type $Mg_xZn_{1-x}O$ without using any doping sources.

6. The structural and optical properties of AZO films with different Al composition have been measured. The low resistivity, high light transmission, and high surface roughness of the AZO films can be achieved by setting a proper Al-doping concentration. Thereafter, the AZO film was applied to GaN-based LEDs as the transparent conducting layer and the results were satisfactory.

As mentioned earlier, we have intensively studied the ZnO-based materials incorporating buffer layers, dopants, and growth techniques and achieved many remarkable results.

Nevertheless, there still exists a good many technical difficulties in the research of ZnO. The most serious obstacle presently, which limits ZnO's potential for certain applications, is the ability to achieve a high, reliable, and reproducible p-type ZnO. Until now, the cause of these problems is still under investigation. Low impurity solubility, excessive acceptor ionization energy, and possible compensating mechanism are three main factors making p-type doping of ZnO difficult. Obviously, these problems must be appropriately overcome before ZnO can be of practical use.

Acknowledgments

This work has been carried out under the support of several funds from National Science Council and Ministry of Economic Affairs, Taiwan, Republic of China (NSC 95-2221-E-005-131-MY3, 97-EC-17-A-07-S1-097, NSC97-2221-E-002-026 and the ATU plan). We would like to thank the coworkers P. R. Lin, T. N. Chen, T. E. Yu, and S. L. Ou at NCHU (Taichung, Taiwan) for some of the results used in this chapter. We are also pleased to acknowledge the helpful discussions with Prof. Z. C. Feng at NTU (Taipei, Taiwan) on characterization results of $Mg_xZn_{1-x}O$ thin films grown on sapphire substrates.

References

1. Ü. Özgür, Ya. I. Alivov, C. Liu et al., A comprehensive review of ZnO materials and devices, *J. Appl. Phys.* **98**, 041301 (2005).
2. T. Sekiguchi, S. Miyashita, K. Obara, T. Shishido, and N. Sakagami, Hydrothermal growth of ZnO single crystals and their optical characterization, *J. Cryst. Growth* **214/215**, 72 (2000).
3. M. Shiloh and J. Gutman, Growth of ZnO single crystals by chemical vapour transport, *J. Cryst. Growth* **11**, 105 (1971).
4. J.-M. Ntep, S. S. Hassani, A. Lusson, A. Tromson-Carli, D. Ballutaud, G. Didier, and R. Triboulet, ZnO growth by chemical vapour transport, *J. Cryst. Growth* **207**, 30 (1999).
5. J. Nause, ZnO broadens the spectrum, *III-Vs Rev.* **12**, 28 (1999).
6. R. J. Lad, P. D. Funkenbusch, and C. R. Aita, Postdeposition annealing behavior of RF sputtered ZnO films, *J. Vac. Sci. Technol.* **17**, 808 (1980).
7. M. A. L. Johnson, S. Fujita, W. H. Rowland, W. C. Hughes, J. W. Cook, and J. F. Schetzina, MBE growth and properties of ZnO on sapphire and SiC substrates, *J. Electron. Mater.* **21**, 157 (1992).
8. W. C. Shih and M. S. Wu, Growth of ZnO films on GaAs substrates with a SiO_2 buffer layer by RF planar magnetron sputtering for surface a coustic wave applications, *J. Cryst. Growth* **137**, 319 (1994).
9. D. K. Hwang, K. H. Bang, M. C. Jeong, and J. M. Myoung, Effects of RF power variation on properties of ZnO thin films and electrical properties of p-n homojunction, *J. Cryst. Growth* **254**, 449 (2003).
10. H. J. Ko, Y. F. Chen, Z. Zhu, T. Hanada, and T. Yao, Biexciton emission from high-quality ZnO films grown on epitaxial GaN by plasma-assisted molecular-beam epitaxy, *J. Cryst. Growth* **208**, 389 (2000).
11. A. Ohtomo, K. Tamura, K. Saikusa, K. Takahashi, T. Makino, Y. Segawa, H. Koinuma, and M. Kawasaki, Single crystalline ZnO films grown on lattice-matched $ScAlMgO_4(0001)$ substrates, *Appl. Phys. Lett.* **75**, 2635 (1999).

12. Y. Zhang, G. Du, D. Liu, X. Wang, Y. Ma, J. Wang, J. Yin, X. Yang, X. Hou, and S. Yang, Crystal growth of undoped ZnO films on Si substrates under different sputtering conditions, *J. Cryst. Growth* **243**, 439 (2002).

13. S. H. Jeong, I. S. Kim, S. S. Kim, J. K. Kim, and B. T. Lee, Homo-buffer layer effects and single crystalline ZnO hetero-epitaxy on *c*-plane sapphire by a conventional RF magnetron sputtering, *Semicond. Sci. Technol.* **19**, L29 (2004).

14. S. H. K. Park, C. S. Hwang, H. S. Kwack, J. H. Lee, and H. Y. Chu, Characteristics of ZnO thin films by means of plasma-enhanced atomic layer deposition, *Electrochem. Solid State Lett.* **9**, G299 (2006).

15. V. Vaithianathan, B. T. Lee, and S. S. Kim, Preparation of As-doped p-type ZnO films using a Zn_3As_2/ZnO target with pulsed laser deposition, *Appl. Phys. Lett.* **86**, 062101 (2005).

16. C. H. Chia, T. Makino, K. Tamura, Y. Segawa, A. Ohtomo, and H. Koinuma, Confinement-enhanced biexciton binding energy in ZnO/ZnMgO multiple quantum wells, *Appl. Phys. Lett.* **82**, 1848 (2003).

17. S. W. Kim, Sz. Fujita, and Sg. Fujita, Self-organized ZnO quantum dots on SiO_2/Si substrates by metalorganic chemical vapor deposition, *Appl. Phys. Lett.* **81**, 5036 (2002).

18. S. Muthukumar, H. Sheng, J. Zhong, Z. Zhang, N. W. Emanetoglu, and Y. Lu, Selective MOCVD growth of ZnO nanotips, *IEEE Trans. Nanotechnol.* **2**, 50 (2003).

19. J. D. Ye, S. L. Gu, W. Liu, S. M. Zhu, R. Zhang, Y. Shi, Y. D. Zheng, X. W. Sun, G. Q. Lo, and D. L. Kwong, Competitive adsorption and two-site occupation effects in metal-organic chemical vapor deposition of ZnO, *Appl. Phys. Lett.* **90**, 174107 (2007).

20. J. Grabowska, A. Meaney, K. K. Nanda, J. P. Mosnier, M. O. Henry, J. R. Duclère, and E. McGlynn, Surface excitonic emission and quenching effects in ZnO nanowire/nanowall systems: Limiting effects on device potential, *Phys. Rev. B* **71**, 115439 (2005).

21. C. C. Wu, D. S. Wuu, P. R. Li, T. N. Chen, and R. H. Horng, Effects of growth conditions on structural properties of ZnO nanostructures on sapphire substrate by metal-organic chemical vapor deposition, *Nanoscale Res. Lett.* **4**, 377 (2009).

22. N. Ohashi, K. Kataoka, T. Ohgaki, T. Miyagi, H. Haneda, and K. Morinaga, Synthesis of zinc oxide varistors with a breakdown voltage of three volts using an intergranular glass phase in the bismuth-boron-oxide system, *Appl. Phys. Lett.* **83**, 4857 (2003).

23. Q. Wan, C. L. Lin, X. B. Yu, and T. H. Wang, Room-temperature hydrogen storage characteristics of ZnO nanowires, *Appl. Phys. Lett.* **84**, 124 (2004).

24. S. J. Young, L. W. Ji, S. J. Chang, and Y. K. Su, ZnO metal-semiconductor-metal ultraviolet sensors with various contact electrodes, *J. Cryst. Growth* **293**, 43 (2006).

25. H. Y. Tsai, Characteristics of ZnO thin film deposited by ion beam sputter, *J. Mater. Process. Technol.* **192/193**, 55 (2007).

26. C. C. Wu, D. S. Wuu, T. N. Chen, T. E. Yu, P. R. Lin, R. H. Horng, and H. Y. Lai, Growth and characterization of epitaxial ZnO nanowall networks using metal organic chemical vapor deposition, *Jpn. J. Appl. Phys.* **47**, 746 (2008).

27. J. Y. Lao, J. Y. Huang, D. Z. Wang, Z. F. Ren, D. Steeves, B. Kimball, and W. Porter, ZnO nanowalls, *Appl. Phys. A* **78**, 539 (2004).

28. J. S. Jeong, J. Y. Lee, J. H. Cho, C. J. Lee, S. J. An, G. C. Yi, and R. Gronsky, Growth behaviour of well-aligned ZnO nanowires on a Si substrate at low temperature and their optical properties, *Nanotechnology* **16**, 2455 (2005).

29. K. Tang, G. K. L. Wong, P. Zu, Z. M. Kawasaki, A. Ohtomo, H. Koinuma, and Y. Segawa, Room-temperature ultraviolet laser emission from self-assembled ZnO microcrystallite thin films, *Appl. Phys. Lett.* **72**, 3270 (1998).

30. S. W. Jung, W. I. Park, H. D. Cheong, G. C. Yi, H. M. Jang, S. Hong, and T. Joo, Time-resolved and time-integrated photoluminescence in ZnO epilayers grown on Al_2O_3(0001) by metalorganic vapor phase epitaxy, *Appl. Phys. Lett.* **80**, 1924 (2002).

31. C. Li, G. Fang, Q. Fu, F. Su, G. Li, X. Wu, and X. Zhao, Effect of substrate temperature on the growth and photoluminescence properties of vertically aligned ZnO nanostructures, *J. Cryst. Growth* **292**, 19 (2006).

32. C. C. Wu, D. S. Wuu, P. R. Lin, T. N. Chen, and R. H. Horng, Realization and manipulation of ZnO nanorod arrays on sapphire substrates using a catalyst-free metalorganic chemical vapor deposition technique, *J. Nanosci. Nanotechnol.* **10**, 3001–3011 (2010).

33. D. C. Kim, B. H. Kong, H. K. Cho, D. J. Park, and J. Y. Lee, Effects of buffer layer thickness on growth and properties of ZnO nanorods grown by metalorganic chemical vapour deposition, *Nanotechnology* **18**, 015603 (2007).

34. C. C. Wu, D. S. Wuu, T. N. Chen, T. E. Yu, P. R. Lin, R. H. Horng, and S. Sun, Characteristics of ZnO nanowall structures grown on GaN template using organometallic chemical vapor deposition, *J. Nanosci. Nanotechnol.* **8**, 3851–3856 (2008).

35. C. C. Wu, D. S. Wuu, P. R. Lin, T. N. Chen, and R. H. Horng, Three-step growth of well-aligned ZnO nanotube arrays by self-catalyzed metalorganic chemical vapor deposition method, *Cryst. Growth Des.* **9**, 4555–4561 (2009).

36. X. Kong, X. Sun, X. Li, and Y. Li, Catalytic growth of ZnO nanotubes, *Mater. Chem. Phys.* **82**, 997–1001 (2003).

37. T. J. Hsueh, S. J. Chang, C. L. Hsu, Y. R. Lin, and I. C. Chen, ZnO nanotube ethanol gas sensors, *J. Electrochem. Soc.* **155**, K152–K155 (2008).

38. S. C. Hung, P. J. Huang, C. E. Chan, W. Y. Uen, F. Ren, S. J. Pearton, T. N. Yang, C. C. Chiang, S. M. Lan, and G. C. Chi, Surface morphology and optical properties of ZnO epilayers grown on Si(111) by metal organic chemical vapor deposition, *Appl. Surf. Sci.* **255**, 3016–3018 (2008).

39. D. J. Lee, J. Y. Park, Y. S. Yun, Y. S. Hong, J. H. Moon, B. T. Lee, and S. S. Kim, Comparative studies on the growth behavior of ZnO nanorods by metalorganic chemical vapor deposition depending on the type of substrates, *J. Cryst. Growth* **276**, 458 (2005).

40. L. Shi, Y. Wu, and Q. Li, Shape-selective synthesis and optical properties of highly ordered one-dimensional ZnS nanostructures, *Cryst. Growth Des.* **9**, 2214–2219 (2009).

41. L. Shi, Y. Wu, and Q. Li, Controlled fabrication of ZnSe arrays of well-aligned nanorods, nanowires, and nanobelts with a facile template-free route, *J. Phys. Chem. C* **113**, 1795–1799 (2009).

42. L. Shi, Y. M. Xu, S. K. Hark, Y. Liu, S. Wang, L.-M. Peng, K. W. Wong, and Q. Li, Optical and electrical performance of SnO_2 capped ZnO nanowire arrays, *Nano Lett.* **7**, 3559–3563 (2007).

43. Y. Chen, Y. Pu, L. Wang et al., Influence of nitrogen annealing on structural and photoluminescent properties of ZnO thin film grown on c-Al_2O_3 by atmospheric pressure MOCVD, *Mater. Sci. Semicond. Process.* **8**, 491–496 (2005).

44. B. Cao, W. Cai, and H. Zeng, Temperature-dependent shifts of three emission bands for ZnO nanoneedle arrays, *Appl. Phys. Lett.* **88**, 161101 (2006).

45. D. Yu, L. Hu, J. Li, H. Hu, H. Zhang, Z. Zhao, and Q. Fu, Catalyst-free synthesis of ZnO nanorod arrays on InP (001) substrate by pulsed laser deposition, *Mater. Lett.* **62**, 4063–4065 (2008).

46. M. Pan, W. E. Fenwick, M. Strassburg et al., Metal-organic chemical vapor deposition of ZnO, *J. Cryst. Growth* **287**, 688 (2006).

47. S. W. Kim, S. Fujita, M. S. Yi, and D. H. Yoon, Catalyst-free synthesis of ZnO nanowall networks on Si_3N_4/Si substrates by metalorganic chemical vapor deposition, *Appl. Phys. Lett.* **88**, 253114 (2006).

48. Y. Ma, G. T. Du, T. P. Yang et al., Effect of the oxygen partial pressure on the properties of ZnO thin films grown by metalorganic vapor phase epitaxy, *J. Cryst. Growth* **255**, 303 (2003).

49. H. S. Kang, J. S. Kang, J. W. Kim, and S. Y. Lee, Annealing effect on the property of ultraviolet and green emissions of ZnO thin films, *J. Appl. Phys.* **95**, 1246 (2004).

50. H. T. Ng, J. Li, M. K. Smith et al., Growth of epitaxial nanowires at the junctions of nanowalls, *Science* **300**, 1249 (2003).

51. S. R. Hejazi, H. R. Madaah Hosseini, and M. Sasani Ghamsari, The role of reactants and droplet interfaces on nucleation and growth of ZnO nanorods synthesized by vapor-liquid-solid (VLS) mechanism, *J. Alloy. Comp.* **455**, 353–357 (2008).

52. P. X. Gao, C. S. Lao, Y. Ding, and Z. L. Wang, Metal/semiconductor core/shell nanodisks and nanotubes, *Adv. Funct. Mater.* **16**, 53 (2006).

53. Z. Yin, N. Chen, R. Dai, L. Liu, X. Zhang, X. Wang, J. Wu, and C. Chai, On the formation of well-aligned ZnO nanowall networks by catalyst-free thermal evaporation method, *J. Cryst. Growth* **305**, 296 (2007).

54. M. Wei, D. Zhi, and J. L. MacManus-Driscoll, Self-catalysed growth of zinc oxide nanowires, *Nanotechnology* **16**, 1364 (2005).

55. A. Umar, E.-K. Suh, and Y. B. Hahn, Non-catalytic growth of high aspect-ratio ZnO nanowires by thermal evaporation, *Solid State Commun.* **139**, 447 (2006).

56. S. L. Mensah, V. K. Kayastha, I. N. Ivanov, D. B. Geohegan, and Y. K. Yap, Formation of single crystalline ZnO nanotubes without catalysts and templates, *Appl. Phys. Lett.* **90**, 113108 (2007).

57. R. C. Wang, C. P. Liu, J. L. Huang, and S.-J. Chen, ZnO hexagonal arrays of nanowires grown on nanorods, *Appl. Phys. Lett.* **86**, 251104 (2005).

58. H. Wang, Z. P. Zhang, X. N. Wang, Q. Mo, Y. Wang, J. H. Zhu, H. B. Wang, F. J. Yang, and Y. Jiang, Selective growth of vertical-aligned ZnO nanorod arrays on Si substrate by catalyst-free thermal evaporation, *Nanoscale Res. Lett.* **3**, 309–314 (2008).

59. C. C. Wu, D. S. Wuu, P. R. Lin, T. N. Chen, and R. H. Horng, Repeated growing and annealing towards ZnO film by metal-organic CVD, *Chem. Vap. Deposition* **15**, 234–241 (2009).

60. A. B. M. A. Ashrafi, Y. Segawa, K. Shin, J. Yoo, and T. Yao, Nucleation and growth modes of ZnO deposited on 6H-SiC substrates, *Appl. Surf. Sci.* **249**, 139 (2005).

61. A. Janotti and C. G. Van de Walle, New insights into the role of native point defects in ZnO, *J. Cryst. Growth* **287**, 58 (2006).

62. J. Dai, H. Su, L. Wang, Y. Pu, W. Fang, and F. Jiang, Properties of ZnO films grown on (0001) sapphire substrate using H_2O and N_2O as O precursors by atmospheric pressure MOCVD, *J. Cryst. Growth* **290**, 426 (2006).

63. F. K. Shan, G. X. Liu, W. J. Lee, and B. C. Shin, The role of oxygen vacancies in epitaxial-deposited ZnO thin films, *J. Appl. Phys.* **101**, 053106 (2007).

64. H. Zhang, Z. G. Wang, X. P. Peng et al., Modification of ZnO films under high energy Xe-ion irradiations, *Nucl. Instrum. Methods B* **266**, 2863 (2008).

65. G. W. Tomlins, J. L. Routbort, and T. O. Mason, Zinc self-diffusion, electrical properties, and defect structure of undoped, single crystal zinc oxide, *J. Appl. Phys.* **87**, 117 (2000).

66. L. P. Dai, H. Deng, F. Y. Mao, and J. D. Zang, The recent advances of research on p-type ZnO thin film, *J. Mater. Sci. Mater. Electron.* **19**, 727 (2008).

67. A. Murai, D. B. Thompson, H. Masui, N. Fellows, U. K. Mishra, S. Nakamura, and S. P. DenBaars, Hexagonal pyramid shaped light-emitting diodes based on ZnO and GaN direct wafer bonding, *Appl. Phys. Lett.* **89**, 171116 (2006).

68. C. C. Wu, P. R. Lin, T. N. Chen, T. E. Yu, and D. S. Wuu, Growth of Al-doped ZnO window layer for GaN LED applications, *Chin. J. Lumin.* **29**, 508–512 (2008).

69. R. K. Shukla, A. Srivastava, A. Srivastava, and K. C. Dubey, Growth of transparent conducting nanocrystalline Al doped ZnO thin films by pulsed laser deposition, *J. Cryst. Growth* **294**, 427 (2006).

70. W. D. Kingery, H. K. Bowen, and D. R. Unlmann (eds.), *Introduction to Ceramics*, John Wiley, New York, 1976.

71. X. Zi-qiang, D. Hong, L. Yan, and C. Hang, Al-doping effects on structure, electrical and optical properties of c-axis-orientated ZnO:Al thin films, *Mater. Sci. Semicond. Process.* **9**, 132 (2006).

72. E. Burstein, Anomalous optical absorption limit in InSb, *Phys. Rev.* **93**, 632–633 (1954).

73. T. S. Moss, The interpretation of the properties of indium antimonide, *Proc. Phys. Soc. Lond. Sect. B* **67**, 775 (1954).

74. A. Tsukazaki, A. Ohtomo, T. Onuma et al., Repeated temperature modulation epitaxy for p-type doping and light-emitting diode based on ZnO, *Nat. Mater.* **4**, 42–46 (2005).

75. A. I. Belogorokhov, A. Y. Polyakov, N. B. Smirnov, A. V. Govorkov, E. A. Kozhukhova, H. S. Kim, D. P. Norton, and S. J. Pearton, Lattice vibrational properties of ZnMgO grown by pulsed laser deposition, *Appl. Phys. Lett.* **90**, 192110 (2007).

76. C. C. Wu, D. S. Wuu, P. R. Lin, T. N. Chen, R. H. Horng, S. L. Ou, Y. L. Tu, C. C. Wei, and Z. C. Feng, Characterization of $Mg_xZn_{1-x}O$ thin films grown on sapphire substrates by metalorganic chemical vapor deposition, *Thin Solid Films* 519, 1966–1970 (2011).

77. L. Bergman, J. L. Morrison, X. B. Chen, J. Huso, and H. Hoeck, Ultraviolet photoluminescence and Raman properties of MgZnO nanopowders, *Appl. Phys. Lett.* **88**, 023103 (2006).

78. D. J. Nagaraju and S. B. Krupanidh, Investigations on multimagnetron sputtered $Zn_{1-x}Mg_xO$ thin films through metal-ferroelectric-semiconductor configuration, *J. Appl. Phys.* **104**, 043510 (2008).

79. Z. G. Ju, C. X. Shan, D. Y. Jiang, J. Y. Zhang, B. Yao, D. X. Zhao, D. Z. Shen, and X. W. Fan, $Mg_xZn_{1-x}O$-based photodetectors covering the whole solar-blind spectrum range, *Appl. Phys. Lett.* **93**, 173505 (2008).

80. D. C. Kim, B. H. Kong, C. H. Ahn, and H. K. Cho, Characteristics improvement of metalorganic chemical vapor deposition grown MgZnO films by MgO buffer layers, *Thin Solid Films* **518**, 1185 (2009).

81. Z. P. Wei, B. Yao, Z. Z. Zhang et al., Formation of *p*-type MgZnO by nitrogen doping, *Appl. Phys. Lett.* **89**, 102104 (2006).

82. C. H. Kuo, C. L. Yeh, P. H. Chen, W. C. Lai, C. J. Tun, J. K. Sheu, and G. C. Chi, Low operation voltage of nitride-based LEDs with Al-doped ZnO transparent contact layer, *Electrochem. Solid-State Lett.* **11**, H269–H271 (2008).

83. T. Minami, S. Takata, and T. Kakumu, New multicomponent transparent conducting oxide films for transparent electrodes of flat panel displays, *J. Vac. Sci. Technol. A* **14**, 1689 (1996).

84. E. Fortunato, L. Raniero, L. Silva et al., Highly stable transparent and conducting gallium-doped zinc oxide thin films for photovoltaic applications, *Sol. Energy Mater. Sol. Cells* **92**, 1605 (2008).

85. Reported by Nanomarkets, Zinc oxide (ZnO)-material, properties, applications and market opportunities of zinc oxide by nanomarkets, December 18, 2008, from http://www.azom.com/details.asp?ArticleID=4522.

86. S. J. Jiao, Z. Z. Zhang, Y. M. Lu et al., ZnO p-n junction light-emitting diodes fabricated on sapphire substrates, *Appl. Phys. Lett.* **88**, 031911 (2006).

87. Y. Sun and J. A. Rogers, Inorganic semiconductors for flexible electronics, *Adv. Mater.* **19**, 1897–1916 (2007).

88. J. H. Kim, B. D. Ahn, C. H. Lee, K. A. Jeon, H. S. Kang, and S. Y. Lee, Characteristics of transparent ZnO based thin film transistors with amorphous HfO_2 gate insulators and Ga doped ZnO electrodes, *Thin Solid Films* **516**, 1529–1532 (2008).

89. S. Roy and S. Basu, Improved zinc oxide film for gas sensor applications, *Bull. Mater. Sci.* **25**, 513–515 (2002).

90. S. J. Pearton, D. P. Norton, Y. W. Heo, L. C. Tien, M. P. Ivill, Y. Li, B. S. Kang, F. Ren, J. Kelly, and A. F. Hebard, ZnO spintronics and nanowire devices, *J. Electron. Mater.* **35**, 862–868 (2006).

91. S. A. Morin, F. F. Amos, and S. Jin, Biomimetic assembly of zinc oxide nanorods onto flexible polymers, *J. Am. Chem. Soc.* **129**, 13776–13777 (2007).

2

Solution-Grown n-Type ZnO Nanostructures Synthesis, Microstructure, and Doping

Rodrigo Noriega, Saahil Mehra, and Alberto Salleo

CONTENTS

2.1 Introduction..59
2.2 Synthesis..61
2.3 Doping..67
2.4 Structural Characterization..67
2.5 Electronic Characterization..71
2.6 Charge Transport in Nanowires and Nanowire Ensembles...............................73
2.7 Optical Properties of Nanowire Meshes 74
2.8 Device Applications...76
2.9 Conclusion 77
References..79

2.1 Introduction

Transparent conducting oxides (TCOs) have been the focus of intense research in recent years [1,2], due to their applications in electronics and photovoltaics. These wide band gap materials have been used as the active channel material in field-effect transistors [3,4]; as the electron-accepting material in dye-sensitized solar cells [5]; and, when doped, as transparent electrodes for transistors, organic light emitting diodes, solar cells, and electrochromic windows [6]. The most widely used TCO is indium tin oxide (ITO). However, due to high processing costs and indium scarcity there has been a field-wide push to find suitable alternatives.

Zinc oxide (ZnO) is an extensively studied material [7] and well suited to replace ITO, due to its low cost and marginal toxicity combined with its competitive electrical and optical properties. Recent research has focused on tailoring the properties of these materials by modifying their morphology at the nanoscale [5,8]. ZnO has a wurtzite crystal structure comprised of a series of alternating planes of tetrahedrally coordinated O^{2-} and Zn^{2+} ions stacked along the *c*-axis, with characteristic polar surfaces that give rise to a number of different nanostructures including rods, wires, belts, springs, and tubes [8].

Without any intentional doping, ZnO is naturally *n*-type due to oxygen vacancies and can be extrinsically doped by introducing heteroatoms (such as Al, Ga, In, B, or Sn) into the ZnO crystal lattice, reaching carrier concentrations as high as 1.5×10^{21} cm^{-3} [7].

Electron mobility in ZnO can reach values of $200 \, cm^2 \, V^{-1} \, s^{-1}$ for ZnO single crystals at low doping levels, with mobility generally decreasing at higher doping levels. However, for heavily doped ZnO films obtained with techniques such as magnetron sputtering, pulsed laser deposition (PLD), and metalorganic chemical vapor deposition (MOCVD), the mobility varies between 10 and $60 \, cm^2 \, V^{-1} \, s^{-1}$ depending on the microstructure of the material [7]. The lowest resistivity achieved in ZnO is on the order of $10^{-4} \, \Omega \, cm$ [7], comparable to ITO. For applications in optoelectronic devices such as UV lasers, the large exciton binding energy of ZnO (60 meV) is a desirable property because it favors the radiative recombination of excitons, which are not thermally dissociated at room temperature [9,10].

The synthesis routes vary widely and so do their results. Films are usually obtained with magnetron sputtering, MOCVD, PLD, and sol–gel techniques. Nanostructured ZnO can also be synthesized via a variety of methods, each with different advantages and applicability ranges. MOCVD processes [8] have the capability to produce the highest diversity of nanostructures, depending on the growth conditions, and also show very high homogeneity within the same batch, as well as high dopant incorporation; however, they are not compatible with flexible substrates due to the high temperatures required (e.g., MOCVD reactions usually operate above 300°C and typically around 600°C [11,12]). PLD has also been used to create nanostructured ZnO films with promising results [13].

From the processing perspective, the idea of solution-processable TCOs is a very attractive one, due to the inherent scalability of the technique, the lack of complicated or expensive synthesis steps, and the potential to achieve high throughputs. Furthermore, solution-processable materials are amenable to incorporation into large-area flexible substrates due to the ability to use low temperature and high-throughput deposition techniques such as gravure printing and spray coating. Examples of solution-based syntheses are hydrothermal growth, thermal decomposition in organic solvents, and electrochemical deposition. In this chapter, we focus on the study of highly doped ZnO nanostructures grown in organic solvents at moderate temperatures (300°C).

The growth of doped nanostructures in solution, however, is not an easy task. For example, doping in aqueous solutions is difficult due to the formation of aquo-ions of the dopant species, and these stable, inert complexes are not incorporated into the host crystal. A strategy to overcome this problem is to apply an electric potential to the growth substrate to favor doping over metal–aquo complex formation [14]. Another option is to use nonaqueous, high boiling point organic solvents. The use of organic solvents prevents the formation of inert metal–aquo complexes and favors dopant incorporation, with the additional benefit of the possibility to achieve higher reaction temperatures. The incorporation of dopants into the crystal lattice, however, is not necessarily site-selective and typical reactions yield a mixture of substitutional and interstitial impurities [15].

Another issue with solution growth is the larger degree of size and shape dispersity among the resulting structures when compared to other growth techniques such as MOCVD or PLD. Such dispersity is evidenced by the variations in size, morphology, and dopant incorporation efficiency within a single synthesis, as well as between nominally identical runs. A reason for such variations is found in the fact that most of the solution-growth methods operate out of thermal equilibrium. A comprehensive understanding of the growth mechanism is lacking in order to achieve a better control of the morphology and impurity incorporation, as well as repeatability.

2.2 Synthesis

Two well-established routes to synthesize high aspect ratio wurtzite ZnO nanowires from solution are based on the decomposition of zinc salts in either aqueous or organic media. Aqueous-based growth of ZnO nanostructures can be performed hydrothermally [16] or by using chemical bath deposition on a textured seed substrate [17,18]. Colloidal synthetic routes to ZnO nanostructures usually involve the thermal decomposition of the zinc precursor in the presence of a growth-directing surfactant and a high boiling point organic solvent. While large-scale synthesis of oriented nanowire arrays has been demonstrated using aqueous methods, the dependence of these methods on a seed layer and substrate for arrayed nanowire growth makes them less amenable to the fabrication of randomly oriented nanowire mesh films for device applications. Synthesis in organic solvents is advantageous in that it allows substrate-independent, template-free synthesis of nanostructures as well as higher doping efficiencies for electronic or dilute magnetic semiconductor (DMS) applications [14].

Vayssieres synthesized arrays of high aspect ratio ZnO nanostructures via an aqueous, bottom-up approach [16]. Using the hydrothermal method, zinc nitrate was decomposed in the presence of methenamine, a cyclic tertiary amine known to direct one-dimensional (1D) growth of ZnO nanowires. One-dimensional growth of ZnO crystals has been reported under a variety of experimental conditions, with wire growth being observed in a range of pH from slightly acidic to basic conditions (pH range: 5–12) and a temperature range from 50°C to 200°C. Nanostructure growth is dependent on the hydrolysis of the zinc precursors to create zinc hydroxyl species, but basic conditions are necessary for ZnO growth because the zinc cation complexes need a higher pH to hydrolyze [14]. Methenamine has shown to be the most effective additive for aqueous growth of ZnO nanowires, although its role in the chemical reaction is still under debate [14].

Colloidal synthesis of ZnO nanostructures in high boiling point organic solvents is an alternative route toward high aspect ratio ZnO nanowires. A low degree of polydispersity and an understanding of morphological control in solution-grown ZnO nanostructures are critical for enabling practical applications. Achieving monodisperse synthetic outcomes requires control of reactant concentrations throughout the growth solution to narrow the size distribution. At high precursor concentrations, the critical size for stable nanoparticles is usually much smaller than the distribution of nanoparticles present in the synthesis, thus encouraging growth of all nanoparticles—with smaller particles growing faster—and narrowing the resulting size distribution [19]. Another strategy used toward achieving monodisperse distributions of nanoparticles is to have nucleation occur over as small of a time window as possible. In similar II–VI systems, strategies such as hot injection of precursors into the solvent are employed to minimize the time over which crystallites nucleate, promoting monodispersity [20].

Morphological control of colloidally grown ZnO nanostructures is dependent on many variables, including the relative molar ratios of the precursors, solvent choice, and surfactant choice. Surfactants typically consist of a coordinating head group and a long alkyl chain to provide a dynamic organic capping layer that selectively adsorbs onto different crystal surfaces. O'Brien and coworkers synthesized ZnO nanostructures by thermally decomposing zinc acetate using oleic acid as a surfactant in the presence of a high boiling point organic solvent [21]. Depending on the solvent used trioctylamine, 1-hexadecanol,

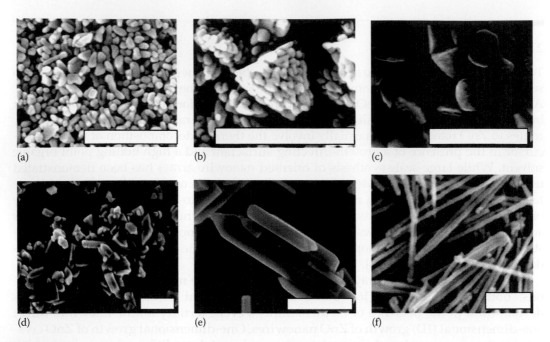

FIGURE 2.1

SEM micrographs of ZnO nanostructures colloidally grown under different conditions in 1-hexadecanol (a–c) and trioctylamine (d–f). All scale bars: 500 nm.

or 1-octadecene—and its coordinating power, a variety of different morphologies were obtained. Scanning electron micrographs (Figure 2.1) denote some of the further synthetic outcomes observed within this synthesis by additional experimentation with the process variables of the ternary reaction system.

Trioctylamine, a tertiary amine with three branched alkyl chains, exhibits a strong coordinating power and enables growth of high aspect ratio structures. This effect is analogous to that observed when the high aspect ratio nanowires are grown in aqueous environments under the presence of methenamine. It has been argued that trioctylamine could act as a ligand during the synthesis as well, strongly coordinating to (11–20) planes and inhibiting radial growth [21]. To further explore the effects of the solvent alkyl chain length on the aspect ratio of the ZnO nanostructures obtained, O'Brien and coworkers synthesized ZnO nanostructures using tertiary amine solvents with three different alkyl chain lengths. They observed the highest aspect ratio structures (aspect ratio: 20–38) with the shortest alkyl chain length solvent, trihexylamine, and lowest aspect ratio structures with tridodecylamine. The tertiary amine is proposed to stabilize the polar surfaces and quench radial growth, and the lower aspect ratios observed with tridodecylamine was attributed to steric hindrance from the longer alkyl chains attached to the amine [22].

The coordinating power of a solvent is defined as its ability to form stable intermediate complexes with the metal ions in question. A more stable intermediate will result in slower growth kinetics, allowing the reaction more time to approach thermodynamic equilibrium. Unstable complexes make the growth faster, which could lead to kinetically dominated growth, where intermediates might get trapped in an energetically unfavorable site by the next wave of incoming particles adsorbing on the crystal surface.

Furthermore, Peng and coworkers developed a theory characterizing the colloidal growth of II–VI nanostructures by relating the degree of monomer supersaturation to the driving force for anisotropic growth and defined concentration-dependent growth regimes [23]. They found growth to be a kinetically limited process, where high precursor concentrations drove anisotropic one-dimensional growth and lower concentrations resulted in isotropic growth of nanostructures. Higher precursor concentrations have also been shown to encourage the narrowing of the size distribution of particles, encouraging monodispersity [19,24].

The presence of surfactant plays an important role in the morphology of solution-grown ZnO nanostructures. For example, adding oleic acid as a surfactant during the growth of ZnO nanowires in trioctylamine directly affects the aspect ratio of the resulting nanowires. A series of synthesis reactions was stopped after different growth times, and the dimensions of the obtained nanowires were determined by analyzing approximately 100 nanowires per synthesis using a scanning electron microscope (SEM) in order to obtain statistically significant results (Figure 2.2). The length of the nanowires continually grows in both cases (although significantly more slowly when no surfactant is present), but the width of the wires behaves differently. When surfactant is used the diameter of the nanowires hardly changes (40–60 nm), but when no surfactant is added the diameter of the wires grows steadily, being as large as 150 nm after 1 h. As a result, short syntheses are practically unaffected by the presence of oleic acid, while in longer syntheses oleic acid significantly increases the aspect ratio of the nanowires [25].

From a thermodynamic perspective, Wulff's rule predicts the equilibrium structure of a crystal to be dictated by the relative surface energies of the different facets. While it is not possible to predict the equilibrium structure for a given set of reaction conditions since the growth is likely an out of equilibrium process, the relative surface energies of each of the crystallographic faces play a large role in driving the reaction and growth of the ZnO nanostructures. Furthermore, the choice of solvent and surfactant may alter the relative surface energies of crystallographic facets due to preferential adsorption to different planes as well as affect the stability of solvent–metal complexes during the synthesis and the resulting growth kinetics [23,26].

The one-dimensional growth of ZnO nanostructures occurs along the close-packed [0001] direction in ZnO nanowires, owing to the electrostatic destabilization (and resulting

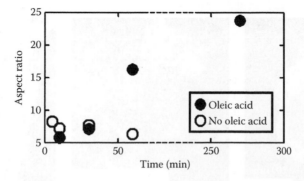

FIGURE 2.2
Aspect ratio of undoped wires with (solid symbols) and without (open symbols) oleic acid, showing the effect of surfactant in the morphology of ZnO nanoparticles. (Reproduced from Noriega, R. et al., Transport and structural characterization of solution-processable doped ZnO nanowires, *Proc. SPIE*, 74110N-1, 2009. With permission from The International Society for Optical Engineering.)

higher surface energies) of the (0001) planes. The highly anisotropic, *c*-axis oriented growth has been rationalized by considering the wurtzite ZnO stacking sequence of alternating planes of tetrahedrally coordinated O^{-2} anions and Zn^{+2} cations along the *c*-direction. In the [000–1] direction, Zn^{+2} cations have three dangling bonds, resulting in minimal charge compensation from O^{-2} anions and significant electrostatic destabilization of cationic planes growing along this direction. As a result of the relative instability of these faces, anisotropic growth along the *c*-axis has been observed with ZnO in this reaction system [14,21,27].

While high aspect ratio nanostructures can be synthesized using template-free methods in this colloidal approach, the ZnO nanowires grown using this mechanism are observed to grow from a hexagonal platelet that serves as the base for nanowire growth (Figure 2.3). These seed-grown nanowire bunches contain various micron-length nanowires grown attached to the base plate, which are not optimal for use as randomly oriented nanowire mesh films due to large inhomogeneities in films that result from nanowire bunching. In the context of electronic applications of nanostructured films, an ideal nanowire mesh film would be a sparse, interconnected array of randomly oriented high aspect ratio nanowires. The bundling of these nanowires presents a synthetic and a processing challenge for deposition of nanowire mesh films. Aggregates of nanowires would create nonuniform surfaces and make it difficult to deposit high-quality films on top of these nanostructured meshes as well as decrease their electrical performance due to the lower degree of interconnectivity.

From a synthetic point of view, a fundamental understanding of the growth mechanism is critical to being able to design stand-alone nanowires that can be integrated into electronic devices. While the influence of surfactants on the synthetic outcome is understood, there is no clear understanding of reaction pathways and how the initial chemical precursors transform to yield the final reaction products [26,28]. The crystallization mechanisms may be complex and difficult to elucidate, often suggesting unconventional crystal growth mechanisms. For instance, Marks and coworkers grew ZnO nanowire arrays in aqueous solution and proposed a ripening-based mechanism for nanowire growth where surface roughness features serve as nucleation sites for thin nanorods, which eventually coalesced to form each individual nanowire (Figure 2.4a–c) [18]. Weller and coworkers synthesized ZnO nanorods by first preparing spherical nanoparticles and evaporating the solvent to increase the nanoparticle concentration. This process induced oriented

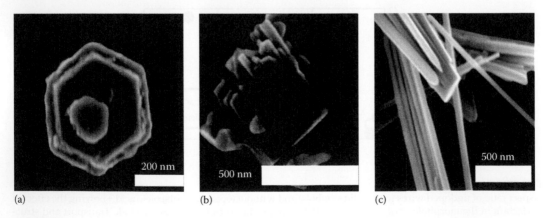

(a)   (b)   (c)

FIGURE 2.3
SEM micrographs taken at different times during the synthesis showing the initial formation of a hexagonal base plate (a), growth of nanowires from initial platelet (b), and final nanowire-on-platelet structures (c).

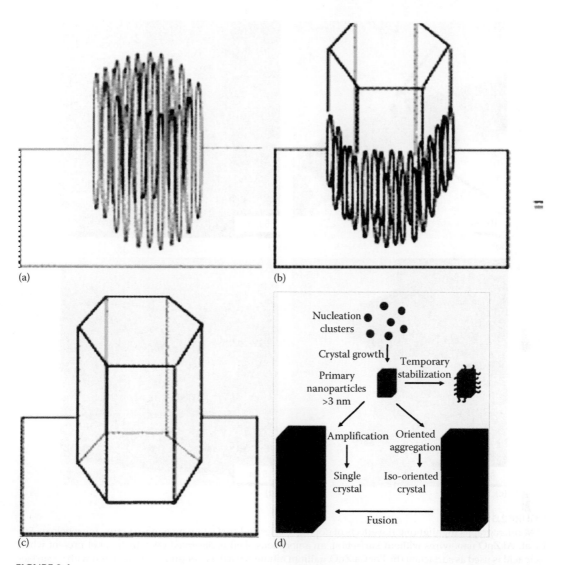

FIGURE 2.4

Proposed growth mechanism for ZnO nanowires in aqueous solution by Li and coworkers. (a) Nanorods grow as a bundle but independently, (b) onset of coalescence of single nanowires (NWs), and (c) eventual growth of a single nanowire. (a–c: Reproduced with permission From Li, Q. et al., Fabrication of ZnO nanorods and nanotubes in aqueous solutions, *Chem. Mater.*, 17(5), 1001–1006. Copyright 2005 American Chemical Society.) The oriented attachment growth mechanism observed by Weller and coworkers is outlined in (d). (Reproduced from Niederberger, M. and Colfen, H., Oriented attachment and mesocrystals: Non-classical crystalization mechanisms based on nanoparticle assembly, *Phys. Chem. Chem. Phys.*, 8(28), 3271–3287, 2006. With permission from The Owner Societies.)

attachment of nanoparticles into chain-like structures, which eventually coarsened into ZnO nanorods (Figure 2.4d) [28–30].

The progress made in the field of solution-grown nanomaterials has resulted in significant work being focused on enhancing the synthetic understanding and enabling rational syntheses. Research efforts are currently ongoing to provide an explanation of the growth mechanism and move toward rational syntheses of ZnO nanostructures.

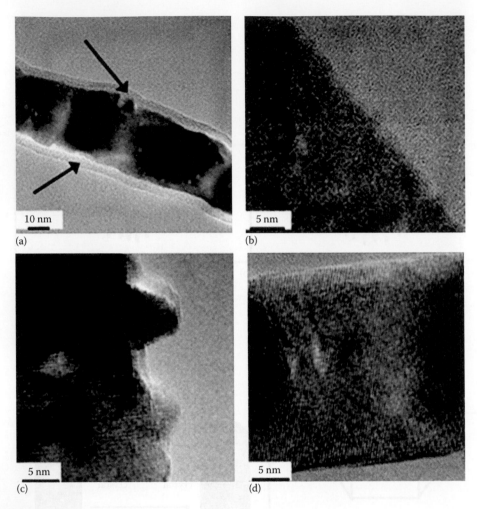

FIGURE 2.5

TEM micrographs highlighting the effects of doping species and surfactant on nanostructure morphology. For 1% at. Al:ZnO nanowires without surfactant, an amorphous shell is observed (a), which is not present when oleic acid is used as surfactant (b). For Ga:ZnO, gallium nitrate is used as a doping precursor and with no surfactant one observes surface instabilities (c), which are absent when surfactant is used (d). (Reproduced from Goris, L. et al., Intrinsic and doped zinc oxide nanowires for transparent electrode fabrication via low-temperature solution synthesis, *J. Electron. Mater.*, 38(4), 586–595, 2009. With permission from The Minerals Society.)

An additional difficulty to doping in solution during growth is the introduction of additional chemical species to the synthesis reaction. These additional species can modify the morphology of the nanostructures (Figure 2.5). The addition of aluminum acetate in order to attain Al:ZnO in a zinc acetate/trioctylamine reaction results in an amorphous shell surrounding a crystalline core when surfactants are not present, but the inclusion of oleic acid as a surfactant prevents the formation of these amorphous shells. If gallium nitrate is used as a doping salt in the same reaction, anisotropic growth is observed even in the absence of surfactants, and high concentrations of $Ga(NO_3)_3$ cause surface instabilities on the otherwise smooth nanowires. These surface instabilities are a crystalline continuation of the nanowire as can be seen by the uninterrupted lattice planes. Once more, the addition of oleic acid prevents the formation of these surface instabilities [27,31].

2.3 Doping

Doping can be achieved by disrupting the crystal structure of an otherwise intrinsic semiconductor. These disruptions can be in the form of impurities or native defects in the lattice. Depending on the ionization state of the dopant, positive or negative charge can be added to the pure, perfect crystal, leading to a shift on the Fermi level of the semiconductor and achieving the desired *n*- or *p*-type behavior. In ZnO, the most common dopant heteroatoms are Al, Ga, In, B, or H. Native defects, primarily oxygen vacancies (V_O), have been posed as donor defects in ZnO.

Group III elements (B, Al, Ga, In) are donor impurities when they substitute Zn. The effectiveness of doping will depend on the ease of incorporation of these heteroatoms into zinc substitutional sites in the ZnO lattice (X_{Zn}). In the case of hydrogen, recent density functional theory calculations have shown that it can create donor sites when incorporated into interstitial sites during growth. Hydrogen is commonly present in most growth environments and is thus a more likely explanation for the *n*-type character of nominally intrinsic ZnO, in contrast to the deeper donor sites created by oxygen vacancies.

The possibility of native point defects acting as doping sites also raises the question of compensating defects. Even if extrinsic dopant incorporation is successfully achieved, the creation of native defects can balance their effect. This has been discussed by Zunger et al. [32–35] as a reason for the "undopability" of certain II–V semiconductors such as *p*-ZnO, *p*-ZnS, and *n*-ZnTe.

The failure to dope has been attributed to three main reasons by Zunger and coworkers [33]. The first one is the use of insoluble dopants such that the desired impurity atom cannot be introduced into the host crystal because of limited solubility, including cases of formation of secondary phases, impurity segregation or precipitation, and incorporation of the impurities on the wrong site in the host lattice. A second pathway for doping failure is the creation of deep donor (or acceptor) sites, as opposed to shallow impurity levels, which prevent the dopant atoms from being ionized at room temperature. The last mechanism is the one mentioned in the last paragraph, where the dopant atom is both soluble and ionized, but as it produces free carriers it favors the spontaneous creation of an opposite-charged native defect that compensates the effect of the dopant.

Given that the electrical properties of these materials are determined by their composition and structure, structural characterization is very important, for which sensitive spectroscopic techniques are very useful, as outlined in the following section.

2.4 Structural Characterization

The importance of structural characterization of doped ZnO has long been a topic of interest [15,36–38]. Since the local coordination environment and oxidation state of an atom are closely related, the electrical properties of the resulting material can be understood by studying the local environment of dopant atoms. Techniques such as nuclear magnetic resonance (NMR) and near-edge x-ray absorption fine structure (NEXAFS) are useful for determining the local coordination environment of dopant atoms.

In the case of solution-grown Al-doped ZnO nanowires, NMR studies (Figure 2.6) show the incorporation of Al atoms into a wide variety of sites, some acting as effective donors

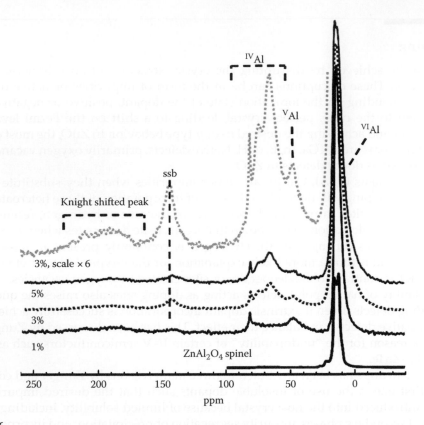

FIGURE 2.6
^{27}Al NMR spectra for doped ZnO nanowires, showing the signals from different local coordination environments. The dopant concentration in the growth solution (1%–5% at. Al:Zn) is shown next to the curves, displaced vertically for clarity. The spectrum of a $ZnAl_2O_4$ spinel is shown for comparison. (Reproduced with permission from Noriega, R. et al., Probing the electrical properties of highly-doped Al:ZnO nanowire ensembles, *J. Appl. Phys.*, 107(7), 074312–074317. Copyright 2010, American Institute of Physics.)

with the remaining inactive Al creating defects that do not contribute to the free electron density and instead hinder charge transport by acting as scattering centers [15]. The ^{27}Al NMR spectra of Al:ZnO nanowires are dominated by two partially overlapping components (Figure 2.6) corresponding mostly to tetrahedral (^{IV}Al) and octahedral (^{VI}Al) sites, as well as a Knight-shifted signal. The data also suggest a small contribution from $ZnAl_2O_4$ spinel and from ^{V}Al. A very narrow peak at 81.5 ppm suggests a small amount of Al in a highly symmetrical ^{IV}Al site. The observed ^{VI}Al peak overlaps with that of α-alumina but the formation of this phase is unlikely at the low temperatures of synthesis (300°C). The ^{IV}Al and ^{VI}Al peaks have different widths and positions from those typical of disordered crystalline phases such as γ-alumina, indicating that these phases could make at most only a minor contribution. These spectra also showed a low, broad peak centered around 180 ppm, contributing approximately 3%–7% of the total signal. This frequency is far outside the known range of normal chemical shifts in oxides, and must, therefore, be due to either the magnetic effects of unpaired electron spins, or, more likely, to a Knight shift from the conduction electrons. In fact, a previous low-resolution NMR study of Al-doped ZnO reported an ^{27}Al peak that was Knight shifted by about 200 ppm [39]. These results illustrate that aluminum as a dopant in solution-grown ZnO nanowires appears in various coordination environments.

Yuhas et al. have shown a detailed structural analysis that allows for a direct correlation between the local atomic structure of cobalt-doped ZnO and its magnetic properties [38]. Using a combination of scanning transmission x-ray microscopy (STXM) and NEXAFS, they reported that magnetic dopants are homogeneously distributed within single crystalline, single phase ZnO nanowires and adopt a highly uniform local coordination environment. These transition-metal-doped nanowires exhibit weak ferromagnetism, correlated with the incorporation of Co dopants. The STXM experiments consist of imaging the transmission of monochromatic x-rays at energies above and below the absorption edge of a particular element, in this case the $L_{2,3}$ edges of Co. The difference in optical density is then related to the presence of the element in question (Figure 2.7).

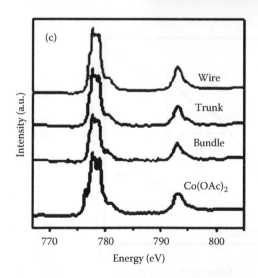

FIGURE 2.7
(a) STXM image of $Zn_{1-x}Co_xO$ nanowire bundles ($x = 0.0671$). Transmission image obtained at Co L_3 edge (778 eV). (b) Co elemental map, obtained from difference of optical density (OD) images recorded at 778 and 765 eV on area A. (c) NEXAFS spectra of different regions of the sample, corresponding to the body of a nanowire, a trunk out of which nanowires grow, a bundle of wires, and the cobalt precursor as a comparison. (Reproduced with permission from Yuhas, B.D. et al., Probing the local coordination environment for transition metal dopants in zinc oxide nanowires, *Nano Lett.*, 7(4), 905–909. Copyright 2007 American Chemical Society.)

NEXAFS spectra were recorded for the same sample. The peak positions and the line shape of the NEXAFS spectra depend on the local electronic structure of the studied atom, providing information on its valence state and the site symmetry, revealing that the Co cations in their ZnO nanowires are divalent. A tetrahedral coordination was confirmed by electronic absorption or photoluminescence spectroscopy, as well as EXAFS measurements.

The work of Yuhas et al. [38] highlights the importance of x-ray spectroscopic techniques as tools to simultaneously investigate the structure and composition of nanostructures. Another powerful technique is anomalous x-ray diffraction (AXRD), in which the intensity of a given *hkl* Bragg peak is monitored as a function of the photon energy of the incident x-ray beam. When the photon energy crosses an absorption edge of an element present in the crystal lattice, the scattering strength of the atoms that absorb will change sharply, reducing the intensity of the scattered beam. This measurement allows determination of the incorporation of dopant atoms into the substitutional sites in the crystal lattice, because only atoms in lattice sites contribute to the diffraction peaks. The ability to tune the energy of the incident x-ray beam and maintain a high intensity and brightness require the use of synchrotron radiation for these experiments.

In the case of Ga-doped ZnO, the incorporation of Ga into Zn sites in the lattice can be observed with AXRD. Since the K edges of Zn and Ga are relatively close together, and Zn is a major component in the samples, the signal of Ga incorporation is partially masked from the post-edge signal of Zn. A typical AXRD spectrum for a nominal 5% at. Ga:ZnO sample of solution-grown nanowires is shown in Figure 2.8. The area of the (002) peak has been calculated for XRD patterns taken for energies in the 10,250–10,500 eV range. The dip at 10,360 eV corresponds to the K edge of Ga, and the background signal also marks the fluorescence from Ga atoms in the sample. The fluorescence signal is related to all Ga atoms in the sample, while the AXRD signal represents only those incorporated into

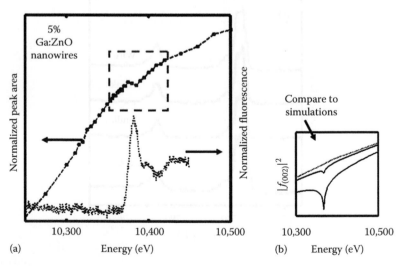

FIGURE 2.8
Anomalous XRD measurements on ZnO nanowires grown out of a 5% at. Ga:Zn solution. (a) Normalized area of the (002) Bragg peak as a function of photon energy, as well as the background fluorescence signal at each energy. Both the dip in the peak area and the rise in the fluorescence signal correspond to the position of the Ga K edge. (b) Results of structure factor simulations for 0%, 1%, and 10% Ga incorporation in ZnO (top to bottom) with the higher incorporation corresponding to a deeper dip.

ZnO as Ga_{Zn}. As a comparison, the simulated signal for samples with 0%, 1%, and 10% at. Ga incorporation is also shown.

2.5 Electronic Characterization

Adequate characterization of the electrical properties of *n*-ZnO requires quantitative determination of the density and mobility of the charge carriers. The carrier concentration and mobility in semiconducting thin films are typically obtained using Hall effect measurements, whose application is problematic for extracting electrical characteristics from quasi-1D structures such as nanowires. The mobility μ_e and free carrier density N_e in flat semiconducting films can also be measured using polarized light ellipsometry. Ellipsometry experiments have the advantage of being a noncontact technique, therefore, avoiding the unintentional modification of the sample. However, values extracted from ellipsometry depend on the accurate measurement of the change in the polarization state of light reflected from the sample surface. Because nanostructured films are rough and scatter light effectively, the validity of ellipsometry experiments is compromised. Fourier-transform infrared spectroscopy (FTIR) or attenuated total reflection FTIR has also been used to analyze the free carrier absorption in ZnO films. For highly scattering films, however, the detected signal attenuation is not necessarily correlated with the absorption of the sample alone. These issues pose a need for a better way to measure the electrical properties of an ensemble of nanostructures.

The large free-carrier concentrations in highly doped ZnO turn it into a Drude metal [40], resulting in a prominent absorption feature in the near infrared, which is due to free charge absorption and thus related to the density (N_e) and mobility (μ) of charges in the material. The Drude model predicts that the free charge density of a metal determines a cut-off frequency, the plasma frequency ω_p, such that for $\omega > \omega_p$ the material is practically transparent, and for radiation with $\omega < \omega_p$ the absorption coefficient and the reflectivity of the material are very high. There are two key parameters in this model: the plasma frequency and carrier scattering time.

$$\omega_p \equiv \sqrt{\frac{N_e e^2}{m_e^* \varepsilon_0}} \quad \text{and} \quad \tau \equiv \frac{\mu m_e^*}{e},$$

where
    m_e^* is the effective electron mass at the bottom of the conduction band
    ε_0 is the permittivity of vacuum

For ZnO with 1 at. % of active dopants, the free carrier absorption is strong for wavelengths longer than 860 nm. Thus, measuring the optical properties of an ensemble of nanostructures can provide insight into their electrical properties. A powerful technique to do so is photothermal deflection spectroscopy (PDS). PDS has a remarkable sensitivity to low signals and is not dependent on light transmission through the sample. This is important for measuring weak absorption from highly scattering nanostructures such as ensembles of ZnO nanostructures.

Using the Drude model to analyze the optical absorption of solution-grown Al:ZnO nanowires with aluminum concentrations in the 0%–4% at. range, one can obtain mean values for the mobility of 27–14 cm^2 V^{-1} s^{-1} and free carrier densities of 8.8–11.6 × 10^{19} cm^{-3} [15]. A similar approach was used to measure the interfacial charge in transparent, non-scattering ZnO field-effect transistors [41].

Since the creation of defects is a common phenomenon in highly doped ZnO nanostructures, their study is of paramount importance in understanding their electrical properties. Defect formation can be monitored by measuring the sub-band gap optical absorption of these samples. The increase in the sub-band gap absorption for an ensemble of undoped ZnO nanoparticles as a result of annealing in air at 200°C is shown in Figure 2.9. The marked increase in sub-band gap absorption corresponds to the creation of a defect state with a peak absorption close to ~3 eV. The nature of this defect state is not clear and is the focus of current investigation.

There are many other characterization techniques that are useful for studying doped nanostructures, such as photoluminescence, conventional x-ray diffraction, Auger electron spectroscopy, or transmission electron microscopy (TEM) coupled with composition sensitive analysis. For example, the presence of secondary phases can be detected by conventional XRD experiments, as long as the secondary phase is crystalline and above the detection limit of the diffractometer. Gradients in dopant incorporation within individual nanostructures can be observed with techniques like Auger electron spectroscopy or a TEM equipped with energy dispersive spectroscopy (EDS). The combination of first-principles calculations and photoluminescence experiments has been used to study the defect chemistry of ZnO films and powders [42,43], with various defects (e.g., V_O, O_{Zn}) being reported as the cause of the green luminescence in ZnO.

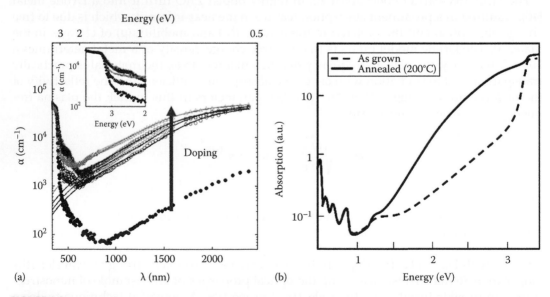

FIGURE 2.9

Absorption spectra of ZnO nanowire films. (a) Increase in IR absorption associated with free charge (from the bottom up: undoped, 0.68 at. % Al, 1.58 at. % Al, 2.34 at. % Al, and 3.98 at. % Al) and the corresponding Drude model fits (solid lines). Inset: absorption coefficient vs. photon energy in the band edge and sub-band gap regions. (Reproduced with permission from Noriega, R. et al., Probing the electrical properties of highly-doped Al:ZnO nanowire ensembles, *J. Appl. Phys.*, 107(7), 074312–074317. Copyright 2010, American Institute of Physics.) (b) Defect-related absorption in the sub-band gap region for undoped ZnO nanoparticles and its increase with moderate annealing temperature.

2.6 Charge Transport in Nanowires and Nanowire Ensembles

In order to use these materials in electronic devices, one must first study the properties and limitations of charge transport in them. It is not only important to understand their electronic properties as isolated nanostructures, but also of networks made of them. As discussed previously, it is possible to study the electrical properties of nanostructures using indirect techniques such as optical absorption measurements; however, it is also important to compare these results to direct measurements of individual nanostructures. In the case of nanowires, one can deposit metallic contacts to form two- or four-point measurements of their current–voltage characteristics. It must be kept in mind that there are additional complications when depositing contacts on top of nanostructures, such as unintentional doping and/or damage caused by the ion or electron beam, as well as the effect of contact resistance, which must be taken into account during measurement interpretation.

Using single-nanowire devices, the doping of ZnO nanowires can be observed in Figure 2.10, where two devices made of intrinsic and doped wires are compared. The intrinsic wire clearly displays semiconducting characteristics, with a current on/off ratio of $\sim 10^6$, a mobility close to $28 \, cm^2 \, V^{-1} \, s^{-1}$, and an off-current on the order of 10^{-13} A. The doped nanowire device (5% at. Ga), on the other hand, shows ohmic behavior with very low gate dependence, as expected for a highly doped semiconductor. If we neglect the resistance of the contacts, the Ga-doped nanowire shown in Figure 2.10 has a resistivity close to $3.6 \times 10^{-3} \, \Omega \, cm$, which agrees with previous reports [7,15].

When making interconnected mats of nanostructures, it is not only important to study charge transport along individual nanostructures, but also the transfer from one nanostructure to another. One of the tools that provides useful information in this regard is electrical impedance spectroscopy (EIS), where the impedance response of a sample is measured at a variety of driving frequencies. This response is then modeled with what is called a "brick layer model" [44] in order to decouple the contributions from different components of the sample. In this case, one of the loops of the Nyquist plot (Figure 2.11) is assigned to the intra-wire contributions and the other is related to the inter-wire connectivity.

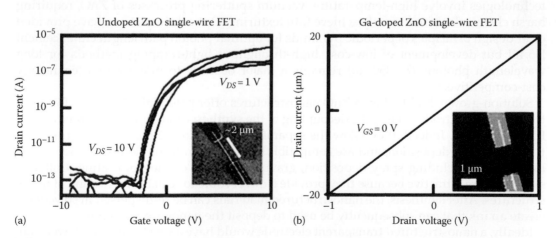

FIGURE 2.10

(a) Transfer curves of an undoped ZnO single-wire FET with drain-source voltages $V_{DS} = 1, 10 \, V$ (inset: SEM image of the device). (Reproduced with permission from Noriega, R. et al., Probing the electrical properties of highly-doped Al:ZnO nanowire ensembles, *J. Appl. Phys.*, 107(7), 074312–074317. Copyright 2010, American Institute of Physics.) (b) Similar measurement on a doped (1% at. Ga) ZnO nanowire, showing ohmic behavior as a result of doping.

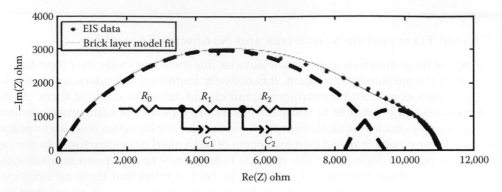

FIGURE 2.11

Nyquist plot of the impedance response for a 1% at. Ga:ZnO nanowire sample, showing the two distinct loops resulting from the different contributions of the intra- and inter-wire charge transport. This can be modeled with a brick layer model.

2.7 Optical Properties of Nanowire Meshes

Thin-film solar technologies have the potential to dramatically decrease solar energy costs due to lower materials usage and high-throughput manufacturing. While the transparent conducting electrode plays a key role in thin-film devices as well as display industries, the future of established ITO is severely limited by insufficient natural abundance, prompting research into replacement materials. With current materials capabilities, transparent electrodes could become serious barriers to thin-film solar cell commercialization. Another key layer in thin-film photovoltaic devices is the back electrode, where texturing techniques are typically employed to increase photon path lengths and offset absorption losses due to the use of less material. Established deposition technologies involve high-temperature, vacuum sputtering processes of ZnO, requiring harsh chemicals to inefficiently achieve film texturing. Optical simulations have provided clear design principles of periodic pyramids for optimal photon path length enhancement [45,46] but development of low-cost, high-throughput light-trapping methods for long wavelength photons ($\lambda > 850\,nm$) remains a major obstacle to making this technology cost-competitive.

Solution-grown, highly doped ZnO nanostructures offer potential advantages for novel transparent electrodes due to the decoupling of the synthesis and deposition processes for film fabrication. Indeed, they allow the separate optimization of the nanostructure syntheses and film deposition, and are compatible with roll-to-roll solution-based deposition mechanisms including spray deposition, gravure printing, or die-slot coating. Colloidal syntheses are attractive because they provide a direct route to scalable fabrication of nanostructures. After synthesis, the nanostructured materials can be re-dispersed in solvents to create an ink that can subsequently be used to deposit the nanowire mesh films.

Ideally, a nanostructured transparent electrode would have a sheet resistance lower than $10\,\Omega/\mathrm{sq}$, a transmission in the visible part of the spectrum greater than 90%, and scatter the majority of the transmitted light at large angles. The figure of merit for the scattering efficiency of a film is the haze factor, which is the fraction of transmitted light that is scattered significantly upon passing through the nanostructured film. Typically in sputtered

ZnO or low-pressure chemical vapor deposition (LPCVD) ZnO used for thin-film devices, the haze drops significantly in the IR to below 10% [47]. The focus of novel texturing techniques for transparent electrodes for these devices is on enhancing the long wavelength path lengths and maximizing the scattering of sub-band gap light.

The haze factors measured for spray-deposited Ga:ZnO nanowire films compared to standard LPCVD-deposited ZnO are shown in Figure 2.12. What is immediately apparent from the shape of the curves for nanostructured films is the smaller rate of decrease in scattering throughout the visible into the IR portion of the spectrum. In comparison, the ZnO vapor-deposited films exhibit a pronounced shoulder around 500 nm resulting

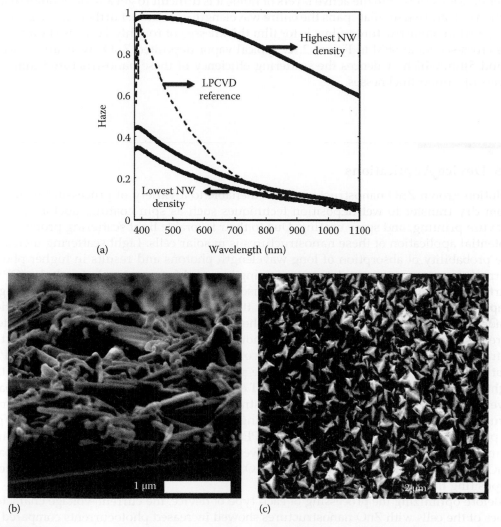

(a)

(b)

(c)

FIGURE 2.12
Scattering properties of spray-deposited nanowire mesh films on glass at different NW densities. The reference curve is for LPCVD ZnO [47] typically used for a-Si PV fabrication. (b) Cross-sectional SEM image of a spray-deposited Ga:ZnO nanowire mesh film at the highest density. (c) SEM of a typical LPCVD-deposited film with randomly distributed pyramids and a 300 nm average lateral feature size. LPCVD optical data. (a and c: Reproduced from Klingshirn, C.F. et al., *Zinc Oxide: From Fundamental Properties towards Nouel Applications*, Springer, Heidelberg, Germany, 2010.)

in a decreased scattering efficiency at longer wavelengths. For the lower nanowire densities, the haze factor in the visible is lower than that of the LPCVD reference. These are the wavelengths where the active layers absorb strongly and where a large haze factor is less critical. For the highest nanowire density the haze factor falls to ~60% at 1100 nm compared to 10% for LPCVD ZnO, indicating the effectiveness of the nanostructured films in scattering long wavelength light. The relatively higher scattering observed across the spectrum with nanostructured ZnO films and the slower dropoff in haze can be explained by considering that the random film meshes exhibit periodicities on the order of the light wavelength throughout the thickness of the films in addition to the surface morphology. In contrast, in the vapor-deposited ZnO, where the textured surface is used to enhance scattering in the active layers of films, it is difficult to get a wide enough distribution of feature sizes that spans the entire wavelength spectrum. Further, the larger haze observed in nanostructured films is for film thicknesses of roughly 1 μm, as shown from the cross-sectional SEM in Figure 2.12. Typical vapor-deposited ZnO layers are between 2 and 5 μm, which evidences the scattering efficiency of these nanostructured films for layers of similar thicknesses.

2.8 Device Applications

Solution-grown ZnO nanostructures are amenable to a variety of processing methods, from dry transfer to wet deposition techniques such as spin coating, doctor blading, gravure printing, and spray coating. Due to their favorable light scattering properties, a potential application of these nanostructures is in solar cells. Light scattering increases the probability of absorption of long wavelength photons and results in higher photocurrent. Low-temperature solution processing is useful for organic photovoltaics (OPV), particularly polymer-based devices. The usefulness of ZnO nanostructures as a light trapping element in OPV was studied for the model system of a bulk heterojunction made with the polythiophene P3HT and the C_{60} fullerene PCBM. The device architecture is shown in Figure 2.13, where the conducting polymer PEDOT:PSS is used for the extraction of holes. The modified devices incorporated ZnO nanowires into a PEDOT:PSS matrix in order to maintain the conductive properties of PEDOT:PSS coupled with the light scattering from ZnO. The ZnO/PEDOT composite was obtained by spinning a base layer of PEDOT onto an ITO-on-glass substrate, then transferring a film of ZnO nanostructures on top of the PEDOT by stamping using a hydraulic press to apply 4500 psi of pressure, followed by a second layer of PEDOT to infiltrate the ZnO porous film and planarize it [48]. The ZnO nanostructures were incorporated in two formats: using bare ZnO nanowires films, and covering the ZnO nanowires with PEDOT in solution before depositing the ZnO films.

Before optimizing the processing of each type of device, the current–voltage characteristics of the cells with ZnO nanostructures showed increased photocurrents compared to the control devices, but poorer charge extraction characteristics as evidenced by reduced fill factors and increasing currents in reverse bias. Optimizing each architecture leads to a smaller (but consistent across batches) difference between the control devices and the ones incorporating the composite electrodes. Devices with thin ZnO layers provide better charge extraction due to increased PEDOT infiltration, but lower light scattering because of the smaller contrast of refraction index between ZnO and PEDOT as compared to the

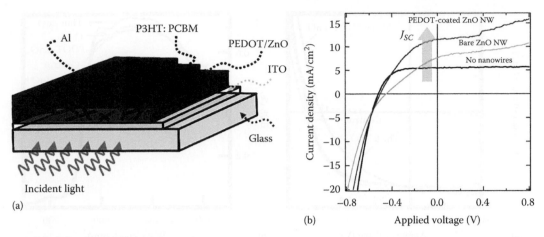

FIGURE 2.13
(a) Sketch of the device architecture incorporating ZnO nanowires into a PEDOT/ZnO composite electrode in order to enhance light scattering and increase photocurrent. (b) Current–voltage curves for devices without ZnO, with bare ZnO nanowires embedded in PEDOT, and pre-coated nanowires. The devices with scattering electrodes showed increased photocurrent.

TABLE 2.1

Device Characteristics of OPV Devices with ZnO/PEDOT Electrodes

	η_{PCE} (%)	J_{SC} (mA/cm²)	V_{oc} (mV)	Fill Factor
Control (no ZnO)	3.3	10.2	590	0.55
Thin ZnO (bare wires)	3.5	10.8	610	0.53
PEDOT-coated ZnO wires	2.2	10.5	510	0.42
Thick ZnO (bare wires)	1.7	8.93	540	0.35

ZnO–air interface. Thicker films have improved scattering properties, but limited charge extraction due to the higher resistance of ZnO nanostructure mats (Figure 2.14, Table 2.1).

Further optimization work is being performed to find the optimal tradeoff between charge collection from nanowire mesh films and maximized absorption due to light scattering. Improved electrical properties of nanostructured ZnO films would lead to increased device performance, allowing one film to perform both functions of light scattering and charge extraction at the same time. This ability could open up applications in other types of devices such as dye-sensitized solar cells or amorphous silicon photovoltaics. The improvement of synthesis routes and dopant incorporation in low-cost, scalable solution techniques will make a broader array of large-area devices based on nanostructures (e.g., UV photodetectors, UV lasers, field-effect transistors, sensors, memory devices, and piezoelectric actuators) a reality.

2.9 Conclusion

Solution-based syntheses provide a low-cost, scalable route to fabrication of ZnO nanostructures for a variety of optoelectronic applications. Colloidal synthesis has shown to

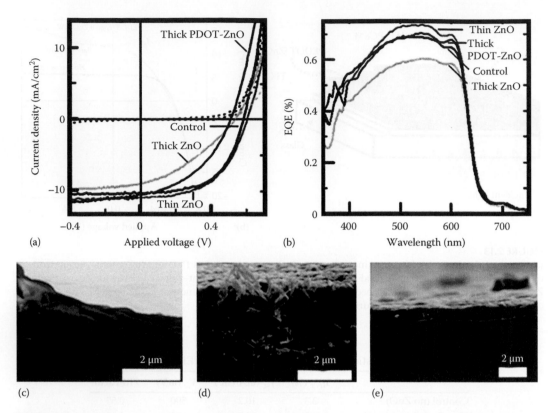

FIGURE 2.14

Device characteristics (a and b) and film morphologies (c–e) for optimized devices with ZnO/PEDOT electrodes. The curves correspond to a thin ZnO layer embedded in PEDOT (c), a thick ZnO layer sandwiched in PEDOT (d), and a layer of PEDOT-coated ZnO nanowires (e). The current–voltage curves are shown in (a) and the external quantum efficiencies in (b).

be a promising route to achieving monodisperse, highly doped nanostructures. ZnO nanostructures grown in high boiling point organic solvents exhibit reasonably high carrier densities when doped with Al or Ga, and post-processing techniques can further improve their electrical characteristics. A variety of characterization techniques are currently being employed to explore the nature of the electronic properties of these nanostructured mats, including x-ray techniques and NMR to characterize the local coordination environment of dopant ions, sensitive spectroscopic techniques to extensively characterize sub-band gap absorption and trap levels, single nanowire electrical measurements, and impedance spectroscopy to analyze the resistive bottlenecks in nanowire mats. The ongoing synthesis and characterization work on ZnO nanostructures will enable the development of optimized ZnO nanowire syntheses for electronic applications.

The separation of the nanostructure synthesis and film deposition allows each of the processes to be individually optimized and enables continuous film deposition instead of batch processing. Colloidally synthesized nanoparticles can be redispersed in a variety of solvents for use in wet coating techniques. Spray-deposited ZnO nanowire films have been shown to exhibit favorable optical properties for thin-film photovoltaic devices by effectively scattering broadband light, thus improving the probability of photons being absorbed within the photovoltaic active layer. Preliminary work fabricating bulk heterojunction photovoltaic devices with ZnO nanostructures has yielded encouraging results, showing increases in the short circuit current due to improved absorption.

Solution-grown ZnO nanowires and randomly oriented mesh films have shown promise for use in optoelectronic devices as substitutes for traditional transparent electrodes. Although nanowire synthesis and doping is not yet optimized for use in these devices, it remains an active field of research and is rapidly progressing toward developing device-quality nanomaterials. Nanowire mesh films have shown promising optical properties for photovoltaic devices, and recent efforts to fabricate devices with ZnO nanostructured films have shown favorable results. As nanomaterials synthesis research continues moving toward rational design rules for nanostructures and device fabrication with nanostructured films becomes optimized, we anticipate a variety of useful applications for all solution-processed nanostructured ZnO in the near future.

References

1. Ginley, D.S. and C. Bright, Transparent conducting oxides. *MRS Bulletin*, **25**(8): 15–18 (2000).
2. Hecht, D.S., L. Hu, and G. Irvin, Emerging transparent electrodes based on thin films of carbon nanotubes, graphene, and metallic nanostructures. *Advanced Materials*, **23**: 1482–1513 (2011).
3. Ju, S. et al., High performance ZnO nanowire field effect transistors with organic gate nanodielectrics: Effects of metal contacts and ozone treatment. *Nanotechnology*, **18**(15): 7 (2007).
4. Pal, B.N. et al., Solution-deposited zinc oxide and zinc oxide/pentacene bilayer transistors: High mobility n-channel, ambipolar and nonvolatile devices (vol 18, pg no 1832, 2008). *Advanced Functional Materials*, **18**(13): 1876–1876 (2008).
5. Law, M. et al., Nanowire dye-sensitized solar cells. *Nature Materials*, **4**(6): 455–459 (2005).
6. Lewis, B.G. and D.C. Paine, Applications and processing of transparent conducting oxides. *MRS Bulletin*, **25**(8): 22–27 (2000).
7. Ellmer, K., Resistivity of polycrystalline zinc oxide films: Current status and physical limit. *Journal of Physics D: Applied Physics*, **21**: 3097–3108 (2001).
8. Wang, Z.L., Zinc oxide nanostructures: Growth, properties and applications. *Journal of Physics-Condensed Matter*, **16**(25): R829–R858 (2004).
9. Huang, M.H. et al., Room-temperature ultraviolet nanowire nanolasers. *Science*, **292**(5523): 1897–1899 (2001).
10. Bagnall, D.M. et al., Optically pumped lasing of ZnO at room temperature. *Applied Physics Letters*, **70**(17): 2230–2232 (1997).
11. Klingshirn, C.F. et al., *Zinc Oxide: From Fundamental Properties towards Novel Applications*. Springer, Heidelberg, Germany (2010).
12. Ren, F., *ZnO Based Thin Films, Nano-Wires, and Nano-Belts for Photonic and Electronic Devices and Sensors*. The Electrochemical Society, Pennington, NJ (2008).
13. Wang, X., C.J. Summers, and Z.L. Wang, Large-scale hexagonal-patterned growth of aligned ZnO nanorods for nano-optoelectronics and nanosensor arrays. *Nano Letters*, **4**(3): 423–426 (2004).
14. Greene, L.E. et al., Solution-grown zinc oxide nanowires. *Inorganic Chemistry*, **45**(19): 7535–7543 (2006).
15. Noriega, R. et al., Probing the electrical properties of highly-doped Al: ZnO nanowire ensembles. *Journal of Applied Physics*, **107**(7): 074312–074317 (2010).
16. Vayssieres, L., Growth of arrayed nanorods and nanowires of ZnO from aqueous solutions. *Advanced Materials*, **15**(5): 464–466 (2003).
17. Greene, L.E. et al., Low-temperature wafer-scale production of ZnO nanowire arrays. *Angewandte Chemie International Edition*, **42**(26): 3031–3034 (2003).

18. Li, Q. et al., Fabrication of ZnO nanorods and nanotubes in aqueous solutions. *Chemistry of Materials*, **17**(5): 1001–1006 (2005).
19. Yin, Y. and A.P. Alivisatos, Colloidal nanocrystal synthesis and the organic-inorganic interface. *Nature*, **437**(7059): 664–670 (2005).
20. Peng, X. et al., Shape control of CdSe nanocrystals. *Nature*, **404**(6773): 59–61 (2000).
21. Andelman, T. et al., Morphological control and photoluminescence of zinc oxide nanocrystals. *The Journal of Physical Chemistry B*, **109**(30): 14314–14318 (2005).
22. Andelman, T., Y. Gong, G. Neumark, and S. O'Brien, Diameter control and photoluminescence of ZnO nanorods from trialkylamines. *Journal of Nanomaterials*, 73824 (2007).
23. Peng, X., Mechanisms for the shape-control and shape-evolution of colloidal semiconductor nanocrystals. *Advanced Materials*, **15**(5): 459–463 (2003).
24. Reiss, H., The growth of uniform colloidal dispersions. *The Journal of Chemical Physics*, **19**(4): 482–487 (1951).
25. Noriega, R. et al., Transport and structural characterization of solution-processable doped ZnO nanowires. *Proceedings of SPIE*, 74110N-1–74110N-6 (2009).
26. Garnweitner, G. and M. Niederberger, Nonaqueous and surfactant-free synthesis routes to metal oxide nanoparticles. *Journal of the American Ceramic Society*, **89**(6): 1801–1808 (2006).
27. Goris, L. et al., Intrinsic and doped zinc oxide nanowires for transparent electrode fabrication via low-temperature solution synthesis. *Journal of Electronic Materials*, **38**(4): 586–595 (2009).
28. Pinna, N. and M. Niederberger, Surfactant-free nonaqueous synthesis of metal oxide nano-structures. *Angewandte Chemie International Edition*, **47**(29): 5292–5304 (2008).
29. Niederberger, M. and H. Colfen, Oriented attachment and mesocrystals: Non-classical crystal-lization mechanisms based on nanoparticle assembly. *Physical Chemistry Chemical Physics*, **8**(28): 3271–3287 (2006).
30. Pacholski, C., A. Kornowski, and H. Weller, Self-assembly of ZnO: From nanodots to nanorods. *Angewandte Chemie International Edition*, **41**(7): 1188–1191 (2002).
31. Kusinski, G.J. et al., Transmission electron microscopy of solution-processed, intrinsic and Al-doped ZnO nanowires for transparent electrode fabrication. *Journal of Microscopy*, **237**(3): 443–449.
32. Zunger, A., Practical doping principles. *Applied Physics Letters*, **83**(1): 57–59 (2003).
33. Zhang, S.B., S.H. Wei, and A. Zunger, Overcoming doping bottlenecks in semiconductors and wide-gap materials. *Physica B: Condensed Matter*, **273–274**: 976–980 (1999).
34. Zhang, S.B., S.H. Wei, and A. Zunger, Microscopic origin of the phenomenological equilibrium "Doping Limit Rule" in n-type III-V semiconductors. *Physical Review Letters*, **84**(6): 1232 (2000).
35. Zhang, S.B., S.-H. Wei, and A. Zunger, A phenomenological model for systematization and prediction of doping limits in II–VI and I–III–VI[sub 2] compounds. *Journal of Applied Physics*, **83**(6): 3192–3196 (1998).
36. Brehm, J.U., M. Winterer, and H. Hahn, Synthesis and local structure of doped nanocrystalline zinc oxides. *Journal of Applied Physics*, **100**(6): 064311 (2006).
37. Singh, A.V. et al., Doping mechanism in aluminum doped zinc oxide films. *Journal of Applied Physics*, **95**(7): 3640–3643 (2004).
38. Yuhas, B.D. et al., Probing the local coordination environment for transition metal dopants in zinc oxide nanowires. *Nano Letters*, **7**(4): 905–909 (2007).
39. Roberts, N. et al., Al-27 and Ga-69 impurity nuclear magnetic resonance in ZnO:Al and ZnO:Ga. *Physical Review B*, **57**(10): 5734–5741 (1998).
40. Dexheimer, S.L., *Terahertz Spectroscopy: Principles and Applications*. CRC Press, Boca Raton, FL (2007).
41. Kim, J. et al., Infrared spectroscopy of the interface charge in a ZnO field-effect transistor. *Applied Physics Letters*, **93**(24): 241902-1–241902-3 (2008).
42. Lin, B., Z. Fu, and Y. Jia, Green luminescent center in undoped zinc oxide films deposited on silicon substrates. *Applied Physics Letters*, **79**(7): 943–945 (2001).
43. Vanheusden, K. et al., Mechanisms behind green photoluminescence in ZnO phosphor powders. *Journal of Applied Physics*, **79**(10): 7983–7990 (1996).

44. Fleig, J., The grain boundary impedance of random microstructures: Numerical simulations and implications for the analysis of experimental data. *Solid State Ionics*, **150**(1–2): 181–193 (2002).

45. Dewan, R. et al., Light trapping in thin-film silicon solar cells with submicron surface texture. *Optics Express*, **17**(25): 23058–23065 (2009).

46. Haase, C. and H. Stiebig, Thin-film silicon solar cells with efficient periodic light trapping texture. *Applied Physics Letters*, **91**(6): 061116–061123 (2007).

47. Soderstrom, T. et al., ZnO transparent conductive oxide for thin film silicon solar cells. *Proceedings of SPIE*, San Francisco, CA (2010).

48. Gaynor, W. et al., Smooth nanowire/polymer composite transparent electrodes. *Advanced Materials*, **23**: 2905–2910 (2011).

3

Heteroepitaxy of ZnO on SiC as a Route toward Nanoscale p–n Junction

Volodymyr Khranovskyy and Rositza Yakimova

CONTENTS

3.1 Introduction 83
3.2 Development of the ZnO/SiC Heterostructures: From Films to Nanostructures 86
3.3 Heteroepitaxial ZnO Nanohexagons on p-Type SiC 106
    3.3.1 Preparation of p-Type SiC Substrates of High Structural
        Quality via Sublimation Epitaxy 106
    3.3.2 Growth of ZnO Nanohexagons on 4H:SiC 107
    3.3.3 Detailed Structural Characterization of ZnO Nanohexagons Grown
        on 4H:SiC with Emphasis on the Heterointerface 112
    3.3.4 Micro-Photoluminescence Investigations of ZnO Seeding Layer and
        Hexagonal Nanopillars Grown on p-Type SiC Substrates 120
3.4 Conclusions 125
Acknowledgments 126
References 126

3.1 Introduction

ZnO has become interesting as a semiconductor material for device fabrication during the last decade. Due to its direct wide band gap (3.37 eV at RT), high exciton binding energy (60 meV at RT), and optical transparency for visible light, ZnO is a prospective material for micro-, opto-, and transparent electronics [1–6].

Particularly, ZnO is promising for optoelectronics as a material for light-emitting devices—light-emitting diodes and laser diodes (LED and LD, respectively) [1,3,4]. Having a band gap of 3.37 eV, ZnO should emit in the range of 366–380 nm (A-UV radiation), but commonly the emission spectra of ZnO consist of two luminescence bands. The narrow peak of the near-band-edge excitonic emission is typically accompanied by a broad visible emission, so-called green-yellow luminescence, centered at $\lambda = 550$ nm [7,8]. The origin of the visible luminescence is still under debate, but it is most probably related to Zn atoms as interstitials (Zn_i) and/or oxygen vacancies (V_o) [9].

Decreasing the size of the ZnO functional elements (nanosized p–n junction) can be a solution toward obtaining monoemission spectra due to a better structural quality and improved stoichiometry, which is a common problem in II–VI compounds. Moreover, the nanopillar structures may relieve strain and thus accommodate the large lattice mismatch at heterointerface, which can enhance the radiative recombination efficiency and thus the internal quantum efficiency of LED devices [10]. Due to the large sidewall surface of nanopillars, light extraction efficiency can also be improved [11]. In addition, the inclined threading dislocations may cease in the nanopillar's side surface and decrease dislocation density, which can reduce the nonradiative recombination rate and thus improve the device performance significantly [12].

However, a persistent problem is to obtain p-type material: Usually the intrinsic (as-grown) ZnO is of n-type conductivity, which is difficult to overcome due to a self-compensation effect. Earlier, there were several reports claiming a successful p-type ZnO preparation, e.g., [13–17]. However, the low hole mobility [14], time instability [15], nonreproducibility [16], and/or impractical growth techniques [17] of the obtained material brings doubts whether the p-type problem is as by today confidently solved. Thus, the realization of long-term stable and reproducible p-type ZnO is still a challenge due to the deep level of the valence band and possible point defects that may act as donors [18,19]. Despite a number of publications about successful p-type doping by N, P, As, and Sb [20–24] and even observed electroluminescence (EL) [25–27], at present no ZnO LED is commercially available.

Many research groups pursued investigation of LEDs using alternative p-type materials as a substrate. Such devices are also called hybrid LEDs [28]. Additional expected advantage for using the p–n heterojunctions is that the devices based on p–n heterojunction are expected to show higher efficiency than homojunction devices [29,30], because the energy barrier formed at the junction interface due to the large band offset may effectively decreases the confined carriers [31].

One of the most promising p-type substrates for hetero n–p junction is GaN. ZnO and GaN possess similar physical properties, including the same wurtzite structure, and similar band energy gap and close lattice constants. The lattice mismatch between ZnO and GaN is as small as 1.9%. Furthermore, the GaN technology is well developed since it is the material for LEDs in the green and blue spectral range. At present, GaN is already used for practical applications for visible-range light-emitting diodes and laser diodes [32–35]. However, the EL spectra of n-ZnO/p-GaN LEDs reported in the literature were mostly polychromatic, showing the emission bands at different energies ranging from yellow/green to the near band edge in the blue/UV spectral range. This suggests that the origin of the emission is until now not completely understood and a detailed investigation is required.

Based on the rather intense study of n-ZnO/p-GaN heterostructures, the next issues have to be highlighted in order to take into account the future heterostructure-based ZnO devices. The p-type material is intended to be a hole injection layer into ZnO. However, due to the fact that the valence band edge of ZnO is very low in energy compared to other conventional semiconductors, an efficient hole injection is rather difficult, due to the unfavorable band alignment. In the first reports concerning the n-ZnO/p-GaN LEDs, the origin of the light emission was even unclear—was it from ZnO or from GaN, mostly being even below the band gap of both materials. Despite some reports about successful electroluminescence (EL) from these LEDs, the light emission mechanism is still under discussion. But all reports have in common that the band alignment is type II with the valence band maximum in GaN and conduction band minimum in ZnO. A LED based on a p-GaN layer with vertically grown n-ZnO on top was reported [34]. A yellow emission around 2 eV was detected in the EL spectrum for a reverse bias of 3 V. This was explained as defect availability in the heterojunction interface.

Certainly, the quality of the heterojunctions interface is of extreme importance for LED performance. However, the drawback for n-ZnO/p-AlGaN LED is an ultrathin Ga_2O_3 layer, which may be formed at the interface with lots of interface states, which act as nonradiative recombination centers [28]. Application of the ZnO nanowires grown on p-GaN allowed the realization of the high-brightness LED. The EL spectrum of such nanoheterojunctions for voltages applied in forward direction ranging from 10 to 35 V showed a peak centered at 400–440 nm wavelengths depending on the bias. A weak UV peak located at 370 nm could also be detected for voltages higher than 25 V [36]. Thus, based on the GaN experience, one can conclude that decrease of the heterojunctions area favors the LED performance.

SiC is a recognized wide band gap semiconductor suitable for high-power, high-frequency, and high-temperature applications. It has huge potential as a luminescent material being able to emit in the visible spectral range from blue to yellow [37]. SiC has excellent physical and electrical properties such as wide band gap, high electrical breakdown field, high thermal conductivity, high surface hardness, and high saturated electron velocity [37]. Furthermore, SiC is suitable as a surface-corrosion-resistant coating [38]. Recently, amorphous SiC, porous crystalline SiC, nanometer SiC, and so on were investigated, and great progress has been made in these fields [39,40].

For ZnO SiC is an attractive substrate material, mainly due to the similarity of the crystal structures and small lattice and thermal mismatch ($a_{6H\text{-}SiC} = 0.308$ nm, $TEC_{6H\text{-}SiC} = 4.3 \times 10^{-6}/K$, while $a_{ZnO} = 0.3252$ nm, $TEC_{ZnO} = 6.51 \times 10^{-6}/K$), giving a misfit in the basal plane ~5.4%. ZnO was reported grown on different SiC substrates, which can be divided into hexagonal type (6H:SiC, 4H:SiC) and the cubic structure (3C:SiC). The properties of ZnO along with the SiC main polytypes are shown in Table 3.1. Particularly, among the many SiC polytypes, 4H-SiC is considered as the most attractive for device applications due to its wider band gap (~3.2 eV) and higher and more isotropic mobility than the other polytypes. However, relative to work carried out with 6H:SiC substrates, little effort has been spent on attempts to grow epitaxial ZnO films on 4H:SiC [42].

Recently, there have been a number of reports on ZnO/6H:SiC heterostructure fabrication and characterization; however, efficient UV LEDs have not been proven [43–45]. Structural/interfacial defects were suggested to be responsible for the enhancement of the defect emission and the quenching of the NBE emission. p-4H:SiC can be obtained via a respective p-type doping of the homoepitaxial SiC layer [46]. However, growth of thick SiC layers by itself is challenging since the substrate quality affects epitaxial growth. Macroscopic imperfections from the substrate are usually reproduced by the layer at the first stages of epitaxy. These defects may continue or evolve with increasing layer thickness and deteriorate the surface morphology. Via sublimation epitaxy, growth rates of several tens of microns per hour or more are easily obtained and thick

TABLE 3.1

Properties of ZnO in Comparison to Different SiC Polytypes

		SiC		
	ZnO	**6H:SiC**	**4H:SiC**	**3C:SiC**
E_g (eV)	3.36	3.02		2.3
Lattice constant (Å)	$a = 3.25$	$a = 3.08$	$a = 3.073$	$a = 4.36$
	$c = 5.21$	$c = 15.12$	$c = 10.053$	
Coefficient of thermal expansion ($10^{-6}/K$)	2.9	3–5		3.3
Coefficient of thermal conductivity (W/(cm·K))	0.6	3.6–4.9	3.7–4.9	3.2–4.9
Relative dielectric coefficient	8.1	9.7		9.7
Breakdown field strength 10^6 V/cm		3		1.54

Source: Zhu, J. et al., *Thin Solid Films*, 478, 218, 2005.

layers (>50 μm) can be grown [37]. These issues are important for the quality of the heterointerface and must be taken into account during the n-ZnO/p-4H:SiC heterojunctions design.

Researchers developing LEDs based on n-ZnO/p-4H:SiC heterojunctions experienced problems similar to the ones peculiar for GaN usage. Morkoç et al. [46] reported the fabrication of thin-film-based heterojunction diode (n-ZnO/p-SiC). Based on measured electro-optical properties, they have concluded that electron injection takes place mainly from the ZnO side. As a result, the emission properties of the diode were driven by the luminescence of p-SiC and evidenced a high density of interface states. Similarly, other authors observed that the electron injection from n-ZnO dominates over the hole injection from p-SiC. Thus, due to the smaller energy barrier for electron, electroluminescence (EL) usually comes from the defect-related states in p-SiC in the n-ZnO/p-SiC heterojunction LEDs [42,47]. Finally, the vital importance of the injection layer for the monochromatic EL spectra has been confirmed by Yuen et al. and Shih et al. [47,48].

To ensure high device performance, the optical quality of the ZnO by itself should be high, i.e., without visible deep-level emission (DLE). The commonly observed DLE from ZnO films/nanostructures may be a serious obstacle for the effective UV electroluminescence [49,50]. The DLE can be overcome by approaching a good stoichiometry of ZnO or respective passivation (e.g., by H) [51].

The importance of the interface structural quality was reported by Bano et al. [52]. The low structural quality of the ZnO nanorods, emerging from the SiC interface resulted in a visible defect emission, observed from electroluminescence. Therefore, the quality of the heterointerface is of primary importance, since it may be an origin of extended defects acting as nonradiative recombination centers. Thus, fabrication of aligned nanosized ZnO/SiC heterojunctions of high structural and interface quality may lead to a successful LED realization. More details about the ZnO films and nanostructures growth, leading toward LED fabrication will be described in the next part.

Here, we present a review of the recent results concerning the growth and investigation of ZnO on SiC substrates, eventually leading to LED performance. Also, original data concerning the epitaxial growth of self-aligned n-type ZnO hexagonal nanorods (HEX) on p-type 4H-SiC substrates and comprehensive investigation of their structural and optical properties are presented.

3.2 Development of the ZnO/SiC Heterostructures: From Films to Nanostructures

Initially, ZnO films were deposited onto SiC substrates and a number of studies of their structural and optical properties were reported [53–56]. In general, it looks that the growth of ZnO on SiC experienced the evolution from polycrystalline to highly textured epitaxial films. For the first time the growth of ZnO layers on 6H-SiC(0001) substrates was reported by Ashrafi in 2004 [53].

Authors attempted to bring in focus the SiC substrates for the ZnO epitaxy, via analyzing the experimental and/or theoretical data. Figure 3.1 demonstrates that the SiC substrate is closely matched to the ZnO aligned with other oxides [53]. As one can see, the ZnO in both wurtzite and zinc blende crystal structures is well matched with the SiC—6H:SiC and 3C:SiC, respectively.

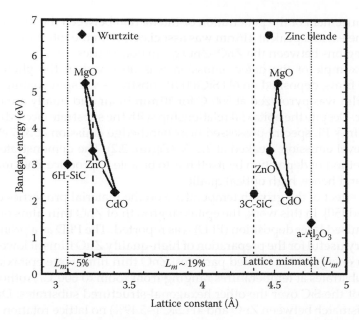

FIGURE 3.1

Lattice constants of wurtzite and zinc blende ZnO, MgO, CdO, and SiC materials. (Reprinted from Ashrafi, A., Zhang, B., Binh, N., Wakatsuki, K., and Segawa, Y., High-quality ZnO layers grown on 6H-SiC substrates by metalorganic chemical vapor deposition, *Jpn. J. Appl. Phys.*, 43, 1114, 2004. With permission from APEX/JJAP.)

Based on experimental results, according to the X-ray diffraction (XRD) measurement, c-axis-oriented growth of ZnO layers on SiC(0001) substrates was obtained. The growth temperature of 450°C was found to be optimal for the MOCVD growth based on the structural quality. As it was observed, the x-ray rocking curve of the ZnO layers grown on SiC possessed a smaller full width at half maximum in comparison to that of the same layers grown on the Al_2O_3 substrate. This positioned the SiC as a suitable substrate and provoked a further interest in the ZnO/SiC heterostructures. The obtained heterostructures demonstrated advanced luminescent properties: A distinct free-exciton emission was dominantly observed even at room temperature (RT) while the donor-bound-exciton peaks disappeared at around 120 K. In addition, no deep-level emission was observed even at RT. Authors explained the results obtained as due to the higher quality of the ZnO layers grown on SiC as a result of the smaller lattice mismatch of ~5% as well as the c-texture orientation in ZnO/ SiC sample geometry.

Later, the influence of the temperature on the interface quality and photoluminescence properties was investigated for the ZnO/SiC heterostructures [54]. Multilayered ZnO/ SiC system was investigated and the as-deposited films were annealed in the temperature range 600°C–1000°C under nitrogen ambient. Based on the ZnO crystal structure, the increase of annealing temperature to ~800°C resulted in the enhancement of the crystal quality of ZnO. However, with further temperature increase up to 1000°C, the crystal quality of ZnO degrades. This was explained because of the ZnO layers and SiC layers interpenetration and thus degradation of the interface between ZnO and SiC.

The optical properties of the samples were found to be strongly dependent on the annealing temperature: The intensities of the PL peaks increase remarkably with its increase. Furthermore, the NBE emission had a tendency of redshift from 3.27 eV for the sample annealed at 600°C to 3.13 eV for the sample annealed at 1000°C. Authors explained this as due to the enhancement of the penetration or interaction between the ZnO layers and SiC. The origin of the existed peak at 416 nm was associated with the interface traps existing in the depletion regions between the ZnO–ZnO grain boundaries.

A classical example of the defect emission was observed in the photoluminescence spectra of ZnO films, deposited on 6H:SiC(0001) substrates via spin-coating pyrolysis [49]. As-deposited film was pyrolyzed at 500°C for 10 min in air and finally annealed at 800°C for 30 min in air. Despite the epitaxial relationship with the substrate (according to the XRD analysis), the film's PL spectra possessed near-band-edge emission at 3.27 eV followed by a broad deep-level emission, peaked at 2.2 eV (Figure 3.2). The demonstrated results suggest that the method of deposition by itself has to be able to produce ZnO material of high stoichiometry and hence, high optical quality.

Recently, the effect of the growth temperature on the material properties of ZnO on SiC was investigated [50]. In this work, the epitaxial growth of ZnO thin films on 4H:SiC(0001) substrates by pulsed laser deposition (PLD) was reported. The PLD as a prominent growth technique is very useful for the preparation of high-quality ZnO films at lower temperature due to the high energy of the ablated particles. ZnO thin films were epitaxially grown on 4H:SiC(0001) substrates at temperatures ranging from room to 600°C. Authors underlined the advantage of the SiC over the other hexagonal-structured substrates: Due to the very small lattice mismatch between ZnO and 4H:SiC (~5.49%) no lattice rotation was observed at all substrate temperatures (Figure 3.3).

Authors pointed out the effect of the substrate temperature on the properties of the ZnO films on 4H:SiC substrates: The crystalline quality of the films was improved, and surfaces became denser and smoother as the substrate temperature increased.

Illustration of the substrate temperature effect on the quality of the heteroepitaxy of ZnO films on 4H:SiC is shown in Figure 3.4. The ZnO films deposited at 400°C revealed sharp and narrow peaks, while room temperature deposited films possess broader and weaker XRD reflections. However, it has to be noted that the epitaxial growth is achieved at both temperatures.

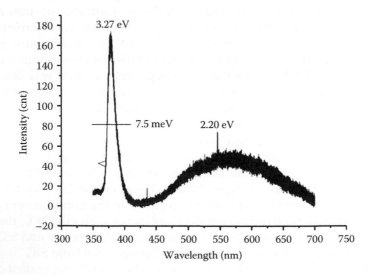

FIGURE 3.2
Typical photoluminescence spectra of the ZnO films, prepared by spin-coating pyrolysis. (Reprinted from *Appl. Surf. Sci.*, 253, Jeon, Y.-S., Kim, D.-M., and Hwang, K.-S., Epitaxially grown ZnO thin films on 6H-SiC(0001) substrates prepared by spin coating-pyrolysis, 7016. Copyright 2007, with permission from Elsevier.)

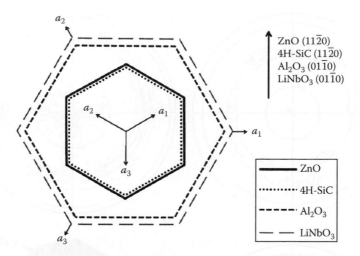

FIGURE 3.3
Schematic diagram of in-plane relationships between ZnO and different hexagonal-structured substrates. (Reprinted from *J. Alloys Compd.*, 489, Kim, J.-H., Cho, D.-H., Lee, W., Moon, B.-M., Bahng, W., Kim, S.-C., Kim, N.-K., and Koo, S.-M., Structural and optical properties of epitaxial ZnO thin films on 4H–SiC (0001) substrates prepared by pulsed laser deposition, 179. Copyright 2010, with permission from Elsevier.)

The authors [50] observed that the PL properties depend strongly on the substrate temperature (Figure 3.5). The PL measurements revealed a strong near-band-edge (NBE) ultraviolet (UV) emission and a weak deep-level (DLE) blue-green band emission at the optimal substrate temperature 400°C. Further increase of the substrate temperature caused enhancement of the DLE intensity, which authors explained by the formation of the point defects including oxygen vacancies.

Recently the interface of ZnO/SiC heterostructures was brought to light by [55]. Authors emphasized a possible formation of an oxide layer on the ZnO/SiC interface upon annealing at 800°C. Since SiC has a large band gap and high physical and chemical stability, resulting in a high temperature resistance, it seems that O atoms diffused during annealing are unlikely to break the strong bonds between Si and C and form a simple oxide layer. According to the author's conclusion, during thermal annealing only a very thin oxide layer forms using the outermost Si atoms of the Si face at the surface. This layer is so thin that it has no influence on degradation of the structural properties.

The reported achievements in the material development of the ZnO/SiC heterostructures consequently yielded the device performances [47–49]. Eventually, the number of real devices has been constructed based on ZnO/SiC, evidencing the vitality of this heterostructures for the industry.

One of the earliest successful realizations of ZnO-film-based LED was reported by Yuen et al. [47]. They demonstrated the realization of *n*-ZnO:Al/*p*-4H:SiC heterojunction LEDs. The *p*-doped single-side polished 4H:SiC wafer was chosen as a substrate and the hole injection layer. The expected hole concentration in the wafers was $\sim 1 \times 10^{19}$ cm^{-3} and carrier mobility was ~ 120 cm^2/(V·s). The SiC substrate was polished in order to provide the abrupt interface; then thin ZnO:Al films were deposited over by filtered cathodic vacuum arc technique. The active layer was ZnO:Al of a concentration 3%, resulting in a concentration of 1×10^{20} cm^{-3} and carrier mobility ~ 7.2 cm^2/(V·s). As a transparent injection layer, the ZnO:Al (7%) was used, having the carrier concentration $\sim 1 \times 10^{21}$ cm^{-3} and a band gap ~ 0.12 eV wider then active layer. The prepared structure underwent rapid thermal

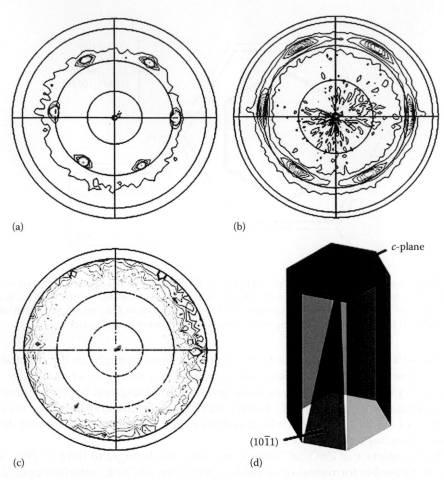

FIGURE 3.4

XRD pole figures of (a) ZnO (10$\bar{1}$1) at 400°C, (b) ZnO (10$\bar{1}$1) at room temperature, and (c) 4H:SiC (10$\bar{1}$1). (d) Schematic of the hexagonal unit cell. (Reprinted from *J. Alloys Compd.*, 489, Kim, J.-H., Cho, D.-H., Lee, W., Moon, B.-M., Bahng, W., Kim, S.-C., Kim, N.-K., and Koo, S.-M., Structural and optical properties of epitaxial ZnO thin films on 4H–SiC (0001) substrates prepared by pulsed laser deposition, 179. Copyright 2010, with permission from Elsevier.)

annealing (RTA) for 5 min in N_2 at 800°C in order to improve the contacts ohmicity, but also supposed to eliminate the DLE emission from ZnO film. The electrical characterization showed a rectifying behavior with a turn-on voltage 3.8 V. It was noted that if the electron and hole densities in *n*-ZnO:Al (3%) films and *p*-SiC are greater than 1×10^{19} cm^{-3}, then the turn-on voltage should be closer to the corresponding band gap energy. The band gap energy of ZnO and *p*-SiC is ~3.3 and 3.28 eV, so the measured turn-on voltage was explained as due to the high ohmic resistance from the Al/Ti metal contacts deposited on the *p*-SiC substrate. The prepared LED demonstrated the EL spectra with a UV emission peaked at 385 nm and the intensity of the UV peak increases with increase in biased voltage. This is stated as indication that the ZnO:Al films exhibited effective near-band-edge radiative recombination and the influence of the heterointerface defects was negligible.

The SiC as a *p*-type substrate was used by Leong and Yu [56]. Authors demonstrated UV random lasing from *p*-SiC(4H)/*i*-ZnO-SiO$_2$ nanocomposite/*n*-ZnO:Al heterojunctions

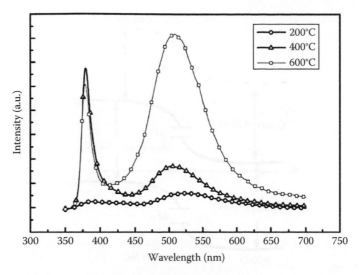

FIGURE 3.5

Photoluminescence spectra of the ZnO thin films deposited by PLD at different substrate temperatures. (Reprinted from *J. Alloys Compd.*, 489, Kim, J.-H., Cho, D.-H., Lee, W., Moon, B.-M., Bahng, W., Kim, S.-C., Kim, N.-K., and Koo, S.-M., Structural and optical properties of epitaxial ZnO thin films on 4H–SiC (0001) substrates prepared by pulsed laser deposition, 179. Copyright 2010, with permission from Elsevier.)

diodes. However, in this case the idea was that the sandwiching of ZnO nanoparticles between the electron- and hole-injection layers may enhance the injection efficiency of the carriers so that the nanoparticles can attain a high optical gain under electrical excitation. Indeed, the UV emission of the laser diodes originated from the ZnO-SiO$_2$ nanocomposite layer and the use of ZnO powder did improve the electrical-to-optical conversion efficiency of the heterojunction.

El-Shaer et al. reported the fabrication of diode, based on ZnO films on *p*-SiC [42]. The ZnO layer was deposited on 8° offcut 4H:SiC substrate. The SiC substrate was covered by *p*-4H:SiC epilayer of 10 μm thickness, having a hole concentration 5×10^{18} cm^{-3}. Authors revealed the single crystalline nature of ZnO layers, according to XRD and RHEED analysis. The grown ZnO layer had a carrier concentration 6×10^{17} cm^{-3} and mobility as high as 55 cm^2/(V·s). The obtained structures demonstrated diode-like behavior with a turn-on voltage about 2 V. Furthermore, authors interpreted their results in term of "ideal" heterojunction band diagram constructed according to the Anderson model (Figure 3.6). The electron affinity of ZnO was 4.35 eV and the electron affinity of 4H-SiC was 4.05 eV. The band gap energies of ZnO and 4H-SiC were 3.3 and 3.2 eV, respectively. Based on the diagram, the energetic barrier for electrons was $\Delta E_C = 0.3$ eV, while the energetic barrier for holes was $\Delta E_V = 0.4$ eV. It was concluded that since ΔE_V is higher than ΔE_C, the electron injection from *n*-ZnO to *p*-SiC was supposed to occur, unlike the hole injection from *p*-SiC to *n*-ZnO. These data suggest additionally the importance of application of both the injection and active ZnO layers for effective UV emission.

This, however, was successfully fulfilled by Shih et al. in 2010 [48]. The authors, being aware of the fact that the electron injection from *n*-ZnO dominates over the hole injection from *p*-SiC, compared the two type of heterostructures: *n*-ZnO/*p*-SiC and *n*-ZnO/*i*-ZnO/*p*-SiC. It was confirmed that inserting an undoped *i*-ZnO layer between *n*-ZnO and *p*-SiC leads to the hole and electron injection from *p*-SiC and *n*-ZnO into the *i*-ZnO layer. For the regular *n*-ZnO/*p*-SiC heterojunction LEDs, the electroluminescence (EL) came from

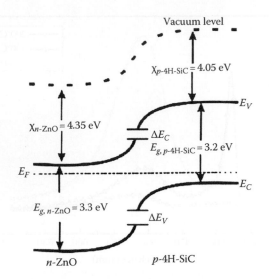

FIGURE 3.6
Schematic energy band diagram of the *n*-ZnO/*p*-4H:SiC heterostructures. (Reprinted from *Superlattices Microstruct.*, 42, El-Shaer, A., Bakin, A., Schlenker, E., Mofor, A.C., Wagner, G., Reshanov, S.A., and Waag, A., Fabrication and characterization of *n*-ZnO on *p*-SiC heterojunction diodes on 4H:SiC substrates, 387. Copyright 2007, with permission from Elsevier.)

the defect-related states in *p*-SiC, while the efficient UV EL at 393 nm from the *n*-ZnO/*i*-ZnO/*p*-SiC heterojunction LED has been observed at room temperature.

Very recently, the significance of the ZnO/SiC heterointerface was additionally reported for the photoelectric conversion. Earlier, the ZnO/Si heterojunction solar cell was reported as a candidate for silicon mono junctions and low-cost solar cells. In such heterostructures, ZnO film was used not only as a solar cell window but also as a *p–n* heterojunction partner [57]. Unfortunately, the photoelectric conversion efficiency of such a heterojunctions was very low. Particularly for this application, the growth of ZnO on cubic SiC gives an enormous prospective. The 3C:SiC has the band gap 2.3 eV, which is larger than that of Si (1.1 eV) and smaller than that of ZnO (3.37 eV). Thus, via designing heterostructures of ZnO/3C:SiC/Si, the utilized range of solar spectrum can be extended. Therefore, such multi-heterostructures will have higher conversation efficiency than the conventional ZnO/Si heterojunctions [58]. In other words, ZnO/3C:SiC/Si heterojunctions can utilize solar energy across a wide spectrum. Wu et al. [58] have shown that the photoelectric conversion efficiency of the *n*-ZnO/*n*-SiC/*p*-Si heterojunction is about four times higher than that of the *n*-ZnO/*p*-Si heterojunction. For this type of application, the quality of the heterointerface is important as well as for the LED.

Phan and Chung reported the deposition of zinc thin films on a polycrystalline 3C–SiC buffer layer using RF magnetron sputtering and a sol–gel method [59]. The 3C–SiC buffer layer was suggested to improve the optical and piezoelectric properties of the ZnO film. Authors demonstrated less residual stress due to a close lattice mismatch of the ZnO and SiC layers.

Zhu et al. reported that by using SiC buffer layer ZnO epitaxial films were deposited on Si (111) substrate [41]. Here, SiC was chosen as a buffer layer to reduce the lattice mismatch between ZnO and Si. According to the measurement results, it was found that the ZnO/SiC/Si (111) films have much stronger ultraviolet emission and better crystal quality than that of the ZnO/Si (111) films. These results prove that the SiC buffer layer is useful for

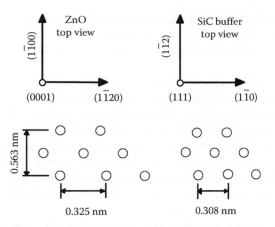

FIGURE 3.7
Schematic depicting the top view of the arrangement of the atoms in the ZnO film and SiC. (Reprinted from *Thin Solid Films*, 478, Zhu, J., Lin, B., Sun, X., Yao, R., Shi, C., and Fu, Z., Heteroepitaxy of ZnO film on Si(111) substrate using a 3C–SiC buffer layer, 218. Copyright 2005, with permission from Elsevier.)

modulating the lattice mismatch between ZnO and Si and improving the photoelectric properties of the ZnO films. Authors proposed the atomic arrangement of ZnO on the SiC surface (Figure 3.7).

Diode based on n-ZnO/p-4H:SiC heterostructures was recently reported in Ref. [60]. The ZnO film has been epitaxially deposited onto the SiC substrate, providing the sixfold symmetry of (10–11) plane. The turn-on voltage of the diode was 3 V, what was considered as reasonable for ZnO and SiC wide band gap. The prepared heterostructure was found to be sensitive for the UV light (254 and 365 nm) and therefore it was concluded that n-ZnO/p-4H:SiC heterojunctions diodes are promising candidates for UV photodetector applications.

Reported device performance initiated a number of papers, devoted to properties optimization of the used materials [61,62,64,66,67]. The experimental results were significantly proven followed by a certain number of works, creating the theoretical basic of ZnO growth on SiC. Particularly, significant attention has been paid to the materials interface quality and strain development on the heterointerface of ZnO/SiC. In terms of ZnO-based LED structure, it is important to achieve lower threshold current density and higher differential quantum efficiency. These parameters strongly depend on the control of biaxial strain and stress kinetics of the materials system.

In principle, the biaxial strain appears in heterostructure due to two reasons: lattice (δ) and thermal (σ) mismatches. The δ appears while the lattice does not match with the underlayer, for example, for ZnO/Al$_2$O$_3$ heterosystem ($\delta = 18\%$, that fully relaxes when the epilayer thickness becomes larger than several nanometers of critical thickness h_c) [61]. On the other hand, the σ occurs while the postdeposition substrate temperature cools down to room temperature; a compressive biaxial stress is induced along the out of plane because of the different thermal expansion coefficients of the materials. To minimize these driving forces in ZnO materials system, lattice-matched and/or closely matched substrates can play a vital role. Here, the SiC as a substrate for ZnO has to be considered.

Availability of the strain strongly affects the device performance. It is noted that the strain induces, for example, dark-line defects and piezoelectric devices which tend to shorten the device lifetime by containing a huge density of misfit dislocations (MDs) and threading dislocations, including cracking [62]. These defects, in principle, led to the failure of lasers and breakdown of *p–n* junctions [63]. To avoid/reduce these drawbacks, especially for the ZnO epitaxy, SiC may work out since the δ and σ are 5% and <1%, respectively.

The basics of ZnO on SiC heteroepitaxy were reported by Ashrafi et al. in Ref. [64]. In this study, the epitaxial ZnO layers were grown directly on the 6H-SiC substrates by MOCVD. Authors, considering the in-plane lattice parameter of SiC (*a*: 3.081 Å) substrate with respect to the ZnO epilayer (*a*: 3.246 Å), calculated the lattice mismatch f_m using the relation $f_m = [(a_1 - a_s)/a_s] \times 100\%$ where a_1 is the epilayer and a_s is the substrate lattice constant. From here, the f_m value of ZnO/SiC heterostructure was estimated to be +5%. Here, sign "+" means that the ZnO/SiC interface induces compressive strain in ZnO. Figure 3.8 presents the basic schematic model of a lattice deformation mechanism for the ZnO/SiC interface.

Furthermore, authors observed that the strain value changes with the thickness of ZnO layer on SiC—the strain relaxation occurs. In order to clarify the strain relaxation in the ZnO/SiC heterostructure, the biaxial strain $\varepsilon_{xx} = \varepsilon_{yy} \neq \varepsilon_{zz}$ along the *a* and *c* axes were estimated by $\varepsilon_{xx} = (a - a_0)/a_0$ and $\varepsilon_{zz} = (c - c_0)/c_0$, where a_0 and c_0 were, respectively, the lattice constants in the unstrained crystal and taken as $a_0 = 3.246$ Å and $c_0 = 5.205$ Å [65]. Thus, the authors estimated the strain along *a* and *c* axis as a function of ZnO layer thickness (Figure 3.9a).

The biaxial strain relaxation rate as a function of the ZnO layer thickness is shown in Figure 3.9b. The relaxation rate was calculated as $r_r = ((\varepsilon_c - \varepsilon_r)/\varepsilon_c \cdot 100\%)$, where ε_c was the strain when the ZnO layers grow coherently on the underlying SiC substrate and ε_r is the measured strain of epitaxial ZnO. As the authors pointed out, the ZnO grows coherently only until reaching the layer thickness 6 nm. With the increase of layer thickness the stress was accounted at a maximum from ≥7 to ~230 nm under the compressive strain and decreased slowly with the increase of the ZnO layer thickness, typically from ≥230 nm to 1.5 μm. It was noted that the maximum strain relaxation was estimated to be ~94% already for the layer thickness around 1.5 μm. This remaining strain in the ZnO/SiC heterostructure was explained as due to the thermal expansion coefficients, which were not accounted for in this study. The thermal expansion coefficients of ZnO and SiC materials

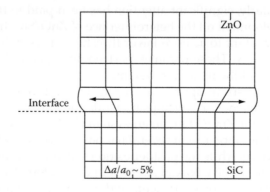

FIGURE 3.8
Schematic model of a lattice deformation mechanism for the ZnO/SiC interface. (Reprinted with permission from Ashrafi, A., Binh, N., Zhang, B., and Segawa, Y., Strain relaxation and its effect in exciton resonance energies of epitaxial ZnO layers grown on 6H:SiC substrates, *Appl. Phys. Lett.*, 84, 2814. Copyright 2004, American Institute of Physics.)

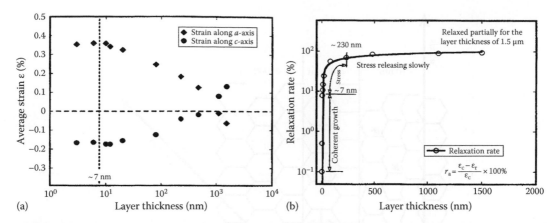

FIGURE 3.9
Evolution of the in-plane and out-of-plane strains (a) and the biaxial strain relaxation rate (b) with the layer thickness for the ZnO/SiC heterostructures. (Reprinted with permission from Ashrafi, A., Binh, N., Zhang, B., and Segawa, Y., Strain relaxation and its effect in exciton resonance energies of epitaxial ZnO layers grown on 6H:SiC substrates, *Appl. Phys. Lett.*, 84, 2814. Copyright 2004, American Institute of Physics.)

were reported to be 4.75 and 4.70×10^{-6}/K, respectively, providing ~1% thermal expansion coefficients mismatch.

Strain is important for the optoelectronic properties of ZnO. For example, it has been demonstrated earlier that the strain relaxation is affected by growth conditions, VI/II molar ratios, and layer thickness, and it is also adhered that the structural and optical properties are strongly correlated with the degrees of the strain. For example, the threshold power density in stimulated emission decreases with an increase of the strain [66]. Authors in [65] also investigated the thickness-dependent strain relaxation effect on exciton resonance energies of ZnO layers. Thus, the shift of the main peaks on the photoluminescence spectra was admitted with the strain development. The three PL peaks assigned to be due to free A-exciton (A), neutral donor bond exciton (D^0X), and neutral acceptor-bound (A^0X) excitons change linearly as a function of strain in the ZnO/SiC. The slope $\Delta E/\varepsilon_{zz}$ was estimated to be 13.1 and 14.6 eV, corresponding to the A and D^0X excitons resonance energies for the ZnO/SiC samples. It was admitted that the observed shift is relatively smaller than that for the ZnO grown on other materials—ZnO/Al$_2$O$_3$ or even GaN (14.2 and 16.4 eV for A exciton peak, respectively). This indication of the relatively smaller strain effect in ZnO exciton resonance energies than in GaN may be noticed as an additional advantage of the ZnO for optoelectronics.

However, the previous model (Figure 3.8) was too primitive to describe the strain in the ZnO/SiC heterostructures. Therefore, later a more detailed strain relaxation model was proposed (Figure 3.10) [61]. Considering the experimental observation of lattice relaxation as a function of ZnO layer thickness, the following relation of a and c axis strains was observed: It was found that at the extreme points the a axis length elongates to the maximum, while the c axis length gets compressed to the minimum. Furthermore, the biaxial strain relaxation was reported to be anomalous. It was suggested that since δ in the ZnO/SiC heterostructure is around 5%, this is responsible for the appearance of misfit dislocations (MDs) (marked by arrows in Figure 3.10) for the ZnO layer of thickness up to 10 nm. At this stage, the lattice relaxes by plastic deformation and induces MDs. For the ZnO layer thickness of up to 100 nm, the lattice relaxation was suggested to take place both by

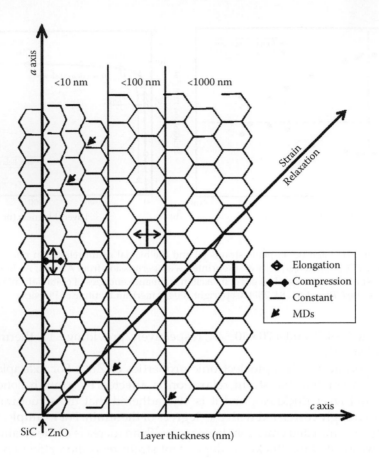

FIGURE 3.10
Schematic model of strain relaxation in ZnO/SiC heterostructures. (Reprinted with permission from Ashrafi, A. and Segawa, Y., Anomalous lattice relaxation mechanics in ZnO/SiC heterostructures, *J. Appl. Phys.*, 103, 093527. Copyright 2008, American Institute of Physics.)

plastic and elastic deformations by keeping in balance. Until now, no anomalous zones were observed here. However, after the complete relaxation, it was demonstrated, that only elastic deformation is dominant and again induces anomalous lattice relaxation due to imbalance of charge distribution.

The strain values were reported to range from −0.3 to +0.1 and −0.22 to +0.3 for the c axis and a axis lengths, respectively. As it was explained by Ashrafi et al., this anomaly in the lattice parameter relaxation is the result of biaxial strain kinetics and/or associated with complex materials engineering and chemistry. Since ZnO belongs to the space group $P_63mc = C^4_{6v}$, this material grows along the c axis and relaxes faster perpendicular to the substrate. Since the δ and σ in the ZnO/SiC heterostructure are even larger, the strain relaxation would imply elimination of dislocations for a few nanometer thicknesses and terminated by big crystallite sizes. Considering these properties, the isotropic spin-orbit coupling, band structure, and electronic properties of ZnO/SiC heterostructures have considerably been modified [61].

The biaxial strain relaxation kinetics has also been theoretically assessed by considering the MDs at the heterointerfaces and thereafter into the thin films [61]. The energy cost E_{MD} required for creating a unit length MD at the heterointerface was given by

$$E_{MD} = \frac{\ln(1 + h/b) x\mu b^2 (1 - \nu/4)}{4\pi(1 - \nu)} \tag{3.1}$$

where

μ is the shear modulus at the interface between the film and the substrate ($\mu \approx (C_{11} - C_{12})$) for the slip system ⟨011⟩ {111}

ν is the Poisson ratio

b is the Burgers vector

For comparison, the theoretical calculation of MD formation energy as a function of ZnO layer thickness for GaN/SiC heterostructure has been fitted theoretically and plotted along with ZnO/SiC in Figure 3.11. When the ZnO layer thickness is ~200 nm, the E_{MD} is 8 meV, but it is 11 meV while it approaches 1000 nm. Based on the Figure 3.11, one can conclude, that the possibility of the formation of dislocations in ZnO/SiC is relatively higher than that of the GaN/SiC heterostructures. The fitted parameter was found to be $b = 2.253$; absolutely, the a axis length of ZnO. It was found that the MDs are easier to form in the ZnO/SiC heterostructure than in the ZnO/Al₂O₃ and GaN/SiC materials systems. There is a common tendency to easily form the MDs in II–VI than in the III–V materials. On the other hand, there are systems such as GaN/AlN, with lower δ (2.4%), which are known for exhibiting a clear SK transition with the formation of coherent islands.

The anomalous strain development was followed by its respective influence on the ZnO photoluminescence properties. Authors observed the influence of the strain in the ZnO/SiC heterostructures on the observed dominant 3.3762, 3.3638, and 3.330 eV peaks originated from free-exciton (FX), neutral-donor bound exciton (D^0X), and neutral acceptor-bound exciton (A^0X) bands, respectively. It was noted that the FX_A band is

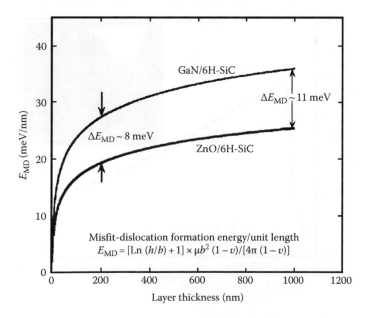

FIGURE 3.11
Theoretical calculation of MD formation energy E_{MD} as a function of ZnO layer thickness. (Reprinted with permission from Ashrafi, A. and Segawa, Y., Anomalous lattice relaxation mechanics in ZnO/SiC heterostructures, *J. Appl. Phys.*, 103, 093527. Copyright 2008, American Institute of Physics.)

observed only for the layers of the thickness more than 6 nm and more than 1000 nm. This FX_A band dominance can be initially attributed to the defect-free ZnO layers since critical layer thickness h_c has been experimentally assigned to be 4–6 nm. In addition, there were no FX_A bands for the layer thickness of more than 10–80 nm, due to defects possibly contributed by δ, as has been clarified in TEM and secondary ion-mass spectroscopy [61]. It was underlined, that in principle, the photon energy does not change due to the cell volume, especially for the wurtzite structure. However, authors observed that with the increase in the ZnO layer thickness, the FX_A, D^0X, and A^0X bands were blueshifted by the energies of 2.5, 4.8, and 5.8 meV, respectively. This blueshifting is consistent with the systematic biaxial strain relaxation as a function of ZnO layer thickness [61]. To estimate the nonradiative defect density, time-resolved photoluminescence (TRPL) study of free excitonic emission peaks in the ZnO epilayers on SiC were performed at room temperature (Figure 3.12).

The TRPL decay spectra was fitted using a biexponential line shape $I(t) = I_1 \exp(-t/\tau_R) + I_2 \exp(-t/\tau_{NR})$, where $I(t)$ is the PL intensity at time t and I_1 represents the initial intensity of the I component. The observed fast decay constant (τ_R) was explained as probably representing the effective radiative lifetime of localized excitons of 450 ps. On the other hand, the slow one (τ_{NR}) is too long to be accounted for in any intrinsic radiative process. The longer time constant in the two-exponential fits was attributed to the radiative lifetime of the exciton, although the origin of the short time constant was not certain. Two possibilities were admitted for the nonradiative processes in origin: Defect concentration changes over the sample in such a way that the concentration increases linearly over the growth time, and surface states may cause nonradiative decay. It was additionally reported that the optical band structure of ZnO/SiC heterostructures was modified considerably by the biaxial strain incorporated. In principle, the c/a ratio reduction counteracts the increase and/or

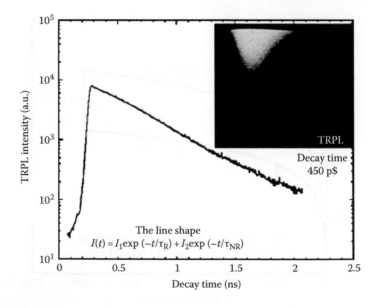

FIGURE 3.12

Typical TRPL spectra of ZnO layer deposited on SiC substrate taken at room temperature. The inset shows the TRPL image taken from the sample at room temperature. (Reprinted with permission from Ashrafi, A.B.M.A., Segawa, Y., Shin, K., and Yao, T., Strain effects in ZnO layers deposited on 6H-SiC, *J. Appl. Phys.*, 100, 063523. Copyright 2006, American Institute of Physics.)

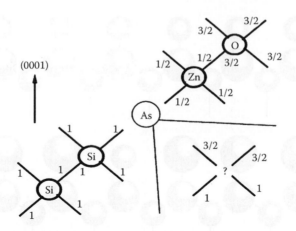

FIGURE 3.13
Sketch of the possible imbalanced interface chemistry in ZnO/SiC heterostructures. (Reprinted with permission from Ashrafi, A., Segawa, Y., Shin, K., and Yao, T., Nucleation and interface chemistry of ZnO deposited on 6H-SiC, *Phys. Rev. B*, 72, 155302. Copyright 2005, American Institute of Physics.)

fluctuations in the optical bands due to the volume compression. In a more detailed study of the ZnO/SiC heterostructure [64], it was suggested that an imbalanced charge distribution can be expected on the interface (Figure 3.13).

It was suggested, that in the ZnO, each Zn (O) atom contributes 1/2 (3/2) electron to each of the four bonds surrounding it, and in bulk Si, each atom contributes exactly one electron per bond. At a (0001) interface between the Si and Zn of ZnO, therefore, the interface bonds with each reach a total of 3/2 electrons instead of the two electrons required. Therefore, the imbalanced electron distribution and dangling atomic bonds are responsible for the columnar growth modes and related defects. It is also assumed, that the SiC substrate surface contains nanoscale fluctuations and scratches, which play an important role in the formation of defects in ZnO epilayers. The important conclusion of this study is that as the authors suggested As or related materials may satisfy the misbalancing problem for this and similar II–VI/IV material system.

The nucleation and interface chemistry of ZnO/SiC heterostructure have been described in Ref. [67]. It was shown that associated ZnO growth modes depend on the strain energy and surface energy that favor growing islands in nucleation. Complex interface chemistry accommodated in the ZnO/SiC heterointerface was found to be responsible for the formation of misfit dislocations and dangling atomic bonds, together with impurity complex contributed from the host material, SiC substrate, and metal-organic precursors. It was additionally concluded that higher solubility of C atoms than N may play an informative role in ZnO epitaxy for conductivity type's conversion.

An observation of oxidized or Zn passivated modes of interface arrangements was performed. The as-received SiC substrates were dominated by oxide layer as shown in Figure 3.14. It was as a result of flowing oxygen on the thermally treated substrates prior the deposition. Under the proper chemical and thermal treatments, the oxide layer desorbs with bare Si atoms. By DEZn irradiation after the thermal treatment, however, on the bare Si atoms, the Si-faced SiC surface is passivated by dangling Zn bonds represented in Figure 3.14. As a result, under the thermal treatment as well as the subsequent DEZn exposition establishes the Zn–Si bonds in the heterointerfaces and hinders the formation of oxidation in nucleation prior to ZnO deposition.

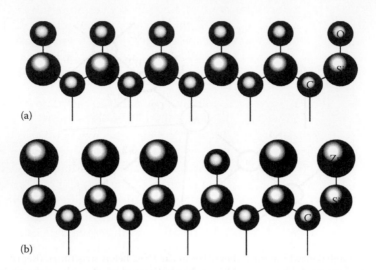

(a)

(b)

FIGURE 3.14
Schematic representation of atomic SiC surfaces with (a) partial oxidation or (b) Zn passivation. (Reprinted with permission from Ashrafi, A., Segawa, Y., Shin, K., and Yao, T., *Phys. Rev. B*, 72, 155302, 2005. Copyright 2005, American Physical Society.)

The dynamics of the zinc and the oxygen adatom supplied as atomic zinc and atomic oxygen on the 6H-SiC(0001) surface was investigated using the first-principles calculation in [68]. The result suggested that the on top site is unstable for the zinc and the oxygen on the 6H-SiC(0001) surface. However, the oxygen is stable at the on top site near a Zn adatom. The calculation showed that the optimized growth condition for the growth of ZnO on 6H:SiC(0001) is Zn-polarity ZnO crystal grown under stoichiometric growth condition.

For practical applications, it is necessary to know the valence (VBO) and conduction band offset values and the band lineup of ZnO/SiC heterojunctions. The ZnO/6H-SiC band offsets, in particular, have been a subject of extensive studies due to the technological importance in optoelectronics. Alivov et al. [69] have reported the conduction band offset (CBO) of ZnO/6H-SiC heterojunction diodes by current-voltage and deep-level transient spectroscopy measurements in the range of 1.10–1.25 eV. Ashrafi [70] addressed the band alignment of epitaxial ZnO/6H-SiC heterojunction, based on XPS measurements.

Taking into account the band offset values, the wurtzite ZnO/6H-SiC heterojunction band lineup is found to be the staggered-type alignment, as illustrated in Figure 3.15. The reported ΔE_V and ΔE_C values for the ZnO epilayers deposited on different substrates are shown in Table 3.2. It was concluded, that the obtained experimental values of ΔE_V and ΔE_C will provide the further understanding in the fundamental properties of ZnO/6H-SiC heterojunction, and will be useful in the design, modeling, and analysis of optoelectronic devices.

The development of nanotechnology as a general trend resulted in research devoted to low-dimensional ZnO/SiC system. The formation of heterostructures is believed to be of importance in tailoring the physical properties of one-dimensional (1D) nanostructures. A number of papers reported the growth of randomly oriented nanosized ZnO growth on SiC [75,76]. Despite the high structural quality revealed, the functionality of such devices was out of reality, due to complicated technological processing of the grown structures (contacts deposition, etc.).

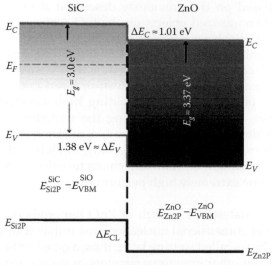

Staggered band alignment

FIGURE 3.15

Energy band diagram of ZnO/6H-SiC heterojunction. A type-II heterojunction is aligned in the staggered arrangement. (Reprinted from *Surf. Sci.*, 604, Ashrafi, A., Band offsets at ZnO/SiC heterojunctions: Heterointerface in band alignment, L63. Copyright 2010, with permission from Elsevier.)

TABLE 3.2

Valence and Conduction Band Offsets of ZnO Heterostructures Deposited on SiC in Comparison to Different Substrates

Heterojunction	ΔE_V (eV)	ΔE_C (eV)	Band Alignment	References
ZnO/6H-SiC	1.38 ± 0.28	1.01 ± 0.28	Type-II	[70]
ZnO/6H-SiC	—	1.1–1.25	—	[71]
ZnO/4H-SiC	1.61 ± 0.23	1.50 ± 0.23	Type-II	[72]
ZnO/p⁺-Si	2.55	0.30	—	[73]
ZnO/GaAs	2.39 ± 0.23	-0.44 ± 0.23	Type-I	[74]
GaN/6H-SiC	1.02 ± 0.10	0.62 ± 0.10	Type-II	[75]

Source: Reprinted from *Surf. Sci.*, 604, Ashrafi, A., Band offsets at ZnO/SiC heterojunctions: Heterointerface in band alignment, L63. Copyright 2010, with permission from Elsevier.

Bae et al. in 2004 reported the heterostructures of ZnO nanorods on SiC nanowires, prepared by thermal CVD [75]. The prepared heterostructures possessed abrupt interface, but their luminescence properties were mainly governed by defect emission in the visible range.

Zhang et al. reported the rose-like zinc oxide nanoflowers synthesized on SiC substrate via CVD [76]. It was shown that the products deposited on different zones of the substrate show different morphologies such as micro-plates, micro-palmerworms, and self-assembled nanosheets. At the same time, the room-temperature photoluminescence spectrum of the rose-like nanoflowers reveals a strong green emission at ~491 nm corresponding to defect-related emission.

At the same time, based on the previously described data, some groups reported ordered growth of ZnO nanosized objects on the SiC surface [77–79]. ZnO nanostructures such as nanorods (NRs) have attracted considerable attention owing to their large surface area, good crystal quality, and unique photonic properties. Additional advantage is that the nanorods can release the strain/stress due to substrate mismatch through a relatively large surface area. Eventually, the well-ordered ZnO nanostructures (nanorods, nanopillars) were obtained on SiC, providing high structural quality, advanced luminescence properties, and therefore making the nanodevices based on ZnO/SiC heterostructures an achievable goal. As a result, a number of real devices, based on ZnO/SiC heterostructures, were obtained. Especially, LEDs based on nanosized n-ZnO/p-SiC were designed, revealing the promising characteristics [79,80]. Possible nanoscale devices will not only show extremely high performance, but shall also be small and less energy consuming.

Mofor et al. reported catalyst-free growth of ZnO nanorods on 6H-SiC [77]. Authors pointed out that the three dimensional nucleation was initiated directly on the substrate without the growth of the so-called wetting layer, thus, a quasi-self-organized growth was reported. After optimizing other growth parameters in several experiments at a growth temperature of 800°C, nanorods with widths of 70–90 nm and length of up to 4 μm with a density of about 3×10^9 cm^{-2} were obtained (Figure 3.16). The room temperature photoluminescence (PL) from the nanorods was dominated by the free exciton PL peaks at 3.26 eV having a full width at half maximum of 95 meV. Authors concluded that the grown ZnO nanorods were of very high optical quality.

Zhao et al. reported ordered ZnO nanostructures grown on 4H:SiC substrates via vapor–liquid–solid (VLS) growth [78]. Authors observed the conventional photoluminescence spectra with the peak at 380 nm, which was related to the free exciton (FE) and the intense emission band close to 520 nm which was related to DLE (Figure 3.17).

However, despite the defects availability, stimulated emission was observed over the threshold excitation power. From the PL spectra, the multiple mode was obvious, similar to when ZnO nanorods become lasing. Although the different lengths and different diameters of ZnO nanorods induced lasing mode variation, the observed multiple lasing modes were mainly due to the cavity modes since the spacing of the main modes is

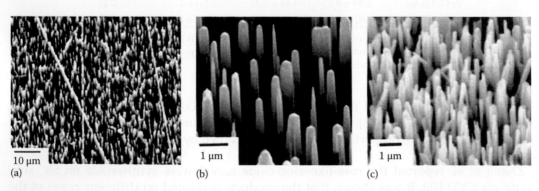

FIGURE 3.16
SEM images of ZnO nanorods on 6H-SiC substrate grown at 750°C and 10 mbar (a, b) and at 750°C and 25 mbar (c). Growth rate is higher in "lines" cause by the abrasive effect of SiC polishing. (From Mofor, A., Bakin, A., Elshaer, A., Fuhrmann, D., Bertram, F., Hangleiter, A., Christen, J., and Waag, A.: Catalyst-free vapor-phase transport growth of vertically aligned ZnO nanorods on 6H-SiC and (11–20)Al$_2$O$_3$. *Phys. Status Solidi C*. 2006. 3. 1046. Copyright Wiley-VCH Verlag GmbH & Co. KGaA. Reprinted with permission.)

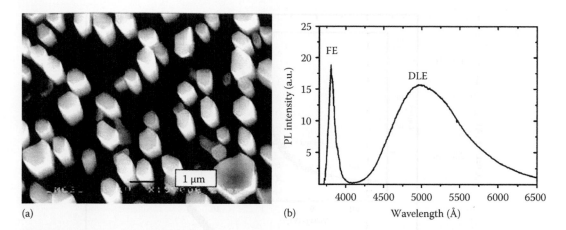

FIGURE 3.17
SEM picture of the ZnO nanorods grown on a 4H:SiC(0001) substrate (a) and their respective PL properties (b). (Reprinted with kind permission from Springer Science+Business Media: *Appl. Phys. A*, Growth of ZnO nanostructures by vapor–liquid–solid method, 88, 2007, 27, Zhao, Q., Klason, P., and Willander, M.)

approximately the same. Because the grown ZnO nanorods were vertically located on the substrate, they formed a good laser cavity. The threshold of excitation power was observed at about 30 kW/cm², which is relatively low [78].

Bano et al. in 2009 reported the study of luminescent centers in ZnO nanorods catalytically grown on *p*-4H:SiC [79]. The electrical characteristics of the prepared heterojunction were examined. Figure 3.18 represents a typical *I–V* characteristic for the ZnO nanorods/*p*-SiC heterojunction at room temperature (RT).

The ideality factor was found to be in the range 3–4 for all diodes investigated. The higher value of the ideality factor indicates that the transport mechanism is no longer dominated by the thermionic emission. It was concluded that the nonideal behavior is often attributed to defect states in the band gap of the semiconductor or at the interface providing other current transport mechanisms such as through structural defects, surface contamination, barrier tunneling, or generation recombination in the space charge region and variations in interface composition.

It was reported that the violet, blue, and green-yellow emission luminescence lies in the deep band emission (420–700 nm) which is a result of superposition of different defect bands emitting in different wavelengths. Moreover, a strong white light emission was observed, being a superposition due to different defects in ZnO NRs. Based on the statement, that in wide band gap semiconductors, the broadband luminescence is related to the transitions from donor states to the deep acceptor states, authors developed the band diagram of the three observed luminescence process (Figure 3.19).

Authors pointed out that the previously published heterojunction LEDs based on ZnO thin films grown on SiC were as a consequence of the high density of interface states. Controversially, their results were explained by the much better interface at the *p–n* heterojunction. This was because the strain/stress due to lattice mismatch in the case of ZnO nanorods were released through their large surface area. Authors concluded that a better *p–n* heterojunction interface and a better ZnO quality are critical for the high efficiency of LEDs based on ZnO/SiC heterostructures.

The same group has recently demonstrated the local luminescence characterization of ZnO NRs via cathodoluminescence probing and has obtained detailed luminescence

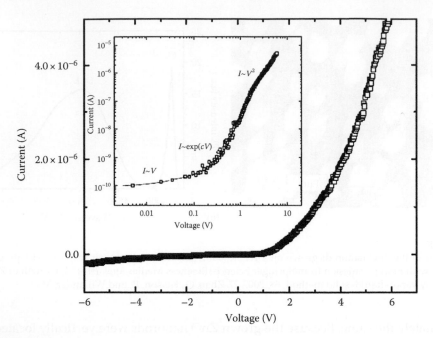

FIGURE 3.18
Typical *I–V* characteristic, obtained for the ZnO nanorods/*p*-SiC heterojunction at room temperature (RT). (Reprinted from Bano, N., Hussain, I., Nur, O., Willander, M., Klason, P., and Henry, A., Study of luminescent centers in ZnO nanorods catalytically grown on 4H-*p*-SiC, *Semicond. Sci. Technol.*, 24, 125015, 2009. With permission from IOP.)

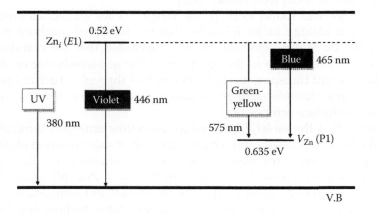

FIGURE 3.19
Schematic band diagram of the three observed luminescence processes. (Reprinted from Ashrafi, A., Zhang, B., Binh, N., Wakatsuki, K., and Segawa, Y., High-quality ZnO layers grown on 6H-SiC substrates by metalorganic chemical vapor deposition, *Jpn. J. Appl. Phys.*, 43, 1114, 2004. With permission from IOP.)

information about ZnO NRs. As it can be seen (Figure 3.20), the CL spectra of SiC have very weak emission at 390 and 540 nm but the CL spectra of ZnO NRs show stronger emissions at 380 nm and DBE (420–720 nm) centered at 521 nm. The figure confirms that most of the emission is originating from ZnO NRs, and the SiC has a minute contribution to the emission. The EL measurement of the ZnO NR-based heterojunction LED was carried out

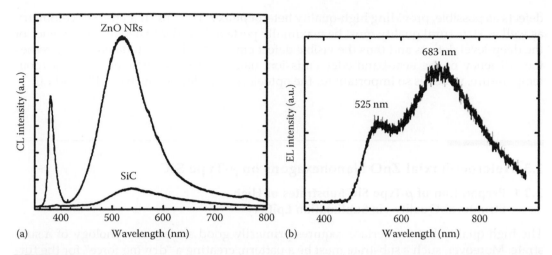

(a)   Wavelength (nm)   (b)   Wavelength (nm)

FIGURE 3.20
Room temperature cathodoluminescence spectra of ZnO nanorods and SiC (a) and the electroluminescence of ZnO/SiC heterostructures (b). (Reprinted from *J. Lumin.*, 130, Bano, N., Hussain, I., Nur, O., Willander, M., Wahab, Q., Henry, A., Kwack, H., and Le Si Dang, D., Depth-resolved cathodoluminescence study of zinc oxide nanorods catalytically grown on *p*-type 4H-SiC, 963. Copyright 2010, with permission from Elsevier.)

at room temperature; the corresponding RT–EL spectrum for forward injection current is shown in Figure 3.20.

The observed EL spectrum showed two peaks: DBE at 525 nm and a red emission at 683 nm. Authors noticed that the peak positions in the EL spectra are different from that in the CL spectra because the sample was heated by electron bombardment during CL measurements. It was concluded that RT–CL spectra DBE peak at 521 nm arises from electron transitions from the energy level of isolated vacancies to the valence band [185].

Mofor et al. in 2006 reported the fabrication of ZnO nanorod-based *p–n* junction on SiC substrate [80]. The growth of *n*-ZnO nanorods on 4H-SiC substrates was investigated on *n*-type 8° off-plane 4H:SiC. Authors suggested that the surface on which the step-flow growth approach is traditionally exploited for homoepitaxy of SiC the terraces would facilitate the 2D ZnO growth. Therefore, different sets of growth parameters that favor the formation of nanocrystals and hence the growth of ZnO nanorods on stepped 8° off-plane 4H-SiC were investigated. Particularly, it was observed that the ZnO nanorods on 8° off-plane 4H-SiC were not as vertically aligned as those grown on bulk substrates. The *I–V* measurements performed on the *p–n* junction between the *n*-ZnO nanorods and the *p*-SiC epilayer exhibited a turn-on voltage of below 2 V and very high breakdown voltage. Authors noted that the crystal quality and optical properties of the grown nanorods create optimism for high-quality optoelectronic devices on a ZnO nanorod basis.

Thus, based on the literature analysis performed, one can conclude that the *n*-ZnO/*p*-SiC system is promising for the next generation of optoelectronics nanodevices and requires comprehensive investigation. Particularly, it is necessary to obtain well-ordered massive of ZnO nanostructures, enabling further processing via conventional planar-based semiconductor techniques (lithography, etching, contacts deposition, etc.). Thus, the "drive force" for control of the location and microstructure of ZnO nanostructures on the SiC surface is necessary to be proposed. An extreme attention has to be paid to the interface quality, since the nano *n*-ZnO/*p*-SiC junctions characteristics are mainly determined by the interface quality. Thus, the abrupt and clear heterointerface is required, containing as low structural defects as possible, providing high-quality heteroepitaxy. The ZnO material stoichiometry as well as structural quality must be maximally perfect, in order to avoid the existence of the deep-level defects and thus the visible defect emission. This simultaneously increases the efficiency of the near-band-edge emission, namely, free excitonic emission at room temperature, which is so important for the optoelectronic devices such as LED and LD.

3.3 Heteroepitaxial ZnO Nanohexagons on *p*-Type SiC

3.3.1 Preparation of *p*-Type SiC Substrates of High Structural Quality via Sublimation Epitaxy

The high-quality heterointerface requires primarily good surface morphology of a substrate. Moreover, such a substrate must be a pattern, creating a "driving force" for the further nanostructures growth. We proposed to use the 4H-SiC layers epitaxially grown on commercially available 2 in. 4H:SiC wafers with standard as-polished surfaces off oriented 8° in the [11–20] direction. Earlier, it was reported, that the significant structural improvement can be obtained via such a technique [37]. The epitaxial growth rate was estimated to be 100 μm/h. The layer thicknesses ranged from 50 to 100 μm. The samples were cleaned accordingly to TL1 and TL2 procedures by solvents and acids prior to growth. The possible oxide on the surface was removed by dipping in hydroflouric acid. The growth approach is based on a polycrystalline SiC source and a SiC substrate separated with a graphite spacer. By applying a temperature gradient with a higher temperature at the source, Si- and C-bearing species are sublimed from the source and deposited on the cooler SiC substrate (Figure 3.21). At low growth pressures, the sublimation of the source is the growth rate determining step [81].

Silicon carbide epitaxy on off-axis substrate surfaces proceeds via step-flow growth [82] and typical morphological features like shallow pits, elongated grooves, and triangular-shaped defects may appear in the layers [83]. The latter two are especially important to avoid since they affect possible device performance [I90]. Disregarding micropipes and screw dislocations (determined by KOH etching) inherited from the substrate, feature-less epilayer surfaces were obtained, via selecting conditions for stable growth within a parameter window for morphological stability [37].

The model of structural improvement was earlier described by M. Syväjärvi et al. [37]. The model describing the structural quality improvement in step-flow growth is given in Figure 3.22. For example, the substrate surface is off oriented in the [11–20] direction

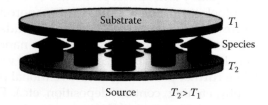

FIGURE 3.21
Schematic view of the growth configuration for sublimation epitaxy of SiC. (Reprinted with permission from Syväjärvi, M., Yakimova, R., Jacobsson, H., and Janzen, E., Structural improvement in sublimation epitaxy of 4H–SiC, *J. Appl. Phys.*, 88, 1407. Copyright 2000, American Institute of Physics.)

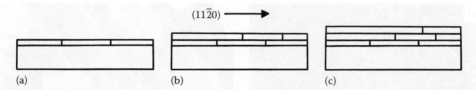

FIGURE 3.22
Illustration of the model for structural quality improvement via step-flow growth, by following respective stages from (a) through (c). (Reprinted with permission from Syväjärvi, M., Yakimova, R., Jacobsson, H., and Janzen, E., Structural improvement in sublimation epitaxy of 4H–SiC, *J. Appl. Phys.*, 88, 1407. Copyright 2000, American Institute of Physics.)

and the substrate contains several misoriented domains (Figure 3.22a) with three domains intersecting the surface. Step-flow growth takes place over the whole substrate. At the edge opposite to the [110] direction domains increase their size in the growth direction by the step-flow growth mechanism. As growth proceeds, these domains partly or completely overtake domains in the down-step direction (Figure 3.22b). In a similar manner, domains in the center of the substrate overtake domains in area to the right (Figure 3.22c). With increasing layer thickness, the domain enlargement at area from the left dominates and finally (in the ideal situation) completely overtakes domains in the down-step direction. In this case the epilayer surface consists of one domain.

From scanning electron microscopy and atomic force microscopy measurements, it was demonstrated that the macrosteps were resolved to be a typical feature on the as-grown epilayer surfaces. The step widths vary from 10 to 70 nm and the step heights are in the range of 1–4 nm. We suggested that the observed homoepitaxial growth mechanism can be applied for heteroepitaxial growth of ZnO on SiC. Thus, the described epilayered SiC substrates were subjected to ZnO growth.

3.3.2 Growth of ZnO Nanohexagons on 4H:SiC

The ZnO nanostructures were grown by atmospheric pressure metalorganic chemical vapor deposition (APMOCVD) using Zn acetylacetonate and oxygen as zinc and oxygen precursors, respectively [84–86]. The substrate temperature was kept at 500°C, the flow rates of Ar, as a buffer gas, and oxygen were 50 and 25 sccm, respectively [84]. More detailed description of the ZnO APMOCVD growth procedure can be found elsewhere [85,86]. The commercial 4H-SiC(0001) substrates were miscut by 8° off the c-axis to [11$\bar{2}$0] and then a p-SiC layer was grown by sublimation epitaxy, as described earlier. The layer thickness and the net acceptor concentrations were ~30 μm and ~5 × 10^{16} cm^{-3}, respectively.

The grown structures were initially characterized in terms of their structural properties by XRD via θ-2θ scans and pole figures using a Siemens D5000 diffractometer, utilizing Cu-Kα radiation ($\lambda = 0.1542$ nm). Scanning electron microscopy (SEM) was used to characterize the microstructure of the prepared sample. Figure 3.23a and b presents SEM images taken from an as-grown sample.

The sample exhibits surface steps along SiC[11$\bar{2}$0] and nanohexagons. Due to anisotropy of the growth rate in wurtzite semiconductors and minimization of the surface energy, vicinal surfaces of SiC show self-ordering phenomena, resulting in step bunching. In our 8° off 4H-SiC the bunched step configurations with an integer number (1, 2, 3, …) of unit cell height are energetically stabilized where the number depends on the growth

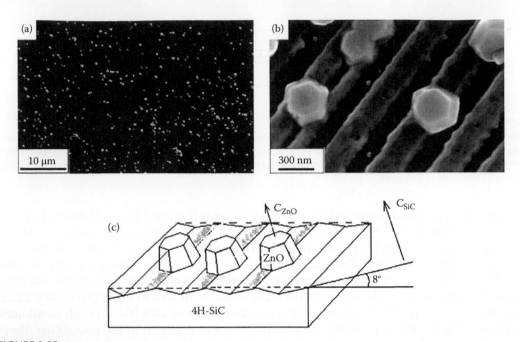

FIGURE 3.23

SEM images of ZnO HEX nanostructures on a stepped SiC surface (a) an overview; (b) a closer view—the light regions are ZnO, the dark area between the stripes reflects the steps; and (c) schematic representation of ZnO growth on 4H-SiC vicinal surface. The shadowed areas indicate ZnO film on the SiC terraces, corresponding to the bright areas in (b). (Reprinted from *J. Cryst. Growth*, 312, Khranovskyy, V., Tsiaoussis, I., Yazdi, G.R., Hultman, L., and Yakimova, R., Heteroepitaxial ZnO nanohexagons on *p*-type SiC, 327. Copyright 2010, with permission from Elsevier.)

conditions [I38]. Here, the 4H-SiC utilized has 20 nm bunched step height. The nanohexagons of alleged ZnO composition in Figure 3.23a and b have an *a*-plane-faceted hexagonal shape ($1\bar{2}00$) facets with a top *c*-plane facet reflecting the crystal symmetry of ZnO and suggesting a single-crystal nature. These hexagons have a characteristic azimuthal diameter of 200–350 nm and height of ~200 nm. Figure 3.23c is a sketch illustrating the orientation relationship between the ZnO nanohexagons and the SiC substrate. The nanohexagons are aligned along the ZnO *c*-axis with an 8° tilt corresponding to the substrate miscut. It should be mentioned that during the XRD measurements it was possible to identify the signal from ZnO HEX only after subtraction of 8° from the angle of the incident x-ray beam, assuming 8° misorientation from the *c*-axis toward [$11\bar{2}0$] direction. This is actually the step-flow direction during the growth of the SiC epitaxial layer. This shows that the ZnO nanohexagons were grown not perpendicularly to the substrate plane, but to the (0001) plane of SiC, i.e., epitaxially onto the SiC substrate.

The detailed microstructure study of the ZnO nanorods was carried out using conventional and high-resolution TEM (HRTEM). For the cross-section TEM (XTEM) specimen preparation, two strips of the specimen were cut and glued face to face, then were mechanically thinned down to 25 μm. Subsequently the specimens were thinned to electron transparency by Ar ion milling with energy of 4 kV, at a low incident angle of 4° in order to avoid amorphization artifacts from the argon ions. For the conventional characterization a TEM JEM 120 CX was used, while for the HRTEM investigation a JEM 2011 having 0.194 nm point-to-point resolution was utilized. Figure 3.24 shows cross-sectional TEM micrographs from the ZnO/SiC heterointerface taken along the 4H-SiC[$11\bar{2}0$] zone axis (along the step

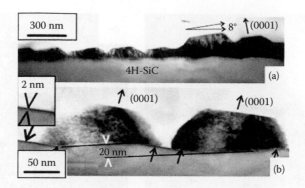

FIGURE 3.24
XTEM images of the ZnO/SiC heterostructures (a) and the enlarged images of ZnO hexagons (b). Inset represents the thin ZnO layer between the hexagons. (Reprinted from *J. Cryst. Growth*, 312, Khranovskyy, V., Tsiaoussis, I., Yazdi, G.R., Hultman, L., and Yakimova, R., Heteroepitaxial ZnO nanohexagons on *p*-type SiC, 327. Copyright 2010, with permission from Elsevier.)

edges as an overview and a higher magnification image). While a few crystallites touch each other, as seen in SEM (Figure 3.23a), the apparent proximity between the crystallites in Figure 3.24 is an artifact because of the projective nature of the TEM technique and sample thickness of about 200 nm.

The crystallites have a trapezoidal shape with flat planes on top forming an angle of 8° with respect of the miscut substrate. This confirms that the growth of ZnO follows the *c*-axis of 4H-SiC. The nanocrystallites nucleate at the step edge and expand laterally and vertically, ultimately overgrowing the step, over the adjacent terraces, as it will be later shown in detail. The XTEM image also indicates the presence of a few monolayer-thick ZnO film on top of the SiC terraces in between the ZnO hexagons. The SEM images from the ZnO-deposited samples also reveal ~20 nm wide longitudinal stripes along the direction expected for the terraces of the miscut SiC substrate, however, with a stronger contrast (bright in Figure 3.23a and b) than for the virgin substrate. A schematic representation is depicted in Figure 3.23c. The EDX and CL analyses presented imply that there is a Stranski–Krastanov (3D-island on 2D-layer) growth mode for the ZnO on SiC. Thus, the origin of the stripes is a decoration of SiC terraces by ZnO.

Figure 3.25 shows a plan-view TEM image from an as-deposited sample after ion beam thinning from the substrate side. The image confirms the presence of the thin ZnO layer that completely covers the SiC terraces as seen by the Moiré pattern.

In addition, Figure 3.25 reveals the presence of ZnO domains with a mean size of 40 nm, which are slightly misoriented, exhibiting a mosaic structure. The misorientation of the grains is evident from the distortions of the Moiré pattern, which are denoted by arrows in Figure 3.23a. In our case the actual in-plane misorientation of the nanograins in respect to the SiC matrix is 0.3°.

The elemental mapping was made by energy disperse x-ray (EDX) analysis in a Leo 1550 Gemini SEM (at operating voltage ranging from 10 to 20 kV and standard aperture value 30 μm). In Figure 3.26, the EDX images are presented along with the SEM image of the scanned area. The Zn and O signals are clearly correlated with the nanohexagons, but also with the stripes, however, of faint contrast consistent with the mosaic structure of the 2D layer of ZnO not much thicker than a few atomic layers (about 1 nm). The signal of Si was detected over the whole area of scanning which is due to a high penetration of the excitation electron beam, thus probing the SiC substrate.

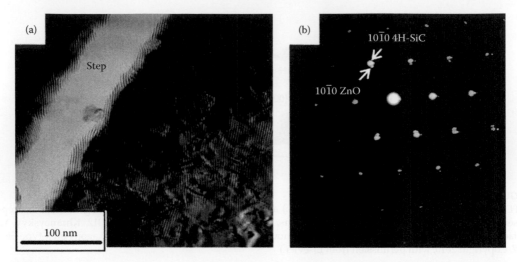

FIGURE 3.25
(a) Plan-view TEM and (b) corresponding selected area electron diffraction pattern from an area of the sample containing thin ZnO layer covering the SiC substrate (apparent from Moiré fringes). (Reprinted from *J. Cryst. Growth*, 312, Khranovskyy, V., Tsiaoussis, I., Yazdi, G.R., Hultman, L., and Yakimova, R., Heteroepitaxial ZnO nanohexagons on *p*-type SiC, 327. Copyright 2010, with permission from Elsevier.)

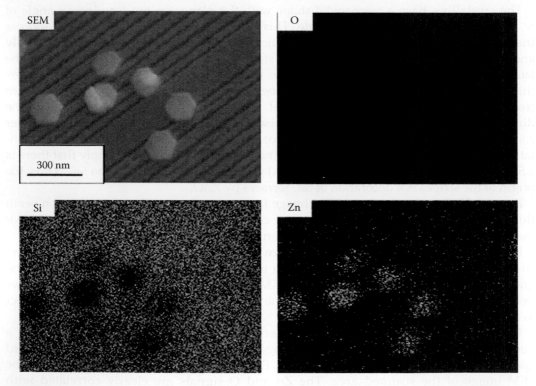

FIGURE 3.26
SEM image and corresponding EDX elemental maps of Si, Zn, and O from a sample with ZnO deposited on a SiC substrate. (Reprinted from *J. Cryst. Growth*, 312, Khranovskyy, V., Tsiaoussis, I., Yazdi, G.R., Hultman, L., and Yakimova, R., Heteroepitaxial ZnO nanohexagons on *p*-type SiC, 327. Copyright 2010, with permission from Elsevier.)

Nevertheless, the area where the ZnO HEX nanorods are located displayed a lower intensity of the Si signal.

The XRD analysis (θ-2θ pattern) of the ZnO/4H-SiC heterostructure exhibits only (0002) and (0004) diffraction peaks characteristic of ZnO, which evidences a good structural quality. In order to determine the epitaxial relationship of ZnO nanohexagons and SiC substrate, we performed pole figure XRD measurements (Φ scan). Experimental studies were carried out using a Philips X′Pert high-resolution x-ray diffraction (HRXRD) diffractometer operating with Cu Kα1 anode at a fixed voltage and current of 45 kV and 40 mA, respectively. The HRXRD was operated in triple axis mode which combines two crystal (Ge 220) four reflection monochromator and three reflection analyzer crystal. The orientation of the samples was aligned by using symmetric (0002) crystallographic planes. Pole figures are widely used to examine epitaxial relationship or in-plane orientation of ZnO films [87,88].

For the $\{11\bar{2}\lambda\}$ family of planes, sixfold peaks are expected, which reflects the hexagonal structure of the unit cell. Figure 3.27 shows the Φ scans of the ZnO hexagons $\{11\bar{2}2\}$ family of planes together with the Φ scan of the $\{11\bar{2}6\}$ family of planes of the SiC substrate. The obvious sixfold symmetry unambiguously demonstrates that the ZnO heterostructures containing the hexagons are grown epitaxially. The ZnO crystallites revealed the perfect epitaxial relation with the substrate having $[0001]_{ZnO}//[0001]_{4H\text{-}SiC}$ and $[10\bar{1}0]_{ZnO}//[10\bar{1}0]_{4H\text{-}SiC}$.

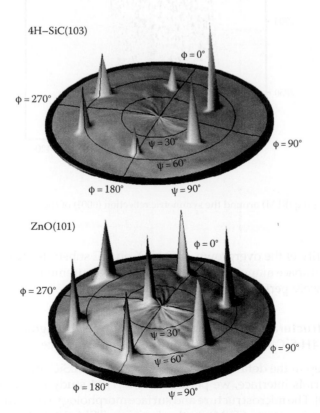

FIGURE 3.27
Pole figure of the ZnO nanohexagons grown on 4H:SiC(0001) substrate.

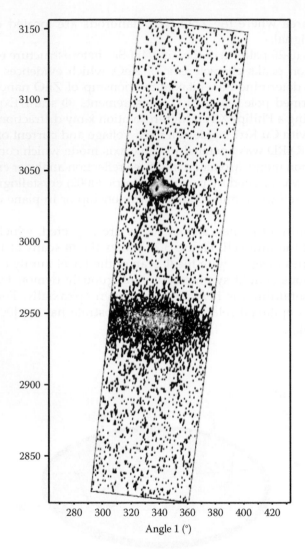

FIGURE 3.28
Reciprocal lattice mapping (RLM) around the symmetric reflection (0001) of the ZnO nanohexagons grown on 4H:SiC substrate.

The crystal quality of the overgrown ZnO HEX on SiC substrates was additionally evaluated by reciprocal space mapping (RLM). The RLM measurements around the symmetric (0002) reflections were performed, as shown in Figure 3.28.

3.3.3 Detailed Structural Characterization of ZnO Nanohexagons Grown on 4H:SiC with Emphasis on the Heterointerface

In order to get deep in the details of the ZnO growth on 4H:SiC and to emphasize specifically on the materials interface, we performed the TEM study of the prepared ZnO/SiC heterojunction [89]. The microstructure and surface morphology were investigated by transmission electron microscopy (TEM), both plan-view TEM (PVTEM) and cross-sectional TEM (XTEM) observations were performed. For the conventional TEM characterizations a JEOL JEM 120 CX was used, while the high-resolution TEM (HRTEM) analysis was performed using a JEOL JEM 3010 transmission electron microscope with point resolution of 0.14 nm. Scanning electron microscopy (SEM) was performed in a Leo 1550 Gemini SEM (at operating voltage ranging from 10 to 20 kV and standard aperture value 30 μm). Quantitative strain measurements were performed on cross-sectional HRTEM images by using Geometrical Phase Analysis (GPA) [90].

The deposited ZnO nanocrystallites are shown in the low-magnification XTEM micrograph in Figure 3.29. In Figure 3.29b, the individual crystallites are evident.

The crystallites have a trapezoidal shape with their flat planes on top forming an angle of about 8° with respect to the miscut substrate. This reveals that the growth of ZnO follows the c-axis of the miscut 4H-SiC. The nanocrystallites grew between adjacent steps of the substrate, as shown by arrows in Figure 3.29b. The free surface of the 4H-SiC substrate between the hexagonal nanocrystallites is covered by a 2 nm thin layer, as shown at higher magnification in the inset in the Figure 3.29b. The nature of this very thin layer was studied by PVTEM observations and will be discussed in the next paragraph. In some cases threading dislocation semi-loops were observed, as indicated by an arrow in

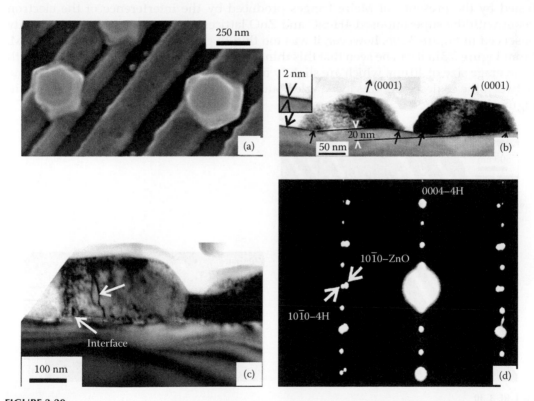

FIGURE 3.29

(a) SEM image of the ZnO hexagonal nanopillars deposited on vicinal 4H-SiC substrate. (b) XTEM micrograph of the deposited ZnO nanocrystallites on 4H-SiC substrate. The arrows indicate that the top plane of the ZnO hexagons is parallel to the SiC terraces. The space between the nanocrystallites is covered by a 2 nm thick film as shown in the inset. (c) A threading dislocation semi-loop is shown by an arrow. (d) SAD pattern taken from the interface of a ZnO nanocrystallite with the 4H-SiC substrate. (Reprinted with permission from Tsiaoussis, I., Khranovskyy, V., Dimitrakopulos, G.P., Stoemenos, J., Yakimova, R., and Pecz, B., Structural characterization of ZnO nanopillars grown by atmospheric-pressure metalorganic chemical vapor deposition on vicinal 4H-SiC and SiO2/Si substrates, *J. Appl. Phys.*, 109, 043507. Copyright 2011, American Institute of Physics.)

Figure 3.29c. The crystallites were in perfect epitaxial relation with the substrate having $[0001]_{ZnO}//[0001]_{4H\text{-}SiC}$ and $[10\bar{1}0]_{ZnO}//[10\bar{1}0]_{4H\text{-}SiC}$ as shown in the selected area diffraction (SAD) pattern of Figure 3.29d. Due to the high misfit between ZnO and 4H-SiC in the basal plane (nominal 5.6%), the $[10\bar{1}0]$ reflections of ZnO and 4H-SiC are well distinguished in Figure 3.29d. However the misfit along the c-axis between the (0002) and (0004) planes of ZnO and 4H-SiC, respectively, is only 3.5%, and in this case the corresponding diffraction spots overlap.

The hexagonal ZnO nanocrystallites were also studied by PVTEM observations along the [0001] zone axis. Well-developed hexagonal nanocrystallites with their $[11\bar{2}0]$ edge pointing toward the step-flow direction are evident. A SAD pattern obtained from the region of the nanocrystallite denoted by the letter A in Figure 3.30a is shown in Figure 3.30b.

The well-aligned diffraction spots from ZnO and 4H-SiC confirm the good epitaxial growth. The faint satellite spots around the main spots are due to double diffraction. The surface between the ZnO nanocrystallites is covered by a large number of small patches. The high magnification PVTEM micrograph in Figure 3.30a reveals that the space between the hexagonal nanocrystallites is covered by a thin ZnO film, as manifested by the presence of Moiré fringes produced by the interference of the electron beam with the superimposed 4H-SiC and ZnO lattices. A 2 nm thick film was already observed in Figure 3.29b; however, it was too thin to give diffraction pattern in XTEM. From Figure 3.31a it can be seen that this thin film consisted of individual domains with an average size of 40 nm, which are slightly misoriented, exhibiting a mosaic structure. Between the coalesced grains low-angle boundaries were formed denoted by arrows in Figure 3.31a.

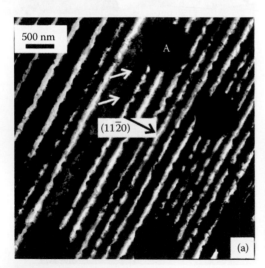

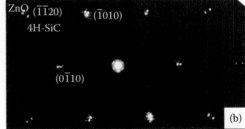

FIGURE 3.30
PVTEM micrographs of the nanocrystallites after thinning the specimen from the substrate side. (a) Hexagonal nanocrystals at low magnification. The $[11\bar{2}0]$ edge of the hexagons points along the step-flow direction. The free SiC surface between the nanocrystallites is covered by small patches denoted by arrows. (b) SAD pattern from the region of the nanocrystal denoted by the letter A in (a). (Reprinted with permission from Tsiaoussis, I., Khranovskyy, V., Dimitrakopulos, G.P., Stoemenos, J., Yakimova, R., and Pecz, B., Structural characterization of ZnO nanopillars grown by atmospheric-pressure metalorganic chemical vapor deposition on vicinal 4H-SiC and SiO2/Si substrates, *J. Appl. Phys.*, 109, 043507. Copyright 2011, American Institute of Physics.)

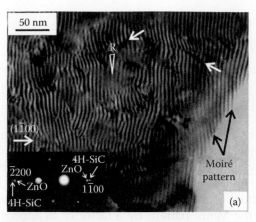
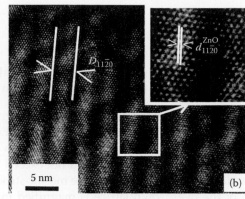

FIGURE 3.31

(a) PVTEM micrograph, taken under near two beam conditions, revealing that the space between the hexagonal nanocrystallites is covered by a very thin ZnO film manifested by the presence of Moiré fringes. Low-angle boundaries are indicated by arrows; the misorientation is evident by the distortion of the Moiré patterns. The corresponding diffraction pattern is shown in the inset. (b) Plan-view HRTEM micrograph taken with the electron beam parallel to the [0001] direction. A net of Moiré fringesis is evident. In the inset ZnO lattice planes are visible at higher magnification. (Reprinted with permission from Tsiaoussis, I., Khranovskyy, V., Dimitrakopulos, G.P., Stoemenos, J., Yakimova, R., and Pecz, B., Structural characterization of ZnO nanopillars grown by atmospheric-pressure metalorganic chemical vapor deposition on vicinal 4H-SiC and SiO2/Si substrates, *J. Appl. Phys.*, 109, 043507. Copyright 2011, American Institute of Physics.)

The grains are slightly misoriented, as is evident from the diffuse ZnO diffraction spots compared to the sharp 4H-SiC ones, as shown in the inset in Figure 3.31a, also from the distortion of the Moiré fringes. The observed Moiré fringes in Figure 3.31a are along the [1$\bar{1}$00] direction and were formed by the interference of the very strong [1$\bar{1}$00] ZnO and SiC reflections, as shown in the inset of Figure 3.31a. Assuming no rotation between the lattices, i.e., Moiré pattern of the translational type [91] the nominal fringe periodicity $D_{1\bar{1}00}$ is given by

$$D_{1\bar{1}00} = \frac{d_{ZnO}d_{SiC}}{d_{ZnO} - d_{SiC}} = \frac{d_{ZnO}}{\delta} \tag{3.2}$$

where

$d_{SiC} = 0.266\,\text{nm}$ and $d_{ZnO} = 0.281\,\text{nm}$ are the (1$\bar{1}$00) interplanar spacings for 4H-SiC and ZnO, respectively

δ is the nominal misfit expressed with respect to the substrate

Therefore, Equation 3.2 gives a nominal periodicity $D_{1\bar{1}00} = 4.99\,\text{nm}$ for an unstrained epilayer lattice. The actual periodicity of the Moiré pattern measured in Figure 3.31a is slightly larger than the nominal value for reasons we will discuss in the next paragraph. It is worth noticing that the Moiré patterns are very sensitive, so that any local strain variation with respect to the matrix lattice results in significant distortion of the Moiré pattern. Hence, the observed distortion of the Moiré patterns is attributed to local changes in the epilayer lattice. According to Equation 3.2 any lattice distortion is magnified in the Moiré pattern by a factor $1/\delta$. Also any rotation of the lattice of the epilayer with respect to the substrate by an angle Q is magnified by the factor $1/\delta$ according to the equation [91]:

$$R = \frac{Q}{\delta} \qquad\qquad (3.3)$$

where
  Q is the actual misorientation of the crystal lattice
  R is the observed rotational misorientation of the Moiré pattern

For example, in Figure 3.31a the Moiré fringes between two adjacent domains are rotated by an angle $R = 13°$. Therefore, according to Equation 3.3, the real misorientation will be Q ~0.7° only, assuming the nominal misfit. In general the calculated misorientation of the ZnO domains in Figure 3.31a did not exceed ±1°. This explains the formation of low-angle boundaries and consequently the mosaic structure of the film. Thus, in our case any distortion of the overgrown lattice is magnified in the Moiré pattern by a factor of 18 approximately. Therefore the Moiré pattern provides a simple direct method for the estimation of the residual elastic strain of the epitaxial film by comparing the observed with the nominal periodicity of Moiré fringes. For this purpose the periodicity of the Moiré was measured in the high-resolution PVTEM micrograph shown in Figure 3.31b, taken with the electron beam parallel to the [0001] direction. A net of Moiré fringes is evident, the mean value $D_{11\bar{2}0}$ of their periodicity along the three $\langle 11\bar{2}0 \rangle$ equivalent directions was deduced by measuring 10 fringes in each direction and was found to be $D_{11\bar{2}0} = 3.692\,\text{nm}$. Considering the SiC substrate undistorted having the nominal spacing $D_{11\bar{2}0} = 0.15407\,\text{nm}$, we have calculated from Equation 3.2 the corresponding $d_{11\bar{2}0}^{ZnO}$ spacing for ZnO, which was 0.1608 nm; this value is significantly smaller than the nominal value 1.6267 nm, revealing that the ZnO film is under compressive strain. Due to this difference the residual misfit is 1.16%. The remaining 4.44% from the total 5.6% misfit is accommodated in a plastic manner by misfit dislocations at the ZnO/SiC interface and dislocations at the low-angle boundaries of the mosaic grains. In the inset of Figure 3.31b the ZnO lattice in the section (0001) is shown at high magnification.

The three-dimensional (3D) island growth on a thin two-dimensional (2D) highly defected crystalline layer is a typical example of the Stranski–Krastanov mode of growth. Namely, after forming a few monolayers, subsequent layer growth is unfavorable and islands are formed on top of this "intermediate" layer [92]. The residual strain in the ZnO nanopillars was estimated from cross-sectional HRTEM observations.

Figure 3.32 illustrates a cross-sectional HRTEM image obtained from the interfacial region of a ZnO nanocrystal on SiC. It is seen that the interface comprises a bunched SiC step of height 9–13 nm. The uncertainty in the measurement of the step height results from the fact that the interface is amorphized at the ZnO side, as indicated at the right-hand side in Figure 3.32. This is an artifact frequently observed in XTEM specimens as a consequence of the ion milling process, due to the significant difference in hardness between substrate and epilayer. Another contributing factor toward amorphization is the strain that is developed at the interface due to the lattice misfit. The same effect has been observed during XTEM specimen preparation of 3C-SiC films epitaxially grown on Si. In this case the amorphous layer was formed at the interface in the Si side. The ZnO nanocrystal exhibits defects emanating from the interface. In particular, a closed domain comprising a local change of the defocus condition is well visible, denoted by "dom" in Figure 3.32, which could be possibly attributed to polarity inversion. An emanating threading dislocation is also visible by faint contrast but it was made better discernible by the strain analysis as shown in Figure 3.32. No detailed analysis of emanating defects was undertaken in this work as this was out of the scope of the present contribution.

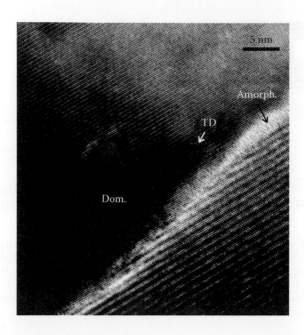

FIGURE 3.32
Cross-sectional HRTEM image of the ZnO/SiC interfacial region showing a bunched SiC step. On the right-hand side the interfacial structure is partially amorphized. A closed domain and an emanating threading dislocation in ZnO are indicated. (Reprinted with permission from Tsiaoussis, I., Khranovskyy, V., Dimitrakopulos, G.P., Stoemenos, J., Yakimova, R., and Pecz, B., Structural characterization of ZnO nanopillars grown by atmospheric-pressure metalorganic chemical vapor deposition on vicinal 4H-SiC and SiO2/Si substrates, *J. Appl. Phys.*, 109, 043507. Copyright 2011, American Institute of Physics.)

Figure 3.33a is a Fourier-filtered image from the interface obtained from part of Figure 3.32, which was taken with the electron beam parallel to the [11$\bar{2}$0] direction. The Bragg image obtained from the interface by using the 1$\bar{1}$00 spatial frequencies of ZnO and 4H-SiC is shown in Figure 3.33b, whereby extra half-planes planes in 4H-SiC are indicated. In this case 29 SiC lattice planes correspond to 28 ZnO planes. Considering the SiC substrate undistorted and taking the spacing $d_{1\bar{1}00}^{SiC} = 0.2669$ nm we find the ZnO spacing $d_{1\bar{1}00}^{ZnO}$ equal to 0.2764 nm, significantly smaller of the nominal spacing 0.281776 nm revealing a residual compressive strain of 1.8% the remaining 3.8% misfit being absorbed by misfit dislocations.

However, these measurements are not very accurate due to the large spacing between the misfit dislocations, large fluctuations in their periodicity, and interfacial amorphization, as shown in Figure 3.32. For this reason a more accurate quantitative analysis was performed by using the GPA method.

The quantitative HRTEM (qHRTEM) analysis was performed by using the GPA method. Three distinct areas of the ZnO/SiC interfacial region were analyzed, starting from the edge of the nanocrystal and going toward the center. For each HRTEM image, 2D lattice strain maps were generated. For such mapping the substrate was assumed to be relaxed, exhibiting the nominal lattice constants $a_{SiC} = 0.3081$ nm and $c_{SiC} = 1.0053$ nm. The lattice strain maps were obtained by selecting the g 1$\bar{1}$00$_{ZnO}$/1$\bar{1}$00$_{SiC}$ and g 0004$_{SiC}$/0002$_{ZnO}$ spatial frequencies in the fast Fourier transform (FFT) of each HRTEM image, using Gaussian masks. Profiles of the variation of the lattice strain along [0001] were then extracted for the two normal strain components ε_{xx}^{lt} and ε_{yy}^{lt}, where the x-direction is [1$\bar{1}$00] and the y-direction is [0001] [i.e., $\varepsilon_{xx}^{lt} = \left(d_{1\bar{1}00}^{ZnO} - d_{1\bar{1}00}^{SiC}\right)/d_{1\bar{1}00}^{SiC}$ and $\varepsilon_{yy}^{lt} = \left(d_{0002}^{ZnO} - d_{0004}^{SiC}\right)/d_{0004}^{SiC}$]. The strain

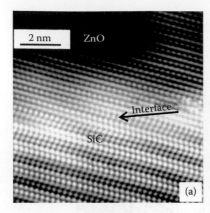
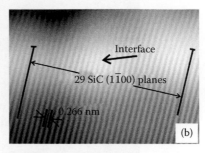

FIGURE 3.33
Cross-sectional HRTEM image along the [11–20] zone axis. (a) Fourier-filtered image from the interface. (b) Bragg-filtered image obtained by using only the 10–10 spatial frequencies. Misfit dislocations are evident at the interface. (Reprinted with permission from Tsiaoussis, I., Khranovskyy, V., Dimitrakopulos, G.P., Stoemenos, J., Yakimova, R., and Pecz, B., Structural characterization of ZnO nanopillars grown by atmospheric-pressure metalorganic chemical vapor deposition on vicinal 4H-SiC and SiO2/Si substrates, *J. Appl. Phys.*, 109, 043507. Copyright 2011, American Institute of Physics.)

profiles extended for more than 20 nm in height into ZnO. The integration width for each profile along the [1$\bar{1}$00] direction, over which the strain was averaged, ranged between 5 and 20 nm. In the following, strain measurements are discussed by illustrating maps obtained from one image, but average values from all three measured specimen areas are reported.

Figure 3.34a illustrates the map of the ε_{xx}^{lt} strain component obtained using a $g/5$ mask size (which corresponds to a spatial resolution of 1.3 nm). In order to enhance clarity, the map is shown superimposed on the HRTEM image. The map shows clearly that there is plastic relaxation of the strain by localization at the interface in the form of misfit dislocations (indicated by arrows in Figure 3.34a). Figure 3.34b illustrates the ε_{xx}^{lt} strain profile along [0001] obtained from the selection box indicated in Figure 3.34a. It is evident from the profile in Figure 3.34b that ε_{xx}^{lt} is less than the nominal value of the misfit $\delta = 5.60\%$. The lattice strain of ZnO along the growth direction was averaged, after discarding the measurements corresponding to the first 5 nm above the interface, where the profile was distorted due to defect emanation and interfacial amorphization. In a similar manner, strain measurements were collected from profiles in the rest two regions, and a total average from over 3000 measurements was extracted. This yielded $\varepsilon_{xx}^{lt} = 4.10\% \pm 0.30\%$, i.e., the lattice strain is 1.50% less than nominal. This corresponds to a residual compressive elastic strain in the basal plane of ZnO equal to $\varepsilon^{el} = -1.42\%$. In Figure 3.34b it is seen that as the thickness of the film increases, the lattice strain ε_{yy}^{lt} slightly increases having a tendency toward the nominal value, and hence the residual strain ε^{el} decreases somewhat. Even so the lattice strain is still much less than the nominal misfit.

Regarding the c lattice parameter, the average lattice strain, ε_{yy}^{lt}, obtained from three images using a $g/5$ mask size around the $0004_{SiC}/0002_{ZnO}$, was $\varepsilon_{yy}^{lt} = 3.61\% \pm 0.44\%$, which is almost the nominal value of 3.67% between $c_{SiC}/2$ and c_{ZnO} (the nominal c lattice constant of ZnO is $c_{ZnO} = 0.52151$ nm). It is interesting to compare the measured value of ε_{yy}^{lt} with the predicted one if we assume a biaxial strain state introduced by the residual elastic strain

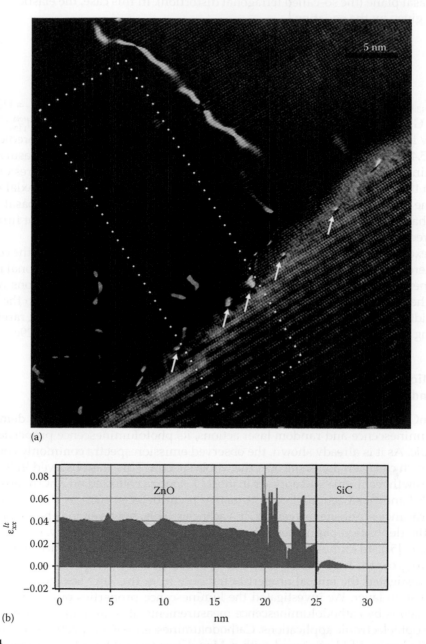

(a)

(b)

FIGURE 3.34
(a) Map of the ε_{xx}^{lt} strain component, obtained using a g/5 mask size which corresponds to a spatial resolution of 1.3 nm. The map is shown superimposed on the HRTEM image. Arrows indicate localized interfacial strain fields corresponding to misfit dislocations. (b) Profile of ε_{xx}^{lt} along [0001] obtained using the selection box indicated in (a). The width of the box along [0$\bar{1}$00] is the employed integration width. (Reprinted with permission from Tsiaoussis, I., Khranovskyy, V., Dimitrakopulos, G.P., Stoemenos, J., Yakimova, R., and Pecz, B., Structural characterization of ZnO nanopillars grown by atmospheric-pressure metalorganic chemical vapor deposition on vicinal 4H-SiC and SiO2/Si substrates, *J. Appl. Phys.*, 109, 043507. Copyright 2011, American Institute of Physics.)

on the basal plane (the so-called tetragonal distortion). In this case, the elastic strain along [0001] is given by the equation

$$\varepsilon_{yy}^{el} = -2 \frac{C_{13}}{C_{33}} \varepsilon_{xx}^{el} \tag{3.4}$$

where C_{13} and C_{33} are elastic constants of ZnO. The employed values were $C_{13} = 118\,\text{GPa}$ and $C_{33} = 211\,\text{GPa}$ [93]. The elastic strain is defined as $\varepsilon_{xx}^{el} = \left(d_{1\bar{1}00}^{ZnO} - d_{1\bar{1}00}^{relaxed\_ZnO} \right) / d_{1\bar{1}00}^{relaxed\_ZnO}$, and similarly for ε_{yy}^{el}. Equation 3.4 yielded $\varepsilon_{yy}^{el} = 1.59\%$ from which we obtain a predicted value of $\varepsilon_{yy}^{lt} = 5.40\%$. This is clearly much larger than both the experimentally measured ε_{yy}^{lt} and the nominal mismatch along the *c*-axis. Hence we conclude that despite the residual elastic strain in the basal plane, the strain state of the ZnO nanocrystals is not biaxial but closer to a plane strain condition. The strain along [0001] is probably relaxed by basal extra half planes that were observed to be inserted at the bunched substrate steps, but further work is required to clarify the exact mechanism.

The hexagonal nanocrystallites are two orders of magnitude thicker than the continuous film. Therefore, as the residual strain reduces with thickness, in the hexagonal nanocrystallites new misfit dislocations will be created. The new misfit dislocations are generated at the free surface and subsequently extend to the interface. Thanks to the relatively small width of the nanocrystallites the generated dislocations extend easily, rarely leaving threading segments in the interior of the crystallites, as shown in Figure 3.29c.

3.3.4 Micro-Photoluminescence Investigations of ZnO Seeding Layer and Hexagonal Nanopillars Grown on *p*-Type SiC Substrates

Since ZnO nanostructures heterojunctions on *p*-type SiC and GaN have demonstrated electroluminescence and random laser actions, its photoluminescence properties became a hot topic. As it is already shown, the observed emission spectra commonly contain DLE band, which can even dominate over the whole spectra. This was observed independently on the growth technique and appears in most PL spectra collected on ZnO nanostructures. The DLE band has been attributed to native point defects, such as the Zn interstitials, oxygen vacancies, antisite oxygen, or Zn vacancies. This, however, may be surmounted by tuning the deposition parameters, maintaining the ZnO stoichiometry, or via respective passivation [51]. To eliminate definitively the green/yellow luminescence, however, a key point is to grow the high-quality ZnO nanostructures.

We investigated the optical properties not only from the ZnO hexagons but also from the nucleation layer. We investigated the luminescence properties of the fabricated ZnO nanohexagons by cathodoluminescence measurements at room temperature in terms of possible optoelectronic applications. Cathodoluminescence (CL) spectra were taken in the Leo 1550 Gemini SEM equipped with a MonoCL system (Oxford Instruments) using a 10 keV electron beam and 30 μm of aperture with 1800 lines mm^{-1} grating. Panchromatic images of ZnO/SiC samples were taken to trace the difference in emission intensity. The CL spectra of the sample demonstrated intense peak of ultraviolet emission at $\lambda = 376\,\text{nm}$ while the visible luminescence was negligible (Figure 3.35a).

The spectral line of the UV emission is very narrow—the full width at half maximum is as low as 12 nm. We assign the luminescence observed to the near-band-edge (NBE) excitonic emission [51]. Moreover, probing over the sample surface with different concentration

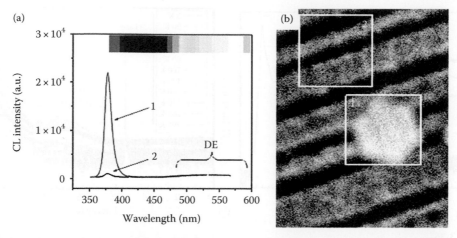

FIGURE 3.35
(a) RT CL spectra taken from ZnO grown on *p*-4H-SiC. The corresponding regions are marked on the panchromatic CL image (b). The spectral range, where a possible defect emission (DE) could be observed is indicated. (Reprinted from *J. Cryst. Growth*, 312, Khranovskyy, V., Tsiaoussis, I., Yazdi, G.R., Hultman, L., and Yakimova, R., Heteroepitaxial ZnO nanohexagons on *p*-type SiC, 327. Copyright 2010, with permission from Elsevier.)

of ZnO nanohexagons displayed different signal intensity, but the characteristic features of the spectra did not change. In order to differentiate the contribution of the emission from nanohexagons and the ZnO stripes, we recorded their emission spectra separately. In Figure 3.35 CL spectra along with the probed regions are shown. Figure 3.35b represents the panchromatic image of the sample. Light is emitted by the whole area covered by ZnO, i.e., CL signal is also observed from the stripes around hexagons, proving their emitting ability and ZnO nature. However, it is evident that the emission intensity from ZnO stripes on top of the SiC terraces is significantly lower than that from ZnO hexagons.

The spectra from both areas display the only peak of NBE emission. The absence of visible emission which is related to point defects in the material suggests a good stoichiometry of both types of ZnO. At the same time, the difference in the emission intensities may be explained by a difference in the concentrations of extended structural defects. It has been reported that structural defects as dislocations are mainly responsible for quenching of the luminescence intensity. The epitaxial interface followed by high structural quality is a prerequisite of obtaining the high light emission efficiency. Since the ZnO nanorods are relaxed heteroepitaxial structures, misfit dislocations are expected, although they are invisible in the TEM images. However, we assumed that they do not affect the CL properties since they are confined at the interface.

We have additionally conducted the temperature-dependent photoluminescence study. The low temperature photoluminescence (LTPL) measurements confirm the absence of the DLE emission in our sample. However, in order to separate the contributions from HNPs and the seeding layer, μ-LTPL mapping and μ-TRPL measurements were performed. This gives a detailed insight into the role of the seeding layer in the recombination process, which has not been established before. Figure 3.36 shows a LTPL spectrum collected at 5 K on a HNP-rich part of the sample. The dominant peak at 3.364 eV originates from D^0X emission with LO phonon replicas on the low energy side. Because the penetration depth of the laser is larger than the thickness (~200 nm) of the ZnO, some admixture with the well-known N-Al donor-acceptor pair (DAP) spectrum of *p*-type 4H-SiC was also observed.

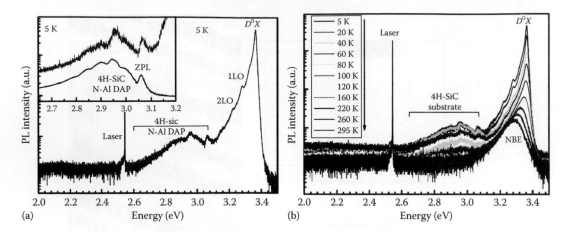

FIGURE 3.36
(a) LTPL spectrum collected from a HNPs-rich area. In the inset, the N-Al DAP spectrum collected from the 4H-SiC reference sample is compared with the corresponding part of the spectrum collected from the ZnO sample. (b) Temperature dependence of the PL spectrum collected from the HNP-rich area in the temperature range 5–295 K.

For clarity, this is ascertained in the inset of Figure 3.36 by comparison with a typical N-Al DAP spectrum collected from a p-type 4H-SiC reference sample. We have to admit that no DLE band in the LTPL spectrum of the HNPs was observed. This was further confirmed by the temperature-dependent PL measurements (Figure 3.36b).

With temperature increase, both the neutral donor and the exciton bound to the neutral donor start to thermally ionize. Although it was not possible to clearly resolve the fine structure of the free exciton peak, it is clear that the D^0X line decreases and broadens gradually, dissociating into a new D^+X and a free exciton recombination line. At room temperature, this results in the NBE emission band at 3.30 eV which mainly originated from the free exciton emission. It was noticed that above 120 K the broad 4H-SiC signal also disappears and only the broad NBE line remains. It should be emphasized that at any temperature, the broad "green/yellow band," which is a common feature in ZnO, does not appear at all. This is clear in the temperature range 5–295 K, but the respective role of HNPs and the seeding layer is still not clear, although both suppose to contribute to the LTPL spectra.

In order to separate the contributions from the seeding layer and ZnO HNPs, micro-LTPL maps were collected at 10 K. The photoluminescence was excited by the fourth harmonic of a continuous wave YAG laser ($\lambda = 266$ nm) focused with a ×20 refractive objective. The spot diameter was ~2 µm and data were collected using 2 µm steps on a 60×60 µm^2 area. Typical spectra behavior is shown in Figure 3.37. Because of the size of the excitation spot (~2 µm) it was not possible to measure the one individual hexagon. However, moving from a (mainly) seeding layer position (upper left) to a HNP-rich one (bottom right) we found that the shape of the LTPL spectrum did not change significantly. No DLE band appears or disappears, only the intensity was found to be changed. As it was expected, the lower intensity is collected on the (mainly) seeding layer and is five times weaker than the one collected on the HNP-rich area. This is in a good agreement with the difference in thickness and the limited coverage by HNPs (even in the areas with more HNPs). The Figure 3.37 represents the LTPL spectra at different points on the surface of ZnO HNPs.

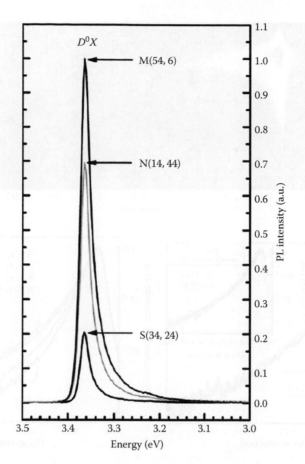

FIGURE 3.37

Micro-LTPL map collected at 10 K on a $60 \times 60\,\mu m^2$ area at three different places. The typical spectra collected from point M(54,6) represent the HNP-rich area, point S(34,24) the seeding layer, and intermediate point is N(14,44).

The TRPL measurements give further understanding of the optical properties, such as free carrier or exciton lifetime, which is an important parameter related to the light-emitting device performance. Earlier, ZnO nanostructures were reported to exhibit a significantly faster decay time compared to ZnO single crystals. For example, it has been reported that exciton lifetime is 65 ps at 4.3 K for ZnO quantum dots [94], compared to 322 ps measured for bulk ZnO [95]. However, in the literature, there is a large dispersion of results for ZnO nanostructures with different recombination times and both single- and biexponential decays have been reported for different nanostructures [96–98]. Reparaz et al. observed that the lifetime of donor bound exciton (D^0X) varies from 67 to 200 ps, depending on the size of the ZnO nanowires [96]. Kwok et al. reported that different morphologies of ZnO nanostructures exhibited very different decay times with biexponential decays, e.g., ZnO nanorods show very fast decays ($\tau_1 = 7\,ps$, $\tau_2 = 44\,ps$) while tetra-pods exhibit comparatively slow lifetimes ($\tau_1 = 91\,ps$, $\tau_2 = 708\,ps$) [97]. Zhao et al. also observed that ZnO nanorods exhibit a very fast (a few tens of picoseconds) biexponential decay [98].

We performed the micro-TRPL measurements in order to study the recombination mechanism of D^0X in more detail. The μ-TRPL was measured by a frequency-tripled

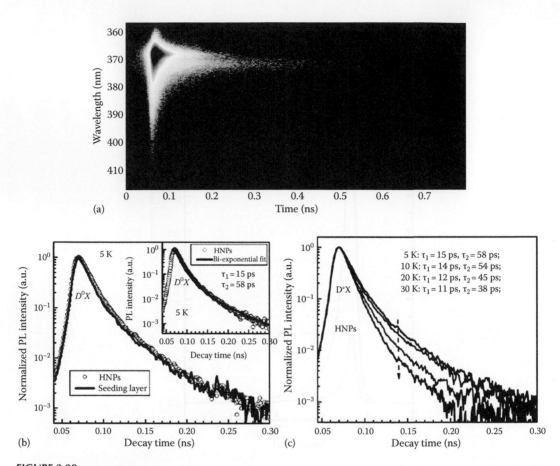

FIGURE 3.38

(a) Typical time dependence of a micro-TRPL spectrum collected at 5K on a HNP-rich area; (b) comparison of decay behaviors of the D^0X emission for a HNP-rich area in comparison to the seeding layer. The inset shows that the decay curve of the D^0X at 5K can be fitted by the same bi-exponential function with $\tau_1 \sim 15$ ps and $\tau_2 \sim 58$ ps; and (c) temperature dependence of the decay curves of the D^0X emission on the HNP-rich area.

sapphire:Ti laser emitting at 266 nm with a 200 fs pulse width and focused with a microscope objective to a diameter of $\sim$2 μm. A streak camera with $\sim$7 ps resolution was used for detection. Figure 3.38a presents the μ-TRPL spectra collected at 5 K on the HNP-rich area. Apart from a higher signal intensity from the HNP-rich area, it was found that the decay curve of D^0X collected on ZnO HNPs is identical to that collected on the seeding layer and can be fitted by a bi-exponential function with $\tau_1 = 15$ ps and $\tau_2 = 58$ ps (as seen in Figure 3.38b). The results are similar to the ones observed in ZnO nanorods by Zhao et al. [98]. Similarly, we attribute the fast component to surface recombination and the slow one to a more specific (bulk-like) process. The result demonstrates clearly that all the optical properties of HNPs including the absence of DLE band start from the ZnO/SiC interface. Generally a measured PL decay time (τ_{PL}) is determined not only by the radiative lifetime (τ_R) but also by the nonradiative lifetime (τ_{NR}), related by $\tau_{PL}^{-1} = \tau_R^{-1} + \tau_{NR}^{-1}$. If the recombination process is dominated by nonradiative recombinations, the lifetime should decrease with increasing temperature since the nonradiative channels play more important role with increasing temperature. This is exactly what we observed (Figure 3.38c), which shows the temperature-dependent TRPL decays collected from the HNP-rich area. It is apparent that the decay time of D^0X is decreasing with increasing temperature. This indicates that the decay time of D^0X is mainly determined by the nonradiative recombination.

An explanation for the observed emission properties can be explained by TEM observations. Accordingly to cross-sectional high-resolution TEM images of the ZnO/SiC interface, the ZnO crystalline planes grow epitaxially on the SiC lattice when crossing the heterointerface. This most probably is the reason for the absence of DLE band, especially in the seeding layer. However, from plan-view TEM, it was also found that the seeding layer exhibits mosaic structures with misoriented domains and low-angle boundaries while threading dislocations appear in the larger HNPs. These defects are generated due to biaxial strain. We believe that the dislocations and domain boundaries revealed by TEM are the same (and unique) source of nonradiative processes, which result in a shorter lifetime in ZnO HNPs and the seeding layer.

3.4 Conclusions

This chapter is focused on the features of epitaxial growth of ZnO on SiC as a route toward fabrication of hybrid light-emitting devices (LED). The features of 2D continuous ZnO layer deposition in comparison to 1D ZnO nanostructures growth on SiC (4H; 6H; 3C polytypes) are investigated. Possessing similar crystal structure, relatively small lattice mismatch (+5%), negligible thermal expansion coefficients mismatch (<1%), and most important accessible p-type conductivity, SiC is considered to be the promising alternative substrate for ZnO heteroepitaxy. It is demonstrated that ZnO grows vertically providing the perfect epitaxial relation with the SiC substrate ($[0001]_{ZnO}//[0001]_{4H\text{-}SiC}$ and $[10\bar{1}0]_{ZnO}//[10\bar{1}0]_{4H\text{-}SiC}$). Lattice and thermal expansion coefficient mismatches induce the strain in the grown ZnO, which was found to be dependent on the layer thickness. The strain relaxation of the overgrown material occurs via the misfit dislocation formation. It was found that the misfit dislocation energy formation in ZnO/SiC heterostructures is lower in comparison to ZnO/GaN. Particular attention is paid to the interface quality of the ZnO/SiC heterostructures and its effect on the structural, optical, and electrical properties of ZnO. It is shown that depending on the initial deposition conditions the oxidized or Zn-passivated ZnO/SiC interface can be obtained. Theoretically it was shown that the optimized growth condition for the growth of ZnO on 6H-SiC(0001) is Zn-polarity ZnO crystal grown under the stoichiometric growth condition. The room temperature luminescence observed from ZnO/SiC is of free excitonic type of nature and is due to band-to-band recombination. The strain induced in the ZnO/SiC interface influences the optical properties of ZnO: The blueshift of main photoluminescence peaks was observed ranged between 2.5 and 5.8 meV. It is shown that the commonly observed defect luminescence from deep levels is an important issue to be avoided for the successful realization of ZnO/SiC-based LED. Via time-resolved photoluminescence measurements the lifetime decay constant was found to be 450 ps, which was suggested to be the effective radiative lifetime of localized excitons. The band alignment of epitaxial ZnO/6H-SiC heterojunction was studied and a type-II heterojunction was found to be aligned in the staggered arrangement. Based on the conducted analysis, it was revealed that the light emission properties of the LED based on ZnO/SiC may be driven by the luminescence of p-SiC and affected by a high density of interface states. This may be due to the electron injection from n-ZnO dominating over the hole injection from p-SiC.

Thus, due to the smaller energy barrier for electron, electroluminescence (EL) may come from the defect-related states in p-SiC in the n-ZnO/p-SiC heterojunction LED. Therefore, the electron injection layer is a vital issue for efficient LED performance. The design of the nanoscaled p–n junction based on ZnO/SiC is discussed. It was experimentally proved that the relaxed 1D ZnO nanostructures on SiC possess improved light emission properties: the highly intense monochromatic emission was demonstrated in the ZnO nanohexagons, which is attributed to the single crystal structure, epitaxial relation, and high-quality heterointerface. However, based on the TRPL analysis, it was observed that the number of defects acting as nonradiative recombination centers is still high in the ZnO/SiC heterostructures and therefore significant efforts has to be applied to decrease their concentration. It is expected that nanosized p–n junction may relieve strain and thus accommodate the lattice mismatch at heterointerface, which can enhance the radiative recombination efficiency and thus the internal quantum efficiency of LED devices. Moreover, the light extraction may be improved. In addition, it is underlined that the inclined threading dislocations may cease in the nanopillar's side surface and decrease dislocation density, which can reduce the nonradiative recombination rate and thus improve the device performance significantly. It is believed that the presented data of ZnO/6H-SiC heterojunction will provide further understanding and will be useful in the design, modeling, and analysis of the consequent optoelectronic devices.

Acknowledgments

Authors are grateful to the personalities, who contributed to the presented results: to Prof. J. Stoemenos and Dr. I Tsiaoussis (Solid State Physics Section, Department of Physics, Aristotle University of Thessaloniki), Prof. B. Pecz (Research Institute for Technical Physics and Materials Science, Hungarian Academy of Sciences), Prof. L Hultman, Prof. P.-O. Holtz, Dr. J. W. Sun, Dr. G. R. Yazdi, Dr. A. Larsson (Department of Physics, Chemistry and Biology [IFM], Linköping University), Prof. J. Camassel (Groupe d'Etudes des Semiconducteurs, Université Montpellier), etc. The financial support from the Swedish Research Council and the VINNEX Centre FunMat is greatly acknowledged.

References

1. A. B. Djurisic, A. M. C. Ng, and X. Y. Chen, ZnO nanostructures for optoelectronics: Material properties and device applications, *Prog. Quant. Electron.* **34**, 191–259 (2010).
2. C. Klingshirn, J. Fallert, H. Zhou, J. Sartor, C. Thiele, F. Maier-Flaig, D. Schneider, and H. Kalt, 65 years of ZnO research—Old and very recent results, *Phys. Status Solidi B* **247**, 1–24 (2010).
3. M. Willander, O. Nur, Q. X. Zhao, L. L. Yang, M. Lorenz, B. Q. Cao et al., Zinc oxide nanorod based photonic devices: Recent progress in growth, light emitting diodes and lasers, *Nanotechnology* **20**, 332001 (2009).
4. Ü. Özgür, Y. I. Alivov, C. Liu, A. Teke, M. A. Reshchikov, S. Doğan, V. Avrutin, S.-J. Cho, and H. Morkoç, A comprehensive review of ZnO materials and devices, *J. Appl. Phys.* **98**, 041301 (2005).
5. D.-K. Hwang, M.-S. Oh, J.-H. Lim, and S.-J. Park, ZnO thin films and light-emitting diodes, *J. Phys. D Appl. Phys.* **40**, R387–R412 (2007).

6. S. Singh, P. Thiyagarajan, K. M. Kant, D. Anita, S. Thirupathiah, N. Rama, B. Tiwari, M. Kottaisamy, and M. S. Ramachandra Rao, Structure, microstructure and physical properties of ZnO based materials in various forms: Bulk, thin film and nano, *J. Phys. D Appl. Phys.* **40**, 6312–6327 (2007).

7. K. Vanheusden, C. H. Seager, W. L. Warren, D. R. Tallant, and J. A. Voigt, Correlation between photoluminescence and oxygen vacancies in ZnO phosphors, *Appl. Phys. Lett.* **68**, 403 (1996).

8. F. Leiter, H. Zhou, F. Henecker, A. Hofstaetter, D. M. Hoffman, and B. K. Meyer, Magnetic resonance experiments on the green emission in undoped ZnO crystals, *Physica B* **908**, 308 (2001).

9. D. C. Look, J. W. Hemsky, and J. R. Sizelove, Residual native shallow donor in ZnO, *Phys. Rev. Lett.* **82**, 2552 (1999).

10. C. J. Neufeld, C. Schaake, M. Grundmann, N. A. Fichtenbaum, S. Keller, and U. K. Mishra, InGaN/GaN nanopillar-array light emitting diodes, *Phys. Status Solidi C* **4**, 1605 (2007).

11. H. M. Kim, Y. H. Cho, H. Lee, S. I. Kim, S. R. Ryu, D. Y. Kim, T. W. Kang, and K. S. Chung, High-brightness light emitting diodes using dislocation-free indium gallium nitride/gallium nitride multiquantum-well nanorod arrays, *Nano Lett.* **4**, 1059 (2004).

12. J. H. Zhu, L. J. Wang, S. M. Zhang, H. Wang, D. G. Zhao, J. J. Zhu, Z. S. Liu, D. S. Jiang, and H. Yang, The fabrication of GaN-based nanopillar light-emitting diodes, *J. Appl. Phys.* **108**, 074302 (2010).

13. M. Joseph, H. Tabata, and T. Kawai, p-type electrical conduction in ZnO thin films by Ga and N codoping, *Jpn. J. Appl. Phys.* **38**, L1205 (1999).

14. K. Minegishi, Y. Koiwai, Y. Kikuchi, K. Yano, M. Kasuga, and A. Shimizu, Growth of p-type zinc oxide films by chemical vapor deposition, *Jpn. J. Appl. Phys.* **36**, L1453 (1997).

15. K. K. Kim, H.-S. Kim, D.-K. Hwang, J.-H. Lim, and S.-J. Park, Realization of p-type ZnO thin films via phosphorus doping and thermal activation of the dopant, *Appl. Phys. Lett.* **83**, 63 (2003).

16. D. C. Look, D. C. Reynolds, C. W. Litton, R. L. Jones, D. B. Eason, and G. Cantwell, Characterization of homoepitaxial p-type ZnO grown by molecular beam epitaxy, *Appl. Phys. Lett.* **81**, 1830 (2002).

17. Y. R. Ryu, T. S. Lee, J. H. Leem, and H. W. White, Fabrication of homostructural ZnO p–n junctions and ohmic contacts to arsenic-doped p-type ZnO, *Appl. Phys. Lett.* **83**, 4032 (2003).

18. J. Li, S.-H. Wei, S.-S. Li, and J.-B. Xia, Design of shallow acceptors in ZnO: First-principles band-structure calculations, *Phys. Rev. B* **74**, 081201 (2006).

19. C. H. Park, S. B. Zhang, and S.-H. Wei, Origin of p-type doping difficulty in ZnO: The impurity perspective, *Phys. Rev. B* **66**, 073202 (2002).

20. J. F. Rommeluere, L. Svob, F. Jomard, J. Mimila-Arroyo, G. Amiri, A. Lusson, V. Sallet, O. Gorochov, P. Galtier, and Y. Marfaing, Nitrogen acceptors in ZnO films grown by metalorganic vapor phase epitaxy, *Phys. Status Solidi C* **1**, 904 (2004).

21. E. Przezdziecka, E. Kaminska, K. P. Korona, E. Dynowska, W. Dobrowolski, R. Jakiela, L. Klopotowski, and J. Kossut, Photoluminescence study of p-type ZnO:Sb prepared by thermal oxidation of the Zn-Sb starting material, *Semicond. Sci. Technol.* **22**, 10 (2007).

22. B. Q. Cao, M. Lorenz, A. Rahm, H. von Wenckstern, C. Czekalla, J. Lenzner, G. Benndorf, and M. Grundmann, Phosphorus acceptor doped ZnO nanowires prepared by pulsed-laser deposition, *Nanotechnology* **18**, 455707 (2007).

23. D. C. Look, G. M. Renlund, R. H. Burgener II, and J. R. Sizelove, As-doped p-type ZnO produced by an evaporation/sputtering process, *Appl. Phys. Lett.* **85**, 5269 (2004).

24. C. H. Zang, D. X. Zhao, Y. Tang, Z. Guo, J. Y. Zhang, D. Z. Shen, and Y. C. Liu, Structural and photoluminescence properties of aligned Sb-doped ZnO nanocolumns synthesized by the hydrothermal method, *Chem. Phys. Lett.* **452**, 148 (2008).

25. J. C. Sun, H. W. Liang, J. Z. Zhao, J. M. Bian, Q. J. Feng, L. Z. Hu, H. Q. Zhang, X. P. Liang, Y. M. Luo, and G. T. Du, Ultraviolet electroluminescence from ZnO-based light-emitting diode with p-ZnO:N/n-GaN:Si heterojunction structure, *Chem. Phys. Lett.* **460**, 548 (2008).

26. Z. Z. Ye, J. G. Lu, Y. Z. Zhang, Y. J. Zeng, L. L. Chen, F. Zhuge et al., ZnO light-emitting diodes fabricated on Si substrates with homobuffer layers, *Appl. Phys. Lett.* **91**, 113503 (2007).

27. J.-H. Lim, C.-K. Kang, K.-K. Kim, I.-K. Park, D.-K. Hwang, and S.-J. Park, UV electroluminescence emission from ZnO light-emitting diodes grown by high-temperature radiofrequency sputtering, *Adv. Mater.* **18**, 2720 (2006).

28. A. Bakin, A. Behrends, A. Waag, H.-J. Lugauer, A. Laubsch, and K. Streubel, ZnO-GaN hybrid heterostructures as potential cost-efficient LED technology, *Proc. IEEE* **98**, 1281 (2010).

29. Z. Qui, K. S. Wong, M. Wu, W. Lin, and H. Xu, Microcavity lasing behavior of oriented hexagonal ZnO nanowhiskers grown by hydrothermal oxidation, *Appl. Phys. Lett.* **84**, 2739 (2004).

30. Q. Wan, Q. H. Li, Y. J. Chen, T. H. Wang, X. L. He, J. P. Li, and C. L. Lin, Fabrication and ethanol sensing characteristics of ZnO nanowire gas sensors, *Appl. Phys. Lett.* **84**, 3654 (2004).

31. Q. H. Li, Q. Wan, Y. X. Liang, and T. H. Wang, Oxygen sensing characteristics of individual ZnO nanowire transistors, *Appl. Phys. Lett.* **84**, 4556 (2004).

32. D. J. Rogers, F. H. Teherani, A. Yasan, K. Minder, P. Kung, and M. Razeghi, Electroluminescence at 375 nm from a ZnO/GaN:Mg/c-Al$_2$O$_3$ heterojunction light emitting diode, *Appl. Phys. Lett.* **88**, 141918 (2006).

33. D. Oh, T. Suzuki, J. J. Kim, H. Makino, T. Hanada, T. Yao, and H. J. Ko, Capacitance-voltage characteristics of ZnO/GaN heterostructures, *Appl. Phys. Lett.* **87**, 162104 (2005).

34. W. I. Park and G. C. Yi, Electroluminescence in n-ZnO nanorod arrays vertically grown on p-GaN, *Adv. Mater.* **16**, 87 (2004).

35. M. C. Jeong, B. Y. Oh, M. H. Ham, and J. M. Myoung, Electroluminescence from ZnO nanowires in n-ZnO film/ZnO nanowire array/p-GaN film heterojunction light-emitting diodes, *Appl. Phys. Lett.* **88**, 202105 (2006).

36. X.-M. Zhang, M.-Y. Lu, Y. Zhang, L.-J. Chen, and Z. L. Wang, Diode using a ZnO-nanowire array grown on p-GaN thin film, *Adv. Mater.* **21**, 1 (2009).

37. M. Syväjärvi, R. Yakimova, H. Jacobsson, and E. Janzen, Structural improvement in sublimation epitaxy of 4H–SiC, *J. Appl. Phys.* **88**, 1407 (2000).

38. Z. Feng, S. Chua, K. Tone, and J. Zhao, Recrystallization of carbon–aluminum ion coimplanted epitaxial silicon carbide—Evidenced by room temperature optical measurements, *Appl. Phys. Lett.* **75**, 472 (1999).

39. J. Fan, X. Wu, F. Kong, T. Qiu, and G. Huang, Luminescent silicon carbide nanocrystallites in 3C-SiC/polystyrene films, *Appl. Phys. Lett.* **86**, 171903 (2005).

40. A. Henry, M. Janson, and E. Janzen, Boron-related luminescence in SiC, *Phys. B* **141**, 340 (2003).

41. J. Zhu, B. Lin, X. Sun, R. Yao, C. Shi, and Z. Fu, Heteroepitaxy of ZnO film on Si (111) substrate using a 3C–SiC buffer layer, *Thin Solid Films* **478**, 218 (2005).

42. A. El-Shaer, A. Bakin, E. Schlenker, A. C. Mofor, G. Wagner, S. A. Reshanov, and A. Waag, Fabrication and characterization of n-ZnO on p-SiC heterojunction diodes on 4H:SiC substrates, *Superlattices Microstruct.* **42**, 387 (2007).

43. S. Jeon, D.-M. Kim, and K.-S. Hwang, Structural and optical properties of epitaxial ZnO thin films on 4H–SiC (0001) substrates prepared by pulsed laser deposition, *Appl. Surf. Sci.* **253**, 7016 (2007).

44. B. Ataev, Y. Alivov, E. Kalinina, V. Mamedov, G. Onushkin, S. Makhmudov, and A. Omaev, Heteroepitaxial ZnO/6H-SiC structures fabricated by chemical vapor deposition, *J. Cryst. Growth* **275**, 2471 (2005).

45. A. Ashrafi, Y. Segawa, K. Shin, J. Yoo, and T. Yao, Nucleation and growth modes of ZnO deposited on 6H–SiC substrates, *Appl. Surf. Sci.* **249**, 139 (2005).

46. Y. Alivov, D. Johnstone, Ü. Özgür, V. Avrutin, Q. Fan, S. Akarca-Biyikli, and H. Morkoç, Electrical and optical properties of n-ZnO/p-SiC heterojunctions, *Jpn. J. Appl. Phys.* **44**, 7281 (2005).

47. C. Yuen, S. Yu, S. Lau, and T. Chen, Fabrication of n-ZnO:Al/p-SiC:4H heterojunction light-emitting diodes by filtered cathodic vacuum arc technique, *Appl. Phys. Lett.* **86**, 241111 (2005).

48. Y. Shih, M. Wu, M. Chen, Y. Chen, J. Jang, and M. Shiojiri, ZnO based heterojunctions light-emitting diodes on p-SiC(4H) grown by atomic layer deposition, *Appl. Phys. B* **98**, 767 (2010).

49. Y.-S. Jeon, D.-M. Kim, and K.-S. Hwang, Epitaxially grown ZnO thin films on 6H-SiC(0001) substrates prepared by spin coating-pyrolysis, *Appl. Surf. Sci.* **253**, 7016 (2007).

50. J.-H. Kim, D.-H. Cho, W. Lee, B.-M. Moon, W. Bahng, S.-C. Kim, N.-K. Kim, and S.-M. Koo, Structural and optical properties of epitaxial ZnO thin films on 4H–SiC (0001) substrates prepared by pulsed laser deposition, *J. Alloys Compd.* **489**, 179 (2010).

51. V. Khranovskyy, G. R. Yazdi, G. Lashkarev, A. Ulyashin, and R. Yakimova, Investigation of ZnO as a perspective material for photonics, *Phys. Status Solidi A* **205**, 144 (2008).

52. N. Bano, I. Hussain, O. Nur, M. Willander, P. Klason, and A. Henry, Study of luminescent centers in ZnO nanorods catalytically grown on 4H-p-SiC, *Semicond. Sci. Technol.* **24**, 125015 (2009).

53. A. Ashrafi, B. Zhang, N. Binh, K. Wakatsuki, and Y. Segawa, High-quality ZnO layers grown on 6H-SiC substrates by metalorganic chemical vapor deposition, *Jpn. J. Appl. Phys.* **43**, 1114 (2004).

54. Z. D. Sha, Y. Yan, W. Qin, X. Wu, and L. Zhuge, Structure and optical properties of the SiC/ZnO five-layer multi-layer on Si (111) substrate with a SiC buffer layer, *J. Phys. D Appl. Phys.* **39**, 3240 (2006).

55. J.-Y. Lee, H.-S. Kim, W.-J. Lee, and K.-R. Ku, Growth and characteristics of annealed ZnO layer on 6H–SiC substrate, *J. Cryst. Growth* **312**, 2393 (2010).

56. E. Leong and S. Yu, UV Random lasing action in p-SiC(4H)/i-ZnO–SiO$_2$ nanocomposite/n-ZnO:Al heterojunction diodes, *Adv. Mater.* **18**, 1685 (2006).

57. W. Zhang, Q. Meng, and S. Zhong, Influence of growth conditions on photovoltaic effect of ZnO/Si heterojunction, *Sol. Energ. Mat. Sol. Cells* **92**, 949 (2008).

58. X. Wu, X. Chen, L. Sun, S. Mao, and Z. Fu, Photoelectric conversion characteristics of ZnO/SiC/Si heterojunctions, *J. Semicond.* **31**, 103002 (2010).

59. D.-Th. Phan and G.-S. Chung, Comparison of ZnO thin films grown on a polycrystalline 3C–SiC buffer layer by RF magnetron sputtering and a sol–gel method, *Appl. Surf. Sci.* **257**, 3285 (2011).

60. J. S. Lee, J. H. Kim, B. M. Moon, W. Bahng, S. C. Kim, N. K. Kim, and S. M. Koo, Epitaxial ZnO/4H:SiC heterojunctions diodes, *Proc of the IEEE* 978-1-4244-3544-9/10 (2010).

61. A. Ashrafi and Y. Segawa, Anomalous lattice relaxation mechanics in ZnO/SiC heterostructures, *J. Appl. Phys.* **103**, 093527 (2008).

62. S.-H. Lim, D. Shindo, H.-B. Kang, and K. Nakamura, Defect structure of epitaxial ZnO films on (0001) sapphire studied by transmission electron microscopy, *J. Vac. Sci. Technol. B* **19**, 506 (2001).

63. R. F. Service, Will UV lasers beat the blues? *Science* **276**, 895 (1997).

64. A. B. M. A. Ashrafi, Y. Segawa, K. Shin, and T. Yao, Strain effects in ZnO layers deposited on 6H-SiC, *J. Appl. Phys.* **100**, 063523 (2006).

65. A. Ashrafi, N. Binh, B. Zhang, and Y. Segawa, Strain relaxation and its effect in exciton resonance energies of epitaxial ZnO layers grown on 6H:SiC substrates, *Appl. Phys. Lett.* **84**, 2814 (2004).

66. D. Reynolds, J. Hoelscher, C. Litton, and T. Collins, Strain splitting of the Γ5 and Γ6 free excitons in GaN, *Appl. Phys. Lett.* **81**, 3792 (2002).

67. A. Ashrafi, Y. Segawa, K. Shin, and T. Yao, Nucleation and interface chemistry of ZnO deposited on 6H-SiC, *Phys. Rev. B* **72**, 155302 (2005).

68. K. Fujiwara, A. Ishii, T. Abe, and K. Ando, Epitaxial growth of ZnO crystal on the Si-terminated 6H-SiC(0001) surface using the first-principles calculation, *e-J. Surf. Sci. Nanotechnol.* **4**, 254–257 (2006).

69. Y. Alivov, J. Nostrand, D. Look, M. Chukichev, and B. Ataev, Observation of 430 nm electroluminescence from ZnO/GaN heterojunction light-emitting diodes, *Appl. Phys. Lett.* **83**, 2943 (2003).

70. A. Ashrafi, Band offsets at ZnO/SiC heterojunctions: Heterointerface in band alignment, *Surf. Sci.* **604**, L63 (2010).

71. Y. Alivov, B. Xiao, Q. Fan, H. Morkoç, and D. Johnstone, Band offset measurements of ZnO/6H-SiC heterostructure system, *Appl. Phys. Lett.* **89**, 152115 (2006).

72. H. Fan, G. Sun, S. Yang, P. Zhang, R. Zhang, H. Wei et al., Valence band offset of ZnO/4H-SiC heterojunction measured by x-ray photoelectron spectroscopy, *Appl. Phys. Lett.* **92**, 192107 (2008).

73. X. Li, B. Zhang, X. Dong, Y. Zhang, X. Xia, W. Zhao, and G. Du, Room temperature electroluminescence from ZnO/Si heterojunction devices grown by metal–organic chemical vapor deposition, *J. Lumin.* **129**, 86 (2009).

74. P. Zhang, X. Liu, R. Zhang, H. Fan, A. Yang, H. Wei, P. Jin, S. Yang, Q. Zhu, and Z. Wang, Valence band offset of ZnO/GaAs heterojunction measured by x-ray photoelectron spectroscopy, *Appl. Phys. Lett.* **92**, 012104 (2008).

75. S. Y. Bae, H. W. Seo, H. C. Choi, and J. Park, Heterostructures of ZnO nanorods with various one dimensional nanostructures *J. Phys. Chem. B* **108**, 12318 (2004).

76. N. Zhang, R. Yi, R. Shi, G. Gao, G. Chen, and X. Liu, Novel rose-like ZnO nanoflowers synthesized by chemical vapor deposition, *Mater. Lett.* **63**, 496 (2009).

77. A. Mofor, A. Bakin, A. Elshaer, D. Fuhrmann, F. Bertram, A. Hangleiter, J. Christen, and A. Waag, Catalyst-free vapor-phase transport growth of vertically aligned ZnO nanorods on 6H-SiC and (11–20)Al$_2$O$_3$, *Phys. Status Solidi C* **3**, 1046 (2006).

78. Q. Zhao, P. Klason, and M. Willander, Growth of ZnO nanostructures by vapor–liquid–solid method, *Appl. Phys. A* **88**, 27 (2007).

79. N. Bano, I. Hussain, O. Nur, M. Willander, Q. Wahab, A. Henry, H. Kwack, and D. Le Si Dang, Depth-resolved cathodoluminescence study of zinc oxide nanorods catalytically grown on p-type 4H-SiC, *J. Lumin.* **130**, 963 (2010).

80. A. Mofor, A. Bakin, U. Chejarla, E. Schlenker, A. El-Shaer, G. Wagner, N. Boukos, A. Travlos, and A. Waag, Fabrication of ZnO nanorod-based p–n heterojunction on SiC substrate, *Superlattices Microstruct.* **42**, 415 (2007).

81. M. Syvajarvi, R. Yakimova, M. Tuominen, A. Kakanakova-Georgieva, M. F. MacMillan, A. Henry, Q. Wahab, and E. Janzen, Growth of 6H and 4H-SiC by sublimation epitaxy, *J. Cryst. Growth* **197**, 155 (1999).

82. T. Kimoto, A. Itoh, and H. Matsunami, Step-controlled epitaxial growth of high-quality sic layers, *Phys. Status Solidi B* **202**, 247 (1997).

83. J. A. Powell and D. J. Larkin, Process-induced morphological defects in epitaxial CVD silicon carbide, *Phys. Status Solidi B* **202**, 529 (1997).

84. V. Khranovskyy, G. R. Yazdi, A. Larson, S. Hussain, P.-O. Holtz, and R. Yakimova, Growth and characterization of ZnO nanostructured material, *J. Opt. Adv. Mater.* **10**, 2629 (2008).

85. V. Khranovskyy, I. Tsiaoussis, L. A. Larsson, P. O. Holtz, and R. Yakimova, Nanointegration of ZnO with Si and SiC, *Phys. B* **404**, 4359 (2009).

86. V. Khranovskyy, I. Tsiaoussis, G. R. Yazdi, L. Hultman, and R. Yakimova, Heteroepitaxial ZnO nanohexagons on p-type SiC, *J. Cryst. Growth* **312**, 327 (2010).

87. C. Gorla, N. Emanetoglu, S. Liang, W. Mayo, Y. Lu, M. Wraback, and H. Shen, Structural, optical, and surface acoustic wave properties of epitaxial ZnO films grown on (012) sapphire by metalorganic chemical vapor deposition, *J. Appl. Phys.* **85**, 2595 (1999).

88. B. Zhang, L. Manh, K. Wakatsuki, T. Ohnishi, M. Lippmaa, N. Usami, M. Kawasaki, and Y. Segawa, Epitaxial growth and polarity of ZnO films on sapphire (0001) substrates by low-pressure metal organic chemical vapor deposition, *Jpn. J. Appl. Phys.* **42**, 2291 (2003).

89. I. Tsiaoussis, V. Khranovskyy, G. P. Dimitrakopulos, J. Stoemenos, R. Yakimova, and B. Pecz, Structural characterization of ZnO nanopillars grown by atmospheric-pressure metalorganic chemical vapor deposition on vicinal 4H-SiC and SiO2/Si substrates, *J. Appl. Phys.* **109**, 043507 (2011).

90. M. J. Hÿtch, J.-L. Putaux, and J.-M. Pénisson, Measurement of the displacement field of dislocations to 0.03 Å by electron microscopy, *Nature* **423**, 270 (2003).

91. P. B. Hirsch, A. Howie, R. B. Nicholson, and D. W. Pashley, *Electron Microscopy of Thin Crystals*, Butterworths, London, U.K., 1965.

92. J. A. Venables, G. D. T. Spiller, and M. Hanbuckenth, Nucleation and growth of thin films, *Rep. Prog. Phys.* **47**, 399 (1984).

93. G. Carlotti, D. Fioretto, G. Socino, and E. Verona, Brillouin scattering determination of the whole set of elastic constants of a single transparent film of hexagonal symmetry, *J. Phys. Condens. Matter* **7**, 9147 (1995).

94. X. H. Zhang, S. J. Chua, A. M. Yong, S. Y. Chow, H. Y. Yang, S. P. Lau, and S. F. Yu, Exciton radiative lifetime in ZnO quantum dots embedded in SiOx matrix, *Appl. Phys. Lett.* **88**, 221903 (2006).

95. D. C. Reynolds, D. C. Look, B. Jogai, J. E. Hoelscher, R. E. Sherriff, M. T. Harris, and M. J. Callahan, Time-resolved photoluminescence lifetime measurements of the Γ5 and Γ6 free excitons in ZnO, *J. Appl. Phys.* **88**, 2152 (2000).

96. J. S. Reparaz, F. Güell, M. R. Wagner, A. Hoffmann, A. Cornet, and J. R. Morante, Size-dependent recombination dynamics in ZnO nanowires, *Appl. Phys. Lett.* **96**, 053105 (2010).

97. W. M. Kwok, A. B. Djurisic, Y. H. Leung, W. K. Chan, and D. L. Phillips, Surface recombination in ZnO nanorods grown by chemical bath deposition, *Appl. Phys. Lett.* **87**, 223111 (2005).

98. Q. X. Zhao, L. L. Yang, M. Willander, B. E. Sernelius, and P. O. Holtz, Surface recombination in ZnO nanorods grown by chemical bath deposition, *J. Appl. Phys.* **104**, 073526 (2008).

4

ZnO-Based Nanostructures

José Ramón Durán Retamal, Cheng-Ying Chen, Kun-Yu Lai, and Jr-Hau He

CONTENTS

4.1 Introduction.. 133
4.2 Growth of ZnO Nanostructures 134
    4.2.1 Vapor–Liquid–Solid Process 134
    4.2.2 Solution-Based Chemical Synthesis 135
    4.2.3 Chemical Vapor Deposition 137
4.3 Transport Properties... 138
4.4 Photoluminescence Properties.. 146
4.5 Photoconductivity Properties 150
4.6 Gas/Chemical Sensor... 154
4.7 Dye-Sensitized Solar Cells 159
4.8 Conclusion 162
Acknowledgments 162
References... 162

4.1 Introduction

A unique group of ZnO nanostructures, owing to three types and a total of 13 fastest growth directions ($\langle 0001 \rangle$, $\langle 01\bar{1}0 \rangle$, and $\langle 2\bar{1}\bar{1}0 \rangle$) together with a pair of polar surfaces (0001), have been synthesized such as nanowires/nanorods (NWs/NRs) [1], nanobelts (NBs) [2], nanoribbons [3], nanoparticles (NPs) [4], 3D nanoarchitectures [5], nanojunction arrays [6], nanosprings, [7], nanorings [8], and more complex nanostructures [9,10]. Moreover, nanostructured ZnO has attracted intensive research efforts due to surface effects, including surface band bending (SBB) [11,12], depletion region [13,14], physisorption/ chemisorption/ionosorption and photodesorption near surfaces [15,16], defects [17–20], surface defects/states [12,14,21–23], and surface roughness [24,25]. The surface effect of ZnO nanostructures are more pronounced than that in thin film and bulk counterparts [11,26] due to the structural uniqueness and the ultrahigh surface-to-volume (S/V) ratio of ZnO nanostructures. In this regard, several efforts have been made with ZnO nanostructures for its versatile applications such as NW field-effect transistors (FETs) [24,27–31], NW-based light emitting diodes (LEDs) [32], gas sensors [33–35], chemical sensors [36,37], photodetectors [38–41], optical switches [39], second harmonics generators [42], solar cells [43,44], logic circuits [45], and biosensor [22]. In order to develop the novel application of ZnO nanostructures utilizing the surface effect, it is very important to understand how the physical properties are affected by shrinking the dimension of ZnO. For example, the room temperature resistivity of an NW is 5–6 orders of magnitude lower than the bulk ZnO single crystal [46,48]. This enormous difference arises from the fabrication methods used to synthesize ZnO nanostructures, where approaches such as vapor transport [49], hydrothermal or solution-based methods [9,50,51], chemical vapor deposition [52,53], metal organic chemical vapor deposition [54], or epitaxial methods [55] induce a high concentration of structural defects such as oxygen vacancies and zinc interstitials [20], leading to n-type ZnO nanostructures rather than insulating ZnO films. Native defects are even more pronounced on the surface than in the core of the nanostructures [9], and depending on either ZnO-O or ZnO-Zn terminated facets [56], the surface defects bring out the upward band bending near the surface [16]. Photoluminescence studies at low temperature or room temperature have been performed extensively in order to elucidate the origin of structural defects (as discussed later), but there is still a lack of consensus. In this chapter, initially we discuss the transport mechanism and surface-related transport properties of ZnO nanostructures, with particular interest in reports of the mobility exceeding state-of-the-art planar devices observed in ZnO NW devices [31]. Afterward, attending to the practical applications of ZnO nanostructures, we examine recent studies on gas/chemical sensing, employing the surface effect, due to their deviation from stoichiometry, and relying on a change of conductivity via electron trapping and detrapping processes at the nanostructure surfaces, thus distinguishing between reducing and oxidizing gases as target species. Furthermore, the photoconductive properties under ultraviolet (UV) illuminations are analyzed in terms of photoconductive gain and response times, underlying the detrimental effects of the surface defects and how the ZnO nanostructures have been tailored in order to overcome such disadvantages. Finally, we summarize the emerging photovoltaic (PV) application of ZnO nanostructures. The ultrahigh S/V ratios of nanostructured devices suggest that the studies on the synthesis and PV properties of various nanostructured ZnO for dye-sensitized solar cells (DSSCs) offer great potential for high efficiency and low-cost solar-cell solutions.

4.2 Growth of ZnO Nanostructures

Most of ZnO nanostructures are grown with the three main methods: vapor–liquid–solid (VLS) process, solution-based chemical synthesis, and chemical vapor deposition (CVD). Depending on the specific growth conditions, the nanostructures can be obtained with a wide range of geometric features, such as NWs, NRs, nanorings, nanocombs, NBs, etc. A common feature of these nanostructures is the very high aspect ratio, which is due to the lowest surface energy of a certain crystalline facet [57]. In this section, detailed chemical/physical processes for each growth method will be described.

4.2.1 Vapor–Liquid–Solid Process

ZnO nanostructures synthesized by VLS processes are produced through the crystallization from the alloys at their supersaturated states. A VLS process generally requires metal droplets (such as Au, Fe, Sn, etc.) acting as the catalyst. The catalyst can be prepared by the deposition on the substrate with a metal thin film of a few tens mm in thickness. When the substrate is heated so that the metal is liquidized into droplets, the reactants in gas phase

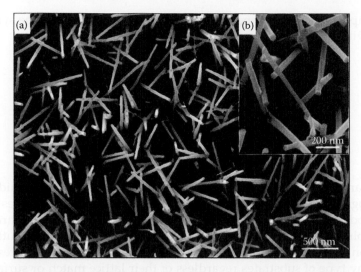

FIGURE 4.1
(a) ZnO nanorods grown by the VLS process using gold as a catalyst. (b) An enlarged image showing gold particles at the tips. (Reproduced from Wang, Z.L., Zinc oxide nanostructures: Growth, properties and applications, *J. Phys. Condens. Matter*, 16, R829, 2004. With permission from Institute of Physics Publishing.)

start to form an alloy with the metal droplets, which serves as the nucleation seeds for the following crystallization of the nanostructures. NW growth is initiated once the metal alloy becomes supersaturated. The growth continues as long as the reactants are available and the ambient temperature is above the eutectic temperature of the metal alloy. The orientation of NWs is usually directed by the metal droplets, which remain on the top of the NW during growth. Therefore, one of the signatures of a VLS process is the solid metal nanoparticles sitting at the NW tips.

Since the synthesis of ZnO nanostructures by VLS technique is a self-organized process, the technique holds the advantage of simplicity. However, the term "simplicity" does not imply the lack of control over the grown structures. Figure 4.1 shows ZnO nanorods and NBs synthesized by VLS processes [57,58]. The different geometrical shapes are governed by different thermodynamic equilibrium conditions during growth. Important parameters include substrate temperature, oxygen partial pressure, and choice of catalyst [59]. It is found that substrate preparations also play an important role in affecting the morphology of the grown nanostructures. For example, Wang et al. reported that the nanorods obtained on Zn-terminated ZnO substrates are very different than those obtained on oxygen-terminated ZnO substrates [60]. These controlled thermal equilibriums result in the growth kinetics favored for the formation of a specific crystalline orientation or facets and thus give rise to a number of nanostructures with different geometries and properties.

4.2.2 Solution-Based Chemical Synthesis

Solution-based chemical synthesis is also referred as the hydrothermal method. The synthesis is carried out in the aqueous solutions where zinc nitrate is generally used as the precursor for Zn. During the reaction, highly water-soluble hexamethylenetetramine (HMT) is often employed as the source of hydroxyl ions, which can be freed through the thermal decomposition of MHT and react with Zn ions to from ZnO. The synthesis can be summarized with the following reactions:

$$(CH_2)_6 N_4 + 6H_2O \leftrightarrow 6HCHO + 4NH_3, \tag{4.1}$$

$$NH_3 + H_2O \leftrightarrow NH_4^+ + OH^-, \tag{4.2}$$

$$2OH^- + Zn_2^+ \rightarrow ZnO_{(s)} + H_2O. \tag{4.3}$$

Alternatively, ammonia can be used to produce hydroxyl ions, and solid ZnO can be obtained through the last two reactions. The growth of ZnO nanostructures by the hydrothermal method is usually initiated on a seed layer, which can be prepared by e-beam evaporation, spin coating of zinc acetate followed by proper annealing, sputtering, pulsed laser deposition, etc. The use of the seed layer allows ZnO nanostructures to be grown on virtually any type of substrates regardless of their lattice match or surface chemistry. The growth of ZnO nanostructures by the hydrothermal method is also a self-organized process, and thus shares the same advantage of simplicity as the VLS method. In addition, since the reactions take place in water solution, the ambiance temperature of a hydrothermal method is below 100°C, which makes the process compatible with many device fabrications.

Similar to the VLS method, hydrothermal synthesis can also produce ZnO nanostructures with different geometrical features by properly controlled growth conditions. Figure 4.2a and b shows ZnO nanorods and nanoflowers, respectively, grown on spin-coated zinc acetate with different rotation speeds. The nanorods in Figure 4.2a are attained by 500 rpm, while the nanoflowers in Figure 4.2b are made by 2000 rpm. The distinct geometries are attributed to the different thickness of the seed layer. It has been found that the thermal decomposition of zinc acetate results in the nucleation seeds orientated along the [0001]

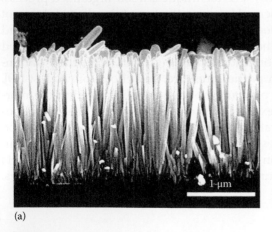

(a)   (b)

FIGURE 4.2
SEM images of ZnO nanorods and nanoflowers grown by hydrothermal processes on the ZnO seed layers spin-coated at (a) 500 and (b) 2000 rpm, respectively. (From Chao, Y.C., Chen, C.Y., Lin, C.A., and He, J.H., Light scattering by nanostructured anti-reflection coatings, *Energy Environ. Sci.*, 4, DOI: 10.1039/c0ee00636j. Reproduced by permission of The Royal Society of Chemistry.)

direction and the alignment toward [0001] increases with the seed layer thickness due to the improved uniformity of the ZnO nanocrystals [61]. Since increasing the rotation speed leads to thinner thickness of the seed layer, the orientation of the nanorods grown with the high rotation speed is less uniformly distributed and thus displays flower-like features.

4.2.3 Chemical Vapor Deposition

CVD technique is widely used in industry for compound semiconductor growth due to its high growth rate and precise control of epitaxial structures (such as chemical compositions, layer thickness, and doping concentration). For ZnO, vapor phase compounds like dimethylzinc or diethylzinc are often employed as the precursor for Zn, while O can be obtained from the sources including O_2, N_2O, H_2O_2, etc. Because of their superior crystal qualities, many ZnO-based quantum wells, the key component in photonic devices, are attained by CVD growth [62,63]. Like the aforementioned two methods, various morphologies of ZnO nanostructures can be achieved through specific growth conditions in CVD processes. Figure 4.3a through d presents the ZnO nanostructures grown by metal-organic

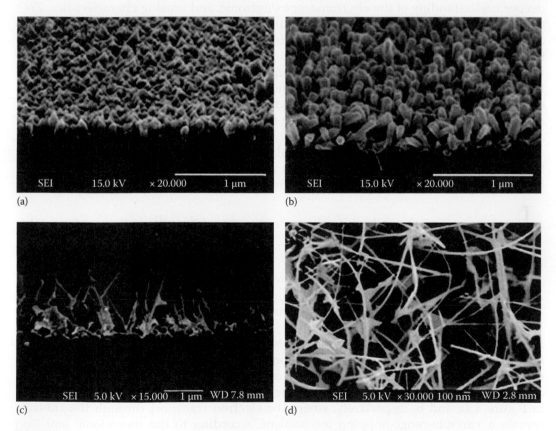

FIGURE 4.3
ZnO nanostructures grown by MOCVD at various substrate temperatures: (a) 240°C, (b) 300°C, (c) 500°C, and (d) 600°C, respectively. (Reproduced with kind permission from Springer Science+Business Media: *J. Mater. Sci. Mater. Electron.*, Morphology control of 1D ZnO nanostructures grown by metal-organic chemical vapor deposition, 19, 2008, 760, Kim, D.C., Kong, B.H., and Cho, H.K.)

chemical vapor deposition (MOCVD) with varied substrate temperatures [54]. It is found that the diameters of the 1-D nanostructures decrease with increasing substrate temperatures. As seen in Figure 4.3a, the low growth temperature of 240°C results in a film-like structure. X-ray diffraction analysis indicated that the film-like structure is polycrystalline with hexagonal tips [8]. As shown in Figure 4.3b through d, increasing the substrate temperature leads to curved nanostructures with improved crystallinity. In addition to the substrate temperature, many other parameters (including reactor pressure, buffer layer preparations, and the use of metal catalysts, etc.) are also reported to have effects on the morphologies of ZnO nanostructures by CVD techniques [54,64,65].

4.3 Transport Properties

The emerging field of nanoelectronics is demanding fundamental studies in the transport mechanism [23,48,66–69] and electrical properties of ZnO nanostructures for a better and deeper understanding of the electronic, optoelectronic, and sensing characteristics. ZnO nanostructures are definitely promising candidates for FET, LED, photodetectors, PV solar cells, chemical sensors, and biosensors, offering enhanced mobilities, high carrier concentrations, fast electron transfer, space charge separation, and greatly improved sensitivities, which are ascribed to the high S/V ratio, confinement effect and surface-related effects. In this regard, the physical, geometrical, and morphological characteristics, such as surface states [11,21], surface roughness [24,25,70], and finite size [21,47], are of vital importance in order to understand their electrical behavior, and determine the transport mechanism for a reliable device fabrication [71,72].

In this regard, Lin et al. [66] and Chiu et al. [48,67] have studied the electronic conduction by measuring the resistivity at different temperatures (from 300 down to 0.25 K) and least-squares fitting the results by the following equation:

$$\rho^{-1}(T) = \rho_1^{-1} \exp^{-(E_1/K_B T)} + \rho_2^{-1} \exp^{-(E_2/K_B T)} + \rho_3^{-1} \exp^{-(E_3/K_B T)} \qquad (4.4)$$

where
ρ_i ($i = 1, 2, 3$) are temperature-independent resistivity parameters
E_i are thermal activation energies describing the electronic conduction in the high (E_1), intermediate (E_2), and low (E_3) temperature regimes

Hence, the electrical properties of the ZnO NW have been categorized into two classes: low-resistance NW with thermal activation transport and high-resistance NW. The semiconducting ZnO NW contacted between two metal electrodes can be modeled as back-to-back Schottky contacts with the corresponding band diagram, as shown in Figure 4.4a and b, respectively, where the electron transport through the contacts reveals a variable-range-hopping mechanism, according to the thermionic emission theory, which defines the current equations of a Schottky junction as follows:

$$J(V, \phi_b) = A^* T^2 \exp\left(-\frac{\phi_b}{k_B T}\right) \exp\left(\frac{qV}{n k_B T}\right) \times \left\{1 - \exp\left(-\frac{qV}{k_B T}\right)\right\} \qquad (4.5)$$

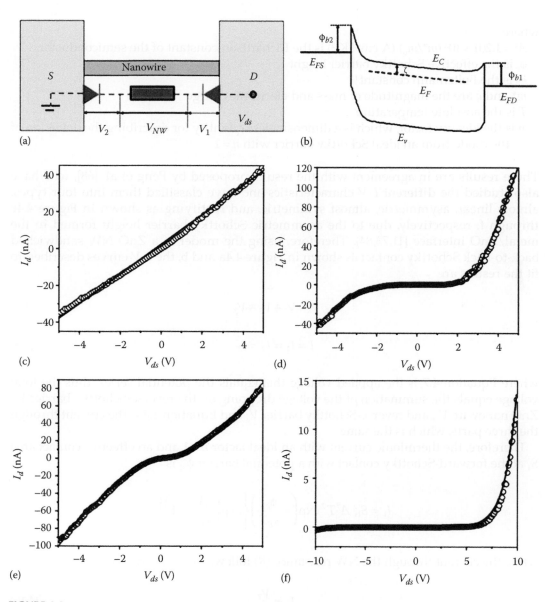

FIGURE 4.4
(a) Metal-semiconductor-metal (MSM) model for a two-terminal semiconducting NW device and its equivalent circuit. (b) Band diagram of the MSM structure. Experimental *I–V* characteristics: (c) almost linear, (d) asymmetric, (e) almost-symmetric, and (f) rectifying. (Zhang, Z., Yao, K., Liu, Y., Jin, C., Liang, X., Chen, Q., and Peng, L.-M.: Quantitative analysis of current–voltage characteristics of semiconducting nanowires: Decoupling of contact effects. *Adv. Funct. Mater.* 2007. 17. 2478. Copyright Wiley-VCH Verlag GmbH & Co. KGaA, Weinheim. Reproduced with permission.)

$$A^* = \frac{4\pi m^* q k_B^2}{h^3} \tag{4.6}$$

where
   $A^* = 1.201 \times 10^2 \; (m^*/m_0)$ (A cm^{-2} K^{-2}) is the Richardson constant of the semiconductor
   ϕ_b is the effective Schottky barrier height
   k_B is the Boltzmann constant
   m^* and q are the magnitude of mass and electronic charge
   T is the absolute temperature
   n is the ideality factor, which is a dimensionless quantity for describing the deviation of
       the diode from an ideal Schottky barrier with $n = 1$

These results are in agreement with the results proposed by Peng et al. [68], who have also studied the different *I–V* characteristics and have classified them into four types: almost linear, asymmetric, almost symmetric, and rectifying as shown in Figure 4.4c through f, respectively, due to the asymmetric Schottky barrier height formed in the metal/ZnO interface [11,73,84]. Therefore, using the model of a ZnO NW sandwiched back-to-back Schottky contact as shown in Figure 4.4a and b, the *I–V* curves described to fit the results are

$$V = V_1 + V_2 + V_3 \tag{4.7}$$

$$I = I_1 = I_2 = I_3 \tag{4.8}$$

where Equation 4.7 is the applied voltage that equals the potential series, thus the total voltage equals the summation of the voltage dropping on the reverse-Schottky barrier V_1, ZnO nanowire V_2, and reverse-Schottky barrier V_3 and Equation 4.8 is the current through the three parts, which is the same.

Therefore, the thermionic current with an ideal factor $n = 1$ and an effective contact area S_1 at the forward-Schottky contact with a potential barrier ϕ_{b1} is

$$I_1 = S_1 \cdot A^* T^2 \exp\left(-\frac{\phi_{b1}}{k_B T}\right)\left\{\exp\left(\frac{q V_1}{k_B T} - 1\right)\right\} \tag{4.9}$$

while the current through the NW resistance (R) follows an ohmic relation

$$I_2 = \frac{V_2}{R} \tag{4.10}$$

and the tunneling current through the reverse-Schottky contact with an effective area S_3 and effective Schottky barrier ϕ_{b2} is

$$I_3 = S_3 \left[j_s(V_3, \phi_{b2}) - j_s(0, \phi_{b2}) \right] \times \exp\left\{ e V_3 \left[\frac{1}{k_B T} - \frac{1}{E_{00}} \tanh\left(\frac{E_{00}}{k_B T}\right) \right] \right\} \tag{4.11}$$

which is a slowly varying function of applied bias and describes the reverse saturation current as follows:

$$j_s = \frac{A^* T^2 (\pi E_{00})^{1/2}}{k_B} \sqrt{(V_3 - \xi) + \frac{\phi_{b2}}{\cosh^2(E_{00}/k_B T)}} \times \exp\left(-\frac{\phi_{b2}}{E_0}\right) \tag{4.12}$$

$$E_0 = E_{00} \coth\left(\frac{q E_{00}}{k_B T}\right) \tag{4.13}$$

$$E_{00} = \frac{h}{2}\left[\frac{N_d}{m_n^* \varepsilon_r \varepsilon_0}\right]^{1/2} \tag{4.14}$$

where
  ξ is the distance between the Fermi energy to the bottom of the conduction band
  E_0 and E_{00} are parameters that can be calculated from Equations 4.13 and 4.14
  N_d is the donor density at the metal/semiconductor interface
  ε_r and ε_0 are the relative and vacuum dielectric constants of the semiconductor, respectively

If the typical parameters assumed by Liao et al. [11] are $\zeta = 0.1$ eV, $N_d = 10^{17}$ cm^{-3}, $m^* = 0.318$ m_0, and $\varepsilon_r = 8.7$ [19], then $A^* = 38.2$ A cm^{-2} K^{-2}, $E_0 = 26.2$ meV, and $E_{00} = 3.52$ meV.

By applying moderate reverse bias, the thickness of the barrier may be reduced, and thus electrons can Fowler–Nordheim tunnel from the contact potential to the ZnO NW, contributing with hundreds of picoampere [23]. According to the output characteristics (I_d–V_d) and transfer characteristics (I_d–V_g) shown in Figure 4.5a and b, respectively, the n-type semiconductor behavior of natively doped ZnO NW is concluded. Moreover, temperature dependence measurements ranging from 180 to 290 K [74] or the Arrhenius plot shown in Figure 4.5c have confirmed the thermionic emission theory assisted by quantum mechanical tunneling.

The conductivity of ZnO nanostructures is typically defined as n-type by electrical experiments, and ascribed to the abundant point defects of the nanostructures [9,20]. Formerly, oxygen vacancies with low formation energies, zinc interstitials with higher formation energies, and hydrogen impurities were considered as shallow donor candidates. Although afterward oxygen vacancies have been considered as deep donors and even further as acceptor defects. Therefore, there is no clear evidence of the role of oxygen vacancies in the conduction mechanism. Meanwhile, zinc vacancies and oxygen interstitials, both with higher energy formation, are regarded as acceptors defects, although oxygen interstitials could have ambipolar behavior.

Recently, the FET electrical characteristics of different 1D ZnO nanostructures [10] such as NW [31,52,75–78], nanorod [79–81], NB [82,83], and nanoribbons [3] have been investigated extensively. Moreover, different device configurations have been successfully tested for a decade under research and practical conditions alike, including suspended NW [78,84–86] and multiple NWs assembled by advanced techniques such as dielectrophoresis (DEP) [87–90] or DEP combined with hot press to ensure good contacts [91] and roll transfer printing [92]. Due to the confined geometry of the low-dimension nanostructures, it is not possible to determinate the electrical properties by traditional thin-film

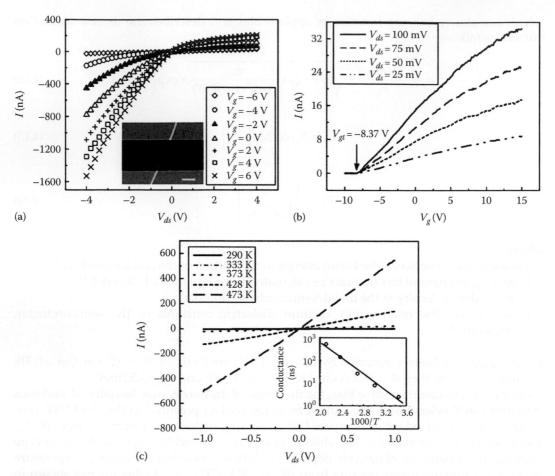

FIGURE 4.5

(a) Room-temperature I_d–V_d curves obtained at different gate voltages. Inset: SEM image of a ZnO NW FET with source and drain electrodes, scale bar is 2 μm. (b) I_d–V_g and transconductance of the ZnO NW FET under 100, 75, 50, and 25 mV bias voltages. (c) I_d–V_d curves obtained at different temperatures at $V_g = 0$ V. Inset: Conductance of NW device vs. inverse temperature. (Reproduced with permission from Fan, Z.Y., Wang, D.W., Chang, P.C., Tseng, W.Y., and Lu, J.G., ZnO nanowire field-effect transistor and oxygen sensing property, *Appl. Phys. Lett.*, 85, 5923. Copyright 2004, American Institute of Physics.)

techniques such as Hall measurements. Alternatively, the measurements of a single ZnO NW-based three-terminal FET are often used to estimate device characteristics, such as carrier concentration, mobility, and ideality factor from *I*–*V* measurements under different gate voltages [24,31,81,93,94]. In the typical FET test structures, a pair of leads patterned on the two ends of the NW serve as the source and drain electrodes, usually Au [83,95], Ti/Au [23,70,74,77,87], Pt [73,96], In/Au [97], and Ni/Au [98]. A substantial improvement in the performance is obtained by using Al instead of Ti/Au [94], as a consequence of the poor wetting of Ti/Au, which results in increased density of surface states at the metal/NW interface. The effective barrier height is estimated by applying the thermionic emission theory [23,68,73,74,77,83,95], consequently Schottky barriers of ~0.57 and 0.3 eV [98], have been reported for Ti/ZnO [70] and Ni/ZnO [98], respectively, although Kim et al. [74] and Liao et al. [23] reported that the single ZnO NW and Ti/Au contact barrier is estimated to be ~30 meV and 0.31 eV, respectively, considering the Arrhenius plot of the two-probe resistance, the thermionic emission conduction, and the Fowler–Nordheim tunneling model. The electrodes are commonly defined by conventional photolithography, by e-beam lithography, by focus ion beam (FIB) [96], or, recently, by self-aligned ink-jet printing technique [99]. Additionally, a weakly capacitively coupled terminal as a gate electrode is employed to control the NW conduction and thus gates the FET device performances such as operation speed, power efficiency, and ON/OFF ratio via applying a transverse electric field [93,100,101].

There are several types of gate configuration for NW-based FET. The back-gate configuration is the most widely used method due to its fabrication simplicity, since a heavily doped p-type or n-type Si substrate with a SiO_2 dielectric layer can serve as a back gate. However, the much larger capacitance of top-gate configuration, owing to the geometric field enhancement around the 1-D nanostructures, has demonstrated to be more effective [102]. In this regard, Kim et al. [88,103] have given evidence by using dual gate and making comparison between both operation modes. A more precise study using scanning probe microscopy (SPM) tips with high spatial resolution have proved the local electronic properties, estimating the potential drop along the NW and the contact resistance [104].

Following the CMOS technology, the commonly deposited gate dielectric layers are SiO_2 [88,103] and Al_2O_3 [102,105] or even air [30], but due to the increasing demand of flexible substrate for display technologies, organic materials are receiving more attention; thus, new strategies to deposit organic material have been developed in order to attend this requirement. In this regard, Janes's group via a self-assembled superlattice (SAS) has interlinked 15 nm of a four layer-by-layer self-assembled nanodielectric (SAND) as organic back-gate insulator [72,94,106,107], achieving low operation voltage such as drain saturation current at $V_{ds} = 0.5$ V, threshold voltage of $V_{th} = -0.4$ V, and mobility of ~196 cm^2/(V s) [106]. Meanwhile, Noh et al. [99] developed a self-aligned ink-jet printing (SAP) technique to deposit poly (3,4-ethylenedioxithiophene) doped with poly-(styrene sulfonate) (PEDOT:PSS) with channel lengths of 50–400 nm on a poly(methyl methacrylate) (PMMA) gate dielectric layer, exhibiting mobilities of 2–4 cm^2/(V s). Moreover, Kwon et al. [108], have investigated the piezoelectric effect induced by bent of the flexible substrate in ZnO NW FET fabricated on poly(ethylene terephthalate) (PET) substrates with a cross-linked poly(4-vinylphenol) (PVP) polymer gate dielectric.

The electron mobility (μ) and the carrier concentration (n) are related to the transconductance ($g_m = dI_{ds}/dV_g$) and can be calculated by the following:

$$\mu = \frac{dI_{ds}}{dV_g} \frac{L^2}{V_{ds} C_g} \tag{4.15}$$

and

$$n = \frac{C_g |V_g - V_{gt}|}{e \pi r^2 L} \tag{4.16}$$

$$C_g = \frac{2\pi \varepsilon_r \varepsilon_0 L}{\cosh^{-1}\left((r+h)/r\right)} \tag{4.17}$$

where typical back-gate to NW capacitance is derived from the metallic cylinder-plane system [31]; thus, C_g is the gate-NW capacitance, h is the thickness of the gate insulating layer, L is the NW channel length, r is the NW radius, ε_r is the dielectric constant of the gate insulating layer, V_{ds} is source-drain voltage, V_g is gate voltage, and V_{gt} is the gate threshold voltage below which current is suppressed to an OFF state [24,93]. Although these equations have been adopted from their planar device counterparts and widely reported, it should be noticed that there is still a lack of accuracy for several reasons: the assumptions of the back-gate capacitance are often incorrect [109], since they do not consider the space surrounding the NW; assume that the NW is electrostatically metallic and infinitely long; the asymmetric contact areas and interface surface states are omitted [68]; the grain boundaries effects are not examined [70]; and the geometry of the NW is idealized, neglecting the NW roughness [25] and the finite size [47]. Furthermore, the influence of the surface chemical dynamics of ZnO needs to be accounted [12].

As we have mentioned previously, the native defects such as oxygen vacancies and zinc interstitials, induce the undoped ZnO n-type semiconductor behavior. It is well known that native defects at the metal-oxide surfaces serve as the binding sites for chemisorption processes, such as the formation of charged oxygen molecule complexes. They also contribute to the scattering and the trapping of carriers [110], thus lowering the carrier concentration. For example, absorbed species such as oxygen can trap electrons (Equation 4.18) and dissociate in O^- ions by further capture of electrons (Equation 4.19); absorbed water molecules by capturing electrons can form hydroxyl groups (Equation 4.20); as well as the presence of zinc ions in the surface can lead to capture of electrons (Equation 4.21); and further absorption of oxygen molecules (Equation 4.22), which form more chemisorbed oxygen species or zinc ions. It should be considered that chemisorbed oxygen species can recombine with photon-generated holes leading to photo desorbed oxygen and detrapped electrons events (Equation 4.23). Such events constitute the basis of the photoconduction mechanism of ZnO nanostructures, as it will be discussed in Section 4.5.

$$O_2 + e^- \rightarrow O_2^- \tag{4.18}$$

$$O_2^- + e^- \rightarrow 2O^- \tag{4.19}$$

$$2H_2O + O_2 + 4e^- \rightarrow 4OH^- \tag{4.20}$$

$$Zn^{+2} + e^- \rightarrow Zn^+ \tag{4.21}$$

$$Zn^+ + O_2 \rightarrow Zn^{2+} + O_2^- \tag{4.22}$$

$$O_2^- + h^+ \rightarrow O_2 \tag{4.23}$$

The width of the surface depletion region caused by adsorbates on the surface of the ZnO NW was estimated to be around ~69 nm [111]. This value can be regarded as a critical radius, since NWs with a radius shorter than the critical radius will be fully depleted. Consequently, the whole trap centers in the NW can trap electrons, suppressing the further injection of electrons from the electrode, until the current is reduced to space-charge-limited current (SCLC) transport. In this circumstance, electrons tunneling from the electrode to those trapped centers and hopping from one trapped state to another are proposed to account for the current in the NW.

Evidence of the instability in the results and thus in the concentration and mobility estimations due to the surface chemical dynamics are demonstrated by repeated measurements of the FET transfer characteristics, which results in a positive shift of the V_{th} associated with the enhancement of oxygen adsorption by capturing electrons from the induced current during the probing [12]. By evaluating the stress bias and stress time, Jane's group [72,94,107] suggest that the ZnO NW–SAND gate insulator interface reduce the electron trapping at the surface, decreasing the off current level and the variation in V_{th}. The evaluation of the transient drain current characteristics and the induced hysteresis was also evaluated by Maeng et al. [112] in terms of oxygen desorption and adsorption rates; meanwhile, Liao et al. [111] indicate that the hysteresis observed in ZnO NW dominated by SCLC transport is due to the capacitance effect of the NW, which produces an additional discharging/charging current flowing on removal/increase of the voltage.

The requirement to understand the surface dynamics of the ZnO NW, bring out the necessity to study the electrical characteristics under different ambient conditions. Under vacuum [77,82,111,113] and N_2 [77,112] conditions, on one side the electron concentration increases due to less oxidized outer surface, and on the other side the mobility decreases due to higher electron–electron scattering near surface. Consequently, the shielding of the gate voltage from the inner NW to the outer metallic surface increases, and the V_{th} is shifted back and hysteresis is reduced. Oppositely, under O_2 [77,82,113], higher electronegative species concentration on the surface is bound, reducing the carrier concentration and increasing the mobility.

This unusual behavior is the SBB [70,114]. It is expected that SBB at the surface of single crystalline NWs is more pronounced than that in the thin film counterpart [11,115]. As a result, the surface treatment on ZnO nanostructures is important to tailor desired electrical characteristics of NW-based devices. Park et al. have reported that ZnO NWs exhibited an electron mobility of $1000 \, cm^2/(V \, s)$ larger than that of thin films ($115–155 \, cm^2/(V \, s)$) after coating the NWs with polyimide [81,116]; meanwhile, Lee et al. [117] reported slight improvements from 5–100 to $40–150 \, cm^2/(V \, s)$ after passivation with ~200 nm of PMMA. Chang et al. [31] have used a SiO_2/Si_3N_4 layer as a passivation layer to enhance electron mobility up to $4120 \, cm^2/(V \, s)$. The passivation of surface states eliminates scattering and reduces trapping centers, leading to the enhanced mobility and on/off ratios. In addition to the passivation layer coating, the post-annealing in ozone [94], O_2 [118], H_2 [119], or N_2 [120] ambient has been applied to compensate the surface states or reconstruct the surface of NWs, resulting in the reduction of the negative surface charges and the width of the depletion regions. For example, Ju et al. [94] have demonstrated that the electron mobility of an ozone-treated single ZnO NW is up to $1175 \, cm^2/(V \, s)$. These results demonstrate that after the surface modification, the ZnO nanostructure-based devices can achieve a faster operation speed than the thin film counterpart. On the other hand, Hu et al. [121] concluded that using hydrogen peroxide with increasing concentration results in a continuous increase in the amount of –OH group at the surfaces of the NW, which can modify the ZnO NW surface and thus the electrical properties in three ways: first, annihilating oxygen vacancies; second, forming hydroxide groups ($H + O_2^- \rightarrow OH^- + e^-$); and third, modifying the surface morphology by etching.

Furthermore, negative oxygen molecule complexes adsorbed on the ZnO surfaces form the depletion regions that decrease the conductivity, influencing the operation modes of NW FETs of transistors (depletion-mode [D-mode] or enhancement-mode [E-mode]) and can be crucially determined by the carrier trapping states at the surfaces. In this regard, Lee's group have intensively studied the adjustment of the operation mode by surface modification, proposing three methods: first, tailoring the surface of the ZnO NW, depending on the NW fabrication method, thus using AZO, Au-GZO, ZnO, Au-ZnO substrates to grow the NW, shorter diameters and rough surface NWs were observed, resulting in

E-mode due to more critical fraction of the surface depletion region in the NW channel, while NWs grown in Au-AZO, GZO, and Au-sapphire substrates showing smooth surface and larger diameter exhibited D-mode [24,122,123]; second, by submerging the ZnO NWs in isopropyl alcohol (IPA) and thus etching the surface to generate surface roughness, converting the original D-mode to E-mode [124]; and third, a high-energy proton irradiation, during a short time, enables to shift the V_{th} toward negative gate voltage (D-mode), while longer irradiation results in E-mode [122]. On the other hand, Janes et al. [94,97] reported that ozone treatment induces positive shift V_{th} hence tuning from D-mode to E-mode. Another approach based on a Ω-shaped Au top contact patterned on the top of the NW changes from D-mode to E-mode [97]. Finally, both D-mode and E-mode FETs are demanded for NW-based integrated circuit electronics with the high logic performance. Jo et al. [123] have reported that the inverter circuits, by combination of both D-mode and E-mode ZnO NW devices, show desired voltage transfer characteristics with a high gain and robust noise margin with less power dissipation, which makes them superior to logic inverters based on single-mode NW FETs.

4.4 Photoluminescence Properties

Photoluminescence (PL) analysis is the most widely applied technique to investigate optical properties of ZnO nanostructures, since it can estimate the tightly bound excitons, the bandgap energy (E_g), and related defect transitions of ZnO. PL measurements have demonstrated highly efficient near-band-edge emission (NBE) at UV regions. Analysis hence becoming a powerful tool to inspect the optical properties of ZnO nanostructures for its applications in light emitters, lasers, solar cells, or photodetectors due to very efficient near-band-edge emission (NBE) at UV regions [13,17–19,21,46,49,53,125–167]. Different nanostructures show some variations of the position and the intensity of the peak in the PL spectra, as well as nanostructures grown by different approaches have led to unestablished consensus in the origin of these emissions. Nevertheless, green emission is commonly observed in vapor deposited sample as well as orange–yellow emission is observed in solution-based samples. RTPL NBE in different ZnO nanostructures have been reported to occur at 373 [126], 378 [127–129], 380 [130–132], 381 [133], 383 [134,135], 384–391 [17], 387.5 [136], 389 [135,137], and 390 nm [138]; a more detailed table has been summarized by Djurišić et al. [9] and reproduced in Table 4.1.

Size-dependent time-resolved PL (TRPL) analysis indicates that the defect densities of various nanostructures with high S/V ratio are varied as compared with bulk ZnO [18,21,139], leading to the fact that the shifts of the NBE for a variety of nanostructures synthesized by different methods are distinct [125]. Furthermore, although the sizes of these ZnO nanostructures are larger than the small size exciton Bohr radius of ZnO (2.34 nm) [9,140], the quantum confinement effects only can be observed in very small nanostructures; thus, blue shift of the NBE in ZnO nanostructures is observed by decreasing the size in ZnO nanocrystals (NCs) [141], nanodots [142], NBs [143], which is indicative of surface effects as well.

Room-temperature PL (RTPL) spectra of ZnO nanostructures typically exhibit NBE at UV regions (UV luminescence) and the visible emission, which corresponds to deep level emission (DLE) as shown in Figure 4.6. Accordingly, the ratio UV-to-visible emission is established to quantify the optical quality. At high excitation intensities, the UV-to-visible

TABLE 4.1

Positions and Proposed Origin of Bound Exciton Lines in ZnO

LTPL Exciton Bound Lines		RTPL Peaks	
Proposed Origin	Position (eV)	Proposed Origin	Position (nm)
Excitons bound to structural defects	3.332	Near- and edge emission	373–390
Surface states	3.31	O_{Zn}	~402 (77 K), ~520
N_o acceptor	3.315, 3.3725, 3.3718, 3.3679	Shallow donor-oxygen vacancy transition	~446
Ionized donor bound exciton	3.3674, 3.3665, 3.3660	Zinc interstitial	~459
Neutral donor bound excitons	3.3614, 3.3604, 3.3600, 3.3593	Oxygen vacancy	~495
H	3.3628, 3.3630	Cu^+/Cu^{2+}	~500/510
Al	3.3608	Surface defects/defect complexes	~510, ~560
Ga	3.3598	Singly ionized vacancy	~510
In	3.3567, 3.3572	Zinc vacancy	~520
Na acceptors	3.356	Oxygen vacancy and zinc interstitials	~520
Li acceptors	3.353, 3.3531, 3.3484	$V_o^{••}$	~540
Neutral acceptor bound exciton	3.3481, 3.3530, 3.3564, 3.3566, 3.3484–3.3614, 3.3562, 3.358	Shallow donor-deep acceptor	~566 (10 K)
Donor	3.3598–3.3693	Oxygen interstitials,	~580,
(neutral or ionized) bound exciton		Li impurities	~626
Rotator states	3.3686, 3.302, 3.3670, 3.3664, 3.3702, 3.3714	Hydroxyl groups	~590
B excitons bound to neutral donor	3.3724, 3.3707, 3.3741, 3.3754, 3.3772	Oxygen related defects, zinc interstitials	~750

Source: Reproduced from *Prog. Quant. Electron.*, 34, Djurišić, A.B., Ng, A.M.C., and Chen, X.Y., ZnO nanostructures for optoelectronics: Material properties and device applications, 191. Copyright 2010, with permission from Elsevier.

Note: Positions and proposed origin of room temperature PL peaks in ZnO.

ratio increases [168]. However, the hypothesis that strong UV emission means good crystalline quality is not correct, because commonly ZnO nanostructures exhibit a large number of defects with ionization energies ranging from ~0.03 to 3.14 eV [19,144], as shown in Figure 4.7. Green emission is the most commonly observed defect emission in ZnO nanostructures [17,53,127–130,134–137,145–148] and often attributed to singly ionized oxygen vacancies [126,130,131,146,148], donor-acceptor transitions [149], recombination at $V_o^{••}$ centers [46,150], zinc vacancies [151,152], and surface defects [153]. Nevertheless, the reality is that the defect origin of the green emission has not been conclusively identified, and that the origin of the oxygen vacancies is still controversial [169]. Formerly, oxygen vacancies were assumed to be shallow donors [170], but recently they have been proposed as a deep donor and the single ionized oxygen vacancies regarded as unstable [171]. Therefore, more robust methods have been attempted, such as polarization-dependent PL spectra from aligned ZnO NRs, which has demonstrated that green emission is originated from the NR surfaces [154], and electron paramagnetic resonance (EPR) has supported the transition between neutral and double-ionized oxygen vacancies [172]. It has been shown that

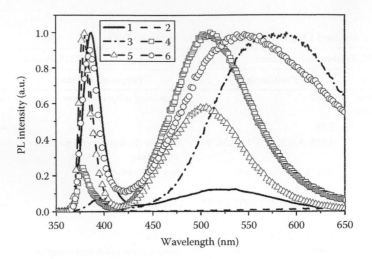

FIGURE 4.6

Room-temperature PL spectra of various nanostructures in the UV range: (1) Tetrapods, (2) needles, (3) nano-rods, (4) shells, (5) highly faceted rods, and (6) ribbons/combs. (From Djurišić, A.B. and Leung, Y.H.: Optical properties of ZnO nanostructures. *Small*. 2006. 2. 944. Copyright Wiley-VCH Verlag GmbH & Co. KGaA, Weinheim. Reproduced with permission.)

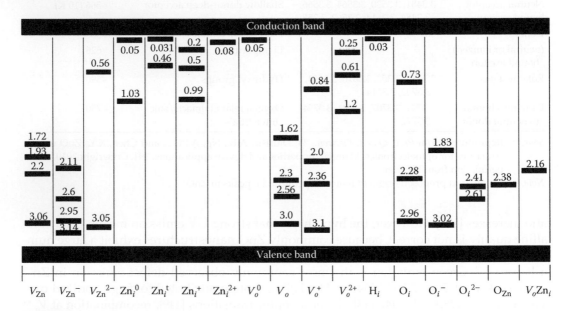

FIGURE 4.7

Energy levels of native defects in ZnO referred to the conduction band minimum. V_{Zn}, V_{Zn}^-, and V_{Zn}^{2-} denote neutral, singly charged, and doubly charged zinc vacancies, respectively. Zn_i^0 and Zn_i^t indicate neutral octahedral zinc interstitial and neutral tetrahedral zinc interstitial, while Zn_i^+ and Zn_i^{2+} indicate singly charged and doubly charged zinc interstitials. V_o^0 and V_o denote neutral oxygen vacancy while V_o^+ and V_o^{2+} denote singly charged and doubly charged oxygen vacancies. H_i and O_i represent hydrogen and oxygen interstitials, respectively, while O_{Zn} indicates antisite oxygen and V_oZn_i denotes a complex of oxygen vacancy and zinc interstitial. (From Kroger, F.A., *The chemistry of Imperfect Crystals*, 2nd edn., North Holland, Amsterdam, the Netherlands, 1974; Reproduced from *Prog. Quant. Electron.*, 34, Djurišić, A.B., Ng, A.M.C., and Chen, X.Y., ZnO nanostructures for optoelectronics: Material properties and device applications, 191. Copyright 2010, with permission from Elsevier.)

coating ZnO nanostructures with the surfactant can suppress the green emission significantly [153]. The thickness of surface recombination layer responsible for visible emission in ZnO NWs is estimated to be ~30nm [21]. In ZnO nanoparticles, the transition from electron near conduction band to a hole trapped at oxygen vacancies has been ascribed to green emission defect [173] and nanoparticles showed that increased particle size results in enhanced UV and visible luminescence intensities [174]. In addition, the intensity of the blue–green defect emission is dependent on the NW diameter; by reducing the NW radius the SV ratio increases and thus the surface recombination increases, leading to higher defect emission to NBE ratio [21], although both, increased and decreased intensities of defect emissions, have been observed with the diameter [49,132]. An important result concerning the surface defect emission was reported by Liao et al. [141], who showed that the green emission came from the nanostructure surface, while the UV emission came from the core, due to the dangling bonds at the surface. More evidence of the green emission has been explained from growth mechanism dependence. ZnO NWs grown by MOVPE on *p*-GaN substrate [55] or vapor transport on Au-coated Al_2O_3 substrate [71] have shown free defect emission, while pulsed-laser ablation [175] and vapor transport on Au catalyst free ZnO film [71] have shown green emission. Calculations of the surface structure, concluded that the surface energy of ZnO-Zn terminated-facet is higher than ZnO-O, thus the ZnO-Zn is more unstable and active, showing *n*-type conduction [56], while oxygen-terminated facets exhibit higher defect emission [176].

Yellow emission attributed to oxygen interstitials is commonly observed in ZnO nanostructures grown by solution methods [127,135,151,155] and with a high dependence on the seed used for nanostructure growth. The defects responsible for the yellow emission are not located at the surface [135,155] and associated with excess oxygen, Li impurities and hydroxyl groups [9]. By annealing the samples in hydrogen/argon ambient, the defect emission is suppressed [127], and at 200°C the hydroxyl groups can be desorbed, resulting in a red shift from yellow to orange–red emission and enhancing the UV-to-visible ratio [129]. By annealing in N_2 and O_2 atmosphere, the NBE due to excitonic transitions sharpen, and DLE is strongly reduced [89].

Various complex nanostructures have been synthesized to tailor the optical properties since the engineering of the ZnO NWs/NRs through surface modification can maximize the benefits provided by nanostructures [156–160]. For example, the enhancement of PL has been observed in ZnO/Er_2O_3 and ZnO/ZnMgO core/shell nanostructures [157,158]. Chen et al. have reported an aqueous chemical method to fabricate well-aligned ZnO/Al_2O_3 core/shell NRs. The Al_2O_3 shell layer resulting in the flat-band effect near the ZnO surfaces leads to a stronger overlap of the wavefunctions of electrons and holes in the ZnO core, further enhancing the NBE [13]. Yu et al. [161] have demonstrated that the PL intensity of ZnO/ZnS nanotetrapods and the lifetime of the UV exciton radiative transition have been increased more than 20 times due to the passivation effect of the core/shell structures, while ZnO/ZnS NWs change the alter the position of the visible maximum in RTPL [177].

In addition to RTPL analysis, low-temperature photoluminescence (LTPL) measurements are often applied to investigate exciton emission in ZnO nanostructures with high efficient recombination. Typically, the low temperature brings up low green emission and the free exciton emission yields information about impurities and radiative decay, which can be associated with optical quality by studying the LTPL dependence and TRPL spectra. The NBE from LTPL consists of several sharp emission lines corresponding to different bound excitons or even biexcitons. In this regard Djurišić et al. [9] reviewed the recent literature and has listed the positions and proposed origins of bound exciton lines in ZnO, such as neutral or ionized donor-bound excitons (excitons bound to H, Al, Ga, In), neutral acceptor-bound excitons relative shallow (60–120 meV), excitons bound to structural defects or neutral donor, among others, overall are labeled from I_0 to I_{11}. Due to the SBB [164,165], ZnO nanostructures have a large fraction of free excitons bound to the near surface defects, namely surface excitons (SX) [162,163], which have higher photon energy than other bound excitons but lower than free excitons. As power intensity of excitation is increased, the saturation behavior in the PL intensity of SX can be observed, which is different from other bound excitons [164]. Hence, when the size of ZnO nanostructures is decreased, the role of SX becomes important [164,166,167]. The LTPL spectra can be altered by fabrication approach, for example, hydrothermal grown NWs shown higher excitons bound to hydrogen [146]. Moreover, post-fabrication treatment such as annealing at N_2 ambient [120], surface treatment with H_2O_2 [121], proton irradiation [122], or surface modification with ZnO/amorphous Al_2O_3 core/shell NWs, which enhance the surface excitonic emission due to reduced SBB, created the high density of SX [159].

4.5 Photoconductivity Properties

UV detectors based on ZnO has attracted great attention due to wide direct bandgap energy (~3.37 eV), high exciton binding energy (~60 meV), and visible transparent properties of ZnO. Formerly, ZnO thin films have shown a slow response time, ranging from a few minutes to several hours and a low optical gain is commonly observed [178]. However, recently tailored ZnO nanostructures have been widely researched as building blocks in nanomaterial-based systems and with the capability to improve the performance, due to its ultrahigh SV ratio and confinement effect. Such tailored nanostructure materials are suitable for high photo-induced gain and fast photoresponse [179], being the perfect candidates for UV photodetectors [180], in aerospace, automotive, and petroleum industries [181]. Special attention has been focused on NW photodetectors with applications in optical chip interconnections, single photon detectors, and image sensors [41].

The photoconduction mechanism of ZnO NW has been extensively reported [41,84,95,182]. An illustration of the process is depicted in the Figure 4.8 and it can be summarized as follows: (1) In the dark, oxygen is adsorbed on the ZnO surface defects capturing electrons from the conduction band, creating a depleted space charge region with low conductivity near the surface (Equations 4.18 through 4.22). (2) Upon turning on the UV illumination above the ZnO bandgap, ZnO NW reacts rapidly photogenerating electron-hole pairs (photoelectrons and photoholes, respectively) in a solid state process. (3) The photoholes migrate toward the NW surface along the SBB, (4) where discharge-absorbed oxygen and hence oxygen molecules (gas) are photodesorbed. (5) As a consequence, the recombination rate of the photoelectrons reduces and the photoelectron's lifetime increases, leading to an accumulation of unpaired photoelectrons in the conduction band due to the spatial charge separation. As a result, the current increases abruptly. Overall, the free carrier density in the bulk ZnO increases, while the surface depletion region (W_{UV}) and the surface barrier height ($\Delta_{\phi b}$) reduce. At higher energy illuminations, the process is maintained with a substantially more efficient absorption near the surface of the NWs [183]. Additionally, the photodesorbed oxygen gives place to new vacancies that are available for readsorption. Therefore, a continuous flow of readsorption and desorption events takes place until the

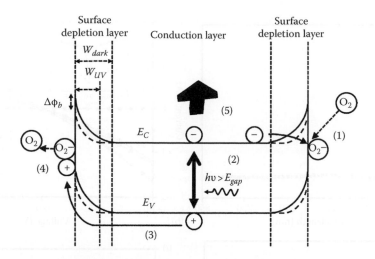

FIGURE 4.8
Schematic of the SBB and the dynamics of the photogenerated electron-hole pairs near the surface of ZnO NW, where E_C is the conduction band energy level, E_V is the valence band energy level, W_{dark} and W_{UV} are the widths of the surface depletion region under dark and UV, respectively, and $\Delta_{\phi b}$ is the decrease in the height of the surface band bending. The photogenerated holes are accelerated by the built-in electrical field in the surface depletion layer and can be trapped by the chemisorbed oxygen ions.

equilibrium between trapping and detrapping events takes place and hence the steady state is reached. As a result, the current rises gradually until the saturation level is reached. Upon turning off the UV illumination, the photogeneration of electron-hole pairs ceases immediately and excess photoelectrons recombine rapidly with the remaining photoholes that were accumulated along the SBB of the ZnO surface, since they were not able to reach the surface and discharge-adsorbed oxygen. Such photoholes can recombine in a fast manner with photoelectrons that have not been collected in the cathode yet, resulting in an abrupt decrease of the photocurrent. However, there are still a large number of free electrons in the ZnO that are captured gradually by reabsorbed oxygen molecules on the NW surface in a slow manner. This slow decay is strongly related with the NW surrounding and surface coating.

An important aspect to remark is that NWs with diameters smaller than the depletion region are fully depleted and minimize the dark current, while larger diameters increase the photoconductivity by hindering photogenerated carrier recombination, thus the photoconduction mechanism under these circumstances is strongly affected [41,178]. The width of the depletion region is an important parameter; thus, some guidelines to calculate it are provided in Section 4.6 by Equations 4.31 and 4.32.

The fundamentals and the theory behind the photoconduction mechanism in ZnO NW have been reported by Soci et al. [41] and Prades et al. [84]. On the other hand, the experimental section typically includes the photocurrent-wavelength dependence, which is a good methodology to observe the visible blind properties and the gain peaks as shown in Figure 4.9a, and the $I-V$ curves under dark and UV illumination determine the ratio I_{photo}/I_{dark} as shown in Figure 4.9b. Moreover by cutting the UV illumination periodically, the optical switching characteristics can be examined as is shown in Figure 4.9c. Typically, the results are presented in terms of the photo-induced gain in order to quantify photoconduction, since it is defined as the number of electrons collected per unit time and the number of absorbed photons per unit time ($G = N_{electron}/N_{photon}$) [38,41,184], and it is a convolution

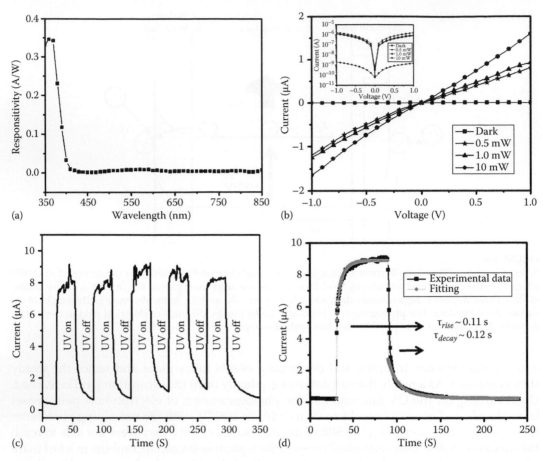

FIGURE 4.9
ZnO NW UV sensor. (a) Photoresponse spectrum as a function of the wavelength of incident light. (b) *I–V* characteristics in dark and under 365nm UV illumination with different power. The inset shows the *I–V* curves in logarithmic scale. (c) Time dependence of the photocurrent growth and decay under periodic illumination of the 365nm UV light on the device. The bias on the device is 1V. (d) Experimental and fitted curve of the photoresponse and photorelaxation process.

of two processes, the ratio of carrier lifetime to carrier transit time and the account of trap saturation. For simplicity, the formula is

$$G = \left(\frac{I_{photo} - I_{dark}}{P_\lambda \times A_{abs}} \right) \times \left(\frac{h\nu}{q} \right) \tag{4.24}$$

where P_λ is the total excitation power intensity (W/m²) absorbed in the NW, and therefore A_{abs} is the absorption area and is calculated as half of the NW surface ($A_{abs} = (2\pi r/2)L$, where (r is the NW radius and L the NW length). The dark current I_{dark} has been subtracted from the photocurrent I_{photo} at the excited wavelength in order to elucidate the real photoconduction enhancement.

On the other hand, the photoresponse has been commonly fitted by a stretched exponential function [185], while the photoresponse and the photocurrent relaxation can be described by either a first-order or second-order exponential function as shown in Figure 4.9d [185]:

$$I = I_0 + A_1 \exp\left(-\frac{t}{\tau_1}\right) + A_2 \exp\left(-\frac{t}{\tau_2}\right) \qquad (4.25)$$

where the constant I_0 represents the steady state photocurrent in the photoresponse or the ultimate dark current in the photorelaxation process. The values of A_1 and A_2 are weighing factors that quantify the relative contribution of each mechanism. The shorter time constant, which means the faster process is dominated by bulk related process and is assigned to τ_1, meanwhile the slower time constant τ_2 is ascribed to surface photoconduction process and is considered as a slower process.

The widely reported surface defects of ZnO NW [186] have been used as an advantage in gas sensors [187], but become a disadvantage in UV photodetectors, since the chemical absorbed species in such surface defects can deplete the NW surface by trapping electrons from the bulk and thus limiting the dark current and photocurrent due to the higher surface recombination velocity [186]. According to the photoconductivity studies of ZnO NWs, the presence of O_2 has an important effect on the photoresponse [11,16,40,80] as shown in Figure 4.10a and c. The photocurrent relaxation time is hours in vacuum but around 8 s in air ambient as shown in Figure 4.10b and d, due to the oxygen-related hole-trapping state effect. Similar results in vacuum has also been proved by the effect of adsorbates on SCLC ZnO NW [111], where it is indicated that desorption of chemisorbed oxygen is more complete in vacuum and leads to incomplete recovery of surface adsorptions in vacuum. It has been reported that the internal photoconductive gain of ZnO NW-based UV photodetectors could be as high as $\sim 10^8$ due to the presence of oxygen-related hole-trapping states at NW surfaces [38,40,96].

In order to overcome the detrimental surface defects, ZnO NW have been functionalized in several ways, such as radial nano-heterostructures [177,188,189], capping layers [190,191], metallic nanoparticles [184,192–195], which form local Schottky contacts, and semiconducting nanoparticles [185,186,196–199] forming nano-junctions. Table 4.2 summarizes several ZnO nanostructures used for UV photodetectors. Some general conclusions about the advantages of the selected nanostructures can be derived from their properties. Heterojunctions compared to homojunctions have the advantage that can tune the photodetection range. Crossed NWs and longitudinal NW–NW junctions have an indefinable and small junction area, as well as capped NWs; therefore, NPs decorating NWs will provide more effective functionalized surface. NPs synthesized by vapor phase approaches avoid the possible contaminants in solution processed approaches. Metallic NPs form localized Schottky effects [184], in contrast, semiconductor NPs are able to form *pn* nano-junction with an improved charge transfer and offer the possibility to tune the photodetection range. Finally, NWs are preferred over thin films due to their enhanced SV ratio.

Furthermore, the photocurrent as a function of the light polarization angle for both UV and visible light has been demonstrated [40]. Polarized photodetection of both UV (365 nm) and visible light shows that the photoconductivity of ZnO NW is maximized when incident light is polarized parallel to the NW axis. This behavior specifies ZnO NW promising application in high contrast polarizer.

Recently, a great attention has been paid on improving the photoresponse, thus by functionalizing the surface with polymers, the photoresponse has been enhanced by five orders of magnitude [200,201]. The reset time of ZnO NW-based photodetector has been

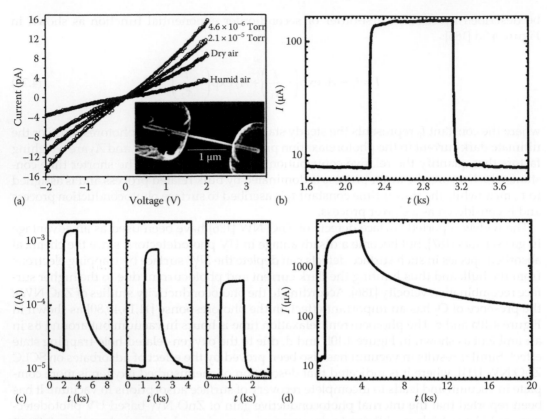

FIGURE 4.10

(a) *I–V* curves of a single ZnO NW at different ambiences as marked. Dots are experimental data and the solid lines are the theoretical fitting. Insert: SEM image of the ZnO NW device. (Reproduced from *Phys. Lett. A*, 367, Liao, Z.-M., Liu, K.-J., Zhang, J.-M., Xu, J., and Yu, D.-P., Effect of surface states on electron transport in individual ZnO nanowires, 207, Copyright 2007, with permission from Elsevier.) Photoresponse of the NW film to UV illumination (b) in air, (c) in various oxygen pressures: 1.9×10^2, 2.0×10^4, and 1.0×10^5 Pa from left to right and (d) under vacuum. (Reproduced with permission from Li, Q.H., Gao, T., Wang, Y.G., and Wang, T.H., Adsorption and desorption of oxygen probed from ZnO nanowire films by photocurrent measurements, *Appl. Phys. Lett.*, 86, 123117. Copyright 2005, American Institute of Physics.)

reduced drastically by surface functionalizing of Ag NPs [193] and CdTe quantum dots [185]. By utilizing a Schottky contact instead of an Ohmic contact in device fabrication, the photoresponse has been improved by four orders of magnitude [84,193]. In addition, by using back-gate FET configuration, the maximum UV photodetection sensitivity ($I_{photo}/I_{dark} \sim 10^6$) occurs at the bottom of the subthreshold swim region [202].

4.6 Gas/Chemical Sensor

ZnO NWs are also applicable on detecting versatile gases due to novel characteristics [37]: (1) the high S/V ratio provides higher surface reactions; (2) the Debye length (λ_D) becoming comparable to the NW radius causes a strong influence in the surface and can result in a better sensitivity and selectivity, since it can modulate the conductivity from highly

TABLE 4.2

Tailored ZnO Nanostructures for UV Photodetection

Nanostructure	Materials	References
Crossed NWs	n-ZnO NW/n-ZnO NW	[203]
Heterojunctions	p-Si substrate/n-ZnO NWs	[204]
Longitudinal homojunctions	p-ZnO/n-ZnO NWs	[205]
Longitudinal heterojunction	p-Si/n-ZnO NW	[206]
	p-NiO/n-ZnO NW	[207,208]
Radial: core/shell	ZnO/ZnS NW	[177]
	ZnO NW/polymer	[200,214]
	ZnO/CuS NW	[189]
	ZnO nanotube	[209]
pn Heterojunction thin films	n-ZnO TF/CdTe QDs	[185]
	n-ZnO nanowall/CdS NPs	[186]
	n-ZnO TF/CdS NPs	[198]
Metal NPs decorated NW	n-ZnO NW/Ag NPs	[193,194]
	n-ZnO NW/Au NPs	[184,195]
	n-ZnO NW/ Pt NPs	[192]
Capped NW	ZnO NWs/NiO cap	[191]
	ZnO NW/CdO capped	[190]
Semiconductor NPs decorated NW	n-ZnO NW/p-HgTe NPs	[196]
	ZnO NW/ZnO NPs	[199]
	SnO$_2$ NW/NiO NPs	[234]
	SnO$_2$ NW/ZnO NPs	
Doped NW	ZnO NW/Cu doped	[210]

NW: nanowire, NP: nanoparticle, TF: thin film.

conductive to fully depleted; (3) the rapid diffusion rate of electrons and holes to the surface improve the adsorption–desorption kinetics; (4) the improved stoichiometry of ZnO NW reduces instability; (5) ZnO NW FET with back-gate configuration can control the surface process electronically; and (6) shrinking the NW size along either its radial or axial direction, more significant quantum effects are expected.

The fundamental mechanism relies on a change of conductivity via electron trapping and detrapping process on NW surfaces. As gas molecules are adsorbed on NW surfaces, charge transfer occurs and modifies the carrier concentration, resulting in a change of conductivity. Therefore, extrinsic ambient conditions, such as temperature, pressure, illumination, target concentration, and intrinsic nanostructure characteristics such as surface defects, surface roughness, surface terminated facet, significantly influence the response for gas sensing [33,211–214].

The target species can be divided into two groups [35,37]: (1) oxidizing gases or electron acceptors, such as NO$_2$ or O$_2$, which tend to capture free electrons in NWs and dissociate into NO$_2^-$ or O$_2^-$, forming a low-conductivity depletion layer near the surfaces. Therefore, the conductivity of n-type ZnO NWs, where the electrons are the major carriers, is decreased in the presence of NO$_2$ and O$_2$ [98,215–217]. The mechanism can be modeled in terms of the oxygen chemisorption and desorption. Oxygen absorption is directly related to the concentration of unoccupied vacancies (N_s) and by trapping free electrons (n) ($O_2^{gas} + e^- \rightarrow O_2^-$) might dissociate into oxygen ions ($O_2^- + e^- \rightarrow 2O^-$), in which the general form $O_{\beta s}^{-\alpha}$, where α, $\beta = \{1,2\}$, accounts for the charge and molecular or atomic nature. Moreover, the surface oxygen coverage (θ) depends on the adsorption/desorption rate constants (k_{ads}, k_{des}) as follows:

$$\frac{\beta}{2} \cdot O_2^{gas} + \alpha \cdot e^- + N_S \rightarrow O_{\beta S}^{-\alpha} \tag{4.26}$$

$$K_{ads} \times N_S \times n \times p_{O_2}^{\beta/2} = k_{des} \times \theta \tag{4.27}$$

(2) On the other hand, reductive gases or electron donors, such as ethanol, H_2, CO, H_2S, and water vapor, react with the charged oxygen molecules on NW surfaces, and thus free electron concentration is increased due to oxygen desorption, leading to an increase of conductivity in *n*-type ZnO NWs [211–213,218,219]. In this case the reaction takes place with ionosorbed oxygen and can be modeled by the absorption of CO as follows [37]:

$$\beta \cdot CO^{gas} + O_{\beta S}^{-\alpha} \rightarrow \beta \cdot CO_2^{gas} + \alpha \cdot e^- \tag{4.28}$$

Generally speaking, the response to target gases in ambient environment is defined as

$$S = \frac{G_1 - G_0}{G_0} \tag{4.29}$$

where G_0 and G_1 are the conductance before and under exposure and are expressed as a function of the carrier concentration (n_0), mobility (μ), NW length (l), and diameter (D):

$$G = n_0 e \mu \frac{\pi(D - 2\omega)^2}{4l} \tag{4.30}$$

Additionally, the width of the surface charge region (ω, width of the depletion region) is related with the Debye length (L_D) as follows:

$$\omega = L_D \left(\frac{eV_S}{k_B T} \right)^{1/2} \tag{4.31}$$

$$L_D = \left(\frac{\varepsilon_r \varepsilon_0 k_B T}{e^2 n_0} \right)^{1/2} \tag{4.32}$$

where
  ε_0 is the absolute dielectric constant
  ε_r is the relative dielectric permittivity of the structure
  k_B is the Boltzmann's constant
  T is the temperature
  V_S is the adsorbate-induced band bending

Size dependence of ZnO NWs for gas sensing has also been reported by Fan et al. [98], indicating that the sensing ability of ZnO NW to oxygen increase by decreasing the diameter, as shown in Figure 4.11. As the NW diameter shrinks, the sensitivity to oxygen sensing

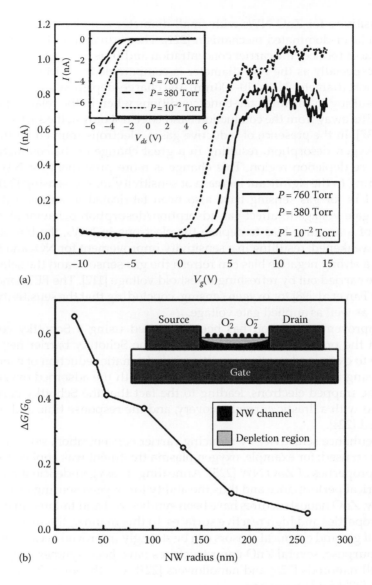

FIGURE 4.11
(a) Transconductance of a ZnO NW FET under different pressures at room temperature, and $V_d = 200$ mV. Inset: $I_d - V_d$ curve of the NW under 760, 380, and 10^{-2} Torr. (b) Ratio of conductance change vs. radius under vacuum and atmosphere. Inset: A schematic of NW FET channel depletion (depicted by gray shading) caused by adsorption of oxygen molecules. (Reproduced with permission from Fan, Z.Y., Wang, D.W., Chang, P.C., Tseng, W.Y., and Lu, J.G., ZnO nanowire field-effect transistor and oxygen sensing property, *Appl. Phys. Lett.*, 85, 5923. Copyright 2004, American Institute of Physics.)

is high due not only to the large S/V ratio but also to high density of oxygen vacancies in NWs [33,98], mostly at the surface of NW and act as adsorption sites for oxygen molecules [81]. The reason for high density of oxygen vacancies in small NWs is that the compressive stress within ZnO NWs increases with the decreasing diameter of NWs and thus induces the increase of the density of oxygen vacancies in the surface layers of the NWs [220]. Consequently, a large quantity of oxygen vacancies and large effective surface areas at the surfaces of NWs with small diameter lead to higher sensitivity to gas molecules. This enhanced sensitivity for ZnO NWs with small diameter can also be explained by a surface depletion layer-dominated mechanism [221]. The formation of the depletion regions near NW surfaces reduces the carrier concentration and thus has a great influence on NW conductivity especially as the NW diameter is comparable to a Debye screening length [221,222]. In short diameter NWs (10–100 nm), the charge-depletion layer encompasses the entire NW, resulting in so-called flat-band conditions wherein the relative position of the Fermi level shifts away from the conduction band edge not only at the surface but throughout the NW [37]. In the presence of reductive gases, electrons trapped on the surface are released by oxygen desorption, resulting in a great change of the conductivity and the width of surface depletion region. This change is more prominent for NWs with small diameter, leading to the significant increase of sensitivity for gas sensing [221,223].

ZnO NW FETs for gas sensing have also been fabricated to enhance the sensitivity via adjusting gate voltage to affect gas adsorption/desorption behavior at NW surfaces. With the aid of a negative bias to deplete the electrons in NWs, the binding of oxygen adsorption is weakened, resulting in a sensitivity improvement for NO_2 and NH_3 sensing. Furthermore, a strong negative bias can refresh the gas sensors, and the selectivity of gas sensing can be carried out by refreshing threshold voltage [112]. The FET concept was also carried out by Fan et al. [98] for oxygen sensing, concluding that the sensitivity depends on NW diameter as well as applied gate voltage.

Another improvement method has been proposed using a Schottky contacted NW gas sensor. In the presence of oxygen molecules, the Schottky barrier height is further increased due to oxygen adsorption, resulting in a dramatic reduction of conductivity. For CO sensing using this method, CO molecules react with the adsorbed oxygen molecules and release the trapped electrons, leading to the fact that the Schottky barrier height is then decreased with a drastic current recovery, and the response time and reset time can also be reduced [224].

In order to enhance sensitivity by reducing carrier concentration, several surface treatments have been used; for example, oxygen plasma treatment was applied to enhance the NH_3 sensing properties of ZnO NW [225]. Annealing in oxygen-deficient atmosphere can tune the electrical performance and thus the ability for oxygen sensing [82].

Additionally, ZnO nanostructures have been synthesized and tailored in order to obtain the desired properties and high reactive surfaces. In this manner, the surface-related sensing behavior of gas and chemical sensors can be strongly improved because of the large SV ratio. In this purpose, several ZnO nanostructures have been synthesized, such as F_2O_3/ZnO core/shell nanorods [226] and nanoflowers [221] for ethanol or ZnO dendrites [227] and tetrapods [228] for H_2S sensing.

In addition to the nanostructured surface modification, the surface decoration with noble metal nanoparticles NPs has attracted increasing attention, due to their surface-related properties [229]. Thus, Au NP-decorated ZnO NWs for CO sensing have been studied. Au NPs on NW surfaces act as a catalyst to reduce the activation energy for oxygen dissociation and spill charged oxygen molecules over NW surfaces due to their availability of free electrons and highly conductive nature, resulting in an excellent gas sensing behavior. This mechanism is also known as spill-over effect [200,221,226]. Due to the large difference between the work function of Au and the electron affinity of ZnO, Au NPs also enhance the surface depletion effect by modulating the surface depletion region and account for the rapid and ultrahigh sensing for ethanol [223,230]. Pt NP-decorated ZnO NWs are also found to be effective to enhance response to H_2S, ethanol, and H_2 [34,231,232]. As shown in Figure 4.12, the response increases with H_2S concentration, and can be further enhanced by Pd NP decoration, as well as for H_2 [233]. On the other hand, decorating with

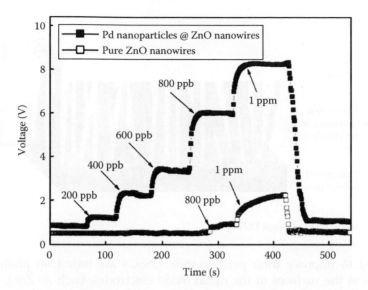

FIGURE 4.12
dynamic responses of sensors with different samples to H₂S. (From Zhang, Y., Xiang, Q., Xu, J.Q., Xu, P.C., Pan, Q.Y., and Li, F., Self-assemblies of Pd nanoparticles on the surfaces of single crystal ZnO nanowires for chemical sensors with enhanced performances, *J. Mater. Chem.*, 19, 4701, 2009. Reproduced by permission of The Royal Society of Chemistry.)

semiconductor NPs, three additives can be considered [234]: (1) induction of more structure defects, by diffusion of additives; (2) accelerated chemical reactions; and (3) surface band structure modifications by the formation of nano-heterojunctions at the NW surface, which couple the effect of the two sensing materials. Hence SnO₂ NWs decorated with NiO and ZnO NPs [234], has been used for H₂S, CO, and CH₄ gases.

Doping the NW, the position of the Fermi level can be shifted, opening the possibility to control the molecular absorption onto its surface. Therefore, The sensitivity of acetone and toluene sensing using InSb-doped and TiO₂-doped ZnO NWs have been improved via reduction of the activation energy of oxygen desorption and thus efficiently desorbs oxygen molecules at the surfaces of ZnO NWs [235,236]. A reduction of activation energy also contributes to the reduction of response times and reset times [222]. The enhanced sensitivity in a bilayered polymer/ZnO gas sensor based on ZnO NBs with a plasma-polymerized acrylonitrile (PP-AN) surface coating has been observed as well [214]. PP-AN can increase the concentration of the adsorption sites for target gases, resulting in a high sensing response. Moreover, it has been reported that desorbing charged oxygen molecules adsorbed at NB surfaces using UV illumination can create more adsorption sites for subsequent oxygen molecules sensing, leading to the improvement of the sensitivity [214].

4.7 Dye-Sensitized Solar Cells

Dye-sensitized solar cells (DSSCs) have been reported by Grätzel and O'Regan in 1991 [237]. Because it is one of the most promising low-cost solar cell, a lot of work has been

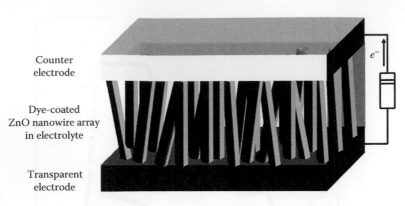

Counter
electrode

Dye-coated
ZnO nanowire array
in electrolyte

Transparent
electrode

FIGURE 4.13
schematic of ZnO nanostructure-based DSSCs.

done in trying to improve their performance. DSSCs are based on photoexcitation of dye molecules at the surfaces of the metal oxide electrodes, such as ZnO. Electrons are injected from the dye molecules into the metal oxide electrodes and the dye is reduced by electron donation from the electrolyte. Therefore, metal oxide electrodes are investigated intensively for achieving a fast electron transport and improving the efficiency of DSSCs [43,238,239]. TiO_2 is the most widely used electrode in DSSCs. However, ZnO is thought to be a potential substitute for TiO_2 due to its high mobility [116].

A dye monolayer on a flat interface for ZnO thin films exhibits only negligible light absorption since the cross section of optical absorption for molecular dyes is typically 2–3 orders of magnitude smaller than their physical cross sections. The use of a nanostructured ZnO significantly enhances the interfacial surface area over the geometric surface area, by up to a 1000-fold for a 10 μm thick film, leading to high absorbance from the many successive monolayers of adsorbed dyes in the optical paths [240]. Figure 4.13 shows the schematic of ZnO nanostructure-based DSSCs. To compare with TiO_2, ZnO can be easily fabricated in various morphologies with a large surface area, which is an essential factor in maximizing dye adsorption. It has been studied that the electron transport of ZnO is related to its morphologies [26,43,241,242,243]. For example, the electron transport in the ZnO NW DSSCs is about two orders of magnitude faster than the ZnO NP DSSCs [241]. However, the short-circuit photocurrent density (I_{sc}) and the energy conversion efficiency (η) of ZnO NP DSSCs (with 22.8 μm in thickness) are superior to those of the ZnO NW DSSCs (with 18–24 μm in length) [43,244]. In general, the I_{sc} is proportional to the amount of the absorbed dyes, which is related to the area of the electrodes of DSSCs. The roughness factor (RF), defined as a ratio of total surface area of electrodes to shadow substrate areas of a cell, is widely used to quantitatively estimate the amount of the absorbed dyes [43]. The RF of 882.35 for ZnO NP DSSCs is much larger than that for ZnO NW DSSCs (RF = 200) [43,244]. If the RF factor is sufficiently large, even a monolayer of dye molecules could absorb most of the incident photons [237]. Various ZnO nanostructures used in DSSCs and their performance are listed in Table 4.3.

The transport in DSSCs is influenced by the surfaces of nanostructured electrodes significantly. For example, some electrons might be captured by the surface state located the ZnO electrode surface, decreasing the η of the cell. It has been reported that surface passivation, which can reduce the surface defects of the ZnO, is another method to enhance the η. After annealing the ZnO electrodes in oxygen ambient, the η increases from 1.03% to 1.59% [245]. Au NPs have been found to enhance the η of ZnO nanoflower-based DSSCs,

TABLE 4.3

Performance of DSSCs Utilizing Different ZnO Nanostructures

Nanostructure	η (%)	V_{oc} (mV)	I_{sc} (mA/cm²)	FF	References
Nanocrystalline film	0.40	550	1.22	0.66	[252]
	2.80	550	9.10	0.57	[253]
NR	1.32	570	7.00	0.33	[254]
	4.70	710	10.70	0.62	[253]
Nanofiber	3.02	570	9.14	0.58	[255]
NW	1.50	710	5.85	0.38	[43]
	0.84	500	3.40	0.49	[44]
NW with branched structure	0.46	620	1.84	0.40	[256]
NW with pillar structure	0.34	690	1.26	0.39	[256]
Nanosphere	1.59	668	5.43	0.44	[245]
	2.60	557	12.30	0.48	[257]
Nanotetrapod	1.02	580	3.76	0.47	[258]
	3.27	614	9.71	0.55	[259]
Nanoflower	1.60	580	8.75	0.32	[246]
	0.30	535	1.10	0.54	[257]
Nanoflower with Au NPs	2.50	500	15.00	0.33	[246]
NP	0.87	670	2.25	0.58	[254]
	6.58	621	18.11	0.59	[245]
	0.75	573	1.20	0.51	[260]
Nanosheet	1.55	593	2.06	0.55	[260]
Mesoporous aerogel thin films	2.40	600	8.32	0.48	[261]
NP aggregate	5.40	595	9.70	0.45	[262]
Plate aggregate	1.90	554	8.40	0.41	[257]
Network structure	1.34	600	3.58	0.62	[263]
Hierarchical structure	6.51	670	10.90	0.48	[264]
Nanosphere/film composite	2.25	718	6.13	0.51	[245]
NW/NP composite	2.20	610	6.30	0.58	[44]
ZnO/TiO₂ core/shell NW	2.27	800	4.78	0.60	[248]
	2.00	705	5.30	0.54	[249]
	2.00	704	5.30	0.53	[250]

η: solar energy conversion efficiency, V_{oc}: open-circuit voltage, I_{sc}: short-current density, FF: fill-factor, NPs: nanoparticles, and NWs: nanowires.

that is, the ZnO nanoflowers decorated with Au NPs showed 2.5% of η, higher than 1.6% for the ZnO nanoflowers [246]. This is because surface defects are eliminated during the Au NPs' formation process, possibly by the electron abstraction from the defects through the reduction of Au ions.

It is known that ZnO surface is unstable in acidic dye of DSSCs, such as N_3 and N719. ZnO surface forms Zn^{2+}/dye complexes in dye environments, leading to a low electron injection efficiency of the dye molecules [247]. The core/shell structure, one of surface passivation methods, provides a good way to protect the ZnO against the harm of the acidic dye. Law et al. [248] have reported that the Al_2O_3 shells acting as a protecting and electron-blocking layer improved the open-circuit voltage (V_{oc}) but greatly decreased the short-circuit current of the ZnO/Al₂O₃ core/shell NW DSSCs. However, TiO₂ shells increased the V_{oc} and fill factor, resulting in the improvement of η up to 2.25% for the ZnO/TiO₂ core/shell NW DSSCs [248]. Similar results were also reported by Wang et al. [249,250] confirming the fact that the shells suppress the recombination by forming an energy barrier and passivating surface recombination centers.

4.8 Conclusion

We have reviewed some of the critical, initial researches designed to determine unique transport, photoluminescence, and photoconductivity properties offered by ZnO nanostructures with ultrahigh S/V ratio. Encouraging progresses on versatile applications of ZnO nanostructures developed on the basis of these distinct physical properties are reviewed as well. We expect that nanostructured ZnO employing the surface effect could open up tremendous potential for the future electronic and optoelectronic devices.

Acknowledgments

A special acknowledgment to Ming-Wei Chen, Jr-Jian Ke, and Chin-An Lin from the *Institute of Photonics and Optoelectronics at the National Taiwan University* for their support, contribution, and collaboration in the redaction of this chapter.

References

1. L. E. Greene, M. Law, D. H. Tan, M. Montano, J. Goldberger, G. Somorjai, and P. Yang, General route to vertical ZnO nanowire arrays using textured ZnO seeds, *Nano Lett.* **5**, 1231 (2005).
2. X. Y. Kong and Z. L. Wang, Spontaneous polarization-induced nanohelixes, nanosprings, and nanorings of piezoelectric nanobelts, *Nano Lett.* **3**, 1625 (2003).
3. L. Wang, K. Chen, and L. Dong, Synthesis of exotic zigzag ZnO nanoribbons and their optical, electrical properties, *J. Phys. Chem. C* **114**, 17358 (2010).
4. E. A. Meulenkamp, Synthesis and growth of ZnO nanoparticles, *J. Phys. Chem. B* **102**, 5566 (1998).
5. J. Shi, S. Grutzik, and X. Wang, Zn cluster drifting effect for the formation of ZnO 3D nanoarchitecture, *ACS Nano* **3**, 1594 (2009).
6. J. Y. Lao, J. G. Wen, and Z. F. Ren, Hierarchical ZnO nanostructures, *Nano Lett.* **2**, 1287 (2002).
7. X. Y. Kong and Z. L. Wang, Polar-surface dominated ZnO nanobelts and the electrostatic energy induced nanohelixes, nanosprings, and nanospirals, *Appl. Phys. Lett.* **84**, 975 (2004).
8. X. Y. Kong, Y. Ding, R. S. Yang, and Z. L. Wang, Single-crystal nanorings formed by epitaxial self-coiling of polar nanobelts, *Science* **303**, 1348 (2004).
9. A. B. Djurišić, A. M. C. Ng, and X. Y. Chen, ZnO nanostructures for optoelectronics: Material properties and device applications, *Prog. Quant. Electron.* **34**, 191 (2010).
10. Z. L. Wang, ZnO nanowire and nanobelt platform for nanotechnology, *Mater. Sci. Eng. R Rep.* **64**, 33 (2009).
11. Z.-M. Liao, K.-J. Liu, J.-M. Zhang, J. Xu, and D.-P. Yu, Effect of surface states on electron transport in individual ZnO nanowires, *Phys. Lett. A* **367**, 207 (2007).

12. J. I. Sohn, W.-K. Hong, M. J. Lee, T. Lee, H. Sirringhaus, D. J. Kang, and M. E. Welland, The influence of surface chemical dynamics on electrical and optical properties of ZnO nanowire field effect transistors, *Nanotechnology* **20**, 505202 (2009).

13. C. Y. Chen, C. A. Lin, and J. H. He, ZnO/Al_2O_3 core–shell nanorod arrays: Growth, structural characterization, and luminescent properties, *Nanotechnology* **20**, 185605 (2009).

14. H. Moormann, D. Kohl, and G. Heiland, Work function and band bending on clean cleaved zinc oxide surfaces, *Surf. Sci.* **80**, 261 (1979).

15. J. Suehiro, N. Nakagawa, S.-I. Hidaka, M. Ueda, K. Imasaka, M. Higashihata, T. Okada, and M. Hara, Dielectrophoretic fabrication and characterization of a ZnO nanowire-based UV photosensor, *Nanotechnology* **17**, 2567 (2006).

16. Q. H. Li, T. Gao, Y. G. Wang, and T. H. Wang, Adsorption and desorption of oxygen probed from ZnO nanowire films by photocurrent measurements, *Appl. Phys. Lett.* **86**, 123117 (2005).

17. Q. Yang, K. Tang, J. Zuo, and Y. Qian, Synthesis and luminescent property of single-crystal ZnO nanobelts by a simple low temperature evaporation route, *Appl. Phys. A* **79**, 1847 (2004).

18. W. Gcpel and U. Lampe, Influence of defects on the electronic structure of zinc oxide surfaces, *Phys. Rev. B* **22**, 6447 (1980).

19. L. Schmidt-Mende and J. L. MacManus-Driscoll, ZnO—Nanostructures, defects, and devices, *Mater. Today* **10**, 40 (2007).

20. A. Janotti and C. G. Van de Walle, Native point defects in ZnO, *Phys. Rev. B* **76**, 165202 (2007).

21. I. Shalish, H. Temkin, and V. Narayanamurti, Size-dependent surface luminescence in ZnO nanowires, *Phys. Rev. B* **69**, 245401 (2004).

22. Z. Zhao, W. Lei, X. Zhang, B. Wang, and H. Jiang, ZnO-based amperometric enzyme biosensors, *Sensors* **10**, 1216 (2010).

23. Z.-M. Liao, C. Hou, Y.-B. Zhou, J. Xu, J.-M. Zhang, and D.-P. Yu, Influence of temperature and illumination on surface barrier of individual ZnO nanowires, *J. Chem. Phys.* **130**, 084708 (2009).

24. W.-K. Hong, G. Jo, S.-S. Kwon, S. Song, and T. Lee, Electrical properties of surface-tailored ZnO nanowire field-effect transistors, *IEEE Trans. Electron Dev.* **55**, 3020 (2008).

25. W.-K. Hong, S. Song, D.-K. Hwang, S.-S. Kwon, G. Jo, S.-J. Park, and T. Lee, Effects of surface roughness on the electrical characteristics of ZnO nanowire field effect transistors, *Appl. Surf. Sci.* **254**, 7559 (2008).

26. C. Look, D. C. Reynolds, J. R. Sizelove, R. L. Jones, C. W. Litton, G. Cantwell, and W. C. Harsch, Electrical properties of bulk ZnO, *Solid State Commun.* **105**, 399 (1998).

27. J. H. He, C. H. Ho, C. W. Wang, Y. Ding, L. J. Chen, and Z. L. Wang, Growth of crossed ZnO nanorod networks induced by polar substrate surface, *Cryst. Growth Des.* **9**, 17 (2009).

28. J. H. He and C. H. Ho, The study of electrical characteristics of heterojunction based on ZnO nanowires using ultrahigh-vacuum conducting atomic force microscopy, *Appl. Phys. Lett.* **91**, 233105 (2007).

29. J. H. He, S. T. Ho, T. B. Wu, L. J. Chen, and Z. L. Wang, Electrical and photoelectrical performances of nano-photodiode based on ZnO nanowires, *Chem. Phys. Lett.* **435**, 119 (2007).

30. S. N. Cha, J. E. Jang, Y. Choi, G. A. J. Amaratunga, G. W. Ho, M. E. Welland, D. G. Hasko, D.-J. Kang, and J. M. Kim, High performance ZnO nanowire field effect transistor using self-aligned nanogap gate electrodes, *Appl. Phys. Lett.* **89**, 263102 (2006).

31. P.-C. Chang, Z. Fan, C.-J. Chien, D. Ronning, C. Stichtenoth, and J. G. Lu, High-performance ZnO nanowire field effect transistors, *Appl. Phys. Lett.* **89**, 133113 (2006).

32. J. Bao, M. A. Zimmler, and F. Capasso, Broadband ZnO single-nanowire light-emitting diode, *Nano Lett.* **6**, 1719 (2006).

33. Z. Fan and J. G. Lu, Chemical sensing with ZnO nanowire, *IEEE Trans. Nanotechnol.* **5**, 393 (2006).

34. T.-J. Hsueh, S.-J. Chang, C.-L. Hsu, Y.-R. Lin, and I.-C. Chen, Highly sensitive ZnO nanowire ethanol sensor with Pd adsorption, *Appl. Phys. Lett.* **91**, 053111 (2007).

35. J. Huang and Q. Wan, Gas sensors based on semiconducting metal oxide one-dimensional nanostructures, *Sensors* **9**, 9903 (2009).

36. P.-C. Chen, G.-Z. Shen, and C.-W. Zhou, Chemical sensors and electronic noses based on 1-D metal oxide nanostructures, *IEEE Trans. Nanotechnol.* **7**, 668 (2008).
37. A. Kolmakov and M. Moskovits, Chemical sensing and catalysis by one-dimensional metal-oxide nanostructures, *ChemInform* **35** (published online), DOI: 10.1002/chin.200441227 (2004).
38. C. Soci, A. Zhang, B. Xiang, S. A. Dayeh, D. P. R. Aplin, J. Park, X. Y. Bao, Y. H. Lo, and D. Wang, ZnO nanowire UV photodetectors with high internal gain, *Nano Lett.* **7**, 1003 (2007).
39. H. Kind, H. Yan, B. Messer, M. Law, and P. Yang, Nanowire ultraviolet photodetectors and optical switches, *Adv. Mater.* **14**, 158 (2002).
40. Z. Fan, P. Chang, E. C. Walter, C. Lin, H. P. Lee, R. M. Penner, and J. G. Lu, Photoluminescence and polarized photodetection of single ZnO nanowires, *Appl. Phys. Lett.* **85**, 6128 (2004).
41. C. Soci, A. Zhang, X.-Y. Bao, H. Kim, Y. Lo, and D. Wang, Nanowire photodetectors, *J. Nanosci. Nanotechnol.* **10**, 1430 (2010).
42. S. W. Chan, R. J. Barille, M. Nunzi, K. H. Tam, Y. H. Leung, W. K. Chan, and A. B. Djurišić, Second harmonic generation in zinc oxide nanorods, *Appl. Phys. B* **84**, 351 (2006).
43. M. Law, L. E. Greene, J. C. Johnson, R. Saykally, and P. Yang, Nanowire dye-sensitized solar cells, *Nat. Mater.* **4**, 455 (2005).
44. C. H. Ku and J. J. Wu, Electron transport properties in ZnO nanowire array/nanoparticle composite dye-sensitized solar cells, *Appl. Phys. Lett.* **91**, 093117 (2007).
45. J. H. He, S. Singamaneni, C. H. Ho, Y. H. Lin, M. E. McConney, and V. V. Tsukruk, A thermal sensor and switch based on a plasma polymer/ZnO suspended nanobelt bimorph structure, *Nanotechnology* **20**, 065502 (2009).
46. A. van Dijken, E. Meulenkamp, D. Vanmaekelbergh, and A. Meijerink, Identification of the transition responsible for the visible emission in ZnO using quantum size effects, *J. Lumin.* **90**, 123 (2000).
47. P.-C. Chang, C.-J. Chien, D. Stichtenoth, C. Ronning, and J. G. Lu, Finite size effect in ZnO nanowires, *Appl. Phys. Lett.* **90**, 113101 (2007).
48. S.-P. Chiu, Y.-H. Lin, and J.-J. Lin, Electrical conduction mechanisms in natively doped ZnO nanowires, *Nanotechnology* **20**, 015203 (2009).
49. M. H. Huang, Y. Wu, H. Feick, N. Tran, E. Weber, and P. Yang, Catalytic growth of zinc oxide nanowires by vapor transport, *Adv. Mater.* **13**, 113 (2001).
50. G.-C. Yi et al., ZnO nanorods: Synthesis, characterization and applications, *Semicond. Sci. Technol.* **20**, S22 (2005).
51. K. H. Tam et al., Defects in ZnO nanorods prepared by a hydrothermal method, *J. Phys. Chem. B* **110**, 20865 (2006).
52. P. Chang, Z. Fan, W. Tseng, D. Wang, W. Chio, J. Hong, and J. G. Lu, ZnO nanowires synthesized by vapor trapping CVD method, *Chem. Mater.* **16**, 5133 (2004).
53. X. Liu, X. Wu, H. Cao, and R. P. H. Chang, Growth mechanism and properties of ZnO nanorods synthesized by plasma-enhanced chemical vapor deposition, *J. Appl. Phys.* **95**, 3141 (2004).
54. D. C. Kim, B. H. Kong, and H. K. Cho, Morphology control of 1D ZnO nanostructures grown by metal-organic chemical vapor deposition, *J. Mater. Sci. Mater. Electron.* **19**, 760 (2008).
55. I. C. Robin et al., Compared optical properties of ZnO heteroepitaxial, homoepitaxial 2D layers and nanowires, *J. Cryst. Growth* **311**, 2172 (2009).
56. Y.-F. Zhang, Z.-Y. Guo, X.-Q. Gao, D.-X. Cao, Y.-X. Dai, and H.-T. Zhao, First-principles of wurtzite ZnO (0001) and (000$\bar{1}$) surface structures, *J. Semicond.* **31**, 082001 (2010).
57. Z. L. Wang, Zinc oxide nanostructures: Growth, properties and applications, *J. Phys. Condens. Matter* **16**, R829 (2004).
58. Z. W. Pan, Z. R. Dai, and Z. L. Wang, Nanobelts of semiconducting oxides, *Science* **291**, 1947 (2001).
59. N. S. Ramgir, D. J. Late, A. B. Bhise, M. A. More, I. S. Mulla, D. S. Joag, and K. Vijayamohanan, ZnO multipods, submicron wires, and spherical structures and their unique field emission behavior, *J. Phys. Chem. B* **110**, 18236 (2006).
60. P. X. Gao and Z. L. Wang, Self-assembled nanowire–nanoribbon junction arrays of ZnO, *J. Phys. Chem. B* **106**, 12653 (2002).

61. Y. C. Chao, C. Y. Chen, C. A. Lin, and J. H. He, Light scattering by nanostructured anti-reflection coatings, *Energy Environ. Sci.* **4** (published on line), DOI: 10.1039/c0ee00636j (2011).

62. T. Gruber, C. Kirchner, R. Kling, F. Reuss, and A. Waag, ZnMgO epilayers and ZnO–ZnMgO quantum wells for optoelectronic applications in the blue and UV spectral region, *Appl. Phys. Lett.* **84**, 5359 (2004).

63. W. I. Park, S. J. An, J. L. Yang, G.-C. Yi, S. Hong, T. Joo, and M. Kim, Photoluminescent properties of ZnO/Zn$_{0.8}$Mg$_{0.2}$O nanorod single-quantum-well structures, *J. Phys. Chem. B* **108**, 15457 (2004).

64. S.-H. Nam, S. H. Jeong, and J.-H. Boo, Growth behavior and characteristics of one dimensional ZnO nanostructures by metalorganic chemical vapor deposition, *J. Nanosci. Nanotechnol.* **11**, 1648 (2011).

65. A. Umar, S. Lee, Y. H. Im, and Y. B. Hahn, Flower-shaped ZnO nanostructures obtained by cyclic feeding chemical vapour deposition: Structural and optical properties, *Nanotechnology* **16**, 2462 (2005).

66. Y.-F. Lin, W.-B. Jian, C. P. Wang, Y.-W. Suen, Z.-Y. Wu, F.-R. Chen, J.-J. Kai, and J.-J. Lin, Contact to ZnO and intrinsic resistances of individual ZnO nanowires with a circular cross section, *Appl. Phys. Lett.* **90**, 223117 (2007).

67. L. T. Tsai, S. P. Chiu, J. G. Lu, and J.-J. Lin, Electrical conduction mechanisms in natively doped ZnO nanowires (II), *Nanotechnology* **21**, 145202 (2010).

68. Z. Zhang, K. Yao, Y. Liu, C. Jin, X. Liang, Q. Chen, and L.-M. Peng, Quantitative analysis of current–voltage characteristics of semiconducting nanowires: Decoupling of contact effects, *Adv. Funct. Mater.* **17**, 2478 (2007).

69. S. M. Sze, D. J. Coleman Jr, and A. Loya, Current transport in metal-semiconductor-metal (MSM) structures, *Solid State Electron.* **14**, 1209 (1971).

70. Y. Yoon, J. Lin, S. J. Pearton, and J. Guo, Role of grain boundaries in ZnO nanowire field-effect transistors, *J. Appl. Phys.* **101**, 024301 (2007).

71. W.-K. Hong, D.-K. Hwang, I.-K. Park, G. Jo, S. Song, S.-J. Park, T. Lee, B.-J. Kim, and E. A. Stach, Realization of highly reproducible ZnO nanowire field effect transistors with n-channel depletion and enhancement modes, *Appl. Phys. Lett.* **90**, 243103 (2007).

72. S. Ju, D. B. Janes, G. Lu, A. Facchetti, and T. J. Marks, Effects of bias stress on ZnO nanowire field-effect transistors fabricated with organic gate nanodielectrics, *Appl. Phys. Lett.* **89**, 193506 (2006).

73. Y. W. Heo, L. C. Tien, D. P. Norton, S. J. Pearton, B. S. Kang, F. Ren, and J. R. LaRoche, Pt/ZnO nanowire Schottky diodes, *Appl. Phys. Lett.* **85**, 3107 (2004).

74. K. Kim, H. Kang, H. Kim, J. Lee, S. Kim, W. Kang, and G.-T. Kim, Contact barriers in a single ZnO nanowire device, *Appl. Phys. A Mater.* **94**, 253 (2009).

75. C. H. Liu, W. C. Yiu, F. C. K. Au, J. X. Ding, C. S. Lee, and S. T. Lee, Electrical properties of zinc oxide nanowires and intramolecular *p–n* junctions, *Appl. Phys. Lett.* **83**, 3168 (2003).

76. Q. H. Li, Q. Wan, Y. X. Liang, and T. H. Wang, Electronic transport through individual ZnO nanowires, *Appl. Phys. Lett.* **84**, 4556 (2004).

77. J. Goldberger, D. J. Sirbuly, M. Law, and P. Yang, ZnO nanowire transistors, *J. Phys. Chem. B* **109**, 9 (2004).

78. F. Xiaojun et al., Fabrication and photoelectrical characteristics of ZnO nanowire field-effect transistors, *J. Semicond.* **30**, 084002 (2009).

79. H. Chik, J. Liang, S. G. Cloutier, N. Kouklin, and J. M. Xu, Periodic array of uniform ZnO nanorods by second-order self-assembly, *Appl. Phys. Lett.* **84**, 3376 (2004).

80. Y. W. Heo, L. C. Tien, D. P. Norton, B. S. Kang, F. Ren, B. P. Gila, and S. J. Pearton, Electrical transport properties of single ZnO nanorods, *Appl. Phys. Lett.* **85**, 2002 (2004).

81. W. I. Park, J. S. Kim, G.-C. Yi, M. H. Bae, and H.-J. Lee, Fabrication and electrical characteristics of high-performance ZnO nanorod field-effect transistors, *Appl. Phys. Lett.* **85**, 5052 (2004).

82. M. S. Arnold, P. Avouris, Z. W. Pan, and Z. L. Wang, Field-effect transistors based on single semiconducting oxide nanobelts, *J. Phys. Chem. B* **107**, 659 (2003).

83. S. Lao, J. Liu, P. Gao, L. Zhang, D. Davidovic, R. Tummala, and Z.-L. Wang, ZnO nanobelt/nanowire Schottky diodes formed by dielectrophoresis alignment across Au electrodes, *Nano Lett.* **6**, 263 (2006).

84. J. Zhou, Y. Gu, Y. Hu, W. Mai, P.-H. Yeh, G. Bao, A. K. Sood, D. L. Polla, and Z. L. Wang, Gigantic enhancement in response and reset time of ZnO UV nanosensor by utilizing Schottky contact and surface functionalization, *Appl. Phys. Lett.* **94**, 191103 (2009).

85. S.-A. Cromar, Properties of suspended ZnO nanowire field-effect transistor (published online http://hdl.handle.net/2142/9690) (2006).

86. L. Ming, H.-Y. Zhang, C.-X. Guo, J.-B. Xu, X.-J. Fu, and C.-P. Feng, Characterization of ZnO nanowire field-effect transistors exposed to ultraviolet radiation, *Chin. Phys. B* **18**, 5020 (2009).

87. D.-I. Suh, S.-Y. Lee, J.-H. Hyung, T.-H. Kim, and S.-K. Lee, Multiple ZnO nanowires field-effect transistors, *J. Phys. Chem. C* **112**, 1276–1281 (2008).

88. D.-J. Kim, J.-H. Hyung, D.-W. Seo, D.-I. Suh, and S.-K. Lee, Dual-gate multiple-channel ZnO nanowire transistors, *J. Electron. Mater.* **39**, 563 (2010).

89. S.-H. Ju, K.-H. Lee, D. B. Janes, J.-Y. Li, R. P. H. Chang, M.-H. Yoon, A. Facchetti, and T. J. Marks, ZnO nanowire field-effect transistors: Ozone-induced threshold voltage shift and multiple nanowire effects, *Nanotechnology, 2006. IEEE-NANO 2006. Sixth IEEE Conference*, West Lafayette, IN, p. 445 (2006).

90. O. Harnack, C. Pacholski, H. Weller, A. Yasuda, and J. M. Wessels, Rectifying behavior of electrically aligned ZnO nanorods, *Nano Lett.* **3**, 1097 (2003).

91. Y.-K. Chang and F. C.-N. Hong, The fabrication of ZnO nanowire field-effect transistors combining dielectrophoresis and hot-pressing, *Nanotechnology* **20**, 235202 (2009).

92. Y.-K. Chang and F. C.-N. Hong, The fabrication of ZnO nanowire field-effect transistors by roll-transfer printing, *Nanotechnology* **20**, 195302 (2009).

93. P.-C. Chang and J. G. Lu, ZnO nanowire field-effect transistors, *IEEE Trans. Electron Dev.* **55**, 2977 (2008).

94. S. Ju, K. Lee, M. H. Yoon, A. Facchetti, T. J. Marks, and D. B. Janes, High performance ZnO nanowire field effect transistors with organic gate nanodielectrics: Effects of metal contacts and ozone treatment, *Nanotechnology* **18**, 155201 (2007).

95. K. Keem, H. Kim, G.-T. Kim, J. S. Lee, B. Min, K. Cho, M.-Y. Sung, and S. Kim, Photocurrent in ZnO nanowires grown from Au electrodes, *Appl. Phys. Lett.* **84**, 4376 (2004).

96. J. H. He, P. H. Chang, C. Y. Chen, and K. D. Tsai, Electrical and optoelectronic characterization of a ZnO nanowire contacted by focused-ion-beam-deposited Pt, *Nanotechnology* **20**, 135701 (2009).

97. W.-Q. Yang, L. Dai, R.-M. Ma, C. Liu, T. Sun, and G.-G. Qin, Back-gate ZnO nanowire field-effect transistors each with a top Omega shaped Au contact, *Appl. Phys. Lett.* **93**, 033102 (2008).

98. Z. Y. Fan, D. W. Wang, P. C. Chang, W. Y. Tseng, and J. G. Lu, ZnO nanowire field-effect transistor and oxygen sensing property, *Appl. Phys. Lett.* **85**, 5923 (2004).

99. Y.-Y. Noh, X. Cheng, H. Sirringhaus, J. I. Sohn, M. E. Welland, and D. J. Kang, Ink-jet printed ZnO nanowire field effect transistors, *Appl. Phys. Lett.* **91**, 043109 (2007).

100. G. J. Wang, E. Polizzi, S. Datta, and M. Lundstrom, Electrostatics of nanowire transistors, *IEEE Trans. Nanotechnol.* **2**, 329 (2003).

101. Y. C. Yeo, Metal gate technology for nanoscale transistors—Material selection and process integration issues, *Thin Solid Films* **34**, 462 (2004).

102. K. Keem, J. Kang, C. Yoon, D. Yeom, D.-Y. Jeong, B.-M. Moon, and S. Kim, A fabrication technique for top-gate ZnO nanowire field-effect transistors by a photolithography process, *Microelectron. Eng.* **84**, 1622 (2007).

103. H.-J. Kim, C.-H. Lee, D.-W. Kim, and G.-C. Yi, Fabrication and electrical characteristics of dual-gate ZnO nanorod metal–oxide semiconductor field-effect transistors, *Nanotechnology* **17**, S327 (2006).

104. Z. Fan and J. G. Lu, Electrical properties of ZnO nanowire field effect transistors characterized with scanning probes, *Appl. Phys. Lett.* **86**, 032111 (2005).

105. Y. W. Heo, L. C. Tien, Y. Kwon, D. P. Norton, S. J. Pearton, B. S. Kang, and F. Ren, Depletion-mode ZnO nanowire field-effect transistor, *Appl. Phys. Lett.* **85**, 2274 (2004).

106. S. Ju, K. Lee, D. B. Janes, M.-H. Yoon, A. Facchetti, and T. J. Marks, Low operating voltage single ZnO nanowire field-effect transistors enabled by self-assembled organic gate nanodielectrics, *Nano Lett.* **5**, 2281 (2005).

107. S. Ju, S. Kim, S. Mohammadi, D. B. Janes, Y.-G. Ha, A. Facchetti, and T. J. Marks, Interface studies of ZnO nanowire transistors using low-frequency noise and temperature-dependent I-V measurements, *Appl. Phys. Lett.* **92**, 022104 (2008).

108. S.-S. Kwon, W.-K. Hong, G. Jo, J. Maeng, T.-W. Kim, S. Song, and T. Lee, Piezoelectric effect on the electronic transport characteristics of ZnO nanowire field-effect transistors on bent flexible substrates, *Adv. Mater.* **20**, 4557 (2008).

109. D. R. Khanal and J. Wu, Gate coupling and charge distribution in nanowire field effect transistors, *Nano Lett.* **7**, 2778 (2007).

110. G. Eranna, B. C. Joshi, D. P. Runthala, and R. P. Gupta, Oxide materials for development of integrated gas sensors—A comprehensive review, *Crit. Rev. Solid State Mater. Sci.* **29**, 111 (2004).

111. Z.-M. Liao, Z.-K. Lv, Y.-B. Zhou, J. Xu, J.-M. Zhang, and D.-P. Yu, The effect of adsorbates on the space–charge-limited current in single ZnO nanowires, *Nanotechnology* **19**, 335204 (2008).

112. J. Maeng, W. Park, M. Choe, G. Jo, Y. H. Kahng, and T. Lee, Transient drain current characteristics of ZnO nanowire field effect transistors, *Appl. Phys. Lett.* **95**, 123101 (2009).

113. S. Song, W.-K. Hong, S.-S. Kwon, and T. Lee, Passivation effects on ZnO nanowire field effect transistors under oxygen, ambient, and vacuum environments, *Appl. Phys. Lett.* **92**, 263109 (2008).

114. F. M. Hossain, J. Nishii, S. Takagi, A. Ohtomo, T. Fukumura, H. Fujioka, H. Ohno, H. Koinuma, and M. J. Kawasaki, Modeling and simulation of polycrystalline ZnO thin-film transistors, *Appl. Phys.* **94**, 7768 (2003).

115. F. Jones, F. Léonard, A. A. Talin, and N. S. Bell, Electrical conduction and photoluminescence properties of solution-grown ZnO nanowires, *J. Appl. Phys.* **102**, 014305 (2007).

116. E. M. Kaidashev, M. Lorenz, H. Wenckstern, A. Rahm, H. C. Semmelhack, K. H. Han, G. Benndorf, C. Bundrsmann, H. Hochmuth, and M. Grundmann, High electron mobility of epitaxial ZnO thin films on *c*-plane sapphire grown by multistep pulsed-laser deposition, *Appl. Phys. Lett.* **82**, 3901 (2003).

117. W.-K. Hong, B.-J. Kim, T.-W. Kim, G. Jo, S. Song, S.-S. Kwon, A. Yoon, E. A. Stach, and T. Lee, Electrical properties of ZnO nanowire field effect transistors by surface passivation, *Colloids Surf. Physicochem. Eng. Aspects* **313–314**, 378 (2008).

118. S. H. Jo, J. Y. Lao, Z. F. Ren, R. A. Farrer, T. Baldacchini, and J. T. Fourkas, Field-emission studies on thin films of zinc oxide nanowires, *Appl. Phys. Lett.* **83**, 4821 (2003).

119. K. J. Kang, C. Yoon, D. Y. Jeong, B. M. Moon, and S. Kim, Enhanced performance of ZnO nanowire field effect transistors by H$_2$ annealing, *Jpn. J. Appl. Phys.* **46**, 6230 (2007).

120. C. Bekeny, T. Voss, B. Hilker, J. Gutowski, R. Hauschild, H. Kalt, B. Postels, A. Bakin, and A. Waag, Influence of ZnO seed crystals and annealing on the optical quality of low-temperature grown ZnO nanorods, *J. Appl. Phys.* **102**, 044908 (2007).

121. Y. Hu, Y. Liu, H. Xu, X. Liang, L.-M. Peng, N. Lam, K. Wong, and Q. Li, Quantitative study on the effect of surface treatments on the electric characteristics of ZnO nanowires, *J. Phys. Chem. C* **112**, 14225 (2008).

122. W.-K. Hong, G. Jo, J. I. Sohn, W. Park, M. Choe, G. Wang, Y. H. Kahng, M. E. Welland, and T. Lee, Tuning of the electronic characteristics of ZnO nanowire field effect transistors by proton irradiation, *ACS Nano* **4**, 811 (2010).

123. G. Jo, W.-K. Hong, J. Maeng, M. Choe, W. Park, and T. Lee, Logic inverters composed of controlled depletion-mode and enhancement-mode ZnO nanowire transistors, *Appl. Phys. Lett.* **94**, 173118 (2009).

124. W. Park, W.-K. Hong, G. Jo, G. Wang, M. Choe, J. Maeng, Y. H. Kahng, and T. Lee, Tuning of operation mode of ZnO nanowire field effect transistors by solvent-driven surface treatment, *Nanotechnology* **20**, 475702 (2009).

125. A. B. Djurišić and Y. H. Leung, Optical properties of ZnO nanostructures, *Small* **2**, 944 (2006).

126. R.-C. Wang, C.-P. Liu, J.-L. Huang, and S.-J. Chen, ZnO symmetric nanosheets integrated with nanowalls, *Appl. Phys. Lett.* **87**, 053103 (2005).

127. L. E. Greene, M. Law, J. Goldberger, F. Kim, J. C. Johnson, Y. Zhang, R. J. Saykally, and P. Yang, Low-temperature wafer-scale production of ZnO nanowire arrays, *Angew. Chem. Int. Ed.* **42**, 3031 (2003).

128. H. J. Fan, R. Scholz, F. M. Kolb, and M. Zacharias, Two-dimensional dendritic ZnO nanowires from oxidation of Zn microcrystals, *Appl. Phys. Lett.* **85**, 4142 (2004).

129. H. J. Fan, R. Scholz, F. M. Kolb, M. Zacharias, U. Gcsele, F. Heyroth, C. Hempel, T. Eisenschmidt, and J. Christen, On the growth mechanism and optical properties of ZnO multi-layer nanosheets, *Appl. Phys. A* **79**, 1895 (2004).

130. X. Q. Meng, D. Z. Shen, J. Y. Zhang, D. X. Zhao, Y. M. Lu, L. Dong, Z. Z. Zhang, Y. C. Liu, and X. W. Fan, The structural and optical properties of ZnO nanorod arrays, *Solid State Commun.* **135**, 179 (2005).

131. L. Wang, X. Zhang, S. Zhao, G. Zhou, Y. Zhou, and J. Qi, Synthesis of well-aligned ZnO nanowires by simple physical vapor deposition on *c*-oriented ZnO thin films without catalysts or additives, *Appl. Phys. Lett.* **86**, 024108 (2005).

132. H. T. Ng, B. Chen, J. Li, J. Han, M. Meyyappan, J. Wu, S. X. Li, and E. E. Haller, Optical properties of single-crystalline ZnO nanowires on *m*-sapphire, *Appl. Phys. Lett.* **82**, 2023 (2003).

133. D. Zhao, C. Andreazza, P. Andreazza, J. Ma, Y. Liu, and D. Shen, Temperature-dependent growth mode and photoluminescence properties of ZnO nanostructures, *Chem. Phys. Lett.* **399**, 522 (2004).

134. L. Huang, S. Wright, S. Yang, D. Shen, B. Gu, and Y. Du, ZnO well-faceted fibers with periodic junctions, *J. Phys. Chem. B* **108**, 19901 (2004).

135. F. Wen, W. Li, J.-H. Moon, and J.-H. Kim, Hydrothermal synthesis of ZnO:Zn with green emission at low temperature with reduction process, *Solid State Commun.* **135**, 34 (2005).

136. F. Wang, Z. Ye, D. Ma, L. Zhu, and F. Zhuge, Rapid synthesis and photoluminescence of novel ZnO nanotetrapods, *J. Cryst. Growth* **274**, 447 (2005).

137. S. Choopun, N. Hongsith, S. Tanunchai, T. Chairuangsri, C. Krua-in, S. Singkarat, T. Vilathong, P. Mangkorntong, and N. Mangkorntong, Single-crystalline ZnO nanobelts by RF sputtering, *J. Cryst. Growth* **282**, 365 (2005).

138. W. Yu, X. Li, and X. Gao, Catalytic synthesis and structural characteristics of high-quality tetrapod-like ZnO nanocrystals by a modified vapor transport process, *Cryst. Growth Des.* **5**, 151 (2005).

139. S. Hong, T. Joo, W. I. Park, Y. H. Jun, and G.-C. Yi, Time-resolved photoluminescence of the size-controlled ZnO nanorods, *Appl. Phys. Lett.* **83**, 4157 (2003).

140. Y. Gu, I. Kuskovsky, M. Yin, S. O'Brien, and G. F. Neumark, Quantum confinement in ZnO nanorods, *Appl. Phys. Lett.* **85**, 3833 (2004).

141. T. B. Hur, Y.-H. Hwang, and H.-K. Kim, Quantum confinement in Volmer–Weber-type self-assembled ZnO nanocrystals, *Appl. Phys. Lett.* **86**, 193113 (2005).

142. Y.-Y. Peng, T.-E. Hsieh, and C.-H. Hsu, Optical characteristics and microstructure of ZnO quantum dots-SiO_2 nanocomposite films prepared by sputtering methods, *Appl. Phys. Lett.* **89**, 211909 (2006).

143. X. Wang, Y. Ding, C. J. Summers, and Z. L. Wang, Large-scale synthesis of six-nanometer-wide ZnO nanobelts, *J. Phys. Chem. B* **108**, 8773 (2004).

144. J. Han, P. Q. Mantas, and A. M. R. Senos, Defect chemistry and electrical characteristics of undoped and Mn-doped ZnO, *J. Eur. Ceram. Soc.* **22**, 49 (2002).

145. M. H. Huang, S. Mao, H. Feick, H. Yan, Y. Wu, H. Kind, E. Weber, R. Russo, and P. Yang, Room-temperature ultraviolet nanowire nanolasers, *Science* **292**, 1897 (2001).

146. Y. Q. Chen, J. Jiang, Z. Y. He, Y. Su, D. Cai, and L. Chen, Growth mechanism and characterization of ZnO microbelts and self-assembled microcombs, *Mater. Lett.* **59**, 3280 (2005).

147. S. M. Abrarov, Sh. U. Yuldashev, T. W. Kim, S. B. Lee, Y. H. Kwon, and T. W. Kang, Effect of photonic band-gap on photoluminescence of ZnO deposited inside the green synthetic opal, *Opt. Commun.* **250**, 111 (2005).

148. Z. Chen, N. Wu, Z. Shan, M. Zhao, S. Li, C. B. Jiang, M. K. Chyu, and S. X. Mao, Effect of N2 flow rate on morphology and structure of ZnO nanocrystals synthesized via vapor deposition, *Scr. Mater.* **52**, 63 (2005).

149. D. C. Reynolds, D. C. Look, and B. Jogai, Fine structure on the green band in ZnO, *J. Appl. Phys.* **89**, 6189 (2001).

150. A. van Dijken, E. Meulenkamp, D. Vanmaekelbergh, and A. Meijerink, The kinetics of the radiative and nonradiative processes in nanocrystalline ZnO particles upon photoexcitation, *J. Phys. Chem. B* **104**, 1715 (2000).

151. Y. W. Heo, D. P. Norton, and S. J. Pearton, Origin of green luminescence in ZnO thin film grown by molecular-beam epitaxy, *J. Appl. Phys.* **98**, 073502 (2005).

152. Q. X. Zhao, P. Klason, M. Willander, H. M. Zhong, W. Lu, and J. H. Yang, Deep-level emissions influenced by O and Zn implantations in ZnO, *Appl. Phys. Lett.* **87**, 211912 (2005).

153. A. B. Djurišić, W. C. H. Choy, V. A. L. Roy, Y. H. Leung, C. Y. Kwong, K. W. Cheah, T. K. Gundu Rao, W. K. Chan, H. F. Lui, and C. Surya, Photoluminescence and electron paramagnetic resonance of ZnO tetrapod structures, *Adv. Funct. Mater.* **14**, 856 (2004).

154. N. E. Hsu, W. K. Hung, and Y. F. Chen, Origin of defect emission identified by polarized luminescence from aligned ZnO nanorods, *J. Appl. Phys.* **96**, 4671 (2004).

155. D. Li, Y. H. Leung, A. B. Djurišić, Z. T. Liu, M. H. Xie, S. L. Shi, S. J. Xu, and W. K. Chan, Different origins of visible luminescence in ZnO nanostructures fabricated by the chemical and evaporation methods, *Appl. Phys. Lett.* **85**, 1601 (2004).

156. C. Hsu, Y. Lin, S. Chang, T. Lin, S. Tsai, and I. Chen, Vertical ZnO/ZnGa$_2$O$_4$ core–shell nanorods grown on ZnO/glass templates by reactive evaporation, *Chem. Phys. Lett.* **411**, 221 (2005).

157. S. Z. Li, C. L. Gan, H. Cai, C. L. Yuan, J. Guo, P. S. Lee, and J. Ma, Enhanced photoluminescence of ZnO/Er$_2$O$_3$ core-shell structure nanorods synthesized by pulsed laser deposition, *Appl. Phys. Lett.* **90**, 263106 (2007).

158. W. I. Park, J. Yoo, D. W. Kim, G. C. Yi, and M. Kim, Fabrication and photoluminescent properties of heteroepitaxial ZnO/Zn$_{0.8}$Mg$_{0.2}$O coaxial nanorod heterostructures, *J. Phys. Chem. B* **110**, 1516 (2006).

159. Y. H. Park, Y. H. Shin, S. J. Noh, Y. Kim, S. S. Lee, C. G. Kim, K. S. An, and C. Y. Park, Optical quenching of NiO/Ni coated ZnO nanowires, *Appl. Phys. Lett.* **91**, 012102 (2007).

160. J. P. Richters, T. Voss, D. S. Kim, R. Scholz, and M. Zacharias, Enhanced surface-excitonic emission in ZnO/Al$_2$O$_3$ core–shell nanowires, *Nanotechnology* **19**, 305202 (2008).

161. L. Yu, X.-F. Yu, Y. Qiu, Y. Chen, and S. Yang, Nonlinear photoluminescence of ZnO/ZnS nano-tetrapods, *Chem. Phys. Lett.* **465**, 272 (2008).

162. V. V. Travnikov, A. Freiberg, and S. F. Savikhin, Surface excitons in ZnO crystals, *J. Lumin.* **47**, 107 (1990).

163. T. Voss and L. Wischmeier, Recombination dynamics of surface-related excitonic states in single ZnO nanowires, *J. Nanosci. Nanotechnol.* **8**, 228 (2008).

164. L. Wischmeier, T. Voss, S. Börner, and W. Schade, Comparison of the optical properties of as-grown ensembles and single ZnO nanowires, *Appl. Phys. A* **84**, 111 (2006).

165. Y. Yang, B. K. Tay, X. W. Sun, J. Y. Sze, Z. J. Han, J. X. Wang, X. H. Zhang, Y. B. Li, and S. Zhang, Quenching of surface-exciton emission from ZnO nanocombs by plasma immersion ion implantation, *Appl. Phys. Lett.* **91**, 071921 (2007).

166. J. Grabowska, A. Meany, K. K. Nanda, J.-P. Mosnier, M. O. Henry, J.-R. Duclère, and E. McGlynn, Surface excitonic emission and quenching effects in ZnO nanowire/nanowall systems: Limiting effects on device potential, *Phys. Rev. B* **71**, 115439 (2005).

167. J. Fallert, R. Hauschild, F. Stelzl, A. Urban, M. Wissinger, H. Zhou, C. Klingshirn, and H. Kalt, Surface-state related luminescence in ZnO nanocrystals, *J. Appl. Phys.* **101**, 073506 (2007).

168. J.-P. Richters, T. Voss, I. Rückmann, and J. Gutowski, Influence of polymer coating on the low-temperature photoluminescence properties of ZnO nanowires, *Appl. Phys. Lett.* **92**, 011103 (2008).

169. M. D. McCluskey and S. J. Jokela, Defects in ZnO, *J. Appl. Phys.* **106**, 071101, 1–13 (2009).

170. K. Vanheusden, C. H. Seager, W. L. Warren, D. R. Tallant, and J. A. Voigt, Correlation between photoluminescence and oxygen vacancies in ZnO phosphors, *Appl. Phys. Lett.* **68**, 403 (1996).

171. A. Janotti and C. G. Van de Walle, Oxygen vacancies in ZnO, *Appl. Phys. Lett.* **87**, 122102 (2005).

172. D. M. Hofmann, D. Pfisterer, J. Sann, B. K. Meyer, R. Tena-Zaera, V. Munoz-Sanjose, T. Frank, and G. Pensl, Properties of the oxygen vacancy in ZnO, *Appl. Phys. A Mater. Sci. Process.* **88**(1), 147 (2007).

173. A. van Dijken, E. A. Meulenkamp, D. Vanmaekelbergh, and A. Meijerink, The kinetics of the radiative and nonradiative processes in nanocrystalline ZnO particles upon photoexcitation, *J. Phys. Chem. B* **104**, 1715 (2000).

174. G. Xiong, U. Pal, and J. G. Serrano, Correlations among size, defects, and photoluminescence in ZnO nanoparticles, *J. Appl. Phys.* **101**, 024317 (2007).

175. K. Sakai, S. Oyama, K. Noguchi, A. Fukuyama, T. Ikari, and T. Okada, Low-temperature photoluminescence of nanostructured ZnO crystal synthesized by pulsed-laser ablation, *Jpn. J. Appl. Phys.* **48**, 085001 (2009).

176. Y. Dong, Z.-Q. Fang, D. C. Look, G. Cantwell, J. Zhang, J. J. Song, and L. J. Brillson, Zn- and O-face polarity effects at ZnO surfaces and metal interfaces, *Appl. Phys. Lett.* **93**, 072111 (2008).

177. A. Bera and D. Basak, Photoluminescence and photoconductivity of ZnS-coated ZnO nanowires, *ACS Appl. Mater. Interfaces* **2**, 408 (2010).

178. J. D. Prades et al., Toward a systematic understanding of photodetectors based on individual metal oxide nanowires, *J. Phys. Chem. C* **112**, 14639 (2008).

179. G. Konstantatos and E. H. Sargent, Nanostructured materials for photon detection, *Nat. Nano* **5**, 391 (2010).

180. E. Monroy et al., Wide-bandgap semiconductor ultraviolet photodetectors, *Semicond. Sci. Technol.* **18**, R33 (2003).

181. M. Razeghi and A. Rogalski, Semiconductor ultraviolet detectors, *J. Appl. Phys.* **79**, 7433 (1996).

182. J. D. Prades, F. Hernandez-Ramirez, R. Jimenez-Diaz, M. Manzanares, T. Andreu, A. Cirera, A. Romano-Rodriguez, and J. R. Morante, The effects of electron–hole separation on the photoconductivity of individual metal oxide nanowires, *Nanotechnology* **19**, 465501 (2008).

183. Y. W. Heo, B. S. Kang, L. C. Tien, D. P. Norton, F. Ren, J. R. La Roche, and S. J. Pearton, UV photoresponse of single ZnO nanowires, *Appl. Phys. A Mater.* **80**, 497 (2005).

184. M.-W. Chen, C.-Y. Chen, D.-H. Lien, Y. Ding, and J.-H. He, Photoconductive enhancement of single ZnO nanowire through localized Schottky effects, *Opt. Express* **18**, 14836 (2010).

185. J. R. S. Aga, D. Jowhar, A. Ueda, Z. Pan, W. E. Collins, R. Mu, K. D. Singer, and J. Shen, Enhanced photoresponse in ZnO nanowires decorated with CdTe quantum dot, *Appl. Phys. Lett.* **91**, 232108 (2007).

186. F. Fang, D. X. Zhao, B. H. Li, Z. Z. Zhang, J. Y. Zhang, and D. Z. Shen, The enhancement of ZnO nanowalls photoconductivity induced by CdS nanoparticle modification, *Appl. Phys. Lett.* **93**, 233115 (2008).

187. C.-Y. Chen, M.-W. Chen, J.-J. Ke, C.-A. Lin, J.-R. Retamal, and J.-H. He, Surface effects on optical and electrical properties of ZnO nanostructures, *Pure Appl. Chem.* **82**, 2055 (2010).

188. A. Bera and D. Basak, Effect of surface capping with poly(vinyl alcohol) on the photocarrier relaxation of ZnO nanowires, *ACS Appl. Mater. Interfaces* **1**, 2066 (2009).

189. X.-M. Qian, J.-X. Fang, and Y.-L. Song, Electrical characteristics of CuS/ZnO PN heterojunction, *Proc. SPIE* **7381**, 738126, DOI: 10.1117/12.835498 (2009).

190. S. Karuppanan, T. Youngjo, S. Minsu, and Y. Kijung, Synthesis and characterization of ZnO nanowire-CdO composite nanostructures, *Nanoscale Res. Lett.* **4**, 1329 (2009).

191. W.-C. Tsai, S.-J. Wang, C.-R. Tseng, R.-M. Ko, and J.-C. Lin, Preparation and optoelectronic properties of NiO/ZnO heterostructure nanowires, *Proc. SPIE* **7356**, 73561F (2009).

192. Y.-H. Lin, Y.-C. Hsueh, C.-C. Wang, J.-M. Wu, T.-P. Perng, and H. C. Shih, Enhancing the photon-sensing properties of ZnO nanowires by atomic layer deposition of platinum, *Electrochem. Solid-State Lett.* **13**, K93 (2010).

193. D. Lin, H. Wu, W. Zhang, H. Li, and W. Pan, Enhanced UV photoresponse from heterostructured Ag-ZnO nanowires, *Appl. Phys. Lett.* **94**, 172103 (2009).

194. Y. Zheng, C. Chen, Y. Zhan, X. Lin, Q. Zheng, K. Wei, and J. Zhu, Photocatalytic activity of Ag/ZnO heterostructure nanocatalyst: Correlation between structure and property, *J. Phys. Chem. C* **112**, 10773 (2008).

195. V. Subramanian, E. E. Wolf, and P. V. Kamat, Green emission to probe photoinduced charging events in ZnO–Au nanoparticles. Charge distribution and fermi-level equilibration, *J. Phys. Chem. B* **107**, 7479 (2003).

196. H. Seong, K. Cho, and S. Kim, A pn heterojunction diode constructed with a n-type ZnO nanowire and a p-type HgTe nanoparticle thin film, *Appl. Phys. Lett.* **94**, 043102 (2009).

197. Y. F. Li et al., Ultraviolet photodiode based on p-$Mg_{0.2}Zn_{0.8}O$/n-ZnO heterojunction with wide response range, *J. Phys. D Appl. Phys.* **42**, 105102 (2009).

198. Y. Wu, T. Tamaki, T. Volotinen, L. Belova, and K. V. Rao, Enhanced photoresponse of inkjet-printed ZnO thin films capped with CdS nanoparticles, *J. Phys. Chem. Lett.* **1**, 89 (2009).

199. H.-J. Seong, J.-G. Yun, J. H. Jun, K.-G. Cho, and S.-S. Kim, The transfer of charge carriers photogenerated in ZnO nanoparticles into a single ZnO nanowire, *Nanotechnology* **20**, 245201 (2009).

200. C. S. Lao, M. C. Park, Q. Kuang, Y. L. Deng, A. K. Sood, D. L. Polla, and Z. L. Wang, Giant enhancement in UV response of ZnO nanobelts by polymer surface-functionalization, *J. Am. Chem. Soc.* **129**, 12096 (2007).

201. J. H. He, Y. H. Lin, M. E. McConney, V. V. Tsukruk, Z. L. Wang, and G. Bao, Enhancing UV photoconductivity of ZnO nanobelt by polyacrylonitrile functionalization, *J. Appl. Phys.* **102**, 084303 (2007).

202. W. Kim and K. S. Chu, ZnO nanowire field-effect transistor as a UV photodetector; optimization for maximum sensitivity, *Phys. Status Solidi A* **206**, 179 (2009).

203. G. Chai, O. Lupan, L. Chow, and H. Heinrich, Crossed zinc oxide nanorods for ultraviolet radiation detection, *Sens. Actuators, A* **150**, 184 (2009).

204. L. Luo, Y. Zhang, S. S. Mao, and L. Lin, Fabrication and characterization of ZnO nanowires based UV photodiodes, *Sens. Actuators A* **127**, 201 (2006).

205. Y. H. Leung, Z. B. He, L. B. Luo, C. H. A. Tsang, N. B. Wong, W. J. Zhang, and S. T. Lee, ZnO nanowires array *p-n* homojunction and its application as a visible-blind ultraviolet photodetector, *Appl. Phys. Lett.* **96**, 053102 (2010).

206. C.-Y. Huang, Y.-J. Yang, J.-Y. Chen, C.-H. Wang, Y.-F. Chen, L.-S. Hong, C.-S. Liu, and C.-Y. Wu, *p*-Si nanowires/SiO_2-ZnO heterojunction photodiodes, *Appl. Phys. Lett.* **97**, 013503 (2010).

207. J.-Y. Wang, C.-Y. Lee, Y.-T. Chen, C.-T. Chen, Y.-L. Chen, C.-F. Lin, and Y.-F. Chen, Double side electroluminescence from *p*-NiO/*n*-ZnO nanowire heterojunctions, *Appl. Phys. Lett.* **95**, 131117 (2009).

208. W.-C. Tsai, S.-J. Wang, C.-R. Tseng, R.-M. Ko, and J.-C. Lin, Preparation of Ni/Zn and NiO/ZnO heterojunction nanowires and their optoelectrical characteristics, *Proc. SPIE* **7356**, 73561D (2009).

209. N. Chantarat et al., Enhanced UV photoresponse in nitrogen plasma ZnO nanotubes, *Nanotechnology* **20**, 395201 (2009).

210. N. Kouklin, Cu-doped ZnO nanowires for efficient and multispectral photodetection applications, *Adv. Mat.* **20**, 2190 (2008).

211. R. K. Joshi, Q. Hu, F. Am, N. Joshi, and A. Kumar, Au decorated zinc oxide nanowires for CO sensing, *J. Phys. Chem. C* **113**, 16199 (2009).

212. T. J. Hsueh, Y. W. Chen, S. J. Chang, S. F. Wang, C. L. Hsu, Y. R. Lin, T. S. Lin, and I. C. Chen, ZnO nanowire-based CO sensors prepared at various temperatures, *J. Electrochem. Soc.* **154**, J393 (2007).

213. L. Liao, H. B. Lu, J. C. Li, C. Liu, D. J. Fu, and Y. L. Liu, He sensitivity of gas sensor based on single ZnO nanowire modulated by helium ion radiation, *Appl. Phys. Lett.* **91**, 173110 (2007).

214. J. H. He, C. H. Ho, and C. Y. Chen, Polymer functionalized ZnO nanobelts as oxygen sensors with a significant response enhancement, *Nanotechnology* **20**, 065503 (2009).

215. E. Comini, C. Baratto, G. Faglia, M. Ferroni, and G. Sberveglieri, Single crystal ZnO nanowires as optical and conductometric chemical sensor, *J. Phys. D* **40**, 7255 (2007).

216. Q. H. Li, Y. X. Liang, Q. Wan, and T. H. Wang, Oxygen sensing characteristics of individual ZnO nanowire transistors, *Appl. Phys. Lett.* **85**, 6389 (2004).

217. C. Y. Lu, S. P. Chang, S. J. Chang, T. J. Hsueh, C. Z. Y. Fan, D. W. Wang, P. C. Chang, W. Y. Tseng, and J. G. Lu, ZnO nanowire field-effect transistor and oxygen sensing property, *Appl. Phys. Lett.* **85**, 5923 (2004); L. Hsu, Y. Z. Chiou, and C. Chen, ZnO nanowire-based oxygen gas sensor, *IEEE Sens. J.* **9**, 485 (2009).

218. E. Comini, G. Faglia, G. Sberveglieri, Z. W. Pan, and Z. L. Wang, Stable and highly sensitive gas sensors based on semiconducting oxide nanobelts, *Appl. Phys. Lett.* **81**, 1869 (2002).

219. G. Kwak and K. J. Yong, Adsorption and reaction of ethanol on ZnO nanowires, *J. Phys. Chem. C* **112**, 3036 (2008).

220. L. Liao, H. B. Lu, J. C. Li, H. He, D. F. Wang, D. J. Fu, and C. Liu, Size dependence of gas sensitivity of ZnO nanorods, *J. Phys. Chem. C* **111**, 1900 (2007).

221. C. C. Li, Z. F. Du, L. M. Li, H. C. Yu, Q. Wan, and T. H. Wang, Surface-depletion controlled gas sensing of ZnO nanorods grown at room temperature, *Appl. Phys. Lett.* **91**, 032101 (2007).

222. Z. Y. Fan and J. G. Lu, Gate-refreshable nanowire chemical sensors, *Appl. Phys. Lett.* **86**, 123510 (2005).

223. C. C. Li, L. M. Li, Z. F. Du, H. C. Yu, Y. Y. Xiang, Y. Li, Y. Cai, and T. H. Wang, Rapid and ultra-high ethanol sensing based on Au-coated ZnO nanorods, *Nanotechnology* **19**, 035501 (2008).

224. T. Y. Wei, P. H. Yeh, S. Y. Lu, and Z. L. Wang, Gigantic enhancement in sensitivity using Schottky contacted nanowire nanosensor, *J. Am. Chem. Soc.* **131**, 17690 (2009).

225. J. B. K. Law and J. T. L. Thong, Improving the NH$_3$ gas sensitivity of ZnO nanowire sensors by reducing the carrier concentration, *Nanotechnology* **19**, 205502 (2008).

226. J. X. Wang, X. W. Sun, S. S. Xie, Y. Yang, H. Y. Chen, G. Q. Lo, and D. L. Kwong, Preferential growth of SnO$_2$ triangular nanoparticles on ZnO nanobelts, *J. Phys. Chem. C* **111**, 7671 (2007).

227. N. Zhang, K. Yu, Q. Li, Z. Q. Zhu, and Q. Wan, Room-temperature high-sensitivity H$_2$S gas sensor based on dendritic ZnO nanostructures with macroscale in appearance, *J. Appl. Phys.* **103**, 104305 (2008).

228. S. K. Gupta, A. Joshi, and M. Kaur, Development of gas sensors using ZnO nanostructures, *J. Chem. Sci.* **122**, 57 (2010).

229. S. J. Chang, T. J. Hsueh, I. C. Chen, and B. R. Huang, Highly sensitive ZnO nanowire CO sensors with the adsorption of Au nanoparticles, *Nanotechnology* **19**, 175502 (2008).

230. E. Wongrat, P. Pimpang, and S. Choopun, Comparative study of ethanol sensor based on gold nanoparticles: ZnO nanostructure and gold: ZnO nanostructure, *Appl. Surf. Sci.* **256**, 968 (2009).

231. Y. Zhang, Q. Xiang, J. Q. Xu, P. C. Xu, Q. Y. Pan, and F. Li, Self-assemblies of Pd nanoparticles on the surfaces of single crystal ZnO nanowires for chemical sensors with enhanced performances, *J. Mater. Chem.* **19**, 4701 (2009).

232. L. C. Tien, P. W. Sadik, D. P. Norton, L. F. Voss, S. J. Pearton, H. T. Wang, B. S. Kang, F. Ren, J. Jun, and J. Lin, Hydrogen sensing at room temperature with Pt-coated ZnO thin films and nanorods, *Appl. Phys. Lett.* **87**, 222106 (2005).

233. H. T. Wang, B. S. Kang, F. Ren, L. C. Tien, P. W. Sadik, D. P. Norton, S. J. Pearton, and J. Lin, Hydrogen-selective sensing at room temperature with ZnO nanorods, *Appl. Phys. Lett.* **86**, 243503 (2005).

234. Q. Kuang, C.-S. Lao, Z. Li, Y.-Z. Liu, Z.-X. Xie, L.-S. Zheng, and Z. L. Wang, Enhancing the photon- and gas-sensing properties of a single SnO$_2$ nanowire based nanodevice by nanoparticle surface functionalization, *J. Phys. Chem. C* **112**, 11539 (2008).

235. Y. Zeng, T. Zhang, L. J. Wang, M. H. Kang, H. T. Fan, R. Wang, and Y. He, Enhanced toluene sensing characteristics of TiO$_2$-doped flowerlike ZnO nanostructures, *Sens. Actuators B* **140**, 73 (2009).

236. N. Kakati, S. H. Jee, S. H. Kim, H. K. Lee, and Y. S. Yoon, Sensitivity enhancement of ZnO nano-rod gas sensors with surface modification by an InSb thin film, *Jpn. J. Appl. Phys.* **48**, 105002 (2009).

237. B. O'Regan and M. Grätzel, A low-cost, high-efficiency solar cell based on dye-sensitized colloidal TiO$_2$ films, *Nature* **353**, 737 (1991).

238. K. Y. Cheung, C. T. Yip, A. B. Djurišić, Y. H. Leung, and W. K. Chan, Long K-doped titania and titanate nanowires on Ti foil and FTO/quartz substrates for solar-cell applications, *Adv. Funct. Mater.* **17**, 555 (2007).

239. K. Sayama, H. Suguhara, and H. Arakawa, Photoelectrochemical properties of a porous Nb$_2$O$_5$ electrode sensitized by a ruthenium dye, *Chem. Mater.* **10**, 3825 (1998).

240. M. D. Archer and A. J. Nozik, *Nanostructured and Photoelectrochemical Systems for Solar Photon Conversion*, World Scientific, Singapore (2008).

241. E. Galoppini, J. Rochford, H. H. Chen, G. Saraf, T. C. Lu, A. Hagfeldt, and G. Boschloo, Fast electron transport in metal organic vapor deposition grown dye-sensitized ZnO nanorod solar cells, *J. Phys. Chem. B* **110**, 16159 (2006).

242. E. A. Meulenkamp, Electron transport in nanoparticulate ZnO films, *J. Phys. Chem. B* **103**, 7831 (1999).

243. Y. S. Yun, J. Y. Park, H. Oh, J. J. Kim, and S. S. Kim, Electrical transport properties of size-tuned ZnO nanorods, *J. Mater. Res.* **21**, 132 (2006).

244. M. Saito and S. Fujihara, Large photocurrent generation in dye-sensitized ZnO solar cells, *Energy Environ. Sci.* **1**, 280 (2008).

245. Y. Zhang, L. Wu, Y. Liu, and E. Xie, Improvements to the hierarchically structured ZnO nanosphere based dye-sensitized solar cells, *J. Phys. D* **42**, 085105 (2009).

246. V. Dhas, S. Muduli, W. Lee, S. H. Han, and S. Ogale, Enhanced conversion efficiency in dye-sensitized solar cells based on ZnO bifunctional nanoflowers loaded with gold nanoparticles, *Appl. Phys. Lett.* **93**, 243108 (2008).

247. K. Keis, J. Lindgren, S. E. Lindquist, and A. Hagfeldt, Studies of the adsorption process of Ru complexes in nanoporous ZnO electrodes, *Langmuir* **16**, 4688 (2000).

248. M. Law, L. E. Greene, A. Radenovic, T. Kuykendall, J. Liphardt, and P. Yang, ZnO–Al$_2$O$_3$ and ZnO–TiO$_2$ core–shell nanowire dye-sensitized solar cells, *J. Phys. Chem. B* **110**, 22652 (2006).

249. M. Wang, C. Huang, Y. Cao, Q. Yu, W. Guo, Q. Huang, Y. Liu, J. Huang, H. Wang, and Z. Deng, The effects of shell characteristics on the current-voltage behaviors of dye-sensitized solar cells based on ZnO/TiO$_2$ core/shell arrays, *Appl. Phys. Lett.* **94**, 263506 (2009).

250. M. Wang et al., Dye-sensitized solar cells based on nanoparticle-decorated ZnO/TiO$_2$ core/ shell nanorod arrays, *J. Phys. D* **42**, 155104 (2009).

251. F. A. Kroger, *The Chemistry of Imperfect Crystals*, 2nd edn., North Holland, Amsterdam, the Netherlands (1974).

252. G. Redmond, D. Fitzmauricem, and M. Grätzel, Visible light sensitization by cis-bis(thiocyanato) bis(2,2'-bipyridyl-4,4'-dicarboxylato)ruthenium(ii) of a transparent nanocrystalline ZnO film prepared by sol-gel techniques, *Chem. Mater.* **6**, 686 (1994).

253. A. R. Rao and V. Dutta, Achievement of 4.7% conversion efficiency in ZnO dye-sensitized solar cells fabricated by spray deposition using hydrothermally synthesized nanoparticles, *Nanotechnology* **19**, 445712 (2008).

254. Z. L. S. Seow, A. S. W. Wong, V. Thavasi, R. Jose, S. Ramakrishna, and G. W. Ho, Controlled synthesis and application of ZnO nanoparticles, nanorods and nanospheres in dye-sensitized solar cells, *Nanotechnology* **20**, 045604 (2009).

255. W. Zhang, R. Zhu, X. Liu, B. Liu, and S. Ramakrishna, Facile construction of nanofibrous ZnO photoelectrode for dye-sensitized solar cell applications, *Appl. Phys. Lett.* **95**, 043304 (2009).

256. D. I. Suh, S. Y. Lee, T. H. Kim, J. M. Chun, E. K. Suh, O. B. Yang, and S. K. Lee, The fabrication and characterization of dye-sensitized solar cells with a branched structure of ZnO nanowires, *Chem. Phys. Lett.* **442**, 348 (2007).

257. M. S. Akhtar, M. A. Khan, M. S. Jeon, and O. B. Yang, Controlled synthesis of various ZnO nanostructured materials by capping agents-assisted hydrothermal method for dye-sensitized solar cells, *Electrochim. Acta* **53**, 7869 (2008).

258. Y. F. Hsu, Y. Y. Xi, C. T. Yip, A. B. Djurišić, and W. K. Chan, Dye-sensitized solar cells using ZnO tetrapods, *J. Appl. Phys.* **103**, 083114 (2008).

259. W. Chen, H. Zhang, I. M. Hsing, and S. Yang, A new photoanode architecture of dye sensitized solar cell based on ZnO nanotetrapods with no need for calcination, *Electrochem. Commun.* **11**, 1057 (2009).

260. A. E. Suliman, Y. T. Tang, and L. Xu, Solar energy mater, *Solar Cells* **91**, 1658 (2007).

261. T. W. Hamann, A. B. F. Martinson, and J. W. Elam, Aerogel templated ZnO dye-sensitized solar cells, *Adv. Mater.* **20**, 1560 (2008).

262. Q. Zhang, T. P. Chou, B. Russo, S. A. Jenekhe, and G. Cao, Aggregation of ZnO nanocrystallites for high conversion efficiency in dye-sensitized solar cells, *Angew. Chem. Int. Ed.* **47**, 2402 (2008).

263. I. D. Kim, J. M. Hong, B. H. Lee, D. Y. Kim, E. K. Jeon, D. K. Choi, and D. J. Yang, Dye-sensitized solar cells using network structure of electrospun ZnO nanofiber mats, *Appl. Phys. Lett.* **91**, 163109 (2007).

264. T. P. Chou, Q. Zhang, G. E. Fryxell, and G. Cao, Hierarchically structured ZnO film for dye-sensitized solar cells with enhanced energy conversion efficiency, *Adv. Mater.* **19**, 2588 (2007).

Part II

ZnO-Based Optoelectronic Devices

5

Light-Emitting Diodes Based on p-GaN/ i-ZnO/n-ZnO Heterojunctions

Yichun Liu and Yanhong Tong

CONTENTS

5.1 Introduction 177
5.2 Basic Principle 179
    5.2.1 *p*-GaN/*n*-ZnO Heterojunction 179
    5.2.2 *p*-GaN/*i*-ZnO/*n*-ZnO Heterojunction 180
    5.2.3 *p*-GaN/*n*-ZnO Heterojunction Based on ZnO Nanostructures 180
5.3 Material Growth and Device Performance 181
    5.3.1 (*p*-GaN Film)/(*n*-ZnO Film) 181
    5.3.2 *p*-GaN/*i*-ZnO/*n*-ZnO 186
    5.3.3 (*p*-GaN film)/(*n*-ZnO Nanostructures) 196
        5.3.3.1 (*p*-GaN Film)/(Single *n*-ZnO Nanowire)............................ 196
        5.3.3.2 (*p*-GaN Film)/(*n*-ZnO Nanoscale Array)........................... 197
        5.3.3.3 (*p*-GaN Film)/(ZnO Nanowire Array)/(*n*-ZnO Film)........ 210
        5.3.3.4 (ZnO Nanowire Array)/(GaN Film)/(ZnO Nanowire Array).......... 211
5.4 Conclusion and Outlook 212
Acknowledgments 214
References.. 214

5.1 Introduction

ZnO-based *p-n* homojunction light emitting diodes (LEDs) have gained great progress in recent years [1–5]. However, their commercial applications face a big challenge due to the lack of a growth technique for realizing reproducible, stable, and high-quality *p*-type ZnO [6]. Although N is known to be the most likely genuine *p*-type dopant based on electronic structure and atomic size considerations, recent reports demonstrate that N cannot lead to *p*-type conductivity in ZnO [7,8]. An alternative approach is to fabricate heterostructured ZnO-based devices using other *p*-type materials to substitute the *p*-ZnO.

The first ZnO-based heterojunction LED was demonstrated in 1967 by Drapak in which Cu_2O was used as the *p*-type material [9]. The other materials, for example, Si [10–14], SiC [14–16], GaN [17–22], AlGaN [23], NiO [24], CdTe [25], $ZnRu_2O_4$ [26], and $SrCu_2O_2$ [27,28], were also selected for the fabrication of the heterojunctions. Among these *p*-type materials, GaN is considered to be the most promising one because of the following: (1) The growth technique of high-quality *p*-GaN is very mature and it is commercially available. The fabrication of heterojunction LEDs based on *n*-ZnO and

TABLE 5.1

Comparison of Crystal Structure of ZnO, GaN, 6H-SiC, and c-Al$_2$O$_3$

Formula	Name	Symmetry	Point Group	Space Group	Stacking Sequence	Lattice Constant (Å)	E_g (eV)
ZnO	Zinc oxide	Hexagonal	6 mm	P6$_3$mc(186)	AB(HH)	$a=3.253$ $c=5.213$	3.37
GaN	Gallium nitride	Hexagonal	6 mm	P6$_3$mc(186)	AB(HH)	$a=3.1891$ $c=5.1855$	3.40
6H-SiC	6H silicon carbide	Hexagonal	6 mm	P6$_3$mc(186)	ABCACB(CCHCCH)	$a=3.081$ $c=15.117$	3.00
c-Al$_2$O$_3$	Sapphire corundum	Trigonal	3 m	R3C(167)	—	$a=4.758$ $c=12.991$	7.30

Source: Özgür, Ü. et al., *J. Appl. Phys.*, 98, 041301, 2005.

p-GaN is facile. (2) GaN has a similar crystal structure as ZnO. The crystal structure parameters of ZnO, GaN, 6H-SiC, and c-Al$_2$O$_3$ are shown in Table 5.1. The structural relationship of the two semiconductors forming the heterojunctions is one of the most important factors in realizing high-quality heterostructure devices. The lattice mismatch increases the number of interface defects with the detrimental effects, for example, creating the nonradiative center and decreasing the emission efficiency. GaN and ZnO have the same wurtzite crystal structure, same stacking sequence (HH), and similar lattice parameters with small lattice mismatch at 1.8%. The similar crystal structure of GaN and ZnO facilitates the growth of high-quality heterojunctions, obtaining exciton-related emission, and improving light emission efficiency. (3) In general, *p-n* heterojunction devices show a lower efficiency than homojunction devices due to an energy barrier formed at the junction interface. GaN has similar room-temperature bandgap (E_g) to ZnO (GaN: 3.40 eV, ZnO: 3.37 eV). Together with similar crystal structure, *p*-GaN/*n*-ZnO heterostructures can effectively substitute the ZnO-based homojunction and at the same time retain similar physical properties with the homojunction. These advantages are favorable for the successful formation of the high-quality heterojunction LED devices.

The fabricated *p*-GaN/*n*-ZnO heterojunction LEDs commonly presented electroluminescence (EL) on the GaN side rather than on the ZnO side. Such hybrid devices have high threshold, have low luminescence efficiency, and cannot work at high temperature because of the small excitonic binding energy of GaN at 26 meV. It is hoped that radiative recombination occurs in the ZnO layer, so that the LED devices can benefit from the robust excitonic emission of ZnO because of its large exciton binding energy at 60 meV. Such devices will possess all the advantages of the ZnO-based homojunction devices, that is, the UV emittance with high brightness, low threshold currents, high emission efficiency, and good high-temperature device performance.

With this aim, an intrinsic ZnO layer has been tried to be inserted between the *p*-GaN layer and the *n*-ZnO layer to form a *p*-GaN/*i*-ZnO/*n*-ZnO heterojunction, where the holes transported from the *p*-GaN side and the electrons from the *n*-ZnO side are recombined in the intrinsic region, and the photons with the energy corresponding to the bandgap of the intrinsic layer are emitted. Some techniques have been developed to grow *p*-GaN/*i*-ZnO/*n*-ZnO heterojunction LEDs, and new device configurations have been designed to optimize device performance. In addition, a rich variety of the ZnO nanostructures stimulates the development of the nanoscale *p*-GaN/*n*-ZnO heterojunction LEDs. Employing the ZnO nanostructures to substitute the conventional ZnO film for the fabrication of the *p*-GaN/*n*-ZnO heterojunction possibly provides another way to realize the ZnO related excitonic emission. This chapter mainly focuses on the issues of *p*-GaN/*n*-ZnO and *p*-GaN/*i*-ZnO/*n*-ZnO heterojunctions and their devices based on ZnO nanostructures. Their basic principle, fabrication methods, and EL properties are discussed.

5.2 Basic Principle

5.2.1 *p*-GaN/*n*-ZnO Heterojunction

When two different semiconductors are brought into physical contact, their energy band alignment changes. To date, none of the theoretical approaches can reliably predict the band offsets of all semiconductor heterostructure combinations. With the aim of simplification, the Anderson model, a simple way to analyze the energy band alignment in heterojunction devices [29], is used to discuss the energy band diagrams of *p*-GaN/*n*-ZnO. The size effect, defect states, and other perturbations, which may be the result of imperfect crystal lattice matches, are ignored in the Anderson model. According to the Anderson model, when two semiconductors are brought in contact and equilibrium is maintained, their energy band offsets, including the valence band offset and the conductive band offset, are unchanged. Figure 5.1a through c shows the energy band diagrams of *p*-GaN, *n*-ZnO, and *p*-GaN/ *n*-ZnO heterojunction, respectively. The electron affinity (χ) for ZnO and GaN are 4.20 and 4.35 eV, and the bandgaps E_g of GaN and ZnO at room-temperature are about 3.40 and 3.37 eV, respectively. The energy discontinuities for the conduction and valence bands are estimated to be $\Delta E_c = \chi_{ZnO} - \chi_{GaN} = 0.12$ eV and $\Delta E_v = E_{gZnO} - E_{gGaN} + \Delta E_c = 0.15$ eV. Similar values between ΔE_c and ΔE_v mean that in the *p*-GaN/*n*-ZnO heterojunction, the barrier height the electron needs to overcome to diffuse toward the *p* region is almost equivalent to that which the hole needs to overcome to diffuse toward the *n* region. In that case, the EL will be mainly determined by the differences of the carrier mobility and the carrier concentration between *p*-GaN and *n*-ZnO.

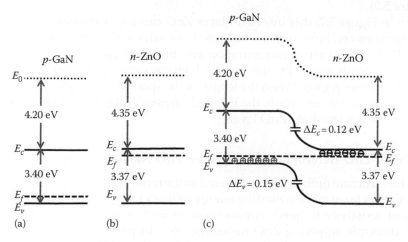

FIGURE 5.1
Energy band diagrams of (a) *p*-GaN, (b) *n*-ZnO, and (c) *p*-GaN/*n*-ZnO.

Generally, the electron concentration of the n-ZnO is over the order of 10^{18} cm^{-3} and the hole concentration of the p-GaN is on the order of 10^{17} cm^{-3}. In addition, the electron mobility is far higher than the hole mobility. Relatively higher carrier concentration of the n-ZnO combined with the higher mobility as compared with the p-GaN, allows the electrons to be injected into the p-GaN region more easily. As a result, the EL of the heterojunctions usually occurs in the GaN side rather than the ZnO side. The emission origin of the EL is usually defined by comparing the position and shape of the EL emission peak with that of the photoluminescence (PL) results of ZnO and GaN.

The EL peak position and intensity are related to the quality of the p-GaN layer, the quality of n-ZnO layer, and the interface quality between GaN and ZnO. The defect states and interface states act as the nanoradiative recombination center, resulting in weak emission from the p-GaN/n-ZnO heterojunction devices. In fact, it is also very difficult to obtain the ultraviolet (UV) EL related to ZnO exciton emission from the heterostructured p-GaN/ n-ZnO LEDs. Some researchers have tried to insert a high-resistance intermediate layer with a thickness of 20–150 nm to p-GaN/n-ZnO in order to improve device performance. The high-resistance materials include wide band-gap materials such as MgO [30] and AlN [31]. These wide band-gap materials have good stability. They can increase the barrier height of the electron blocker and facilitate realization of emission from the n-ZnO side. An intrinsic ZnO (i-ZnO) is also used as the intermediate layer, which has good lattice match with the heterojunction. In general, the heterojunctions are fabricated by growing n-ZnO onto p-GaN. In comparison with other high-resistance materials, applying i-ZnO between p-GaN and n-ZnO facilitates obtaining high-quality n-ZnO, minimizing the number of the defects, and decreasing the capture of the carriers. Therefore, an exciton emission related to ZnO possibly occurs in such a high quality heterojunction. The p-GaN/i-ZnO/n-ZnO is introduced and discussed in Sections 5.2.2 and 5.3.2.

5.2.2 p-GaN/i-ZnO/n-ZnO Heterojunction

The p-i-n heterostructures are known to show high-efficiency LED performance, where the p and n layers provide the hole and electron carriers, respectively [32]. The energy band diagram inferred by the Anderson model is drawn to determine the reasons of the improved emission properties of the heterojunction with insertion of an intrinsic ZnO layer (Figure 5.2).

As shown in Figure 5.2, this inserted i-layer ZnO can lower the energy barriers for the supply of electrons and holes into the intrinsic layer, which plays the role of an active layer. i-ZnO usually has low carrier concentration and high resistivity. Therefore, external forward voltage is mostly applied to this layer, which corresponds to widen the space charge region, that is, barrier region. When the length of the space charge region is longer than the electron and hole diffusion length, the injected carriers will be confined to this layer and recombine in the intermediate ZnO layer.

5.2.3 p-GaN/n-ZnO Heterojunction Based on ZnO Nanostructures

The ZnO nanostructure probably is considered as the most promising one in low-dimension systems due to its large exciton binding energy (60 meV), high electromechanical coupling constant, and resistivity to harsh environment as well as the relatively easy fabrication process. In principle, applying ZnO nanostructures, in particular, one-dimensional (1D) ZnO nanorods or nanowires, as n layer for the fabrication of the heterojunction LEDs has many merits including (1) the confinement effect of 1D ZnO nanostructures facilitates the

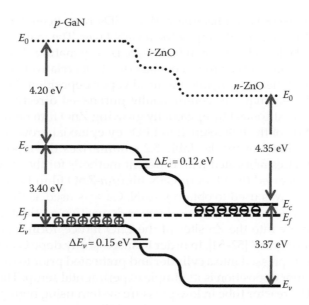

FIGURE 5.2
Energy band diagram of *n*-ZnO/*i*-ZnO/*p*-GaN.

electronic transport and can enhance radiative recombination of carriers [33]. (2) Applying ZnO nanostructure gives the nanosized contacts at the interface between ZnO and GaN, which possibly provides a way to improve the carrier injection rate [34,35]. (3) The ZnO nanostructures generally have the structure of a single crystal. The high crystal quality can minimize the number of defects and decrease electron diffusion. (4) ZnO has self organization growth property. Its nanostructure is more easily realized than the film. For example, the well-oriented high-quality ZnO nanorod array even can be obtained in an aqueous solution using a mixture of zinc nitrate $(Zn(NO_3)_2)$ and hexa-methylene-tetramine (HMT) at a reaction temperature of 80°C–90°C [36], while the high-quality films are generally realized at high temperature of over 500°C. Because of these advantages, ZnO nanostructures are ideal for fabricating the heterojunctions and developing new generation nanodevices with high performance.

5.3 Material Growth and Device Performance

5.3.1 (*p*-GaN Film)/(*n*-ZnO Film)

Lattice mismatch is one of the most important factors for the growth of high-quality films. It will cause extended defects with detrimental effects such as creation of nonradiative center, making the use of ZnO for optoelectronic applications a challenge. GaN has similar structural properties as ZnO. Compared with the sapphire substrate (a large lattice mismatch with ZnO at 18%), GaN is a closely lattice-matched material to ZnO with a lattice mismatch of 1.8%. Therefore, GaN is an appropriate material for the epitaxial growth of ZnO. In the early reports about GaN film/ZnO film structure, researchers applied *n*-GaN as the buffer layer with the aim of growth of high-quality ZnO [37–39].

The (*p*-GaN film)/(*n*-ZnO film) heterojunction LEDs have been constructed mostly by depositing *n*-type ZnO layer on the top of the *p*-GaN layer. The main reason is that *p*-GaN is commercially available and its growth technique is very mature. The *c*-plane sapphire is a common used substrate for the growth of GaN due to its relatively lower cost. The GaN layer is fabricated by molecular organic chemical vapor deposition (MOCVD), molecular-beam epitaxy (MBE) technique, or commercially purchased directly. ZnO and GaN are brought into the physical contact by epitaxially growing ZnO film on the top of GaN film. To date, the fabrication of the heterojunction LEDs by epitaxial growth is the cleanest and most reproducible method available. Table 5.2 summarizes the reported (*p*-GaN film)/(*n*-ZnO film) LED devices fabricated with different methods for the growth of ZnO layer.

Alivov et al. first reported the EL of (*p*-GaN film)/(*n*-ZnO film) LED in 2003 [40,41]. Mg doped GaN (GaN:Mg) was used to obtain *p*-GaN. Ga was used as the donor dopant species for the growth of *n*-ZnO by a chemical vapor deposition method, in which Ga incorporates substitutionally into the Zn sites of the ZnO lattice, forming a donor level with activation energy of ~50 meV [52–54]. In order to successfully dope Ga into ZnO, a mixture of ZnO and Ga_2O_3 was pressed into a cylinder and preheated prior to the growth of *n*-ZnO. The advantage of vapor deposition is its simple experimental setup. The growth of *n*-ZnO was carried out in the reactor tube in low-pressure system using high-purity hydrogen as a carrier gas. The free carrier concentration and mobility were measured to be $p = 3.5 \times 10^{17}$ cm^{-3} and $\mu_p = 10\,cm^2\,V^{-1}\,s^{-1}$ for *p*-GaN film, and $n = 4.5 \times 10^{18}\,cm^{-3}$, and $\mu_n = 40\,cm^2\,V^{-1}\,s^{-1}$

TABLE 5.2

Growth Method of ZnO, Carrier Concentration, Contact Electrodes, and EL Emission Color Peaks of (*p*-GaN Film)/(*n*-ZnO Film) Heterojunction LEDs

Method	Carrier Concentration GaN/ZnO ($cm^2\,V^{-1}\,s^{-1}$)	Contact GaN/ZnO	Emission (Color) (nm)	References
CVD	$3.5 \times 10^{17}/4.5 \times 10^{18}$	Au/In	430 (blue–violet)	[40,41]
Vapor cooling	$3.7 \times 10^{17}/1.7 \times 10^{20}$	(Ni/Au)/(Al/Pt)	432 (blue–violet)	[42]
ALD	$3 \times 10^{17}/2 \times 10^{18}$	Ni/–	386 (UV), 420 (blue–violet)	[43]
ALD	$2.6 \times 10^{17}/2 \times 10^{18}$	(NiO/Au)/Al	391 (UV), 425 (blue–violet)	[44]
MBE	$1.98 \times 10^{17}/3.96 \times 10^{18}$	(Ni/Au)/(Al//Au)	570 (yellowish-green)	[45]
MBE	$10^{17}/1.2 \times 10^{18}$	(Ni/Au)/(Ti/Au)	389, 570 (yellowish-green)	[18]
MBE	$1.19 \times 10^{17}/2.56 \times 10^{18}$	(Ni/Au)/In	375 (UV), 415 (blue–violet), 525 (green)	[46]
MOCVD	$7.36 \times 10^{17}/2.22 \times 10^{18}$	In/In	415 (blue–violet)	[47]
PLD	$5 \times 10^{17}/2 \times 10^{17}$	(Ni/Au)/(Ti/Au)	375 (UV)	[48]
Laser ablation	$10^{18}/3.36 \times 10^{17}$	(Ni/Au)/ITO	365.5 (UV), 384 (UV)	[20]
RF sputtering	$10^{18}/10^{15}$	(Ni/Au)/(Cr/Ni/Au)	400–700 (yellowish)	[49]
RF sputtering	$10^{18}/10^{17}$	(Ni/Au)/(Cr/Ni/Au)	400 (blue–violet)	[49]
RF sputtering	$10^{18}/10^{15} - 10^{17}$	(Ni/Au)/(Cr/Ni/Au)	381 (UV)	[50]
Solution growth	$6 \times 10^{17}/3 \times 10^{20}$	(Ni/Au)/(Ti/Au)	393 (UV), 529.5 (green)	[51]

CVD, chemical vapor deposition; ALD, atomic layer deposition; MBE, molecular beam epitaxy; MOCVD, molecular organic chemical vapor deposition; PLD, pulse laser deposition; ECD, electrochemical deposition; RF, radio-frequency.

for *n*-ZnO film, respectively. The cathodoluminescence (CL) spectrum of the *n*-ZnO film shows an intense UV emission at 390 nm with a full width at half maximum (FWHM) of 118 meV, and a weak broad green band centered at 510 nm. The CL spectrum of *p*-GaN presents a weak near-band-edge (NBE) emission at 383 nm and a strong broad band at 430 nm. The emission band of *p*-GaN at 430 nm is generally attributed to transitions from the conduction band or shallow donors to deep Mg acceptor levels [55,56]. The EL spectrum of the forward-biased LED indicates a broad emission band centered at about 430 nm and a tail that extends to longer wavelengths. Compared with the PL results of *n*-ZnO film and *p*-GaN film, the EL peak therefore is attributed to transitions from the conduction band or shallow donors to deep Mg acceptor level of *p*-GaN. It was caused by the electron injection from the *n*-ZnO side of the heterojunction to the *p*-GaN side. The EL spectra of the *p*-GaN/*n*-ZnO heterostructure LED were measured at different injection currents. The shape of the EL spectrum does not change significantly with the injection current. The FWHM of the peak exhibits a slight increase. Significantly, no shifting of the peak emission is observed with the increasing of the injection current. The light-current characteristics have been obtained from a direct measurement of the peak emission intensity at 430 nm. The light-current curve shows a superlinear dependence at low currents (<10 mA) with a slope 1.9 ($L \sim I_m$, $m = 1.9$), which becomes sublinear ($m = 0.85$) at higher currents (>10 mA). The change in the slope is believed to correspond to the current value at which the intensity of blue–violet emission at 430 nm is saturated. However, it can also be caused by Auger recombination at high injection current densities and by heating effects or series resistances.

Another vapor deposition method called the vapor cooling condensation technique was used by Lee et al. [42], in which the material sources, including 0.85 g of ZnO powder and 0.12 g of In, in a tungsten boat were sublimated by heating and then condensed on the cooled GaN:Mg substrate by liquid nitrogen. The electron concentration and mobility of the deposited ZnO:In are 1.7×10^{20} cm^{-3} and 3.7 cm^2 V^{-1} s^{-1}, respectively. The EL shows a similar result with that of Alivov et al., that is, a broad emission band at 432 nm, which is attributed to the defect-related emission of GaN.

Chen et al. used the atomic layer deposition (ALD) method to grow a 540 nm thick ZnO layer followed by a rapid thermal annealing in N$_2$ ambient at 950°C for 5 min for the fabrication of the *p*-GaN/*n*-ZnO heterostructure LEDs [43]. ALD is a layer–layer growth technique with atomic-layer accuracy. It offers many advantages, including easy and accurate thickness control, excellent conformality, abrupt interfaces, high uniformity over a large area, good reproducibility, dense and pinhole-free structures, and low growth temperatures. The ZnO layer fabricated with this method shows high (0002) orientation, almost a perfect single crystal with a very few dislocations, and good epitaxial relation with GaN. Transmission electron microscopy (TEM) and high-angle annular dark-field scanning TEM images reveal that a 4–5 nm thick interfacial layer between *p*-GaN and *n*-ZnO was formed by the diffusion of Mg atoms from *p*-GaN during rapid thermal annealing. The PL shows a strong NBE UV emission at 378 nm with negligible defect-related band in the visible region. The EL spectra of the heterostructure consist of UV emission at 391 nm and blue emission at 425 nm, as shown in Figure 5.3, which respectively are attributed to NBE emission of ZnO and the defect emission of GaN. The emission intensity is enhanced with increasing injection current, and ZnO-related UV emission becomes dominant when the current is higher than 12.5 mA (in the inset of Figure 5.3). It shows a competition process between the emission peaks of GaN and ZnO. This competition process is believed to be related with the differences in the carrier concentration and light emission efficiency in *p*-GaN and *n*-ZnO, and the interface states. As mentioned in Section 5.2.1, the higher

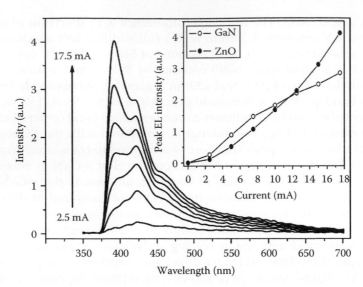

FIGURE 5.3
RT EL spectra of ALD-grown *p*-GaN/*n*-ZnO heterojunction LED at various injection currents. The inset shows EL peak intensities of *n*-ZnO and *p*-GaN as a function of injection current. (After Chen, H.C., Chen, M.J., and Cheng, Y.C., Amplified spontaneous emission from ZnO in *n*-ZnO/*p*-GaN heterojunction light-emitting diodes with an external-feedback reflector, *IEEE Photon. Technol. Lett.*, 22, 248. © 2010. With permission from IEEE.)

carrier concentration in *n*-ZnO makes the amount of electrons injected from *n*-ZnO much larger than that of the holes injected from *p*-GaN at forward bias. On the other hand, the carriers may be trapped by GaN/ZnO interface states when they transport across the interfacial layer. At a low injection current of 2.5 mA, the interface states may capture both the holes and electrons. Because of higher electron concentration and higher injection current, the remaining electrons will be injected from *n*-ZnO into *p*-GaN resulting in defect emission of *p*-GaN. With increasing injection current, the holes injected from *p*-GaN not only fill up the interface states but also enter into the *n*-ZnO layer. Because of the higher emission efficiency of ZnO than GaN, the EL from *n*-ZnO increases rapidly and eventually dominates over the emission from *p*-GaN.

By depositing metal Al on the back of the sapphire substrate as an external-feedback reflector, Chen et al. found a significant enhancement of the EL intensity from ZnO, decrease of the FWHM of the emission peak, and superlinear increase in ZnO EL intensity with the injection current [44]. As shown in Figure 5.4, it is clearly seen that the EL intensity of the device with Al back reflector (Device A) is considerably greater than that of devices without the Al deposition at the back of the sapphire (Devices B and C). It has been found that the UV EL peak intensity of the heterostructure (from ZnO) with an Al back reflector (Device A) increases rapidly with the injection current, showing a remarkable superlinear dependence on the injection current compared with those heterostructures without Al back reflector (Devices B and C). These results suggest that the external optical feedback from the polished back surface with an Al reflector stimulates light emission from the ZnO epilayer.

Until now, it is still very difficult to obtain the dominant UV emission from ZnO layer of the *p*-GaN/*n*-ZnO film heterojunction devices. Similar to the results reported by Chen et al., some research groups obtained radiative recombination emissions in both *n*-ZnO and *p*-GaN sides, with the MBE technique [46], and the MOCVD method [47].

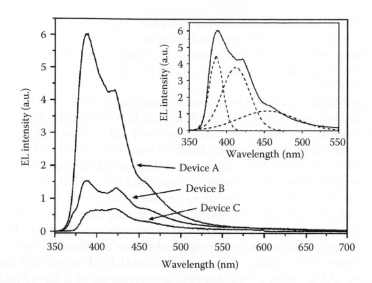

FIGURE 5.4

RT EL spectra of *p*-GaN/*n*-ZnO LEDs at a pulsed current of 800 mA. The inset shows that Gaussian functions (dashed lines) were used to fit the EL spectra to extract peak EL intensities and FWHMs of ZnO and GaN. Device A: a double-sided polished sapphire substrate with aluminum back reflector. Device B: a double-sided polished sapphire substrate without aluminum at the back. Device C: rough sapphire substrate without aluminum at the back. (After Chen, H., Chen, M., Wu, M., Li, W., Tsai, H., Yang, J., Kuan, H., and Shiojiri, M., UV electroluminescence and structure of *n*-ZnO/*p*-GaN heterojunction LEDs grown by atomic layer deposition, *IEEE J. Quantum Electron.* 46, 265. © 2010. With permission from IEEE.)

However, most methods present the defect-related emission from ZnO or/and GaN, or the band-edge emission from GaN.

Rogers et al. fabricated a *p-n* heterojunction LED from GaN/ZnO by using pulse laser deposition (PLD) for the growth of ZnO and obtained UV EL emission from the ZnO side of the heterojunction under high injection current [48]. In the PLD method, a stream of high-power laser pulses is used to evaporate the target materials. The ablated species are condensed on the substrate placed opposite to the target. The main advantages of PLD are its ability to create high-energy source particles, permitting high-quality film growth at low substrate temperatures, and operating in high ambient gas pressure of 10^{-5}–10^{-1} Torr. Rogers et al. grew ZnO thin films using a Coherent KrF excimer laser (248 nm). During growth the pulse repetition rate varied between 1 and 50 Hz. The laser spot was focused on the high purity target to give a fluence of up to about 4 J cm^{-2}. High purity O_2 ranging between 10^{-6} and 10^{-3} Torr was introduced for the growth of ZnO. X-ray diffraction (XRD) confirms that the growth of ZnO with PLD gives the high quality epitaxy of ZnO on GaN. The electron concentration of ZnO film is about 2×10^{17} cm^{-3}, and the GaN film has a similar carrier concentration at 5×10^{17} cm^{-3}. RT PL of the heterojunction shows an intense emission at 375 nm, indicating a NBE excitonic emission from a ZnO layer with low dislocation/defect density. The RT EL spectrum from the *p*-GaN/*n*-ZnO heterostructure under forward bias consists of one UV peak centered at 375 nm with FWHM of about 25 nm. This EL emission is located at the same position as that observed in the PL spectrum, which suggests that radiative recombination occurs in the ZnO layer. Therefore, the EL emission is attributed to the excitonic emission of ZnO. These results show that it is possible to obtain significant hole injection from GaN into the intrinsically doped ZnO. The increased injection current from 500 to 875 mA causes slight redshift of the EL emission peak probably due to thermal effect, and EL power shows an almost linear increase with the forward drive current. LEDs ranging from 0.05 to 0.3 mm^2 in size have similar outputs. It indicates that there is a good lateral homogeneity of the layer/device properties. The LED devices show light conversion efficiency of about 0.00005%.

Magnetron sputtering is a useful technique to grow oxide films on a large scale and for easy doping control of various dopants [56]. Kim et al. fabricated the *p*-GaN/*n*-ZnO heterostructure by depositing ZnO film with radio frequency (RF) magnetron sputtering [49,50]. They found that the EL result under the forward bias was dependent on the annealing conditions, including the annealing ambient and the annealing time. An optimal annealing condition can realize the NBE emission of the *n*-ZnO layer at 381 nm, where the annealing time is limited at 0.5 h. With increasing annealing time, the EL peak is broadened and a strong deep-level emission appears. When the annealing time is fixed at 2 h, the heterostructure annealed in N$_2$ ambient gives blue and violet light emission centered at ~400 nm, while the heterostructure annealed in air ambient shows a yellowish light emission with the broad band from 400 to 700 nm. The cross-sectional TEM image and the Auger electron spectroscopy (AES) investigation suggest the formation of a Ga–O mixed region at the interface of GaN and ZnO when the heterojunction was annealed in air ambient. This finding reveals that the O$_2$ in the ambient can diffuse into the interface of GaN and ZnO and react with Ga. The formation of the interface state depresses the radiative recombination process, resulting in the degradation of the EL emission when annealed in air ambient for a longer time.

5.3.2 *p*-GaN/*i*-ZnO/*n*-ZnO

In *p*-GaN/*i*-ZnO/*n*-ZnO heterojunction, *i*-ZnO layer has higher resistance than the *n*-ZnO layer. It is realized by using the different deposition techniques, different O$_2$ atmosphere, or different doping methods. Table 5.3 summarizes the growth method, doping for the growth of *n*-ZnO and *i*-ZnO layers, contact electrodes, and emission color of *p*-GaN/*i*-ZnO/*n*-ZnO heterojunction LEDs. The basic structure for *p*-GaN/*i*-ZnO/*n*-ZnO is to sandwich the high-resistance *i*-ZnO layer between *p*-GaN and *n*-ZnO. In general, sapphire is used as the substrate material. Ni/Au was deposited onto the GaN, and Ti/Au, In, Al, Ni, or indium-tin-oxide (ITO) was deposited on ZnO as the contact electrodes.

Lu et al. fabricated the *p*-GaN/*i*-ZnO/*n*-ZnO heterojunction by plasma-assisted MBE [57]. They used nitrogen-doped ZnO (ZnO:N) as the *i*-layer and undoped ZnO as the *n* layer. Prior to the growth of *i*-ZnO, GaN was pre-treated for 30 min at 800°C in order to clean the surface and reduce the surface potential barrier. N$_2$ was introduced into the chamber to deposit ZnO:N. Sequentially, the undoped ZnO ($n = 2 \times 10^{18}$ cm^{-3}, $\rho = 0.6 \Omega$ cm) was deposited on the *i*-ZnO. Al and Ni/Au electrodes were made by thermal evaporation for the ohmic contact for ZnO and GaN film, respectively. Both heterojunctions show diode rectifying behaviors. Compared with the *p*-GaN/*n*-ZnO heterojunction, *p*-GaN/*i*-ZnO/*n*-ZnO heterojunction presents much bigger series resistance and much smaller leakage current. It indicates that the insertion of *i*-ZnO increases series resistance and improves the electrical properties of the heterojunction. The EL spectra of *p*-GaN/*i*-ZnO/*n*-ZnO and *p*-GaN/*n*-ZnO heterojunctions at RT are shown in Figure 5.5. A very broad emission band centered at 430 nm is observed for *p*-GaN/*n*-ZnO heterojunction, which originates from GaN. In comparison, the EL peak of the *p*-GaN/*i*-ZnO/*n*-ZnO heterojunction blueshifts to 400 nm, and shows a narrower FWHM. It is believed to originate from *i*-layer because N doping can cause redshift of NBE emission [66].

TABLE 5.3

Growth Method, Doping for the Growth of *n*-ZnO Layer and *i*-ZnO Layer, Contact Electrodes, and EL Emission Color of the *p*-GaN/*i*-ZnO/*n*-ZnO Heterojunction LEDs

Growth Method (*i*-ZnO/*n*-ZnO)	Doping (*i*/*n*)	Contact GaN/ZnO	Emission (Color) (nm)	References
MBE	N/–	(Ni/Au)/Al	400 (blue–violet)	[57]
RF magnetron sputtering	–/Ga	(Ni/Au)/(Ti/Au)	650 (orange–red)	[58]
RF magnetron sputtering	–/Ga	(Ni/Au)/(Ti/Au)	405 (blue–violet)	[59]
RF magnetron sputtering	–/Ga	(Ni/Au)/(Ti/Au)	430 (blue), 440 (blue), 480 (blue)	[60]
EBE/RF magnetron sputtering	—	(Ni/Au)/In	386 (UV), 403 (blue–violet), 590 (orange)	[61]
Vapor cooling condensation	–/In	(Ni/Au)/(Al/Pt)	385 (UV)	[42]
Vapor cooling condensation	–/In	(Ni/Au)/ITO	386 (UV)	[62,63]
FCVA	–/Al	Au/Ni	387 (UV)	[64]
PLD	—	(Ni/Au)/In	410 (blue–violet) 500–680 (yellow)	[65]

MBE, molecular beam epitaxy; RF, radio-frequency; EBE, electron beam evaporation; FCVA, filtered cathodic vacuum arc; PLD, pulse laser deposition.

Cho et al. showed an EL peak at 405 nm from the *p*-GaN/*i*-ZnO/*n*-ZnO heterojunction [59]. The heterojunction was fabricated by magnetron sputtering. The undoped ZnO target was used as ZnO source for the growth of *i*-ZnO, and 1 wt% Ga doped ZnO target was used to grow *n*-ZnO. A mixture of high purity Ar and O_2 at a ratio of 1:2 was used as ambient gas. The ZnO film was grown at 750°C. The EL spectrum of *p*-GaN/*n*-ZnO LEDs exhibits a broad deep level related emission band centered at 530 nm (Figure 5.6a) while the *p*-GaN/*i*-ZnO/*n*-ZnO heterojunction shows strong near-UV emission centered at 405 nm with absence of visible emission (Figure 5.6b). The results show that improved device performance is achieved with insertion of a thin intrinsic layer between *n*-ZnO and *p*-GaN. The EL peak position and shape of the *p*-GaN/*i*-ZnO/*n*-ZnO heterojunction is similar to that reported by Lu et al. [57]. Cho et al. attributed this peak to *i*-ZnO layer. The redshift of the EL peak compared with the PL peak of the *i*-ZnO layer is believed to be related to the heating effect of the device.

Liu et al. used magnetron sputtering to fabricate *i*-ZnO and electron beam evaporation to fabricate *n*-ZnO [61]. It is the earliest report on inserting *i*-ZnO layer to improve the EL performance of *p*-GaN/*n*-ZnO. The high-resistance *i*-ZnO with thickness of 30 nm was realized in an oxygen-rich environment by RF reactive magnetron sputtering. An ultrapure metallic Zn disk was used as the sputtering target. Ultrapure Ar and O_2 gas mixtures with flow-rate ratio of 2:1 were introduced into the growth chamber. The substrate temperature and RF power density were controlled at 300°C and 2.0 W cm^{-2} during film deposition, respectively. The working pressure was kept at 1.0 Pa. The oxygen-rich atmosphere decreases the number of oxygen vacancy and interstitial zinc, which are the main defects for the formation of *n*-ZnO. Next, an *n*-ZnO layer was grown by electron beam evaporating ZnO ceramic target. Higher evaporation power density (195 W cm^{-2}) and substrate temperature (500°C) were used for obtaining *n*-ZnO with higher electron

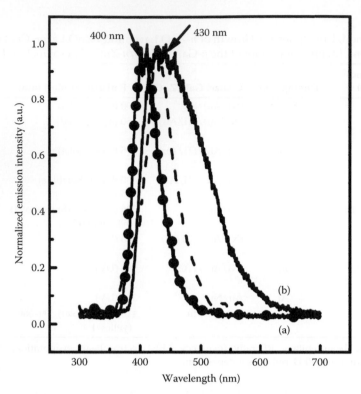

FIGURE 5.5

RT EL spectra of (a) *p*-GaN/*i*-ZnO/*n*-ZnO and (b) *p*-GaN/*n*-ZnO heterojunctions fabricated by plasma-assisted MBE technique. The dashed line presents the PL of *p*-type GaN. (After Jiao, S.J., Lu, Y.M., Shen, D.Z., Zhang, Z.Z., Li, B.H., Zhang, J.Y., Yao, B., Liu, Y.C., and Fan, X.W.: Ultraviolet electroluminescence of ZnO based heterojunction light emitting diode. *Phys. Status Solidi C.* 2006. 3. 972. Copyright Wiley-VCH Verlag GmbH & Co. KGaA. Used with permission.)

concentration. The RT PL spectra of *n*-ZnO, *i*-ZnO, and *p*-GaN are given in Figure 5.7, and RT EL spectra of *p-i-n* and *p-n* heterojunctions LEDs are shown in Figure 5.8. Compared with the EL results of *p*-GaN/*n*-ZnO, two new emissions, including a narrow UV EL peak at 386 nm and a very broad, weak emission band at 590 nm, are observed in the EL spectrum of *p*-GaN/*i*-ZnO/*n*-ZnO heterojunction. Both emissions are located at almost the same positions as those in PL spectra of *i*-ZnO (Figure 5.7). Thus, the EL emission peaks located at 386 and 590 nm are believed to originate from the active *i*-ZnO region, and are attributed to UV NBE emission related to ZnO exciton recombination and deep-level emission related to native defects, respectively. The relatively weak emission at 403 nm detected in the EL spectrum of the *p-i-n* heterojunction is attributed to *p*-GaN layer. The possible reason is that a small quantity of electrons tunnel through *i*-ZnO layer, and are injected into the *p*-GaN region. The observation of the EL related to ZnO confirms that the radiative recombination can be partly confined in *i*-ZnO region by constructing the *p-i-n* heterojunction.

Lee et al. grew *p*-GaN/*n*-ZnO and *p*-GaN/*i*-ZnO/*n*-ZnO film LED devices by vapor cooling condensation technique [42]. By putting 0.85 g of ZnO powder on the tungsten boat, a 100 nm thick ZnO film was deposited on cooled GaN:Mg in the vapor cooling condensation system that was used as *i*-ZnO. The mixture of 0.85 g of ZnO powder and 0.12 g of In

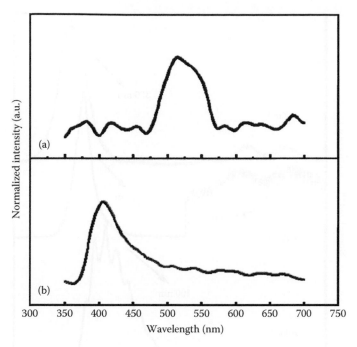

FIGURE 5.6
RT EL spectra from (a) *p*-GaN/*n*-ZnO and (b) *p*-GaN/*i*-ZnO/*n*-ZnO fabricated by magnetron sputtering. (After *Thin Solid Films*, 517, Han, W.S., Kim, Y.Y., Kong, B.H., and Cho, H.K., Ultraviolet light emitting diode with *n*-ZnO:Ga/*i*-ZnO/*p*-GaN:Mg heterojunction, 5106. Copyright 2009, with permission from Elsevier.)

were put together on the tungsten boat for the growth of 300 nm thick *n*-ZnO. The electron concentration of *i*-ZnO and *n*-ZnO films was obtained as 7.6×10^{15} and 1.7×10^{20} cm^{-3}, respectively. Both types of heterojunction LEDs showed the RT EL. As shown in Figure 5.9, *p*-GaN/*n*-ZnO had a broad emission band centered at 432 nm while *p*-GaN/*i*-ZnO/*n*-ZnO had an UV emission at 385 nm. From correlations with PL spectra it was inferred that the former was due to defect-related radiative recombination in GaN, while the latter was due to radiative recombination in *i*-ZnO layer.

Further, Lee et al. applied a similar method to obtain *p*-GaN/*i*-ZnO/*n*-ZnO nanostructured LED devices [62,63]. The *p*-GaN film/*i*-ZnO nanorod array/*n*-ZnO:In nanorod array LED was realized by growing *i*-ZnO and *n*-ZnO nanorod arrays on a *p*-GaN layer with an anodic alumina membrane (AAM) template using a vapor cooling condensation method. As shown in Figure 5.10, double-layer GaN film, which consisted of an undoped GaN buffer layer and a Mg-doped GaN layer, was first thermally annealed at 750°C for 30 min in N$_2$ ambient. The hole concentration and hole mobility of GaN were measured to be 3.7×10^{17} cm^{-3} and 11.2 cm V^{-1} s^{-1}, respectively. The Ni/Au electrode was deposited on the perimeter of *p*-GaN using a standard photolithographic technique and lift-off process, followed by thermal annealing at 500°C for 10 min in an air ambient in order to form ohmic contact with GaN. The area enclosed by Ni/Au square strips was about 7.2 mm^2. An AAM template with a pore diameter of 200 nm was put on top of GaN with the aim of deposition of ZnO nanorods. In the vapor cooling condensation system used here, the source materials was put on tungsten boats and then heated up to 500°C at the step of 25°C min^{-1}. At this temperature, metal In began to melt. After 30 min, the shutter was opened and then temperature was increased continuously. The ZnO:In vapor gas started to evaporate gradually when the

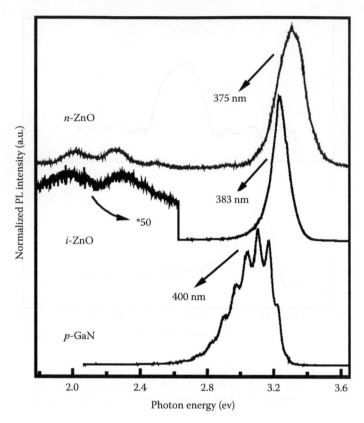

FIGURE 5.7

RT PL spectra of *n*-ZnO, *i*-ZnO, and *p*-GaN. (With kind permission from Springer Science+Business Media: *Appl. Phys. B*, Ultraviolet electroluminescence from *p*-GaN/*i*-ZnO/*n*-ZnO heterojunction light-emitting diodes, 80, 2005, 871, Xu, H.Y., Liu, Y.C., Liu, Y.X., Xu, C.S., Shao, C.L., and Mu, R.)

temperature was around at 1050°C and were then condensed and deposited on the cooled substrate by liquid nitrogen. The nanorod arrays could thus be deposited on the *p*-GaN layer. The electron concentration and mobility for *i*-ZnO and *n*-ZnO are 7.6×10^{15} cm^{-3} and 6.0 cm V^{-1} s^{-1}, and 1.7×10^{20} cm^{-3} and 3.1 cm V^{-1} s^{-1}, respectively. For comparison, a *p*-GaN film/*n*-ZnO:In nanorod array LED was fabricated. Figure 5.11 shows the RT EL spectra of *p-n* and *p-i-n* nanorod heterostructured LEDs operated at 15 and 35 μA, respectively. The EL spectra of the *p-i-n* nanorod devices present a similar rule with *p-n* devices at a different injection current. The *p-n* nanorod heterostructured LEDs reveal a broad emission band centered at 430 nm, while the *p*-GaN/*i*-ZnO/*n*-ZnO nanostructured LEDs reveal NBE exciton emission at 386 nm from *i*-ZnO. In addition, *p-i-n* nanorod devices present a broad emission band at 415 nm, which is attributed to the defect emission of *i*-ZnO. When the injection current of *p-i-n* nanorod heterostructured LEDs increases, the increase in the rate of NBE emission intensity is faster than that of the defects, as shown in Figure 5.11. Therefore, the NBE emission exhibits the dominant role in the *p*-GaN film/*i*-ZnO nanorod array/*n*-ZnO:In nanorod array LED when operated at a higher injection current.

Yu et al. fabricated a *p*-GaN/*i*-ZnO/*n*-ZnO heterojunction LED with a 3 μm wide rib waveguide, which was different from conventional heterojunction device structure. It is proposed that the *p-i-n* rib waveguide structure can achieve more effective radiative

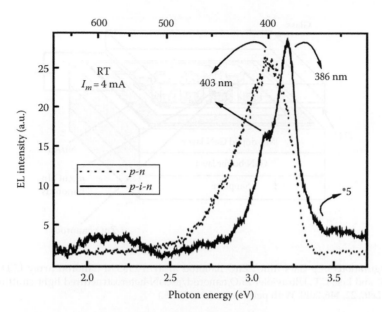

FIGURE 5.8
RT EL spectra of *p*-GaN/*n*-ZnO and *p*-GaN/*i*-ZnO/*n*-ZnO heterojunctions LEDs by magnetron sputtering and electron beam evaporation deposition. (With kind permission from Springer Science+Business Media: *Appl. Phys. B*, Ultraviolet electroluminescence from *p*-GaN/*i*-ZnO/*n*-ZnO heterojunction light-emitting diodes, 80, 2005, 871, Xu, H.Y., Liu, Y.C., Liu, Y.X., Xu, C.S., Shao, C.L., and Mu, R.)

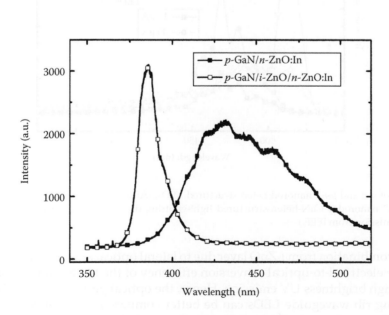

FIGURE 5.9
RT EL spectra of *p*-GaN/*n*-ZnO and *p*-GaN/*i*-ZnO/*n*-ZnO heterojunction LEDs fabricated by vapor cooling condensation technique. (Used with permission from Chuang, R.W., Wu, R.X., Lai, L.W., and Lee, C.T., ZnO on GaN heterojunction light emitting diode grown by vapor cooling condensation technique, *Appl. Phys. Lett.*, 91, 231113. Copyright 2007, American Institute of Physics.)

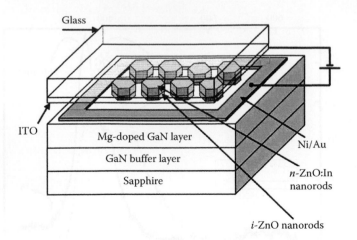

FIGURE 5.10
Schematic configurations of *p*-GaN film/*i*-ZnO nanorod array/*n*-ZnO:In nanorod array LED. (After Yan, J., Chen, C., Yen, S., and Lee, C.T., Ultraviolet ZnO nanorod/*p*-GaN-heterostructured light-emitting diodes, *IEEE Photon. Technol. Lett.*, 22, 146, 2010. With permission from IEEE.)

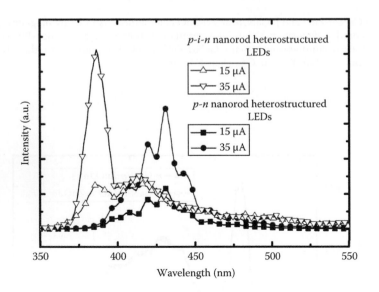

FIGURE 5.11
RT EL spectra of *p-n* and *p-i-n* nanorod heterostructured LEDs. (After Yan, J., Chen, C., Yen, S., and Lee, C.T., Ultraviolet ZnO nanorod/*p*-GaN-heterostructured light-emitting diodes, *IEEE Photon. Technol. Lett.*, 22, 146, 2010. With permission from IEEE.)

excitonic recombination from *i*-ZnO layer due to lateral optical and electrical confinement, can improve electrical-to-optical conversion efficiency of the LEDs, and can obtain directional and high brightness UV emission. Hence, the optical performance of the proposed edge-emitting rib waveguide LEDs can be better compared with conventional surface-emitting LEDs. As shown in Figure 5.12, in order to fabricate *p-i-n* rib waveguide LED, a 150 nm thick undoped ZnO was used as *i*-ZnO and 100 nm thick Al doped ZnO was used as *n*-ZnO. The *i*-ZnO film was first deposited onto half of the *p*-GaN:Mg/sapphire substrate by using filtered cathodic vacuum arc (FCVA) deposition technique at the substrate

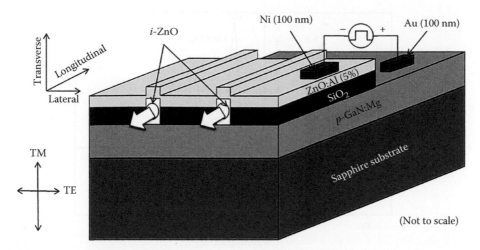

FIGURE 5.12

Schematic of *p*-GaN/*i*-ZnO/*n*-ZnO heterojunction LED with rib waveguide. (After Liang, H.K. et al., *Opt. Express*, 15, 3687, 2010. With permission from Optical Society of America.)

temperature of ~150°C and oxygen partial pressure of 2×10^{-5} Torr. Subsequently, a 3 μm wide and 0.8 μm thick line-mask with 500 μm separation was coated onto the surface of the *i*-ZnO film by photolithography technique. The unmasked *i*-ZnO layer was then completely removed by ion beam sputtering with an etching rate of ~10 nm min^{-1} for 15 min. Then, a 120 nm thick of SiO$_2$ cladding layer was deposited onto the sample by electron beam evaporation with substrate temperature of 50°C. After the deposition, a lift-off process was carried out in acetone to remove excess SiO$_2$ layer attached on the surface of *i*-ZnO. Finally, a layer of ZnO:Al was deposited onto the sample by FCVA technique. The carrier concentration of *i*-ZnO and *n*-ZnO films was found to be about 10^{18} and 10^{21} cm^{-3}, respectively. The metal contacts were formed by depositing the Au film onto the *p*-GaN substrate and the Ni film on the *n*-ZnO layer. Figure 5.13 shows the EL spectra of heterojunction LED with the rib waveguide at different bias voltages, and the inset shows the corresponding PL spectra of *i*-ZnO and *p*-GaN layers. All the EL spectra exhibit only the UV emission with peak at 387 nm. The heterojunction LED without rib waveguide exhibits similar EL spectra but weaker intensity than that with a rib waveguide. Together with the PL spectra of the *i*-ZnO layer it indicates that radiative excitonic recombination occurs in the *i*-ZnO layer of the heterojunction LED, and proves that the proposed *p*-*i*-*n* heterojunction confines radiative excitonic recombination in the *i*-ZnO layer.

From the experimental results mentioned earlier, emission of ZnO layer is much more easily to realize in *p*-GaN/*i*-ZnO/*n*-ZnO heterojunction LED devices compared with *p*-GaN/*n*-ZnO heterojunction devices. The emission in the heterojunctions mainly shows UV, blue, and blue–yellow color. Liu et al. recently reported a white light emission resembling conventional YAG:Ge phosphor white LED [65]. They applied a PLD method to fabricate ZnO. Ablation was performed using a YAG pulsed laser ($\lambda = 355$ nm, $ts = 20$ ns; 10 Hz). The substrate temperature was maintained at 500°C to ensure that the resulting ZnO films would have the desired amount of deep-level emission. The growth of *n*-ZnO and *i*-ZnO was controlled by adjustment of oxygen pressure at 0.02 and 20 Pa, respectively. The lower oxygen pressure produced more oxygen vacancies that served as donors, leading to higher electron concentration. In the 20 nm thick *i*-ZnO case, the EL shows a blue–violet emission peak at 410 nm and a broad yellow emission band from 500 to 680 nm (Figure 5.14). The CIE

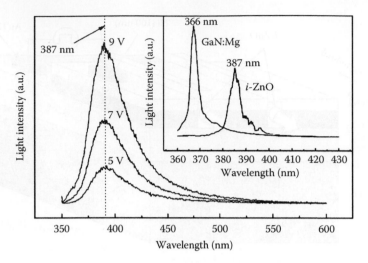

FIGURE 5.13
EL spectra of the heterojunction LED with rib waveguide. The inset shows the PL spectra observed from *p*-GaN and *i*-ZnO layers. (After Liang, H.K. et al., *Opt. Express*, 15, 3687, 2010. With permission from Optical Society of America.)

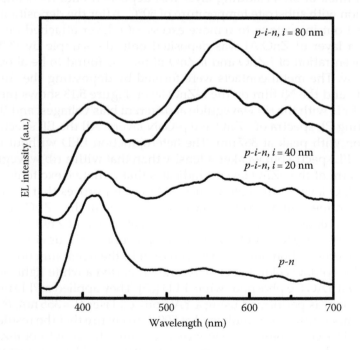

FIGURE 5.14
EL spectra of *p*-GaN/*n*-ZnO and *p*-GaN/*i*-ZnO/*n*-ZnO heterojunctions with *i*-layer thickness of 20, 40, and 80 nm. (With kind permission from Springer Science+Business Media: *Appl. Phys. B*, A new approach to white emitting diodes of *p*-GaN/*i*-ZnO/*n*-ZnO heterojunctions, 92, 2008, 185, Zhao, L., Xu, C.S., Liu, Y.X., Shao, C.L., Li, X.H., and Liu, Y.C.)

chromaticity coordinates for heterojunctions with *i*-ZnO thickness of 20 nm is (0.34, 0.32), which is very close the ideal white light value (0.33, 0.33). The commixing deep-level ZnO emission and blue GaN emission presents a white emission for the possible application in white LED.

As for the *p*-GaN/*i*-ZnO/*n*-ZnO heterojunction, the thickness of *i*-ZnO layer is one key factor to affect EL properties. Liu et al. investigated the dependence of the EL spectra of the heterojunction on *i*-layer thickness. As shown in Figure 5.14, when the thickness of *i*-ZnO increases, the intensity of yellow emission (centered at 560 nm) increases relative to that of blue–violet (centered at 410 nm), suggesting a straightforward method for modulating emitted color through control of *i*-ZnO layer thickness. The white EL is observed when the *i*-layer is 20 nm thick, while yellow emission is dominant when the *i*-layer is 80 nm thick. For *p*-GaN/*n*-ZnO heterojunctions, due to lower carrier concentration and mobility of *p*-GaN compared with those of *n*-ZnO, and similar conduction and valence band offsets values as calculated previously, carrier recombination occurs mainly in the *p*-GaN layer, giving blue–violet emission. In *p*-GaN/*i*-ZnO/*n*-ZnO heterojunctions, *i*-ZnO has the lowest carrier concentration and mobility of the three layers. Therefore, electrons from *n*-ZnO and holes from *p*-GaN mostly inject into the *i*-ZnO layer, where recombination occurs, giving rise to yellow deep-level emission. Also, some electrons from *n*-ZnO tunnel through *i*-ZnO into the *p*-GaN layer (and some electrons from *i*-ZnO also inject into the *p*-GaN layer), where they recombine with holes to give blue–violet emission. As *i*-ZnO thickness increases, however, fewer electrons tunnel through, instead recombining in *i*-ZnO layer, resulting in increase in yellow emission intensity relative to blue–violet emission.

The effect of injection current on EL in a *p*-GaN/*i*-ZnO/*n*-ZnO heterojunction was measured at injection currents ranging from 5 to 100 mA. As shown in Figure 5.15a, the EL

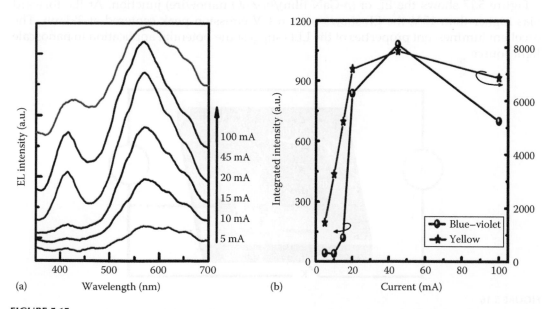

(a)   Wavelength (nm)   (b)   Current (mA)

FIGURE 5.15
(a) EL spectra of a *p*-GaN/*i*-ZnO/*n*-ZnO heterojunction with injection currents ranging from 5 to 100 mA and (b) EL integrated intensities of blue–violet and yellow emissions at various injection currents. (With kind permission from Springer Science+Business Media: *Appl. Phys. B*, A new approach to white emitting diodes of *p*-GaN/*i*-ZnO/*n*-ZnO heterojunctions, 92, 2008, 185, Zhao, L., Xu, C.S., Liu, Y.X., Shao, C.L., Li, X.H., and Liu, Y.C.)

lineshapes depend significantly on the injection current. The integrated intensity of blue–violet emission and yellow emission calculated at different injection currents are shown in Figure 5.15b. For injection currents below 10 mA, yellow emission dominates and increases in intensity with increasing current, while blue–violet emission was negligible. In this range, as described earlier, n-ZnO electrons and p-GaN holes are injected into i-ZnO and combine radiatively there (yellow), while a very small amount of i-ZnO electrons are injected into p-GaN and combine radiatively (blue–violet). Between 10 and 20 mA, both blue–violet and yellow emission intensity increased dramatically. In this range, more carriers are injected into i-ZnO and recombine there, while more electrons in n-ZnO tunnel through i-ZnO into p-GaN, radiatively recombining with holes there. When the injection current is above 50 mA, heating effect became prominent, which decreases the efficiency of radiative recombination.

5.3.3 (p-GaN film)/(n-ZnO Nanostructures)

5.3.3.1 (p-GaN Film)/(Single n-ZnO Nanowire)

The application of (p-GaN film)/(n-ZnO nanowire) devices for LEDs has also sparked considerable interest because they may be able to act as nanoscale solid-state light sources. Liao et al. produced (p-GaN film)/(single n-ZnO nanowire) heterojunctions for the first time. As shown in Figure 5.16, Ni/Au ohmic contacts were first prepared on GaN:Mg film. A 70 nm thick Al_2O_3 insulating film was then deposited by electron beam evaporation onto half of the GaN film with a shadow mask. A single ZnO nanowire was subsequently placed across the boundary between Al_2O_3 and GaN films. A platinum electrode was deposited onto the ZnO nanowire on the Al_2O_3 side by focused ion beam (FIB) induced deposition technique.

Figure 5.17 shows the EL of (p-GaN film)/(n-ZnO nanowire) junction. At the forward bias higher than 15 V, the EL shows a main UV emission peak centered at 391 nm. The excellent luminescent properties of the LED suggest the potential application in nanoscale light source.

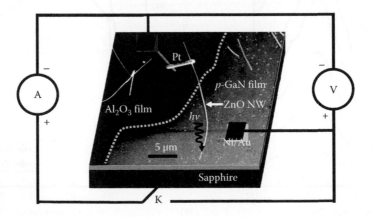

FIGURE 5.16
SEM image of (p-GaN film)/(single n-ZnO nanowire) heterojunction and the schematic test circuit with the denoted positive connection. (From Bie, Y., Liao, Z., Wang, P., Zhou, Y., Han, X., Ye, Y., Zhao, Q., Wu, X., Dai, L., Xu, J., Sang, L., Deng, J., Laurent, K., Wang, Y., and Yu, D.: Single ZnO nanowire p-type GaN heterojunctions for photovoltaic devices and UV light-emitting diodes. *Adv. Mater.* 2010. 22. 4284. Copyright Wiley-VCH Verlag GmbH & Co. KGaA. Used with permission.)

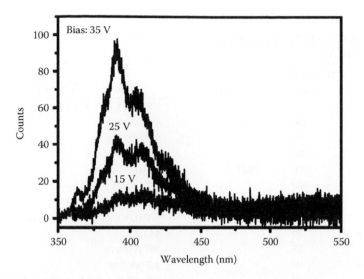

FIGURE 5.17

EL of the (*p*-GaN film)/(single *n*-ZnO nanowire) heterojunction at different forward biases from 15 to 35 V. (From Bie, Y., Liao, Z., Wang, P., Zhou, Y., Han, X., Ye, Y., Zhao, Q., Wu, X., Dai, L., Xu, J., Sang, L., Deng, J., Laurent, K., Wang, Y., and Yu, D.: Single ZnO nanowire *p*-type GaN heterojunctions for photovoltaic devices and UV light-emitting diodes. *Adv. Mater.* 2010. 22. 4284. Copyright Wiley-VCH Verlag GmbH & Co. KGaA. Used with permission.)

5.3.3.2 (p-GaN Film)/(n-ZnO Nanoscale Array)

Due to the small lattice mismatch between GaN and ZnO, GaN can directly serve as a template for epitaxial growth of high-quality ZnO nanostructures. A ZnO nanorod or nanowire array normal to the GaN film is preferred for the formation of heterojunction LED, which is more easily fabricated than the film and generally presents single-crystal structure. Here, we mainly focus on growth of ZnO nanostructures for (*p*-GaN film)/ (*n*-ZnO nanoscale array) heterojunction LED. Table 5.4 summarizes the growth method and growth temperature of ZnO nanostructures, the fabrication technique of the seed layer prior to the growth of ZnO nanostructures, and EL peak for heterojunction LEDs. The growth of ZnO nanostructure for *p*-GaN/*n*-ZnO heterojunction LEDs mainly includes the vapor deposition in a horizontal tube [68–71] and solution-based techniques [75–84]. The other methods, for example, PLD [72], MOCVD [73,74], and MOVPE [35], are also used for the fabrication of the ZnO nanostructure. The vapor deposition is one of the most extensively used methods for the growth of ZnO nanostructures. In a typical vapor deposition technique, the source material, generally, metal Zn, or ZnO, was sublimated at the high-temperature zone of the horizontal quartz tube and then deposited in the low-temperature zone. High pure source materials are easily used in vapor method. In addition, the sublimation process in the vapor deposition is favorable to further improve the purity of the deposited materials. Therefore, vapor-deposited ZnO nanostructures are expected to have better crystal quality compared to those prepared by solution-based techniques.

Zhang et al. grew a ZnO nanowire array on a GaN film using vapor deposition method via a vapor–liquid–solid (VLS) process [68,87]. A 2 nm thick gold layer was first deposited onto GaN substrate by plasma sputtering as catalyst for the growth of the ZnO nanowires. The pure Zn powders were used as the reaction source and were loaded in an alumina boat located at the center of an alumina tube, which was placed in a single-zone tube furnace. Ar was used as carrier gas at a flow rate of 49 sccm with additional 2% (1 sccm) O_2 to

TABLE 5.4

Growth Method, Growth Temperature, Seed Layer, and EL Emission Color of (*p*-GaN Film)/
(*n*-ZnO Nanoscale Array)

Method	Temperature (°C)	Seed	Emission (Color) (nm)	References
Vapor deposition	950	—	370 (UV), 400–440 (blue)	[68]
	450	—	405 (violet)	[69]
	550	—	415–417 (blue)	[70]
	600	—	400–700 (white)	[71]
PLD	1000	PLD	380 (UV–violet)	[72]
MOCVD	500	—	432 (blue)	[73]
	500	—	386 (UV–violet)	[74]
MOVPE	400–500	—	370 (UV), 443 (blue), 560 (yellow)	[35]
Electrodeposition	80	—	440 (blue)	[75]
	85	—	397 (violet)	[76,77]
Aqueous solution	90	—	407 (violet)	[78]
	90	—	385 (UV), 415 (blue), 530 (yellow)	[79]
	90	—	365 (UV), 530 (yellow)	[79]
	130	—	Violet	[80]
	90	Dip coating	415 (blue)	[81]
	90	Drop	UV, violet, blue, or yellow	[82]
	85	Drop	440 (blue)	[75]
	85	Electrodeposition	440 (blue), 600 (yellow)	[75]
	90–96	Spin coating	400–900 (white)	[14,83–85]
	90	Spin coating	398 (blue–violet)	[86]

PLD, pulse laser deposition; MOCVD, molecular organic chemical vapor deposition; MOVPE, metal organic vapor phase epitaxy.

facilitate the reaction. The temperature of the source materials was increased to 950°C at a rate of 50°C min^{-1}, and remained for 40 min under a pressure of 10 Torr. Then the furnace was turned off, and the tube was cooled in air to RT under an Ar flow. The obtained nanowires have a uniaxial orientation with an epitaxial orientation with respect to the GaN substrate. The grown ZnO nanowires are about 100–150 nm in diameter and about 5 mm in lengths. They form fairly uniform arrays on GaN. Each nanowire grows up independently from the GaN film.

Based on the nanowire array, the LED device was fabricated by the following procedure. The poly(methyl methacrylate) (PMMA) was spin coated on the ZnO nanowires array with a rotation rate of 4000 rpm for 2 min to form a smooth surface, which can effectively avoid leak current and successfully realize subsequent deposition of the contact electrode for the EL measurement. The PMMA coated on the surface of the exposed ZnO nanowires was removed with oxygen plasma etching by flowing 30 cm^{-3} min^{-1} O_2 in an 8×10^{-3} Torr vacuum with etching power 50 W for 2 h in order to obtain electrode contact. After that, a 300 nm thick ITO film was deposited by RF magnetron sputtering in a vacuum of 5×10^{-3} Torr. Finally, two copper wires were attached on the surface of the exposed GaN and ITO films using silver paste as electrodes. The schematic image of the device is shown in Figure 5.18.

The EL spectrum of the (*p*-GaN film)/(*n*-ZnO nanowires) LED device was measured under various forward bias voltages. As shown in Figure 5.19, the EL spectrum shows

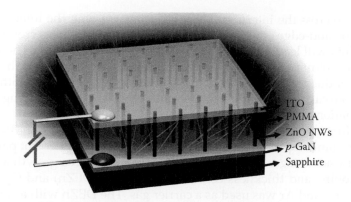

FIGURE 5.18
Schematic illustration of the (*p*-GaN film)/(*n*-ZnO nanowires) heterojunction LED device. (From Zhang, X.M., Lu, M.Y., Zhang, Y., Chen, L.J., and Wang, Z.L.: Fabrication of a high-brightness blue-light-emitting diode using a ZnO-nanowire array grown on *p*-GaN thin film. *Adv. Mater.* 2009. 21. 2767. Copyright Wiley-VCH Verlag GmbH & Co. KGaA. Used with permission.)

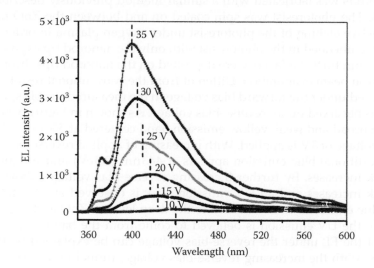

FIGURE 5.19
EL spectrum of the (*p*-GaN film)/(*n*-ZnO nanowires) LED device under various forward bias voltages, showing broad emission peaks from UV to blue. (From Zhang, X.M., Lu, M.Y., Zhang, Y., Chen, L.J., and Wang, Z.L.: Fabrication of a high-brightness blue-light-emitting diode using a ZnO-nanowire array grown on *p*-GaN thin film. *Adv. Mater.* 2009. 21. 2767. Copyright Wiley-VCH Verlag GmbH & Co. KGaA. Used with permission.)

broad emission peaks centered at around 370 nm and at 400–440 nm. When the forward bias increases from 10 to 35 V, the emission peak at about 440 nm is significantly enhanced and shifts to 400 nm. Zhang et al. proposed that the blueshift of the peak is related to the changes of the emission mechanism and could be explained using the band structure model. Under the small applied bias, the electrons on the conduction band of ZnO and the hole on the valence band of GaN will recombine directly at the interface between ZnO and GaN, producing an emission at about 440 nm. When the applied bias value is over the threshold voltage (10 V), parts of the carriers, including the electrons and holes, will obtain enough energy to cross the interface. These electrons crossing the interface from ZnO to GaN will lead to band-edge emission of GaN (387 nm), and the holes crossing the interface from GaN to ZnO will lead to band-edge emission of ZnO (368 nm). The broad emission band from 380 to 500 nm combines the band-edge emission of GaN and the interface emission from the conduction band of ZnO to the valence band of GaN. With the increasing applied bias, more carriers contribute to the band-edge emission, resulting in the blueshift of the recombination peak at about 440 nm.

Park and Yi first grew vertically-aligned n-ZnO nanorod arrays on 3 μm thick p-GaN layer by metal organic vapor phase epitaxy (MOVPE) [35]. MOVPE is of particular interest since it has many advantages such as the feasibility of large area growth as well as simple and accurate doping and thickness control. Diethyl-zinc (DEZn) and O_2 were employed as reactant sources, and Ar was used as a carrier gas. The DEZn with a carrier gas flowed for 30 s prior to ZnO nanorod growth in order to prevent GaN surface oxidation. A typical growth temperature was in the range of 400°C–500°C. Prior to ZnO nanorod growth, very thin ZnO buffer layers were grown at a low temperature. In this growth, no metal catalyst is coated on the substrates. For the heterojunction device fabrication, 10 nm thick Pt and 50 nm thick Au were deposited on p-GaN to form ohmic contact. The top electrode on the ZnO nanorods was fabricated with a similar method previously described by Zhang in Figure 5.18. The photoresist was spin coated on and between the ZnO nanorods, followed by selective etching of the photoresist under oxygen plasma in order to produce a nanorod array embedded in the photoresist with only the nanorod tip exposed. The 10 nm thick Ti and 50 nm thick Au layers were deposited on the nanorod tips through a shadow mask by electron beam evaporation. Different from the conventional results, EL emission was not observed under the forward bias voltage in the measured visible spectrum range. However, EL is observed under reverse-bias voltage. As shown in Figure 5.20, the EL spectrum shows a broad and weak yellow emission band centered at 560 nm (2.21 eV) when a reverse-bias voltage of 3 V is applied. With increasing of applied reverse-bias voltage from 3 to 4 V, an additional blue emission appears at 443 nm (2.8 eV) and the intensity of the emission peak increases. By further increasing reverse-bias voltages above 4 V, the blue emission peak increases and an UV emission peak is observed at 370 nm (3.35 eV). The yellow and blue emissions are attributed to the deep level in ZnO and Mg acceptor level in GaN, while the UV emission is believed to come from the band-edge emission. The observation of the EL under the reverse-bias voltage can be explained by the tunneling of the carriers. With the increasing reverse-bias voltage, tunneling probability increases, resulting in the enhanced EL intensity.

Based on the well-aligned ZnO nanobottles fabricated by vapor deposition, Chen et al. applied similar device configuration to create the (p-GaN film)/(ZnO nanobottle) heterojunctions [69]. The ZnO nanobottles were grown by using high purity Zn foil (99.99%) as precursor and high-purity Ar (99.9%) as carrier gas. The Zn foil was placed in the alumina boat at the center of a tube furnace. The reaction chamber was sealed and pumped to keep at a pressure of 200 Torr. Subsequently, the temperature of the furnace was raised to 450°C and maintained at this temperature for 30 min. The main difference between this growth process and the conventional one is that the GaN-sapphire substrate used to collect the synthesized products was located above the Zn foil with the GaN surface placed facing upwards and the sapphire surface placed facing down. Based on this special upside-down arrangement, the bottle-like ZnO nanoscaled array with an excellent crystalline quality is obtained. With the substrate placed facing upwards, however, only the ZnO nanowires are obtained. After the growth, the sample was then spin-coated with electron resist of ZEP520 for subsequent EL measurement. Finally, In was thermally

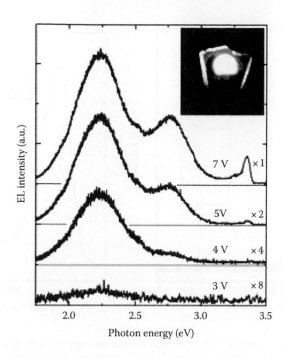

FIGURE 5.20
RT EL spectra of a *p*-GaN/*n*-ZnO heterojunction device at a different bias. The inset is a photograph of light emission from the EL device at a bias voltage of 5 V. (From Park, W.I. and Yi, G.C.: Electroluminescence in *n*-ZnO nanorod arrays vertically grown on *p*-GaN. *Adv. Mater.* 2004. 16. 87. Copyright Wiley-VCH Verlag GmbH & Co. KGaA. Used with permission.)

heated onto the *p*-GaN thin film and Ti/Au were sputtered onto the exposed ZnO nano-bottles to form electrode contacts.

Figure 5.21a shows the EL spectra of the (*p*-GaN)/(*n*-ZnO nanobottle) heterojunction generated from 0 to 50 V. The dominant peak is centered at 405 nm and the defect emission is very weak. The EL peak position at 405 nm is lower than the band-edge emission of the ZnO nanobottles (377 nm), but larger than the emission from *p*-type GaN film (~412 nm). It is attributed to the transition from the conduction band edge of ZnO to the acceptor level of GaN. Figure 5.21b shows the EL spectrum of (*p*-GaN)/(*n*-ZnO nanowire) heterojunction under bias voltage of 30 V. The ZnO-nanowire-based heterojunction exhibits a blue emission at 426 nm and the defect emission band centered at ~570 nm. The EL at 426 nm is attributed to the transition from the donor level in ZnO to the acceptor level in GaN. For the (*p*-GaN)/(*n*-ZnO nanobottle) heterojunction devices, the well-aligned ZnO nanobottles on a GaN thin film provide a chance to realize the selective angle emission as a result of the waveguided emission, which arises from the special geometry of the bottle structures. For the waveguide measurement, the (*p*-GaN)/(*n*-ZnO nanobottle) heterojunction device was placed on the stage of a rotator to identify the collecting angle between the normal direction of the device and the normal direction of the collecting lens. The slit of the spectrometer entrance was tuned to ensure that the EL from the device could be entirely collected. The EL spectra were measured from +90° to −90° at the step of 10°. Figure 5.22a shows the angle-distribution EL spectra collected from the dominant peak at 405 nm with all intensities background subtracted (530 units) and normalized to the highest intensity. The two-petal-like angle distribution displays a wide-angle emission with a maximum

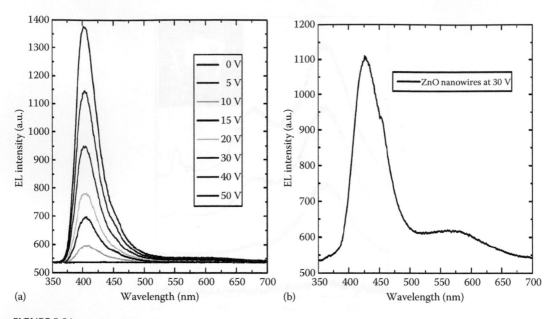

FIGURE 5.21

EL of (a) (*p*-GaN)/(*n*-ZnO nanobottle) heterojunction at different applied voltages and (b) (*p*-GaN)/(*n*-ZnO nanowire) at 30 V. (From Fu, H., Cheng, C., Wang, C., Lin, T., and Chen, Y.: Selective angle electroluminescence of light-emitting diodes based on nanostructured ZnO/GaN heterojunctions. *Adv. Funct. Mater.* 2009. 19. 3471. Copyright Wiley-VCH Verlag GmbH & Co. KGaA. Used with permission.)

extraction of light at ±50°. This result is different from the well-aligned ZnO nanowires devices, which extract light with maximum intensity in the normal direction of the device (Figure 5.22b). It is believed that the bottle-like ZnO nanostructure with a hexagonal cap plays an important role in the two-petals like angle-distribution light emission.

The nanobottle is composed of a long bottle and a hexagonal cap with an upside down cone. The hexagonal cap serves as a good reflector and prevents emitted light from passing out of the vertical direction. In addition, it guides the light extracted from the side wall. Because the extracted light is normal to the surface of the side wall, it has an angle of around 50° with respect to the vertical direction. This angle is in good agreement with the angle of the maximum extracted intensity as shown in Figure 5.22a. In comparison, the (*p*-GaN)/(*n*-ZnO nanowire) devices show droplet-like angle distribution of the emission. As shown in Figure 5.22b, the maximum intensity of light extraction occurs in the normal direction of the device. The different results between nanobottle and nanowire based devices confirm that the morphology and the crystalline quality of ZnO nanostructure do play a significant role in determining the optical properties. The heterojunction LEDs based on the vertically aligned ZnO nanobottles presented here provide an excellent alternative for the design of a selective angle light source, which will be very useful for the future development of high-brightness solid-state emitters.

In comparison with the vapor method, the solution-based growth has important advantages in low cost, low temperature, and simple process. Most of the efforts for (*p*-GaN)/(*n*-ZnO nanostructures) heterojunction LED fabrication have been focus on solution based method, as shown in Table 5.4. Two facile solution methods, aqueous chemical growth and electrodeposition, have been extensively applied to obtain ZnO nanorod array on GaN. The aqueous chemical growth is the most commonly used method, which first was described

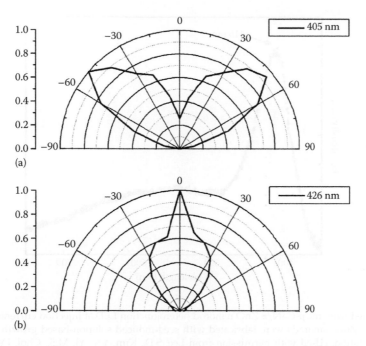

FIGURE 5.22

Angular distributions of EL spectra of (a) (*p*-GaN)/(*n*-ZnO nanobottle) heterojunction and (b) (*p*-GaN)/(*n*-ZnO nanowire) from +90° to −90° with intensities normalized. (From Fu, H., Cheng, C., Wang, C., Lin, T., and Chen, Y.: Selective angle electroluminescence of light-emitting diodes based on nanostructured ZnO/GaN heterojunctions. *Adv. Funct. Mater.* 2009. 19. 3471. Copyright Wiley-VCH Verlag GmbH & Co. KGaA. Used with permission.)

by Vayssieres et al. [36]. The equimolar zinc nitrate [$Zn(NO_3)_2$] and HMT was mixed and heated at low temperature of 80°C–96°C. This method combined with the preprepared seed layer can produce well-oriented ZnO nanorod arrays. For example, Kim et al. first fabricated ZnO seed layer on *p*-GaN substrate by dipping the substrate into 5 mM of zinc acetate [$Zn(C_2H_3O_2)_2$] dissolved in deionized water at 90°C [81]. To control the seeds, the dipping time ranged from 5 to 30 min. Prior to growth of nanorods, the seed-coated samples were then annealed for 20 min at 100°C. The ZnO nanorods were grown in an aqueous solution of 25 mM of $Zn(NO_3)_2$ and 25 mM of HMT for 4 h at 90°C. In order to fabricate the *p*-GaN heterojunction/*n*-ZnO nanorod LED, a 500 nm thick SiO_2 layer was filled into the free space between ZnO nanorods by plasma-enhanced chemical vapor deposition, and a 200 nm thick ITO layer was sputtered onto the top of ZnO nanorods. Ti/Au and Ni/Au were deposited on ITO and *p*-GaN, respectively.

They found that the density and size of the grown nanorods and microrods depended significantly on ZnO seed density. Uniformly distributed ZnO seed nanoparticles of smaller diameters lead to the formation of vertically aligned ZnO nanorods with relatively uniform diameters during the main growth. More dippings lead to nonuniform distribution of agglomerated seed nanoparticles with larger diameters, resulting in the formation of microrods as well as nanorods during the main growth. Synchrotron x-ray scattering measurements show an epitaxial relationship between the ZnO nanorods and *p*-GaN thin film. RT EL spectra show a dominant emission centered at 415 nm and the broad emission band covering the range from 485 to 750 nm (Figure 5.23). Differences in the shape of the emission spectrum are not noticeable in the EL spectra in different injection currents. This result indicates that such heterojunction LEDs fabricated from solution

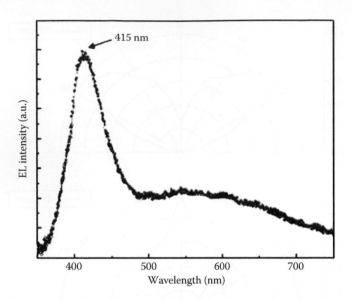

FIGURE 5.23
EL spectra obtained from the *p*-GaN/*n*-ZnO nanorod heterojunction LED at injection currents of 40 mA under forward bias. The ZnO nanorods were fabricated with seed-induced solution-based growth in Zn(NO$_3$)$_2$ and HMT aqueous solution. (Used with permission from Lee, S.D., Kim, Y.S., Yi, M.S., Choi, J.Y., and Kim, S.W., Morphology control and electroluminescence of ZnO nanorod/GaN heterojunctions prepared using aqueous solution, *J. Phys. Chem. C*, 113, 8954. Copyright 2009 American Chemical Society.)

are a promising approach for realizing large-area light-emitting sources with high external quantum efficiency.

Zhao et al. applied a similar solution method to fabricate *n*-ZnO microcrystalline film on *p*-GaN. The equimolar Zn(CH$_3$COO)$_2$·2H$_2$O and HMT were used as reactant sources [78]. The reaction kettle was put into an oven and maintained at 90°C for 16 h for each experiment. The ZnO microcrystalline film was obtained through repeated growth for six times under the same conditions. Their (*p*-GaN film)/(*n*-ZnO microcrystalline film) heterojunction realized the electrically pumped single-mode lasing emission at 407 nm with FWHM of 0.7 nm when the injection current exceeds 36 mA, as shown in Figure 5.24. It is believed that the excellent crystal quality and the ideal interface facilitate efficient carrier accumulation, and larger refractive index of ZnO than that of air makes the smooth top surface of ZnO serve as a mirror that defines optical microcavity.

Zhao et al. further found that vertical ZnO nanorod arrays with nominal diameters of 100–500 nm and lengths of 3 mm could be epitaxially grown on *p*-GaN thin film [79]. The (*p*-GaN film)/(*n*-ZnO nanorod) device shows typical diode rectifying behavior, which confirms the formation of *p-n* junction. Figure 5.25 shows the EL of the heterojunction before and after hydrogen treatment. Hydrogen treatment on the LED was carried out under atmospheric pressure with a hydrogen flow rate of 2000 ppm at RT for 30 min. The results revealed that such a device is sensitive to the surrounding atmosphere. In air, the EL spectrum of the device is broad and consists of peaks centered at 385, 415, and 530 nm. The emissions at 385 and 530 nm are attributed to ZnO, and the emission at 415 nm is attributed to GaN. Hydrogen-treated (*p*-GaN film)/(*n*-ZnO nanorod) device shows a dominant peak at 365 nm, which is believed to originate from GaN. The additional hydrogen shallow donors increase effective electron concentration in ZnO nanorods and the EL recombination zone changes from ZnO nanorods to GaN film, which can visually change the color from cyan to violet.

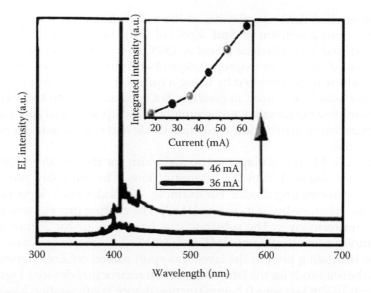

FIGURE 5.24
Single mode lasing emission for the *p*-GaN film/*n*-ZnO microcrystals heterojunction fabricated by using the aqueous solution of $Zn(CH_3COO)_2$ and HMT as the reaction solution. The lasing effect is evident when the injection current exceeds 36mA; the integrated EL intensity as a function of the injection current is shown in the inset. (Used with permission from Guo, Z., Zhao, D.X., Liu, Y.C., Shen, D.Z., Yao, B., Zhang, Z., and Li, B., Electrically pumped single-mode lasing emission of self-assembled *n*-ZnO microcrystalline film/*p*-GaN heterojunction diode, *J. Phys. Chem. C*, 114, 15499. Copyright 2010 American Chemical Society.)

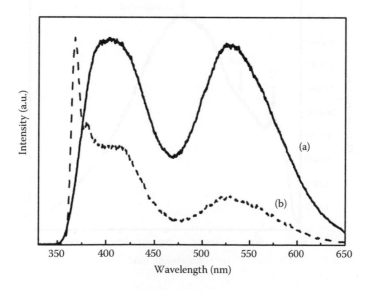

FIGURE 5.25
RT EL spectra for a *p*-GaN/*n*-ZnO nanorod heterojunction LED before (a) and after (b) hydrogen absorption at 10mA. (After Fang, F., Zhao, D.X., Li, B.H., Zhang, Z., and Shen, D.Z., Hydrogen effects on the electroluminescence of *n*-ZnO nanorod/*p*-GaN film heterojunction light-emitting diodes, *Phys. Chem. Chem. Phys.*, 12, 6759, 2010. With permission of The Royal Society of Chemistry.)

Mastro et al. applied ZnO nanoflowers as the *n*-ZnO layer [80]. The ZnO nanoflowers were fabricated from a solution of 5 mL $Zn(CH_3COO)_2$ in water (0.001 M), 10 mL NaOH in water (0.01 M), and 1 mL ethylene glycol at 130°C in a covered Teflon beaker. The ZnO nanoflowers were dispersed in isopropyl alcohol and dropped onto a *p*-GaN thin film. An ac electric field was then applied by using a function generator to produce the dielectrophoretic force, which was used to position ZnO nanoflowers on the *p*-type GaN. The (*p*-type GaN film)/(*n*-ZnO nanoflower) heterojunction displays rectifying current–voltage behavior characteristic of a pristine *p-n* junction diode and emit violet light under forward bias above 4.7–5.5 V.

Although the UV EL in ZnO layer is the main aim for the *p*-GaN/*n*-ZnO, white light emission has been obtained from the heterojunctions. This probably provides a way to fabricate white-light-emitting diodes. For example, Willander et al. deposited a thin ZnO seed layer on the substrate by spin coating [14,83]. The coating process was repeated three times to get a uniform layer. The substrate was heated in air for 20 min at 250°C. And then the equimolar concentration of HMT and $Zn(NO_3)_2$ were use to obtain oriented ZnO nanorods. The insulating photoresist layer was spun coated on ZnO nanorods to fill the gaps between the nanorods for the fabrication of heterostructure devices. Figure 5.26 shows the (*p*-GaN film)/(*n*-ZnO nanorod) heterojunction device configuration based on the ZnO nanorods. The EL result of the heterojunction (Figure 5.26) shows weak and sharp violet–blue emission peak at 457 nm, the green emission peak at 531 nm, and a broad orange–red band centered at 660 nm. The combination of three emission lines composed of violet–blue, green, and orange–red peaks showed very bright nearly white light EL emission.

By further suspending the ZnO nanorods upside down in an aqueous solution of potassium chloride (KCl) in a 5 M concentration at 95°C for 10 h, Willander et al. fabricated

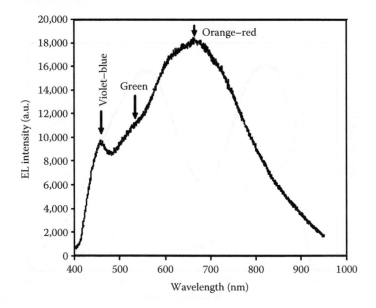

FIGURE 5.26

EL spectrum for the (*p*-GaN film)/(*n*-ZnO nanorod) LED fabricated with $Zn(NO_3)_2$ and HMT aqueous solution as the reaction solution. (After Alvi, N.H., Riaz, M., Tzamalis, G., Nur, O., and Willander, M., Fabrication and characterization of high-brightness light emitting diodes based on *n*-ZnO nanorods grown by a low-temperature chemical method on *p*-4H-SiC and *p*-GaN, *Semicond. Sci. Technol.*, 25, 065004, 2010. With permission from Institute of Physics.)

the well aligned ZnO nanotubes [85]. The ZnO nanorods can be successfully etched into the nanotubes due to the difference in stability between the polar and nonpolar planes of ZnO nanorods. The nonpolar planes (corresponding to the side surfaces of the nanorods) are the most stable planes by exhibiting lower surface energy than polar planes (correspond to top surface), which are metastable. The etching rate is faster along the polar plane than nonpolar planes. The diameter of the ZnO nanotubes is in 100–150 nm, the wall thickness is 25–40 nm, and the average length of the nanotubes is about 1.7 µm. The etching depth of ZnO nanotubes has been intentionally kept between 700 and 900 nm to protect the *p-n* junction of the heterostructure. The EL spectrum of *p*-GaN film/ZnO nanotubes exhibits relatively weak violet–blue emission centered at around 450, and a broad emission band centered at 560 nm, respectively. Their combination produces a strong white light, which are attributed to a lot of surface states and bulk defects.

The electrodeposition is another cost-efficiency solution based method for the fabrication of the heterojunction LEDs. It is also catalyst- and seed-layer-free growth technique of ZnO nanorod arrays. The method is clean, using only zinc ions and molecular oxygen as the growth precursor dissolved in an aqueous bath. The electrodeposition process involves an exchange of electrons between the substrate (the electrode) and the solution and this insures a very good electrical contact between the deposit and the substrate. Moreover, it was demonstrated recently that the method is outstanding for growing high-quality ZnO nanowires after optimizing various growth parameters such as the bath temperature, the supporting salt, the precursor concentration, and so on [88]. Djuriśić et al. applied $Zn(NO_3)_2$ of 3.8 mM and HMT of 6 mM in deionized water of 30 mL as the reaction solution to electrodeposit ZnO nanorods [75]. A platinum foil was used as the anode, and *p*-GaN film was used as the cathode. The solution was heated up to 80°C, and then 10 V voltage was applied, and both temperature and voltage were maintained constant for 30 min. They found that the interface is one key to affect the emission properties of the heterojunctions. It has been proposed that energy band alignment across GaN/ZnO interface will changed with change of interface quality [14]. Djuriśić et al. found the heterojunction based on the electrodeposited ZnO nanorods (ED) and the heterojunction based on an electrodeposited-seed ZnO nanorods (ED-HT) have similar *I–V* curves and EL at reverse bias even though the ZnO nanorods are grown with different methods [75]. This confirms that the recombination processes in GaN/ZnO heterojunctions are mainly affected by interface quality. The interface states could also be affected by the presence of a seed layer and/or its preparation method.

Lupan et al. demonstrated the electrochemical growth of ZnO nanowire arrays on *p*-type GaN (0001) single crystalline thin films supported on sapphire [76,77]. The cleaned GaN substrates were mounted on a static working electrode (WE) with a copper support. The aqueous solution of $ZnCl_2$ and KCl saturated with O_2 were used as a supporting electrolyte. The device based on ZnO nanowires is schematically shown in Figure 5.27. The nanowire device deposited with 0.1 mM $ZnCl_2$ exhibited a rectifying behavior with a forward current onset at 3 V, and a unique UV-light emission centered at 397 nm. The annealing is found to be able to reduce the density of point defects in ZnO and improve the quality of ZnO surface and interface of GaN/ZnO and hence can effectively improve device performance. As shown in Figure 5.28, the EL intensity of the heterojunction increased dramatically after annealing.

Further, they applied 0.2 mM $ZnCl_2$ to fabricate the ZnO based heterojunction. Similar to the previous device with 0.1 mM $ZnCl_2$ as the reaction solution, the device presents a UV emission at 397 nm combined with a low forward-voltage emission threshold of 4.4 V and

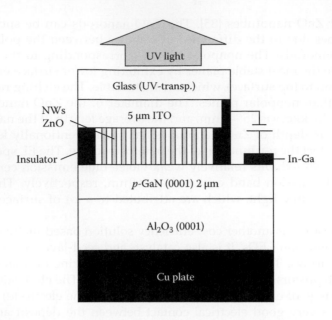

FIGURE 5.27
Schematic image of the (*p*-GaN film)/(*n*-ZnO nanowire) heterojunction LED fabricated by electrodeposition in aqueous solution of ZnCl$_2$ and KCl. (From Lupan, O., Pauporté, T., and Viana, B.: Low-voltage UV-electroluminescence from ZnO-nanowire array/*p*-GaN light-emitting diodes. *Adv. Mater.* 2010. 22. 3298. Copyright Wiley-VCH Verlag GmbH & Co. KGaA. Used with permission.)

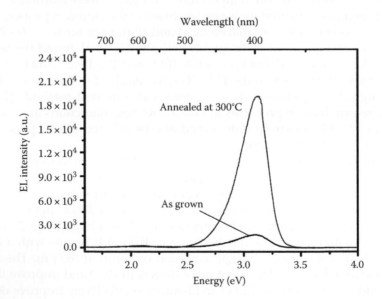

FIGURE 5.28
RT EL spectra of the LED under 6.5 V bias for as grown and thermal annealed devices where the ZnO nanowires were grown by the electrodeposition with KCl and 0.1 mol ZnCl$_2$. (Used with permission from Lupan, O., Pauporté, T., Viana, B., Tiginyanu, I.M., Ursaki, V.V., and Cortés, R., Epitaxial electrodeposition of ZnO nanowire arrays on *p*-GaN for efficient UV-light-emitting diode fabrication, *ACS Appl. Mater. Interface*, 2, 2083. Copyright 2010 American Chemical Society.)

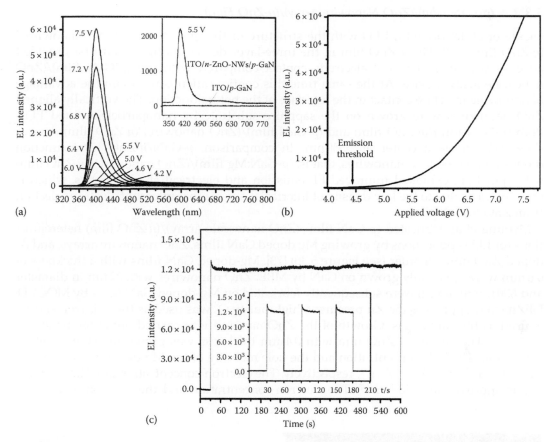

FIGURE 5.29
(a) RT EL spectra of the heterojunction LED based on the electrodeposited ZnO under different forward bias voltages. The inset shows the comparison of EL spectra from the heterojunction LED structure and an ITO/GaN/(In/Ga) structure. (b) Relationship between the RT EL intensity and the forward bias voltage of the fabricated LED structure. (c) Time dependence (10 min) of the RT EL of the fabricated LED under a dc bias of 6.2 V. The inset shows the RT EL time dependence under a dc bias of 6.2 V for three switch-on/switch-off pulses (30 s between each switch). (From Lupan, O., Pauporté, T., and Viana, B.: Low-voltage UV-electroluminescence from ZnO-nanowire array/*p*-GaN light-emitting diodes. *Adv. Mater.* 2010. 22. 3298. Copyright Wiley-VCH Verlag GmbH & Co. KGaA. Used with permission.)

a high brightness above 5–6 V. As shown in Figure 5.29a, EL emission rapidly increased with the applied forward bias. The peak shape of EL is similar to the PL of ZnO. Therefore, the UV EL emission most likely results from the radiative recombinations in ZnO. The inset in Figure 5.29a illustrates the compared EL emission results of *p*-GaN/*n*-ZnO heterojunction and ITO/GaN/(In/Ga) structure, which confirms that EL emission comes from *p*-GaN/*n*-ZnO heterojunction devices. Figure 5.29b shows rapidly increased EL emission with the applied voltage. The EL emission becomes intense between 5 and 7 V (0.5–1.5 mA). The remarkable low emission threshold suggests high-quality interface between GaN/ZnO with a very low density of defects. The time response of EL intensity is shown in Figure 5.29c. It shows good stability of EL emission with the rapid and reproducible response of emission intensity. These results suggest that the device is promising for the fabrication of a nanoscaled UV LEDs, and the developed electrodeposition procedure is effective to produce such excellent interfaces.

5.3.3.3 (p-GaN Film)/(ZnO Nanowire Array)/(n-ZnO Film)

Okada et al. fabricated a LED with the structure of the (p-GaN film)/(ZnO nanowire)/(n-ZnO film) [72]. The n-ZnO film in the three-layer device configuration first provided an easy way to realize the electrode contact compared with the (p-GaN film)/(ZnO nanoscale array) device. At the same time this configuration can remain the advantage of nanoscale interface contact in the interface of the heterojunction. The vertically aligned ZnO nanowires were grown on the sapphire substrate by nanoparticle-assisted PLD. Both (p-GaN film)/(n-ZnO film) and (p-GaN film)/(ZnO nanowire)/(n-ZnO film) devices show UV emission center at ~380 nm. In comparison, p-GaN/n-ZnO heterojunction diodes with embedded nanowires, that is, (p-GaN:Mg film)/(ZnO nanowire)/(n-ZnO film) heterojunction, exhibited improved EL emission and electrical characteristics, which is believed to be due to the low density of interfacial defects that interrupt electron injection from ZnO to GaN films.

Myoung et al. fabricated (p-GaN film)/(ZnO nanoscale array)/(n-ZnO film) heterojunctions for LED applications by growing Mg doped GaN films, ZnO nanowire arrays, and Al doped ZnO films in order (see Figure 5.30) [73]. Mg-doped GaN films with a thickness of 0.6 mm were epitaxially grown on GaN by MBE. ZnO nanowires with 32 nm in diameter and 600 nm in length were subsequently fabricated on Mg-doped GaN films by MOCVD. DEZn was supplied as the Zn precursor, high-purity O_2 was used as the oxidizer, and Ar is used as the carrier gas. Growth of the ZnO nanorod was carried out at temperatures of 500°C. The Al-doped ZnO film with 0.4 mm thickness was grown by RF magnetron sputtering. The hole concentration and the hole mobility of the Mg doped GaN films are 1.2×10^{18} cm^{-3} and 15 cm^2 V^{-1} s^{-1}, respectively. The electron concentration of n-ZnO nanowires is approximately 10^{12} cm^{-3}. The electron concentration and the electron mobility of

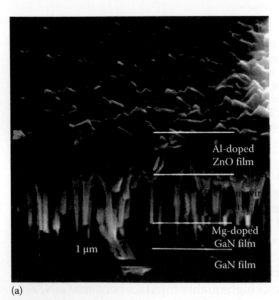

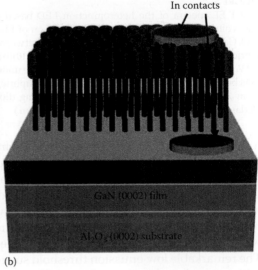

(a)   (b)

FIGURE 5.30
(a) SEM image and (b) schematic illustration of the (Mg-doped GaN film)/(ZnO nanowire array)/(Al-doped ZnO film) structures for nanometer-sized GaN/ZnO heterojunction diode applications. (From Jeong, M.C., Oh, B.Y., Ham, M.H., Lee, S.W., and Myoung, J.M.: ZnO-nanowire- inserted GaN/ZnO heterojunction light-emitting diodes. *Small.* 2007. 3. 568. Copyright Wiley-VCH Verlag GmbH & Co. KGaA. Used with permission.)

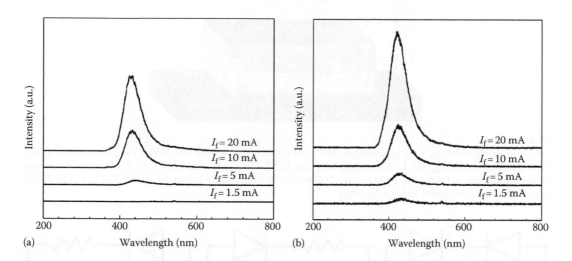

FIGURE 5.31

EL spectra of (a) film-based GaN/ZnO heterojunction LED and (b) (Mg-doped GaN film)/(ZnO nanowire array)/(Al-doped ZnO film) LED. (From Jeong, M.C., Oh, B.Y., Ham, M.H., Lee, S.W., and Myoung, J.M.: ZnO-nanowire- inserted GaN/ZnO heterojunction light-emitting diodes. *Small.* 2007. 3. 568. Copyright Wiley-VCH Verlag GmbH & Co. KGaA. Used with permission.)

ZnO films were 9.7×10^{18} cm^{-3} and 6.9 cm^2 V^{-1} s^{-1}, respectively. The Al-doped n-ZnO films not only facilitate the deposition of the metal for the formation of ohmic contacts but also supply electrons into the heterojunctions for radiative recombination. By comparing the film-based GaN/ZnO heterojunction diodes and ZnO-nanowire-inserted GaN/ZnO heterojunction diodes, the EL results in Figure 5.31 show that emission from the nanowire-inserted heterojunction diodes is stronger than that of film-based heterojunction diodes. For (p-GaN film)/(ZnO nanowire array)/(n-ZnO film) heterojunctions diodes, EL emission centered at 432 nm at the low forward current of 1.5 mA blueshifts to 421 nm, and EL intensity increases with the increasing forward current up to 20 mA. It is expected that the relatively strong intensity and blueshift of the EL peak of the nanosized heterojunction diodes compared to those of film-based diodes are due to the low density of interfacial defects, which interrupts electron injection from ZnO to GaN films.

5.3.3.4 *(ZnO Nanowire Array)/(GaN Film)/(ZnO Nanowire Array)*

Hsu et al. fabricated an (ZnO nanowire array)/(GaN film)/(ZnO nanowire array) n-p-n heterojunction LED based on ZnO nanowire array. The vertically aligned ZnO nanowires were grown on p-GaN substrate in a thermal furnace by chemical vapor deposition method [70]. High-purity Zn metal powder was used as the zinc vapor source. Ar (54.4 sccm) and O$_2$ (0.8 sccm) were continuously introduced into the thermal furnace at a temperature of 600°C, and the growth process was processed for 50 min. The pressure in the thermal furnace was controlled with a mechanical pump and maintained at 10 Torr. The obtained nanowires are 0.7 μm in length and 50–300 nm in diameter. Figure 5.32 displays the schematic image of the n-p-n heterojunction LED. A 0.25 cm wide ditch was etched into the ITO film to separate it into two electrodes. The (p-GaN)/(n-ZnO nanowire array) was put on the prepared ITO/glass substrate with the ZnO nanowires facing downward. The tips of the ZnO nanowires were directly in contact with the ITO, forming an n-p-n heterojunction structure for an LED. This LED was then annealed at 200°C for 10 min, followed by a

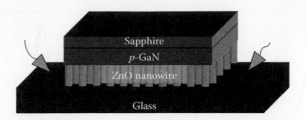

FIGURE 5.32

ZnO nanowires forming a *n-p-n* heterojunction LED on the *p*-GaN substrate. (With permission from Chen, C.H., Chang, S.J., Chang, S.P., Li, M.J., Chen, I.C., Hsueh, T.J., Hsu, A.D., and Hsu, C.L., Fabrication of a white-light-emitting diode by doping gallium into ZnO nanowire on a *p*-GaN substrate, *J. Phys. Chem. C*, 114, 12422. Copyright 2010 American Chemical Society.)

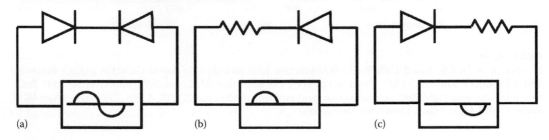

FIGURE 5.33

(a) Equivalent circuit of the *n-p-n* heterojunction structure, (b) at positive bias, and (c) at negative bias. (With permission from Chen, C.H., Chang, S.J., Chang, S.P., Li, M.J., Chen, I.C., Hsueh, T.J., Hsu, A.D., and Hsu, C.L., Fabrication of a white-light-emitting diode by doping gallium into ZnO nanowire on a *p*-GaN substrate, *J. Phys. Chem. C*, 114, 12422. Copyright 2010 American Chemical Society.)

flip-chip package process to form a good ohmic contact. Figure 5.33a displays the equivalent circuit of the *n-p-n* heterojunction structure. When a positive bias is applied to the right side, this side of the *n-p* junction is forward biased, and this junction acts as a variable resistor, while the left side *p-n* junction becomes an LED (Figure 5.33b). When a negative bias is applied to the right side, the right side *n-p* junction becomes an LED and the left side *p-n* junction becomes a resistor (Figure 5.33c). Figure 5.34 presents the EL spectrum of the heterostructure LED at 20 mA. The peak wavelength and FWHM are 417 and 50 nm, respectively. It is attributed to a defect-related radiative recombination in the *p*-GaN region when the carriers are injected under forward bias.

Ga-doped ZnO (ZnO:Ga) nanowires were also used to fabricate the *n-p-n* heterojunction [71]. The Ga metal ball (99.99% purity) was placed in the alumina boat 5 cm upstream of the Zn powder. The EL shows broad emission band centered at about 500 nm with the FWHM of 200 nm (Figure 5.34). It shows a white light LED was successfully fabricated by doping Ga into ZnO nanowire and forming an *n-p-n* heterojunction on an ITO/glass substrate.

5.4　Conclusion and Outlook

p-GaN/*n*-ZnO and *p*-GaN/*i*-ZnO/*n*-ZnO heterojunction LEDs offer promising potential in UV, blue, and white light emission. Various technological approaches to the device

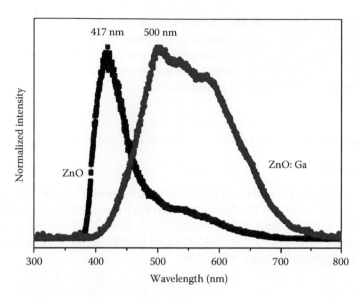

FIGURE 5.34
EL spectra of the *n-p-n* LED based on ZnO and ZnO:Ga nanowires at a DC current of 20 mA. (With permission from Chen, C.H., Chang, S.J., Chang, S.P., Li, M.J., Chen, I.C., Hsueh, T.J., Hsu, A.D., and Hsu, C.L., Fabrication of a white-light-emitting diode by doping gallium into ZnO nanowire on a *p*-GaN substrate, *J. Phys. Chem. C*, 114, 12422. Copyright 2010 American Chemical Society.)

fabrication are being tested, and the ZnO nanostructure, in particular, the ZnO nanowire and nanorod array, have been introduced to the heterojunction LEDs for enhanced performance and integration levels. Some new configurations have been proposed to obtain the ideal EL from the ZnO layer. For example, using the multi-quantum well to substitute the *i*-ZnO between *p*-GaN and *n*-ZnO possibly provides a more effective way to realize high-efficiency UV emission. The quantum well can confine the injection carriers in the active region and thus increases the carrier density in the recombination zone [46,89,90]. In addition, the confinement effect of the quantum well facilitates obtaining the shorter-wavelength emission; employing a *p-n-n* (or *n-p-p*) type double heterostructure is a very effective means to achieve the low-threshold laser diodes owing to better carrier and optical confinement in the active regions of the devices [91]; applying the SiO_2-ZnO nanocomposite to substitute the ZnO layer accomplishes a role of the current blocking layer because of large energy barrier for electron injection from *n*-ZnO into *p*-GaN, and also causes increase in the light extraction efficiency from *n*-ZnO due to its low refractive index, resulting in a significant UV EL from *n*-ZnO at a low forward bias current [92]. These typical methods pave the way for the development of the *p*-GaN/*n*-ZnO-based heterojunction LED performance.

Despite this, encouraging progress has been made in the research, while there are still a number of important issues that need to be further investigated before these heterojunctions can be transitioned to commercial use for the stated applications. The main aim for *p*-GaN/*n*-ZnO is to obtain UV emission. Presently, a broad emission spectrum extending into the visible range is frequently generated in the *p*-GaN/*n*-ZnO heterojunction LED. The reproducibility and light emission efficiency remains to be the main issue. The device-to-device reproducibility is low, and the LEDs poorly control the emission spectra. The task is made more difficult by the highly successful GaN that competes for similar applications. However, the heterojunction possesses advantages over GaN in UV solid state light emission. In the case of heterojunction, this field is still in a state of infancy and much more effort is needed to pave the way for the realization of the effective exciton emission in the ZnO side. Further studies are required to propose the effective method for practical applications. The related parameters to the emission intensity and emission lifetime of the LEDs need be considered. The new techniques and reasonable structure design need to be developed for the goal of commercial applications.

Acknowledgments

The authors acknowledge financial support from the National Nature Science Foundation of China (no. 10974027, 51103018), the National High Technology Research and Development Program of China (863 program) (no. 2006AA03Z311), National Basic Research Program of China (973 program) (grant no. 2012CB933703) and the National Natural Science Fund for Distinguished Young Scholars (no. 50725205).

References

1. A. Tsukazaki, T. Onuma, M. Ohtani, T. Makino, M. Sumiya, K. Ohtani et al., Repeated temperature modulation epitaxy for p-type doping and light-emitting diode based on ZnO, *Nat. Mater.* **4**, 42 (2005).
2. H. S. Kim, F. Lugo, S. J. Pearton, D. P. Norton, Y. L. Wang, and F. Ren, Dielectric passivation effects on ZnO light emitting diodes, *Appl. Phys. Lett.* **92**, 112101 (2008).
3. X. W. Sun, B. Ling, J. L. Zhao, S. T. Tan, Y. Yang, Y. Q. Shen, Z. L. Dong, and X. C. Li, Ultraviolet emission from a ZnO rod homojunction light-emitting diode, *Appl. Phys. Lett.* **95**, 133124 (2009).
4. Y. Choi, J. Kang, D. Hwang, and S. Park, Recent advances in ZnO-based light-emitting diodes, *IEEE Trans. Electron Devices* **57**, 26 (2010).
5. D. Hwang, M. Oh, J. Lim, and S. Park, ZnO thin films and light-emitting diodes, *J. Phys. D Appl. Phys.* **40**, R387 (2007).
6. Ü. Özgür, Y. I. Alivov, C. Liu, A. Teke, M. A. Reshchikov, S. Doğan, V. Avrutin, S. J. Cho, and H. Morkoc, A comprehensive review of ZnO materials and devices, *J. Appl. Phys.* **98**, 041301 (2005).
7. Y. Cui and F. Bruneval, p-type doping and codoping of ZnO based on nitrogen is ineffective: An ab initio clue, *Appl. Phys. Lett.* **97**, 042108 (2010).
8. J. L. Lyons, A. Janotti, and C. G. Van de Walle, Why nitrogen cannot lead to p-type conductivity in ZnO, *Appl. Phys. Lett.* **95**, 252105 (2009).
9. I. T. Drapak, Visible luminescence of a. ZnO–Cu$_2$O heterojunction, *Semiconductors* **2**, 624 (1968).
10. X. D. Chen, C. C. Ling, S. Fung, C. D. Beling, Y. F. Mei, R. K. Y. Fu, G. G. Siu, and P. K. Chu, Current transport studies of ZnO/p-Si heterostructures grown by plasma immersion ion implantation and deposition, *Appl. Phys. Lett.* **88**, 132104 (2006).
11. G. Xiong, J. Wilkinson, S. Tuzemen, K. B. Ucer, and R. T. Williams, Proceeding of laser and laser applications, *Proc. SPIE* **256**, 4644 (2002).
12. I. S. Jeong, J. H. Kim, and S. Im, Ultraviolet-enhanced photodiode employing n-ZnO/p-Si structure, *Appl. Phys. Lett.* **83**, 2946 (2003).
13. S. E. Nikitin, Yu. A. Nikolaev, I. K. Polushina, V. Yu. Rud, Yu. V. Rud, and E. I. Terukov, Photoelectric phenomena in ZnO:Al-p-Si heterostructures, *Semiconductors* **37**, 1291 (2003).
14. N. H. Alvi, M. Riaz, G. Tzamalis, O. Nur, and M. Willander, Junction temperature in n-ZnO nanorods/(p-4H-SiC, p-GaN, and p-Si) heterojunction light emitting diodes, *Solid State Electrochem.* **54**, 536 (2010).

15. Y. A. Alivov, Ü. Özgur, S. Dogan, D. Johnstone, V. Avrutin, N. Onojima, C. Liu, J. Xie, Q. Fan, and H. Morkoc, Photoresponse of n-ZnO/p-SiC heterojunction diodes grown by plasma-assisted molecular-beam epitaxy, *Appl. Phys. Lett.* **86**, 241108 (2005).
16. C. Yuen, S. F. Yu, S. P. Lau, Rusli, and T. P. Chen, Fabrication of n-ZnO:Al/p-SiC(4H) heterojunction light-emitting diodes by filtered cathodic vacuum arc technique, *Appl. Phys. Lett.* **86**, 241111 (2005).
17. H. Asll, E. Gür, K. Çmar, and C. Coşkun, Electrochemical growth of n-ZnO onto the p-type GaN substrate: p-n heterojunction characteristics, *Appl. Phys. Lett.* **94**, 253501 (2009).
18. Q. Qin, L. Guo, Z. Zhou, H. Chen, X. Du, Z. Mei, J. Jia, Q. Xue, and J. Zhou, Electroluminescence of an n-ZnO/p-GaN heterojunction under forward and reverse biases, *Chin. Phys. Lett.* **22**, 2298 (2005).
19. Y. I. Alivov, J. E. Van Nostrand, D. C. Look, M. V. Chukichev, and B. M. Ataev, Annealing effect on the optical response and interdiffusion of *n*-ZnO/p-Si (111) heterojunction grown by atomic layer deposition, *Appl. Phys. Lett.* **97**, 181915 (2010).
20. Q. X. Yu, B. Xu, Q. H. Wu, Y. Liao, G. Z. Wang, R. C. Fang, H. Y. Lee, C. T. Lee, and C. Lee, Optical properties of ZnO/GaN heterostructure and its near-ultraviolet light-emitting diode, *Appl. Phys. Lett.* **83**, 4713 (2003).
21. A. Bakin, A. Behrends, A. Waag, H. Lugauer, A. Laubsch, and K. Streubel, ZnO-GaN hybrid heterostructures as potential cost-efficient LED technology, *Proc. IEEE* **98**, 1281 (2010).
22. P. Kung and M. Razeghi, Electroluminescence at 375 nm from a ZnO/GaN:Mg/c-Al₂O₃ hetero-junction light emitting diode, *Opto-Electron. Rev.* **8**, 3 (2000).
23. Y. I. Alivov, E. V. Kalinina, A. E. Cherenkov, D. C. Look, B. M. Ataev, A. K. Omaev, M. V. Chukichev, and D. M. Bagnall, Fabrication and characterization of n-ZnO/p-AlGaN hetero-junction lightemitting diodes on 6H-SiC substrates, *Appl. Phys. Lett.* **83**, 4719 (2003).
24. H. Ohta, M. Hirano, K. Nakahara, H. Maruta, T. Tanabe, M. Kamiya, T. Kamiya, and H. Hosono, Fabrication and photoresponse of a pn-heterojunction diode composed of transparent oxide semiconductors, p-NiO and n-ZnO, *Appl. Phys. Lett.* **83**, 1029 (2003).
25. J. A. Aranovich, D. Golmayo, A. L. Fahrenbruch, and R. H. Bube, Photovoltaic properties of ZnO/CdTe heterojunctions prepared by spray pyrolysis, *J. Appl. Phys.* **51**, 4260 (1980).
26. H. Ohta, H. Mizoguchi, M. Hirano, S. Narushima, T. Kamiya, and H. Hosono, Fabrication and characterization of heteroepitaxial p-n junction diode composed of wide-gap oxide semicon-ductors p-ZnRh₂O₄/n-ZnO, *Appl. Phys. Lett.* **82**, 823 (2003).
27. H. Ohta, K. Kawamura, M. Orita, M. Hirano, N. Sarukura, and H. Hosono, Current injection emission from a transparent p-n junction composed of p-SrCu₂O₂/n-ZnO, *Appl. Phys. Lett.* **77**, 475 (2000).
28. H. Ohta, M. Orita, M. Hirano, and H. Hosono, Surface morphology and crystal quality of low resistive indium tin oxide grown on yttria-stabilized zirconia, *J. Appl. Phys.* **89**, 5720 (2001).
29. D. Qiao, L. S. Yu, S. S. Lau, J. M. Redwing, J. Y. Lin, and X. Jiang, Dependence of Ni/AlGaN Schottky barrier height on Al mole fraction, *J. Appl. Phys.* **87**, 801 (2000).
30. H. Zhu, C. X. Shan, B. Yao, B. H. Li, J. Y. Zhang, Z. Z. Zhang et al., Ultralow-threshold laser real-ized in zinc oxide, *Adv. Mater.* **21**, 1613 (2009).
31. J. B. You, X. W. Zhang, S. G. Zhang, J. X. Wang, Z. G. Yin, H. R. Tan et al., Improved electrolu-minescence from n-ZnO/AlN/p-GaN heterojunction light-emitting diodes, *Appl. Phys. Lett.* **96**, 201102 (2010).
32. A. Tsukazaki, M. Kubota, A. Ohtomo, T. Onuma, K. Ohtani, H. Ohno, S. F. Chichibu, and M. Kawasaki, Blue light-emitting diode based on ZnO, *Jpn. J. Appl. Phys.* **44**, L643 (2005).
33. Z. L. Wang, Zinc oxide nanostructures: Growth, properties and applications, *J. Phys. Condens. Matter* **16**, R829 (2004).
34. G. D. J. Smit, S. Rogge, and T. M. Klapwijk, Schottky nanocontacts on ZnO nanorod arrays, *Appl. Phys. Lett.* **82**, 4358 (2003).
35. W. I. Park and G. C. Yi, Electroluminescence in n-ZnO nanorod arrays vertically grown on p-GaN, *Adv. Mater.* **16**, 87 (2004).
36. L. Vayssieres, K. Keis, A. Hagfeldt, and S. Lindquist, Three-dimensional array of highly oriented crystalline ZnO microtubes, *J. Phys. Chem. B* **105**, 3350 (2001).

37. R. D. Vispute, V. Talyansky, S. Choopun, R. P. Sharma, T. Venkatesan, M. He et al., Heteroepitaxy of ZnO on GaN and its implications for fabrication of hybrid optoelectronic devices, *Appl. Phys. Lett.* **73**, 348 (1998).

38. S. K. Hong, H. J. Ko, Y. Chen, and T. J. Yao, Interface engineering in ZnO epitaxy, *J. Cryst. Growth* **209**, 537 (2001).

39. M. V. Chukichev, B. M. Ataev, V. V. Mamedov, Y. I. Alivov, and I. I. Khodos, Cathodoluminescence of ZnO/GaN/α-Al$_2$O$_3$ heteroepitaxial structures grown by chemical vapor deposition, *Semiconductors* **36**, 977 (2002).

40. Y. I. Alivov, J. E. Van Nostrand, D. C. Look, M. V. Chukichev, and B. M. Ataev, Observation of 430 nm electroluminescence from ZnO/GaN heterojunction light-emitting diodes, *Appl. Phys. Lett.* **83**, 2943 (2003).

41. B. M. Ataev, Y. I. Alivov, and V. A. Nikitenko, n-ZnO/p-GaN/α-Al$_2$O$_3$ heterojunction as a promising blue light emitting system, *J. Optoelectron. Adv. Mater.* **5**, 899 (2003).

42. R. W. Chuang, R. X. Wu, L. W. Lai, and C. T. Lee, ZnO on GaN heterojunction light emitting diode grown by vapor cooling condensation technique, *Appl. Phys. Lett.* **91**, 231113 (2007).

43. H. C. Chen, M. J. Chen, and Y. C. Cheng, Amplified spontaneous emission from ZnO in n-ZnO/p-GaN heterojunction light-emitting diodes with an external-feedback reflector, *IEEE Photon. Technol. Lett.* **22**, 248 (2010).

44. H. Chen, M. Chen, M. Wu, W. Li, H. Tsai, J. Yang, H. Kuan, and M. Shiojiri, UV electroluminescence and structure of n-ZnO/p-GaN heterojunction LEDs grown by atomic layer deposition, *IEEE J. Quantum Electron.* **46**, 265 (2010).

45. S. P. Chang, R. W. Chuang, S. J. Chang, Y. Z. Chiouand, and C. Y. Lu, MBE n-ZnO/MOCVD p-GaN heterojunction light-emitting diode, *Thin Solid Films* **517**, 5054 (2009).

46. J. W. Sun, Y. M. Lu, Y. C. Liu, D. Z. Shen, Z. Z. Zhang, B. H. Li, J. Y. Zhang, B. Yao, D. X. Zhao, and X. W. Fan, Excitonic electroluminescence from ZnO-based heterojunction light emitting diodes, *J. Phys. D Appl. Phys.* **41**, 155103 (2008).

47. T. P. Yang, H. C. Zhu, J. M. Bian, J. C. Sun, X. Dong, B. L. Zhang, H. W. Liang, X. P. Li, Y. G. Cui, and G. T. Du, Room temperature electroluminescence from the n-ZnO/p-GaN heterojunction device grown by MOCVD, *Mater. Res. Bull.* **43**, 3614 (2008).

48. D. J. Rogers, F. H. Teherani, A. Yasan, K. Minder, P. Kung, and M. Razeghi, Electroluminescence at 375 nm from a ZnO/GaN:Mg/c-Al$_2$O$_3$ heterojunction light emitting diode, *Appl. Phys. Lett.* **88**, 141918 (2006).

49. J. Lee, J. Lee, H. Kim, C. Lee, H. Ahn, H. Cho, Y. Kim, B. Kong, and H. S. Lee, A study on the origin of emission of the annealed n-ZnO/p-GaN heterostructure LED, *Thin Solid Films* **517**, 5157 (2009).

50. J. Y. Lee, J. H. Lee, and H. S. Kim, Effect of the interface on n-ZnO/p-GaN heterojunction light emitting diodes, *J. Korean Phys. Soc.* **55**, 1568 (2009).

51. H. Q. Le, S. K. Lim, G. K. L. Goh, S. J. Chua, N. S. S. Ang, and W. Liu, Solution epitaxy of gallium-doped ZnO on p-GaN for heterojunction light-emitting diodes, *Appl. Phys. B* **100**, 705 (2010).

52. H. Kato, M. Sano, K. Miyamoto, and T. Yao, Growth and characterization of Ga-doped ZnO layers on a-plane sapphire substrates grown by molecular beam epitaxy, *J. Cryst. Growth* **237**, 538 (2002).

53. B. M. Ataev, A. M. Bagamadova, A. M. Djabrailov, V. V. Mamedov, and R. A. Rabadanov, Highly conductive and transparent Ga-doped epitaxial ZnO films on sapphire by CVD, *Thin Solid Films* **260**, 19 (1995).

54. V. A. Nikitenko, Luminescence and EPR of zinc oxide, *J. Appl. Spectrosc.* **57**, 783 (1993).

55. S. Nakamura, T. Mukai, and M. Senon, High-power GaN p-n junction blue-light-emitting diodes, *J. Appl. Phys.* **30**, L1998 (1991).

56. M. A. Khan, Q. Chen, R. A. Skogman, and J. N. Kuznia, Violet-blue GaN homojunction light emitting diodes with rapid thermal annealed p-type layers, *Appl. Phys. Lett.* **66**, 2046 (1995).

57. S. J. Jiao, Y. M. Lu, D. Z. Shen, Z. Z. Zhang, B. H. Li, J. Y. Zhang, B. Yao, Y. C. Liu, and X. W. Fan, Ultraviolet electroluminescence of ZnO based heterojunction light emitting diode, *Phys. Status Solidi C* **3**, 972 (2006).

58. B. H. Kong, W. S. Han, Y. Y. Kim, H. K. Cho, B. H. Kong, W. S. Han, Y. Y. Kim, H. K. Cho, and J. H. Kim, Heterojunction light emitting diodes fabricated with different n-layer oxide structures on p-GaN layers by magnetron sputtering, *Appl. Surf. Sci.* **256**, 4972 (2010).

59. W. S. Han, Y. Y. Kim, B. H. Kong, and H. K. Cho, Ultraviolet light emitting diode with n-ZnO:Ga/i-ZnO/p-GaN:Mg heterojunction, *Thin Solid Films* **517**, 5106, (2009).

60. S. Hwang, T. H. Chung, and B. T. Lee, Study on the interfacial layer in ZnO/GaN heterostructure light-emitting diode, *Mater. Sci. Eng. B* **157**, 32 (2009).

61. H. Y. Xu, Y. C. Liu, Y. X. Liu, C. S. Xu, C. L. Shao, and R. Mu, Ultraviolet electroluminescence from p-GaN/i-ZnO/n-ZnO heterojunction light-emitting diodes, *Appl. Phys. B* **80**, 871 (2005).

62. J. Yan, C. Chen, S. Yen, and C. T. Lee, Ultraviolet ZnO nanorod/p-GaN-heterostructured light-emitting diodes, *IEEE Photon. Technol. Lett.* **22**, 146 (2010).

63. H. Lee, C. Lee, and J. Yan, Emission mechanisms of passivated single n-ZnO:In/i-ZnO/p-GaN-heterostructured nanorod light-emitting diodes, *Appl. Phys. Lett.* **97**, 111111 (2010).

64. H. K. Liang, S. F. Yu, and H. Yang, Edge-emitting ultraviolet n-ZnO:Al/i-ZnO/p-GaN heterojunction light-emitting diode with a rib waveguide, *Opt. Express* **15**, 3687 (2010).

65. L. Zhao, C. S. Xu, Y. X. Liu, C. L. Shao, X. H. Li, and Y. C. Liu, A new approach to white emitting diodes of p-GaN/i-ZnO/n-ZnO heterojunctions, *Appl. Phys. B* **92**, 185 (2008).

66. M. Fyutsuhara, K. Yoshioka, and O. Takai, Optical properties of zinc oxynitride thin films, *Thin Solid Films* **317**, 322 (1998).

67. Y. Bie, Z. Liao, P. Wang, Y. Zhou, X. Han, Y. Ye et al., Single ZnO nanowire p-type GaN heterojunctions for photovoltaic devices and UV light-emitting diodes, *Adv. Mater.* **22**, 4284 (2010).

68. X. M. Zhang, M. Y. Lu, Y. Zhang, L. J. Chen, and Z. L. Wang, Fabrication of a high-brightness blue-light-emitting diode using a ZnO-nanowire array grown on p-GaN thin film, *Adv. Mater.* **21**, 2767 (2009).

69. H. Fu, C. Cheng, C. Wang, T. Lin, and Y. Chen, Selective angle electroluminescence of light-emitting diodes based on nanostructured ZnO/GaN heterojunctions, *Adv. Funct. Mater.* **19**, 3471 (2009).

70. C. H. Chen, S. J. Chang, and S. P. Chang, Electroluminescence from n-ZnO nanowires/p-GaN heterostructure light-emitting diodes, *Appl. Phys. Lett.* **95**, 223101 (2009).

71. C. H. Chen, S. J. Chang, S. P. Chang, M. J. Li, I. C. Chen, T. J. Hsueh, A. D. Hsu, and C. L. Hsu, Fabrication of a white-light-emitting diode by doping gallium into ZnO nanowire on a p-GaN substrate, *J. Phys. Chem. C* **114**, 12422 (2010).

72. R. Guo, J. Nishimura, M. Matsumoto, M. Higashihata, D. Nakamura, and T. Okada, Electroluminescence from ZnO nanowire-based p-GaN/n-ZnO heterojunction light-emitting diodes, *Appl. Phys. B* **94**, 33 (2009).

73. M. C. Jeong, B. Y. Oh, M. H. Ham, S. W. Lee, and J. M. Myoung, ZnO-nanowire- inserted GaN/ZnO heterojunction light-emitting diodes, *Small* **3**, 568 (2007).

74. M. C. Jeong, B. Y. Oh, M. H. Ham, and J. M. Myong, Electroluminescence from ZnO nanowires in n-ZnO film/ZnO nanowire array/p-GaN film heterojunction, *Appl. Phys. Lett.* **88**, 202105 (2006).

75. A. M. C. Ng, X. Y. Chen, F. Fang, Y. F. Hsu, A. B. Djurišić, C. C. Ling et al., Solution-based growth of ZnO nanorods for light-emitting devices: Hydrothermal vs. electrodeposition, *Appl. Phys. B* **100**, 851 (2010).

76. O. Lupan, T. Pauporté, B. Viana, I. M. Tiginyanu, V. V. Ursaki, and R. Cortés, Epitaxial electrodeposition of ZnO nanowire arrays on p-GaN for efficient UV-light-emitting diode fabrication, *ACS Appl. Mater. Interfaces* **2**, 2083 (2010).

77. O. Lupan, T. Pauporté, and B. Viana, Low-voltage UV-electroluminescence from ZnO-nanowire array/p-GaN light-emitting diodes, *Adv. Mater.* **22**, 3298 (2010).

78. Z. Guo, D. X. Zhao, Y. C. Liu, D. Z. Shen, B. Yao, Z. Zhang, and B. Li, Electrically pumped single-mode lasing emission of self-assembled n-ZnO microcrystalline film/p-GaN heterojunction diode, *J. Phys. Chem. C* **114**, 15499 (2010).

79. F. Fang, D. X. Zhao, B. H. Li, Z. Zhang, and D. Z. Shen, Hydrogen effects on the electroluminescence of n-ZnO nanorod/p-GaN film heterojunction light-emitting diodes, *Phys. Chem. Chem. Phys.* **12**, 6759 (2010).

80. J. Ahn, M. A. Mastro, and J. Hite, Electroluminescence from ZnO nanoflowers/GaN thin film p-n heterojunction, *Appl. Phys. Lett.* **97**, 082111 (2010).

81. S. D. Lee, Y. S. Kim, M. S. Yi, J. Y. Choi, and S. W. Kim, Morphology control and electroluminescence of ZnO nanorod/GaN heterojunctions prepared using aqueous solution, *J. Phys. Chem. C* **113**, 8954 (2009).

82. A. M. C. Ng, Y. Y. Xi, Y. F. Hsu, A. B. Djurišić, W. K. Chan, S. Gwo et al., GaN/ZnO nanorod light emitting diodes with different emission spectra, *Nanotechnology* **20**, 445201 (2009).

83. S. Kishwar, K. Hasan, and G. Tzamalis, Electro-optical and cathodoluminescence properties of low temperature grown ZnO nanorods/p-GaN white light emitting diodes, *Phys. Status Solidi A* **207**, 67 (2010).

84. N. H. Alvi, M. Riaz, G. Tzamalis, O. Nur, and M. Willander, Fabrication and characterization of high-brightness light emitting diodes based on n-ZnO nanorods grown by a low-temperature chemical method on p-4H-SiC and p-GaN, *Semicond. Sci. Technol.* **25**, 065004 (2010).

85. J. R. Sadaf, M. Q. Israr, S. Kishwa, O. Nur, and M. Willander, White electroluminescence using ZnO nanotubes/GaN heterostructure light-emitting diode, *Nanoscale Res. Lett.* **5**, 957, (2010).

86. S. Dalui, C. Lin, H. Lee, S. Yen, Y. Lee, and C. Lee, Electroluminescence from solution grown n-ZnO nanorod/p-GaN-heterostructured light emitting diodes, *J. Electrochem. Soc.* **157**, H516 (2010).

87. X. Wang, J. Song, P. Li, J. H. Ryou, R. D. Dupuis, C. J. Summers, and Z. L. Wang, Growth of uniformly aligned ZnO nanowire heterojunction arrays on GaN, AlN, and $Al_{0.5}Ga_{0.5}N$ substrates, *J. Am. Chem. Soc.* **127**, 7920 (2005).

88. T. Pauporté, E. Jouanno, F. Pellé, B. Viana, and P. Ashehoug, Key growth parameters for the electrodeposition of ZnO films with an intense UV-light emission at room temperature, *J. Phys. Chem. C* **113**, 10422 (2009).

89. A. Osinsky, J. W. Dong, M. Z. Kauser, B. Hertog, A. M. Dabiran, P. P. Chow, S. J. Pearton, O. Lopatiuk, and L. Chernyak, MgZnO/AlGaN heterostructure light emitting diodes, *Appl. Phys. Lett.* **85**, 4272 (2004).

90. C. Bayram, F. H. Teherani, D. J. Rogers, and M. Razeghi, A hybrid green light-emitting diode comprised of n-ZnO/(InGaN/GaN) multi-quantum-wells/p-GaN, *Appl. Phys. Lett.* **93**, 08111 (2008).

91. Y. I. Alivov, Ü. Özgür, S. Doğan, C. Liu, X. Moon, X. Gu, V. Avrutin, Y. Fu, and H. Morkoc, Forward-current electroluminescence from GaN/ZnO double heterostructure diode, *Solid State Electron.* **49**, 1693 (2005).

92. M. Chen, Y. Shih, M. Wu, H. Chen, H. Tsai, W. Li, J. Yang, H. Kuan, and M. Shiojiri, Structure and ultraviolet electroluminescence of n-ZnO/SiO₂-ZnO nanocomposite/p-GaN heterostructure light-emitting diodes, *IEEE Trans. Electron Devices* **57**, 2195 (2010).

6

Light-Emitting Diodes Based on n-ZnO/n-Si(GaAs) Isotype Heterojunctions

S.T. Tan, J.L. Zhao, and X.W. Sun

CONTENTS

6.1 Introduction 219
6.2 n-ZnO/SiO$_x$/n-Si and n-ZnO/SiO$_x$/p-Si Heterostructured LEDs 220
6.3 Epitaxially Grown n-ZnO/MgO/TiN/n-Si(111) Heterostructured
     Light-Emitting Diode 224
6.4 n-ZnO/n-GaAs Heterostructured White Light-Emitting Diode 229
6.5 Summary 233
References 235

6.1 Introduction

ZnO, with a strong exciton binding energy (60 meV) and wide direct bandgap (3.37 eV), is a promising candidate for short-wavelength optoelectronic devices.[1] The capability of fabricating ZnO at a lower temperature (500°C–800°C) and the feasibility of wet-chemical etching make the integration of ZnO with Si technology highly feasible. It has motivated tremendous research activities on ZnO-based materials in the past few decades.[2,3] However, reliable and high-quality p-type ZnO films are difficult to achieve and have stumbled many researchers. The lack of p-type ZnO films has always been a major challenge for fabricating p-n homojunction optoelectronic devices.[4,5] As an alternative solution, a lot of effort has been devoted to ZnO-based heterojunction light-emitting diodes (LEDs) by growing n-type ZnO films on top of a variety of p-type layers, such as GaN, Si, SiC, SrCu$_2$O$_2$, and CuGaS$_2$.[6–12] Among the attempts, ZnO/Si heterostructured LEDs have stirred up considerable interest in integrating ZnO onto Si technology. Although most of the works on ZnO/Si heterostructured LEDs adopt a n-ZnO/p-Si heterostructure, n-ZnO/n-Si isotype heterostructure has also been found promising for optoelectronic applications.[13–19] However, reports on n-ZnO/n-Si isotype heterostructure for LED application are rare. In this chapter, n-ZnO/SiO$_x$/(n,p)-Si, n-ZnO/MgO/TiN/n$^+$-Si and n-ZnO/n-GaAs heterostructured LEDs were repeatedly fabricated and investigated.[16–19] The electroluminescence (EL) and current transport properties of these heterostructured diodes shall be discussed.

6.2 *n*-ZnO/SiO$_x$/*n*-Si and *n*-ZnO/SiO$_x$/*p*-Si Heterostructured LEDs

ZnO film was deposited by a homemade shower-head injector MOCVD system on *n*- and *p*-Si(100) wafers (both have carrier density of ~10^{18} cm^{-3}). The ZnO film was grown at 500°C with a chamber pressure of ~100 Torr. Dimethylzinc and oxygen were used as precursors and the molar ratio of oxygen to zinc was controlled at 2000. The thickness of the ZnO film is about 900 nm. After the deposition of the ZnO film, Au electrodes were deposited on the surface of the ZnO film and the backside of Si wafers by direct-current magnetron sputtering. The Au electrodes on ZnO were patterned into squares (1.5 × 1.5 mm^2) using a shadow mask. The ZnO films exhibit *n*-type conduction, with electron concentration of ~10^{17} cm^{-3}, Hall mobility of ~0.67 cm^2/V s, and resistivity of ~57 Ω cm from Hall effect measurement. The Si wafers were precleaned by sequential ultrasonic baths of acetone and ethanol, resulting in a thin layer of SiO$_x$. Transmission electron microscopy (TEM) was employed to examine the interface of the diode. In the cross-sectional TEM, a well-defined SiO$_x$ layer with a thickness of ~3.7 nm was observed on Si surface, as shown in Figure 6.1. The SiO$_x$ layer has no obvious crystallites and it functions as a carrier blocking layer for light-emitting diodes.

The typical current–voltage (*I–V*) characteristics of the ZnO/Si heterostructured diodes are shown in Figure 6.2. Ohmic contact was established for Au/*n*-ZnO, as shown in the inset of Figure 6.2. *n*-ZnO/*p*-Si heterostructured diodes show diode-like rectification characteristics with a low breakdown voltage (~3 V) at reverse bias. The space charge layer is mainly formed in *n*-ZnO due to high doping density in *p*-Si. Thus, electrons may tunnel from the *p*-Si valence band (VB) to the ZnO conduction band at reverse bias, which will lead to tunneling breakdown. On the other hand, the *I–V* curve of ZnO/*n*-Si

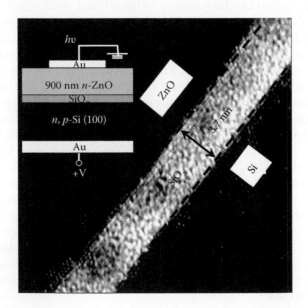

FIGURE 6.1
Schematic diagram of a ZnO/Si heterostructured light-emitting diode and the cross-sectional TEM image of the diode. (Reprinted with permission from Tan, S.T., Sun, X.W., Zhao, J.L., Iwan, S., Cen, Z.H., Chen, T.P., Ye, J.D., Lo, G.Q., Kwong, D.L., and Teo, K.L., Ultraviolet and visible electroluminescence from *n*-ZnO/SiO$_x$/(n,p)-Si heterostructured light-emitting diodes, *Appl. Phys. Lett.*, 93, 013506. Copyright 2008, American Institute of Physics.)

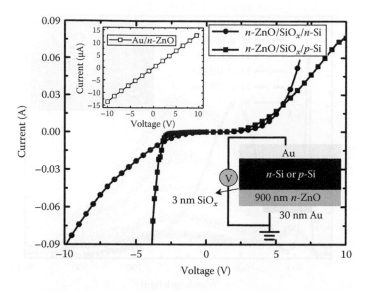

FIGURE 6.2
Current–voltage characteristics of ZnO/Si heterostructured diodes. Top left inset shows the ohmic character-istic of Au/*n*-ZnO contact. Bottom right inset is the bias scheme of the diodes. (Reprinted with permission from Tan, S.T., Sun, X.W., Zhao, J.L., Iwan, S., Cen, Z.H., Chen, T.P., Ye, J.D., Lo, G.Q., Kwong, D.L., and Teo, K.L., Ultraviolet and visible electroluminescence from *n*-ZnO/SiO$_x$/(n,p)-Si heterostructured light-emitting diodes, *Appl. Phys. Lett.*, 93, 013506. Copyright 2008, American Institute of Physics.)

heterostructured diodes shows rectifying characteristics at both positive and negative bias. The native SiO$_x$ at the ZnO/Si interface acts as a double Schottky barrier for both *n* layers and the junction can be considered as a series of two back-to-back Schottky diodes.[20]

Figure 6.3a shows the EL spectra with current injection from 40 to 138 mA as well as the emission photo of ZnO/*n*-Si isotype diode. The emission occurs only when Si is positively biased. Orange-color emission (centered at ~600 nm), which originated from defects, was clearly observed by naked eyes. Other than the defect-related emission, there is an ultraviolet (UV) emission peaking at 390 nm. The UV emission originated from the ZnO excitonic emission, which is red-shifted by about 10 nm compared with the PL peak. Structural defects, such as grain boundary in polycrystalline ZnO film can form band-tail states near band edge. The density of injected nonequilibrium carriers by electric current is far lower than that excited by laser in PL measurement. The low density of injected car-riers will mainly fill the band-tail states, leading to the red-shift of near-band-edge (NBE) emission. In contrast to ZnO/Si isotype diodes, ZnO/Si *p-n* diodes exhibit whitish EL with a positive bias on Si, as shown in Figure 6.3b. It can be seen that the EL spectrum covers a wide range from 360 to 850 nm and centered at ~520 nm.

The observation of EL in these devices is interesting yet complicated. The observation of excitonic and defect-related emissions of ZnO indicates that electron-hole pairs are cre-ated in ZnO layer. The possible mechanism responsible for generating holes in ZnO is discussed in the following. The energy band diagrams of both types of diodes at equi-librium and under bias that derived from Anderson model are shown in Figure 6.4. The electron affinities of ZnO (χ_{ZnO}) and Si (χ_{Si}) were taken as 4.35 and 4.05 eV, respectively,[21] while the bandgaps of ZnO and Si were taken as 3.37 and 1.12 eV, respectively. The band-gap discontinuity at conduction band and VB is therefore equal to $\Delta E_c = \chi_{ZnO} - \chi_{Si} = 0.3$ eV

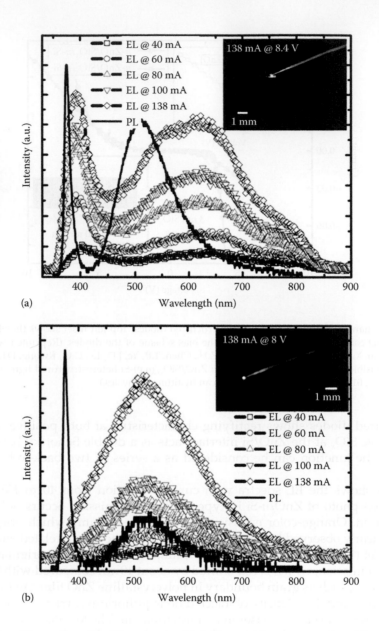

FIGURE 6.3

Room temperature EL spectra with different current injection and room temperature PL spectrum for the (a) n-ZnO/SiO$_x$/n-Si diode and (b) n-ZnO/SiO$_x$/p-Si diode. Insets show the EL photos of the diodes in dark at the current injection of 138 mA. (Reprinted with permission from Tan, S.T., Sun, X.W., Zhao, J.L., Iwan, S., Cen, Z.H., Chen, T.P., Ye, J.D., Lo, G.Q., Kwong, D.L., and Teo, K.L., Ultraviolet and visible electroluminescence from n-ZnO/SiO$_x$/(n,p)-Si heterostructured light-emitting diodes, *Appl. Phys. Lett.*, 93, 013506. Copyright 2008, American Institute of Physics.)

and $\Delta E_v = E_{g\text{ZnO}} - E_{g\text{Si}} + \Delta E_c = 2.55$ eV. For ZnO/Si isotype diode, the bands of Si near the Si/SiO$_x$ interface will bend upward with a positive bias applied on Si (Figure 6.4a and b). It can be seen from Figure 6.4b that, under the influence of electric field, the band bending of Si/SiO$_x$ interface will gradually induce an inversion layer for n-ZnO/SiO$_x$/n-Si diodes, which is responsible for the hole injection. Besides, the VB of Si is getting closer to the VB

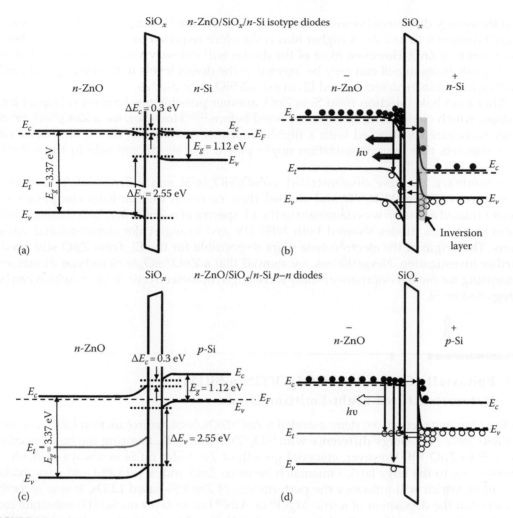

FIGURE 6.4
Energy band diagrams of the *n*-ZnO/SiO$_x$/*n*-Si isotype diodes under (a) thermal equilibrium and (b) positive bias at Si side and *n*-ZnO/SiO$_x$/*p*-Si *p-n* diodes under (c) thermal equilibrium and (d) positive bias at Si side. (Reprinted with permission from Tan, S.T., Sun, X.W., Zhao, J.L., Iwan, S., Cen, Z.H., Chen, T.P., Ye, J.D., Lo, G.Q., Kwong, D.L., and Teo, K.L., Ultraviolet and visible electroluminescence from *n*-ZnO/SiO$_x$/(n,p)-Si heterostructured light-emitting diodes, *Appl. Phys. Lett.*, 93, 013506. Copyright 2008, American Institute of Physics.)

of ZnO under sufficient bias (Figure 6.4b). As a result, accumulated holes in the inversion layer could tunnel through the barrier into the VB and defect band of ZnO and recombine with the electrons in ZnO conduction band that blocked by SiO$_x$, resulting in UV and orange-color emission, respectively. However, this speculation is accurate only if the thin oxide at the interface can sustain certain voltage and hence facilitate the creation of inversion layer. Figure 6.4c and d show the energy band diagram of the *n*-ZnO/SiO$_x$/*p*-Si *p-n* diodes at equilibrium and under forward bias, respectively. The holes could be injected from *p*-Si into *n*-ZnO by means of tunneling through the thin SiO$_x$ layer (Figure 6.4d). Besides, the E_v of Si moves to lower energy under positive bias and becomes closer to the deep defect level of ZnO. Therefore, it is possible that the holes can be injected from E_v of Si into ZnO under sufficient bias. However, it is notable from the energy band diagram that the energy difference between the VB of *p*-Si and *n*-ZnO is larger than that of *n*-Si and *n*-ZnO (Figure 6.4b and d). A higher bias is therefore required to effectively inject holes from *p*-Si to *n*-ZnO. However, most of the diodes will not survive under such high bias. As a result, holes in *p*-Si can only be injected to the defect levels in the bandgap of ZnO, leading to dominant defect-related EL in *n*-ZnO/SiO$_x$/*p*-Si diodes.

Other than hole injection from Si to ZnO, another possible mechanism is impact ionization, which could result in EL as reported before.[22,23] However, for a ZnO/SiO$_x$/*n*$^+$-Si heterostructure constructed with a highly doped Si (~10^{19} cm^{-3}), no EL was detected. This suggests that impact ionization might play a less significant role in the *n*-ZnO/SiO$_x$/*n*-Si LEDs.

In summary, we have demonstrated *n*-ZnO/SiO$_x$/*n*-Si and *n*-ZnO/SiO$_x$/*p*-Si heterostructured LEDs by MOCVD and studied their emission mechanisms comparatively. Defect-related emission was dominant in the EL spectra of *n*-ZnO/SiO$_x$/*p*-Si diodes, while *n*-ZnO/SiO$_x$/*n*-Si diodes showed both NBE UV and orange-color defect-related emissions. The origin of the electron-hole pairs responsible for the EL from ZnO still needs further investigation. Nevertheless, we showed that *n*-ZnO/SiO$_x$/*n*-Si isotype diodes are promising for the development of short-wavelength optoelectronic devices, which can be integrated on Si.

6.3 Epitaxially Grown *n*-ZnO/MgO/TiN/*n*-Si(111) Heterostructured Light-Emitting Diode

In the previous section, we demonstrated a ZnO/SiOx/*n*-Si heterojunction LED. The *n*-Si exhibits lower VB energy difference with SiO$_x$/ZnO than *p*-Si, favoring the hole injection from Si to ZnO VB. However, epitaxial growth of ZnO film on Si is always difficult to achieve due to the large lattice mismatch between ZnO and Si (–15.4%) and easy oxidation of Si, which will influence the performance of ZnO/Si-based LEDs. It was recently shown that the deposition of a thin MgO[24] or AlN[25] buffer layer on Si (111) substrate can significantly improve the epitaxial quality of ZnO film. Moreover, these dielectric buffer layers can also act as the carrier blocker for the ZnO/Si heterostructured LED, providing a promising way for the development of low-cost Si-integrated ZnO-based LEDs. In this section, we shall report the epitaxial growth of ZnO (0002) on Si (111) substrate using an MgO/TiN buffer. The obtained *n*-ZnO/MgO/TiN/*n*-Si heterostructure was further employed to fabricate LED and strong EL was observed.

Pulsed laser deposition (PLD) was employed to deposit MgO/TiN buffer layers on *n*$^+$-Si (111) substrate (resistivity ~10^{-3} Ω cm). Before they were loaded into the growth chamber, Si wafers were cleaned in acetone and ethanol, then etched by HF solution for 3 min, and rinsed in deionized water. A 248 nm KrF excimer laser with a pulse repetition of 3 Hz was used to ablate the TiN and MgO sintered targets sequentially. The targets were rotated at a suitable speed to ensure uniform ablation. The growth chamber was evacuated to the order of 10^{-7} Torr, a thin layer of TiN (~2 nm) was then deposited on the Si (111). The ultra-thin TiN layer was employed to improve the mismatch between MgO and Si and avoid the oxidation of the Si surface during the MgO growth that followed. The MgO layer (~20 nm) was then grown *in situ* on TiN at oxygen ambient with a pressure of 1×10^{-4} Torr. After the deposition of MgO/TiN buffers by PLD, ZnO films were further grown by a homemade shower-head injector MOCVD system on MgO/TiN buffered Si. The ZnO film was grown at 600°C with a chamber pressure of ~25 Torr. Dimethylzinc and oxygen were used as precursors and the molar ratio of oxygen to zinc was controlled at 2000. The thickness of ZnO film was about 900 nm. To obtain the ZnO/MgO/TiN/Si-based LED, Au electrodes were deposited on the surface of ZnO film and the backside of Si wafers by direct-current magnetron sputtering. The Au electrodes on ZnO were patterned into circular pads with diameter of 1 mm using a shadow mask.

The epitaxial growth of MgO(111)/TiN on Si (111) by PLD was confirmed by *in situ* reflective high-energy electron diffraction (not shown here). Figure 6.5 shows the cross-sectional high-resolution transmission electron microscopy (HRTEM) of the ZnO/MgO/TiN/Si heterostructure. A well-defined MgO layer below ZnO and a thin TiN layer between MgO and Si can be found in HRTEM. Some amorphous phase is observed in TiN layer, which is caused by the oxidation of TiN during the growth of MgO layers. Although the MgO/TiN/Si interface was degraded, the top layer of MgO can maintain a good crystal quality, which serves as a good template for ZnO epitaxy, as shown in the inset of Figure 6.5.

The crystal quality of ZnO film was further investigated by HRXRD on a Philips double-axis high-resolution diffractometer at 45 kV and 40 mA. Figure 6.6a shows a $\theta/2\theta$ scan of XRD, in which only a sharp diffraction peak of (0002) can be found, indicating that the ZnO film is strictly grown along c axis. To further analyze the in-plane orientation of the ZnO film, Φ-scan was performed on the ZnO ($10\bar{1}1$) and Si (220) planes, as shown in Figure 6.6b. Six sharp Φ-scan peaks at 60° intervals confirm that the epitaxial ZnO exhibits a single-domain wurtzite structure with hexagonal symmetry. Moreover, Φ-scan also indicates that the surface lattice of ZnO (0002) and Si (111) are exactly overlapped without any rotation, with an in-plane epitaxial relationship as $\langle 10\bar{1}0 \rangle_{ZnO} \| \langle 11\bar{2} \rangle_{Si}$. The ω-scan rocking curve of ZnO (0002) shows a symmetric Gauss peak, with the full width at half maximum (FWHM) of 0.8°, as illustrated in Figure 6.6c.

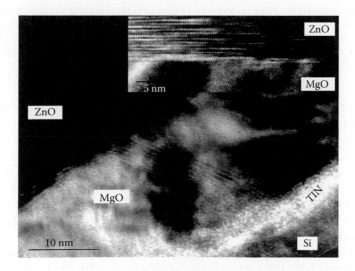

FIGURE 6.5
Cross-sectional HRTEM image of the ZnO/MgO/TiN/Si heterostructure. The inset shows an enlarged HRTEM image of ZnO/MgO interface. (Reprinted with permission from Sun, X.W., Zhao, J.L., Tan, S.T., Tan, L.H., Tung, C.H., Lo, G.Q., Kwong, D.L., Zhang, Y.W., Li, X.M., and Teo, K.L., Epitaxially grown *n*-ZnO/MgO/TiN/*n*⁺-Si(111) heterostructured light-emitting diode, *Appl. Phys. Lett.*, 92, 111113. Copyright 2008, American Institute of Physics.)

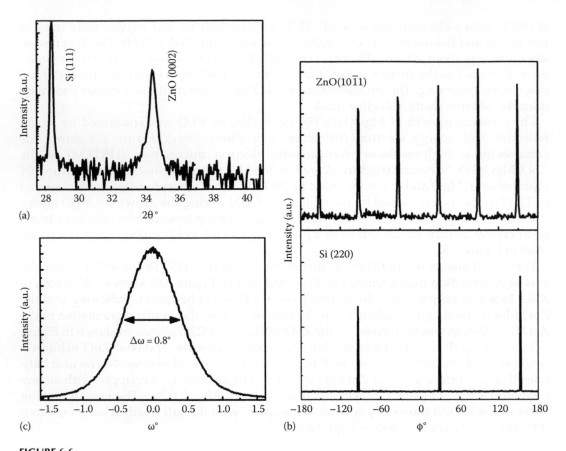

FIGURE 6.6

XRD spectrum of the epitaxial ZnO film on MgO/TiN buffered Si (111): (a) θ/2θ scan, (b) Φ scans for ZnO (10$\bar{1}$1) and Si (220), and (c) ω-scan rocking curve of ZnO (0002). (Reprinted with permission from Sun, X.W., Zhao, J.L., Tan, S.T., Tan, L.H., Tung, C.H., Lo, G.Q., Kwong, D.L., Zhang, Y.W., Li, X.M., and Teo, K.L., Epitaxially grown *n*-ZnO/MgO/TiN/*n*+-Si(111) heterostructured light-emitting diode, *Appl. Phys. Lett.*, 92, 111113. Copyright 2008, American Institute of Physics.)

The LED was further fabricated based on the epitaxial *n*-ZnO/MgO/TiN/*n*+-Si heterostructure. The typical *I–V* characteristics of the diode are shown in Figure 6.7. Ohmic contact was established for Au/*n*-ZnO, as shown in the inset of Figure 6.7. The *I–V* curve shows typical characteristics of a back-to-back diode. The band diagram of the diode is also shown in Figure 6.7. Since the TiN layer is very thin, we neglected its effects in the diagram. The MgO acts as a double Schottky barrier for both *n*-ZnO and *n*+-Si and the heterojunction can be considered as a series of two back-to-back Schottky diodes. This back-to-back diode *I–V* behavior has also been demonstrated in ZnO/SiO$_x$/*n*+-Si-based LEDs in our previous studies.

Figure 6.8a shows the photoluminescence spectrum from the ZnO/MgO/TiN/Si heterostructure. A narrow NBE UV emission can be observed, with the FWHM of only 88 meV, indicating good optical quality for the epitaxial ZnO film on the MgO buffer. Moreover, a strong wide green band is also observed in our PL, which is the deep-level emission (DLE) that has been commonly attributed to the oxygen vacancy.[26,27] Figure 6.8b illustrates the EL photo and spectra of the heterostructured LED at room temperature under

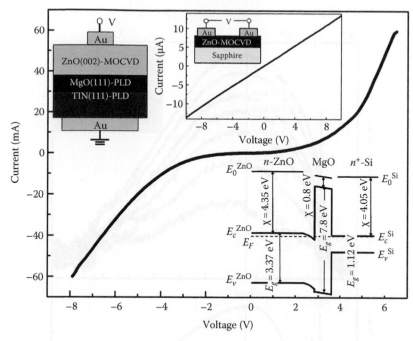

FIGURE 6.7
Current–voltage characteristic of the *n*-ZnO/MgO/TiN/*n*⁺-Si diode. The top left inset illustrates the diode structure, the top right inset shows the ohmic contact between Au and *n*-ZnO, and the bottom right inset shows the band diagram of the ZnO/MgO/Si. (Reprinted with permission from Sun, X.W., Zhao, J.L., Tan, S.T., Tan, L.H., Tung, C.H., Lo, G.Q., Kwong, D.L., Zhang, Y.W., Li, X.M., and Teo, K.L., Epitaxially grown *n*-ZnO/MgO/TiN/*n*⁺-Si(111) heterostructured light-emitting diode, *Appl. Phys. Lett.*, 92, 111113. Copyright 2008, American Institute of Physics.)

injection current from 40 to 192 mA. The output light from the LED is clearly observed by naked eyes at room temperature, with a yellow-whitish color, when a positive voltage was applied on the back electrode of Si substrates. However, no light can be observed when a positive voltage was applied on the top electrode of ZnO film. A wide emission band ranging from 350 to 850 nm, centered at about 530 nm, can be detected in the EL spectra. The EL mechanism is similar with that in ZnO/SiO$_x$/Si-based LEDs as proposed in our previous studies. Under sufficient positive bias on the Si/MgO/ZnO diode, the energy bands of Si near interface bend upward, inducing an inversion layer with holes accumulation. The holes accumulated in Si/MgO interface are swept to ZnO side through the MgO under the electric field by tunneling. Meanwhile, the electrons are injected into ZnO from Au and accumulate at ZnO/MgO interface due to the blockage of the MgO barrier. Therefore, EL originates from the recombination between the injected electrons and holes in ZnO layer near ZnO/MgO interface. Due to the large VB offset between ZnO and Si, the holes are difficult to inject from the VB of Si into the VB of ZnO and most of the holes are injected into the deep level in the bandgap of ZnO. Furthermore, more structural defects such as misfit dislocations exist in the ZnO layer near the interface compared to the sound top ZnO layer, resulting in a weak NBE emission and strong and wide DLE emission. The measured output light intensity (*L*) as a function of injection current (*I*) is depicted in Figure 6.9. The *L–I* curve can be fitted using the well-known power law of $L \sim I^m$,[28] where *m* accounts for the influence of nonradiative defects on the characteristics

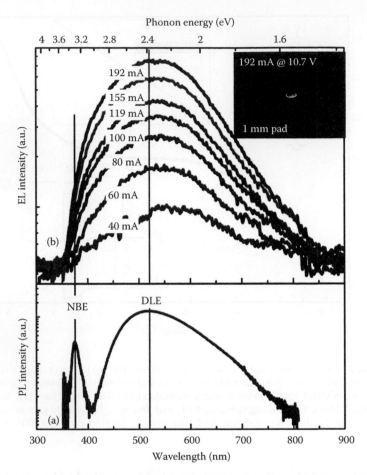

FIGURE 6.8

(a) Room-temperature PL spectrum from the ZnO film on MgO/TiN buffered Si. (b) Room-temperature EL spectra from the *n*-ZnO/MgO/TiN/*n*⁺-Si LED at various injection currents. Inset in (b) shows the photo of emitting LED at 192 mA and 10.7 V. (Reprinted with permission from Sun, X.W., Zhao, J.L., Tan, S.T., Tan, L.H., Tung, C.H., Lo, G.Q., Kwong, D.L., Zhang, Y.W., Li, X.M., and Teo, K.L., Epitaxially grown *n*-ZnO/MgO/TiN/*n*⁺-Si(111) heterostructured light-emitting diode, *Appl. Phys. Lett.*, 92, 111113. Copyright 2008, American Institute of Physics.)

of light emission. For our LED with injection current from 40 to 192 mA, the *L–I* exhibits a linear behavior with *m* = 1.02, indicating the limited influence of nonradiative defects. Moreover, previous studies also show that *m* will be lower than 1 at high injection current, due to the limited emission efficiency restricted by Auger recombination and heating effects. However, our LED can keep the linear *L–I* relationship up to 192 mA (current density 24 A/cm²), indicating the good stability of the LED based on good quality epitaxial ZnO/MgO/TiN/Si heterostructure.

In summary, we have demonstrated an *n*-ZnO/MgO/TiN/*n*⁺-Si heterostructured LED using pulsed laser deposition (buffer) and metal-organic chemical-vapor deposition (ZnO). Epitaxial growth of ZnO on Si (111) was achieved by introducing an MgO/TiN buffer. Distinct room temperature EL with a wide emission band ranging from 350 to 850 nm can be observed from the diode when a positive voltage is applied on the Si substrate. Such epitaxial ZnO/Si-based heterostructured diode is promising for the development of low-cost and high-performance optoelectronic devices integrated on Si.

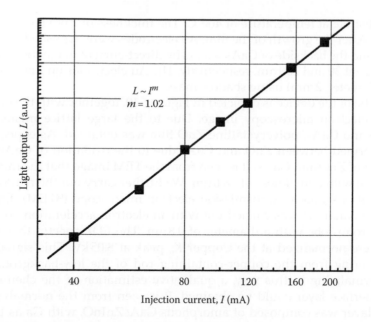

FIGURE 6.9
Light output intensity as a function of injection current for the *n*-ZnO/MgO/TiN/*n*⁺-Si LED. (Reprinted with permission from Sun, X.W., Zhao, J.L., Tan, S.T., Tan, L.H., Tung, C.H., Lo, G.Q., Kwong, D.L., Zhang, Y.W., Li, X.M., and Teo, K.L., Epitaxially grown *n*-ZnO/MgO/TiN/*n*⁺-Si(111) heterostructured light-emitting diode, *Appl. Phys. Lett.*, 92, 111113. Copyright 2008, American Institute of Physics.)

6.4 *n*-ZnO/*n*-GaAs Heterostructured White Light-Emitting Diode

In previous section, we have demonstrated ZnO/Si heterostructured LEDs by using MOCVD and ultrasonic spray pyrolysis (USP) technique. Room-temperature EL was observed in the *n*-ZnO/*n*-Si and *n*-ZnO//MgO/TiN/*n*⁺-Si LEDs when *n*-Si was positively biased. The field-induced inversion layer in *n*-Si was proposed to be responsible for hole injection and hence photon emission. However, due to the indirect bandgap of the Si, recombination of electron-hole pairs at the Si site will generate phonons. In this section, ZnO has been deposited on GaAs to form a *n*-ZnO/*n*-GaAs heterostructured diode. Both the emissions from ZnO and GaAs have been observed in the devices.

GaAs and its related compound has been applied as infra-red and red LEDs from 1990s. With the UV/blue and green-band emission nature of ZnO, the integration of ZnO on GaAs technology would be expected to demonstrate LEDs that emit red, green, and blue light, forming white-light LEDs. As a result, the integration of ZnO onto GaAs will definitely accelerate the development of ZnO especially as a light emitter.[29,30] In this section, the fabrication, characterization, and EL mechanism of the *n*-ZnO/*n*-GaAs heterostructured diodes shall be discussed.

The indium doped ZnO thin film used in this study was deposited by USP on *n*-GaAs(100) wafers (carrier density of ~10^{18} cm⁻³). An aqueous solution of zinc acetate ($Zn(CH_3COO)_2 \cdot 2H_2O$) and indium nitrate ($In(NO_3)_3 \cdot 2H_2O$) in a ratio of 20:1 was chosen as the precursor to deposit indium doped ZnO layer. The aerosol of the precursor solution was generated by a commercial ultrasonic nebulizer (frequency, 1.65 MHz), and transported to the heated substrate at temperature of 400°C. The thickness of ZnO layer is controlled at about 200 nm. After the growth of the film, Au electrodes were deposited on the surface of the ZnO film and the backside of GaAs wafers by direct-current magnetic sputtering, with thickness of about 30 and 200 nm, respectively. The Au electrodes on ZnO were patterned into circles (diameter, 2 mm) using shadow mask.

The schematic of the device is depicted in Figure 6.10, together with the cross-sectional transmission electron microscopy image. Due to the large lattice mismatch (~42.4%) between ZnO and GaAs, polycrystalline ZnO film was obtained. As expected, an interface layer is formed between ZnO and GaAs due to the oxidation of GaAs surface and inter-diffusion of ZnO and GaAs. It is seen from the TEM image that the interface layer is amorphous-like with a thickness of ~8.6 nm. We further carry out the nanoscale analysis with an FEI Titan scanning transmission electron microscope (STEM). Energy dispersive x-ray (EDX) analysis was carried out with an electron acceleration voltage of 80 kV, using an electron probe with a diameter of 0.8 nm. The EDX spectra that are shown in Figure 6.11 were normalized at the Copper K_α peak at 8.05 kV. This signal arises from spurious scattering from the copper-containing rod of the low-background specimen holder. By normalizing in this way, a qualitative estimation of the chemical composition of the interface layer could be obtained. It is seen from the normalized EDX that the interface layer was composed of amorphous GaAsZnInO, with Ga as the dominant element. Electron energy-loss spectroscopy (EELS) was further employed to probe the band structure of the interface layer. Figure 6.11b shows the EEL spectra that were taken using a monochrome STEM probe with a diameter of 1 nm. The energy resolution of the measurement was better than 1.7 eV, being the full width at 0.1% of the maximum zero-loss peak intensity. This energy resolution should be just sufficient to resolve low-energy transitions, such as the GaAs bandgap at 1.4 eV, in the low-loss EEL spectrum.

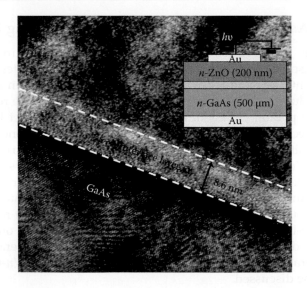

FIGURE 6.10

Schematic diagram of a *n*-ZnO/*n*-GaAs heterostructured light-emitting diode and the cross-sectional TEM image of the diode. (Reprinted with permission from Tan, S.T., Zhao, J.L., Iwan, S., Sun, X.W., Tang, X.H., Ye, J.D., Bosman, M., Tang, L.J., Lo, G.Q., and Teo, K.L., *n*-ZnO/*n*-GaAs heterostructured white light-emitting diode: Nanoscale interface analysis and electroluminescence studies, *IEEE Trans. Electron Devices*, 57, 129. © 2010, IEEE.)

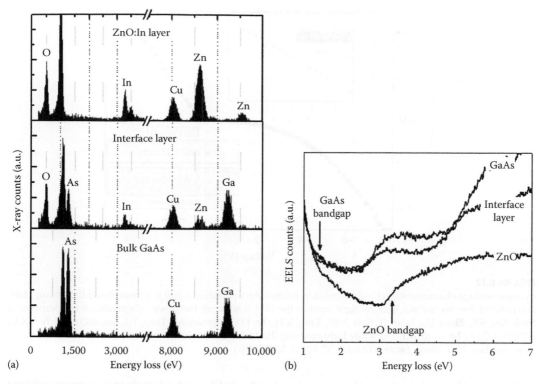

FIGURE 6.11
(a) Normalized EDX spectra at the interface of *n*-ZnO/*n*-GaAs. (b) Low-loss EEL spectra of *n*-ZnO/*n*-GaAs heterostructured diode. (Reprinted with permission from Tan, S.T., Zhao, J.L., Iwan, S., Sun, X.W., Tang, X.H., Ye, J.D., Bosman, M., Tang, L.J., Lo, G.Q., and Teo, K.L., *n*-ZnO/*n*-GaAs heterostructured white light-emitting diode: Nanoscale interface analysis and electroluminescence studies, *IEEE Trans. Electron Devices*, 57, 129. © 2010, IEEE.)

Measurements on the GaAs and ZnO were taken more than 15 nm from the interface to avoid signals from adjacent layers. For the interface layer, the probe was positioned in its center, but due to the delocalized nature of the low-loss signal, it cannot be avoided that a small part of the signal also comes from the ZnO and GaAs. Figure 6.11b displays the EELS results, with a decreasing tail of the zero-loss peak at the low energy-loss part of the spectra. For small bandgap materials, the identification of the bandgap could be challenging due to the difficulty of separating zero-loss and inelastic-loss signals. As a result, the bandgap of the GaAs and interface layer could not be distinctively measured in our EELS measurement. Nevertheless, it is seen from Figure 6.11b that the GaAs spectrum shows a relatively high EELS signal above 1.4 eV, which is arising from electron transitions across its bandgap. The interface layer EEL spectrum shows a similar trend as the GaAs EEL spectrum, suggesting that the interface layer has an electronic structure more closely related to GaAs than ZnO.

The typical *I–V* characteristic of the *n*-ZnO/*n*-GaAs heterostructured LEDs is shown in Figure 6.12. Ohmic contact was established for Au/*n*-ZnO, as shown in Figure 6.12 inset. The two Schottky junctions at both side of the interface layer, as depicted in the band diagram, can be considered as a series of two back-to-back Schottky diodes. The *I–V* of the *n*-ZnO/*n*-GaAs diode, showing nonlinear behavior at both bias schemes, is in fact reflecting the *I–V* behavior of the back-to-back Schottky diode, as shown in Figure 6.12.

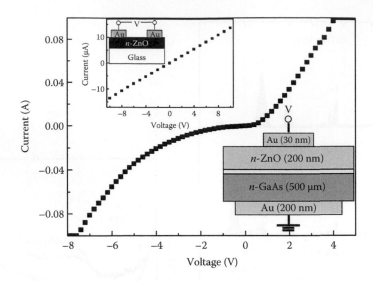

FIGURE 6.12

Current–voltage characteristics of *n*-ZnO/*n*-GaAs heterostructured diode. Top left inset shows the Ohmic characteristic of the Au/*n*-ZnO. Bottom right inset is the bias scheme of the diode. (Reprinted with permission from Tan, S.T., Zhao, J.L., Iwan, S., Sun, X.W., Tang, X.H., Ye, J.D., Bosman, M., Tang, L.J., Lo, G.Q., and Teo, K.L., *n*-ZnO/*n*-GaAs heterostructured white light-emitting diode: Nanoscale interface analysis and electroluminescence studies, *IEEE Trans. Electron Devices*, 57, 129. © 2010, IEEE.)

Figure 6.13a and b show the room-temperature EL and photoluminescence (PL) of the *n*-ZnO/*n*-GaAs heterostructured LED, respectively. The room-temperature PL was obtained using 325 nm He–Cd laser as the quantum confinement effect of 2-dimensional hole gas induced upon the *n*-GaAs being positively biased, as shown in Figure 6.14b. Since both ZnO and GaAs-related EL emissions were observed upon biased, the recombination process should be taking place at the ZnO/GaAs interface. The possible mechanisms that are responsible for the EL upon positively biased *n*-GaAs could be (1) induced inversion layer in *n*-GaAs, resulting in a 2-dimensional hole gas near the interface and (2) impact ionization that facilitates the electron-hole pair generation.

On the other hand, it is also worth mentioning that a thin *p*-type ZnO:As layer could be formed at the interface by As out-diffusion from GaAs substrate.[29] Although the ZnO layer was doped by In, the out-diffusion of As (there might be a small amount of Ga as well) from the GaAs substrate could possibly induce the formation of V–III–V acceptor complex by In and As co-doping.[31] If the *p*-ZnO layer is formed at the interface, the device structure would be *n*-ZnO/*p*-ZnO/*i*-GaAsZnInO/*n*-GaAs, which consists of two back-to-back connected diodes (a *p*-*n* diode and a Schottky diode). In this case, the holes would be injected from induced inversion layer in *n*-GaAs (tunneling and/or impact ionization) into the *p*-ZnO layer. The recombination process that occurs at the interface of ZnO *p*-*n* junction would then give rise to NBE and DLE emission. While the NIR emission is a result from the recombination process that occurs at *p*-ZnO/*i*-GaAsZnInO/*n*-GaAs Schottky interface.

Figure 6.15 shows the measured output light intensity (*L*) of ZnO- and GaAs-related emission as a function of injection current (*I*). The *L*–*I* curve can be fitted using the well-known power law of $L \sim I^m$, where *m* accounts for the influence of nonradiative defects on the characteristics of light emission.[24] It can be seen in Figure 6.15 that the *L*–*I* curve of ZnO-related emission exhibits superlinear behavior, with *m* = 1.42 extracted from the fitting. The superlinear *L*–*I* curve of ZnO-related emission indicates the presence of nonradiative

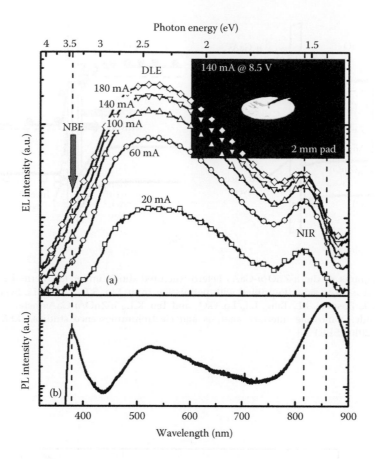

FIGURE 6.13

(a) Room-temperature EL spectra from the *n*-ZnO/*n*-GaAs LED at various injection currents. (b) Room-temperature PL spectrum from the ZnO film on GaAs. Inset in (a) shows the photo of emitting LED at 140 mA and 8.5 V. (Reprinted with permission from Tan, S.T., Zhao, J.L., Iwan, S., Sun, X.W., Tang, X.H., Ye, J.D., Bosman, M., Tang, L.J., Lo, G.Q., and Teo, K.L., *n*-ZnO/*n*-GaAs heterostructured white light-emitting diode: Nanoscale interface analysis and electroluminescence studies, *IEEE Trans. Electron Devices*, 57, 129. © 2010, IEEE.)

recombination centers, which provide a recombination path to the electron-hole pairs. On the other hand, the *L–I* curve of GaAs-related emission exhibits linear behavior with $m = 1.07$, indicating limited occurrence of nonradiative recombination in the electron-hole pairs recombination. It is also noted that the EL intensity saturated at the injection current is higher than 160 mA. The saturation of EL intensity is due to the limitation of the electrical-to-optical conversion efficiency caused by Auger recombination or by heating effects at high current level.

6.5 Summary

In summary, we have demonstrated *n*-ZnO/*n*-GaAs heterostructured LEDs by a simple and low-cost spray pyrolysis method. Both white and near-infrared EL can be achieved at room temperature when the *n*-GaAs is positively biased. The demonstration of the

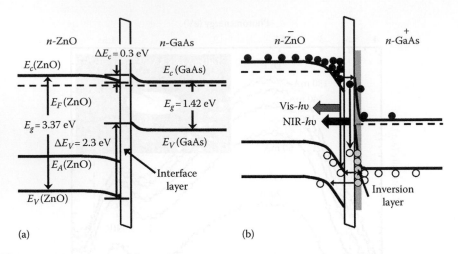

(a)   (b)

FIGURE 6.14
Energy band diagrams of the n-ZnO/n-GaAs heterostructured diode under (a) thermal equilibrium and (b) positive bias at n-GaAs side. (Reprinted with permission from Tan, S.T., Zhao, J.L., Iwan, S., Sun, X.W., Tang, X.H., Ye, J.D., Bosman, M., Tang, L.J., Lo, G.Q., and Teo, K.L., n-ZnO/n-GaAs heterostructured white light-emitting diode: Nanoscale interface analysis and electroluminescence studies, *IEEE Trans. Electron Devices*, 57, 129. © 2010, IEEE.)

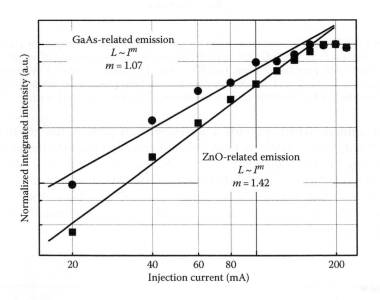

FIGURE 6.15
Light output intensity as a function of injection current for the n-ZnO/n-GaAs LED. (Reprinted with permission from Tan, S.T., Zhao, J.L., Iwan, S., Sun, X.W., Tang, X.H., Ye, J.D., Bosman, M., Tang, L.J., Lo, G.Q., and Teo, K.L., n-ZnO/n-GaAs heterostructured white light-emitting diode: Nanoscale interface analysis and electroluminescence studies, *IEEE Trans. Electron Devices*, 57, 129. © 2010, IEEE.)

ZnO/GaAs heterostructured LEDs are promising for the development of solid-state lighting especially white LED, which could be achieved by integrating the ZnO onto III–V compound semiconductors.

References

1. X.W. Sun and H.S. Kwok, Optical properties of epitaxially grown zinc oxide films on sapphire by pulsed laser deposition, *J. Appl. Phys.* **86**, 408 (1999).
2. S.J. Pearton, D.P. Norton, K. Ip, Y.W. Heo, and T. Steine, Recent progress in processing and properties of ZnO, *Prog. Mater. Sci.* **50**, 293 (2005).
3. Ü. Özgür, Ya.I. Alivov, C. Liu, A. Teke, M.A. Reshchikov, S. Doğan, V. Avrutin, S.J. Cho, and H. Morkoç, A comprehensive review of ZnO materials and devices, *J. Appl. Phys.* **98**, 041301 (2005).
4. C.H. Park, S.B. Zhang, and S.-H. Wei, Origin of *p*-type doping difficulty in ZnO: The impurity perspective, *Phys. Rev. B* **66**, 073202 (2002).
5. S.T. Tan, X.W. Sun, Z.G. Yu, P. Wu, G.Q. Lo, and D.L. Kwong, *p*-Type conduction in unintentional carbon-doped ZnO thin films, *Appl. Phys. Lett.* **91**, 072101 (2007).
6. Ya.I. Alivov, E.V. Kalinina, A.E. Cherenkov, D.C. Look, B.M. Ataev, A.K. Omaev, M.V. Chukichev, and D.M. Bagnall, Fabrication and characterization of *n*-ZnO/*p*-AlGaN heterojunction light-emitting diodes on 6H-SiC substrates, *Appl. Phys. Lett.* **83**, 4719 (2003).
7. J.D. Ye, S.L. Gu, S.M. Zhu, W. Liu, S.M. Liu, R. Zhang, Y. Shi, and Y.D. Zheng, Electroluminescent and transport mechanisms of *n*-ZnO/*p*-Si heterojunctions, *Appl. Phys. Lett.* **88**, 182112 (2006).
8. P. Chen, X. Ma, and D. Yang, Ultraviolet electroluminescence from ZnO/*p*-Si heterojunctions, *J. Appl. Phys.* **101**, 053103 (2007).
9. H. Sun, Q.-F. Zhang, and J.-L. Wu, Electroluminescence from ZnO nanorods with an *n*-ZnO/*p*-Si heterojunction structure, *Nanotechnology* **17**, 2271 (2006).
10. C. Yuen, S.F. Yu, S.P. Lau, Rusli, and T.P. Chen, Fabrication of *n*-ZnO:Al/*p*-SiC(4H) heterojunction light-emitting diodes by filtered cathodic vacuum arc technique, *Appl. Phys. Lett.* **86**, 241111 (2005).
11. H. Ohta, K. Kawamura, M. Orita, M. Hirano, N. Sarukura, and H. Hosono, Current injection emission from a transparent *p*–*n* junction composed of *p*-SrCu₂O₂/*n*-ZnO, *Appl. Phys. Lett.* **77**, 475 (2000).
12. S.F. Chichibu, T. Ohmori, N. Shibata, T. Koyama, and T. Onuma, Greenish-white electroluminescence from *p*-type CuGaS₂ heterojunction diodes using *n*-type ZnO as an electron injector, *Appl. Phys. Lett.* **85**, 4403 (2004).
13. T.L. Tansley and S.J.T. Owen, Conductivity of Si-ZnO *p*-*n* and *n*-*n* heterojunctions, *J. Appl. Phys.* **55**, 454 (1984).
14. H. Kobayashi, H. Mori, T. Ishida, and Y. Nakato, Zinc oxide/*n*-Si junction solar cells produced by spray—Pyrolysis method, *J. Appl. Phys.* **77**, 1301 (1995).
15. J.L. Zhao, X.M. Li, A. Krtschil, A. Krost, W.D. Yu, Y.W. Zhang, Y.F. Gu, and X.D. Gao, Study on anomalous high *p*-type conductivity in ZnO films on silicon substrate prepared by ultrasonic spray pyrolysis, *Appl. Phys. Lett.* **90**, 062118 (2007).
16. S.T. Tan, X.W. Sun, J.L. Zhao, S. Iwan, Z.H. Cen, T.P. Chen, J.D. Ye, G.Q. Lo, D.L. Kwong, and K.L. Teo, Ultraviolet and visible electroluminescence from n-ZnO/SiOₓ/(n,p)-Si heterostructured light-emitting diodes, *Appl. Phys. Lett.* **93**, 013506 (2008).
17. X.W. Sun, J.L. Zhao, S.T. Tan, L.H. Tan, C.H. Tung, G.Q. Lo, D.L. Kwong, Y.W. Zhang, X.M. Li, and K.L. Teo, Epitaxially grown *n*-ZnO/MgO/TiN/*n*⁺-Si(111) heterostructured light-emitting diode, *Appl. Phys. Lett.* **92**, 111113 (2008).

18. J.L. Zhao, X.W. Sun, S.T. Tan, G.Q. Lo, D.L. Kwong, and Z.H. Cen, Realization of n-$Zn_{1-x}Mg_xO$/i-ZnO/SiO_x/n^+-Si heterostructured n-i-n light-emitting diodes by low-cost ultrasonic spray pyrolysis, *Appl. Phys. Lett.* **91**, 263501 (2007).

19. S.T. Tan, J.L. Zhao, S. Iwan, X.W. Sun, X.H. Tang, J.D. Ye, M. Bosman, L.J. Tang, G.Q. Lo, and K.L. Teo, n-ZnO/n-GaAs heterostructured white light-emitting diode: Nanoscale interface analysis and electroluminescence studies, *IEEE Trans. Electron Devices* **57**, 129 (2010).

20. C.V. Opdorp and H.K.J. Kanerva, Current-voltage characteristics and capacitance of isotype heterojunctions, *Solid-State Electron.* **10**, 401 (1967).

21. J.A. Aranovich, D. Golmayo, A.L. Fahrenbuch, and R.H. Bube, Photovoltaic properties of ZnO/CdTe heterojunctions prepared by spray pyrolysis, *J. Appl. Phys.* **51**, 4260 (1980).

22. Y.I. Alivov, D.C. Look, B.M. Ataev, M.V. Chukichev, V.V. Mamedov, V.I. Zinenko, Y.A. Agafonov, and A.N. Pustovit, Fabrication of ZnO-based metal–insulator–semiconductor diodes by ion implantation, *Solid-State Electron.* **48**, 2343 (2004).

23. D.K. Hwang, M.S. Oh, J.H. Lim, Y.S. Choi, and S.J. Park, ZnO-based light-emitting metal-insulator-semiconductor diodes, *Appl. Phys. Lett.* **91**, 121113 (2007).

24. X.N. Wang, Y. Wang, Z.X. Mei, J. Dong, Z.Q. Zeng, H.T. Yuan et al., Low-temperature interface engineering for high-quality ZnO epitaxy on Si(111) substrate, *Appl. Phys. Lett.* **90**, 151912 (2007).

25. F. Jiang, C. Zheng, L. Wang, W. Fang, Y. Pu, and J. Dai, The growth and properties of ZnO film on Si(111) substrate with an AlN buffer by AP-MOCVD, *J. Lumin.* **122–123**, 905 (2007).

26. K. Vanheusden, C.H. Seager, W.L. Warren, D.R. Tallant, and J.A. Voigt, Correlation between photoluminescence and oxygen vacancies in ZnO phosphors, *Appl. Phys. Lett.* **68**, 403 (1996).

27. S.T. Tan, B.J. Chen, X.W. Sun, M.B. Yu, X.H. Zhang, and S.J. Chua, Realization of intrinsic p-type ZnO thin films by metal organic chemical vapor deposition, *J. Electron. Mater.* **34**, 1172 (2005).

28. I. Martil, E. Redondo, and A. Ojeda, Influence of defects on the electrical and optical characteristics of blue light-emitting diodes based on III–V nitrides, *J. Appl. Phys.* **81**, 2442 (1997).

29. J.C. Sun, J.Z. Zhao, H.W. Liang, J.M. Bian, L.Z. Hu, H.Q. Zhang, X.P. Liang, W.F. Liu, and G.T. Du, Realization of ultraviolet electroluminescence from ZnO homojunction with n-ZnO/p-ZnO:As/GaAs structure, *Appl. Phys. Lett.* **90**, 121128 (2007).

30. G.T. Du, Y.G. Cui, X.C. Xia, X.P. Li, H.C. Zhu, B.L. Zhang, Y.T. Zhang, and Y. Ma, Visual-infrared electroluminescence emission from ZnO/GaAs heterojunctions grown by metal-organic chemical vapor deposition, *Appl. Phys. Lett.* **90**, 243504 (2007).

31. T. Yamamoto and H. Katayama-Yoshida, Solution using a codoping method to unipolarity for the fabrication of p-type ZnO, *Jpn. J. Appl. Phys.* **38**, L166 (1999).

7

ZnO Thin Films, Quantum Dots, and Light-Emitting Diodes Grown by Atomic Layer Deposition

Miin-Jang Chen, Jer-Ren Yang, and Makoto Shiojiri

CONTENTS

7.1 Introduction to ZnO 237
7.2 Introduction to ALD 238
7.3 High-Quality ZnO Epilayers on *c*-Sapphire Substrates 239
7.4 Low-Threshold Stimulated Emission in Highly Oriented ZnO Thin
     Films on Amorphous Glass Substrates 244
7.5 ZnO Quantum Dots Embedded in a SiO$_2$ Nanoparticle Layer 250
7.6 UV Electroluminescence from *n*-ZnO/*p*-GaN Heterojunction LEDs 254
7.7 UV Electroluminescence from *n*-ZnO/*i*-ZnO/*p*-SiC Heterojunction LEDs 261
7.8 Long-Term Stable *p*-Type ZnO Thin Films 266
7.9 Summary 271
Acknowledgments 272
References 272

7.1 Introduction to ZnO

Zinc oxide (ZnO) has attracted considerable attention as a promising material for ultraviolet (UV) photonic devices such as light-emitting diodes and lasers. ZnO also has numerous applications in transparent electronics, piezoelectric devices, gas sensors, solar cells, spin electronics, etc. [1,2]. The large excitonic binding energy (60 meV) of ZnO suggests that it has excellent UV light emission properties. ZnO also has several other advantages: amenability to conventional wet chemical etching, relatively low material costs, long-term stability, environmental friendliness, biocompatibility, and excellent radiation resistance.

Many techniques have been applied to prepare high-quality ZnO thin films, including molecular beam epitaxy (MBE) [3–7], metal-organic chemical vapor deposition (MOCVD) [8–11], chemical vapor deposition (CVD) [12,13], pulsed laser deposition (PLD) [14], radio-frequency magnetron sputtering (RFMS) [15–17], and filtered cathodic vacuum arc technique [18,19]. Lim et al. reported that ZnO thin films grown on the (0001) sapphire (*c*-Al$_2$O$_3$) at 200°C–275°C by electron cyclotron resonance-assisted MBE were basically single crystals, which comprised subgrains 15–150 nm in size and accompanied by threading dislocations [5]. Generally, a lot of columnar grains were embedded in the ZnO thin films grown by MOCVD [8–11]. PLD and RFMS may be the most popular techniques for preparing ZnO thin films, but they usually require a high deposition temperature (above 350°C or more).

Atomic layer deposition (ALD) is another noteworthy method of growing high-quality ZnO thin films [20–22]. Recently, Wójcik et al. have examined how to control the preferred orientation of polycrystalline ZnO thin films grown by ALD on the c-Al$_2$O$_3$, lime glass, and (001) and (111) Si substrates at 230°C–400°C [23]. They have also reported that the ALD ZnO thin films grown on Si substrates at low temperatures (90°C–200°C) are high-quality polycrystalline with surface roughness of 1–4 nm [24,25].

7.2 Introduction to ALD

ALD is a thin-film deposition technique for preparing high-quality oxides with atomic-layer accuracy. The most distinguishing feature of ALD is that each precursor is alternately pulsed into the reaction chamber, as shown schematically in Figure 7.1. When the first precursor is introduced into the chamber, the precursor molecules adsorb on the substrate surface by chemical absorption, resulting in the saturative absorption of one monolayer of precursor on the surface. Between the precursor pulses, the chamber is purged with an inert gas to remove all the excess precursors and by-products. As the next precursor is dosed in, it reacts with the precursor previously adsorbed on the surface, producing one monolayer of solid product and gaseous by-products. In contrast to conventional CVD, the chemical reactions in ALD proceed only at the substrate surface, resulting in self-limiting and layer-by-layer growth. The self-limiting or self-terminating reaction also suggests that precise control of precursor homogeneity is not necessary. The only requirement is that sufficient precursor molecules are needed to cover the adsorption sites on the surface. In addition, ALD can provide conformal

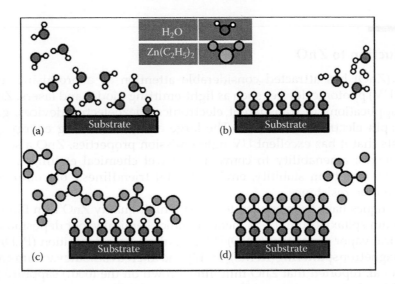

FIGURE 7.1
Schematic diagram of an ALD cycle (1–4) for depositing one monolayer of ZnO. (a) First step: introduction of the first gaseous precursor to chemisorb on the substrate, (b) second step: inert gas purge to remove the excess unsorbed precursor, (c) third step: introduction of the second precursor to react with the first one and produce the desired thin film on the substrate, and (d) last step: inert gas purge to expel the gaseous reaction by-products.

coatings on complex 3-D structures due to the diffusion of vapor-phase precursors into porous structures as well as the surface-controlled self-limiting growth. The advantages of ALD are summarized as follows [20,21]:

1. Precise and easy thickness control with one monolayer accuracy
2. Accurate composition control and facile doping for achieving high doping concentrations
3. Stoichiometric films with low defect density
4. High uniformity over a large area, leading to large-area and large-batch capacity
5. Excellent conformality and good step coverage
6. Low deposition temperature
7. Good reproducibility

In this review article, ZnO was deposited by ALD using alternating pulses of $Zn(C_2H_5)_2$ (DEZn, the Zn precursor) and H_2O (the oxygen precursor) vapor in a N_2 carrier gas flow:

A. $Surface\text{-}OH + Zn(C_2H_5)_2 \rightarrow Surface\text{-}O\text{-}Zn\text{-}C_2H_5 + C_2H_6\uparrow$
B. $Surface\text{-}O\text{-}Zn\text{-}C_2H_5 + H_2O \rightarrow ZnO + Surface\text{-}OH + C_2H_6\uparrow$

The deposition for one monolayer of ZnO by ALD is shown schematically in Figure 7.1. The ALD process consisted of a number of identical cycles and repeated reactions in an ABAB... sequence (AB is called one ALD cycle). The thickness of the ZnO thin films can be controlled accurately by the number of ALD cycles.

7.3 High-Quality ZnO Epilayers on *c*-Sapphire Substrates [26]

A ZnO epilayer was deposited on the $c\text{-}Al_2O_3$ substrate at a low deposition temperature of 180°C by ALD. Each ALD cycle deposited ~0.18 nm thickness of ZnO. After the deposition, the films were treated by rapid thermal annealing (RTA) at 950°C for 5 min in a nitrogen atmosphere to improve the crystal quality. The crystalline structure of the ZnO epilayer was characterized by X-ray diffraction (XRD) and transmission electron microscopy (TEM). A PANalytical X'Pert PRO diffractometer with a wavelength of 1.54 Å was used for the XRD measurement. TEM images were recorded using a Tecnai 30 electron microscope equipped with a lens of $Cs = 1.2$ mm and operated at 300 keV. The specimens for TEM observation were prepared by mechanical polishing and ion milling with a Gatan Precision Ion Polish System Model-691.

Figure 7.2 shows the XRD pattern of the ZnO epilayer, indicating that it was preferentially oriented with respect to the $c\text{-}Al_2O_3$ substrate with the relation of $(0001)ZnO//(0001)Al_2O_3$. The XRD θ–2θ scan (the inset of Figure 7.2) exhibited a sharp ZnO (0002) peak with a full-width at half-maximum (FWHM) of 0.06°, revealing high crystallinity with the c-axis orientation of the ZnO epilayer. Figure 7.3 shows a cross-sectional TEM image of the ZnO epilayer about 830 nm thick on the $c\text{-}Al_2O_3$ substrate. The highest ZnO growth rate along the c-axis usually results in a columnar morphology [4–11]. However, the columnar structures were not observed in the ZnO epilayer grown by ALD and followed by RTA treatment. This can be attributed to layer-by-layer growth and low deposition temperature

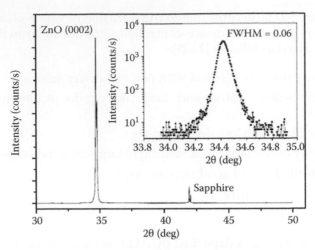

FIGURE 7.2
XRD pattern of the ZnO epilayer grown on the c-Al$_2$O$_3$ substrate by ALD and treated by post-deposition RTA.

FIGURE 7.3
Cross-sectional TEM image of the ZnO epilayer grown on the c-Al$_2$O$_3$ substrate by ALD and treated by post-deposition RTA.

of 180°C in ALD. The RTA treatment recrystallized the granular particles with strained lattices in the as-deposited ZnO thin film to a highly crystalline crystal with the preferred c-axis orientation.

Figure 7.4 displays a high-resolution (HR) TEM image of an area including the ZnO and c-Al$_2$O$_3$ substrate. The insets present the computer diffractograms (fast Fourier transform images) (a through c) and (d) of the areas indicated by (a through c) and of the whole area in the HRTEM image, respectively. Diffractogram (c) reveals that the ZnO layer is oriented with the [0$\bar{1}$10] axis parallel to the incident beam and the [0001] axis normal to the film surface. The interfacial area (b) exhibiting brighter contrast can be also identified as ZnO crystal with almost the same orientation with the crystal in (c). The difference in contrast between (c) and (b) might be caused by very small difference in the diffraction condition. Thus, the dark area (c) and the bright area (b) are almost similar in the crystal structure, although they are not the same because the lattices of interfacial area (b) are more

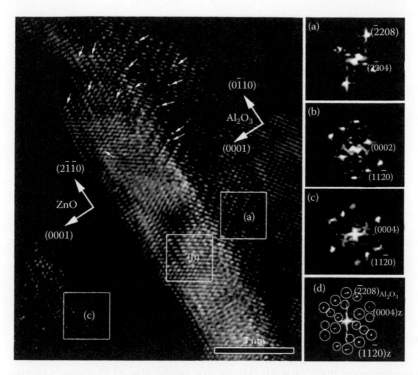

FIGURE 7.4
HRTEM image of an area including the ZnO/Al₂O₃ interface. Computer diffractograms are of the small areas indicated by (a through c) and the whole area (d). Indexes in diffractogram (a) correspond to the lattice fringes of Al₂O₃ crystal and indexes in diffractograms (b) and (c) correspond to the lattice fringes of ZnO crystal. In the diffractogram (d), the spots enclosed by large circles come from ZnO and the spots enclosed by small circles originate from Al₂O₃.

distorted under the influence of the c-Al₂O₃ substrate lattice as seen in the HRTEM image. The Al₂O₃ crystal including the area (a) is nearly oriented with the $[2\bar{1}\bar{1}0]$ axis parallel to the incident beam and the [0001] axis normal to the film surface, which indicates that the ZnO crystal was grown in the epitaxial relation with respect to the c-Al₂O₃: [0001]ZnO// [0001]Al₂O₃ and $[0\bar{1}10]$ZnO//$[2\bar{1}\bar{1}0]$Al₂O₃. The lattice of the ZnO crystal in the interfacial area is conjugated rather homogeneously and coherently with the lattice of c-Al₂O₃ crystal, having very few threading dislocations (also see Figure 7.3). It should be noted that the threading dislocations used to occur in the ZnO crystals grown by other techniques due to a lattice misfit as large as 18% between ZnO and c-Al₂O₃ [5]. The ZnO thin film grown on c-Al₂O₃ by ALD might relax the misfit by distorting its lattice in the interfacial layer. Consequently, the ZnO crystal developed to the highly crystalline crystal as seen in Figure 7.3, in a short annealing time of 5 min. Figure 7.5 shows an HRTEM image of the ZnO epilayer, whose area is indicated by the square in Figure 7.3, revealing that the crystal is a nearly perfect ZnO crystal although some lattice distortion still appears as the traces of coalescence growth between the granular particles in the as-deposited ZnO thin film.

The spots in the diffractogram of the whole area in Figure 7.4d, can be indexed as the spots from the ZnO and Al₂O₃ lattice fringes. We cannot recognize any other crystal in our specimen although Wang et al. identified Al₂ZnO₄ phase in a TEM image at the interface between ZnO and Al₂O₃ [14]. The Al₂ZnO₄ phase might be formed by the reaction of ZnO and Al₂O₃ during PLD at a high growth temperature of 600°C and a long

FIGURE 7.5
HRTEM image of the ZnO epilayer enclosed by a square in Figure 7.3.

annealing treatment at 900°C for 2h [14]. In this study, the ALD growth temperature was as low as 180°C and the post-deposition annealing treatment was short (950°C for 5min). However, we can see extraordinarily bright spots (some examples of which are indicated by arrowheads in Figure 7.4) in the lattice fringes in the HRTEM image, which may be ascribed to the substituted Al atoms in the interfacial ZnO layer and the Al vacancies in the Al_2O_3 crystal near the interface. Therefore, it is reasonable to assume that the Al atoms might diffuse from the substrate to the ZnO thin film during the RTA treatment.

The high crystallinity shown in the XRD and TEM measurements suggests excellent light emission characteristics of the ZnO epilayer grown by ALD. A continuous-wave He-Cd laser ($\lambda = 325$ nm) and a fourth harmonic Q-switched Nd-YAG laser ($\lambda = 266$ nm, pulse width ~10ns, repetition rate = 15Hz) were used as the excitation sources to measure the spontaneous emission and stimulated emission of the ZnO epilayer at room temperature, respectively. The spontaneous emission and stimulated emission photoluminescence (PL) spectra were measured in a standard backscattering configuration by collecting the light emission from top surface of the sample. Figure 7.6 shows the room-temperature spontaneous emission PL spectrum of the ZnO epilayer, which exhibited a strong near-band-edge (NBE) emission at 3.28eV and was nearly free of the defect-related band. The inset of Figure 7.6 presents the same PL spectrum in logarithm scale, indicating a very weak defect-related band at 2.57eV associated with the oxygen vacancy (V_O) [1]. It can be observed that the intensity of the NBE emission (3.28eV) is about 2 orders of magnitude larger than that of the defect-related band (2.57eV).

Figure 7.7 shows the stimulated emission PL spectra of the ZnO epilayer at various excitation intensities from 14.9 to 339.5kW/cm². At low excitation intensity, the spectral peak around 3.26eV can be ascribed to the radiative recombination of free excitons [3,27]. A spectral peak appeared at 3.17eV as the excitation intensity increased. This spectral peak comes from the radiative recombination associated with exciton–exciton scattering, in which one exciton recombines to emit a photon with the energy given by [28]

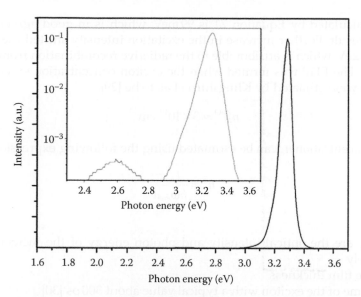

FIGURE 7.6
Spontaneous emission PL spectrum of the ZnO epilayer pumped by a CW He-Cd laser at room temperature.
The inset is the plot of the same PL spectrum in the logarithm scale.

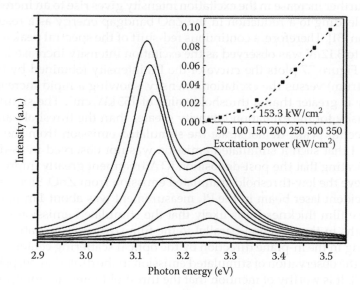

FIGURE 7.7
Stimulated emission PL spectrum of the ZnO epilayer pumped by a pulsed Nd-YAG laser at room temperature.
The inset shows the characteristic of the PL intensity as a function of the excitation intensity.

$$E_n = E_{ex} - E_B^{ex}\left(1 - \frac{1}{n^2}\right) - \frac{3}{2}k_BT \qquad (7.1)$$

and the other exciton is scattered to a higher energy state ($n > 2$). In Equation 7.1, E_{ex} is the
free-exciton energy, E_B^{ex} is the binding energy of the exciton (60 meV), n is the quantum
number of the excited exciton, and k_BT is thermal energy. The difference between E_{ex} and

E_n ($n \gg 2$) as calculated by Equation 7.1 is 99 meV, which is in good agreement with the experimental result. Further increase in the excitation intensity shifted the spectral peak from 3.16 to 3.12 eV, which is attributable to the radiative recombination from electron-hole plasma (EHP). The EHP was formed when the exciton concentration exceeded the Mott density, which was estimated by Klingshirn et al. to be [29]

$$n_M^{ZnO} \approx 3 \times 10^{17} \text{ cm}^{-3} \tag{7.2}$$

The exciton concentration n_p can be estimated using the following expression [29]:

$$n_p = \frac{I_{exc}\tau}{\hbar\omega_{exc}d} \tag{7.3}$$

where
  I_{exc} and $\hbar\omega_{exc}$ are the optical intensity and photon energy of the incident laser beam, respectively
  d denotes the film thickness
  τ is the lifetime of the exciton with a typical value about 300 ps [30]

According to Equation 7.3, the exciton concentration at the excitation intensity of 210.9 kW/cm² was about 1.03×10^{18} cm⁻³, which is higher than the Mott density given by Equation 7.2. Further increase in the excitation intensity gives rise to an increase in the EHP concentration, leading to a reduction in the ZnO bandgap energy as a result of bandgap renormalization [31]. Therefore, a continuous red-shift of the spectral peak associated with EHP from 3.16 to 3.12 eV was observed as the excitation intensity increased.

The inset of Figure 7.7 plots the curve of the PL intensity (obtained by the integration of the PL spectrum) versus the excitation intensity, showing a rapid increase as the excitation intensity is greater than a threshold value of 153 kW/cm². The significant increase in the PL intensity for the excitation intensity greater than the threshold and the appearance of the spectral peaks demonstrate the stimulated emission from the ZnO epilayer. It may be noted that such a stimulated emission was not observed in as-deposited ZnO thin films, indicating that the post-deposition RTA treatment greatly improves the crystal quality to achieve the low-threshold stimulated emission from ZnO. Since the diameter of the focused incident laser beam in the PL measurement was about 1 mm, which is much greater than the film thickness, it is likely that the stimulated emission results from the closed loop paths in the direction parallel to the film surface. The scattering of the photons propagating in the in-plane direction by the optical scattering centers and crystalline grains leads to the observation of stimulated emission in the direction perpendicular to the film surface [27]. It is worthy of mention that the threshold intensity for optically pumped stimulated emission of this ZnO epilayer is much smaller than that of GaN (800 kW/cm²) [32,33], suggesting high optical quality of the ZnO epilayer grown by ALD.

7.4 Low-Threshold Stimulated Emission in Highly Oriented ZnO Thin Films on Amorphous Glass Substrates [34]

Among various substrates on which ZnO thin films were deposited, glass substrates have been extensively used due to their high transparency and low costs. For example, depositing

ZnO on glass substrates as an active layer for transparent thin film transistors was widely studied. However, there are few study reports on ZnO photonic devices grown on glass substrates because of poor crystallinity due to lack of epitaxy of ZnO with amorphous glass substrates. It has been demonstrated in Ref. [34] that highly oriented ZnO thin films can be prepared on amorphous Corning 1737 glass substrate by ALD technique. Low-threshold stimulated emission was also observed in ZnO thin films, indicating good optical and crystal quality of the ALD-grown ZnO thin films on amorphous glass substrates.

Figure 7.8 shows a schema of the layer structure in ZnO thin films prepared on amorphous Corning 1737 glass substrate. A ZnO buffer layer was deposited on the glass substrate to facilitate growth of the ZnO main layer. The buffer layer was deposited by 500-cycle ALD at a high temperature of 300°C and treated by post-deposition RTA at 800°C for 15 min in N_2 ambient to improve crystal quality. Afterward, 1000 ALD cycles were applied to grow the ZnO main layer on the buffer layer at a low temperature of 150°C. The growth rate of ZnO at 150°C was about 0.2 nm for each ALD cycle. It is known that the temperature range between 120°C and 180°C is the so-called ALD window of ZnO [35], in which the deposition proceeds with self-limiting mode. As the temperature is above 200°C, precursor molecules are desorbed from the substrate surface and the growth rate becomes much less than one monolayer per ALD cycle. The ALD rate of ZnO dropped to about 0.14 nm/cycle at 250°C and 0.1 nm/cycle at 300°C.

Figure 7.9 shows the XRD patterns of the ZnO layers deposited on the Corning 1737 glass substrate by 500-cycle ALD at different temperatures. It reveals that the exclusively (0001) oriented growth of ZnO, which has the *c*-axis normal to the substrate surface, occurs at the growth temperature of 300°C. In the ZnO thin films deposited below 300°C, crystallites with other orientations also grow; the ($10\bar{1}1$)-oriented crystallites at 250°C and the ($10\bar{1}0$)-oriented crystallites at 175°C and 125°C, respectively. This result is almost consistent with those reported in the literature using (100)-oriented Si [36], SiO_2 on Si [37], and glass [35] as substrates.

Figure 7.10 displays the surface morphologies of these ZnO buffer layers observed by atomic force microscopy (AFM). Taking into account the aforementioned XRD measurement, the AFM images can be explained as follows. The surface energy of the ($10\bar{1}0$) plane is so low that the ZnO crystal generally forms in the hexagonal prism of the first order, elongating normal to the (0001) plane with the highest surface energy. Therefore, ZnO crystallites or embryos might have the form of a hexagonal prism. When the ALD temperature is as low as 125°C, ZnO nucleates and exposes the (0001) or ($10\bar{1}0$) faces easily

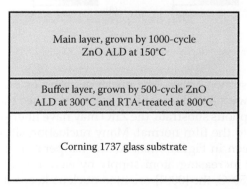

FIGURE 7.8
Schematic diagram of the ZnO layers grown by ALD on the Corning 1737 glass substrate.

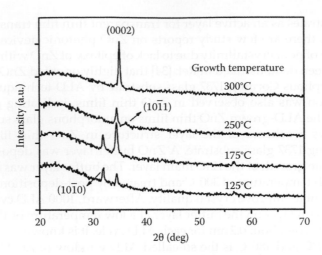

FIGURE 7.9
XRD patterns of the ZnO buffer layers deposited at the temperatures from 125°C to 300°C.

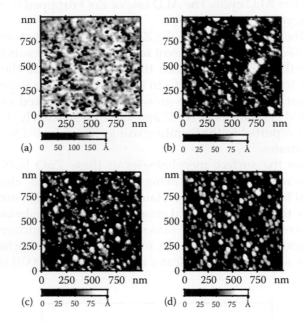

FIGURE 7.10
AFM images of the ZnO buffer layers deposited at (a) 125°C, (b) 175°C, (c) 250°C, and (d) 300°C.

contacting with the substrate surface. The nuclei or embryos can hardly develop due to low migration of atoms with low thermal kinetic energy. Since there is no constant lattice matching with the amorphous substrate, the ZnO may have fiber structures with the [0001] and [10$\bar{1}$0] axes parallel to the film normal. Many nucleation sites and less growth make the surface smooth as seen in Figure 7.10a. As the temperature rises to 175°C, the nuclei can grow as a result of increasing atom supply by easy migration to larger hexagonal crystallites. At around 250°C, the [0001]-oriented nucleus keeps its shape and the embryo with *c*-axis parallel to the substrate surface may grow to a crystallite (with the (10$\bar{1}$1)-orientation) like a leaning tower of hexagonal prism, through twining by the same growth mechanism as a ZnO tetrapod grows [38]. The desorption of precursors is effective at these temperatures. When the deposition temperature reaches 300°C, embryos easily nucleate and grow to the hexagonal prisms by abundant migration of adsorbed atoms with a high kinetic energy. The grain size could reach 50 nm as seen in Figure 7.10d. The grown particles are well-oriented with the *c*-axis normal to the substrate surface although they have the fiber structure as a whole.

Since both the AFM images and XRD patterns indicate that the ZnO grown at 300°C has a preferable crystalline quality, this deposition temperature was chosen to prepare the ZnO buffer layer on the Corning 1737 glass substrate. Afterward, the ZnO buffer layer was annealed at 500°C–800°C for 2 min in N_2 atmosphere by RTA to further improve crystallization. The upper limit of the annealing temperature, 800°C, was fixed taking into account the softening point of the Corning 1737 glass at 975°C. Figure 7.11 shows the XRD patterns of the as-deposited and post-annealed ZnO buffer layers at different RTA temperatures. All the as-deposited and post-annealed films present only the (0002) reflection at 34.7°, indicating the exclusive *c*-axis orientation of the buffer layers. The intensity of the (0002) peak apparently increases for the specimens treated by the post-deposition RTA as compared with the as-deposited film. The FWHM of the (0002) peak is 0.417° for the as-deposited film. The FWHM decreases with the annealing temperature, from 0.326° at 500°C to 0.222° at 800°C. The XRD measurement indicates that the post-deposition RTA treatment significantly improves the crystal quality of the ZnO buffer layer.

On the basis of the high (0001) orientation of the ZnO buffer layer treated by RTA at 800°C, the ZnO main layer was subsequently grown by 1000-cycle ALD at 150°C on this buffer layer. Figure 7.12 shows the spontaneous emission PL spectra of the ZnO main layer and buffer layer, where the spectrum of main layer was measured from the main/buffer/substrate structure, and the spectrum of buffer layer was measured from the buffer/substrate structure, respectively. Two spectral peaks at 432 and 463 nm in the PL spectrum of the ZnO buffer layer are attributed to light emission from the Corning 1737 glass, owing

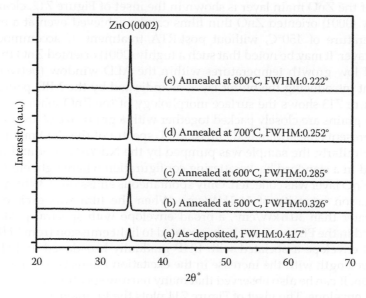

FIGURE 7.11
XRD patterns of the as-deposited and post-annealed ZnO buffer layers treated by RTA at the temperatures from 500°C to 800°C in N_2 ambient.

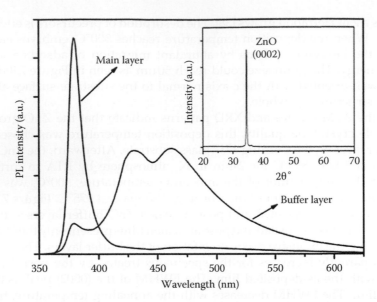

FIGURE 7.12
Room-temperature PL spectra of the ZnO buffer and main layers. Inset shows the corresponding XRD pattern of the ZnO main layer.

to penetration of the incident laser beam through the thin ZnO buffer layer into the glass substrate. As compared with the buffer layer, the PL of the main layer is composed of a dominant spectral peak corresponding to the ZnO NBE emission at 380 nm and a very weak defect-related band around 475 nm associated with the oxygen vacancy (V_O). The disappearance of the spectral peaks at 432 and 463 nm from the glass substrate may be due to optical absorption of the pumping laser beam by the thick ZnO main layer. The XRD pattern of the ZnO main layer is shown in the inset of Figure 7.12, clearly indicating that the highly (0001) oriented ZnO thin films can be achieved even at a extremely low growth temperature of 150°C, without post-RTA treatment, if accommodated with a proper buffer layer. It may be noted that such a highly (0001) oriented ZnO thin film *cannot* be obtained at low growth temperatures within the ALD window (between 120°C and 180°C), without introducing the buffer layer, as indicated by the XRD patterns shown in Figure 7.9. Figure 7.13 shows the surface morphology of the ZnO main layer. It is clearly seen that large grains are closely packed together with a grain size of around 100 nm.

Figure 7.14 presents the stimulated emission PL spectra of the ZnO main layer at room temperature. Similarly, the sample was pumped by the Nd-YAG laser and the PL spectra were measured in a standard backscattering configuration where light emission from top surface of the ZnO layer was collected. Only spontaneous emission at 380 nm was observed at a low excitation intensity of 14.2 kW/cm². When the film was excited at excitation intensities greater than 30.6 kW/cm², a broad envelope with spectral peak around 391–395 nm emerged in the PL spectra, which is related to light emission from EHP in ZnO. The broad spectral envelope associated with EHP gradually broadened and shifted toward the longer wavelength with the increase in the excitation intensity as a result of bandgap renormalization. It can be also observed that many narrow spectral peaks appeared on the broad spectral envelope. The inset of Figure 7.14 plots the PL intensity as a function of the excitation intensity, revealing a rapid increase in the PL intensity as the excitation intensity exceeded a threshold value. The presence of the broad spectral envelope together with

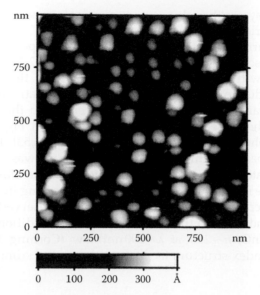

FIGURE 7.13
AFM image of the ZnO main layer.

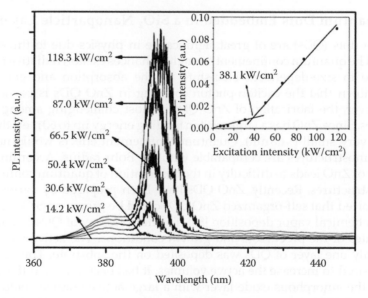

FIGURE 7.14
Room-temperature PL spectra of the ZnO main layer pumped by a pulsed Nd:YAG laser at various excitation intensities. Inset show the PL intensity as a function of the excitation intensity.

the narrow spectral peaks and significant increase of the PL intensity for the excitation intensity greater than the threshold indicate that the onset of stimulated emission takes place at a threshold intensity as low as $38.1\,kW/cm^2$ in the ZnO main layer [14,39–51]. It is worth noting that the stimulated emission *cannot* be observed in the ZnO layer deposited by ALD at 150°C *directly* on the glass substrate without introducing the buffer layer. Hence the ZnO thin film of a high optical quality was fabricated on the amorphous glass substrate by a trick of the buffer layer, although it was not of a single crystal but had a fiber structure composed of highly *c*-oriented columnar crystallites.

The origin of the narrow spectral peaks on the broad spectral envelope shown in Figure 7.14 may be ascribed to random lasing action [52,53]. Unlike conventional semiconductor lasers with well-defined resonant cavities, the random-laser cavities were self-constructed due to the constructive closed-loop paths *via* multiple scattering between crystalline grains and optical scattering centers in the ZnO main layer. The boundaries between the close-packed grains, as shown in Figure 7.13, may account for the scattering of photons to form the closed loops [52,53]. Because the diameter of the incident laser beam in this measurement (~1 mm) was much larger than the film thickness, stimulated emission may originate from the closed-loop paths in the direction parallel to the film surface. The photons propagating in the in-plane direction were scattered by the crystalline grains, leading to the observed stimulated emission in the direction perpendicular to the film surface [27]. In addition, a planar waveguide with low (air)-high (ZnO, $n \sim 2.37$ at $\lambda \sim 390$ nm)-low (Corning 1737 glass, $n \sim 1.53$ at $\lambda \sim 390$ nm) refractive index structure facilitates the onset of stimulated emission in the ZnO thin film.

7.5 ZnO Quantum Dots Embedded in a SiO$_2$ Nanoparticle Layer

ZnO quantum dots (QDs) are of great significance in physics due to three-dimensional confinement. The quantum confinement strongly enhances exciton radiative recombination and gives rise to size-dependent blue-shifts in the absorption and emission spectra. It has been shown that the exciton-photon coupling in ZnO QDs is particularly strong [55–57]. However, the fabrication of ZnO QDs is most challenging among various ZnO nanostructures. Since ZnO has a large exciton binding energy up to 60 meV, the Bohr radius of excitons is very small (~1.8 nm). Quantum confinement effects would not occur until the sizes of nanostructures are comparable with the Bohr radius. Accordingly, the small exciton radius of ZnO leads to difficulty in the observation of quantum confinement effects in ZnO nanostructures. Recently, ZnO QDs have been prepared by various techniques. Kim et al. reported that self-organized ZnO QDs could be grown on SiO$_2$/Si substrates by metalorganic chemical vapor deposition [58]. The morphology of QDs strongly depended on the flow rates of the precursors and growth conditions. The active volume was quite small since only one layer of QDs was deposited on the substrate. Thus the multilayers of QDs are desired to increase the active volumes. It has been reported that the ZnO QDs embedded in the amorphous oxide layer with a large active volume could be fabricated [59,60]. The thin ZnO film was grown on the SiO$_2$ surface and followed by high-temperature thermal treatment. ZnO was then diffused into the SiO$_2$ layer and formed small and homogenous QDs embedded in the oxide films.

A new approach has been demonstrated to fabricate the ZnO QDs embedded in amorphous silicon oxide, as schematically shown in Figure 7.15 [54]. SiO$_2$ nanoparticles with diameters about 10 nm were initially dispersed in an isopropyl alcohol solution. By carefully controlling the concentrations of SiO$_2$ nanoparticles in the solution and the spin coating condition, the oxide layer with a thickness in the range from 30 to 300 nm was easily and quickly deposited on the Si substrate. Figure 7.16 shows the surface morphology of the SiO$_2$ nanoparticle layer coated on the Si wafer. It is seen that the SiO$_2$ particles were

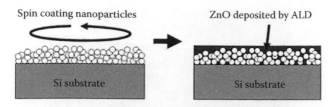

FIGURE 7.15
Schema of the fabrication of ZnO quantum dots. ZnO was deposited in the voids between the SiO$_2$ nanoparticles.

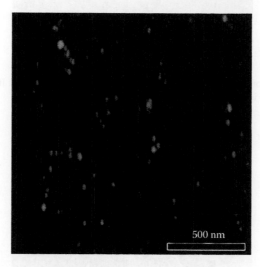

FIGURE 7.16
AFM image of the surface of SiO$_2$ nanoparticles layer.

stacked closely with diameters about 8–10 nm and very small voids between the particles. Afterward, 67 ALD cycles were applied to deposit ZnO at a temperature of 180°C. The multiple-pulses scheme, i.e., one ALD cycle consisting of 30 H$_2$O pulses followed by 30 DEZn pulses, were used to increase the infiltration of the precursors into the small voids between the SiO$_2$ nanoparticles. Thus the precursors can infiltrate into the small voids between the SiO$_2$ nanoparticles to produce ZnO QDs. In order to improve the crystallinity of the ZnO QDs, the sample was subsequently treated by a RTA for 7 min in an N$_2$ ambience at 700°C.

Figure 7.17 shows a cross-sectional TEM image of the SiO$_2$ layer embedded with ZnO. It is seen that the thickness of this layer is about 150 nm. The bright thin layer (~7 nm thick) covering the top of SiO$_2$ layer was confirmed by X-ray analysis to be Zn$_2$SiO$_4$ formed by inter-diffusion of atoms between the SiO$_2$ and ZnO during the RTA treatment. Fringes appearing in a particle in a high-resolution TEM image are reproduced in Figure 7.18. The FFT image or computer diffractogram of the fringe (inset in Figure 7.18) identifies the particle as a ZnO crystal (wurtzite-type) with the [2$\bar{1}$10] axis parallel to the incident electron beam. These crystalline ZnO nanoparticles can be regarded as QDs, as described in the following PL measurement. Figure 7.19 reveals that ZnO QDs were formed uniformly in the voids between the SiO$_2$ nanoparticles. It may be noted that the ZnO QDs can be embedded at the bottom of the film, indicating that the precursors can penetrate through the

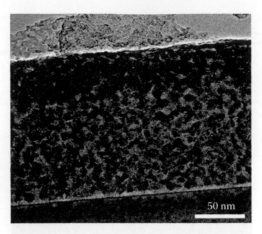

FIGURE 7.17
Cross-sectional TEM image of the SiO_2 layer embedded with ZnO quantum dots. It is epoxy resin used for the specimen preparation which is seen on the top of the SiO_2 layer.

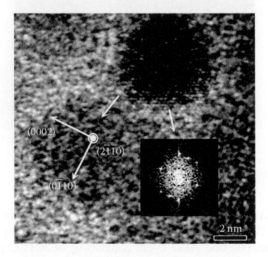

FIGURE 7.18
HRTEM image of the ZnO quantum dots embedded in the SiO_2 layer.

thick, stacked SiO_2-nanoparticle layer to fill the small voids near the substrate. Since the ZnO particles appear with darker contrast in Figures 7.18 and 7.19, we can estimate roughly the size of the QDs to be about 3–8 nm.

The PL spectrum of the ZnO QDs excited by the Nd-YAG laser was measured to investigate the quantum confinement effect. Figure 7.20 shows the room-temperature PL spectrum of the ZnO QDs. This figure also includes the PL spectrum of a ZnO thin film for comparison. The ZnO thin film was grown with 200 nm in thickness on a c-Al_2O_3 substrate by ALD and followed by post-deposition RTA at 950°C for 5 min in N_2 atmosphere. The PL spectrum of the ZnO thin film is dominated by the NBE emission at 3.26 eV. As for the ZnO QDs, the PL spectrum exhibits a spectral peak around 3.34 eV, indicating a remarkable blue-shift about 80 meV compared to the PL spectrum of the ZnO thin film. The blue-shift of the PL spectrum is attributable to the quantum confinement effect in the ZnO dots, which can be described as [59–61].

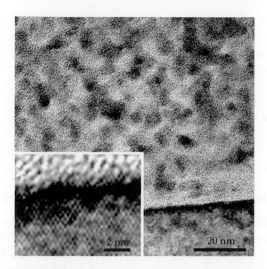

FIGURE 7.19
Cross-sectional HRTEM TEM image of the SiO$_2$ layer embedded with ZnO quantum dots. Inset is an enlarged image of the interface between the Si substrate (where the lattice fringes are seen) and the SiO$_2$ layer.

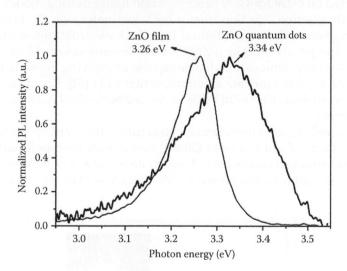

FIGURE 7.20
Normalized PL spectra of ZnO quantum dots embedded in the SiO layer and the ZnO thin film at room temperature.

$$\Delta E = \frac{\pi^2 \hbar^2}{2R^2}\left(\frac{1}{m_e^*} + \frac{1}{m_h^*}\right) - 0.248 E_{Ryd}^* \qquad (7.3a)$$

where

$\hbar$ and R are reduced Plank's constant and the radius of ZnO QDs
E_{Ryd}^* is the exciton binding energy with a value of 60 meV for ZnO

Taking $m_e^* = 0.24 m_0$ and $m_h^* = 1.8 m_0$ for electron and hole effective masses in ZnO [59,60], and using $\Delta E = 80$ meV according to the blue-shift of the PL spectrum, the radius of ZnO QDs is calculated to be about 4.3 nm, which is somewhat larger than the dot diameters (3–8 nm) observed from the TEM images shown in Figures 7.17 through 7.19. This difference is attributed to the assumption of the infinite potential barrier in the theoretical calculation. Actually, the finite potential barrier of the surrounding SiO_2 reduces the quantum confinement effect. It may be noted that the FWHM of the PL spectrum of ZnO QDs (~210 meV) is much larger than that of ZnO thin film (~110 meV). This discrepancy may be due to the non-uniform sizes of the ZnO QDs embedded in the SiO_2 layer.

7.6 UV Electroluminescence from *n*-ZnO/*p*-GaN Heterojunction LEDs [62]

Although ZnO technology has considerably been progressing, it is still difficult to prepare reproducible and stable *p*-type ZnO thin films with high carrier concentration and low resistivity. As a result, growing high-quality *n*-type ZnO thin films on good lattice-matching *p*-type materials is an alternative approach to fabricating the ZnO-based LEDs. Since ZnO and GaN have the same crystal structure (wurtzite) with a small in-plane lattice mismatch (1.8%), *p*-type GaN would substitute for *p*-ZnO. Actually, several investigators have been reported on *n*-ZnO/*p*-GaN heterojunction light-emitting diodes (LEDs) [63–68]. However, since the electron injection from *n*-ZnO dominates over the hole injection from *p*-GaN due to higher carrier concentration in *n*-ZnO, electroluminescence (EL) usually originates from the *p*-GaN side in *n*-ZnO/*p*-GaN heterojunction LEDs [63,65–67]. The ALD technique has been demonstrated to be capable of growing high-quality *n*-type ZnO epilayer on *p*-type GaN to fabricate a heterojunction LED [62]. Room temperature EL at 391 nm was observed from the *n*-ZnO side in the *n*-ZnO/*p*-GaN heterojunction LED at a low injection current.

The inset in Figure 7.21 shows the schematic structure of the *n*-ZnO/*p*-GaN heterojunction LED [62]. First, a thin GaN buffer layer (30 nm) and a thick undoped-GaN layer (2.5 μm) were successively deposited on the *c*-Al_2O_3 substrate by MOCVD. Next, a Mg-doped GaN (GaN:Mg) layer (300 nm) was also grown by MOCVD. The Hall effect measurement with

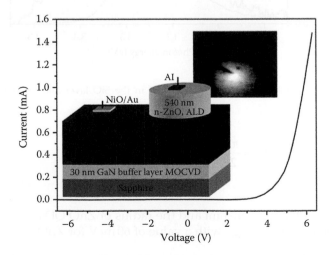

FIGURE 7.21
I–V curve of the *n*-ZnO/*p*-GaN heterojunction LED. The insets show the device structure and the EL image.

the van der Pauw configuration reveals that the hole concentration and mobility in this *p*-type GaN:Mg layer are approximately 2.6×10^{17} cm^{-3} and 11.9 cm^2/(V s), respectively. Afterward, a 540 nm thick ZnO layer was grown at 180°C on *p*-GaN by ALD. After the deposition, the ZnO layer was treated at 950°C for 5 min in N$_2$ ambient by RTA. The electron concentration and mobility in this annealed un-doped ZnO layer are 2×10^{18} cm^{-3} and 22.1 cm^2/(V s), showing intrinsic *n*-type conductivity. A circular ZnO mesa with 1 mm diameter was fabricated by masking the surface and chemically etching ZnO away using a diluted HCl solution. The Ohmic contact to *p*-GaN was made by thermal evaporation of Ni/Au, followed by annealing at 500°C for 30 min in O$_2$ ambient. Finally an Al layer with radius of 100 μm was deposited on top of the *n*-ZnO layer to form the *n*-Ohmic contact.

Figure 7.21 shows the current versus voltage (*I–V*) characteristics of the *n*-ZnO/*p*-GaN heterojunction LED at room temperature. A rectifying diode-like behavior is clearly observed. In Figure 7.22 the normalized PL spectra of the *n*-ZnO and *p*-GaN layers under the same excitation intensity are shown. The PL spectrum of *n*-ZnO comprises a strong excitonic NBE emission at 378 nm and negligible defect-related bands in the visible region. In contrast, the PL spectrum of *p*-GaN reveals a deep-level emission around 430 nm, which can be ascribed to the transitions from the conduction band or unidentified shallow donors to the Mg acceptor levels [69]. The NBE emission from *p*-GaN was not observed. The PL intensity of *n*-ZnO is significantly (21 times) greater than that of *p*-GaN layer, indicating higher light emission efficiency in ZnO due to its larger excitonic binding energy.

Figure 7.23 shows the room-temperature EL spectra of the *n*-ZnO/*p*-GaN heterojunction LED under DC forward bias. The EL image of this LED is shown in the inset of Figure 7.21. At a low injection current of 2.5 mA, the spectral peak of the EL is located at 425 nm. Emission at 391 nm apparently increases with the injection current and becomes dominant when the current exceeds 12.5 mA. A comparison between EL and PL spectra reveals that the spectral peak around 425 nm is attributed to light emission from the Mg acceptor levels in *p*-GaN, and the EL at 391 nm originates from the *n*-ZnO layer. Unlike the *n*-ZnO/*p*-GaN heterojunction LEDs in [63,65–67], where the EL originated only from either the *n*-ZnO or *p*-GaN side, the EL spectra (Figure 7.23) show the superposition

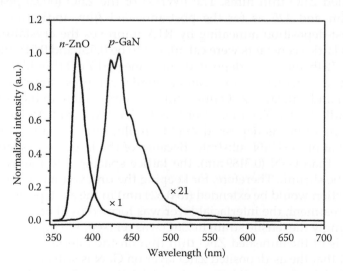

FIGURE 7.22
Room-temperature PL spectra of the *p*-GaN layer and the *n*-ZnO layer grown on the *p*-GaN.

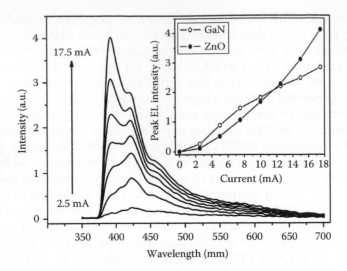

FIGURE 7.23
Room-temperature EL spectra of the *n*-ZnO/*p*-GaN heterojunction LED at various DC injection currents. The inset shows the EL intensities of the *n*-ZnO and *p*-GaN versus the injection current.

of light emissions form both the *n*-ZnO and *p*-GaN layers. Inset of Figure 7.23 depicts the intensities of EL spectral peak corresponding to the *n*-ZnO ($\lambda = 391$ nm) and *p*-GaN ($\lambda = 425$ nm) as a function of the injection current. Competition between the emissions at 391 and 425 nm might result from the differences in the carrier concentration and light emission efficiency in *n*-ZnO and *p*-GaN, as well as the ZnO/GaN interface states, which will be discussed later.

Figure 7.24a shows the XRD patterns of the as-deposited and post-annealed ZnO thin films grown by ALD on *p*-GaN. The intensity of the ZnO (0002) XRD peak for the post-annealed ZnO thin film is four times larger than that for the as-deposited film. Figure 7.24b displays the X-ray rocking curve (XRC) curves of the as-deposited and post-annealed ZnO thin films. The FWHM of the ZnO (0002) peak is 680 as for as-deposited film and 425 as for the post-annealed film, indicating that the high-temperature post-deposition annealing by RTA improves the crystallinity. The corresponding *c*-axis lattice constants were calculated to be 0.524 and 0.518 nm from the XRC peaks in Figure 7.24b, for the as-deposited and annealed ZnO thin films, respectively. It seems to indicate a larger tensile strain and a smaller compress strain along the *c*-axis in the as-deposited and annealed ZnO thin films, as compared with the lattice constant in *c*-axis of pure bulk ZnO (0.5206 nm, according to ASTEM card CAS Number 1314-13-2). The lattice spacing of the as-deposited ZnO thin film might be almost the same in the basal plane as that of the GaN substrate. Because of the larger *a*-axis lattice constant of ZnO (0.3252 nm) than GaN (0.3189 nm), the lattice spacing *a* in the as-deposited ZnO thin film might be shrunk. Therefore, for keeping the unit volume of the ZnO the lattice along the *c*-direction would be extended (to 0.524 nm) in the as-deposited ZnO thin film. After the RTA treatment, the interfacial layer was formed (will be shown in the following) and accordingly the tensile strain in ZnO was relaxed. We think the intrinsic lattice constant in *c*-axis of the annealed ZnO thin film on the interfacial layer is 0.518 nm. It should be noted that the as-deposited ZnO layer on GaN is still crystalline and preferably (0001)-oriented due to a small lattice mismatch between ZnO and GaN, which results in almost a perfect crystal by coalescence during the RTA treatment.

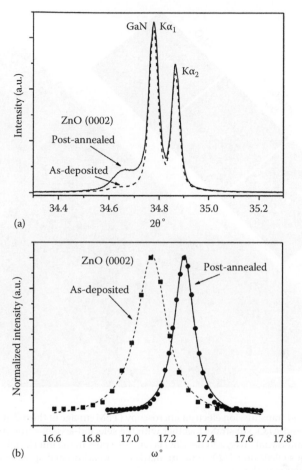

FIGURE 7.24

(a) XRD patterns and (b) XRC of the as-deposited and post-annealed ZnO films grown by ALD on *p*-GaN.

Figure 7.25 shows a cross-sectional TEM image of the *n*-ZnO/*p*-GaN heterojunction LED on the *c*-Al$_2$O$_3$ substrate. The inset selected-area electron diffraction patterns reveal that the ZnO grows in completely parallel orientation, $[0001][11\bar{2}0]_{ZnO}//[0001][11\bar{2}0]_{GaN}$, with the GaN, which was deposited with a good epitaxial relation, $[0001][11\bar{2}0]_{GaN}//[0001][1\bar{1}00]_{Al2O3}$, on the *c*-Al$_2O_3$ substrate. The *n*-ZnO, *p*-GaN and underlying non-doped GaN layers are a little different in contrast and can been identified in the image, accordingly. The threading dislocations, which were generated on the interface between the GaN buffer layer and the *c*-Al$_2$O$_3$ substrate, run along the *c*-axis of the GaN layers. Some of the dislocations still penetrate through the ZnO layer and reach the vacuum. The others however disappear at the interface between the *n*-ZnO and *p*-GaN or at the interface between the *p*-GaN and undoped GaN layers. Hence, the TEM image reveals that the ZnO layer is almost a perfect single crystal with very few survival threading dislocations, which can be ascribed to the layer-by-layer growth of ALD and the recrystallization during the post-deposition RTA treatment. Therefore, the crystal quality of the ZnO epilayer seems much superior to that of ZnO layers prepared by other techniques such as electron cyclotron resonance-assisted MBE [5,70]. The formation of high-quality crystalline ZnO is surely one of requirements for efficient UV EL from ZnO.

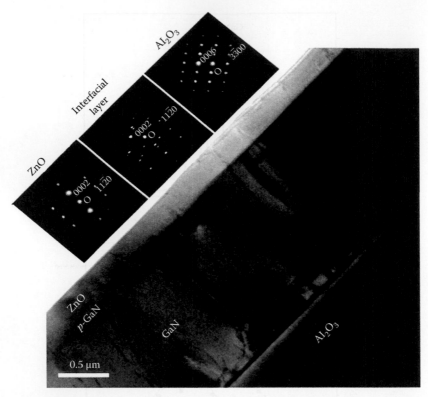

FIGURE 7.25
Cross-sectional bright-field transmission electron microscopy (TEM) image of the n-ZnO/p-GaN heterojunction on the underlying GaN and Al_2O_3 substrate, together with the selected-area electron diffraction patterns from the Al_2O_3 substrate, the interfacial layer, and the ZnO layer. The spots in the interfacial layer comprise the spots from the ZnO layer (whose 0002 and 11$\bar{2}$0 spots are marked) and additional spots from an unknown crystal formed with an orientation with the c-axis.

The details of the interface between the p-ZnO and n-GaN layers were measured to deliberate on the EL mechanism. Figure 7.26a shows the dislocations were annihilated at the interface. Furthermore, it indicates the presence of an interfacial layer along the (0001) interface between the p-ZnO and n-GaN layers. The thickness of the interfacial layer is about 4–5 nm as seen in HRTEM images in Figure 7.26b and c. The interface is composed of crystals apparently different from the adjoining GaN and ZnO crystals where well-ordered lattice fringes, particularly of their (0001) planes, are seen. This can be seen in the electron diffraction pattern from the interfacial layer as shown in Figure 7.25. As compared with the diffraction pattern from ZnO it includes additional spots from an unknown crystal formed with an orientation with the c-axis. We have not yet characterized the phase in the interfacial layer. Since the ALD can make an abrupt interface over a large area, the interfacial layer might be formed rather during the post-deposition RTA treatment than by ALD. Such high-temperature annealing may lead to the diffusion of the Mg atoms in the p-GaN layer toward the interface between n-ZnO and p-GaN, and consequently the interfacial layer between ZnO and GaN was formed. Such diffusion cannot proceed uniformly over the interface as is usual with a heat treatment. An evidence of diffusion of Mg atoms can be seen as bright contrast in the p-GaN layer in Figure 7.27, which is a high-angle annular dark-field

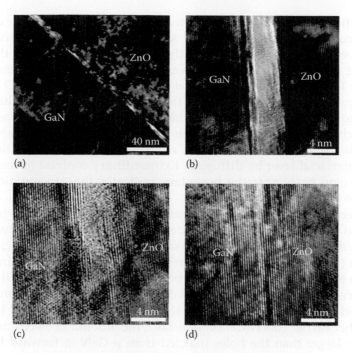

(a) (b)

(c) (d)

FIGURE 7.26
(a) Cross-sectional TEM image of an area in the *n*-ZnO/*p*-GaN heterojunction. (b through d) HRTEM images of the interfacial layer between ZnO and GaN.

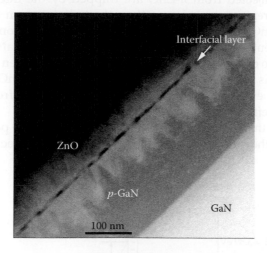

FIGURE 7.27
High-angle annular dark-field scanning transmission electron microscopy image of the *n*-ZnO/*p*-GaN heterojunction on the GaN layer.

(HAADF) scanning transmission electron microscopy (STEM) image of the *n*-ZnO/*p*-GaN on the GaN layer. The HAADF-STEM images are mainly formed by thermal diffuse scattering of electrons or incoherent imaging of elastically scattered electrons and provide the Z-contrast, approximately proportional to the square of the atomic number [70,71]. Therefore, contrast of the *p*-GaN darker than the underlying Ga[31]N[7] is ascribed to the doped Mg[12] atoms, and the parts exhibiting brighter contrast in the *p*-GaN layer are poor in concentration of Mg atoms. However, the brighter contrast was not observed in the lower portion of the *p*-GaN layer, indicating the preferential diffusion of Mg atoms toward ZnO, not toward the underlying GaN. This result may be ascribed to the formation of a new phase at the interface between ZnO and GaN. The phase is a compound including Mg atoms, though its crystal structure has not yet been established. The demand of Mg atoms by the new phase gives rise to the supply of Mg from *p*-GaN, which is greater than the demand by the underlying GaN layer. Figure 7.26d, in which the ZnO and GaN layers contact almost directly, also confirms the formation of interfacial layer by diffusion. Extraordinary contrast of the (0001) fringes appearing in some parts near the boundary indicates the precipitation of diffused atoms along the (0001) plane.

The band diagram of the *n*-ZnO/*p*-GaN heterojunction including this interfacial layer was constructed using the Anderson model, as illustrated in Figure 7.28, which is similar to one presented by Lee and Kim [72]. As shown in Figure 7.26a and c, the interfacial layer is so thin (4–5 nm) as to allow the carriers to tunnel through, as indicated by the forward bias in the *I–V* curve. Competition between light emissions at 391 and 425 nm as shown in Figure 7.23 might result from the differences in the carrier concentration and light emission efficiency in *n*-ZnO and *p*-GaN, as well as the ZnO/GaN interface states. Because of higher carrier concentration in *n*-ZnO, the amount of electrons injected from *n*-ZnO is much larger than the holes injected from *p*-GaN at forward bias [63,65–67]. When the carriers transport across the interfacial layer, they may be trapped at the ZnO/GaN interface states [72]. At a low injection current of 2.5 mA, the interface states may capture a large amount of the holes injected from *p*-GaN. However, only a small portion of electrons injected from *n*-ZnO are trapped by the interface states since the electron injection current is much larger than the hole injection current. Accordingly, light emission from ZnO is not obvious and the EL from GaN dominates the spectrum. As the injection current increases, the holes injected from *p*-GaN not only fill up the interface states but also enter into the *n*-ZnO layer. As a result, even although the electron injection into *p*-GaN is higher than the hole injection into *n*-ZnO, the EL form *n*-ZnO increases rapidly and eventually dominates over the emission from *p*-GaN due to the higher light emission efficiency in ZnO.

On the other hand, the EL peak at 391 nm longer than the PL peak of ZnO at 378 nm can be attributed to the presence of the interfacial layer between *n*-ZnO and *p*-GaN.

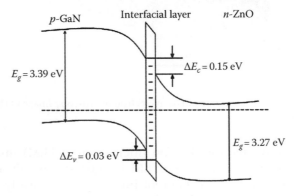

FIGURE 7.28
Band diagram of the *n*-ZnO/*p*-GaN heterojunction without applying voltage.

The electrons and holes may accumulate in ZnO and GaN near the interface, respectively, due to the wide bandgap energy of the interfacial layer. The accumulated electrons might result in a slight decrease in the ZnO bandgap energy due to bandgap renormalization. These electrons recombine with the holes supplied from p-GaN to produce the EL from ZnO. Because of the valence band offset ΔE_v (shown schematically in Figure 7.28) as well as the bandgap renormalization, the photon energy of EL peak is smaller than the ZnO bandgap. Actually, a similar phenomenon with an EL peak at 394 nm has been observed from the n-ZnO/p-GaN heterojunction LED with a thin SiO_2 layer inserted between ZnO and GaN [73]. The UV EL from ZnO at a DC low injection current about 10 mA indicates a high optical quality of the ZnO epilayer grown by AKD and treated by the post-deposition RTA process.

7.7 UV Electroluminescence from n-ZnO/i-ZnO/p-SiC Heterojunction LEDs [74]

In addition to p-type GaN, commercially available 4H-SiC is also a good p-type substrate for ZnO-based heterojunction LEDs because of the same crystal structure (wurtzite) and a small in-plane lattice mismatch (~5%) with ZnO [75]. However, since the electron injection from n-ZnO dominates over the hole injection from p-SiC due to the smaller energy barrier for electron, the EL usually comes from the defect-related states in p-SiC in the n-ZnO/p-SiC heterojunction LEDs [76,77]. The ALD technique has been applied to grow high-quality ZnO thin films on p-type 4H-SiC substrates to fabricate n-ZnO/p-SiC and n-ZnO/i-ZnO/p-SiC heterojunction LEDs [74]. It has been demonstrated that inserting an undoped i-ZnO layer between n-ZnO and p-SiC leads to the hole and electron injection from p-SiC and n-ZnO into i-ZnO layer. Significant UV EL at 393 nm from the n-ZnO/i-ZnO/p-SiC heterojunction LED has been accomplished at room temperature.

Figure 7.29 shows schematic structures of n-ZnO/i-ZnO/p-SiC and n-ZnO/p-SiC heterojunction LEDs. Commercially available p-type 4H-SiC with a hole concentration of 6×10^{18} cm^{-3} was used as the substrate and hole injection layer. For the n-ZnO/i-ZnO/p-SiC structure shown in Figure 7.29a, a 25 nm thick i-ZnO layer was grown on the p-SiC substrate using the ALD technique. The substrate temperature was 300°C during the growth of this

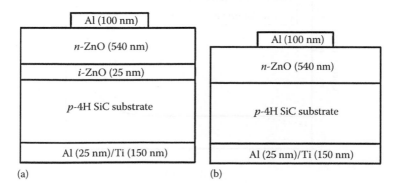

FIGURE 7.29
Schematic diagrams of (a) the n-ZnO/i-ZnO/p-SiC and (b) n-ZnO/p-SiC heterojunction LEDs.

i-ZnO layer. Afterward, an *n*-type Al-doped ZnO (ZnO:Al) layer with thickness about 540 nm was deposited on the *i*-ZnO layer at 180°C. Al(CH$_3$)$_3$ (Trimethylaluminum, TMA) served as the precursor for the Al-doping. The Al-doping percentage was 2%, i.e., one Al-doping ALD cycle was followed by 49 ZnO ALD cycles, and the Al-doping ALD cycles were uniformly distributed over all the ALD cycles. After ZnO deposition, the back of the SiC substrate was deposited with 25 nm of Al, followed by 150 nm of Ti. The sample was then annealed at 850°C for 5 min in N$_2$ atmosphere by RTA. This RTA treatment not only improved the Ohmic contacts between the *p*-SiC and Al/Ti alloy, but also enhanced the crystal quality of ZnO. The XRD pattern of the ZnO layer reveals a sharp ZnO (0002) peak at 34.7°, indicating that the ZnO layer has high crystallinity with *c*-axis orientation. A circular ZnO mesa with 1 mm in diameter was fabricated by masking the surface and chemically etching away the ZnO using a dilute HCl solution. Finally, a 100 nm thick Al layer with a radius of 100 μm was thermally evaporated on top of *n*-ZnO for the *n*-type Ohmic contact. The *n*-ZnO/*p*-SiC heterojunction LEDs were also fabricated for comparison, and all the fabrication conditions were the same as those of the *n*-ZnO/*i*-ZnO/*p*-SiC structure except that no *i*-ZnO layer was grown, as shown in Figure 7.29b.

Figure 7.30 shows the *I–V* characteristics of *n*-ZnO/*i*-ZnO/*p*-SiC and *n*-ZnO/*p*-SiC heterojunction LEDs at room temperature. Both devices exhibited rectifying, diode-like behavior. Clearly, the undoped *i*-ZnO layer in the *n*-ZnO/*i*-ZnO/*p*-SiC structure results in a smaller forward-bias injection current and a smaller reverse-bias leakage current than the *n*-ZnO/*p*-SiC LED.

Figure 7.31 shows the PL spectra of the *n*-ZnO, *i*-ZnO layers and *p*-SiC substrate, where the PL intensities were normalized to that of *n*-ZnO. The samples were put on the same holder without changing any set up of the optical configuration, i.e., the PL spectra were measured under the same optical alignment and excitation intensity. It is clearly shown that the PL intensities of *n*-ZnO and *i*-ZnO layers are much greater than that of the SiC substrate due to direct bandgap of ZnO. The lower PL intensity of *i*-ZnO than that of the *n*-ZnO layer is ascribed to the smaller thickness of *i*-ZnO. The PL spectra of *n*-ZnO and *i*-ZnO layers are similar, and consist of a strong NBE UV emission associated with the radiative recombination of free excitons. The defect-related emission in ZnO is negligible, indicating that the ZnO thin films grown by ALD have a good optical quality. As compared with *i*-ZnO layer, the PL spectrum of *n*-ZnO shows a slight blue-shift related

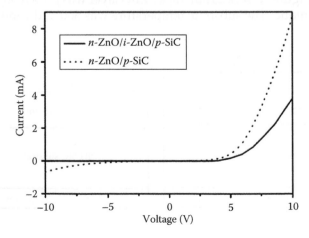

FIGURE 7.30
I–V characteristics of the *n*-ZnO/*i*-ZnO/*p*-SiC and *n*-ZnO/*p*-SiC heterojunction LEDs at room-temperature.

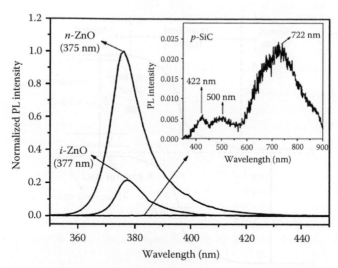

FIGURE 7.31
Room-temperature PL spectra of the *n*-ZnO, *i*-ZnO, and *p*-SiC.

to the well-known Burstein-Moss effect [78]. As shown in the inset of Figure 7.31, the PL spectrum of *p*-SiC comprises a blue luminescence at 422 nm, a green luminescence near 500 nm, and a broad deep-level band around 722 nm. The blue and green luminescence are considered to originate from radiative recombination of donor-acceptor pairs (DAP) associated with the shallow nitrogen donors, aluminum acceptors, and deep boron acceptors in 4H-SiC, respectively [79]. The broad defect-related emission around 722 nm may come from the transition between conduction band and the deep levels associated with the vacancies in 4H-SiC [80,81].

Figure 7.32 shows the EL spectra of the *n*-ZnO/*p*-SiC heterojunction LED at various injection currents. The devices were electrically pumped by pulsed currents from 300 to 600 mA with a pulse duration of 1 ms and a repetition rate of 64 Hz. Comparing these

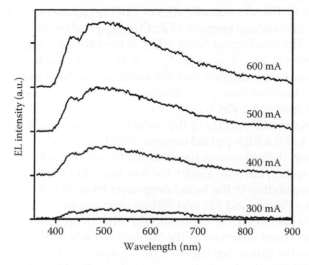

FIGURE 7.32
Room-temperature EL spectra of the *n*-ZnO/*p*-SiC heterojunction LED at various injection currents.

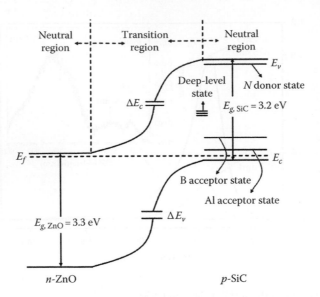

FIGURE 7.33
Band diagram of n-ZnO/p-SiC heterojunction at thermal equilibrium.

EL spectra with the PL spectra shown in Figure 7.31 indicates that the DAP emissions in p-SiC are responsible for these EL spectra, and the light emission from n-ZnO/p-SiC heterojunction mainly originates from the p-SiC side. Figure 7.33 shows that the band diagram of the n-ZnO/p-SiC(4H) heterojunction at thermal equilibrium, which was constructed using the Anderson model and has been presented in [77]. The Al and B acceptor levels and deep-level states related to the vacancies in 4H-SiC, together with the Fermi level, are also indicated in Figure 7.33. Hall effect measurement revealed that the carrier concentrations in n-ZnO and p-SiC are 2×10^{18} cm^{-3} and 6×10^{18} cm^{-3}, respectively, indicating that the separation between conduction band and Fermi level in n-ZnO is about 25 meV [29,82], and the difference between Fermi level and valence band in p-SiC is around 40 meV [83,84]. The electron affinities of ZnO (χ_{ZnO}) and 4H-SiC (χ_{SiC}) are 4.35 and 4.05 eV [85,86], and the bandgap energies of ZnO ($E_{g,ZnO}$) and SiC ($E_{g,SiC}$) are taken as 3.3 and 3.2 eV, respectively. The conduction band offset is therefore $\Delta E_c = \chi_{ZnO} - \chi_{SiC} = 0.3$ eV, while the valence band offset is $\Delta E_v = E_{g,ZnO} + \Delta \tilde{E}_c E_{g,SiC} = 0.4$ eV. The smaller energy barrier ΔE_c for electrons than ΔE_v for holes suggests that the electron injection from n-ZnO into p-SiC is greater than the hole injection from p-SiC into n-ZnO [76,77]. As a result, dominant EL from p-SiC was observed from the n-ZnO/p-SiC heterojunction LED.

Figure 7.34 displays the EL spectra of the n-ZnO/p-SiC heterojunction LED under a low DC current of 25 mA and a high pulsed current of 600 mA, together with the PL spectrum of p-SiC. Considerable difference can be observed between the EL spectra at the low DC and high pulsed injection current. Under the low injection current, the device exhibited light emission corresponding to the broad deep-level band in p-SiC around 722 nm, while the EL from DAP in p-SiC around 422 and 500 nm dominated at the high injection current. Actually, competition between the EL from DAP and deep-level states at high and low injection currents had been observed in the SiC homojunction devices [87].

In order to explain the difference between the PL spectrum of p-SiC and the EL spectra of the n-ZnO/p-SiC LED, the band diagram of the n-ZnO/p-SiC heterojunction under forward bias is shown in Figure 7.35. Under low injection, it is known that the forward-bias

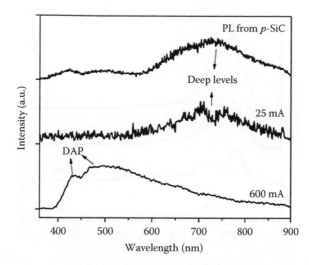

FIGURE 7.34

PL spectrum of the *p*-SiC substrate, EL spectrum of the *n*-ZnO/*p*-SiC heterojunction under a low DC current of 25 mA, and EL spectrum of the *n*-ZnO/*p*-SiC heterojunction under a high pulsed current of 600 mA.

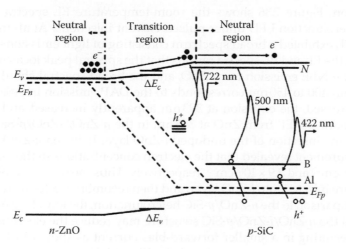

FIGURE 7.35

Band diagram of *n*-ZnO/*p*-SiC heterojunction under forward bias.

current in a *pn* junction is dominated by the carrier recombination in the depletion region. As shown in Figure 7.35, the Fermi-level in the depletion region is close to the middle of bandgap, and thus the deep-level states in *p*-SiC are near the Fermi level. It is also well known that carrier capture and emission occur primarily in the states close to the Fermi level, indicating that those states near the Fermi level constitute the most active recombination centers [88–90]. Therefore, electrons make the transition to the deep-level states, producing the light emission around 722 nm under low injection current. As the injection current increases, carrier diffusion across the junction and recombination in the neutral region dominate the forward-bias current. Since the Fermi level in the neutral region of *p*-SiC is close to the valence band, carrier recombination associated with the Al and B acceptor levels prevails over that related to the deep-level states, thus resulting in the dominant EL from DAP at 422 and 500 nm under high injection current.

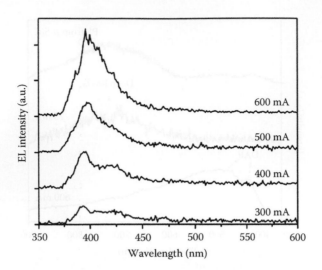

FIGURE 7.36
Room-temperature EL spectra of the n-ZnO/i-ZnO/p-SiC heterojunction LED at various injection currents.

For comparison, Figure 7.36 shows the room-temperature EL spectra of the n-ZnO/i-ZnO/p-SiC heterojunction LED under pulsed current injections. At an injection current of 300 mA, the EL exhibited a broad spectrum consisting of light emissions from ZnO and SiC. Comparing the EL and PL spectra reveals that the spectral peak located at 393 nm may be ascribed to the NBE emission from ZnO, and the broad spectral shoulder around the wavelengths from 400 to 450 nm corresponds to the DAP emission in p-SiC. As the injection current increased, the emission at 393 nm apparently increased and became dominant. The dominant UV EL from ZnO at 393 nm in the n-ZnO/i-ZnO/p-SiC structure can be attributed to the insertion of the undoped i-ZnO layer between n-ZnO and p-SiC. The Hall effect measurement revealed that the electron concentrations in the n-ZnO and i-ZnO layers are 2×10^{18} cm^{-3} and 2.9×10^{17} cm^{-3}, respectively. Thus the holes in p-SiC and the electrons in n-ZnO are injected into i-ZnO layer, and then recombine radiatively to generate UV emission. In comparison to the n-ZnO/p-SiC heterojunction, the low electron concentration of i-ZnO layer in the n-ZnO/i-ZnO/p-SiC structure may reduce the electron injection from ZnO into SiC, resulting in a smaller forward-bias current as shown in Figure 7.30. As a result, the SiC emission was suppressed and significant UV EL from ZnO was achieved in the n-ZnO/i-ZnO/p-SiC heterojunction LED. The EL intensity of the ZnO NBE emission at 393 nm from the n-ZnO/i-ZnO/p-SiC LED was greater than that from the SiC DAP emissions at 422 and 500 nm from the n-ZnO/p-SiC device, attributable to the greater carrier recombination rate in ZnO due to its direct bandgap.

7.8 Long-Term Stable p-Type ZnO Thin Films [91]

A major obstacle to the development of ZnO-based photonic devices is the difficulty in fabricating stable p-type ZnO with high hole concentration. Native crystallographic defects of interstitial Zn (Zn$_i$), oxygen vacancy (V_O), etc. [92,93] and unintentional incorporation of foreign impurities (such as hydrogen [94]) tend to behave as donors, leading to intrinsic *n*-type conductivity of ZnO. Hence, the achievement of stable *p*-type ZnO with high hole concentration is rather challenging due to compensation of the acceptors by a large amount of electrons supplied by the native defects. Nevertheless, researchers have proposed several ways to realize *p*-type ZnO using PLD [95–98], RFMS [99–104], MBE [105,106], MOCVD [107–109], and ALD [110,111]. Long-term stable *p*-type ZnO thin films prepared by ALD on semi-insulating GaAs substrates have been demonstrated [91]. Under appropriate post-deposition RTA treatment, arsenic atoms diffused from GaAs to ZnO and the As-related acceptors were activated. A high hole concentration of 3.44×10^{20} cm^{-3} and long-term stability up to 180 days in this *p*-type arsenic-doped ZnO thin film was achieved.

High-quality ZnO thin films were deposited on commercially available semi-insulating (100) GaAs substrates by the ALD technique. 1100 ALD cycles were applied to grow the ZnO thin film at 180°C and the resulting film thickness was about 200 nm. After the deposition of ZnO, the samples were treated by RTA at 500°C, 550°C, 600°C, 650°C, and 700°C in oxygen atmosphere for 3 min, so as to improve the crystalline quality, facilitate thermal diffusion of arsenic atoms from the GaAs substrate into ZnO thin film, and activate the As-related acceptors.

Figure 7.37a through c show the dependence of carrier concentration, mobility, and resistivity on the temperature of post-deposition RTA treatment. Ecopia HMS-3000 Hall-effect measurement system in the van der Pauw configuration was used to measure the resistivity, carrier density, mobility, the type of conductivity of the films. The samples were cut into the square size about 0.7×0.7 cm^2 for the Hall-effect measurements. The four corners of the specimens were soldered by small indium blobs for the Ohmic contacts. The label "AS" on the horizontal axis of Figure 7.37 denotes the as-deposited ZnO sample. In Figure 7.37a, the as-deposited ZnO thin film shows intrinsic *n*-type conductivity with an electron concentration of 3.10×10^{19} cm^{-3}. The high electron concentration in the as-deposited ZnO thin film might result from the decomposition of the DEZn at the ALD temperature of 180°C, resulting in a Zn-rich composition and a lot of oxygen vacancies (which are generally regarded as the donors) in the ZnO thin films [112–115]. For the specimens treated by the post-RTA process, the electron concentration decreased to 1.71×10^{17} cm^{-3} at the RTA temperature of 500°C and further dropped to 1.98×10^{16} cm^{-3} at 550°C, which was 2–3 orders of magnitude lower than that of the as-deposited film. *n*-type conductivity was converted to *p*-type as the RTA temperature was greater than 600°C. One can observe that a hole concentration as high as 3.44×10^{20} cm^{-3} was obtained at the RTA temperature of 700°C, which was about two orders of magnitude higher than that at 600°C (4.32×10^{18} cm^{-3}). In addition, the type of conductivity in the ZnO thin films can also be confirmed by the polarity of Seebeck voltage which is generated by a thermal electromotive force due to a temperature gradient across the material, which is known as the Seebeck test. In the Seebeck test, heat was applied to the indium contact on the sample by touching a heated soldering iron on it. A multimeter was used to measure the polarity of voltage difference between the heated and un-heated contacts, and then the conduction type can be determined. The Seebeck test exhibited a positive voltage at the cold end of the samples treated by RTA at the temperatures from 600°C to 700°C, clearly indicating the *p*-type conductivity. Such significant decrease in the electron concentration and increase in the hole concentration in the ZnO films treated by the post-RTA process are mainly attributed to the following mechanisms: (1) Ambiance of oxygen gas during the post-RTA treatment might effectively convert the Zn-rich composition to ZnO [114,115] and reduce the oxygen vacancies [102]. Accordingly, the electron concentration decreased. (2) Diffusion of arsenic atoms into the ZnO thin

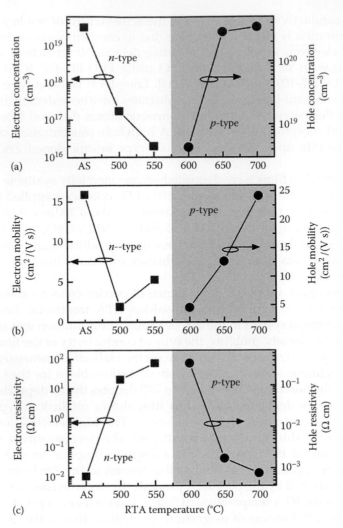

FIGURE 7.37

(a) Carrier concentration, (b) mobility, and (c) resistivity of the ZnO films as a function of the post-RTA temperature.

films from GaAs substrates together with activation of As-related acceptors (As$_{Zn}$-2V$_{Zn}$) in ZnO by the RTA treatment [95,99–101]. For the RTA treatments at 500°C and 550°C, the amount of As-related acceptors was insufficient to compensate for the plentiful background electrons in ZnO. Thus the conductivity was not converted to *p*-type yet. When the RTA temperature further increased, more arsenic atoms gained enough thermal energy to diffuse from GaAs into ZnO. As a result, transformation from *n*-type to *p*-type conductivity was observed in the ZnO thin films when the RTA temperature was greater than 600°C. The X-ray photoelectron spectroscopy (XPS) spectra of the as-deposited and RTA-treated (at 700°C) ZnO thin films, measured by the Thermo-VG Scientific XPS system, are presented in Figure 7.38. As compared with the as-deposited film, an arsenic 2p core-level peak at binding energy of 1325 eV was observed in the RTA-treated samples. It may verify the diffusion of arsenic atoms into the ZnO thin films by the RTA treatment.

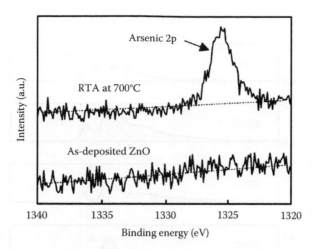

FIGURE 7.38
XPS spectra of the as-deposited and RTA-treated (at 700°C) samples.

It can be seen in Figure 7.37b that for the sample treated by the post-RTA process at 500°C, the electron mobility decreased to 1.89 cm^2/(V s) as compared with that of the as-deposited sample (15.87 cm^2/(V s)). Arsenic diffusion into the ZnO thin films may be responsible for the decrease of the electron mobility. A slight increase in electron mobility was observed at the RTA temperature of 550°C. The hole mobility significantly increased from 4.53 to 24.2 cm^2/(V s) as the RTA temperature increased from 600°C to 700°C, which may be attributed to the improvement in crystalline quality of the ZnO thin films by the RTA treatment. In comparison with the as-deposited sample, significant increase in the resistivity at the RTA temperatures of 500°C and 550°C, as shown in Figure 7.37c, results from the drastic drop of the electron concentration. On the other hand, considerable increment of the hole concentration and mobility leads to the remarkable decrease in the resistivity with the increase of the RTA temperature from 600°C to 700°C, achieving a p-type ZnO thin films with resistivity as low as 7.51×10^{-4} Ω cm. This p-type conductivity in ZnO was further confirmed by remove of the ZnO thin films using wet chemical etching. The Hall effect measurement on the remaining GaAs substrate exhibited a very weak p-type conductivity with resistivity as high as 1.6×10^3 Ω cm, ascribed to the diffusion of zinc atoms from ZnO into the semi-insulating GaAs substrate. Therefore, it can be deduced that the measured p-type resistivity as low as 7.51×10^{-4} Ω cm mainly originate from the post-annealed ZnO thin film, not from the GaAs substrate.

Figure 7.39 shows the low-temperature PL spectra of the RTA-treated samples at 14 K. For the low-temperature PL measurement, samples were placed into a closed-cycle helium refrigerator to keep the temperature at 14 K. A 30 mW continuous-wave He-Cd laser ($\lambda = 325$ nm) was used as the excitation source. Two remarkable spectral peaks appear in the spectra. The spectral peak at 3.32 eV can be identified as the transition from free electrons to the acceptor state (FA) [95]. Another spectral peak around 3.36–3.37 eV is attributed to the superposition of radiative recombinations from acceptor-bound excitations (A^oX) at 3.36 eV and donor-bound excitons (D^oX) at 3.37 eV [95]. With increase in the RTA temperature, it is clearly seen that the spectral peak around 3.36–3.37 eV gradually shifted from D^oX to A^oX, suggesting the decrease in the electron concentration and increase in the hole concentration due to the compensation of the donor states by a great amount of holes. In addition, the PL intensity of A^oX spectral peak increased significantly with the RTA

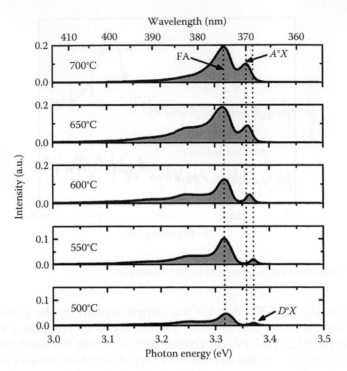

FIGURE 7.39
PL spectra of the RTA-treated ZnO films at a low temperature of 14 K.

temperature, indicating the considerable increase of the amount of acceptors due to the diffusion of arsenic atoms and activation of As-related acceptors in the ZnO thin films. The results are in good agreement with the electrical characteristics of ZnO thin films as shown in Figure 7.37.

Table 7.1 shows the hole concentration, mobility, and resistivity of the samples treated by the post RTA process at 700°C at the first day, as well as 30, 60, 90, 120, 150, and 180 days later. One can clearly see that all the electrical characteristics of the samples are stable with time. Noticeable degradation in the *p*-type conductivity of the ZnO thin film was not observed.

TABLE 7.1

Hole Concentration, Mobility, and Resistivity Were
Measured at the 1st Day and 30, 60, 90, 120, and 180
Days Later of the Sample Treated by RTA at 700°C

Time (Days)	Concentration $(10^{20} \text{ cm}^{-3})$	Mobility $(\text{cm}^2/(\text{V s}))$	Resistivity $(10^{-4}\ \Omega \text{ cm})$
1	3.44	24.2	7.51
30	3.24	25.6	7.52
60	3.24	25.6	7.52
90	3.07	26.8	7.58
120	3.02	27.2	7.61
150	3.05	26.8	7.63
180	3.04	26.6	7.71

7.9 Summary

ALD has been demonstrated as one of the noteworthy techniques to prepare high-quality ZnO thin films, QDs, and LEDs. A high-quality ZnO epilayer was grown on the c-Al_2O_3 substrate by ALD and treated by the post-deposition RTA treatment. The layer-by-layer growth and low deposition temperature of ALD prevent the formation of columnar structures in the ZnO epilayer. Obvious threading dislocations were not observed in the ZnO epilayer even though the lattice mismatch between ZnO and c-Al_2O_3 is up to 18%. Optically pumped stimulated emission with a low threshold value was achieved at room temperature. The high crystalline quality and low-threshold stimulated emission indicate that the ALD technique is appropriate for preparing high-quality ZnO epilayers.

Highly oriented ZnO thin films accompanied by stimulated emission have been prepared by ALD on amorphous glass substrates. By introducing a buffer layer which was deposited at a high temperature of 300°C and followed by post-deposition RTA treatment at 800°C, ZnO thin films with the (0001) preferred orientation could be grown by ALD at a low deposition temperature of 150°C on the buffer layer. The high optical quality of the ZnO thin films was demonstrated by the predominant NBE spontaneous emission at 380 nm with a negligible defect-related band and the low-threshold stimulated emission at room temperature.

The highly conformal ALD technique was also applied to deposit ZnO into the voids between the SiO_2 nanoparticles, even in the SiO_2 nanoparticle layer as thick as 150 nm or above. The ZnO QDs with diameter in the range of 3–8 nm were uniformly embedded in the SiO_2 matrix. Quantum confinement effect of the ZnO QDs was well manifested by a significant blue-shift of the PL spectrum at room temperature. This method can easily prepare ZnO QDs embedded in the SiO_2 layer with a thickness greater than 100 nm, providing a large active volume for optoelectronic applications.

The n-ZnO/p-GaN heterojunction LEDs have been fabricated by the growth of n-type ZnO epilayer using ALD on p-type GaN as well as the post-deposition RTA treatment. TEM observations revealed that the ZnO grows in good epitaxial relation with GaN and to almost a perfect single crystal with very few dislocations. It has been found that the threading dislocations from the underlying layers reduce at the interface between GaN and ZnO, ascribed to the layer-by-layer growth of ALD and the recrystallization during the RTA treatment. At a low injection current, light emission from the Mg acceptor levels in p-GaN is obvious. Increase in the forward bias gives rise to dominant emission from n-ZnO. Competition between the ELs from ZnO and GaN is elucidated to result from the interfacial layer as well as the differences in the light emission efficiency and carrier concentration in n-ZnO and p-GaN layers. The achievement of UV EL at a low DC injection current from ZnO indicates that the ZnO epilayers grown by the ALD technique are effectually applicable to the next-generation short-wavelength photonic devices.

The ALD technique was also applied to grow n-type Al-doped ZnO (n-ZnO) and undoped ZnO (i-ZnO) layers on the p-type 4H-SiC substrate, to fabricate the n-ZnO/ p-SiC and n-ZnO/i-ZnO/p-SiC heterojunction LEDs. The EL from the n-ZnO/p-SiC LED originated from the radiative recombination of DAP in SiC due to the predominant electron injection from n-ZnO into p-SiC. On the other hand, the n-ZnO/i-ZnO/p-SiC LED exhibited dominant UV emission at 393 nm from ZnO. This difference can be attributed to the insertion of the undoped i-ZnO layer between n-ZnO and p-SiC, leading to the injection of holes from p-SiC and electrons from n-ZnO into i-ZnO layer and thus the generation of UV EL from ZnO.

Long-term stable p-type ZnO thin films were prepared on semi-insulating GaAs substrates using ALD and the post-deposition RTA treatment in the oxygen ambient. The RTA treatment facilitates the diffusion of arsenic atoms from GaAs into ZnO and the activation of As-related acceptors, resulting in significant decrease in the electron concentration and increase in the hole concentration. A high-quality p-type arsenic-doped ZnO thin film was achieved with a hole concentration as high as 3.44×10^{20} cm^{-3}, resistivity as low as 7.51×10^{-4} Ω cm, and long-term stability up to 180 days.

All of these results indicate that ALD is a very promising technique for preparing high-quality ZnO thin films and QDs for the UV LEDs and lasers in the near future.

Acknowledgments

This work was financially supported by the National Science Council in Taiwan. We would like to acknowledge the great contribution of our collaborators, particularly, coauthors of our papers cited in references, to the present investigations.

References

1. Ü. Özgür, Ya.I. Alivov, C. Liu, A. Teke, M.A. Reshchikov, S. Doğan, V. Avrutin, S.-J. Cho, and H. Morkoç, A comprehensive review of ZnO materials and devices, *J. Appl. Phys.* **98**, 041301 (2005).
2. C. Jagadish and S.J. Pearton, *Zinc Oxide Bulk, Thin Films and Nanostructures, Processing, Properties and Application*, Elsevier, Oxford, U.K. (2006).
3. Y. Chen, D. Bagnall, and T. Yao, ZnO as a novel photonic material for the UV region, *Mater. Sci. Eng. B* **75**, 190 (2000).
4. J.M. Yuk, J.Y. Lee, J.H. Jung, T.W. Kim, D.I. Son, and W.K. Choi, Initial formation mechanisms of the supersaturation region and the columnar structure in ZnO thin films grown on n-Si (001) substrates, *Appl. Phys. Lett.* **90**, 031907 (2007).
5. S.H. Lim, D. Shindo, H.B. Kang, and K. Nakamura, Defect structure of epitaxial ZnO films on (0001) sapphire studied by transmission electron microscopy, *J. Vac. Sci. Technol. B* **19**, 506 (2001).
6. A. Setiawan, Z. Vashaei, M.W. Cho, T. Yao, H. Kato, M. Sano, K. Miyamoto, I. Yonenaga, and H.J. Ko, Characteristics of dislocations in ZnO layers grown by plasma-assisted molecular beam epitaxy under different Zn/O flux ratios, *J. Appl. Phys.* **96**, 3763 (2004).
7. S.H. Lim, J. Washburn, Z.L. Weber, and D. Shindo, Transmission electron microscopy of threading dislocations in ZnO films grown on sapphire, *J. Vac. Sci. Technol. A* **19**, 2601 (2001).
8. V. Sallet, C. Thiandoume, J.F. Rommeluere, A. Lusson, A. Rivière, J.P. Rivière, O. Gorochov, R. Triboulet, and V. Muñoz-Sanjosé, Some aspects of the MOCVD growth of ZnO on sapphire using *tert*-butanol, *Mater. Lett.* **53**, 126 (2002).
9. C.M. Hou, K.K. Huang, Z.M. Gao, X.S. Li, S.H. Feng, Y.T. Zhang, and G.T. Du, Structural and optical properties of ZnO films with different thicknesses grown on sapphire by MOCVD, *Chem. Res. Chinese. U.* **22**, 552 (2006).
10. C.Y. Liu, B.P. Zhang, N.T. Binh, and Y. Segawa, Third-harmonic generation from ZnO films deposited by MOCVD, *Appl. Phys. B* **79**, 83 (2004).
11. W.Y. Shiao, C.Y. Chi, S.C. Chin, C.F. Huang, T.Y. Tang, Y.C. Lu et al., Comparison of nanostructure characteristics of ZnO grown on GaN and sapphire, *J. Appl. Phys.* **99**, 054301 (2006).
12. J. Hu and R.G. Gordon, Textured aluminum-doped zinc oxide thin films from atmospheric pressure chemical-vapor deposition, *J. Appl. Phys.* **71**, 880 (1992).

13. K. Minegishi, Y. Koiwai,Y. Kikuchi, K. Yano, M. Kasuga, and A. Shimizu, Growth of *p*-type zinc oxide films by chemical vapor deposition, *Jpn. J. Appl. Phys.* **36**, L1453 (1997).

14. Y.G. Wang, N. Ohashi, Y. Wada, I. Sakaguchi, T. Ohgaki, and H. Haneda, Lowering of stimulated emission threshold of zinc oxide by doping with thermally diffused aluminum supplied from sapphire substrate, *J. Appl. Phys.* **100**, 023524 (2006).

15. P.F. Carcia, R.S. McLean, M.H. Reilly, and G. Nunes, Transparent ZnO thin-film transistor fabricated by RF magnetron sputtering, *Appl. Phys. Lett.* **82**, 1117 (2003).

16. L.Y. Chen, W.H. Chen, J.J. Wang, F.C.N. Hong, and Y.K. Su, Hydrogen-doped high conductivity ZnO films deposited by radio-frequency magnetron sputtering, *Appl. Phys. Lett.* **85**, 5628 (2004).

17. H.H. Hsieh and C.C. Wu, Scaling behavior of ZnO transparent thin-film transistors, *Appl. Phys. Lett.* **89**, 041109 (2006).

18. C. Yuen, S.F. Yu, E.S.P. Leong, H.Y. Yang, S.P. Lau, and H.H. Hng, Formation conditions of random laser cavities in annealed ZnO epilayers, *IEEE J. Quant. Electron.* **41**, 970 (2005).

19. Y.G. Wang, S.P. Lau, H.W. Lee, S.F. Yu, B.K. Tay, X.H. Zhang, K.Y. Tse, and H.H. Hng, Comprehensive study of ZnO films prepared by filtered cathodic vacuum arc at room temperature, *J. Appl. Phys.* **94**, 1597 (2003).

20. M. Ritala and M. Leskelä, Atomic layer epitaxy—A valuable tool for nanotechnology? *Nanotechnology* **10**, 19 (1999).

21. L. Niinisto, J. Päiväsaari, J. Niinistö, M. Putkonen, and M. Nieminen, Advanced electronic and optoelectronic materials by atomic layer deposition: An overview with special emphasis on recent progress in processing of high-*k* dielectrics and other oxide materials, *Phys. Status Solidi A* **201**, 1443 (2004).

22. J. Lim, K. Shin, H.W. Kim, and C. Lee, Photoluminescence studies of ZnO thin films grown by atomic layer epitaxy, *J. Lumin.* **109**, 181 (2004).

23. A. Wójcik, M. Godlewski, E. Guziewicz, R. Minikayev, and W. Paszkowicz, Controlling of preferential growth mode of ZnO thin films grown by atomic layer deposition, *J. Cryst. Growth* **310**, 284 (2008).

24. E. Guziewicz, I.A. Kowalik, M. Godlewski, K. Kopalko, V. Osinniy, A. Wójcik, S. Yatsunenko, E. Łusakowska, W. Paszkowicz, and M. Guziewicz, Extremely low temperature growth of ZnO by atomic layer deposition, *J. Appl. Phys.* **103**, 033515 (2008).

25. I.A. Kowalik, E. Guziewicz, K. Kopalko, S. Yatsunenko, M. Godlewski, A. Wójcik et al., Extra-low temperature growth of ZnO by atomic layer deposition with diethylzinc precursor, *Acta Phys. Pol. A* **112**, 401 (2007).

26. H.C. Chen, M.J. Chen, T.C. Liu, J.R. Yang, and M. Shiojiri, Structure and stimulated emission of a high-quality zinc oxide epilayer grown by atomic layer deposition on the sapphire substrate, *Thin Solid Films* **519**, 536 (2010).

27. X.Q. Zhang, I. Suemune, H. Kumano, J. Wang, and S.H. Huang, Surface-emitting stimulated emission in high-quality ZnO thin films, *J. Appl. Phys.* **96**, 3733 (2004).

28. C. Klingshirn, The luminescence of ZnO under high one- and two-quantum excitation, *Phys. Status Solidi B* **71**, 547 (1975).

29. C. Klingshirn, R. Hauschild, J. Fallert, and H. Kalt, Room-temperature stimulated emission of ZnO: Alternatives to excitonic lasing, *Phys. Rev. B* **75**, 115203 (2007).

30. M.H. Huang, S. Mao, H. Feick, H. Yan, Y. Wu, H. Kind, E. Weber, R. Russo, and P. Yang, Room-temperature ultraviolet nanowire nanolasers, *Science* **292**, 1897 (2001).

31. P. Zu, Z.K. Tang, G.K.L. Wong, M. Kawasaki, A. Ohtomo, H. Koinuma, and Y. Segawa, Ultraviolet spontaneous and stimulated emissions from ZnO microcrystallite thin films at room temperature, *Solid State Commun.* **103**, 459 (1997).

32. X.H. Yang, T.J. Schmidt, W. Shan, J.J. Song, and B. Goldenberg, Above room temperature near ultraviolet lasing from an optically pumped GaN film grown on sapphire, *Appl. Phys. Lett.* **66**, 1 (1995).

33. D. Wiesmann, I. Brener, L. Pfeiffer, M.A. Khan, and C.J. Sun, Gain spectra and stimulated emission in epitaxial (In,Al) GaN thin films, *Appl. Phys. Lett.* **69**, 3384 (1996).

34. Y.T. Shih, C.Y. Chiu, C.W. Chang, J.R. Yang, M. Shiojiri, and M.J. Chen, Stimulated emission in highly (0001)-oriented ZnO films grown by atomic layer deposition on the amorphous glass substrates, *J. Electrochem. Soc.* **157**, H879–H883 (2010).
35. S.J. Lim, S. Kwon, and H. Kim, ZnO thin films prepared by atomic layer deposition and RF sputtering as an active layer for thin film transistor, *Thin Solid Films* **516**, 1523 (2008).
36. S.H.K. Park, C.S. Hwang, H.S. Kwack, J.H. Lee, and H.Y. Chu, Characteristics of ZnO thin films by means of plasma-enhanced atomic layer deposition, *Electrochem. Solid-State Lett.* **9**, G299 (2006).
37. S.K. Kim, C.S. Hwang, S.H.K. Park, and S.J. Yun, Comparison between ZnO films grown by atomic layer deposition using H_2O or O_3 as oxidant, *Thin Solid Films* **478**, 103 (2005).
38. K. Nishio, T. Isshiki, M. Kitano, and M. Shiojiri, Structure and growth mechanism of tetrapod-like ZnO particles, *Philos. Mag. A* **76**, 889 (1997).
39. Z.K. Tang, M. Kawasaki, A. Ohtomo, H. Koinuma, and Y. Segawa, Self-assembled ZnO nano-crystals and exciton lasing at room temperature, *J. Crys. Growth* **287**, 169 (2006).
40. M. Kawasaki, A. Ohtomo, I. Ohkubo, H. Koinuma, Z.K. Tang, P. Yu, G.K.L. Wong, B.P. Zhang, and Y. Segawa, Excitonic ultraviolet laser emission at room temperature from naturally made cavity in ZnO nanocrystal thin films, *Mater. Sci. Eng. B* **56**, 239 (1998).
41. A. Ohtomo, M. Kawasaki, Y. Sakurai, Y. Yoshida, H. Koinuma, P. Yu, Z.K. Tang, G.K.L. Wong, and Y. Segawa, Room temperature ultraviolet laser emission from ZnO nanocrystal thin films grown by laser MBE, *Mater. Sci. Eng. B* **54**, 24 (1998).
42. Y. Segawa, A. Ohtomo, M. Kawasaki, H. Koinuma, Z.K. Tang, P. Yu, and G.K.L. Wong, Growth of ZnO thin film by laser MBE: Lasing of exciton at room temperature, *Phys. Status Solidi B* **202**, 669 (1997).
43. G. Tobin, E. McGlynn, M.O. Henry, J.P. Mosnier, E.D. Posada, and J.G. Lunney, Effects of excitonic diffusion on stimulated emission in nanocrystalline ZnO, *Appl. Phys. Lett.* **88**, 071919 (2006).
44. X.H. Zhang, S.J. Chua, A.M. Yong, H.D. Li, S.F. Yu, and S.P. Lau, Exciton related stimulated emission in ZnO polycrystalline thin film deposited by filtered cathodic vacuum arc technique, *Appl. Phys. Lett.* **88**, 191112 (2006).
45. D.M. Bagnall, Y.F. Chen, Z. Zhu, T. Yao, M.Y. Shen, and T. Goto, High temperature excitonic stimulated emission from ZnO epitaxial layers, *Appl. Phys. Lett.* **73**, 1038 (1998).
46. H.D. Li, S.F. Yu, S.P. Lau, E.S.P. Leong, H.Y. Yang, T.P. Chen, A.P. Abiyasa, and C.Y. Ng, High-temperature lasing characteristics of ZnO epilayers, *Adv. Mater.* **18**, 771 (2006).
47. Y. Chen, N.T. Tuan, Y. Segawa, H.-J. Ko, S.-K. Hong, and T. Yao, Stimulated emission and optical gain in ZnO epilayers grown by plasma-assisted molecular-beam epitaxy with buffers, *Appl. Phys. Lett.* **78**, 1469 (2001).
48. Ü. Özgür, A. Teke, C. Liu, S.-J. Cho, H. Morkoç, and H.O. Everitt, Stimulated emission and time-resolved photoluminescence in rf-sputtered ZnO thin films, *Appl. Phys. Lett.* **84**, 3223 (2004).
49. C.Y. Liu, B.P. Zhang, N.T. Binh, K. Wakatsuki, and Y. Segawa, Stimulated ultraviolet emission from ZnO films annealed at high temperature, *Physica B* **381**, 20 (2006).
50. R.P. Wang, H. Muto, X. Gang, P. Jin, and M. Tazawa, Ultraviolet lasing with low excitation intensity in deep-level emission free ZnO films, *J. Cryst. Growth* **282**, 359 (2005).
51. G. Tobin, E. McGlynn, M.O. Henry, J.P. Mosnier, J.G. Lunney, D. O'Mahony, and E. Deposada, Ultraviolet stimulated emission from bulk and polycrystalline ZnO thin films with varying grain sizes, *Physica B* **340**, 245 (2003).
52. F.A. Pinheiro and L.C. Sampaio, Lasing threshold of diffusive random lasers in three dimensions, *Phys. Rev. A* **73**, 013826 (2006).
53. H. Cao, J.Y. Xu, S.H. Chang, and S.T. Ho, Transition from amplified spontaneous emission to laser action in strongly scattering media, *Phys. Rev. E* **61**, 1985 (2000).
54. M.K. Wu, Y.T. Shih, M.J. Chen, J.R. Yang, and M. Shiojiri, ZnO quantum dots embedded in a SiO_2 nanoparticle layer grown by atomic layer deposition, *Phys. Status Solidi RRL* **3**, 88 (2009).
55. B. Gil and A.V. Kavokin, Giant exciton-light coupling in ZnO quantum dots, *Appl. Phys. Lett.* **81**, 748 (2002).
56. K.F. Lin, H.M. Cheng, H.C. Hsu, and W.F. Hsieh, Band gap engineering and spatial confinement of optical phonon in ZnO quantum dots, *Appl. Phys. Lett.* **88**, 263117 (2006).

57. H.M. Cheng, K.F. Lin, H.C. Hsu, and W.F. Hsieh, Size dependence of photoluminescence and resonant Raman scattering from ZnO quantum dots, *Appl. Phys. Lett.* **88**, 261909 (2006).
58. S.W. Kim, S. Fujita, and S. Fujita, Self-organized ZnO quantum dots on SiO_2/Si substrates by metalorganic chemical vapor deposition, *Appl. Phys. Lett.* **81**, 5036 (2002).
59. K.K. Kim, N. Koguchi, Y.W. Ok, T.Y. Seong, and S.J. Park, Fabrication of ZnO quantum dots embedded in an amorphous oxide layer, *Appl. Phys. Lett.* **84**, 3810 (2004).
60. X.H. Zhang, S.J. Chua, A.M. Yong, and S.Y. Chow, Exciton radiative lifetime in ZnO quantum dots embedded in SiO_x matrix, *Appl. Phys. Lett.* **88**. 221903 (2006).
61. S.T. Tan, X.W. Sun, X.H. Zhang, B.J. Chen, S.J. Chua, A. Yong, Z.L. Dong, and X. Hu, Zinc oxide quantum dots embedded films by metal organic chemical vapor deposition, *J. Cryst. Growth* **290**, 518 (2006).
62. H.C. Chen, M.J. Chen, M.K. Wu, W.C. Li, H.L. Tsai, J.R. Yang, H. Kuan, and M. Shiojiri, UV electroluminescence and structure of *n*-ZnO/*p*-GaN heterojunction LEDs grown by atomic layer deposition, *IEEE J. Quant. Electron.* **46**, 265 (2010).
63. Y.I. Alivov, J.E. Van Nostrand, D.C. Look, M.V. Chukichev, and B.M. Ataev, Observation of 430 nm electroluminescence from ZnO/GaN heterojunction light-emitting diodes, *Appl. Phys. Lett.* **83**, 2943 (2003).
64. D.J. Rogers, F.H. Teherani, A. Yasan, K. Minder, P. Kung, and M. Razeghi, Electroluminescence at 375 nm from a ZnO/GaN:Mg/*c*-Al_2O_3 heterojunction light emitting diode, *Appl. Phys. Lett.* **88**, 141918 (2006).
65. S.J. Jiao, Y.M. Lu, D.Z. Shen, Z.Z. Zhang, B.H. Li, J.Y. Zhang, B. Yao, Y.C. Liu, and X.W. Fan, Ultraviolet electroluminescence of ZnO based heterojunction light emitting diode, *Phys. Status Solidi. C* **3**, 972 (2006).
66. H.Y. Xu, Y.C. Liu, Y.X. Liu, C.S. Xu, C.L. Shao, and R. Mu, Ultraviolet electroluminescence from *p*-GaN/*i*-ZnO/*n*-ZnO heterojunction light-emitting diodes, *Appl. Phys. B* **80**, 871 (2005).
67. R.W. Chuang, R.X. Wu, L.W. Lai, and C.T. Lee, ZnO-on-GaN heterojunction light-emitting diode grown by vapor cooling condensation technique, *Appl. Phys. Lett.* **91**, 231113 (2007).
68. H.S. Yang, S.Y. Han, Y.W. Heo, K.H. Baik, D.P. Norton, S.J. Pearton et al., Fabrication of hybrid *n*-ZnMgO/*n*-ZnO/*p*-AlGaN/*p*-GaN light-emitting diodes, *Jpn. J. Appl. Phys.* **44**, 7296 (2005).
69. U. Kaufmann, M. Kunzer, M. Maier, H. Obloh, A. Ramakrishnan, B. Santic, and P. Schlotter, Nature of the 2.8 eV photoluminescence band in Mg doped GaN, *Appl. Phys. Lett.* **72**, 1326 (1998).
70. S.J. Pennycook and P.D. Nellist, Z-contrast scanning transmission electron microscopy, in *Impact of Electron and Scanning Probe Microscopy on Materials Research*, eds D.G. Rickerby et al., Kluwer Academic Publishers, Dordrecht, the Netherlands, pp. 161–207, 1999.
71. M. Shiojiri and H. Saijo, Imaging of high-angle annular dark-field scanning transmission electron microscopy and observations of GaN-based violet laser diodes, *J. Microsc.* **223**, 172 (2006).
72. S. Lee and D.Y. Kim, Characteristics of ZnO/GaN heterostructure formed on GaN substrate by sputtering deposition of ZnO, *Mater. Sci. Eng. B* **137**, 80 (2007).
73. C.P. Chen, M.Y. Ke, C.C. Liu, Y.J. Chang, F.H. Yang, and J.J. Huang, Observation of 394 nm electroluminescence from low-temperature sputtered *n*-ZnO/SiO_2 thin films on top of the *p*-GaN heterostructure, *Appl. Phys. Lett.* **91**, 091107 (2007).
74. Y.T. Shih, M.K. Wu, M.J. Chen, Y.C. Cheng, J.R. Yang, and M. Shiojiri, ZnO-based heterojunction light-emitting diodes on *p*-SiC(4H) grown by atomic layer deposition, *Appl. Phys. B* **98**, 767 (2010).
75. C. Yuen, S.F. Yu, S.P. Lau, Rusli, and T.P. Chen, Fabrication of *n*-ZnO:Al/*p*-SiC(4H) heterojunction light-emitting diodes by filtered cathodic vacuum arc technique, *Appl. Phys. Lett.* **86**, 241111 (2005).
76. Y.I. Alivov, D. Johnstone, Ü. Özgür, V. Avrutin, Q. Fan, S.S. Akarca-Biyikli, and H. Morkoç, Electrical and optical properties of *n*-ZnO/*p*-SiC heterojunctions, *Jpn. J. Appl. Phys.* **44**, 7281 (2005).
77. A. El-Shaer, A. Bakin, E. Schlenker, A.C. Mofor, G. Wagner, S.A. Reshanov, and A. Waag, Fabrication and characterization of *n*-ZnO on *p*-SiC heterojunction diodes on 4H-SiC substrates, *Superlattices Microstruct.* **42**, 387 (2007).
78. J.G. Lu, S. Fujita, T. Kawaharamura, H. Nishinaka, Y. Kamada, T. Ohshima et al., Carrier concentration dependence of band gap shift in *n*-type ZnO:Al films, *J. Appl. Phys.* **101**, 083705 (2007).

79. A. Kakanakova-Georgieva, R. Yakimova, M.K. Linnarsson, and E. Janzén, Site-occupying behavior of boron in compensated *p*-type 4H-SiC grown by sublimation epitaxy, *J. Appl. Phys.* **91**, 3471 (2002).

80. Y. Negoro, T. Kimoto, and H. Matsunami, Stability of deep centers in 4H-SiC epitaxial layers during thermal annealing, *Appl. Phys. Lett.* **85**, 1716 (2004).

81. F. Fabbri, D. Natalini, A. Cavallini, T. Sekiguchi, R. Nipoti, and F. Moscatelli, Comparison between cathodoluminescence spectroscopy and capacitance transient spectroscopy on Al$^+$ ion implanted 4H-SiC p$^+$/n diodes, *Superlattices Microstruct.* **45**, 383 (2009).

82. S.A. Studenikin, N. Golego, and M. Cocivera, Carrier mobility and density contributions to photoconductivity transients in polycrystalline ZnO films, *J. Appl. Phys.* **87**, 2413 (2000).

83. H. Matsuura, M. Komeda, S. Kagamihara, H. Iwata, R. Ishihara, T. Hatakeyama, T. Watanabe, K. Kojima, T. Shinohe, and K. Arai, Dependence of acceptor levels and hole mobility on acceptor density and temperature in Al-doped *p*-type 4H-SiC epilayers, *J. Appl. Phys.* **96**, 2708 (2004).

84. C. Hemmingsson, N.T. Son, O. Kordina, J.P. Bergman, E. Janzén, J.L. Lindström, S. Savage, and N. Nordell, Deep level defects in electron-irradiated *4H* SiC epitaxial layers, *J. Appl. Phys.* **81**, 6155 (1997).

85. J.A. Aranovich, D.G. Golmayo, A.L. Fahrenbruch, and R.H. Bube, Photovoltaic properties of ZnO/CdTe heterojunctions prepared by spray pyrolysis, *J. Appl. Phys.* **51**, 4260 (1980).

86. T.V. Blank, Y.A. Goldberg, E.V. Kalinina, O.V. Konstantinov, A.O. Konstantinov, and A. Hallén, Temperature dependence of the photoelectric conversion quantum efficiency of 4H-SiC Schottky UV photodetectors, *Semicond. Sci. Technol.* **20**, 710 (2005).

87. M. Yoganathan, W.J. Choyke, R.P. Devaty, and P.G. Neudeck, Free to bound transition-related electroluminescence in 3C and 6H SiC p$^+$-n junctions at room temperature, *J. Appl. Phys.* **80**, 1763 (1996).

88. T. Hori, *Gate Dielectrics and MOS ULSIs: Principles, Technologies, and Application*, Springer, Berlin, Germany (1997).

89. E.H. Nicollian and J.R. Brews, *MOS Physics and Technology*, Wiley, New York (1982).

90. D.K. Schroder, *Semiconductor Material and Device Characterization*, 2nd edn., Wiley, New York (1998).

91. Y.C. Cheng, Y.S. Kuo, Y.H. Li, J.J. Shyue, and M.J. Chen, Stable *p*-type ZnO films grown by atomic layer deposition on GaAs substrates and treated by post-deposition rapid thermal annealing, *Thin Solid Films* **519**, 5558 (2011).

92. D.C. Look, J.W. Hemsky, and J.R. Sizelove, Residual native shallow donor in ZnO, *Phys. Rev. Lett.* **82**, 2552 (1999).

93. S.B. Zhang, S.H. Wei, and A. Zunger, Intrinsic *n*-type versus *p*-type doping asymmetry and the defect physics of ZnO, *Phys. Rev. B* **63**, 075205 (2001).

94. C.G. Van de Walle, Hydrogen as a cause of doping in zinc oxide, *Phys. Rev. Lett.* **85**, 1012 (2000).

95. Y.R. Ryu, S. Zhu, D.C. Look, J.M. Wrobel, H.M. Jeong, and H.W. White, Synthesis of *p*-type ZnO films, *J. Cryst. Growth* **216**, 330 (2000).

96. Y.R. Ryu, T.S. Lee, J.A. Lubguban, H.W. White, Y.S. Park, and C.J. Youn, ZnO devices: Photodiodes and *p*-type field-effect transistors, *Appl. Phys. Lett.* **87**, 153504 (2005).

97. Y.W. Heo, Y.W. Kwon, Y. Li, S.J. Pearton, and D.P. Norton, *p*-type behavior in phosphorus-doped (Zn,Mg)O device structures, *Appl. Phys. Lett.* **84**, 3474 (2004).

98. K. Lord, T.M. Williams, D. Hunter, K. Zhang, J. Dadson, and A.K. Pradhan, Effects of As and Mn doping on microstructure and electrical conduction in ZnO films, *Appl. Phys. Lett.* **88**, 262105 (2006).

99. G. Braunstein, A. Muraviev, H. Saxena, N. Dhere, V. Richter, and R. Kalish, *p* type doping of zinc oxide by arsenic ion implantation, *Appl. Phys. Lett.* **87**, 192103 (2005).

100. H.S. Kang, G.H. Kim, D.L. Kim, H.W. Chang, B.D. Ahn, and S.Y. Lee, Investigation on the *p*-type formation mechanism of arsenic doped *p*-type ZnO thin film, *Appl. Phys. Lett.* **89**, 181103 (2006).

101. J.C. Fan, C.Y. Zhu, S. Fung, Y.C. Zhong, K.S. Wong, Z. Xie et al., Arsenic doped *p*-type zinc oxide films grown by radio frequency magnetron sputtering, *J. Appl. Phys.* **106**, 073709 (2009).

102. A.V. Singh, R.M. Mehra, A. Wakahara, and Yoshida, *p*-type conduction in codoped ZnO thin films, *J. Appl. Phys.* **93**, 396 (2003).

103. K.K. Kim, H.S. Kim, D.K. Hwang, J.H. Lim, and S.J. Park, Realization of *p*-type ZnO thin films via phosphorus doping and thermal activation of the dopant, *Appl. Phys. Lett.* **83**, 63 (2003).

104. C.C. Lin, S.Y. Chen, S.Y. Cheng, and H.Y. Lee, Properties of nitrogen-implanted *p*-type ZnO films grown on Si$_3$N$_4$/Si by radio-frequency magnetron sputtering, *Appl. Phys. Lett.* **84**, 5040 (2004).

105. D.C. Look, D.C. Reynolds, C.W. Litton, R.L. Jones, D.B. Eason, and G. Cantwell, Characterization of homoepitaxial *p*-type ZnO grown by molecular beam epitaxy, *Appl. Phys. Lett.* **81**, 1830 (2002).

106. A. Tsukazaki, A. Ohtomo, M. Ohtani, T. Makino, M. Sumiya, K. Ohtani et al., Repeated temperature modulation epitaxy for *p*-type doping and light-emitting diode based on ZnO, *Nat. Mater.* **4**, 42 (2005).

107. T.M. Barnes, K. Olson, and C.A. Wolden, On the formation and stability of *p*-type conductivity in nitrogen-doped zinc oxide, *Appl. Phys. Lett.* **86**, 112112 (2005).

108. J.C. Sun, J.Z. Zhao, H.W. Liang, J.M. Bian, L.Z. Hu, H.Q. Zhang, X.P. Liang, W.F. Liu, and G.T. Du, Realization of ultraviolet electroluminescence from ZnO homojunction with *n*-ZnO/*p*-ZnO:As/GaAs structure, *Appl. Phys. Lett.* **90**, 121128 (2007).

109. A. Krtschil, A. Dadgar, N. Oleynik, J. Bläsing, A. Diez, and A. Krost, Local *p*-type conductivity in zinc oxide dual-doped with nitrogen and arsenic, *Appl. Phys. Lett.* **87**, 262105 (2005).

110. L. Dunlop, A. Kursumovic, and J.L. MacManus-Driscoll, Reproducible growth of *p*-type ZnO:N using a modified atomic layer deposition process combined with dark annealing, *Appl. Phys. Lett.* **93**, 172111 (2008).

111. C. Lee, S.Y. Park, J. Lim, and H.W. Kim, Growth of *p*-type ZnO thin films by using an atomic layer epitaxy technique and NH$_3$ as a doping source, *Mater. Lett.* **61**, 2495 (2007).

112. D.M. King, X.H. Liang, C.S. Carney, L.F. Hakim, P. Li, and A.W. Weimer, Atomic layer deposition of UV-absorbing ZnO films on SiO$_2$ and TiO$_2$ nanoparticles using a fluidized bed reactor, *Adv. Funct. Mater.* **18**, 607 (2008).

113. D.M. King, J. Li, X.H. Liang, S.I. Johnson, M.M. Channel, and A.W. Weimer, Crystal phase evolution in quantum confined ZnO domains on particles via atomic layer deposition, *Cryst. Growth Des.* **9**, 2828 (2009).

114. D.M. King, X.H. Liang, P. Li, S.M. George, and A.W. Weimer, Low-temperature atomic layer deposition of ZnO films on particles in a fluidized bed reactor, *Thin Solid Films* **516**, 8517 (2008).

115. H. Dumont, A. Marbeuf, J.-E. Bourée, and O. Gorochov, Pyrolysis pathways and kinetics of thermal decomposition of diethylzinc and diethyltellurium studied by mass spectrometry, *J. Mater. Chem.* **3**, 1075 (1993).

8

Hybrid Light-Emitting Diodes
Based on ZnO Nanowires

Jinzhang Liu

CONTENTS

8.1 Introduction...279
8.2 Hybrid LEDs Based on Vertically Aligned ZnO Nanowires 280
    8.2.1 Heterostructure of *n*-ZnO-Nanowire/*p*-GaN...280
    8.2.2 Heterojunction of *n*-ZnO-Nanowire/*p*-Si..284
    8.2.3 Heterojunction of *n*-ZnO-Nanowire/*p*-SiC 287
    8.2.4 Heterojunction between *n*-Type ZnO Nanowires and *p*-Type
        Organic Semiconductors...287
        8.2.4.1 Structure of Conductive-Glass/ZnO-Nanowire/
            Organic-Semiconductor/Metal..288
        8.2.4.2 Structure of Glass/PEDOT:PSS/Polymer-Semiconductor/
            ZnO-Nanowire/Metal..291
        8.2.4.3 Bendable LEDs Based on ZnO Nanowires and Organic
            Semiconductors 293
8.3 Hybrid LEDs Based on *n*-ZnO Nanowires Laid on *p*-Si 297
8.4 Hybrid Organic–Inorganic LEDs Based on Free-Standing ZnO Nanowire Films.....301
    8.4.1 Fabrication of the Nanowire Film 301
    8.4.2 Structure of *n*-ZnO-Nanowire/*p*-Si..302
    8.4.3 Structure of *n*-ZnO-Nanowire-Film/*p*-Organic-Semiconductor 303
8.5 Conclusion 305
References..306

8.1 Introduction

ZnO is a wide band-gap semiconductor that has several desirable properties for optoelectronic devices. With its large exciton binding energy of ~60 meV, ZnO is a promising candidate for high stability, room-temperature luminescent and lasing devices [1]. Ultraviolet light-emitting diodes (LEDs) based on ZnO homojunctions have been reported [2,3], while preparing stable *p*-type ZnO is still a challenge. An alternative way is to use other *p*-type semiconductors, either inorganic or organic, to form heterojunctions with the naturally *n*-type ZnO. The crystal structure of wurtzite ZnO can be described as Zn and O atomic layers alternately stacked along the [0001] direction. Because of the fastest growth rate over the polar (0001) facet, ZnO crystals tend to grow into one-dimensional structures, such as nanowires and nanobelts. Since the first report of ZnO nanobelts in 2001 [4], ZnO nanostructures have been particularly studied for their potential applications in nano-sized devices. Various growth methods have been developed for growing ZnO nanostructures, such as chemical vapor deposition (CVD), metal-organic CVD (MOCVD), aqueous growth, and electrodeposition [5]. Based on the successful synthesis of ZnO nanowires/nanorods, various types of hybrid LEDs were made. Inorganic *p*-type semiconductors, such as GaN, Si, and SiC, have been used as substrates to grown ZnO nanorods/nanowires for making LEDs. GaN is an ideal material that matches ZnO not only in the crystal structure but also in the energy band levels. However, to prepare Mg-doped *p*-GaN films via epitaxial growth is still costly. Compared with the inorganic semiconductors, the organic semiconductors are inexpensive and have many options to select, for a large variety of *p*-type polymer or small-molecule semiconductors are now commercially available. The organic semiconductor has the limitation of durability and environmental stability. Many polymer semiconductors are susceptible to damage by humidity or mere exposure to oxygen in the air. Also, the carrier mobilities of polymer semiconductors are generally lower than the inorganic semiconductors. However, the combination of polymer semiconductors and ZnO nanostructures opens the way for making flexible LEDs. There are few reports on the hybrid LEDs based on ZnO/polymer heterojunctions, some of them showed the characteristic UV electroluminescence (EL) of ZnO.

This chapter reports recent progress of the hybrid LEDs based on ZnO nanowires and other inorganic/organic semiconductors. We provide an overview of the ZnO-nanowire-based hybrid LEDs from the perspectives of the device configuration, growth methods of ZnO nanowires, and the selection of *p*-type semiconductors. Also, the device performances and remaining issues are presented.

8.2 Hybrid LEDs Based on Vertically Aligned ZnO Nanowires

8.2.1 Heterostructure of *n*-ZnO-Nanowire/*p*-GaN

The low-lattice constant misfit of ~1.9% between wurtzite ZnO and GaN facilitates the vertical growth of ZnO nanowires on the GaN substrate [6,7]. The GaN film is conventionally grown onto a single-crystal sapphire substrate by epitaxial method. By using Mg-doped *p*-type GaN film as substrate for growing ZnO nanowires/nanorods, heterojunctions are spontaneously formed at the interface between ZnO and GaN. In 2004, Park and Yi for the first time reported hybrid LEDs based on ZnO nanorods grown onto *p*-GaN by MOCVD [6]. Diethyl-zinc and O_2 were employed as reactant sources. The structure of the device is illustrated in Figure 8.1a. The free space between the individual ZnO nanowires was filled with photoresist polymer, followed by selective etching under oxygen plasma to expose the nanowire tips. Onto the nanowire arrays, Ti and Au layers were deposited in order as electrodes. EL measurements revealed visible emissions and relatively low UV emission in their devices. Figure 8.1b shows the energy band diagram of the *n*-ZnO/*p*-GaN heterojunction in equilibrium. It can be seen that the holes from *p*-GaN encounter a small energy barrier to be transferred to the valence band (E_V) of ZnO. However, the crystal defects in ZnO usually lead to visible emissions, which are dominant in the EL spectra. Since 2004, other research groups have reported hybrid LEDs based on the structure of *n*-ZnO-nanowire/*p*-GaN. The devices varied in synthesis routes of ZnO nanowires, insulating materials filling the space between nanowires and the top electrode materials. Zhang et al. grew ZnO nanowires onto *p*-GaN in a tube furnace by thermal evaporation

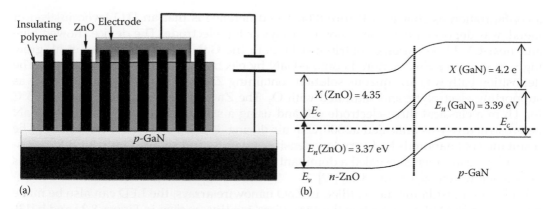

FIGURE 8.1
(a) Schematic of the LED based on vertically aligned ZnO nanowires grown onto p-GaN. (b) Energy band diagram of the n-ZnO/p-GaN heterojunction.

of Zn powder in oxygen atmosphere [8]. After filling the spaces between nanowires with poly(methyl methacrylate) (PMMA) and then exposing the nanowires' top tips by oxygen plasma etching, an ITO film was deposited onto the nanowire arrays. EL measurements revealed weak UV emission and relatively strong visible emissions, as seen in Figure 8.2a. Further work in their group used well-ordered ZnO nanowires grown onto p-GaN by a hydrothermal method combined with e-beam lithography technique [9]. For growing patterned ZnO nanowires, first, periodic holes in a layer of PMMA coated onto p-GaN were made by the e-beam lithography process. GaN surface is selectively exposed at the holes in PMMA layer for ZnO nucleation and epitaxial growth. Second, the PMMA-coated GaN substrate was put into the nutrient solution containing 5 mM 1:1 ratio of zinc nitrate and hexamethylenetetramine at 70°C–95°C for ~24 h [10], after which patterned ZnO nanowires were obtained. Based on the ZnO nanowire arrays, the device was fabricated with

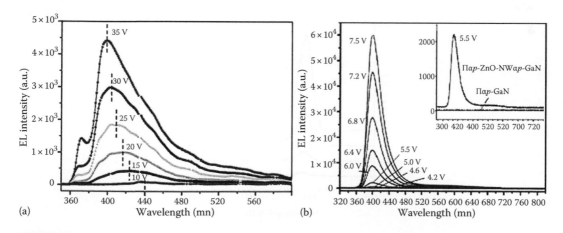

FIGURE 8.2
Room-temperature EL spectra of n-ZnO-nanowire/p-GaN heterostructure LEDs with the ZnO nanowires grown by (a) CVD method and (b) electrodeposition. (From Zhang, X.M., Lu, M.Y., Zhang, Y., Chen, L.-J., and Wang, Z.L.: Fabrication of a high-brightness blue-light-emitting diode using a ZnO nanowire array grown on p-GaN thin film. *Adv. Mater.* 2009. 21. 1; Lupan, O., Pauporte, T., and Viana, B.: Low-voltage UV-electroluminescence from ZnO-nanowire-array/p-GaN light-emitting diodes. *Adv. Mater.* 2010. 22. 3298. Copyright Wiley-VCH Verlag GmbH & Co. KGaA. Reproduced with permission.)

a configuration as shown in Figure 8.1a. The difference is that an ITO layer, instead of metal, was deposited onto the nanowire arrays as top electrode. The device showed blue and near-UV EL emissions, contributed by both the GaN and ZnO. Lupan et al. grew high-crystallinity ZnO nanorods onto *p*-GaN by electrochemical deposition [11,12]. The deposition bath was an aqueous solution containing $ZnCl_2$ (0.2 mM), plus KCl (0.1 M) as supporting electrolyte, and saturated with O_2. The ZnO nanowires were grown at 85°C for 1 h in a classical three-electrode cell and using a static electrode. The *n*-ZnO/*p*-GaN heterostructure was annealed at 300°C in air during 30 min before LED fabrication. The resultant ZnO nanorods had very low density of defects and showed only NEB PL emission. EL measurements revealed a dominant UV emission at 397 nm when the device was biased at low voltage (Figure 8.2b) [12].

Without an insulating media filled in ZnO nanowire arrays, the LED can also be made by coating a ZnO film connecting the nanowires' top tips, as seen in Figure 8.3a and b [13]. In that work, the ZnO nanowires were grown onto p^+-GaN by MOCVD. A 0.4 μm thick

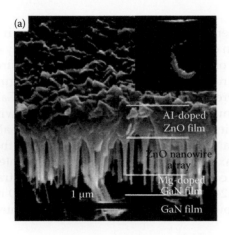
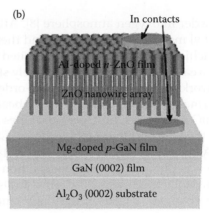

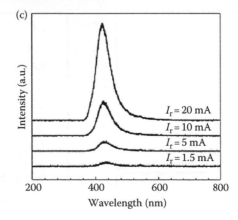

FIGURE 8.3
(a) Tilted-view SEM image and (b) schematic illustration of the *p*-GaN/ZnO-nanowire-array/Al-doped-ZnO-film structures for GaN/ZnO heterojunction diode applications. The inset in (a) is a photograph of blue-light emission of the heterojunction diodes observed through a microscope at the forward current of 10 mA. (c) EL spectra of the LEDs. (From Jeong, M.C., Oh, B.Y., Ham, M.H., Lee, S.W., and Myoung, J.M.: ZnO-nanowire-inserted GaN/ZnO heterojunction light-emitting diodes. *Small*. 2007. 3. 568. Copyright Wiley-VCH Verlag GmbH & Co. KGaA. Reproduced with permission.)

Al-doped ZnO film was deposited onto the nanowire arrays by RF magnetron sputtering. Figure 8.3a shows the titled-view SEM image of a device. The inset in Figure 8.3a is an optical photograph of light emission from the heterojunction diodes observed from a microscope at the forward current of 10 mA. Figure 8.3b schematically illustrates the device structure. The EL spectra of the device under various forward current are shown in Figure 8.3c. It was concluded that the blue EL emission was from Mg-doped GaN.

It is demonstrated that growing defect-free ZnO nanowires is a key factor to achieve dominant UV EL from the hybrid LEDs. While for the n-ZnO/p-GaN case, the hole mobility of GaN also plays an important role.

The crystal defects in ZnO can be utilized to emit white light from n-ZnO/p-GaN heterostructure LEDs. White-light LEDs are being intensively studied aiming to high efficiency illuminators. The white-light LEDs in market are normally based on GaN homojunctions combined with phosphors. Light emissions related to the defects in ZnO mostly sit within the visible range. Chen et al. reported the white-light LEDs based on Ga-doped ZnO nanowires grown onto p-GaN [14]. The nanowires were grown by a thermal CVD, with Zn and Ga powders as precursors. During the nanowires' growth, argon and oxygen were introduced into the furnace at a constant temperature of 600°C. For comparison, undoped ZnO nanowires were also grown with this method without using Ga. To make a device based on the nanowire arrays, an ITO-coated glass substrate was etched to form a 0.25 cm wide ditch dividing the ITO film into two parts. The ZnO nanowire arrays on GaN were placed on the ITO/glass substrate with the nanowires' tips directly contacting with ITO, as shown in Figure 8.4a. Therefore, it is an n-p-n heterojunction when the separated ITO films are used as electrodes. This device was then annealed at 200°C for 10 min. Afterward, spin-on-glass (SOG) was used to bind the LED. This flip-chip package process formed a good ohmic contact and did not deposit any material on the ZnO nanowire tips. Figure 8.4c shows the tilted view SEM image of the ZnO nanowires vertically grown onto GaN substrate. The inset is an x-ray diffraction pattern of the ZnO nanowire arrays. The diffraction peaks of the sample at 34.4°, 34.8°, and 41.9° were indexed to hexagonal wurtzite ZnO, GaN, and sapphire, respectively. Figure 8.4e shows the data of energy-dispersive x-ray (EDX) spectrum of the nanowires, confirming the Ga-doping in ZnO. Figure 8.4b displays the room-temperature PL spectra of the ZnO nanowire, ZnO nanowire doped with Ga, and the p-GaN film. The PL spectrum of undoped ZnO nanowires shows an ultraviolet emission peak at 378 nm, which is attributed to the near-band-edge (NBE) emission of ZnO. The PL spectrum of Ga-doped ZnO nanowires shows two peaks, one at 384 nm and the other at 500 nm. The Ga-doping results in the red-shift of the NBE emission peak of ZnO. The visible emission of ZnO is related to crystal defects. Figure 8.4d presents the EL spectra from the ZnO:Ga LED and ZnO LED at 20 mA. The peak from the ZnO LED at 417 nm is attributed to radiative recombination at the defects that unexpectedly form when carriers are injected at forward bias. The shift in wavelength and broadening of the EL band from the ZnO:Ga LED are attributed to the doping of Ga into the ZnO nanowire and the unexpected defects between sapphire and p-GaN. The broad EL band of the ZnO:Ga LED renders white light to the naked eye.

Though ordered ZnO nanowires can be synthesized by different methods, controlling the defects in ZnO is a big issue for LED application. Growing ordered ZnO nanowires onto GaN can be easily achieved, especially by using the chemical-bath-based growth methods that are promising for low-cost and large-scale synthesis, while the fabrication processes of such hybrid LEDs are still complicated compared with the commercially available GaN homojunction LEDs. In regard to the ZnO nanowire, it can be a waveguide confining lights to propagate along the axial direction. Also, optically pumped lasing had been reported

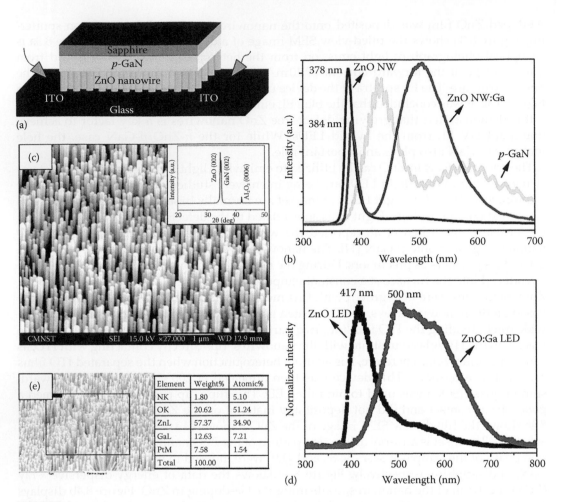

FIGURE 8.4
(a) Scheme of the heterojunction LED based on Ga-doped ZnO nanowire arrays grown onto *p*-GaN. (b) Room-temperature PL spectra of ZnO nanowire, ZnO nanowire doped with Ga, and *p*-GaN film. (c) A SEM image of the ZnO nanowire arrays grown on GaN. (d) EL spectra of ZnO LED and ZnO:Ga LED at a DC current of 20 mA. EDX analysis of ZnO:Ga nanowires grown on the *p*-GaN substrate. (Reproduced with permission from Chen, C.H., Chang, S.J., Chang, S.P., Li, M.J., Chen, I.C., Hsueh, T.J., Hsu, A.D., and Hsu, C.L., Fabrication of a white-light-emitting diode by doping gallium into ZnO nanowire on *p*-GaN substrate, *J. Phys. Chem. C*, 114, 12422. Copyright 2010 American Chemical Society.)

from single ZnO nanowires. The advantages of ZnO nanowire that surpass the bulk film should be stressed in further studies on the *n*-ZnO-nanowire/*p*-GaN LEDs. For the hybrid LEDs based on well-faceted ZnO nanowires with high crystallinity, it is possible to realize electrically driven lasing from ZnO nanowire cavities.

8.2.2 Heterojunction of *n*-ZnO-Nanowire/*p*-Si

Among the *p*-type inorganic semiconductors, Si is an important material that is inexpensive compared with GaN and SiC, though both materials match ZnO well in lattice constant. Devices based on the *n*-ZnO/*p*-Si thin-film heterojunctions have been widely studied, including applications as photodiodes and LEDs [15–17]. Much work has been devoted to the growth of ZnO nanowires/nanorods onto Si, such as by using metal catalysts, or ZnO seed layers. Hybrid LEDs based on ZnO nanowires grown onto p-Si were fabricated, with the device configuration similar to that in Figure 8.1a. In 2006, Sun et al. reported the UV EL from the heterojunction between ZnO nanorods and p-type Si [18]. The ZnO nanorods were hydrothermally grown onto heavily doped p-Si in an aqueous solution containing 0.05 M zinc nitrate and 0.05 M methenamine. Because of the large lattice mismatch (~40%) between ZnO and Si, the ZnO nanorods grown onto Si are generally tilted. To make an LED, the space between nanorods was filled with photoresist, which was subsequently thinned to expose the nanorod tips. A layer of indium was deposited onto the nanorod arrays as electrode. Figure 8.5a shows the room temperature of the device at different voltages. The UV peak at 387 nm is related to the exciton recombination of ZnO. The broad band around 535 nm is attributed to the transition from the shallow donor level of oxygen vacancies to the valence band. As seen from the energy band diagram in Figure 8.5b, there is a large barrier about 2.55 eV for the holes injected from the valence band of Si to that of ZnO. Therefore, the defect levels within the bang gap of ZnO could easily accept the holes from Si, impairing the UV EL emission of ZnO.

To transfer holes from the p-Si to the n-ZnO an interlayer of other p-type semiconductor can be used. The LEDs with a $CuAlO_2$ interlayer between ZnO nanorods and p-Si were fabricated by Ling et al., as illustrated in Figure 8.6a [19]. The p-type $CuAlO_2$ films were deposited onto heavily-doped p^+-Si substrates by direct-current magnetron sputtering at room temperature. The postgrowth annealing of the films was conducted in air for 30 min ranging from 900°C to 1050°C followed by the deposition of a thin ZnO film (~30-nm thick) by sputtering, which served as the seed layer for ZnO nanorods growth. Vertically aligned ZnO nanorod arrays were grown on the prepared substrates by vapor-phase transport method at 600°C–700°C. SOG was chosen as the spacer layer to fill up the interspace of ZnO nanorods by spin coating. Au electrodes were also prepared by sputtering at both sides of the devices. The top cathode electrodes were patterned into circular shape using a shadow mask, and the bottom anode electrodes were deposited directly on the back side of the Si wafers without any patterning. Figure 8.6b shows the cross-sectional view SEM image of the ZnO nanorod arrays grown on a ZnO-buffered

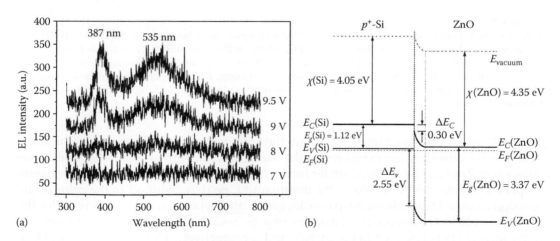

FIGURE 8.5
(a) EL spectra and (b) energy-band diagram of the n-ZnO-nanorod/p-Si heterojunction. (From Sun, H., Zhang, Q.F., and Wu, J.L., Electroluminescence from ZnO nanorods with an n-ZnO/p-Si heterojunction structure, *Nanotechnology*, 17, 2271, 2006. Reproduced with permission from the Institute of Physics.)

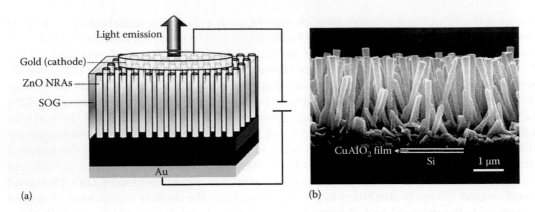

FIGURE 8.6

(a) Schematic of the p^+-Si/p-CuAlO$_2$/n-ZnO nanorod LED. (b) Cross-sectional SEM image of the ZnO nanorods grown on a ZnO coated p-CuAlO$_2$/p^+-Si substrate. (Reproduced with permission from Ling, B., Zhao, J.L., Sun, X.W., Tan, S.T., Kyaw, A.K.K., Divayana, Y., and Dong, Z.L., Color tunable light-emitting diodes based on p^+-Si/p-CuAlO$_2$/n-ZnO nanorod array heterojunctions, *Appl. Phys. Lett.*, 97, 013101. Copyright 2010, American Institute of Physics.)

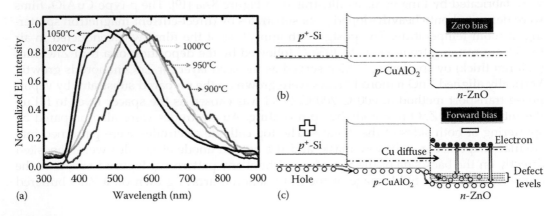

FIGURE 8.7

(a) Room-temperature EL spectra of p^+-Si/p-CuAlO$_2$/n-ZnO nanorod array heterostructured LEDs with CuAlO$_2$ film annealed at different temperatures, that is, 900°C, 1000°C, 1020°C, and 1050°C. (b) and (c) Schematic band diagram of the LED under zero bias and forward bias, respectively. (Reproduced with permission from Ling, B., Zhao, J.L., Sun, X.W., Tan, S.T., Kyaw, A.K.K., Divayana, Y., and Dong, Z.L., Color tunable light-emitting diodes based on p^+-Si/p-CuAlO$_2$/n-ZnO nanorod array heterojunctions, *Appl. Phys. Lett.*, 97, 013101. Copyright 2010, American Institute of Physics.)

p-CuAlO2/p^+-Si substrate. Figure 8.7a shows the room temperature EL spectra of the LEDs with various annealing temperatures for the CuAlO2 layer. Both excitonic and deep-level emission of ZnO can be detected under forward bias. With the decrease of the annealing temperature from 1050°C to 900°C, the dominant EL peak red-shifted as the color of the emission changed from bluish purple, cyan, green, and yellow to red emissions under the same bias of 20 V. The energy band diagram of p^+-Si/p-CuAlO2/n-ZnO at equilibrium and under forward bias is shown in Figure 8.7b and c, respectively. The band gaps of ZnO, Si, and CuAlO$_2$ are 3.37, 1.12, and 4.2 eV, respectively. Under forward bias, the holes are first injected from the valence band of Si into that of CuAlO$_2$ and then injected into the valence band or Cu-related deep defect level bands in ZnO, resulting in UV and visible emissions from ZnO, respectively. The gradual red-shift in the EL spectra is related to the band gap reduction of ZnO, according to the previous reports both in theoretical calculation and experimental work [20,21].

By inserting a *p*-type intermedium between *p*-Si and *n*-ZnO for efficiently injection holes into ZnO is a promising way for making UV LEDs based on ZnO. It leaves many rooms for this study, that is, selection of proper *p*-type semiconductors with high hole mobilities and deep E_V levels close to that of ZnO.

8.2.3 Heterojunction of *n*-ZnO-Nanowire/*p*-SiC

4H-SiC is a good candidate for ZnO-based heterostructure devices because of its wurtzite crystal structure matching and relatively small lattice mismatch of ~2% with ZnO. 4H-SiC is also a wide-band-gap semiconductor ($E_g = 3.23$ eV), suitable for achieving UV EL from the *n*-ZnO/*p*-SiC (4H) heterojunction. By growing ordered ZnO nanorods onto *p*-4H-SiC, Willander et al. had investigated the LEDs based on the *n*-ZnO/*p*-SiC heterojunctions, with the device configuration similar to that in Figure 8.1a [22–25]. The nanowires were grown by either the chemical-bath method [24] or the vapor-phase epitaxy method [25]. Figure 8.8a shows a SEM image of the vertically aligned ZnO nanorods grown onto *p*-SiC by a carbothermal reduction method. The device fabrication follows the scheme in Figure 8.8b. A thin layer of Ni/Al was deposited onto the *p*-SiC substrate, followed by the rapid annealing at 900°C for 3 min in Ar atmosphere for establishing an ohmic contact. After filling PMMA into the interspace between nanorods, the Al contacts of a diameter of 0.5 mm were evaporated onto a group of rods. Figure 8.8c shows an equilibrium energy band diagram of *p*-*n* heterojunction of *p*-4H-SiC and *n*-ZnO. The conduction band offset (ΔE_C) for electrons is $\Delta E_C = \chi$ZnO $- \chi$SiC $= 0.3$ eV, where χ is the electron affinity. While the valance band offset (ΔE_V) for holes is $\Delta E_V = E_g$(ZnO) $+ \Delta E_C - E_g$(SiC) $= 0.4$ eV. ΔE_V has a higher value than ΔE_C, which means that electron injection from *n*-ZnO to *p*-SiC is larger than hole injection from *p*-SiC to *n*-ZnO.

Figure 8.9a shows the room-temperature EL spectra of the LED with ZnO nanorods grown onto *p*-SiC by the CVD method. All the spectra at different forward injection currents show three peaks at 425 (violet), 527 (green), and 683 nm (red), which compose white color to the naked eye. These peaks are due to different deep level defects in the ZnO nanorods. Figure 8.9b shows the color coordinates, with a color rendering index of 93 and a color temperature of around 14,000 K. Figure 8.9c shows an optical photograph of the *n*-ZnO-nanorod/*p*-SiC heterojunction LED. In operation, the device shows bright white light as seen in Figure 8.9d.

8.2.4 Heterojunction between *n*-Type ZnO Nanowires and *p*-Type Organic Semiconductors

With the development of organic LEDs (OLEDs), various *p*-type organic semiconductors have been synthesized, which provide alternatives for making hybrid LEDs with *n*-ZnO. In comparison with the inorganic *p*-type semiconductors (such as GaN and SiC), the organic semiconductors are inexpensive and have many options for selection. Moreover, the adoption of organic semiconductors opens ways for making bendable light-emitting displays. Based on the ordered ZnO nanowires, two types of hybrid LEDs with organic semiconductors have been designed, as illustrated in Figure 8.10a and b. For the type in Figure 8.10a, organic semiconductors are applied onto the top of nanowire arrays. For the other type in Figure 8.10b, organic layers were first coated onto a substrate, followed by the vertical

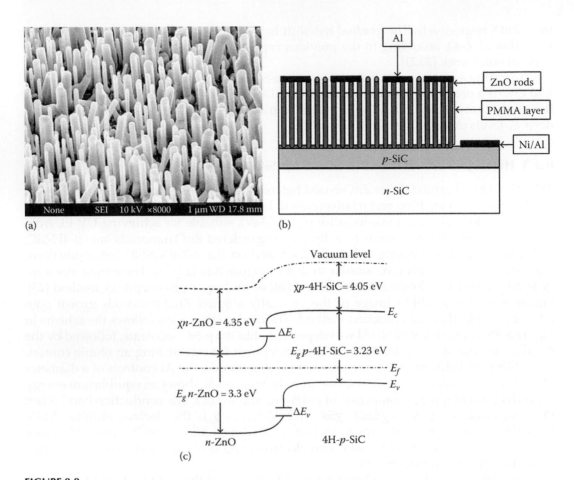

FIGURE 8.8
(a) SEM image of ZnO nanorods grown on 4H-SiC substrate. (b) Schematic illustration of ZnO/*p*-SiC hetero-structure device. (c) Energy-band diagram of *p-n* heterojunction of 4H-*p*-SiC and *n*-ZnO.

growth of ZnO nanowires from chemical bath at low temperature (<100°C). Recent works of hybrid LEDs based on the two structures in Figure 8.10a and b are reviewed in the following.

8.2.4.1 *Structure of Conductive-Glass/ZnO-Nanowire/Organic-Semiconductor/Metal*

For the type in Figure 8.10a, Könenkamp et al. reported for the first time the hybrid LEDs based on ZnO nanowires and poly(3,4-ethylenedioxythiophene):poly(styrenesulfonate) (PEDOT:PSS) [26,27]. In their work, free-standing ZnO nanorods were grown onto a fluorine-doped tin oxide (FTO) glass substrate by electrodeposition from aqueous solution. Polystyrene (PS) was spun coated onto the nanorod arrays to fill up the interspace between nanorods, followed by the coating of PEDOT:PSS. Gold was deposited onto the PEDOT:PSS layer as electrode. Though the visible EL emission related to crystal defects in ZnO primarily appeared, the UV EL ascribed to near-band-edge emission of ZnO was observed from the ZnO/PEDOT:PSS heterojunctions [26]. However, the UV EL from ZnO can be enhanced by annealing the device at 300°C in air [27]. Figure 8.11a shows the SEM image of ZnO nanowires electrodeposited onto a FTO glass substrate. Figure 8.11b emphasizes the nanowires embedded in PS that have undergone oxygen plasma etching to reduce the

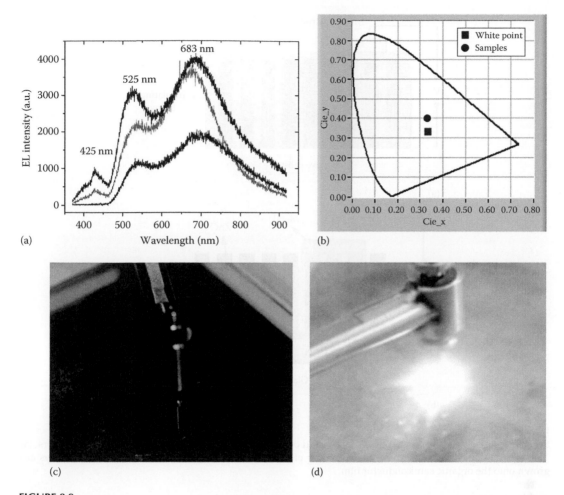

(a)

(b)

(c)

(d)

FIGURE 8.9
(a) EL spectra from the ZnO nanorods/p-4H-SiC heterojunction LEDs at different bias, where the ZnO nanorods were grown by the vapor–liquid–solid approach. (b) Color coordinate measurement of the ZnO nanorod/p-4H-SiC LED. (c) and (d) Typical photograph of the device before and in operation. (From Willander, M., Nur, O., Zhao, Q.X., Yang, L.L., Lorenz, M., Cao, B.Q. et al., Zinc oxide nanorod based photonic devices: Recent progress in growth, light emitting diodes and lasers, *Nanotechnology*, 20, 332001, 2009. Reproduced with permission from the Institute of Physics.)

coverage of nanowire tips. The room-temperature EL spectra of such a device annealed at 300°C for 2 h in air are shown in Figure 8.11c. The UV emission peak at 393 nm is ascribed to the bound-exciton transition.

Polymer semiconductors except PEDOT:PSS are dissolved in nonpolar organic solvent for spin coating. Therefore, the PMMA or PS filled in nanowire arrays would be dissolved when coating polymers from nonpolar solvent for making contacts with the nanowire tips. Differing from the polymer films made by spin coating or brushing, small-molecular organic semiconductors can be deposited by thermal evaporation in vacuum. Sun et al. had deposited N,N'-di(naphtha-2yl)-N,N'-diphenyl-benzidine (NPB), a small molecule organic semiconductor used for hole transport, onto the ZnO nanowire arrays by thermal evaporation [28]. The ZnO nanowires were grown onto ITO glass substrate by hydrothermal method. Before depositing NPB, PMMA was spun coated onto the nanowire arrays,

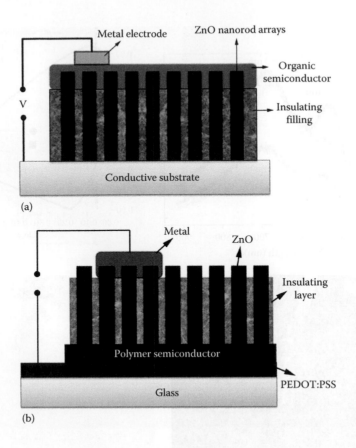

FIGURE 8.10
Schematic illustrations of two types of LEDs based on ZnO nanowire arrays and organic semiconductors. (a) The organic semiconductor layer is coated on top of the nanowire arrays. (b) The ZnO nanowire arrays are grown onto the organic semiconductor film.

which was subsequently etched by oxygen plasma to slightly expose the nanowire tips. The device structure is illustrated in Figure 8.12a. The light absorption spectrum and the PL spectrum of NPB and the PL spectrum of as-grown ZnO nanowires are shown in Figure 8.12b. Figure 8.12c shows the EL spectra of a device with 200 nm thick NPB under different currents. It is notable that an UV emission at 342 nm is revealed, showing a blue-shift compared with either the PL emission peak of NBP or the UV emission ZnO. The broad visible band in EL spectra is attributed to the transition between defect levels in ZnO. Figure 8.12d shows the EL spectra of a device with 300 nm thick NPB, in which the UV emission at 342 nm appears as well. However, in the case of ZnO/PEDOT:PSS heterojunction, the UV EL emission of ZnO shows a bit red-shift to the UV PL peak of ZnO in literatures [27]. The blue-shifted UV emission from ZnO is ascribed to the conduction band filling by electron accumulation at the ZnO/NPB interface, which can be explained by the energy diagrams in Figure 8.13a and b. As can be seen from the energy levels that the LUMO (lowest unoccupied molecular orbit) of NPB (2.1 eV) is much higher than that of PEDOT:PSS (3.3 eV), so the electrons from the conduction band of ZnO are accumulated at the interface, resulting in the band-filling effect as well as the blue-shift of UV EL emission (Figure 8.13b). However, for the PEDOT:PSS/ZnO heterojunction, electrons from ZnO are more freely to be transferred to the LUMO of PEDOT:PSS, leading to no band-filling effect.

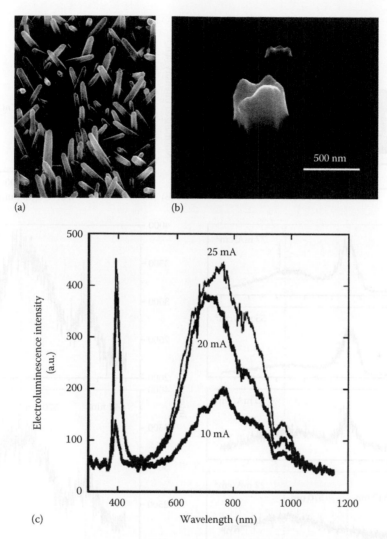

(a)

(b)

(c)

FIGURE 8.11

(a) SEM image of the ZnO nanorods grown onto FTO substrate by electrodeposition. (b) SEM image of a ZnO nanorod embedded in a PS layer that was subsequently plasma-etched to reduce the coverage of the nanorod tip. (c) EL spectra from the ZnO-nanorods/PEDOT: PSS heterojunction annealed at 300°C for 2 h in air. Current values refer to the driving current for a device area of 0.3 cm². (Reproduced with permission from Könenkamp, R., Word, R.C., and Godinez, M., Ultraviolet electroluminescence from ZnO/polymer heterojunction light-emitting diodes, *Nano Lett.*, 5, 2005. Copyright 2005 American Chemical Society.)

8.2.4.2 *Structure of Glass/PEDOT:PSS/Polymer-Semiconductor/ZnO-Nanowire/Metal*

As illustrated in Figure 8.10b, this approach is to grow ZnO nanowires onto a layer of *p*-type organic semiconductor that is precoated onto a substrate. Based on this structure, a group led by M. Willander had studied the hybrid LEDs by using various *p*-type polymer semiconductors [29,30]. To make such a device, the PEDOT:PSS is first coated onto a glass substrate and baked to form a uniform film with thickness of 60 nm. Over the PEDOT:PSS layer, *p*-type polymer semiconductors are spun coated. Either multiple layers or a single layer consisting of two different polymer semiconductors are used for hole transport. ZnO

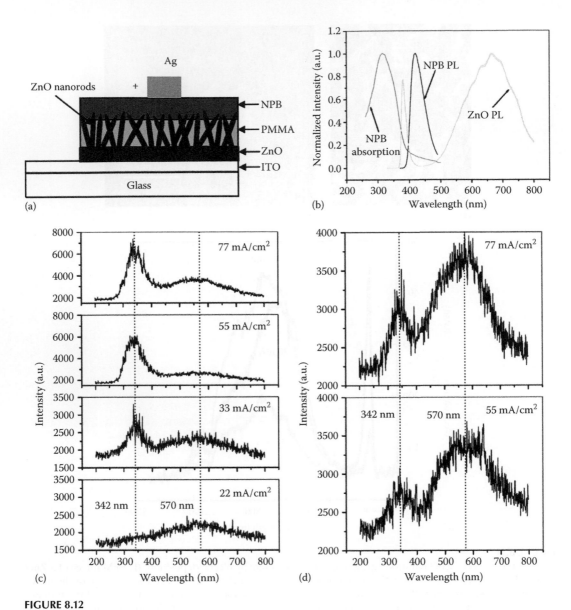

FIGURE 8.12
(a) Schematic diagram of the inorganic/organic heterostructure LED. (b) PL spectrum of the ZnO nanorods and absorption and PL spectrum of NPB. (c) and (d) EL spectra of the ZnO/NPB heterostructure LEDs, with different NPB thicknesses of (c) 200 nm and (d) 300 nm. (Reproduced with permission from Sun, X.W., Huang, J.Z., Wang, J.X., and Xu, Z., A ZnO nanorod inorganic/organic heterostructure light-emitting diode emitting at 342 nm, *Nano Lett.*, 8, 1219. Copyright 2008 American Chemical Society.)

nanowire arrays are grown onto the hole-transport layer by an aqueous chemical solution method. The growth temperature is below 100°C, so the polymer layer(s) is unaffected. Before depositing Au electrode onto ZnO nanowire arrays, a photoresist polymer is used to fill the spaces between nanowires and then partially etched by oxygen plasma to expose the nanowire tips. Figure 8.14 shows a cross-sectional SEM image of a typical device based on the ZnO nanorods and polymer semiconductors. Because of the different

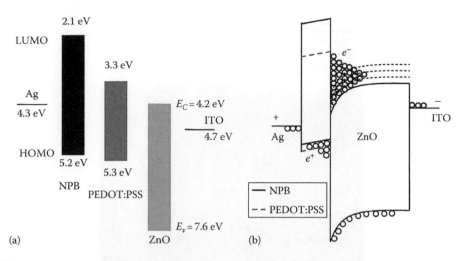

FIGURE 8.13
(a) Energy band diagram of ZnO, NPB, PEDOT:PSS, Ag, and ITO. (b) Energy band diagram of the ZnO/NPB heterostructure LED under a positive bias. (Reproduced with permission from Sun, X.W., Huang, J.Z., Wang, J.X., and Xu, Z., A ZnO nanorod inorganic/organic heterostructure light-emitting diode emitting at 342 nm, *Nano Lett.*, 8, 1219. Copyright 2008 American Chemical Society.)

energy levels of HOMO and LUMO, as well as the carrier mobilities, of the organic semiconductors, various polymers had been used for the making LEDs with the structure of glass/PEDOT:PSS/polymers/ZnO/Au, such as poly(9,9-dioctylfluorene) (PFO), 4.4′-bis[*N*-(1-napthyl)-*N*-phenyl-amino]biphenyl (NPD), poly(*N*-vinylcarbazole) (PVK), and poly(9,9-dioctyl-fluorene-co-*N*(4-butylphenyl)diphenylamine) (TFB). However, in EL measurements, the characteristic UV emission of ZnO failed to appear. Instead, the visible emissions contributed by both the polymer semiconductors and crystal defects in ZnO were observed. Figure 8.15a and b shows the current–voltage (*I–V*) curves of two devices, with the structures of PEDOT:PSS/NPD-PFO/ZnO and PEDOT:PSS/PVK-TFB/ZnO, respectively. The insets are corresponding energy band diagrams of the devices. The EL spectra in Figure 8.15b and c correspond to the heterojunctions of NPD-PFO/ZnO and PVK-TFB/ZnO, respectively. The observed emission covers wavelengths ranging from a peak centered at ∼425 nm up to a peak centered at 525 nm for both configurations. In Figure 8.15c, the first sharp high-intensity peak centered at 458 belongs to the PFO polymer. The second peak at 524 nm, which lies within the ZnO green emission band, is due to Zn-vacancies. The third peak centered at 495 nm, observed in both Figure 8.15c and d, originates from the oxygen vacancies in ZnO. In Figure 8.15d, the blue emission at 455 nm corresponds to the emission of PVK polymer. The other two peaks at 495 and 520 nm are related to the defects in ZnO.

8.2.4.3 Bendable LEDs Based on ZnO Nanowires and Organic Semiconductors

There is a growing interest and investment in the development of flexible electronic devices for foldable and unbreakable displays. The OLEDs based on bendable or even stretchable substrates have been studied, taking advantage of the high flexibility of organic semiconductors. Based on the successful fabrication of LEDs on glass substrates by combining polymer semiconductors and ZnO nanowires, flexible LEDs were also studied by using bendable substrates.

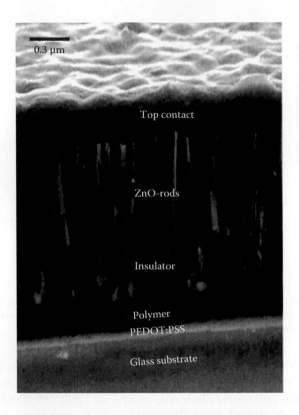

FIGURE 8.14
Cross-sectional SEM of a ZnO/polymer multi-layer LED fabricated on glass substrate. (From Willander, M., Nur, O., Zhao, Q.X., Yang, L.L., Lorenz, M., Cao, B.Q. et al., Zinc oxide nanorod based photonic devices: Recent progress in growth, light emitting diodes and lasers, *Nanotechnology*, 20, 332001, 2009. Reproduced with permission from the Institute of Physics.)

Nadarajah et al. had grown ZnO nanowires onto ITO-coated PET substrate by electrodeposition for making flexible LEDs by utilizing PEDOT:PSS [31]. The electrodeposition was carried out at 80°C in an aqueous electrolyte containing 3 mM $ZnCl_2$ and 0.1 M KCl. For growing Al-doped ZnO nanowires, 5 μM $AlCl_3$ was added into the electrolyte. Oxygen bubbling and magnetic stirring were provided during the electrodeposition to produce an oxygen-saturated electrolyte solution. To show that the ZnO nanowires grown onto a flexible substrate remain attached when the substrate is bended, Figure 8.16a shows the SEM image of ZnO nanowires on a bent Au film with small curvature radius of <10 μm. The fabrication process of hybrid LEDs based on ZnO nanowires onto plastic substrates is similar as that described above for the device structure in Figure 8.10a. The device structure of a bendable LED is schematically shown in Figure 8.16c. PS is used as an insulating material filling the free space between nanowires. After reducing the thickness of the PS film by oxygen plasma etching to expose the nanowire tips, a top contact consisting of a thin PEDOT:PSS layer and an evaporated Au film are provided to serve as the hole injection anode in the LED. Figure 8.16b shows the EL spectra for undoped and Al-doped nanowires on a flexible ITO-coated substrate in comparison to spectra obtained on planar FTO-coated glass substrates. The spectra from flexible devices have slightly less intensity, but the spectral distribution is qualitatively similar to those on solid substrates. The UV emission of ZnO failed to appear, leaving a broad emission band covering the range 500–1100 nm with a peak that is slightly red-shifted from the spectra

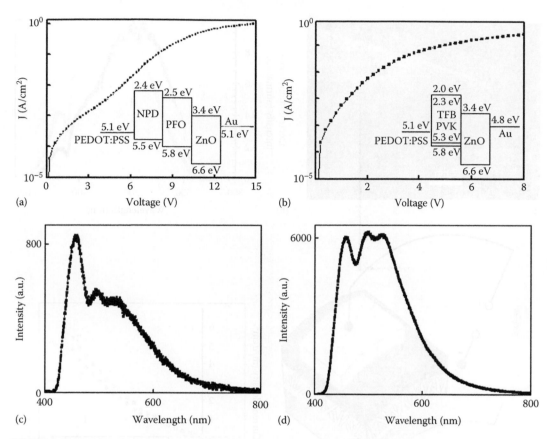

FIGURE 8.15

(a) and (b) Current density–voltage characteristics of NPD-PFO/ZnO structure and PVK-TFB/ZnO structure, respectively. The inset shows a schematic diagram of the energy band diagram for: NPD-PFO and PVK-TFB structures in combination with ZnO. (c) and (d) EL spectra of the hybrid LEDs corresponding to (a) and (b), respectively. The measurement conditions are (c) 18 V and 0.12 mA and (d) 14 V and 0.10 mA. (From Willander, M., Nur, O., Zhao, Q.X., Yang, L.L., Lorenz, M., Cao, B.Q. et al., Zinc oxide nanorod based photonic devices: Recent progress in growth, light emitting diodes and lasers, *Nanotechnology*, 20, 332001, 2009. Reproduced with permission from the Institute of Physics.)

obtained on the glass substrate. Figure 8.16d shows the *I–V* for the flexible LEDs. The diode characteristics and the rectification are mainly produced at the back contact, which is a n⁻-i-p⁺ heterojunction with the layer sequence, ZnO/PS/PEDOT:PSS/Au.

Base on the structure in Figure 8.10b, Zainelabdin et al. had fabricated flexible LEDs by using plastic substrate coated with PEDOT:PSS, as schematically shown in Figure 8.17a [32]. The TFB hole transporting solution was spin-coated onto the substrate, followed by baking at 75°C for 10 min. The PFO, a blue luminescent polymer, was then spin-coated onto the TFB layer and subsequently baked at 75°C for 15 min. ZnO nanorods were grown onto the TFB layer by a hydrothermal method. Prior to the contact deposition, a photoresist was spin coated to insulate the ZnO nanorods from each other and then cured at 90°C for 2 min. Oxygen-reactive ion etching was then employed to partially etch the upper part of the photoresist that covered ZnO nanorod tips. Ohmic contact was achieved by thermally evaporating Au/Ti layers on top of the device. Silver paste was used to act as a bottom contact to the PEDOT:PSS substrate. A photograph of the

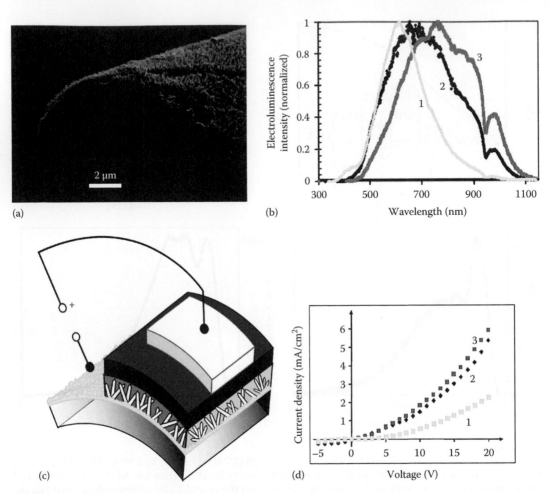

FIGURE 8.16
SEM image of the ZnO nanowires on a bent Au with curvature radius of <10 μm. (b) Normalized EL spectra for various LED structures. (c) Design scheme for a flexible LED structure consisting of vertically oriented single crystalline nanowires grown on a polymeric ITO-coated substrate. The top contact consists of *p*-type polymer and an evaporated Au layer. Light is emitting though the transparent polymer. (d) Electrical characteristics of various LED structures. Curves 1: Undoped nanowires grown on SnO$_2$-coated glass substrates. Curves 2: Al-doped nanowires grown on flexible ITO-coated PET foils. Curves 3: Undoped nanowires grown on flexible ITO-coated PET foils. (Reproduced with permission from Nadarajah, A., Word, R.C., Meiss, J., and Könenkamp, P., Flexible inorganic nanowire light-emitting diode, *Nano Lett.*, 8, 534. Copyright 2008 American Chemical Society.)

flexible LEDs is shown in Figure 8.17c. The *I–V* curve of the device with a structure Ag/PEDOT:PSS/TFB/PFO/ZnO/Ti/Au is shown in Figure 8.17b. Clear diode behavior can be found. Figure 8.17d shows the EL spectra of a device under constant bias of 20 V but at different temperatures. The intrinsic white light is covering the whole visible region from 420 to 800 nm as a broad peak centered at ~560 nm. The intermixing of the blue light generated by PFO layer with the green and red/orange bands produced by the defect-related emissions of ZnO have led to the observed broad band. The emission intensity was suppressed to ~80% by changing the device temperature from 20°C to 60°C. The flexible device was bent at large angles (>60°) and still remained its electrooptical characteristics.

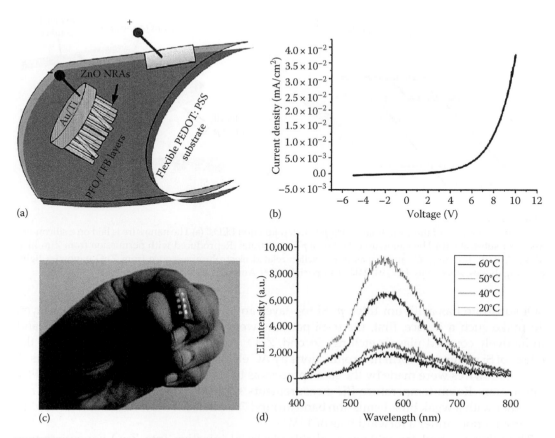

FIGURE 8.17
(a) Schematic diagram of the flexible white LED showing the different parts of the device. (b) *I–V* characteristics of the device. (c) A digital photograph of the flexible PEDOT:PSS substrate containing 20 LEDs bent at large angle of around 60°. (d) EL spectra of the device at different ambient temperatures at a bias voltage of 20 V, showing the depreciation of the intrinsic white light intensity with increasing the ambient temperature. (Reproduced with permission from Zainelabdin, A., Zaman, S., Amin, G., Nur, O., Zhao, Q.X., Yang, L.L., Lorenz, M., Cao, B.Q. et al., Zinc oxide nanorods based photonic devices: Recent progress in growth, light emitting diodes and lasers, *Nanotechnology*, 20, 332001. Copyright 2009 Institute of Physics.)

8.3 Hybrid LEDs Based on *n*-ZnO Nanowires Laid on *p*-Si

The electrical and optoelectronic properties of individual ZnO nanowires have been studied by the fabrication of single-nanowire devices with lithography technique. A conventional way is to use a lithography process to make metal contacts onto a nanowire laid on a planar substrate. For example, field-effect transistors (FETs) based on ZnO nanowires on SiO$_2$-coated Si substrate have been studied [33]. Building on the FETs of ZnO nanowires, some researchers fabricated the *n*-ZnO-nanowires/*p*-Si heterostructure LEDs to study the EL properties of single nanowires. In 2006, Yang et al. reported the ultraviolet EL from single ZnO nanowires laid on *p*-Si [34]. The ZnO nanowires were synthesized within a tube furnace by a carbothermal reduction method. The *n*-ZnO-nanowire/*p$^+$*-Si heterojunctions were fabricated on silicon-on-insulator (SOI) substrates, as illustrated in Figure 8.18a. Each

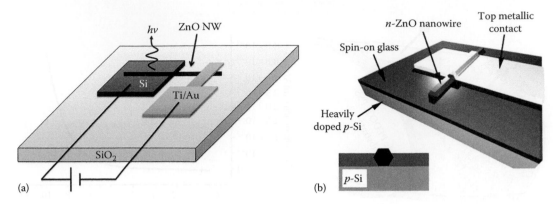

FIGURE 8.18
Schematic structures of the *n*-ZnO-nanowire/*p*-Si heterojunction LEDs. (a) The nanowire is laid on a silicon-on-insulator substrate. (b) The nanowire is laid on a *p*-Si substrate. (Reproduced with permission from Zimmler, M.A., Voss, T., Ronning, C., and Capasso, F., Exciton-related electroluminescence from ZnO nanowire light-emitting diodes, *Appl. Phys. Lett.*, 94, 241120. Copyright 2009, American Institute of Physics.)

SOI substrate has a 100 nm thick p^+-Si top layer and a 380 nm thick SiO_2 insulator layer. To make such a device, first, the p^+-Si patterns were fabricated by UV lithography and inductively coupled plasma etching. Second, ZnO nanowires were assembled across the edges of Si patterns to form heterojunctions by AC electric fields. Finally, ohmic contacts to ZnO nanowires were made by the same process as that used in ZnO single-nanowire FET fabrication. Room-temperature EL measurements showed a sharp peak around 382 nm together with a weak broad emission band around 700 nm from the *n*-ZnO-nanowire/p^+-Si heterojunction under a forward bias of 10 V.

The other method for achieving reliable electrical injection into ZnO nanowires was proposed by Capasso et al., as seen in Figure 8.18b [35,36]. The ZnO nanowires laid on an Si substrate are imbedded in an insulating layer, either PMMA or SOG, with the nanowires partially exposed to let the top sides contact with top metal electrodes. For the device fabrication, first, the nanowires grown by the vapor–liquid–solid growth technique in a horizontal tube furnace are collected and then dissolved in ethanol. Second, the nanowires are randomly dispersed on a heavily doped *p*-type Si substrate by dropping the nanowire suspension. A thin film of PMMA or hydrogen silsesquioxane (HSQ, known as "spin-on glass") can be spun coated onto the substrate as the insulating spacer. If using PMMA, the insulating film is selectively exposed to e-beam by a focused ion beam system. Therefore, after washing with acetone, the nanowire is embedded in a patch of cross-linked PMMA film, with its top surface exposed. After an O_2 plasma cleaning step to remove any organic residue on the exposed surface of the nanowire, Ti/Au is deposited to form the top contact. A rapid thermal annealing at ~300°C for 60 s in He–H_2 is necessary for increasing the forward current and EL intensity [35]. If using HSQ, after cross-linking it with an e-beam writer, the film is thinned down by reactive ion etching using CF_4 gas, which results in the top surfaces of the nanowires being exposed but not the Si substrate [36]. Broad EL emission band had been observed from the *n*-ZnO-nanowire/*p*-Si heterojunctions with PMMA as insulating layer. Figure 8.19a through c shows the EL spectra of three different devices, measured at room temperature in ambient air. Figure 8.19d is a PL spectrum showing the visible emission centered at 556 nm. Due to abundant defects, the characteristic UV emission of ZnO is not emerged in the PL spectrum. In the EL spectra, the NBE emission of ZnO at 380 nm slightly appears together with the relatively strong and broad subband-gap

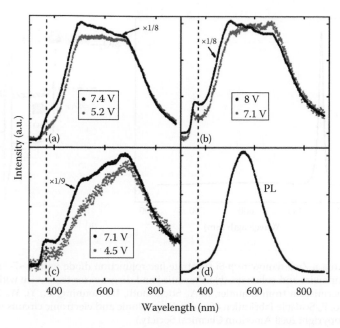

FIGURE 8.19

(a through c) Room-temperature EL spectra of three different LEDs made by laying ZnO nanowires on *p*-Si. (d) PL spectrum of a typical single ZnO nanowire. (Reproduced with permission from Bao, J., Zimmler, M.A., and Capasso, F., Broadband ZnO single-nanowire light-emitting diode, *Nano Lett.*, 6, 1719. Copyright 2006 American Chemical Society.)

emissions. However, in an extension work of their group, strong UV EL was achieved [36]. On the *p*-Si substrate, an oxide layer of ~7–8 nm in thickness was intentionally deposited by plasma-enhanced CVD. Also, they used ZnO nanowires with less crystal defects and HSQ instead of PMMA as the insulating layer. Figure 8.20a shows the EL spectrum of such a device measured at room temperature, revealing the dominant UV emission and negligible deep-level emission. Figure 8.20b shows the device configuration and energy-band diagram. It is noted that an oxide layer between ZnO and Si blocks the electrons from ZnO to Si, promoting the recombination of electrons and holes tunneled across the oxide layer from *p*-Si.

The low-temperature EL spectra from a heterojunction of *n*-ZnO-nanowire/*p*-Si, of which the device structure is illustrated in Figure 8.18b, were studied by Zimmler et al. [37]. The nanowires were grown by a carbothermal reduction method and were typically 10–40 μm long and 100–350 nm in diameter. Figure 8.21 shows the PL and EL spectra of a single-nanowire LED, with the intensity on a logarithmic scale. In general, the PL and EL spectra present similar structure, with the EL spectrum shifted to lower energies by about 12 meV and also exhibiting slightly broader emission lines. At 6 K, the spectra (Figure 8.21a and c) reveal clear and distinct exciton-related emission lines where in both cases, neutral donor bound exciton (D^0X) recombination processes dominate the emission. Other emission lines on the low-energy side of the main D^0X band are known to be related to structural defects or to the longitudinal optical (LO) phonon replicas of the free and bound excitons [38,39]. Both the PL and EL spectra also show an additional emission band on the high-energy side of the FX emission. In the PL spectrum, this band is centered at 3.42 eV. In the EL spectrum, however, it is centered at around 3.44 eV and also exhibits a multipeak structure. For excitonic emission in ZnO, two different processes can be responsible for

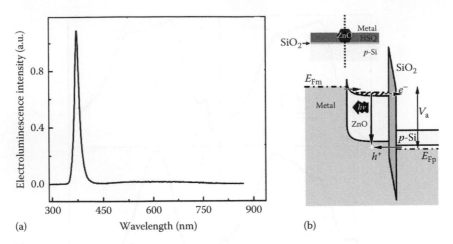

(a)   (b)

FIGURE 8.20
(a) EL spectrum from a *n*-ZnO-nanowire/*p*-Si substrate heterojunction diode with a ~7–8 nm SiO₂ insulating layer between the nanowire and substrate. (b) Schematic band diagram for the device with an SiO₂ thin film. (Reproduced with permission from Zimmler, M.A., Stichtenoth, D., Ronning, C., Yi, W., Narayanamurti, V., Voss, T., and Capasso, F., Scalable fabrication of nanowire photonic and electronic circuits using spin-on glass, *Nano Lett.*, 8, 1695. Copyright 2008 American Chemical Society.)

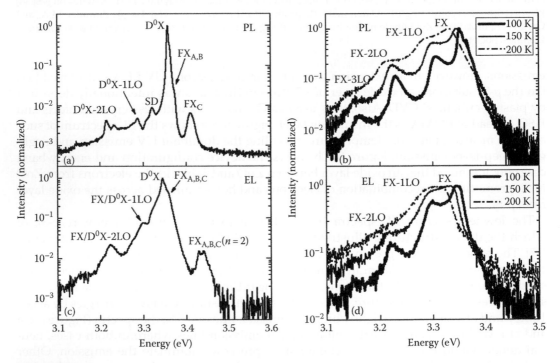

FIGURE 8.21
PL and EL spectra of an *n*-ZnO-nanowire/*p*-Si LED recorded at (a and c) 6 K and (b and d) 100–200 K. The PL was measured under exciton with a HeCd laser (325 nm), while for the EL a bias of 6 V (50 μA current) was applied. At 6 K, the main emission lines are assigned to the neutral donor-bound exciton (D⁰X) and the FX as well as their LO phonon replicas (1 and 2 LO). (Reproduced with permission from Zimmler, M.A., Voss, T., Ronning, C., and Capasso, F., Exciton-related electroluminescence from ZnO nanowire light-emitting diodes, *Appl. Phys. Lett.*, 94, 241120. Copyright 2009, American Institute of Physics.)

these emission bands: the multipeak structure at around 3.44 eV in the EL spectrum suggests emission from excited exciton states ($n = 2$), which can have contributions from the A, B, and C excitons [40]. On the other hand, the emission at 3.42 eV in the PL spectrum can be attributed to the radiative recombination of the C exciton in its ground state. The PL and EL spectra of the single nanowire LED at higher temperatures reveal a continuous increase of the intensities of the FX lines together with their LO phonon replica, as can be seen from the results shown in Figure 8.21b and d. The PL emission at temperatures above 150 K is clearly dominated by the FX emission. At 200 K and above, the FX band merges with its LO phonon replica, resulting in an asymmetric shape of the PL emission. In contrast, in the EL spectra at $T > 150$ K, the FX band shows a comparable intensity with its one LO side band, which is distinctly more pronounced than in the PL spectra. This results in a slight red-shift of the overall EL emission band at 200 K when compared to the PL emission taken at the nominally same temperature.

Therefore, at $T = 6$ K, D^0X emission dominates the EL, whereas at temperatures above 150 K, the FX, together with its LO phonon replica, is responsible for light emission. The results demonstrate that n-ZnO/p-Si nanowire LED operates with efficient excitonic recombination processes and should provide high internal quantum efficiencies.

8.4 Hybrid Organic–Inorganic LEDs Based on Free-Standing ZnO Nanowire Films

ZnO nanowires are normally grown onto planar substrates by either CVD or chemical-bath growth methods. The CVD method is prior to the liquid-phase growth in producing long and thin ZnO nanowires. Liu et al. had produced ZnO nanowires on a large scale by a vapor-phase reaction method without using any vacuum system [41]. The ZnO nanowires were synthesized within a horizontal quartz tube furnace (inner diameter 5 cm) at atmospheric pressure without using any catalyst. By heating the mixture (2–3 g) of ZnO and graphite powders with weight ratio of 1:1 at 1100°C–1200°C, the vaporized growth species were transported by the 1000-sccm N_2 flow mixed with O_2 of 30 sccm, and the cotton-like white product was deposited at the low temperature region (~200°C to room temperature). Typically, the growth duration was 30 min. To apply these mass-produced ZnO nanowires in LEDs, they fabricated paper-like sheets made up of the nanowires. The free-standing nanowire sheets were combined with inorganic or organic p-type semiconductors for making hybrid LEDs [42].

8.4.1 Fabrication of the Nanowire Film

Figure 8.22a shows an optical photograph of the cotton-like product consisting of ZnO nanowires. Thin sheets of the ZnO nanowires were fabricated by a simple filtration method. First, a ZnO-nanowire suspension solution, concentration of 1 mg/mL, was prepared by ultrasonically dispersing the nanowires in isopropanol. Second, the ZnO-nanowire suspension was vacuum-filtered through a porous anode aluminum oxide (AAO) membrane, diameter of 4.3 cm and pore size of 200 nm, purchased from Whatman Co. Third, a network film of ZnO nanowires on an AAO membrane was dried in air at 100°C for 1 h. Finally, a thin sheet of ZnO nanowires was detached off the membrane filter. A typical free-standing sheet of ZnO nanowires is shown in Figure 8.22b. Two SEM images of the nanowire film

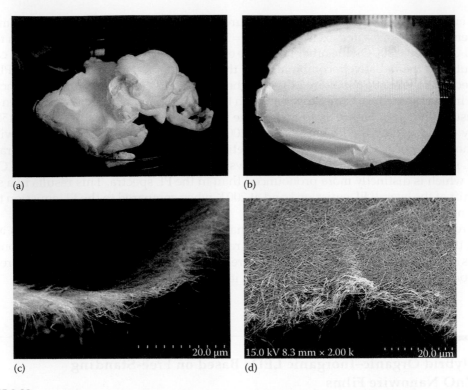

(a)

(b)

(c)

(d)

15.0 kV 8.3 mm × 2.00 k   20.0 μm

FIGURE 8.22
(a and b) Optical photographs of the cotton-like product consisting of ZnO nanowires and the paper-like nanowire film, respectively. (c and d) SEM images of the nanowire film. (From Liu, J., Ahn, Y.H., Park, J.Y., Koh, K.H., and Lee, S., Hybrid light-emitting diodes based on flexible sheets of mass-produced ZnO nanowires, *Nanotechnology*, 20, 445203, 2009. Reproduced with permission from Institute of Physics.)

are shown in Figure 8.22c and d, revealing that the nanowires interlap with each other to form a felty morphology. It is interesting to note that the thin nanowire sheet is flexible and translucent, and it can be cut in any size and shape, using a blade, to be used for hybrid LED fabrications.

8.4.2 Structure of *n*-ZnO-Nanowire/*p*-Si

Figure 8.23a illustrates the process to make a ZnO-nanowire-film/*p*-Si LED. The ITO layer on glass was patterned with a photolithography technique. After squeezing the nanowire film by silicon and ITO substrates, the edge of the device was sealed by epoxy. Note that all the processes were performed in air. Oxygen resides in the felty film. Figure 8.23b shows the *I–V* characteristic of the *n*-ZnO/*p*-Si heterojunction, demonstrating a rectification behavior. Ohmic contact was established for Au/Si by annealing treatment, as shown in the inset of Figure 8.23b. Room-temperature EL spectra of the ZnO-nanowire-film/*p*-Si LED is shown in Figure 8.23c, with the forward driving voltage of 9–15 V. The unique ultra-violet EL peaks show a red-shift and are broader compared with the UV peak at 381 nm in the PL spectrum of as-grown ZnO nanowires, as plotted by the dot–line curve. The NBE emission at ~381 nm appeared both in the PL and EL spectra as marked by the vertical dot line. It is notable that the PL spectrum contains a broad visible emission band, indicating that the mass-produced ZnO nanowires possess abundant defects, such as zinc vacancies

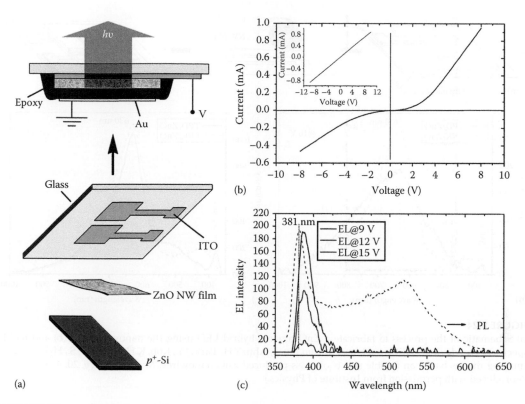

FIGURE 8.23
(a) Illustration for the fabrication process of a ZnO-nanowire-film/p-Si heterostructure LED. (b) I–V curve of the LED. (c) EL spectra of the device at driving biases of 9 V, 12 V, and 15 V. The PL spectrum of ZnO nanowires are shown in dotted line.

(V_{Zn}), oxygen interstitials (O_i), antisite oxygen (O_{Zn}), and oxygen vacancies (V_O). The EL emissions in the range of 390–400 nm can be attributed to electron transitions from conduction band to some deep acceptor levels.

This work demonstrates that the free-standing ZnO nanowire film can be applied to any inorganic p-type semiconductor substrates other than Si, such as GaN and SiC, for making hybrid LEDs.

8.4.3 Structure of n-ZnO-Nanowire-Film/p-Organic-Semiconductor

The EL emissions of LEDs based on the configurations in Figure 8.10 only occur at the top or bottom ends of the nanowires. Based on the free-standing ZnO nanowire sheets, Liu et al. developed a method to make hybrid LEDs by using organic semiconductors, as illustrated in Figure 8.24a [42]. To prepare the substrate, first, the ITO layer on a glass substrate was patterned by using the photolithography technique, followed by the cleaning process including oxygen plasma treatment. A hole-injection material, PEDOT:PSS, was spun coated onto the patterned ITO-glass substrate and dried at 150°C for 20 min in N₂ atmosphere. Onto this layer, the other hole-transport organic layer was spin coated from solution (10 mg/mL in chloroform) at 2000 rpm. Next, a piece of ZnO-nanowire sheet was placed onto the organic layer and pressed down by roller. In the roller-pressing process,

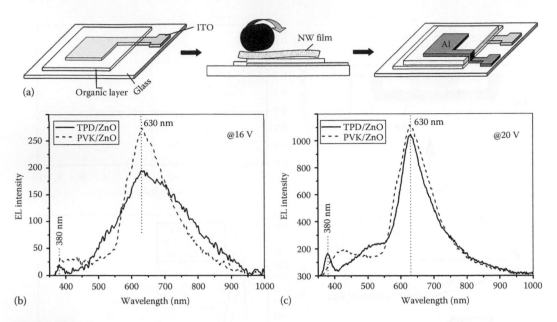

FIGURE 8.24
(a) Schematic of the process to fabricate a ZnO/organic hybrid LED using the nanowire film. (b) and (c) EL spectra of two devices at different biases. (From Liu, J., Ahn, Y.H., Park, J.Y., Koh, K.H., and Lee, S., Hybrid light-emitting diodes based on flexible sheets of mass-produced ZnO nanowires, *Nanotechnology*, 20, 445203, 2009. Reproduced with permission from Institute of Physics.)

the pressure was mechanically controlled, resulting in even adhesion of a ZnO-nanowire sheet to an organic layer. Finally, after evaporating Al to form a cathode onto the ZnO-nanowire sheet, the device was encapsulated by a glass cap in dry N_2 atmosphere. Onto the PEDOT:PSS layer, different polymer semiconductors were used to contact with ZnO nanowires, with the device configuration ITO/PEDOT:PSS/polymer/ZnO/Al. Figure 8.24b and c shows the EL spectra under different biases for two devices with heterojunctions TPD/ZnO and PVK/ZnO, respectively. The small peak at 380 is ascribed to the NBE emission of ZnO. The visible peak at 630 nm is from the electron transition in the band gap between two defect levels. Blue emissions in the region 400–450 nm originate from the polymer semiconductor.

These devices show an identical red emission around 630 nm (1.97 eV) regardless of the different HOMO levels of *p*-type organic semiconductors used in this work, indicating that the exciplexes between ZnO and organic semiconductors can be ruled out causing this emission. Figure 8.25 shows the energy band diagram of the ZnO/polymer heterojunction. The energy band gap of ZnO is 3.37 eV, within which the energy positions of various defects had been calculated or experimentally measured. Generally, V_O and Zn_i are donors, and V_{Zn}, O_i, and O_{Zn} are acceptors. The V_O^+ that is frequently referred to as the origin of green emission from ZnO has the energy level located about 2.45 eV (506 nm) above the valence band maximum [43,44]. Assuming that the red emission is due to the electron transition from the V_O^+ level to an acceptor level, this acceptor level is deduced to be 0.48 eV above the valence-band top. Among the aforementioned acceptor levels, it appears that the V_{Zn}^{2-} level, which was calculated to be 0.51 eV above the valence-band top [45], is closest to this requirement. Therefore, it can be concluded that the red emission centered at 630 nm originates from the electron transition of $V_O^+ \rightarrow V_{Zn}^{2-}$ (see Figure 8.25). The oxygen interstitial has a level

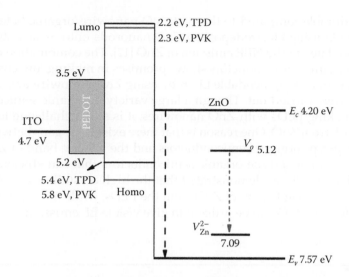

FIGURE 8.25
Energy band diagram of the ZnO/organic heterojunctions accounting for the EL emission from ZnO.

0.66 eV above the valance-band maximum [41]. The $V_O^+ \rightarrow O_i^-$ transition may result in an emission band at 693 nm (1.79 eV), which accidentally corresponds to the right shoulder of the red emission band in the EL spectrum of TPD/ZnO device measured at lower anode bias (Figure 8.24b).

As can be seen from Figure 8.25 that there exist large energy barriers between the HOMO levels of *p*-type organic semiconductors and the valence band of ZnO. Among the currently available *p*-type semiconductors, TPD is a unipolar transporter exhibiting high hole mobility (10^{-3} cm²/(V s)), but the holes from its HOMO level ~5.4 eV encounter a large gap of ~2.2 eV to be injected into the valence band of ZnO. CBP (4,4'-bis(*N*-carbazolyl)-1,1'-biphenyl) has the deepest HOMO level at ~−6.0 eV. However, this material has a bipolar transport character, readily accepting electrons from the conduction band of ZnO to reduce the excitons in ZnO. For efficiently injecting holes into ZnO, it is desirable to develop *p*-type organic materials with deeper HOMO levels and high hole mobilities. This is also requested by the hybrid LEDs with the configurations in Figure 8.10.

8.5 Conclusion

A review of recent progress in the fabrication and performance of hybrid LEDs based on ZnO nanowires is presented. Methods for synthesizing ZnO nanowires and using them to make hybrid LEDs are briefly introduced. The hybrid LEDs can be classified as ZnO/inorganic and ZnO/organic heterojunctions according to the *p*-type semiconductors used to contact with ZnO nanowires. However, in device configuration, there are three types based on vertically aligned nanowires, lying nanowires on planar substrates, and free-standing nanowire sheets. In performance, the ZnO-nanowire/inorganic heterojunction

LEDs are more durable compared to those of ZnO-nanowire/organic heterojunctions. In particular, the LED using electrodeposited ZnO nanorods onto *p*-GaN shows dominant UV EL emission related to the NBE emission of ZnO [12]. The combination of ZnO nanowires and *p*-type organic semiconductors shows promises in making low-cost hybrid LEDs. Also, experiments in making bendable LEDs by using ZnO nanowire-arrays and polymer semiconductors were carried out. Though a large variety of organic semiconductors have been tested in making LEDs with ZnO nanowires, it is still a challenge to obtain strong and unique UV EL from ZnO. One reason is that there exist large gaps between the HOMO levels of current *p*-type organic semiconductors and the valence band of ZnO. Therefore, to synthesize appropriate *p*-type organic semiconductors that can efficiently inject holes into ZnO is in demand. It has demonstrated that the high crystallinity of ZnO nanowires is crucial to achieve strong UV EL of ZnO from the LEDs, while for emitting white lights, some crystal defects in ZnO can be utilized to give visible EL emissions.

References

1. Ü. Özgür, Ya. I. Alivov, C. Liu, A. Teke, M. A. Reshchikov, S. Doğan, U. Avrutin, S.-J. Cho, and H. Morkoc, A comprehensive review of ZnO materials and devices, *J. Appl. Phys.* **98**, 041301 (2005).
2. Z. P. Wei, Y. M. Lu, D. Z. Shen, Z. Z. Zhang, B. Yao, B. H. Li, J. Y. Zhang, D. X. Zhao, X. W. Fan, and Z. K. Tang, Room temperature *p-n* ZnO blue-violet light-emitting diodes, *Appl. Phys. Lett.* **90**, 042113 (2007).
3. J. H. Lim, C. K. Kang, K. K. Kim, I. K. Park, D. K. Hwang, and S. J. Park, UV electroluminescence emission from ZnO light-emitting diodes grown by high-temperature radiofrequency sputtering, *Adv. Mater.* **18**, 2720 (2006).
4. Z. W. Pan, Z. R. Dai, and Z. L. Wang, Nanobelts of semiconducting oxides, *Science* **291**, 1947 (2001).
5. L. Schmidt-Mende and J. L. MacManus-Driscoll, ZnO–nanostructures, defects, and devices, *Mater. Today* **10**, 40 (2007).
6. W. I. Park and G. C. Yi, Electroluminescence in n-ZnO nanorod arrays vertically grown on p-GaN, *Adv. Mater.* **16**, 87 (2004).
7. X. Wang, J. Song, P. Li, J. H. Ryou, R. D. Dupuis, C. J. Summers, and Z. L. Wang, Growth of uniformly aligned ZnO nanowire heterojunction arrays on GaN, AlN, and $Al_{0.5}Ga_{0.5}N$ substrates, *J. Am. Chem. Soc.* **127**, 7920 (2005).
8. X. M. Zhang, M. Y. Lu, Y. Zhang, L.-J. Chen, and Z. L. Wang, Fabrication of a high-brightness blue-light-emitting diode using a ZnO nanowire array grown on p-GaN thin film, *Adv. Mater.* **21**, 1 (2009).
9. S. Xu, C. Xu, Y. Liu, Y. Hu, R. Yang, Q. Yang et al., Ordered nanowire arrays blue/near-UV light-emitting diodes, *Adv. Mater.* **22**, 4749 (2010).
10. S. Xu, Y. Wei, M. Kirkham, J. Liu, W. Mai, D. Daridovic, R. L. Snyder, and Z. L. Wang, Patterned growth of vertically aligned ZnO nanowire arrays on inorganic substrates at low temperature without catalyst, *J. Am. Chem. Soc.* **130**, 14959 (2008).
11. O. Lupan, T. Pauporte, B. Viana, I. M. Tiginyanu, V. V. Ursaki, and R. Cortes, Epitaxial electrodeposition of ZnO nanowire arrays on p-GaN for efficient UV-light-emitting diode fabrication, *ACS Appl. Mater. Inter.* **7**, 2083 (2010).
12. O. Lupan, T. Pauporte, and B. Viana, Low-voltage UV-electroluminescence from ZnO-nanowire-array/p-GaN light-emitting diodes, *Adv. Mater.* **22**, 3298 (2010).
13. M. C. Jeong, B. Y. Oh, M. H. Ham, S. W. Lee, and J. M. Myoung, ZnO-nanowire-inserted GaN/ZnO heterojunction light-emitting diodes, *Small* **3**, 568 (2007).

14. C. H. Chen, S. J. Chang, S. P. Chang, M. J. Li, I. C. Chen, T. J. Hsueh, A. D. Hsu, and C. L. Hsu, Fabrication of a white-light-emitting diode by doping gallium into ZnO nanowire on p-GaN substrate, *J. Phys. Chem. C* **114**, 12422 (2010).

15. J. Y. Lee, Y. S. Choi, J. H. Kim, M. O. Park, and S. Im, Optimizing n-ZnO/p-Si heterojunctions for photodiode applications, *Thin Solid Films* **403–404**, 553 (2002).

16. S. Mridha and D. Basak, Ultraviolet and visible photoresponse properties of n-ZnO/p-Si heterojunctions, *J. Appl. Phys.* **101**, 083102 (2007).

17. S. T. Tan, X. W. Sun, J. L. Zhao, S. Iwan, Z. J. Cen, T. P. Chen, J. D. Ye, G. Q. Lo, D. L. Kwong, and K. L. Teo, Ultraviolet and visible electroluminescence from n-ZnO/(n,p)-Si heterostructured light-emitting diodes, *Appl. Phys. Lett.* **93**, 013506 (2008).

18. H. Sun, Q. F. Zhang, and J. L. Wu, Electroluminescence from ZnO nanorods with an n-ZnO/p-Si heterojunction structure, *Nanotechnology* **17**, 2271 (2006).

19. B. Ling, J. L. Zhao, X. W. Sun, S. T. Tan, A. K. K. Kyaw, Y. Divayana, and Z. L. Dong, Color tunable light-emitting diodes based on p+-Si/p-CuAlO₂/n-ZnO nanorod array heterojunctions, *Appl. Phys. Lett.* **97**, 013101 (2010).

20. Y. Yan, M. Al-jassim, and S. H. Wei, Doping of ZnO by group IB elements, *Appl. Phys. Lett.* **89**, 181912 (2006).

21. K. S. Ahn, T. Deutsch, Y. Yan, C. S. Jiang, C. L. Perkins, J. Turner, and A. Al-Jassim, Synthesis of band-gap-reduced p-type ZnO films by Cu incorporation, *J. Appl. Phys.* **102**, 023517 (2007).

22. M. Willander, O. Nur, Q. X. Zhao, L. L. Yang, M. Lorenz, B. Q. Cao et al., Zinc oxide nanorod based photonic devices: Recent progress in growth, light emitting diodes and lasers, *Nanotechnology* **20**, 332001 (2009).

23. M. Willander, O. Nur, N. Bano, and K. Sultana, Zinc oxide nanorod-based heterostructures on solid and soft substrates for white-light-emitting diode applications, *New J. Phys.* **11**, 125020 (2009).

24. N. H. Alivi, M. Riaz, G. Tzamalis, O. Nur, and M. Willander, Fabrication and characterization of high-brightness light emitting diodes based on ZnO nanorods grown by a low-temperature chemical method on p-4H-SiC and p-GaN, *Semicond. Sci. Technol.* **25**, 065004 (2010).

25. N. Bano, I. Hussain, O. Nur, M. Willander, and P. Klason, Study of radiative defects using current-voltage characteristics in ZnO nanorods catalytically grown on 4H-p-SiC, *J. Nanomater.* **2010**, 817201 (2010).

26. R. Könenkamp, R. C. Word, and C. Schlegel, Vertical nanowire light-emitting diode, *Appl. Phys. Lett.* **85**, 6004 (2004).

27. R. Könenkamp, R. C. Word, and M. Godinez, Ultraviolet electroluminescence from ZnO/polymer heterojunction light-emitting diodes, *Nano Lett.* **5**, 2005 (2005).

28. X. W. Sun, J. Z. Huang, J. X. Wang, and Z. Xu, A ZnO nanorod inorganic/organic heterostructure light-emitting diode emitting at 342 nm, *Nano Lett.* **8**, 1219 (2008).

29. A. Wadeasa, S. L. Beegum, S. Raja, O. Nur, and M. Willander, The demonstration of hybrid n-ZnO nanorod/p-polymer heterojunction light emitting diodes on glass substrates, *Appl. Phys. A* **95**, 807 (2009).

30. A. Wadeasa, O. Nur, and M. Willander, The effect of the interlayer design on the electroluminescence and electrical properties of n-ZnO nanorod/p-type blended polymer hybrid light-emitting diodes, *Nanotechnology* **20**, 065710 (2009).

31. A. Nadarajah, R. C. Word, J. Meiss, and P. Könenkamp, Flexible inorganic nanowire light-emitting diode, *Nano Lett.* **8**, 534 (2008).

32. A. Zainelabdin, S. Zaman, G. Amin, O. Nur, and M. Willander, Stable white light electroluminescence from highly flexible polymer/ZnO nanorods hybrid heterojunction grown at 50°C, *Nanoscale Res. Lett.* **5**, 1442 (2010).

33. J. Goldberger, D. J. Sirbuly, M. Law, and P. Yang, ZnO nanowire transistors, *J. Phys. Chem. B* **109**, 9 (2005).

34. W. Q. Yang, H. B. Huo, L. Dai, R. M. Ma, S. F. Liu, G. Z. Ran, B. Shen, C. L. Lin, and G. G. Qin, Electrical transport and electroluminescence properties of n-ZnO single nanowires, *Nanotechnology* **17**, 4868 (2006).

35. J. Bao, M. A. Zimmler, and F. Capasso, Broadband ZnO single-nanowire light-emitting diode, *Nano Lett.* **6**, 1719 (2006).
36. M. A. Zimmler, D. Stichtenoth, C. Ronning, W. Yi, V. Narayanamurti, T. Voss, and F. Capasso, Scalable fabrication of nanowire photonic and electronic circuits using spin-on glass, *Nano Lett.* **8**, 1695 (2008).
37. M. A. Zimmler, T. Voss, C. Ronning, and F. Capasso, Exciton-related electroluminescence from ZnO nanowire light-emitting diodes, *Appl. Phys. Lett.* **94**, 241120 (2009).
38. B. K. Meyer, H. Alves, D. M. Hofmann, W. Kriegseis, D. Forster, F. Bertram et al., Bound exciton and donor-acceptor pair recombinations in ZnO, *Phys. Status Solidi B* **241**, 231 (2004).
39. M. Schirra, R. Schneider, A. Reiser, G. M. Prinz, M. Feneberg, J. Biskupek, U. Kaiser, C. E. Krill, K. Thonke, and R. Sauer, Stacking fault related 3.31-eV luminescence at 130-meV acceptors in zinc oxide, *Phys. Rev. B* **77**, 125215 (2008).
40. U. Rössler (ed.), *Landolt-Börnstein, Semiconductors, Group III: Condensed Matter*, Vols. III/17b and 22a, Springer, Berlin, Germany (1998).
41. J. Liu, S. Lee, Y. H. Ahn, J. Y. Park, and K. H. Koh, Tailoring the visible photoluminescence of mass-produced ZnO nanowires, *J. Phys. D Appl. Phys.* **42**, 095401 (2009).
42. J. Liu, Y. H. Ahn, J. Y. Park, K. H. Koh, and S. Lee, Hybrid light-emitting diodes based on flexible sheets of mass-produced ZnO nanowires, *Nanotechnology* **20**, 445203 (2009).
43. K. Vanheusden, W. L. Warren, C. H. Seager, D. R. Tallant, J. A. Voigt, and B. E. Gnade, Mechanisms behind green photoluminescence in ZnO phosphor powders, *J. Appl. Phys.* **79**, 7983 (1996).
44. F. Leiter, H. Zhou, F. Henecker, A. Hofstaetter, D. M. Hofmann, and B. K. Meyer, Magnetic resonance experiments on the green emission in undoped ZnO crystals, *Physica B* **308–310**, 908 (2001).
45. A. Janotti and C. G. Van de Walle, New insights into the role of native point defects in ZnO, *J. Cryst. Growth* **287**, 58 (2006).

9

ZnBeMgO Alloys and UV Optoelectronic Applications

Hsin-Ying Lee, Li-Ren Lou, and Ching-Ting Lee

CONTENTS

9.1 Introduction 309
9.2 Synthesis and Characterization 311
    9.2.1 ZnBeO Alloys 311
    9.2.2 ZnBeMgO Alloys 319
9.3 Band Structure Calculation 323
9.4 Optoelectronic Applications 328
9.5 Summary 335
References 335

9.1 Introduction

Significant progresses on ZnO-based semiconductor materials have been achieved in recent decades. Recently, ZnO-based materials have been regarded as promising wide-bandgap semiconductors for application in ultraviolet (UV) region optoelectronic devices because of their excellent optical and electrical properties, such as the wide direct bandgap of 3.37 eV, the high exciton binding energy of 60 meV at room temperature, the good chemical properties, and the low cost. Recently, various applications of ZnO-based optoelectronic devices, such as UV photodetectors and light-emitting diodes (LEDs), have been demonstrated. These progresses have been comprehensively reviewed in some review papers [1–5], and the various aspects have been discussed in detail in the relevant chapters of this book. However, there are still some obstacles needed to be overcome. One of the tasks of great interests is to obtain reliable and stable p-type doping and to suppress the native n-type conductivity of the ZnO-based materials. Many efforts have been devoted to solve this problem, and promising successes have been achieved continuously. For example, the vapor-cooling condensation system was developed and established, which demonstrated to be an ideal technique for growing high-quality intrinsic ZnO films [6]. The associated deposition mechanisms were presented [7]. This technique was successfully used to fabricate various devices, such as the ZnO-on-GaN heterojunction LEDs [6] and the UV photodetectors [8], the n-i-p ZnO-based LEDs [9,10], and the single n-ZnO:In/i-ZnO/p-GaN-heterostructured n-i-p nanorod LEDs [11]. The technique has also been used to deposit intrinsic ZnO film as the gate insulator layer for the AlGaN/GaN metal-oxide-semiconductor high-electron mobility transistors (HEMTs) [12,13].

For practical applications, it is of interest to develop ZnO-based alloys with bandgap energy larger than that of ZnO. First, it is required for producing high-efficiency ZnO UV devices composed of ZnO-based quantum wells (QWs) and superlattices. Most designed semiconductor devices rely on heterostructures for providing carrier and/or optical confinement, such as structures in HEMTs and QWs in LEDs and laser diodes. In other words, bandgap engineering has to be performed. These heterostructures are composed of layers with different materials/compositions, for which the most relevant parameters are the bandgap of each layer and the valence-band offset and conduction-band offset between the individual layers. In order to develop ZnO-based UV optoelectronic devices, it is necessary to find suitable materials with the same crystal structure as ZnO but with wider bandgap with respect to ZnO. This requirement can be achieved by alloying ZnO with proper compounds, in analogy to GaN, which can be alloyed with AlN and/or InN. On the other hand, the bandgap broadening extends the working wavelength toward deeper UV region, for example, of LEDs and photodetectors. It is also beneficial for solar cells when the broader bandgap material is used on solar cells as transparent conducting films to enhance the transmittance in a larger transmission wavelength range, which increases the light absorption in solar cells, and thus improve the conversion efficiency of solar cells.

Ohtomo et al. [14] proposed MgZnO for such ZnO alloys. They reported that the bandgap of $Mg_xZn_{1-x}O$, where x is the atomic fraction, can be increased to 3.99 eV at room temperature as the content of Mg is increased upward to x = 0.33. Superlattices were fabricated by employing ZnO and ZnMgO as alternate layers [15,16]. $Zn_{1-x}Mg_xO$ films exhibit interesting electrical and optical properties and have found wide applications in the fields of optoelectronic and transparent conductive oxide, because the bandgap energy of $Zn_{1-x}Mg_xO$ films is easily controlled by controlling the Mg/(Mg + Zn) atomic ratio. However, crystal phase segregation between the ZnO and the MgO was observed, when the Mg content x exceeded 0.36, due to the different crystal structures and large lattice mismatch between the ZnO and the MgO; specifically, MgO is cubic structure with lattice constant of 4.22 Å, and ZnO is hexagonal structure with a-axis lattice constant of 3.25 Å [17]. To investigate the mechanisms and performances, the $Zn_{1-x}Mg_xO$ films were grown on glass substrates using a sol–gel spin-coating technique [18]. The band-edge luminescence (BEL) intensity of the grown $Zn_{1-x}Mg_xO$ films increased up to 21-fold for $Zn_{0.94}Mg_{0.06}O$ and up to 4-fold for $Zn_{0.958}Mg_{0.042}O$ compared with the $Zn_{0.973}Mg_{0.027}O$ film. The enhanced BEL intensity of the $Zn_{1-x}Mg_xO$ with an increase in the Mg content was related to the change in nonradiative efficiency and the number of oxygen-vacancy-related defects. Furthermore, the increase in bandgap energy and the decrease in oxygen-vacancy-related defects could be resulted by increasing the Mg content in the grown $Zn_{1-x}Mg_xO$ films [18]. To achieve a larger bandgap modulation than that mentioned earlier, Ryu et al. [19] developed another promising candidate of BeZnO alloy alternatively for achieving a larger bandgap modulation than that of ZnO. Considering the large size difference between Zn and Be, which degraded the alloy crystal quality, the investigation was then extended to develop ZnBeMgO alloys. The relevant aspects about ZnMgO alloys have been intensively investigated and comprehensively discussed in the preceding relevant chapters of this book. This chapter devotes to the issue of ZnBeO and ZnBeMgO alloys. Section 9.2 describes the preparation and characterization, including mainly the structure and optical properties, which are fundamental for successful application in technologies. Section 9.3 introduces the theoretical calculation on the band structure of the ZnBeMgO alloys, which is a key point for understanding the bandgap engineering. Finally, the optoelectronic applications of the alloys are described in Section 9.4, followed by a short summary.

9.2 Synthesis and Characterization

9.2.1 ZnBeO Alloys

As mentioned earlier, the component of the ZnO and the MgO in the ZnMgO alloys possesses different structure, in which the ZnO exhibits the wurtzite structure, and the MgO is the rock-salt structure. This situation is different from the behaviors of the GaN families. Especially in the ZnO-based alloys with high content of MgO, the phase separation is expected to be occurred. Only for moderate content of MgO, however, MgZnO alloys assume the wurtzite crystal structure of the parent compound [14]. Consequently, the bandgap of the alloys usually is limited to a value less than 3.99 eV, when Mg substitutes 33 at.% of Zn without changing the hexagonal crystal structure of ZnO. However, the phase segregation between hexagonal ZnO and cubic MgO was observed from the $Mg_xZn_{1-x}O$ alloy with more than 36 at.% of Mg content. Because of the phase segregation and the lattice mismatch between ZnO and MgO, the ZnO and ZnMgO-based heterostructures were not very efficient for developing very reliable ZnO-based devices [15]. To avoid such problems, a new metal oxide of hexagonal BeO was proposed as a promising candidate material due to its good solubility with ZnO and a large bandgap of 10.6 eV [19,20], which indicates a possibility for modulating bandgap to obtain a deep UV region. The bandgap of BeZnO films could be modulated, in principle, from 3.37 to 10.6 eV without phase segregation, because the BeO (10.6 eV) has the same hexagonal wurtzite structure with the ZnO.

The hybrid beam deposition (HBD) method was used to synthesize ZnBeO films on c-Al_2O_3 substrates [19–21]. During the growth of BeZnO films, the c-Al_2O_3 substrates were heated to 500°C. Besides, a polycrystalline ZnO target was evaporated to create Zn and O sources using an electron beam, and Be was separately provided by a thermal evaporation method using Be cell. Because the vapor pressure of Be was very low, the Be effusion cell was operated at a high temperature. The Be concentration in BeZnO could be varied by changing the Be evaporation temperature ranging from 900°C to 1200°C. In order to reduce oxygen deficiency in the BeZnO films, an oxygen plasma source was utilized in connection to the deposition chamber to provide radical oxygen as an additional O source.

The BeZnO alloy films were characterized with an UV–visible infrared spectrometer to study their energy bandgap. Figure 9.1 shows the transmittance spectra measured at room temperature, where the compositional information for the elements of Zn and Be in the grown BeZnO films was obtained by x-ray photoelectron spectroscopy (XPS) and ion sputtering [19]. The results show that the transmission cutoff wavelength shifts continuously to a shorter wavelength with an increase in the Be concentration. The bandgap energy (E_g) of BeZnO is derived from extrapolating the graph of α^2 versus hv-E_g, where α is the absorption coefficient, h is the Planck constant, and v is the photon frequency. Figure 9.2 shows that the bandgap energy of the BeZnO increased as a function of Be concentration [19]. The optical transmission measurements demonstrate that the bandgap energy of the BeZnO can be varied over the range from 3.3 eV (ZnO) to 10.6 eV (BeO).

Figure 9.3 shows the results of x-ray diffraction (XRD) measurements for BeZnO samples with different Be concentrations, controlled by changing Be-evaporator temperatures, such as 1000°C, 1040°C, 1080°C, 1120°C, and 1140°C [20]. The XRD results indicated that c-axis-oriented BeZnO films were epitaxially grown on c-Al_2O_3, and no phase segregation between ZnO and BeO in the BeZnO alloy series was observed. The (114) reflection of

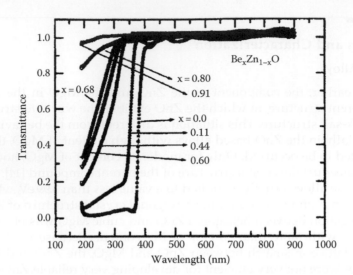

FIGURE 9.1
Transmittance spectra of BeZnO films measured at room temperature. (Reproduced with permission from Ryu, Y.R. et al., *Appl. Phys. Lett.*, 88, 052103, 2006.)

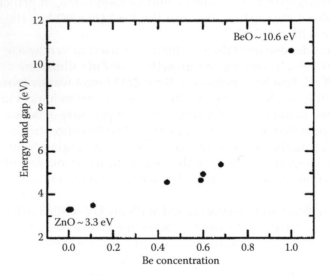

FIGURE 9.2
Energy bandgap of BeZnO as a function of Be concentration in atomic fraction. (Reproduced with permission from Ryu, Y.R. et al., *Appl. Phys. Lett.*, 88, 052103, 2006.)

the BeZnO films shown in Figure 9.3b was investigated to study the in-plane structural changes by varying the Be concentration. A 30° in-plane rotation between the BeZnO films and the Al_2O_3 substrates was observed by x-ray investigation, which might be the result of lattice relaxation due to a large lattice mismatch between them. Because of the poor crystallinity of the BeZnO film deposited at 1140°C, (114) reflection of the BeZnO film was not able to be detected by XRD. This could be attributed to the possible Zn reevaporation from the deposited BeZnO film during the high-temperature deposition procedure.

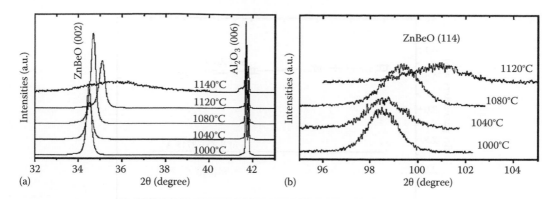

FIGURE 9.3
(a) Normal θ–2θ and (b) asymmetrically scanned XRD patterns of BeZnO films. (Reproduced with permission from Kim, W.J. et al., *J. Appl. Phys.*, 99, 096104, 2006.)

From the XRD results, the a- and c-axes parameters for the BeZnO films as a function of Be-cell temperature were derived and shown in Figure 9.4 [20] in which the known lattice parameters of bulk ZnO and BeO were referred. It is noted that the lattice parameter of the BeZnO films decreased toward that of BeO, when the Be concentration in the BeZnO alloy series increased. It is also noted that the lattice parameter of those BeZnO samples with low Be content (deposited at low temperature) was larger than that of the bulk ZnO. The reason of this unusual phenomenon was attributed to the oxygen vacancies in the film as well as the lattice mismatch between the deposited film and the substrate. The aforementioned results indicated that the Be concentration in the alloy could be varied over the entire range from 0% to 100% without phase segregation observation, unlike the situation happened in the MgZnO alloys.

Yu et al. [22] reported that BeZnO layers were deposited on Al_2O_3 (0001) substrate by simultaneous RF magnetron cosputtering using ZnO target and Be target. The target-substrate distance was set at 150 mm. The base pressure was maintained at 1.3×10^{-8} Torr, and working pressure was maintained at 4 mTorr by a turbo-molecular pump. Prior to the deposition, the surface of the Al_2O_3 substrate was cleaned with acetone and ethanol for 5 min and etched in H_2SO_4:H_3PO_4 = 3:1 for 7 min. It was then rinsed in deionized water. The loaded target was presputtered for 30 min under 50 W RF power to remove any surface contamination on the target. To reduce the lattice mismatch between the Al_2O_3 substrate and the deposited BeZnO film, a 10 Å-thick ZnO buffer layer was first deposited onto the Al_2O_3 substrate. Here, the ZnO buffer layer served as a nucleation layer for the BeZnO growth. The ZnO buffer layer was grown for 5 min under 50 W RF power of ZnO target at a low temperature of 200°C. Thereafter, in order to deposit BeZnO layers, RF power of ZnO target and Be target was set at 150 and 40 W, respectively. The substrate was rotated at 10 rpm during the deposition to obtain a uniform layer thickness. The substrate temperature was maintained at 400°C during the deposition. The layer thickness and growth rate were about 300 nm and 0.33 Å/s, respectively [23].

Figure 9.5 presents the rocking curves of XRD θ–2θ scans for the BeZnO layer [22]. As shown in Figure 9.5, two strong patterns corresponding to the diffraction peaks of BeZnO (0002) and Al_2O_3 (0006) can be observed. The BeZnO (0002) peak shows strong sharp intensity and a full width at half maximum (FWHM) of 702 arcsrc. According to Vegard's law, the Be value of x in the grown $Be_xZn_{1-x}O$ layer turned out to be 0.2. As shown in Figure 9.5,

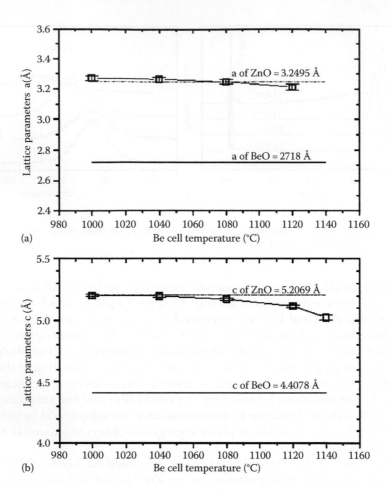

FIGURE 9.4
Be cell temperature dependence of (a) in-plane lattice parameters and (b) surface normal lattice parameters of BeZnO films. (Reproduced with permission from Kim, W.J. et al., *J. Appl. Phys.*, 99, 096104, 2006.)

no other peak related to BeZnO beyond the BeZnO (0002) plane was observed. This indicates that the crystal structure of the BeZnO layer is strongly oriented along the c-axis of the hexagonal structure. According to the cross-sectional TEM micrograph and the selected area electron diffraction (SAED) patterns of the BeZnO layer, it is revealed that there are dislocations located at near the $BeZnO/Al_2O_3$ substrate interface. But, away from the $BeZnO/Al_2O_3$ substrate interface, these dislocation density decreases dramatically [22]. The SAED patterns exhibit visible bright spots corresponding to the crystal planes of the hexagonal BeZnO layer. The results demonstrate that the deposited BeZnO layer is a single crystal grown along (0001) direction. This fact is in agreement with the XRD experimental results, which suggests that the BeZnO layer is crystallized even though the lattice mismatch between the BeZnO layer and the Al_2O_3 substrate is very large.

The XPS results of Be and O elements in the BeZnO layer were studied too [22]. First, the O 1s core level of the O atom exhibits an asymmetric high binding energy shoulder. This XPS band can be divided into two suitable Gaussian components labeled as peaks I and II. It is known that the peak I located at 530.4 eV is attributed to O^{2-} ions surrounded

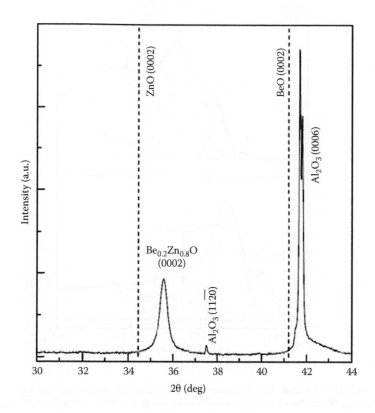

FIGURE 9.5
Rocking curve of XRD ω–2θ scans on the BeZnO layer. (Reproduced with permission from Yu, J.H. et al., *J. Cryst. Growth*, 312, 1683, 2010.)

by four Zn^{2+} ions in the wurtzite ZnO lattice [24]. Thus, this dominant peak I comes from O–Zn bonds, namely, the chemical bonding of Zn–O [25]. On the other hand, the weak peak II centered at 531.7 eV, which is related to the deviation of stoichiometry in ZnO [26], would be attributed to Be–O bonds [25]. The role of Be participating in BeZnO bonding was confirmed through an investigation of the Be 1s core level. As shown in the XPS results of BeZnO layer [21], the Be component corresponding to the Be 1s core level of the Be atom was divided by Gaussian fitting. The derived peak located at 113.2 eV was attributed to the chemical bonding of Be–O [27]. Thus, the finding of Be–O bonding indicated that the Be atoms were well incorporated into the lattice of the BeZnO, as components of the alloy. In other words, the sputtered Be atoms were substituted for the host-lattice site in the ZnO and formed a chemical bonding of Be–O. However, this peak was broadened and of low intensity, which indicated a large fluctuation of Be content in the BeZnO layers.

The optical properties, which are important for the optoelectronic applications, of the grown BeZnO films were studied previously. Jeong et al. [28] reported the photoluminescence (PL) spectra of the $Be_xZn_{1-x}O$ film, deposited by using HBD method, with the Be content ratio x estimated to be 0.11. Figure 9.6a and b shows the PL spectra of the ZnO and BeZnO layers, measured at 5 K, respectively [28]. These PL spectra were resolved into individual peaks using a multi-Gaussian fitting. As shown in Figure 9.6a and b, there are three dominant peaks and a shoulder at the shorter wavelength side.

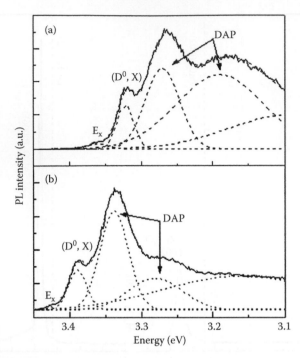

FIGURE 9.6
PL spectra of (a) ZnO and (b) BeZnO layers measured at 5 K. (Here, the dotted line spectra were obtained by a multi-Gaussian fitting.) (Reproduced with permission from Jeong, T.S. et al., *J. Phys. D Appl. Phys.*, 40, 370, 2007.)

However, the resolved spectra of BeZnO (Figure 9.6b) shifted to the shorter-wavelength in comparison with those of ZnO (Figure 9.6a). The shoulder at 3.4267 eV and the peak at 3.3892 eV of the BeZnO were attributed to the emissions of exciton (E_x) and neutral donor bound exciton (D^0, X), respectively. Consequently, the exciton binding energy, $E_{bx}^b = E(E_x) - E(D^0, X)$, was extracted to be 37.5 meV. The other two dominant peaks at 3.3384 and 3.2784 eV were attributed to the donor–acceptor pairs (DAP) and its longitudinal-optical (LO) phonon replica, respectively. Figure 9.7 shows the PL spectra of the BeZnO films measured at various temperatures [28]. The PL spectra show that the position of the emission peak tended to the shorter wavelength region with decreasing temperature, and the emission intensity decreased with an increase in temperature. In particular, the intensity of the (D^0, X) emission increased exponentially, and the associated width of the spectra broadened with an increase in the temperature. It was indicated that the phonons participated in the quenching process [29,30]. Similar results were also observed from the BeZnO films deposited by RF magnetron sputter [22]. Figure 9.8 shows the low temperature (5 K) PL spectrum of the BeZnO layer with the Be compositional ration $x = 0.2$ [22]. As shown in Figure 9.8, two dominant peaks appeared at 3.7313 and 3.6615 eV could be found. Besides, a shoulder peak toward the left-hand side of the dominant peak at 3.7692 eV was observed. This emission was attributed to a free exciton (E_x) transition, but which with a much larger energy in comparison with the corresponding E_x emission of 3.376 eV of the ZnO films measured at 10 K [31]. This result clearly indicated that the ternary BeZnO compound was a wide bandgap material. As shown in Figure 9.8, the emission peak at 3.7313 eV was associated to the neutral donor-bound exciton (D^0, X). Consequently, the binding energy E_{bx}^b of (D^0, X) was derived to be 37.9 meV, which was

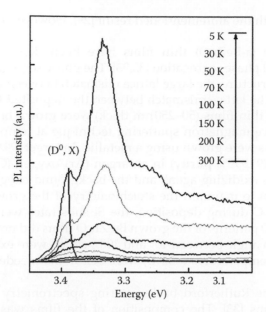

FIGURE 9.7
PL spectra of BeZnO with varying measurement temperatures. (Reproduced with permission from Jeong, T.S. et al., *J. Phys. D Appl. Phys.*, 40, 370, 2007.)

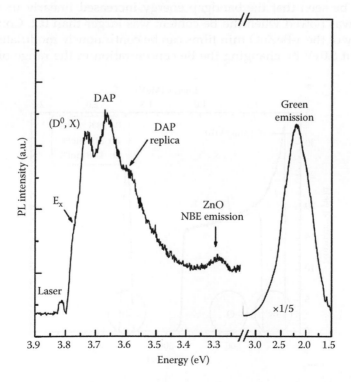

FIGURE 9.8
PL spectrum of the BeZnO layer measured at 5 K. (Reproduced with permission from Yu, J.H. et al., *J. Cryst. Growth*, 312, 1683, 2010.)

quite in agreement with the aforementioned result [28]. However, the origin of the donor is still not clear yet.

Amorphous BeZnO (a-BeZnO) thin films have been deposited for avoiding the problems of the crystal phase segregation [32,33]. The phase segregation is resulting from the different crystal structure, the large lattice mismatch between the components of the crystalline alloy, and the lattice mismatch between the deposited film and the substrate [34]. Various Be_xZn_yO thin films (50–250 nm thick) were grown onto crystalline Si (100) substrates by a reactive magnetron sputtering technique at temperature of <52°C [33]. The Be_xZn_yO thin films were grown using a metallic Zn target (99.999% purity) and four pieces of metallic Be (99.999% purity) in an argon (Ar)/oxygen (O_2) atmosphere. High-purity O_2 was used as oxidizing agent, and the background oxygen pressure was kept at 8×10^{-3} Torr (1.07 Pa) to maintain the stoichiometry of the grown films. To limit the temperature to $T < 52°C$ during deposition, the Si substrates were clamped to a thick copper block. The XRD patterns of the grown Be_xZn_yO films did not show any diffraction peaks at 2θ diffraction angle between 10° and 100°, which were expected because of the fact that the growth temperatures (<52°C) were below that needed to grow crystalline BeZnO films [20,35].

Figure 9.9 shows the Rutherford backscattering spectrometry (RBS) spectra of the a-$Be_{0.286}Zn_{0.286}O_{0.428}$ films [33]. The composition of the films was derived from fitting the experimental RBS spectra with the theoretical curve. The Zn and Be signals in the RBS spectra and the corresponding ion channeling clearly indicated that the Zn was substituted for Be in the amorphous structure and, thus, formed the alloy. Figure 9.10 shows the optical bandgap engineering of the a-Be_xZn_yO films versus the Be concentration x [33]. It could be seen that the bandgap energy increased linearly in the Be content of <0.2, and it was bowed when the Be content was larger than 0.2. Consequently, the bandgap energy of the a-BeZnO thin films can be continuously modulated from 3.35 eV (a-ZnO) to about 8.0 eV by changing the Be concentration in the range of 4.0%–45.05%.

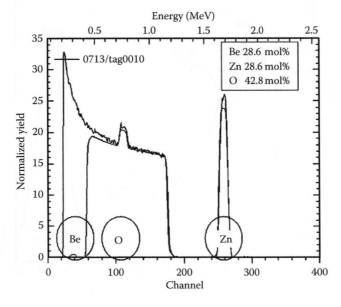

FIGURE 9.9
Rutherford backscattering spectra of amorphous $Be_{0.28}Zn_{0.28}O$ thin film sputtered onto Si (100) substrate at temperature of <52°C. (Reproduced with permission from Khoshman, J.M. et al., *Appl. Phys. Lett.*, 92, 091902, 2008.)

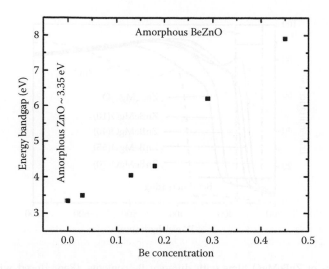

FIGURE 9.10
Bandgap engineering of *a*-BexZnyO alloy films as a function of Be concentration x. (Reproduced with permission from Khoshman, J.M. et al., *Appl. Phys. Lett.*, 92, 091902, 2008.)

The bandgap modulation of amorphous BeZnO films is promising for applications of amorphous ZnO-based devices.

9.2.2 ZnBeMgO Alloys

Although BeZnO alloys have a larger bandgap modulation range in comparison with MgZnO alloys, the large ionic radius difference between Be^{2+} (0.27 Å) and Zn^{2+} (0.60 Å) is still a problem needed to be studied. The crystal quality of the BeZnO is degraded, when the Be content increases and causes a large lattice mismatch. In order to obtain high-quality crystalline ZnO-based films with bandgap modulation to the deeper UV region, it is required to incorporate appropriate amount of Be into ZnMgO to form quaternary alloys [36]. Therefore, the large lattice mismatch of ZnO/BeO and ZnO/MgO is expected to be counteracted by each other [18].

Yang et al. [36] deposited $Zn_{1-x-y}Be_xMg_yO$ films on c-sapphire substrates using a pulsed laser deposition (PLD) system. Two ceramics, $Zn_{0.8}Mg_{0.2}O$ target and BeO target (99.99% purity), were employed. The temperature of the sapphire substrate was 600°C, and the oxygen partial pressure was 1×10^{-2} Pa. A KrF excimer laser (wavelength of 248 nm and energy of 120 mJ/pulse) was applied to ablate these two targets alternately by different numbers of laser pulse for adjusting the Be content in the ZnBeMgO films. To be specific, in each cycle, there were 10, 40, 55, or 70 pulses ablated on BeO target and 25 pulses on $Zn_{0.8}Mg_{0.2}O$ target. The whole growth process included 350 cycles for each deposited film. The obtained samples were labeled as ZnBeMgO(10), ZnBeMgO(40), ZnBeMgO(55), and ZnBeMgO(70), respectively, by the numbers of laser pulse ablated on BeO target per cycle (BeO pulse for short). The estimated composition of the Zn, Be, and Mg in the ZnBeMgO(40) sample was approximate 30%, 25%, and 45%, respectively.

Figure 9.11 shows the transmittance spectra of the ZnBeMgO films with various Be contents [36]. The cutoff wavelength shifted from 335 to 253 nm with an increase in Be content. The bandgap could be estimated using the relationship of $(\alpha h\nu)^2 = (h\nu - E_g)$, where α is the absorption coefficient and hν is the photon energy. The bandgap energy of the

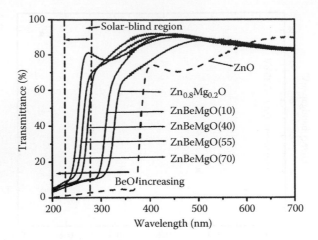

FIGURE 9.11

Transmittance spectra of ZnBeMgO films with different Be contents. (Reproduced with permission from Yang, C. et al., *Appl. Phys. Lett.*, 93, 112114, 2008.)

ZnBeMgO was modulated from 3.7 to 4.9 eV continuously as an increase in the Be content [35]. Hence, the bandgap could be conveniently modulated to a required value by adjusting BeO pulse number. The bandgap energy of the ZnBeMgO(40) could be risen up to 4.5 eV, which was an appropriate value for applications in the solar-blind UV photodetectors as will be discussed later.

Figure 9.12 shows the XRD patterns of (a) ZnMgO and ZnBeMgO(10) and (b) ZnBeMgO films with various Be contents [36]. There are two diffraction peaks at 34.64° and 36.64° detected in the $Zn_{0.8}Mg_{0.2}O$ films, which are assigned to the (002) plane of the hexagonal ZnO and the (111) plane of the cubic MgO, respectively. The phase segregation problem in the ZnMgO films with a high Mg content was generally observed [14,37–39]. However, the impurity phase of MgO was eliminated here in the ZnBeMgO films, related to the incorporation of Be. It could also be derived from Figure 9.12b that the FWHM values of

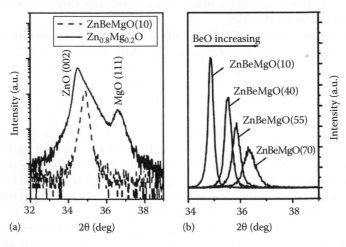

FIGURE 9.12

XRD patterns of (a) ZnMgO and ZnBeMgO(10) and (b) ZnBeMgO films with different Be contents. (Reproduced with permission from Yang, C. et al., *Appl. Phys. Lett.*, 93, 112114, 2008.)

the diffraction peaks of the ZnBeMgO films with appropriate contents of Be were mostly lower than 0.33°, indicating that the high crystal quality of ZnBeMgO films was deposited by multitarget PLD. Obviously, the crystal quality of the ZnBeMgO films was significantly improved compared with that of the ZnMgO films. This result could be demonstrated by the drop in FWHM value of ZnMgO from 0.49° to 0.23° when Be was incorporated.

Comparing with the BeZnO films, the ZnBeMgO films presented a notable advantage in crystal quality. Figure 9.13 exhibited the XRD patterns of the BeZnO and the ZnBeMgO films with the same bandgap value of 4.5 eV [36]. Both the BeZnO and ZnBeMgO films were wurtzite structure without phase segregation. The FWHM value of the ZnBeMgO films was much lower than that of the BeZnO films. This experimental result demonstrated that a higher crystal quality of the ZnBeMgO films was deposited. These XRD results could be explained by the coordination of Be and Mg in the ZnBeMgO films. The large lattice mismatch of ZnO/BeO and ZnO/MgO in the ZnBeMgO films could be counteracted each other in a wide range [17]. However, the excess Be led to a poor crystal quality such as the ZnBeMgO(70) sample. The substitution of Zn by either Be or Mg will result in a lower c-axis parameter value [14,19]. Therefore, the out-of-plane residual compressive stresses in the ZnBeMgO films were related to the content of Be and Mg. Consequently, the coordination of Be and Mg plays an important role in the crystal quality improvement of the ZnBeMgO films. It was clearly demonstrated that by incorporating different amounts of beryllium and magnesium into ZnO, the bandgap energy of the obtained ZnBeMgO films could be modulated from 3.7 to 4.9 eV continuously. The crystal quality of the ZnBeMgO films with proper composition was improved significantly, compared with that of either the ZnMgO films or the BeZnO films. These ZnBeMgO films are promising for fabricating high-efficiency optoelectronic devices such as solar-blind UV detectors.

Yang et al. [40] deposited wurtzite ZnBeMgO films with high magnesium content using the PLD method at various oxygen partial pressures. Experimental results revealed that the crystallinity and the bandgap energy of the ZnBeMgO films were strongly influenced by the oxygen partial pressure. With an increase in the oxygen pressure, the crystalline structure of the ZnBeMgO films changed from cubic rock-salt structure to hexagonal wurtzite structure. The steady decrease in the bandgap energy of the ZnBeMgO films from 4.71 to 4.50 eV was observed when the oxygen partial pressure increased from 2×10^{-4} to 2 Pa.

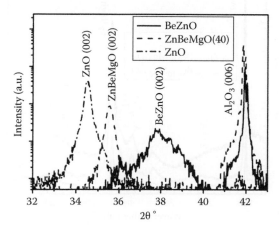

FIGURE 9.13
XRD patterns of BeZnO and ZnBeMgO with the same bandgap value of 4.5 eV. (Reproduced with permission from Yang, C. et al., *Appl. Phys. Lett.*, 93, 112114, 2008.)

Further investigation on the structural and optical properties of the $Zn_{1-x-y}Be_xMg_yO$ ($0 \leq x \leq 0.15$; $0 \leq y \leq 0.20$) powders and thin films were carried out [41]. The $Zn_{1-x-y}Be_xMg_yO$ ($0 \leq x$, $y \leq 0.10$) powders were synthesized by low-energy ball milling using commercial ZnO, BeO, and MgO starting materials. The ZnBeMgO films were grown on (0001)-oriented Al_2O_3 substrates by pulse laser deposition technique using an excimer laser (KrF, 248 nm, 10 Hz repetition frequency) with laser energy of 2.5 J/cm². According to the XRD pattern of the $Zn_{1-x-y}Be_xMg_yO$ ($0 \leq x \leq 0.15$; $0 \leq y \leq 0.20$) powder samples, the lattice volume decreased, and the peak shifted slightly to a higher 2θ value due to the much smaller size of Be^{2+} compared with Zn^{2+} whereas Mg^{2+} and Zn^{2+} ions have similar ionic sizes. Furthermore, the Zn–O bond length decreased because of the smaller ions substitution at the Zn-site [41]. Figure 9.14 shows the unpolarized Raman spectra of the $Zn_{1-x-y}Be_xMg_yO$ ($0 \leq x$, $y \leq 0.10$) samples [41]. The optical mode at the Γ point can be written as

$$\Gamma_{opt} = A_1 + E_1 + 2E_2 + 2B_1 \tag{9.1}$$

where

  B_1 mode is silent in Raman scattering

  A_1 and E_1 modes, both are Raman and infrared active, are polar and split into transverse-optical (TO) component and LO component

  E_2 modes includes E_2 (low) (associated with the motion of Zn sublattices) and E_2 (high) (associated with the motion of oxygen atoms) [42]

For the ZnO samples shown in Figure 9.14, the E_2 (low) and E_2 (high) optical modes are located at 100.2 and 439.4 cm⁻¹. Besides, the $A_1(z)(TO)$, $E_1(x)(TO)$, and $E_1(x)(LO)$ modes located at 381, 413, and 588 cm⁻¹, respectively. In the ZnBeMgO samples, the E_2 (low) peak located at 103.7 cm⁻¹, which was shifted from 100.2 cm⁻¹ of the ZnO samples. It was worth noting that the additional mode located around 556 cm⁻¹ was exhibited for the ZnBeMgO samples. This additional mode named as quasi-LO phonon mode originated

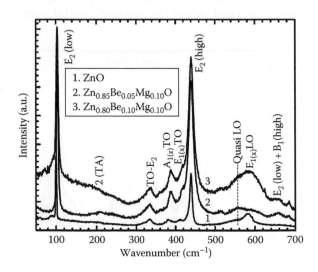

FIGURE 9.14
Room temperature Raman spectra of $Zn_{1-x-y}Be_xMg_yO$ ($0 \leq x$, $y \leq 0.10$) powders. (Reproduced with permission from Panwar, N. et al., *J. Alloy. Compd.*, 509, 1222, 2011.)

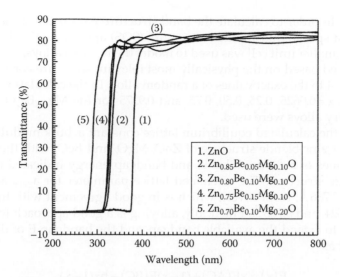

FIGURE 9.15

UV-transmittance spectra of $Zn_{1-x-y}Be_xMg_yO$ ($0 \leq x \leq 0.15$; $0 \leq y \leq 0.20$) thin films. (Reproduced with permission from Panwar, N. et al., *J. Alloy. Compd.*, 509, 1222, 2011.)

from the abundant shallow donor defects including oxygen vacancies and/or Zn interstitials, bounded on the tetrahedral Mg/Be sites. Furthermore, the peak of the mode located at $685\,cm^{-1}$ was attributed to the incorporation of Be/Mg at Zn sites. Figure 9.15 shows the optical transmittance of the ZnO and various ZnBeMgO thin films [41]. Using the experimental results shown in Figure 9.15, a plot of $(\alpha h\nu)^2$ versus the photon energy $h\nu$, where α is the absorption coefficient, h is the Planck constant, and ν is the frequency, could be obtained. From the linear fitting of the $(\alpha h\nu)^2$ versus $h\nu$ curves, the bandgap energy of the ZnO film was 3.37 eV whereas the bandgap energy of the $Zn_{0.7}Be_{0.1}MgO$ film increased up to 4.51 eV.

9.3 Band Structure Calculation

Band structure is a fundamental issue for material study, especially for band engineering study. To obtain a more complete picture of the band structure variation of the $Be_xMg_yZn_{1-x-y}O$ alloy systems, the theoretically calculated band structure has also been carried out for various compositions. Shi and Duan [43] systematically investigated ZnBeO, ZnMgO, and MgBeO alloys, in order to understand their band structure. Their first-principle band structure and total energy calculations were carried out using the density functional theory (DFT) within the local-density approximation (LDA) [44] as implemented in the Vienna *ab initio* simulation package (vasp) code [45]. The electron and core interactions were included using the frozen-core projected augmented wave approach [46]. The Zn 3d electrons were explicitly treated as valence electrons. The whole electron wave function was expanded in plane waves up to a cutoff energy of 400 eV. All the geometries were optimized by minimizing the quantum mechanical forces acting on the atoms. For the Brillouin zone integration, the k points, which were equivalent to the $4 \times 4 \times 4$ Monkhorst–Pack special k-point meshes [47] in the zinc-blende (ZB) Brillouin zone, were used. In order to calculate the band structure parameter of the $A_xB_{1-x}C$ alloys, the more efficient special quasirandom structure (SQS) approach [48,49] was adopted. In this approach, a smaller unit cell was used to model the random alloys, where mixed-atom sites were occupied based on the physically most relevant structure correction functions, which were closest to the exact values of a random alloy. In the calculation, the $2a \times 2a \times 2a$ 64 atoms SQS at $x = 0.0625$, 0.25, 0.50, 0.75, and 0.9375 for the $Mg_{1-x}Be_xO$, $Zn_{1-x}Be_xO$, and $Zn_{1-x}Mg_xO$ ternary alloys were used.

Table 9.1 gives the calculated equilibrium lattice constant a, bulk modulus B, and bandgap at Γ point for zinc-blende structure of ZnO, MgO, and BeO, with the corresponding experimental values of bulk modulus B, and bandgap energy indicated (in brackets) for their ground-state structure. The calculated lattice parameter for a_{ZnO}, a_{MgO}, and a_{BeO} is 4.506, 4.517, and 3.766 Å, respectively, which is in good agreement with the experimental values of 4.47, 4.524, and 3.768 Å. Practically, alloying is a good approach to perform bandgap engineering to extend the available bandgap, and the bandgap E of the alloy $A_xB_{1-x}C$ can be described as

$$E(x) = xE(AC) + (1-x)E(BC) - bx(1-x) \qquad (9.2)$$

where
  b is called the bandgap (optical) bowing parameter
  E(AC) and E(BC) are the bandgap of the binary constituent of AC and BC materials, respectively

For most semiconductor alloys, the bowing parameter b is almost independent of composition x [17]. However, the bowing parameter b of $Ga_{1-x}N_xAs$ and $Zn_{1-x}Te_xO$ alloys is strongly composition-dependent, indicating that these alloys contain some elements with large difference in the size and the atomic orbital energies [54–56]. It is of great interest, therefore, to investigate how the bowing parameter b and the bandgap of the ZnBeO, MgBeO, and ZnMgO vary due to the large difference between the Zn (Mg) and the Be atoms resided in these alloys. Ding et al. [57] calculated the electronic structure of the $Be_xZn_{1-x}O$ by numerical simulation. The calculated results indicated that the bowing parameter b of the BeZnO was much larger than that of AlGaN. It was attributed to

TABLE 9.1

Calculated Lattice Constant a, Bulk Modulus B, and Band gap E_g at Γ for Zinc-Blende Binary Alloys ZnO, MgO, and BeO

	ZnO	MgO	BeO
a (Å)	4.506(4.47[a])	4.517(4.524[b])	3.766(3.768[c])
B (Mbar)	1.562(1.837)	1.554(1.603)	2.354(2.244)
			2.29[c]
E_g (eV)	0.70(3.4)	3.731(7.67)	7.852(10.585)

Source: Reproduced with permission from Shi, H.L. and Duan, Y., *Eur. Phys. J. B*, 66, 439, 2008.

The experimental values of B and Eg suggested in reference [17] are also listed in parentheses.
[a] Experimental value in ref. [50,51].
[b] Theoretical value in ref. [52].
[c] Theoretical value in ref. [53].

the large ionic radius difference between Be^{2+} (0.27 Å) and Zn^{2+} (0.60 Å). The calculated bandgap bowing coefficient of the $Mg_{1-x}Be_xO$, $Zn_{1-x}Be_xO$, and $Zn_{1-x}Mg_xO$ ternary alloys was a function of the composition x. It could be found that the bowing parameter b was large in the ZnBeO and MgBeO alloys. This phenomenon was attributed to the large atomic size difference and the large chemical mismatch between Zn (Mg) and Be. However, compared with ZnBeO and MgBeO, the bowing parameter of the ZnMgO alloys was smaller because of the similar atomic size of the Zn and Mg. Since the atomic size of Zn was slightly larger than that of Mg, the bowing parameter b of the ZnBeO was also a little larger than that of the MgBeO. The bowing parameter was also composition dependent and large at the Be-rich region for the ZnBeO and MgBeO alloys. These results were attributed to the large size and chemical differences between the component atoms. Figure 9.16 shows the calculated bandgap of the ZnBeO, MgBeO, and MgBeO alloys at various concentrations [43]. Although the bandgap of the BeO is larger than that of the MgO, incorporation of a small amount of Be into the MgO matrix can lead to a reduction in the bandgap. This result makes that the bowing parameter is larger than the bandgap difference.

The band offset for the heterostructure is an important issue in the application of resulting devices. Figure 9.17 shows the calculated natural band alignment [43]. For the valence band maximum (VBM) alignment, the VBM of the ZnO is higher than that of the MgO and the BeO. This is because the p-d (p_O-d_{Zn}) repulsion pushes up the VBM, whereas Mg and Be have no active d electrons. According to the common-anion rule [58], the valence band

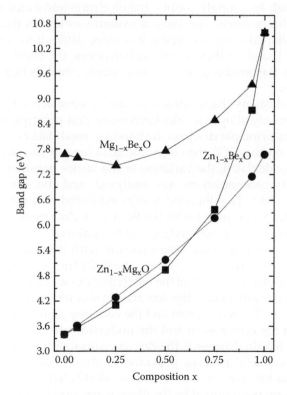

FIGURE 9.16
Predicted bandgaps of $Mg_{1-x}Be_xO$, $Zn_{1-x}Be_xO$, and $Zn_{1-x}Mg_xO$ ternary alloys calculated at various compositions. (Reproduced with permission from Shi, H.L. and Duan, Y., *Eur. Phys. J. B*, 66, 439, 2008.)

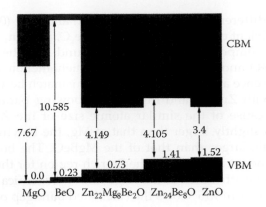

FIGURE 9.17
Calculated natural band alignment of MgO, BeO, $Zn_{22}Mg_8Be_2O_{32}$, $Zn_{24}Be_8O_{32}$, and ZnO alloys. (Reproduced with permission from Shi, H.L. and Duan, Y., *Eur. Phys. J. B*, 66, 439, 2008.)

offset between common-anion systems is small. However, the VBM of the BeO is higher than that of the MgO. This result is attributed to the strong covalency of the BeO alloy, resulting in the kinetic-energy-induced valence band broadening [59]. As shown in Figure 9.17, it can be concluded that the $Zn_{0.75}Be_{0.25}O$ alloy acts as a good barrier material for ZnO-based UV LEDs. The ΔE_C and ΔE_V values are approximately 0.60 and 0.11 eV, respectively, which are large enough to strongly confine the electrons and holes [60]. The quaternary $Zn_{22}Mg_8Be_2O_{32}$ alloy has a conduction band maximum as low as that of ZnO, indicating that it can be n-type doped easily. However, it is more difficult to achieve p-type doping due to the low VBM. Based on these chemical behaviors, the band offset can be tailored to practically perform the band engineering as desired, which is helpful for the design of optoelectronic devices.

Lattice constants, elasticity, band structure, and piezoelectricity of hexagonal wide bandgap $Be_xZn_{1-x}O$ ternary alloys are calculated using first-principle methods [61]. Su et al. [62] reported their first-principle study on the bandgap modulation of Be and Mg codoped ZnO systems. Based on the DFT, the electronic structure of the wurtzite ZnO codoped with Be and Mg was calculated, the variation of the lattice parameter of $Be_xMg_yZn_{1-x-y}O$ with various impurity concentrations was analyzed, and the mechanism for bandgap broadening was discussed. The obtained results exhibited that the Be and Mg codoped systems were more stable compared with the Be-doped ZnO systems. For certain doping concentrations, a solar-blind region bandgap can be achieved.

The ideal ZnO has a hexagonal wurtzite structure with the space group of P63/mc and C6v-4 symmetry. The geometry constants are a = b = 0.3249 nm, c = 0.5206 nm, $\alpha = \beta = 90°$, and $\gamma = 20°$. The 16-atom supercell ($2 \times 2 \times 1$) of the wurtzite ZnO was used in the calculation. The ligand of ZnO was a triangular cone, the side arris is shorter than the underside arris, and the bond length between the center atom and the cone-top atom is slightly longer than the bond length between the center atom and the underside atom [62]. The O^{2-} coordination polyhedron is $O–Zn_4$ tetrahedron, so is the Zn^{2+} coordination polyhedron. The calculation is carried out using CASTEP package provided by the Material Studios 4.1 of Accelrys Company. In the package, the ionic potential is substituted by a pseudo-potential, the electronic wave function is expanded by the plane wave, and the exchange and correlative potentials of electronic–electronic interaction are emendated by the generalized gradient approximation. In the calculation, the energy cutoff is set to be 450 eV, the total energy is converged to less than 2×10^{-5} eV/atom, and a $4 \times 4 \times 4$ k-point mesh in the Brillouin zone is used [62].

The formation energy $E_f(x, y)$ of the alloys is defined as [57,63]:

$$E_f(x,y) = E_{tot}(Be_xMg_yZn_{1-x-y}O) - xE_{tot}(BeO) - yE_{tot}(MgO) - (1-x-y)E_{tot}(ZnO) \quad (9.3)$$

where $E_{tot}(Be_xMg_yZn_{1-x-y}O)$, $E_{tot}(BeO)$, $E_{tot}(MgO)$, and $E_{tot}(ZnO)$ are the total energies of the corresponding material systems. Here, the total system of a quaternary compound is composed of three binary compounds. When $E_f(x, y)$ is a negative value, the quaternary compound is more steady than the binary compound. On the other hand, when $E_f(x, y)$ is positive value, the quaternary compound is less steady than the binary compound. In general, a larger $E_f(x, y)$ is unfavorable for the stability of the total system.

According to the listed optimized lattice parameter and formation energy for different systems [62], the lattice constant and primitive cell volume V_0 of the $Be_xZn_{1-x}O$ alloy decrease with an increase in the Be concentration; for the $Mg_xZn_{1-x}O$ alloy, they increase with an increase in the Mg composition; for the $Be_xMg_yZn_{1-x-y}O$ alloy, their values lie between the values of the $Be_xZn_{1-x}O$ alloy and the $Mg_xZn_{1-x}O$ alloy. However, they are still smaller than those of the ZnO. For the Be concentration of 0.125 and the Mg concentration of 0, 0.125, 0.25, 0.375, and 0.5, the parameter a increases from 3.191 to 3.214, which approaches the value of pure ZnO. It is predicted by Vegard's law that the lattice constant of the alloys changes linearly with the concentration of Be or Mg. For both Be and Mg concentration of 0.25, the parameter a becomes much smaller than that of the $Be_{0.125}Mg_{0.25}Zn_{0.625}O$ alloy, and the formation energy is much larger, that is, the alloy is less stable. Therefore, it is concluded that when the Be concentration is 0.125, the system is relatively stable. In practical applications, it is reasonable to investigate the case of Be concentration of 0.125 and Mg concentration of 0, 0.125, 0.25, 0.375, and 0.5.

As shown in Figure 9.18 [62], the formation energy increases linearly with an increase in the Mg concentration. It is also worth to note that the formation energy of the $Be_xMg_yZn_{1-x-y}O$ system is very small compared with the formation energy of the $Al_xGa_{1-x}N$ system calculated by de Paiva [63], indicating that such $Be_xMg_yZn_{1-x-y}O$ alloys are very stable. This calculation indicates that the bottom of the conduction band is decided by Zn 4s states, and the top of the valence band is decided by O 2p states; the bandgap is decided by O 2p and Zn 4s states. The Zn 4s states at the bottom of the conduction band move to the high-energy region due to the incorporation of Mg and Be, while the O 2p states slightly change to lead the bandgap broadening.

The bandgap of the $Be_xMg_yZn_{1-x-y}O$ (x = 0.125) alloys as a function of Mg concentration is shown in Figure 9.19 [62]. Bandgap of the $Be_xMg_yZn_{1-x-y}O$ alloys increases with an increase in the Mg concentration. The bandgap energy of 2.56 eV for y = 0.5 can be obtained. As pointed out earlier, underestimation of the bandgap is a well-known drawback of using standard DFT calculations, and the results can be corrected by the scissors approximation. Taking pure ZnO as the reference, the scissors correction can approximately be set to be 2.40 eV; thus, the bandgap of the $Be_xMg_yZn_{1-x-y}O$ system can reach 4.96 eV.

From the previous discussion, it is concluded that the $Be_xMg_yZn_{1-x-y}O$ alloys for certain composition ratio are very stable and have a bandgap in deep UV region, which is suitable for the application of optoelectronic devices in UV region, such as solar-blind UV photodetectors.

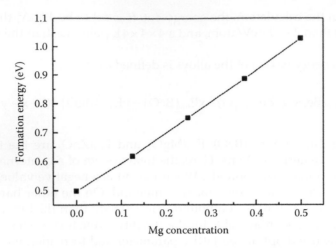

FIGURE 9.18
Formation energies of $Be_xMg_yZn_{1-x-y}O$ (x = 0.125). The formation energy increases linearly with increase in Mg concentration. (Reproduced with permission from Su, X. et al., *Phys. B*, 404, 1794, 2009.)

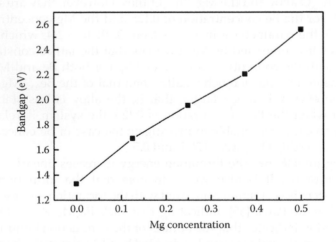

FIGURE 9.19
Bandgap of $Be_xMg_yZn_{1-x-y}O$ (x = 0.125). Bandgap increases as the concentration of Mg increases, it reaches 2.56 eV for y = 0.5. (Reproduced with permission from Su, X. et al., *Phys. B*, 404, 1794, 2009.)

9.4 Optoelectronic Applications

The bandgap of the BeZnO films could be modulated in a much larger range than that of the MgZnO films without phase segregation, because the BeO has the same hexagonal wurtzite structure with ZnO. Therefore, it is more promising for UV optoelectronic applications. However, there are still problems caused by the large ionic radius difference between Be^{2+} and Zn^{2+}. They degrade the resulting crystal quality and generate a large lattice mismatch, especially at a high Be content. In this aspect, quaternary $Be_xMg_yZn_{1-x-y}O$ alloys exhibit to be more promising. These alloy systems have been successfully used to construct various heterostructures and devices.

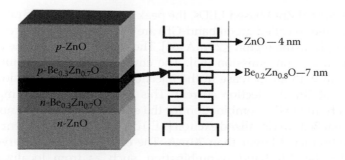

FIGURE 9.20

Schematic illustration of the structure of the ZnO-based UV LED devices that employ a BeZnO/ZnO active layer comprised of MQWs. (Reproduced with permission from Ryu, Y. et al., *Appl. Phys. Lett.*, 88, 241108, 2006.)

Ryu et al. [64] reported ZnO-based UV LEDs. Figure 9.20 shows the schematic configuration of the LEDs, in which the multiple quantum well (MQW) active layer is constructed by the $Be_{0.2}Zn_{0.8}O/ZnO$ [64]. The active layer is composed of seven QWs for which undoped 7 nm thick $Be_{0.2}Zn_{0.8}O$ and 4 nm thick ZnO form barrier layer and well layer, respectively. The p-type ZnO and the $Be_{0.3}Zn_{0.7}O$ layers were formed using As as acceptor dopant, while the n-type ZnO and $Be_{0.3}Zn_{0.7}O$ layers were formed using Ga as donor dopant. The hole concentrations in p-type layers of the ZnO-based UV LEDs are in the mid-10^{17} cm^{-3} range for the ZnO layer and in the upper 10^{16} cm^{-3} range for the $Be_{0.3}Zn_{0.7}O$ layer, while the electron concentrations in n-type layers are in the mid-10^{18} cm^{-3} and low-10^{17} cm^{-3} ranges for the ZnO layer and the $Be_{0.3}Zn_{0.7}O$ layer, respectively, as determined by Hall measurements. All the ZnO and BeZnO layers in the devices were deposited by the HBD method.

For fabricating the LEDs, ohmic contacts were formed on each of the p-type and n-type layers using Ni/Au metals and Ti/Au metals, respectively. The current–voltage (I–V) characteristics of the LEDs are shown in Figure 9.21 [64]. I–V measurements demonstrated the p-n junction characteristics for the ZnO-based structures, featuring low reverse bias current as shown in the insert of Figure 9.21. According to the electroluminescence (EL)

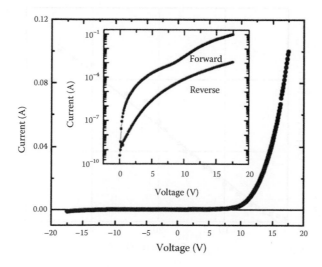

FIGURE 9.21

Current–voltage (I–V) characteristics for a p-n junction ZnO-based LED. (Reproduced with permission from Ryu, Y. et al., *Appl. Phys. Lett.*, 88, 241108, 2006.)

spectra of the fabricated ZnO-based LEDs, the peak located at near 388 nm (bound exciton [BE]) and 550 nm (the so-called green band-GB) are the dominant features at low forward currents, for example, in the region ≤20 mA [64]. The peak at 388 nm is the impurity (donor or acceptor)-bound exciton emission [65]. The emission peak at 550 nm is shifted to a shorter wavelength with an increase in the injection current; the emission peak location is around 520 nm at 50 mA injection current. This behavior indicates impurity-involved emission such as from DAP recombination for the emission peak at 550 nm. The emission peak located at near 363 nm (localized exciton [LE]) becomes the prominent spectral feature at a current injection level larger than 20 mA. The emission peak observed at 363 nm is deduced from the band-to-band recombination, such as from localized-exciton peak in the MQW active layer, as was observed in the GaN-based LEDs [66]. Optical output power (P) for the unpackaged (on-chip) ZnO-based UV LEDs ($500 \times 500 \, \mu m^2$), operating at room temperature in continuous current injection mode, as a function of injection current is depicted in Figure 9.22 [64]. The output power (P) increases in near linear fashion as the injected current (I_f) is increased; more specifically, $P \sim (I_f)^m$, where m is about 1.2. The emitted power of typical packaged LEDs is estimated to be several microwatts at 20 mA injection current. These ZnO-based UV LEDs are potential for applications such as solid-state lighting and antimicrobial lamps.

Figure 9.23 shows the normalized PL spectra of the seven ZnO/BeZnO MQWs measured at room temperature under the excitation of an ArF excimer laser [67]. The laser beam was incident in a direction normal to the sample surface in order to illuminate the entire sample area. The PL spectral emission was detected from the edge of the sample, that is, in a direction perpendicular to the incident beam. When the excitation intensity increases, the PL intensity grows superlinearly with no change in peak position and then transforms into a sharp peak located at 3.21 eV, which becomes the dominant peak finally. Such phenomena are characteristics of stimulated emission and lasing associated with a Fabry–Pérot resonant cavity.

The ZnO/BeZnO-based UV laser diodes were reported previously [67]. The devices were fabricated using the same growth and fabrication methods employed for the devices

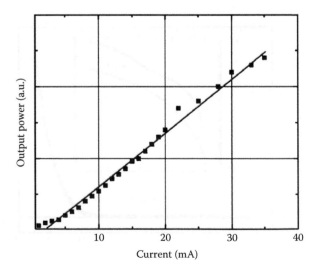

FIGURE 9.22
Integrated output power as a function of device injection current for one of the LEDs. (Reproduced with permission from Ryu, Y. et al., *Appl. Phys. Lett.*, 88, 241108, 2006.)

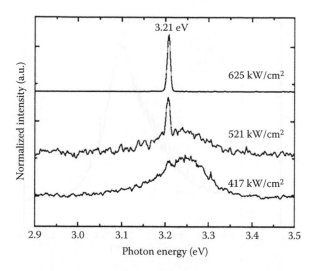

FIGURE 9.23
Normalized PL spectra measured at RT of a ZnO/BeZnO superlattice containing seven QWs excited by an ArF excimer laser at selected excitation intensities to illustrate lasing action. (Reproduced with permission from Ryu, Y.R. et al., *Appl. Phys. Lett.*, 90, 131115, 2007.)

reported earlier [19,21,64]. The devices are composed of a p–n heterojunction structure with a MQW active layer sandwiched between guide-confinement layers. The MQW active layer consisted of three QWs, while the two guide-confinement layers were As-doped p-type ZnO/BeZnO and Ga-doped n-type BeZnO/ZnO films, respectively. The active layer was fabricated with 7 nm thick BeZnO used to form an individual barrier layer and with 4 nm thick undoped ZnO for the well layer. The mixing ratio of Be to Zn atoms in the barrier layers was estimated to be about 0.25. Two Ga-doped BeZnO layers formed the cladding and confinement layers in the n-type layers of the devices. The first layer was a 0.5 μm thick $Be_{0.1}Zn_{0.9}O$ layer deposited on an 1 μm Ga-doped ZnO layer. The second layer was a 0.05 μm thick $Be_{0.2}Zn_{0.8}O$ layer deposited prior to the growth of the MQW active layer. After growing the three periods of the MQWs, in succession, As-doped $Be_{0.3}Zn_{0.7}O$ (0.03 μm)/ $Be_{0.2}Zn_{0.8}O$ (0.5 μm)/$Be_{0.1}Zn_{0.9}O$ (0.7 μm)/ZnO (0.2 μm) layers were deposited on the p-type side of the ZnO/BeZnO-based devices. Both Ga and As concentrations were in the low-10^{19} cm^{-3} range for the devices. The area used for electrical pumping was $300 \times 600\,\mu m^2$. Figure 9.24 shows the EL spectra of the ZnO/BeZnO-based MQW devices measured at room temperature with continuous injection current of 10 and 60 mA [67]. The key features of the EL spectrum were consistently in agreement with those of the PL spectrum obtained for the ZnO/BeZnO superlattice. Excitonic recombinations are the dominant features for the MQW device even at room temperature, and this result can be expected owing to the large exciton binding energy (263 meV) in the ZnO/BeZnO MQWs. As increasing the injection current, the emission rate of the Pn band is much higher than that of the superlinearly grown EX peak, which indicates the phenomenon of stimulated emission [67].

Figure 9.25 exhibits the EL spectra of the ZnO/BeZnO-based MQW devices measured at room temperature in pulsed mode (10% duty cycle) current injection [67]. As increasing the injection current, the presence of sharp Fabry–Pérot-type oscillation became more prominent and caused lasing action. The threshold current density was about 420 A/cm². This low-threshold current density for lasing was attributed to the large exciton binding energy and the presence of exciton–exciton inelastic scattering.

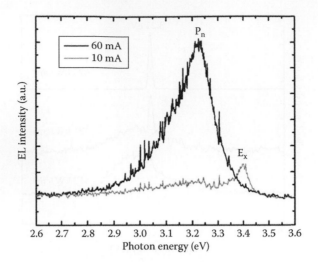

FIGURE 9.24
Electroluminescence (EL) spectra measured at RT with continuous current injection values of 10 and 60 mA for a ZnO/BeZnO-based MQW device. (Reproduced with permission from Ryu, Y.R. et al., *Appl. Phys. Lett.*, 90, 131115, 2007.)

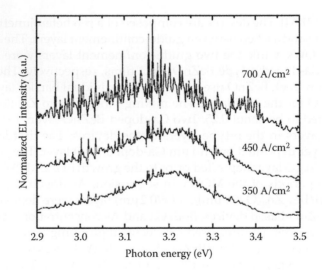

FIGURE 9.25
EL spectra measured at RT in pulse current injection mode (10% duty cycle). Sharp Fabry–Pérot type oscillations become more prominent with increasing current density, indicating lasing action. (Reproduced with permission from Ryu, Y.R. et al., *Appl. Phys. Lett.*, 90, 131115, 2007.)

The solar-blind UV photodetectors have vast potential applications in the fields of solar astronomy, missile plume detection, space-to-space transmission, fire alarms, and combustion monitoring. Since the solar radiation component in the solar-blind spectrum range (220–280 nm) is absorbed by the atmosphere, hence a photodetector working in this range will not be interfered by the solar radiation [68]. For the solar-blind UV detector, its cutoff working wavelength should be shorter than 280 nm, corresponding to the bandgap wider than 4.4 eV. However, high-performance solar-blind ZnO-based photodetectors have not yet been obtained. Moreover, an external (reverse) bias is required to achieve high photoresponsivity for most ZnO/Si photodiodes.

As discussed earlier, the $Zn_{1-x-y}Be_xMg_yO$ alloys have been successfully grown with continuous bandgap modulation from 3.7 to 4.9 eV. Furthermore, the advantages of crystalline and optical properties were demonstrated compared with those of the ZnMgO alloy and the ZnBeO alloy. Based on the quaternary alloy, the zero-biased solar-blind photodetectors composed of ZnBeMgO/Si heterojunction were reported previously [69]. In the photodetectors, the ZnBeMgO film was deposited on p-Si and quartz substrates by PLD at a substrate temperature of 600°C. The oxygen partial pressure was 1×10^{-4} Pa. Ceramic $Zn_{0.7}Be_{0.1}Mg_{0.2}O$ target was ablated by a KrF excimer laser (wavelength of 248 nm, energy of 200 mJ pulse^{-1}, and a repetition rate of 5 Hz). The thickness of the deposited films was about 300 nm. To analyze the spectral response properties of the ZnBeMgO/Si heterojunction, ohmic contacts were made by depositing a regular semitransparent Ti/Pt electrode on the n-ZnBeMgO and an Al/Pt electrode on the reverse surface of p-Si [70]. Furthermore, in order to reduce the specific contact resistance between the metal electrode and the ZnBeMgO layer, an additional highly doped 50 nm thick $Zn_{0.98}Al_{0.02}O$ (AZO) thin film (sheet resistance of about 70 Ω/□) was deposited as a contact layer for comparison. The device structure is illustrated in Figure 9.26, which operates without bias [69]. Figure 9.26 displays the rectifying performance of the n-ZnBeMgO/p-Si heterojunction structure. As shown in Figure 9.26, the threshold voltage is about 0.5 V. The dark current of the photodetectors is approximately 10^{-6} A at a bias of −5 V, which is three orders of magnitude higher than that of GaN-based solar-blind photodetectors. This high dark current is mainly attributed to the defects in the films and interfaces. The dark current can be further decreased by improving the crystalline quality of the ZnBeMgO films. The cutoff wavelength of the transmission spectrum for an individual ZnBeMgO layer is 280 nm, which indicates the solar blindness function [69]. The absolute spectral responsivity of the n-ZnBeMgO/p-Si heterojunction photodetectors without bias is shown in Figure 9.27 [69]. The response curves at 280 nm drop two orders of magnitude and remain at a low response for the entire near UV and visible spectra,

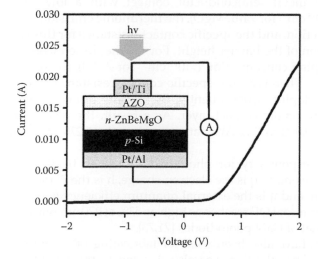

FIGURE 9.26
I–V characteristic of the n-ZnBeMgO/p-Si heterojunction. The inset is the schematic cross section of the device. (Reproduced with permission from Yang, C. et al., *J. Phys. D Appl. Phys.*, 42, 152002, 2009.)

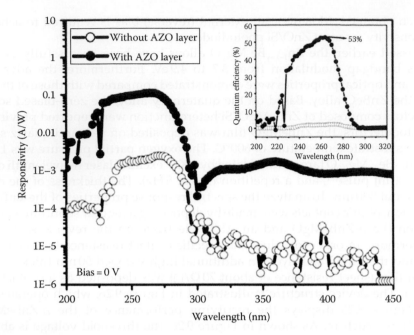

FIGURE 9.27
Measured spectral responsivity curves at zero bias of the n-ZnBeMgO/p-Si heterojunction. The inset is the corresponding external quantum efficiency of the device. (Reproduced with permission from Yang, C. et al., *J. Phys. D Appl. Phys.*, 42, 152002, 2009.)

demonstrating the typical solar-blind photoresponse characteristics. However, because of the high resistivity of the ZnBeMgO layer and the weak UV response, the collection of photogenerated carriers is less efficient without a contact layer [71]. In order to enhance the carrier collection efficiency, an additional AZO film was deposited on the ZnBeMgO layer as a highly doped contact layer. The conductive mechanism could be explained as follows. For a metal–semiconductor contact with a low semiconductor doping concentration ($\sim 10^{13}$ cm^{-3} for ZnBeMgO), the thermionic emission current is the dominant current in this junction, and the specific contact resistance for this case decreases rapidly due to the reduction of the barrier height. For the metal–semiconductor junction with a high impurity doping concentration (>10^{21} cm^{-3} for AZO), the tunnelling process is the dominate function. Therefore, the specific contact resistance is a very strong function of semiconductor doping. Consequently, the heavily doped AZO layer was designed in the device as an ohmic contact layer. As a result, the peak responsivity improved significantly from 0.003 to 0.11 AW^{-1} at a wavelength of 270 nm without bias, as shown in Figure 9.27 [69].

The theoretical responsivity for photodiodes is given by $R(\lambda) = (q\lambda/hc)\eta$ [72], where λ is the photon wavelength, q is the electron charge, h is the Planck constant, c is the light velocity in vacuum, and η is the external quantum efficiency (QE). As shown in the inset of Figure 9.27, the external QE reaches as high as 53% at 270 nm without bias. This QE value is comparable to that of GaN photodiodes [71,73].

ZnBeMgO alloys have also been used for fabricating other optoelectronic and microelectronic devices. The attention was also devoted to the reliable numerical design for optoelectronic devices with ZnO-based alloys [74].

9.5 Summary

For optoelectronic applications of ZnO-based materials, it is required to modulate the bandgap toward deeper UV region. ZnO-based ternary and quaternary alloys, especially the ZnBeO and ZnBeMgO families, have been proposed and studied. For the ZnBeO alloys, the component of the binary oxides of BeO and ZnO possesses the same hexagonal wurtzite structure and good solubility. Therefore, the ZnBeO alloys can achieve a wider bandgap modulation. However, the large size difference between the Be and Zn atoms degrades the crystal quality of the resulting ZnBeO alloys. In this aspect, ZnBeMgO quaternary alloys present advantage over ternary alloys, because the large lattice mismatch of the ZnO/BeO and the ZnO/MgO can be counteracted by each other in this quaternary system.

In this chapter, the preparation of ZnBeO and ZnBeMgO alloys, succeeded in recent years, has been demonstrated. The structural and optical properties of these alloys have also been investigated experimentally and theoretically. The experimental results show that these alloys possessed a larger bandgap modulation range and a good crystal quality. The optoelectronic applications in UV spectral region by using these ZnO-based alloys have also been demonstrated. It is expected that these oxide alloys are promising for applications in UV region.

References

1. R.F. Service, Will UV lasers beat the blues? *Science* **276**, 5314 (1997).
2. Ü. Özgür, Y.I. Alivov, C. Liu, A. Teke, M.A. Reshchikov, S. Dogan, V. Avrutin, S.J. Cho, and H. Morkoç, A comprehensive review of ZnO materials and devices, *J. Appl. Phys.* **98**, 041301 (2005).
3. I.S. Jeong, J.H. Kim, and S. Im, Ultraviolet-enhanced photodiode employing n-ZnO/p-Si structure, *Appl. Phys. Lett.* **83**, 2946 (2003).
4. C.T. Lee, Fabrication methods and luminescent properties of ZnO materials for light-emitting diodes, *Materials* **3**, 2218 (2010).
5. A. Janotti and C.G.V. de Walle, Fundamentals of zinc oxide as a semiconductor, *Rep. Prog. Phys.* **72**, 126501 (2009).
6. R.W. Chuang, R.X. Wu, L.W. Lai, and C.T. Lee, ZnO-on-GaN heterojunction light-emitting diode grown by vapor cooling condensation technique, *Appl. Phys. Lett.* **91**, 231113 (2007).
7. H.Y. Lee, S.D. Xia, W.P. Zhang, L.R. Lou, J.T. Yan, and C.T. Lee, Mechanisms of high quality i-ZnO thin films deposition at low temperature by vapor cooling condensation technique, *J. Appl. Phys.* **108**, 073119 (2010).
8. C.T. Lee, T.S. Lin, and H.Y. Lee, Mechanisms of low noise and high detectivity of p-GaN/i-ZnO/n-ZnO:Al-heterostructured ultraviolet photodetectors, *IEEE Photon. Technol. Lett.* **22**, 1117 (2010).
9. C.T. Lee, Y.H. Lin, L.W. Lai, and L.R. Lou, Mechanism investigation of p-i-n ZnO-based light-emitting diodes, *IEEE Photon. Technol. Lett.* **22**, 30 (2010).
10. C.T. Lee and J.T. Yan, Ultraviolet electroluminescence from ZnO-based n-i-p light-emitting diodes, *IEEE Photon. Technol. Lett.* **23**, 353 (2011).
11. H.Y. Lee, C.T. Lee, and J.T. Yan, Emission mechanisms of passivated single n-ZnO:In/i-ZnO/p-GaN-heterostructured nanorod light-emitting diodes, *Appl. Phys. Lett.* **97**, 111111 (2010).
12. C.T. Lee, Y.L. Chiou, and C.S. Lee, AlGaN/GaN MOS-HEMTs with gate ZnO dielectric layer, *IEEE Electron Device Lett.* **31**, 1220 (2010).

13. Y.L. Chiou, C.S. Lee, and C.T. Lee, AlGaN/GaN metal-oxide-semiconductor high-electron mobility transistors with ZnO gate layer and $(NH_4)_2S_x$ surface treatment, *Appl. Phys. Lett.* **97**, 032107 (2010).

14. A. Ohtomo, M. Kawasaki, T. Koida, K. Masubuchi, H. Koinuma, Y. Sakurai, Y. Yoshida, T. Yasuda, and Y. Segawa, $Mg_xZn_{1-x}O$ as a II–VI widegap semiconductor alloy, *Appl. Phys. Lett.* **72**, 2466 (1998).

15. A. Ohtomo, M. Kawasaki, I. Ohkubo, H. Koinuma, T. Yasuda, and Y. Segawa, Structure and optical properties of $ZnO/Mg_{0.2}Zn_{0.8}O$ superlattices, *Appl. Phys. Lett.* **75**, 980 (1999).

16. A. Ohtomo, K. Tamura, M. Kawasaki, T. Makino, Y. Segawa, Z.K. Tang, G.K.L. Wong, Y. Matsumoto, and H. Koinuma, Room-temperature stimulated emission of excitons in ZnO/(Mg,Zn)O superlattices, *Appl. Phys. Lett.* **77**, 2204 (2000).

17. O. Madelung, *Semiconductors: Data Handbook*, 3rd edn. Springer, New York (2003).

18. Y.J. Lin, P.H. Wu, C.L. Tsai, C.J. Liu, C.T. Lee, H.C. Chang, Z.R. Liu, and K.Y. Jeng, Mechanisms of enhancing band-edge luminescence of $Zn_{1-x}Mg_xO$ prepared by the sol-gel method, *J. Phys. D Appl. Phys.* **41**, 125103 (2008).

19. Y.R. Ryu, T.S. Lee, A. Lubguban, A.B. Corman, H.W. White, J.H. Leem, M.S. Han, Y.S. Park, C.J. Youn, and W.J. Kim, Wide-band gap oxide alloy: BeZnO, *Appl. Phys. Lett.* **88**, 052103 (2006).

20. W.J. Kim, J.H. Leem, M.S. Han, I.W. Park, Y.R. Ryu, and T.S. Lee, Crystalline properties of wide bandgap BeZnO films, *J. Appl. Phys.* **99**, 096104 (2006).

21. Y.R. Ryu, T.S. Lee, and H.W. White, A technique of hybrid beam deposition for synthesis of ZnO and other metal oxides, *J. Cryst. Growth* **261**, 502 (2004).

22. J.H. Yu, J.H. Kim, D.S. Park, T.S. Kim, T.S. Jeong, C.J. Youn, and K.J. Hong, A study on structural formation and optical property of wide band-gap $Be_{0.2}Zn_{0.8}O$ layers grown by RF magnetron co-sputtering system, *J. Cryst. Growth* **312**, 1683 (2010).

23. D.S. Park, J.H. Kim, J.H. Yu, T.S. Jeong, C.J. Youn, and K.J. Hong, Synthesis of a $Be_xZn_{1-x}O$ crystalline film by using RF magnetron co-sputtering, *J. Korean Phys. Soc.* **56**, 70 (2010).

24. C. Lennon, R.B. Tapia, R. Kodama, Y. Chang, S. Sivananthan, and M. Deshpande, Effects of annealing in a partially reducing atmosphere on sputtered Al-doped ZnO thin films, *J. Electron. Mater.* **38**, 1568 (2009).

25. J.F. Moulder, W.F. Stickle, P.E. Sobol, and K.D. Bomben, *Handbook of X-Ray Photoelectron Spectroscopy*, II. Standard XPS spectra of the elements, ed. J. Chastain. Perking-Elmer Corporation, Eden Prairie, MN (1992).

26. M. Chen, X. Wang, Y.H. Yu, Z.L. Pei, X.D. Bai, C. Sun, R.F. Huang, and L.S. Wang, X-ray photoelectron spectroscopy and auger electron spectroscopy studies of Al-doped ZnO films, *Appl. Surf. Sci.* **158**, 134 (2000).

27. C.D. Wagner, J.F. Moulder, L.E. Davis, and W.M. Riggs, *Handbook of X-Ray Photoelectron Spectroscopy*, II. Standard XPS spectra of the elements. ed. G.E. Muilenberg. Perking-Elmer Corporation, Eden Prairie, MN (1979).

28. T.S. Jeong, M.S. Han, J.H. Kim, S.J. Bae, and C.J. Youn, Optical properties of BeZnO layers studied by photoluminescence spectroscopy, *J. Phys. D Appl. Phys.* **40**, 370 (2007).

29. T.S. Jeong, C.J. Youn, M.S. Han, J.W. Yang, and K.Y. Lim, Analysis of Mg-related emissions in p-GaN grown by MOCVD, *J. Cryst. Growth* **259**, 267 (2003).

30. N.Q. Liem, V.X. Quang, D.X. Thanh, J.I. Lee, and D. Kim, Temperature dependence of biexciton luminescence in cubic ZnS single crystals, *Solid State Commun.* **117**, 255 (2001).

31. K. Sakai, K. Noguchi, A. Fukuyam, T. Ikari, and T. Okada, Low-temperature photoluminescence of nanostructured ZnO crystal synthesized by pulsed-laser ablation, *Jpn. J. Appl. Phys.* **48**, 085001 (2009).

32. J.M. Khoshman, D.C. Ingram, and M.E. Kordesch, Growth and optical properties of amorphous $Be_{0.13}Zn_{0.38}O_{0.49}$ thin films prepared by radio frequency magnetron sputtering, *J. Non-Cryst. Solids* **354**, 2783 (2008).

33. J.M. Khoshman, D.C. Ingram, and M.E. Kordesch, Bandgap engineering in amorphous Be_xZn_yO thin films, *Appl. Phys. Lett.* **92**, 091902 (2008).

34. J.M. Khoshman, A. Khan, and M.E. Kordesch, Optical properties of sputtered amorphous beryllium nitride thin films, *J. Appl. Phys.* **101**, 103532 (2007).

35. J.M. Khoshman and M.E. Kordesch, Optical constants and band edge of amorphous zinc oxide thin films, *Thin Solid Films* **515**, 7393 (2007).

36. C. Yang, X.M. Li, Y.F. Gu, W.D. Yu, X.D. Gao, and Y.W. Zhang, ZnO based oxide system with continuous bandgap modulation from 3.7 to 4.9 eV, *Appl. Phys. Lett.* **93**, 112114 (2008).

37. R. Ghosh and D. Basak, Composition dependent ultraviolet photoresponse in $Mg_xZn_{1-x}O$ thin films, *J. Appl. Phys.* **101**, 113111 (2007).

38. X. Zhang, X.M. Li, T.L. Chen, C.Y. Zhang, and W.D. Yu, p-type conduction in wide-gap $Zn_{1-x}Mg_xO$ films grown by ultrasonic spray pyrolysis, *Appl. Phys. Lett.* **87**, 092101 (2005).

39. J.F. Kong, W.Z. Shen, Y.W. Zhang, C. Yang, and X.M. Li, Resonant Raman scattering probe of alloying effect in ZnMgO thin films, *Appl. Phys. Lett.* **92**, 191910 (2008).

40. C. Yang, X.M. Li, X.D. Gao, X. Cao, R. Yang, and Y.Z. Li, Effects of the oxygen pressure on the structural and optical properties of ZnBeMgO films prepared by pulsed laser deposition, *J. Cryst. Growth* **312**, 978 (2010).

41. N. Panwar, J. Liriano, and R.S. Katiyar, Structural and optical analysis of ZnBeMgO powder and thin films, *J. Alloy. Compd.* **509**, 1222 (2011).

42. D.G. Mead and G.R. Wilkinson, The temperature dependence of the Raman effect in some wurtzite type crystals, *J. Raman Spectrosc.* **6**, 123 (1977).

43. H.L. Shi and Y. Duan, Bandgap bowing and p-type doping of (Zn, Mg, Be)O wide-gap semiconductor alloys: A first-principles study, *Eur. Phys. J. B* **66**, 439 (2008).

44. W. Kohn and L.J. Sham, Self-consistent equations including exchange and correlation effects, *Phys. Rev.* **140**, A1133 (1965).

45. G. Kresse and J. Furthmuller, Efficient iterative schemes for ab initio total-energy calculations using a plane-wave basis set, *Phys. Rev. B* **54**, 11169 (1996).

46. G. Kresse and D. Joubert, From ultrasoft pseudopotentials to the projector augmented-wave method, *Phys. Rev. B* **59**, 1758 (1999).

47. H.J. Monkhorst and J.D. Pack, Special points for Brillouin-zone integrations, *Phys. Rev. B* **13**, 5188 (1972).

48. A. Zunger, S.H. Wei, L.G. Ferreira, and J.E. Bernard, Special quasirandom structures, *Phys. Rev. Lett.* **65**, 353 (1990).

49. S.H. Wei, L.G. Ferreira, J.E. Bernard, and A. Zunger, Electronic properties of random alloys: Special quasirandom structures, *Phys. Rev. B* **42**, 9622 (1990).

50. K.J. Chang, S. Froyen, and M.L. Cohen, The electronic band structures for zincblende and wurtzite BeO, *J. Phys. C Solid State Phys.* **16**, 3475 (1983).

51. S. Limpijumnong, S.B. Zhang, S.H. Wei, and C.H. Park, Doping by large-size-mismatched impurities: The microscopic origin of arsenic- or antimony-doped p-type zinc oxide, *Phys. Rev. Lett.* **92**, 155504 (2004).

52. A.B.M.A. Ashrafi, H. Kumano, I. Suemune, Y.W. Ok, and T.Y. Seong, Single-crystalline rocksalt CdO layers grown on GaAs (001)substrates by metalorganic molecular-beam epitaxy, *Appl. Phys. Lett.* **79**, 470 (2001).

53. J.C. Boettger and J.M. Wills, Theoretical structural phase stability of BeO to 1 TPa, *Phys. Rev. B* **54**, 8965 (1996).

54. S.H. Wei and A. Zunger, Giant and composition-dependent optical bowing coefficient in GaAsN alloys, *Phys. Rev. Lett.* **76**, 664 (1996).

55. L. Bellaiche, S.H. Wei, and A. Zunger, Localization and percolation in semiconductor alloys: GaAsN vs GaAsP, *Phys. Rev. B* **54**, 17568 (1996).

56. C.Y. Moon, S.H. Wei, Y.Z. Zhu, and G.D. Chen, Band-gap bowing coefficients in large size-mismatched II-VI alloys: First-principles calculations, *Phys. Rev. B* **74**, 233202 (2006).

57. S.F. Ding, G.H. Fan, S.T. Li, K. Chen, and B. Xiao, Theoretical study of $Be_xZn_{1-x}O$ alloys, *Phys. B* **394**, 127 (2007).

58. S.H. Wei and A. Zunger, Calculated natural band offsets of all II-VI and III-V semiconductors: Chemical trends and the role of cation d orbitals, *Appl. Phys. Lett.* **72**, 2011 (1998).

59. D. Segev and S.H. Wei, Effects of covalency, p-d coupling, and epitaxial strain on the band offsets of II-VI semiconductors, *Phys. Rev. B* **68**, 165336 (2003).

60. M. Kondow, K. Uomi, A. Niwa, T. Kitatani, S. Watahiki, and Y. Yazawa, GaInNAs: A novel material for long-wavelength-range laser diodes with excellent high-temperature performance, *Jpn. J. Appl. Phys.* **35**, 1273 (1996).

61. Y. Duan, H. Shi, and L. Qin, Elasticity, band structure, and piezoelectricity of $Be_xZn_{1-x}O$ alloys, *Phys. Lett. A* **372**, 2930 (2008).

62. X. Su, P. Si, Q. Hou, X. Kong, and W. Cheng, First-principles study on the bandgap modulation of Be and Mg co-doped ZnO systems, *Phys. B* **404**, 1794 (2009).

63. R. de Paiva, J.L.A. Alves, R.A. Nogueira, C. de Oliveira, H.W.L. Alves, L.M.R. Scolfaro, and J.R. Leite, Theoretical study of the $Al_xGa_{1-x}N$ alloys, *Mater. Sci. Eng. B* **93**, 2 (2002).

64. Y. Ryu, T.S. Lee, J.A. Lubguban, H.W. White, B.J. Kim, Y.S. Park, and C.J. Youn, Next generation of oxide photonic devices: ZnO-based ultraviolet light emitting diodes, *Appl. Phys. Lett.* **88**, 241108 (2006).

65. Y.R. Ryu, T.S. Lee, and H.W. White, Properties of arsenic-doped p-type ZnO grown by hybrid beam deposition, *Appl. Phys. Lett.* **83**, 87 (2003).

66. S. Nakamura and G. Fasol, *The Blue Laser Diode: GaN Based Light Emitters and Lasers*. Springer, New York (1997).

67. Y.R. Ryu, J.A. Lubguban, T.S. Lee, H.W. White, T.S. Jeong, C.J. Youn, and B.J. Kim, Excitonic ultraviolet lasing in ZnO-based light emitting devices, *Appl. Phys. Lett.* **90**, 131115 (2007).

68. D. Walker, V. Kumar, K. Mi, P. Sandvik, P. Kung, X.H. Zhang, and M. Razeghi, Solar-blind AlGaN photodiodes with very low cutoff wavelength, *Appl. Phys. Lett.* **76**, 403 (2000).

69. C. Yang, X.M. Li, W.D. Yu, X.D. Gao, X. Cao, and Y.Z. Li, Zero-biased solar-blind photodetector based on ZnBeMgO/Si heterojunction, *J. Phys. D Appl. Phys.* **42**, 152002 (2009).

70. Y.F. Gu, X.M. Li, J.L. Zhao, W.D. Yu, X.D. Gao, and C. Yang, Visible-blind ultra-violet detector based on n-ZnO/p-Si heterojunction fabricated by plasma-assisted pulsed laser deposition, *Solid State Commun.* **143**, 421 (2007).

71. N. Biyikli, O. Aytur, I. Kimukin, T. Tut, and E. Ozbay, Solar-blind AlGaN-based Schottky photodiodes with low noise and high detectivity, *Appl. Phys. Lett.* **81**, 3272 (2002).

72. D. Walker, A. Saxler, P. Kung, X. Zhang, M. Hamilton, J. Diaz, and M. Razeghi, Visible blind GaN p-i-n photodiodes, *Appl. Phys. Lett.* **72**, 3303 (1998).

73. T. Tut, T. Yelboga, E. Ulker, and E. Ozbay, Solar-blind AlGaN-based p-i-n photodetectors with high breakdown voltage and detectivity, *Appl. Phys. Lett.* **92**, 103502 (2008).

74. E. Furno, S. Chiaria, M. Penna, E. Bellotti, and M. Goano, Electronic and optical properties of $Mg_xZn_{1-x}O$ and $Be_xZn_{1-x}O$ quantum wells, *J. Electron. Mater.* **39**, 7 (2010).

10

Ultraviolet ZnO Random Laser Diodes

Siu Fung Yu and Hui Ying Yang

CONTENTS

10.1 Introduction..339
10.2 Random Laser Diodes Using ZnO Clusters as Optical Cavities..............................340
    10.2.1 SiC-ZnO Heterojunction Laser Diodes...340
    10.2.2 GaN-ZnO Heterojunction Laser Diodes...343
10.3 Random Laser Diodes Using Polycrystalline ZnO Thin Film as Optical Cavities.....345
10.4 ZnO-Based Laser Diodes..348
    10.4.1 As-Doped *p*-Type ZnO-Based Materials as Hole-Injection Layer...................349
    10.4.2 Sb Doped *p*-Type ZnO-Based Materials as Hole-Injection Layer 350
10.5 Heterojunction ZnO Laser Diodes with Low Threshold Current............................352
10.6 Advanced Design for Random Laser Diodes..356
    10.6.1 ZnO Polycrystalline Thin-Film Random Lasers with a Ridge
        Waveguide Structure..356
    10.6.2 Directional and Controllable Edge-Emitting ZnO UV Random
        Laser Diodes..357
    10.6.3 High-Power and Single-Mode Operation UV ZnO Laser Diodes...................359
10.7 Discussions and Conclusions...362
    10.7.1 Realization of High Optical Gain under Electrical Excitation.........................362
    10.7.2 Realization of Coherent Optical Feedback to Sustain Lasing Emission........363
    10.7.3 Conclusions..364
Acknowledgment..364
References...365

10.1 Introduction

ZnO, which is a wide bandgap ($\sim$3.37 eV) semiconductor, has a high exciton binding energy of $\sim$60 meV [1]. Furthermore, ZnO is an inexpensive, relatively abundant, chemically stable, easy to prepare, and nontoxic material. Therefore, ZnO has been recognized to be a promising candidate for the fabrication of low-cost ultraviolet (UV) optoelectronic devices. One of the most promising applications of ZnO is to replace indium tin oxide (ITO) in the displays and photovoltaic panels due to the low cost of transparent conductors. In addition, due to the conductive nature of ZnO, it has the potential to be used as the semiconductor materials for making of inexpensive UV optoelectronic devices such as UV laser diodes. Nevertheless, there are not more than 10 original reports on the realization of UV ZnO laser diodes that have been published during the past decade [2–11]. This may be due to two technical challenges that need to be overcome before ZnO can be fully utilized to fabricate UV laser diodes.

One important problem is that there are no stable *p*-type ZnO-based semiconductors. Despite a large number of recent reports on the fabrication of *p*-type ZnO growth by various methods [12,13], no reproducible and high-quality *p*-type ZnO is available. Hence, it is not possible to achieve high-quality *p-n* junction to obtain sufficient optical gain under electrical excitation (i.e., achieve strong amplification of spontaneous emission from via external injection of electrons and holes). Another difficulty is to obtain smooth facets from ZnO films to realize laser cavity (i.e., as if the fabrication of conventional GaAs or InP-based Fabry–Perot lasers). This is because ZnO has a wurtzite crystal structure so that it is impossible to obtain Fabry–Perot cavities by cleaving of ZnO films. As a result, insufficient optical gain (i.e., under electrical excitation) and high scattering loss from ZnO films limited the development of useful UV laser diodes for many potential applications. In the following sections, we will discuss on the current achievement to overcome the difficulties for the realization of ZnO UV laser diodes.

10.2 Random Laser Diodes Using ZnO Clusters as Optical Cavities

There are methods to avoid using *p*-type ZnO. For example, we can use other lattice-match *p*-type semiconductor materials to form heterojunction for the injection of holes. In addition, the difficulty to form cleaved facets in ZnO material can be replaced by using random cavities to support coherent optical feedback inside ZnO medium. Hence, electrically pumped ZnO laser diodes emitted at UV wavelength can still be obtained.

10.2.1 SiC-ZnO Heterojunction Laser Diodes

Figure 10.1 shows the schematic of the ZnO heterostructural *p-i-n* junction diode using *p*-SiC to replace *p*-ZnO for the injection of holes [2]. A *p*-doped single-side polished 4H-SiC substrate (with size of $5 \times 5 \, \text{mm}^2$) is used as the hole-injection layer, and an *n*-doped ZnO:Al

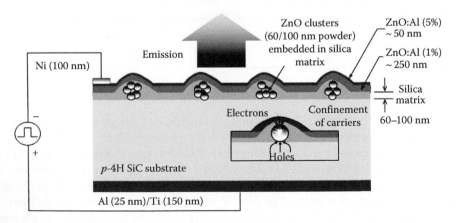

FIGURE 10.1

Schematic of the *p*-SiC(4H)/*i*-ZnO-SiO$_2$ nanocomposite/*n*-ZnO:Al heterojunction diode structure. (After Leong, E.S.P. and Yu, S.F.: UV random lasing action in *p*-SiC(4H)/*i*-ZnO–SiO$_2$ nanocomposite/*n*-ZnO:Al heterojunction diodes. *Adv. Mater.* 2006. 18. 1685–1688. Copyright Wiley-VCH Verlag GmbH & Co. KGaA. Reproduced with permission.)

(1%) layer of thickness ~250 nm is used as the electron-injection layer. An intrinsic layer, which consists of ZnO powder embedded in SiO_2 matrix, is inserted between the *n*- and *p*-injection layers to form the *p-i-n* junction diode. The advantage of sandwiching ZnO powders between the electron and hole-injection layers is the enhancement of injection efficiency of carriers so that the ZnO powders can attain high optical gain under electrical excitation. The intrinsic layer is prepared by the sol–gel technique so that clusters of ZnO powder with stripe pattern can be formed. The thickness of the SiO_2 matrix is selected to be ~60–100 nm so as to optimize the series resistance across the *p-i-n* junction and to direct the injected electrons and holes into the ZnO powder.

Figure 10.2 explains the formation of random laser action inside the intrinsic layer of the diodes. If stimulated recombination occurs inside the ZnO powder (i.e., as indicated in Figure 10.1), light will emit in all directions. Some of the light will travel in open-loop zig-zag paths to result in further amplification and scattering by the ZnO clusters so that incoherent random lasing can be obtained. Coherent random lasing is due to the closed-loop zig-zag paths of light. In addition, the lack of transverse confinement of light (i.e., the refractive indices of SiC substrate and *n*-ZnO:Al layer are higher than that of the ZnO-SiO_2 nanocomposite layer) and corrugated surface increase the emission of light from the surface of the diode but limit the formation of random cavities (i.e., only a few peaks can be excited). Hence, the proposed heterojunction can achieve high optical gain under electrical excitation as well as obtain sufficient optical feedback from the laser medium to sustain lasing oscillation.

Figure 10.3 shows the electroluminance (EL) spectra of the diode with 60 nm ZnO powder. Narrowing of the UV spectra is observed from the diode. The full-width at half-maximum (FWHM) of the UV spectra reduces from 19 to 11 nm with an increase of injection current. The peak wavelength of the emission spectra, however, remains at ~383 nm for different values of injection current. Sharp peaks of linewidth ~0.5 nm also emerged from the emission spectra at high injection current. It is believed that the narrowing of emission spectra is due to incoherent random lasing action, and this is favorable for the diode with denser

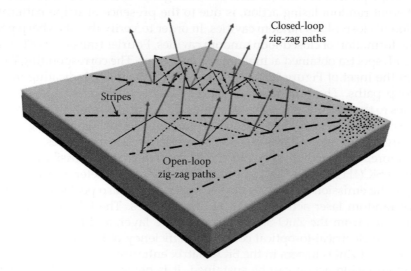

FIGURE 10.2
Possible mechanisms to achieve random laser action inside the diodes. (After Leong, E.S.P. and Yu, S.F.: UV random lasing action in *p*-SiC(4H)/*i*-ZnO–SiO_2 nanocomposite/*n*-ZnO:Al heterojunction diodes. *Adv. Mater.* 2006. 18. 1685–1688. Copyright Wiley-VCH Verlag GmbH & Co. KGaA. Used with permission.)

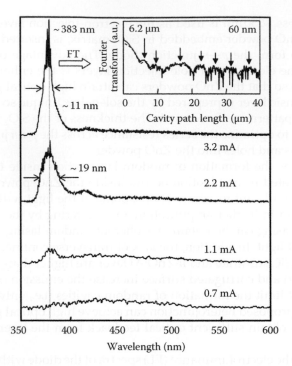

FIGURE 10.3

EL spectra for the *p-i-n* diode with 60 nm ZnO powder. Fourier transform spectrum of the EL emission of the diode with 60 nm ZnO powder biased at 3.2 mA is shown in the inset. (After Leong, E.S.P. and Yu, S.F.: UV random lasing action in *p*-SiC(4H)/*i*-ZnO–SiO₂ nanocomposite/*n*-ZnO:Al heterojunction diodes. *Adv. Mater.* 2006. 18. 1685–1688. Copyright Wiley-VCH Verlag GmbH & Co. KGaA. Used with permission.)

and longer stripe pattern of ZnO clusters. The emergence of sharp peaks, which are attributed to coherent random lasing action, is due to the presence of stripe patterns to allow the formation of closed-loop random cavities. In order to verify that the sharp peaks correspond to the formation of closed-loop random cavities, Fourier transform (FT) is applied to analyze the EL spectra obtained at high injection current. The corresponding FT spectrum as shown in the inset of Figure 10.3b exhibits periodic features, indicating the formation of closed-loop paths. The fundamental (shortest) cavity path formed is ~6.2 μm, which approximates to the spacing of the closest stripes. This value is smaller than the diode with 100 nm ZnO clusters (~7.5 μm) since the stripes are more closely spaced. Higher harmonics indicating larger closed-loop paths can also be formed inside the diodes.

From the aforementioned observations, it has verified that the possibility to realize UV lasing from *p*-SiC(4H)/*i*-ZnO-SiO₂ nanocomposite/*n*-ZnO:Al heterojunction diodes. The narrowing of the emission spectra and the emergence of sharp peaks have confirmed the existence of random laser action inside the laser diodes. The UV emission of the laser diodes originates from the ZnO-SiO₂ nanocomposite layer, and the use of ZnO powder can improve the electrical-to-optical conversion efficiency of the heterojunction. In addition, the patterned ZnO clusters in the SiO₂ matrix enhance the quality of random media so that random lasing action can be sustained. It is believed that if low-index *p*-doped wide-bandgap materials (e.g., *p*-GaN) can be used to replace the *p*-SiC substrate as the hole-injection layer (i.e., in order to obtain transverse confinement of light), strong coherent random lasing can be achieved inside the proposed *p-i-n* laser diodes.

10.2.2 GaN-ZnO Heterojunction Laser Diodes

The aforementioned ZnO laser diode uses p-SiC, which has refractive index higher than that of ZnO clusters in SiO_2 matrix, as the hole-injection layer. However, if higher (lower) refractive-indexed intrinsic (injection) layers are used, transverse confinement of light can be realized to sustain in-plane lasing inside the nanocomposite film. This will improve the optical confinement of the laser diodes. Hence, the fabrication of ZnO-SiO_2 nanocomposite film laser diode is repeated by using p-GaN film, which has lower refractive index than that of p-SiC, as the hole-injection layer.

Figure 10.4 shows the schematic of the proposed n-ZnO:Al/i-ZnO-SiO_2 nanocomposite/ p-GaN heterojunction diode [3]. Sol–gel technique was used to prepare the nanocomposite films in which ZnO nanopowder (with average particle size of 60 nm obtained from Nanophase) is embedded inside the SiO_2 matrix at various molar ratios ranging from 1:1 to 1:10 (ZnO:SiO_2). Half of an $8 \times 5\,mm^2$ (i.e., an area of $4 \times 5\,mm^2$) p-GaN/sapphire substrate (Technologies and Devices International Inc.) was first coated with photoresist (AZ 5214E). After that, the ZnO-SiO_2 sol was spun coated on it at a spin speed of 3500 rpm for 30 s and then thermally polymerized to form a film by baking in a furnace for ~2 h at 110°C. Radial spreading of stripes of ZnO clusters is formed inside the nanocomposite film. After that, a ~200 nm thick ZnO:Al (3%) film followed by a ~100 nm thick ZnO:Al (5%) film was deposited using filtered cathodic vacuum arc (FCVA) technique to form the n-layers, and the photoresist was removed using acetone. Ti (100 nm) and Ni (100 nm) were then e-beam evaporated onto the p- and n-layers, respectively. The cross section scanning electron microscope (SEM) image of the heterojunction diode with molar ratio 1:5 (see the inset of Figure 10.4) shows that the ZnO clusters are embedded within the SiO_2 matrix.

Figure 10.5a shows the EL spectra measured from the edge of the diode with molar ratio 1:5 (denoted as diode A). The diode was biased under electrical pulses (60 ms, 12.6 Hz) at room temperature (RT). The corresponding current–voltage (I–V) curve, which shows a rectifying behavior, is also plotted in the inset of Figure 10.5a. It is observed from Figure 10.5a that at injection current, $I < 4.3$ mA, a broad spontaneous emission spectrum at ~395 nm is

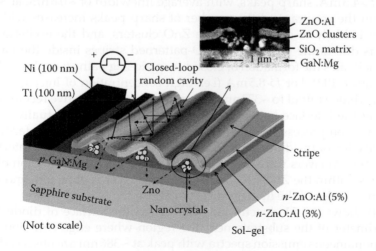

FIGURE 10.4
Schematic of p-GaN/i-ZnO-SiO_2 nanocomposite/n-ZnO heterojunction diode. The formation of the closed-loop random cavity is illustrated in the figure. The inset shows the cross section SEM image of the diode with molar ratio 1:5. (Used with permission from Leong, E.S.P., Yu, S.F., and Lau, S.P., Directional edge-emitting UV random laser diodes, *Appl. Phys. Lett.*, 89, 221109. Copyright 2006, American Institute of Physics.)

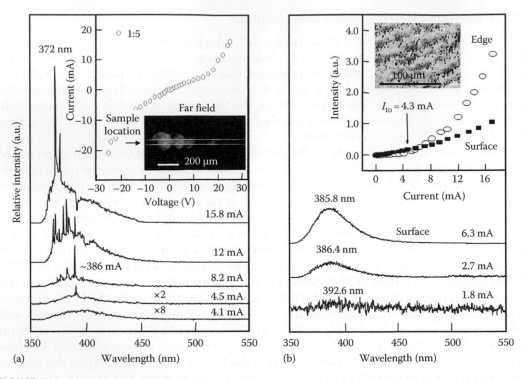

FIGURE 10.5

EL spectra recorded from the (a) edge and (b) surface of the diode. The inset of (a) shows the far-field profile of the diode. The *I–V* curve for diode is also plotted in the inset of (a). The surface morphology of diode A taken at 5× optical magnifications and *L–I* curves for the surface and edge emissions are shown in the inset of (b). (Used with permission from Leong, E.S.P., Yu, S.F., and Lau, S.P., Directional edge-emitting UV random laser diodes, *Appl. Phys. Lett.*, 89, 221109. Copyright 2006, American Institute of Physics.)

recorded. As $I > 4.3$ mA, sharp peaks, with average linewidth of ~0.6 nm, at ~386 nm start to emerge from the spectra, and the number of sharp peaks increases with I. Radiative recombination is likely to occur inside the ZnO clusters, and the excitation of random lasing action is due to the presence of ZnO-patterned stripes inside the nanocomposite layer, which facilitates the formation of closed-loop zig-zag paths (i.e., closed-loop random cavities; see Figure 10.1). For $I > 8.5$ mA (i.e., high concentration of injected electrons), the lasing spectra is blue-shifted to ~372 nm. This indicates that radiative recombination starts to take place at the interface between the *p*- and *i*-layers and eventually dominates the entire recombination process. The inset of Figure 10.5a also shows the far-field emission taken from the edge of the diode at $I = 15.8$ mA. Several spots along the interface between GaN:Mg and ZnO:Al layers were observed. This shows that the formation of closed-loop random cavities within the ZnO-SiO$_2$ nanocomposite layer is obtained, and the emission observed from the edge of the diode is directional.

Figure 10.5b shows the EL spectra measured from the surface of diode A (i.e., taken along the perimeter of the substrate near the region where edge emission is obtained). Only broad spontaneous emission spectra with peak at ~386 nm are observed, and no sign of gain narrowing or lasing peak is recorded with the increase of I. This is because the patterned stripes of ZnO clusters (which have average stripe width and spacing of ~7 and ~16 μm, respectively, see the inset in Figure 10.5b) are not dense enough to support scattering of lasing light to the surface. The light–current (*L–I*) curves of diode A are also plotted in the inset of Figure 10.5b. A kink at I that equals 4.3 mA (i.e., turn-on current, I_{to}) is observed in the L–I curve for the case of edge emission. It is noted that for $I < I_{to}$, both edge and surface emissions are of similar intensity. When $I > I_{to}$, the intensity of edge emission increases exponentially and lasing peaks emerged from the emission spectra. However, the intensity of surface emission is suppressed at $I > I_{to}$. The aforementioned results show that the presence of lower refractive-indexed injection layers enhances the transverse confinement of in-plane scattered light inside the nanocomposite layer so that lasing emission is observed from the edge of diode A. Similar results have been observed for diodes with higher molar ratios (i.e., lower concentration of ZnO nanopowder) and are not repeated here. However, for diodes with lower molar ratio, a different phenomenon is observed. For diodes with lower molar ratio, edge lasing emission is not supported in diodes. This is because strong scattering effect (i.e., small spacing of patterned stripes) will supersede the transverse optical confinement of the diodes. Therefore, lasing is only observed from the surface of the diodes with low molar ratio, and the lasing characteristics are similar to that using p-SiC as the hole-injection layer.

Hence, highly directional random lasing action has been realized from p-GaN/i-ZnO-SiO_2 nanocomposite/n-ZnO:Al heterojunction diode. Improvement in the transverse confinement of light has been achieved through the use of relatively low refractive index of n-ZnO:Al and p-GaN:Mg films as the injection layers. However, the influence of the surface corrugation on surface emission should not be neglected in the design of nanocomposite random laser diodes. Although higher concentration of ZnO nanopowder (low value of x) is found to reduce I_{to} of the laser diodes, the increase of surface roughness deteriorates the transverse confinement of lasing light so that in-plane emission of lasing modes is suppressed. Hence, it is believed that our study will aid in the future design and development of more practical UV random laser diodes.

10.3 Random Laser Diodes Using Polycrystalline ZnO Thin Film as Optical Cavities

Recent investigations have shown that laser ablation can be used to grow ZnO polycrystalline films on amorphous fused silica substrates. ZnO polycrystalline films can also be obtained from the postgrowth annealing of high-crystal-quality ZnO thin films fabricated by FCVA technique [14,15]. The realization of ZnO polycrystalline films is important for the fabrication of ZnO lasers because optical cavities to support lasing action at UV wavelength can be self-generated inside the films without any rigorous control.

The method to grow ZnO polycrystalline films can be realized by postgrowth annealing. Prior to the deposition of ZnO thin films, a SiO_2 buffer layer of thickness around 420 nm was formed on the n-type (100) Si substrate by thermal dry oxidation. A 200 nm thick ZnO film was deposited on the surface of the SiO_2 buffer layer by using the FCVA technique. During the deposition, substrate temperature and oxygen partial pressure were set to 230°C and 2×10^{-4} Torr, respectively. Low temperature deposition of ZnO thin films on the smooth surface (i.e., no preferred lateral growth direction) of amorphous SiO_2 buffer layer, which allows the effective growth of ZnO grains and voids through postgrowth annealing, is the key to realize highly disordered gain media. The growth of ZnO grains (voids) inside the ZnO thin films can be explained as the merger (evaporation) of weakly bonded ZnO grains. Due to the creation of voids, the ZnO grains become loosely in contact, and the corresponding lateral facets are widely exposed. Hence, the lateral facets of the irregular ZnO grains or scatterers can provide a strong optical scattering that forms closed-loop paths of lights (i.e., laser cavities). The main advantage of using postgrowth annealing to form laser cavities is the controllability of the generation of scatterers through the manipulation of annealing temperature and time.

Postgrowth annealing in open air was carried out in a standard Lindberg type furnace using quartz tube reactor. Initially, the samples inside the furnace were heated up from RT to 900°C within 20 min. Then, the samples were annealed for a period of time, T_a, at a constant temperature of 900°C. Finally, the samples were cooled down to RT inside the furnace. This annealing process was used to minimize the built-up of thermal strain/stress at the interfaces of Si/SiO$_2$ and SiO$_2$/ZnO. It must be noted that the postgrowth annealing temperature should be higher than that of the growth temperature in order to achieve effective formation of irregular ZnO grains. Annealing temperature of 900°C was used in the process because scatterers could be generated effectively at this temperature. Figure 10.6 shows the SEM images and x-ray diffraction patterns of the samples before and after annealing. The ZnO grains of the as-grown sample are very small and closely packed. For the sample with $T_a = 1$ min, it is observed that the ZnO grains become loosely in contact with deep slits existing between the grains (i.e., the lateral facets of the ZnO grains are beginning to develop). For $T_a = 2$ h, the size of ZnO grains grows further, and almost all the ZnO grains have developed lateral facets. The corresponding x-ray diffraction patterns ($\theta/2\theta$) show a strong (002) peak at around 34.4°. This indicates that the ZnO thin films have a hexagonal wurtzite structure with their *c*-axis normal to the substrate basal plane. The reduction of FWHM of the (002) peaks implies the increase of the average size of ZnO grains after postgrowth annealing, which is consistent with the observation from the SEM images. Using the method of coherent backscattering, it can be shown that

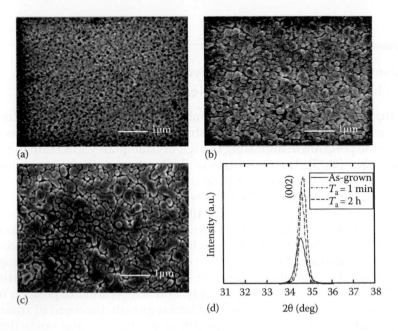

FIGURE 10.6
SEM images of ZnO thin films. (a) As-grown, (b) $T_a = 1$ min, and (c) $T_a = 2$ h. (d) Corresponding x-ray diffraction patterns. (Used with permission from Yu, S.F., Yuen, C., Lau, S.P., and Lee, H.W., Zinc oxide thin-film random lasers on silicon substrate, *Appl. Phys. Lett.*, 84, 3244–3246. Copyright 2004, American Institute of Physics.)

only the annealed samples exhibited scattering mean-free path of 390 nm, which is similar to the emission wavelength of the ZnO thin films. Hence, the formation of irregular ZnO grains can create lateral facets (i.e., laser cavities) so that UV lasing is possible in the ZnO thin-film waveguides.

Using the aforementioned approach, lasing characteristics were demonstrated from a proposed metal-oxide-semiconductor (MOS) structure of Au/SiO$_x$ ($x < 2$)/ZnO polycrystalline film fabricated on a silicon substrate [4]. Figure 10.7a shows the plan-view SEM image of the ZnO-based MOS structure. A ~300 nm thick ZnO film was deposited on the silicon substrate. It is clearly seen that the ZnO grains are distributed randomly but are quite closely packed in the plane of the ZnO film. Figure 10.7b shows the cross-sectional SEM image of the deposited ZnO film, illustrating that the grains with lateral sizes of 50–100 nm are arrayed in the direction perpendicular to the substrate, which is consistent with the fact that the ZnO film is highly *c*-axis oriented. Combining the plan- and cross-sectional views of the ZnO film, it is derived that most of the *c*-axis-oriented ZnO grains grown perpendicularly to the silicon substrate are well faceted at both the end and side surfaces.

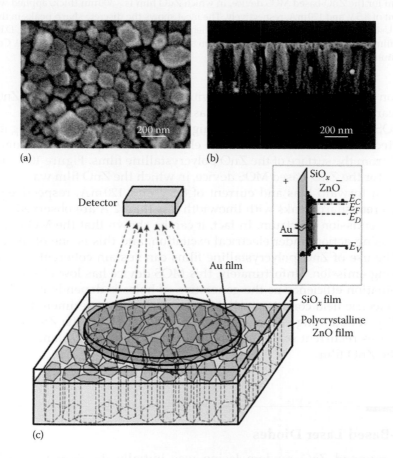

FIGURE 10.7
(a) Plan-view and (b) cross-sectional SEM images of the ~300 nm thick ZnO film deposited on the silicon substrate. (c) Schematic diagram illustrating the formation of in-plane closed-loop random cavities in the *c*-axis oriented ZnO polycrystalline film. The inset shows the schematic energy-band diagram for the forward-biased ZnO-based MOS device. (Used with permission from Ma, X.Y., Chen, P.L., Li, D.S., Zhang, Y.Y., and Yang, D.R., Electrically pumped ZnO film ultraviolet random lasers on silicon substrate, *Appl. Phys. Lett.*, 91, 251109. Copyright 2007, American Institute of Physics.)

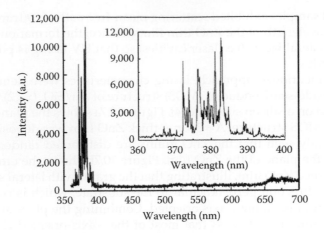

FIGURE 10.8

RT EL spectrum for the ZnO-based MOS device, in which ZnO film is ~300 nm thick, applied with a forward bias and current of 8.5 V and 120 mA, respectively. The inset shows the discrete sharp peaks in the UV band of EL spectrum. (Used with permission from Ma, X.Y., Chen, P.L., Li, D.S., Zhang, Y.Y., and Yang, D.R., Electrically pumped ZnO film ultraviolet random lasers on silicon substrate, *Appl. Phys. Lett.*, 91, 251109. Copyright 2007, American Institute of Physics.)

The formation of the closed-loop random cavities within the *c*-axis-oriented ZnO polycrystalline film can be schematically illustrated as in Figure 10.7c.

For the MOS structure with sufficiently high forward bias, where the negative voltage was connected to the silicon substrate, the electrically pumped UV random lasing can be observed from the surface of the ZnO polycrystalline films. Figure 10.8 shows the RT EL spectrum for the ZnO-based MOS device, in which the ZnO film was ~300 nm thick, applied with a forward bias and current of 8.5 V and 120 mA, respectively. It can be seen that discrete sharp peaks with linewidth less than 2 Å are observed from the UV regime of the emission spectrum. In fact, it can be shown that the MOS device exhibits random lasing action under electrical excitation, and this is one of the examples to prove that the use of ZnO polycrystalline films can sustain coherent optical feedback to excite lasing emission. Unfortunately, this MOS device has low electron-hole radiative recombination efficiency as the generation of holes is dependent of the SiO_x layer under high electric field conditions [4]. Nevertheless, this experiment has verified that the possibility of obtaining electroluminescent UV light from the ZnO polycrystalline film experiences recurrent scattering and interference in the in-plane random cavities formed in the ZnO film.

10.4 ZnO-Based Laser Diodes

Electrically pumped ZnO random lasing was initially developed by using lattice-matched *p*-type semiconductor materials [2,3]. Alternatively, MOS was proposed to achieve injection of holes into the ZnO polycrystalline thin films to obtain random lasing action. Apparently, the fabrication of UV laser diode based on ZnO *p-n* homojunction with high light gain is essential toward high-performance UV lasers. However, the difficulty in the development of electrically pumped ZnO *p-n* homojunction UV lasers is primarily due to the lack of reliable and controllable p-type ZnO. So far, there are only two reports on the realization of UV ZnO laser diodes using p-type ZnO for the injection of holes [5,6].

10.4.1 As-Doped p-Type ZnO-Based Materials as Hole-Injection Layer

First report on the fabrication of ZnO-based UV laser diodes has used p-type As-doped ZnO and As-doped ZnBeO films for the injection of holes [5]. The laser diode consists of a multiple quantum well (MQW) active layer sandwiched between guide-confinement layers. The MQW active layer comprises seven undoped ZnO wells and ZnBeO barriers, while the two guide-confinement layers were As-doped p-type ZnO/ZnOBe and Ga-doped n-type ZnBeO/ZnO films, respectively. Figure 10.9 shows the photoluminance (PL) spectrum of the ZnO/ZnBeO MQW under the excitation by an ArF excimer laser at RT. The laser beam was incident in a direction normal to the surface, and the emission spectra were measured from the edge of the sample (i.e., in a direction perpendicular to the incident beam). Due to the large exciton binding energy of the ZnO/ZnBeO MQW arisen from the strong confinement of excitons inside the QWs, excitons are stable at RT in the ZnO/ZnBeO-based QWs laser under optical excitation. As the excitation intensity increases from 417 to 625 kW/cm^2, the intensity of an emission peak located at 3.21 eV becomes the dominant peak. The corresponding linewidth suddenly reduces from ~10 to ~0.9 nm. This indicated that the sample exhibits excitonic lasing under optical excitation despite there is no optical feedback mechanism inside the sample. Figure 10.10 shows EL spectra of the sample measured at RT under pulsed operation. As the current was increased, sharp peaks are emerged from the spectrum over a width of ~63 nm. The threshold current density for the excitation of sharp peaks is ~420 A/cm^2. The low threshold current densities for lasing of ZnO/ZnBeO-based devices can be attributed to the large exciton binding energy and the presence of exciton–exciton inelastic scattering. However, the nonuniform modal spacing observed from the emission spectra cannot justify that the lasing mechanism of this ZnO/ZnBeO MQW laser is due to the formation of Fabry–Perot cavity.

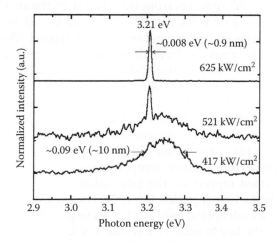

FIGURE 10.9
Normalized PL spectra measured at RT of a ZnO/BeZnO superlattice containing seven QWs excited by an ArF excimer laser at selected excitation intensities to illustrate lasing action. (Used with permission from Ryu, Y.R., Lubguban, J.A., Lee, T.S., White, H.W., Jeong, T.S., Youn, C.J., and Kim, B.J., Excitonic ultraviolet lasing in ZnO-based light emitting devices, *Appl. Phys. Lett.*, 90, 131115. Copyright 2007, American Institute of Physics.)

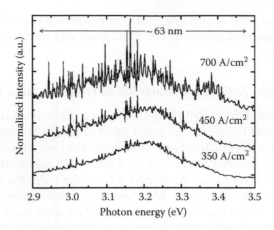

FIGURE 10.10

EL spectra measured at RT in pulse current injection mode (10% duty cycle). Sharp Fabry–Perot-type oscillations become more prominent with increasing current density, indicating lasing action. (Used with permission from Ryu, Y.R., Lubguban, J.A., Lee, T.S., White, H.W., Jeong, T.S., Youn, C.J., and Kim, B.J., Excitonic ultraviolet lasing in ZnO-based light emitting devices, *Appl. Phys. Lett.*, 90, 131115. Copyright 2007, American Institute of Physics.)

10.4.2 Sb Doped *p*-Type ZnO-Based Materials as Hole-Injection Layer

Sb is also proposed as the *p*-type dopant to fabricate *p*-type ZnO for the realization of UV ZnO-based laser diodes [6]. Figure 10.11a shows the schematic of the ZnO-based single quantum well (SQW) laser diode. The laser diode consists of a SQW active layer sandwiched between two guide-confinement layers. The SQW active layer comprises ZnO well and ZnMgO barriers, while the two guide-confinement layers are Sb-doped *p*-type ZnO and Ga-doped *n*-type ZnO films, respectively. Figure 10.11b describes the lasing characteristics of the laser diode under random lasing action. Figure 10.11c and d shows the top and side SEM images, respectively, of the ZnO film deposited on Si substrate.

Figure 10.12a shows the emission spectrum observed from the surface of the laser diode. The laser diode is biased at a dc voltage at RT. It is noted that at low injection current of ~10 mA, a weak spontaneous emission band centered at around 378 nm is appeared from the emission spectrum. This peak can be attributed to free-exciton spontaneous emission. As the injection current increases to ~20 mA, sharp peaks with linewidth less than 0.4 nm are emerged from the single-broad emission spectra around 380 nm. This indicates that the optical gain provided by the SQW is now large enough to enable cavity mode to start lasing. Further increase in the injection current from 20 to 60 mA increases both the number and the intensity of sharp peaks in the emission spectra; see also Figure 10.12b. As the laser is biased at a dc condition, the center of the lasing spectrum redshifts about 4–5 nm with the increase of bias current is due to temperature-induced bandgap variations.

From the aforementioned reports on the fabrication of UV laser diodes using *p*-type ZnO-based materials as the hole-injection layer, the laser diodes can obtain a low threshold current density and a high optical gain at RT. This implies that the realization of *p*-type ZnO-based materials is the key to achieve low-cost UV laser diodes. However, few of the main issues have not yet been addressed in their reports—reliability and stability as well as the reproducibility of their *p*-type ZnO-based materials. To the best of the authors' understanding, the lasing performance of these UV ZnO-based laser diodes may be deteriorated significantly within a few months, and the *p*-type properties of ZnO may disappear.

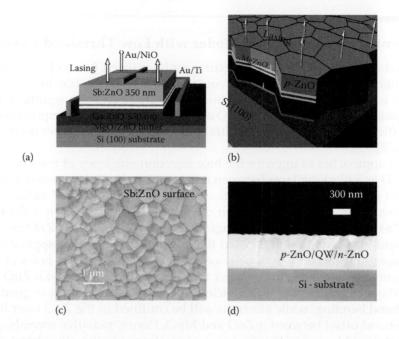

FIGURE 10.11
(a) Schematic of the ZnO laser diode showing a SQW sandwiched between *p*-type and *n*-type layers. Device mesa with Au/NiO and Au/Ti Ohmic contacts is also shown. (b) Schematic of the columnar structures of the ZnO diode "film" on Si substrate. (c) SEM image of sample surface. (d) Cross-sectional SEM image of the ZnO diode. (After Chu, S. et al., *Appl. Phys. Lett.*, 93, 181106, 2008.)

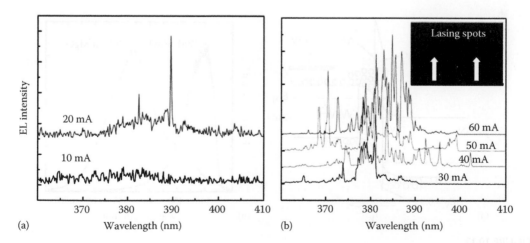

FIGURE 10.12
(a) EL spectra with low injection currents of 10 and 20 mA. Lasing effect is evident when the injection current reaches about 20 mA. (b) EL spectra with higher injection currents from 30 to 60 mA. All spectra are within UV around 380 nm. The spectra were shifted in y scale for clarity. Inset is an optical microscope image of lasing device driven at 30 mA. Arrows indicate isolated lasing spots on the diode surface. (After Chu, S. et al., *Appl. Phys. Lett.*, 93, 181106, 2008.)

10.5 Heterojunction ZnO Laser Diodes with Low Threshold Current

As discussed in Section 10.3, MOS structure can be utilized to inject hole into the ZnO polycrystalline active layer without the requirement of hole-injection layer. In fact, recent publications have shown that other modified MOS structures can significantly improve the hole-injection efficiency into the ZnO active layer, and the corresponding UV ZnO-based laser diodes can demonstrate an ultralow threshold current under normal operation conditions [7,8].

One of the approaches to improve the hole-injection efficiency of the ZnO active layer is to use MgO as a blocking layer between the p-GaN and n-ZnO heterojunction [7]. This is because in p-GaN/n-ZnO heterojunction under forward bias, UV radiative recombination is possible in GaN region due to diffusion of electrons from n-ZnO to p-GaN. Figure 10.13a shows the energy band diagram of the p-GaN/MgO/n-ZnO heterojunction under forward bias condition. It is noted that most of the voltage is applied to the MgO dielectric layer, and its bands will be strongly bended. As a result, holes in the p-GaN layer can tunnel through the thin barrier of MgO and enter into the n-ZnO layer. This is because the effective barrier in the vicinity of valence band offset is greatly reduced due to the band bending, while electrons will be confined in the ZnO layer by the large conduction-band offset between n-ZnO and MgO. Hence, radiative recombination from p-GaN is not possible due to the depletion of electron. On the other hand, extra holes can be injected from the p-GaN layer to the n-ZnO layer so that radiative recombination at the n-ZnO layer has been significantly enhanced. Figure 10.13b shows the EL spectra of the heterojunction with and without a MgO dielectric layer under the same injection current. For the case without MgO layer, the spectrum exhibits a broad peak centered at 530 nm (i.e., related to the deep-level emission in ZnO) and a weak peak at 445 nm (i.e., originates from the GaN layer). For the case with MgO layer, only single peak at ~400 nm can be observed, and the corresponding intensity is almost two orders of magnitude

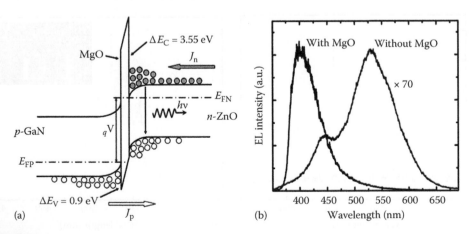

(a) (b)

FIGURE 10.13
(a) Schematic diagram showing the band alignment of the n-ZnO/MgO/p-GaN heterojunction under forward bias. (b) EL spectra of the n-ZnO/p-GaN heterojunction with and without MgO dielectric layer under the same injection current. (From Zhu, H., Shan, C.X., Yao, B., Li, B.H., Zhang, J.Y., Zhang, Z.Z., Zhao, D.X., Shen, D.Z., Fan, X.W., Lu, Y.M., and Tang, Z.K.: Ultralow-threshold laser realized in zinc oxide. *Adv. Mater.* 2009. 21. 1613–1617. Copyright Wiley-VCH Verlag GmbH & Co. KGaA. With permission.)

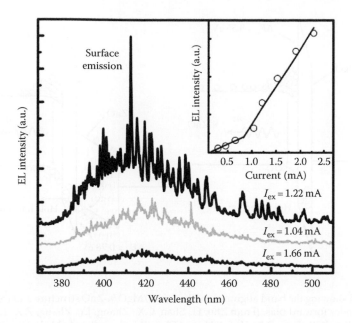

FIGURE 10.14

EL spectra of the heterojunction laser diode under different currents. The inset illustrates the emission intensity of the diode as a function of injection current. (From Zhu, H., Shan, C.X., Yao, B., Li, B.H., Zhang, J.Y., Zhang, Z.Z., Zhao, D.X., Shen, D.Z., Fan, X.W., Lu, Y.M., and Tang, Z.K.: Ultralow-threshold laser realized in zinc oxide. *Adv. Mater.* 2009. 21. 1613–1617. Copyright Wiley-VCH Verlag GmbH & Co. KGaA. With permission.)

higher than that of the heterojunction without MgO layer. This peak can be attributed to the donor–acceptor pair recombination inside the n-ZnO layer.

Figure 10.14 plots the lasing characteristics of the heterojunction diode under forward bias. The EL spectra are measured from the top face of the laser diode at RT. For injection current at ~0.66 mA, a broad spontaneous emission with a FWHM of about 41 nm is observed. For slightly increase of bias current to 1.04 mA, sharp peaks of FWHM ~0.8 nm are emerged from the broad spontaneous emission. Further increase of injection current above 1.22 mA, more sharp peaks appear in a wide spectral range from 380 to 510 nm. The lasing mechanism of such a laser diode is due to random lasing action, and the random cavities are present inside the n-ZnO layer similar to that of the structure of the ZnO polycrystalline as described in Section 10.2. The excitation of sharp peaks started at injection current equals to ~0.8 mA, which is considered to be a low value for the reported threshold current of some UV ZnO-based laser diodes. Hence, this verified that the use of blocking layer between n-ZnO and p-GaN can improve the radiative recombination of carriers inside the ZnO layer.

As discussed in Section 10.3, MOS structure allows the generation of holes from the dielectric oxide layer to contribute for the radiative recombination inside the n-type ZnO semiconductor layer. The use of SiO_2 as the oxide layer to inject holes into the n-ZnO layer has been explored for the fabrication of UV laser diodes. Currently, the use of MgO as the dielectric oxide layer in the MOS structure has also been studied for the realization of UV laser diodes. Figure 10.15a explains the carrier transportation mechanism inside an Au/MgO/ZnO structure under forward bias condition. In fact, the transportation mechanism is similar to that of the Au/SiO_2/n-ZnO structure. As the conduction-band offset between n-ZnO and MgO is large (i.e., ~3.55 eV), electrons are presented to drift

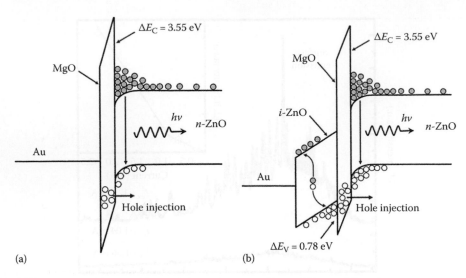

FIGURE 10.15
Schematic diagram showing the band alignment of the (a) Au/MgO/n-ZnO structure and (b) Au/i-ZnO/MgO/n-ZnO structure under forward bias. (From Zhu, H., Shan, C.X., Zhang, J.Y., Zhang, Z.Z., Li, B.H., Zhao, D.X., Yao, B., Shen, D.Z., Fan, X.W., Tang, Z.K., Hou, X.H., and Choy, K.L.: Low-threshold electrically pumped random lasers. *Adv. Mater.* 2010. 22. 1877–1881. Copyright Wiley-VCH Verlag GmbH & Co. KGaA. With permission.)

from the n-ZnO layer to the Au electrode. As a result, electrons will be accumulated at the MgO/n-ZnO interface under forward bias condition. On the other hand, electrons and holes can be generated inside the MgO layer by the impact-ionization process due to the development of high-electric field strength (i.e., electric field in the MgO layer can be in order of 10^6 V/cm at the threshold). Under the forward bias, the generated holes can be injected into the ZnO layer and recombined radiatively with the electrons accumulated at the MgO/n-ZnO interface. If the applied bias is high enough such that the optical gain under electrical excitation is larger than the cavity loss, lasing action can be sustained. Nevertheless, this approach for the generation of hole concentration is still limited by the dielectric oxide layer—it is not an effective way to generate holes. This is because threshold ionization energy of a material is proportional to its bandgap so that the relatively large bandgap of MgO (7.7 eV) is a disadvantage for increasing the generation efficiency of holes. Hence, the laser diodes using MOS structure usually have high threshold conditions.

It is suggested to introduce an insulator ZnO (i-ZnO), which has a smaller bandgap (3.37 eV) than MgO, to reduce the threshold conditions of the MOS laser diodes. This is because the use of i-ZnO can generate holes at a relatively lower electric field [8]. Figure 10.15b explains the transportation mechanism of an Au/i-ZnO/MgO/n-ZnO metal–insulator–oxide–semiconductor structure by using its band alignment profile. Similar to the MOS structure, electron–hole pairs will be generated mainly in the i-ZnO layer at a relatively low electric field, and the MgO layer acts merely as an electron-blocking layer that confines electrons in the ZnO layer. As the valence-band offset between ZnO and MgO is small (0.78 eV), the effective width of the MgO layer in the vicinity of valence-band offset will be greatly reduced due to the band bending of MgO caused by the bias applied. Thus, the generated holes in the i-ZnO layer can tunnel through the MgO barrier and enter into the n-ZnO layer under the forward bias. As the threshold ionization energy of ZnO is lower than that of MgO, it is expected that the threshold conditions of laser diodes using

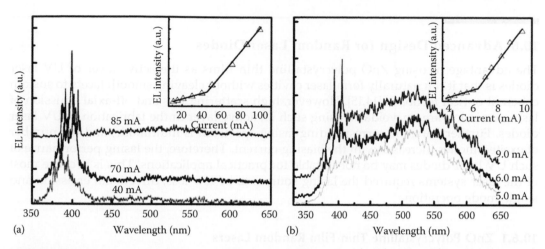

(a) Wavelength (nm)   (b) Wavelength (nm)

FIGURE 10.16

EL spectrum of laser diodes with (a) Au/MgO/n-ZnO and (b) Au/i-ZnO/MgO/n-ZnO structures under different injection currents. The inset shows the integrated intensity of the emission at around 400 nm from these structures as a function of injection current. (From Zhu, H., Shan, C.X., Zhang, J.Y., Zhang, Z.Z., Li, B.H., Zhao, D.X., Yao, B., Shen, D.Z., Fan, X.W., Tang, Z.K., Hou, X.H., and Choy, K.L.: Low-threshold electrically pumped random lasers. *Adv. Mater.* 2010. 22. 1877–1881. Copyright Wiley-VCH Verlag GmbH & Co. KGaA. With permission.)

Au/i-ZnO/MgO/n-ZnO structure can be significantly reduced when compared to that of the laser diode with Au/MgO/n-ZnO structure.

Figure 10.16a shows the lasing characteristics of the Au/MgO/ZnO structure under forward bias. The EL spectra are measured from the top face of the laser diode at RT. For injection current at ~40 mA, a broad EL emission with a FWHM of about 20 nm is observed. For slightly increase of bias current to 70 mA, sharp peaks of FWHM ~0.8 nm are emerged from the broad spontaneous emission. Further increase of injection current above 85 mA, more sharp peaks appear over the emission spectra. The lasing mechanism of such a laser diode is due to random lasing action, and the random cavities are presence inside the n-ZnO layer similar to that for the structure of the ZnO polycrystalline as described in Section 10.2. The dependence of the integrated emission intensity on the injection current is also plotted in the inset of Figure 10.16a. It is clearly shown that the laser diode has a threshold current of ~43 mA. Figure 10.16b shows the room-temperature EL spectra of the Au/i-ZnO/MgO/ZnO structure. For injection current at ~5 mA, a broad EL emission with two broad emissions centered at about 400 and 530 nm can be observed. The emission at around 530 nm arises from the deep-level emission of the i-ZnO layer, while the one at around 400 nm corresponds to the near bandedge emission of the n-ZnO layer. For slightly increase of bias current to 6.0 mA, sharp peaks of FWHM ~0.8 nm are emerged from the broad spontaneous emission. Further increase of injection current above 9 mA, the sharp peaks appear dominant over the emission spectra. The integrated intensity of the emission at around 400 nm versus the injection current is shown in the inset of Figure 10.16b. It is clearly shown that the laser diode has a threshold current of 6.5 mA. From the aforementioned observations, it is noted that the presence of i-ZnO in the MOS structure can significantly reduce the threshold conditions of the laser diodes. In addition, this confirms that the i-ZnO layer indeed increases the generation efficiency of holes in the MOS structure.

10.6 Advanced Design for Random Laser Diodes

The advantage of using ZnO polycrystalline thin films as the active layer of UV laser diodes is that it can naturally form laser cavities without cleaving smooth facets to sustain coherent optical feedback [14,15]. However, high scattering loss and off-axial emission of lasing emission are unavoidable using such an active layer in the fabrication of UV laser diodes. Furthermore, the corresponding lasing spectrum is multimode, and the number of excited modes increases with the biasing current. Therefore, the lasing performance of such UV laser diodes may be not suitable for practical applications. This is because most of the laser systems required the lasing sources to be low-loss, directional emission and single-mode operation.

10.6.1 ZnO Polycrystalline Thin-Film Random Lasers with a Ridge Waveguide Structure

We have examined the high scattering loss and off-axial emission from a ZnO polycrystalline thin film with a ridge waveguide structure [16]. Figure 10.17 shows the typical lasing characteristics of the samples with and without MgO-capped layer on the surface of the ZnO polycrystalline thin film. The samples were excited under optical pumping by a 355 nm Nd:YAG pulsed laser. From the unpolarized light–light curves given in Figure 10.17a, it is observed that a kink (i.e., pump threshold) occurs at ~0.61 MW/cm^2 (~0.69 MW/cm^2) for the sample with (without) MgO-capped layer. The increase in pump intensities above the pump threshold excites more sharp peaks, with linewidth less than 0.4 nm, as observed from the emission spectra in Figure 10.17b. Random lasing action is

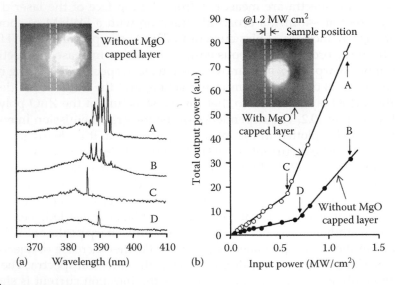

FIGURE 10.17

(a) Light-light curves and (b) emission spectra of the samples with (○) and without (●) MgO-capped layer measured at RT. The inset shows the TE emission far fields of the samples with and without MgO-capped layer at pump intensity of 1.2 MW/cm^2. The dashed lines indicate the location of the sample. (Used with permission from Yuen, C., Yu, S.F., Leong, E.S.P., Yang, H.Y., Lau, S.P., Chen, N.S., and Hng, H.H., Low-loss and directional output ZnO thin-film ridge waveguide random lasers with MgO-capped layer, *Appl. Phys. Lett.*, 86, 031112. Copyright 2005, American Institute of Physics.)

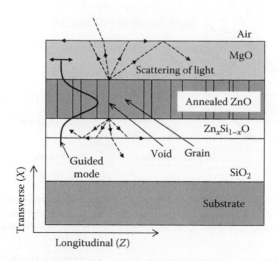

FIGURE 10.18

Schematic cross section of the annealed ZnO ridge waveguide with MgO-capped layer. (Used with permission from Yuen, C., Yu, S.F., Leong, E.S.P., Yang, H.Y., Lau, S.P., Chen, N.S., and Hng, H.H., Low-loss and directional output ZnO thin-film ridge waveguide random lasers with MgO-capped layer, *Appl. Phys. Lett.*, 86, 031112. Copyright 2005, American Institute of Physics.)

responsible for these observations because the facets are too rough to sustain round-trip conditions for Fabry–Perot-type lasing.

The reduction (increment) in pump threshold (output power) implies that the MgO-capped layer has reduced the scattering loss of the random cavities. The TE-polarized far fields of the samples with and without the MgO-capped layer at pump intensity of $\sim$1.2 MW/cm^2 are also shown in the insets of Figure 10.17. Light emitted from facet of the sample without MgO-capped layer exhibits multiple circular spots. On the contrary, the sample with MgO-capped layer shows a single bright-spot emission. Hence, it is shown that the annealed ZnO ridge waveguide random lasers with MgO-capped layer can reduce scattering loss as well as realizing directional UV lasing output. The improvement in lasing characteristics of the ZnO ridge waveguide random lasers can be explained by the trapping of scattering light inside the random cavities, as noted from Figure 10.18. MgO-capped layer traps the light scattered from the ZnO voids to the surface and to the side of the ridge waveguide by total internal reflection. Hence, light emitted from the facets is collimated to a single-spot profile, and the scattering loss of the random cavities is also reduced. It can be shown that the use of MgO-capped layer can significantly reduce the scattering loss of the ZnO polycrystalline thin film by more than 57 cm^{-1}.

10.6.2 Directional and Controllable Edge-Emitting ZnO UV Random Laser Diodes

Based on the investigation of the ZnO polycrystalline thin-film random lasers with a ridge waveguide structure as shown earlier, it is noted that if the ZnO polycrystalline thin film is used as the active layer of the UV laser diodes, introduction of a ridge waveguide enclosed with cladding layers will significantly improve the collimation of the emission beam and reduce the scattering loss of the random cavities. Hence, using the same idea, we can design a low-loss and beam-controllable random laser diode using ZnO polycrystalline thin film [10]. Figure 10.19 shows a schematic of the proposed buried heterojunction rib waveguide laser. A p-GaN:Mg/sapphire substrate with hole concentration of $\sim$5 × 10^{17} cm^{-3}

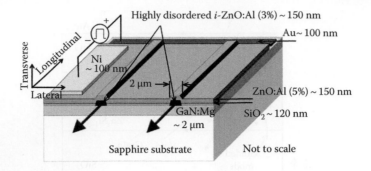

FIGURE 10.19

Schematic of a p-GaN/annealed i-ZnO:Al (3%)/n-ZnO:Al (5%) buried heterojunction rib waveguide laser. (Used with permission from Liang, H.K., Yu, S.F., and Yang, H.Y., Directional and controllable edge-emitting ZnO ultraviolet random laser diodes, *Appl. Phys. Lett.*, 96, 101116. Copyright 2010, American Institute of Physics.)

was used as the hole-injection layer and substrate. A ~150 nm thick ZnO:Al (3%) thin film was deposited onto half of the p-GaN:Mg/sapphire substrate by FCVA technique. During the deposition, substrate temperature and oxygen partial pressure were set to ~150°C and ~2×10^{-5} Torr, respectively. The sample was annealed at 900°C for 30 min in open air to form highly disordered ZnO grains and voids in order to sustain random lasing action. In addition, the use of Al-doped ZnO film as the rib waveguide is to maintain electrically conductive after the annealing process. The annealed i-ZnO:Al (3%) is found to have electron concentration and mobility of ~5×10^{16} cm^{-3} and ~6 cm^2/(Vs), respectively. Subsequently, a line mask (with width, thickness, and separation equal to 2, 0.8, and 500 μm, respectively) was coated onto the surface of the annealed i-ZnO:Al (3%) film by photolithography technique. The unmasked i-ZnO:Al (3%) layer was then completely removed by ion-beam sputtering with an etching rate of ~10 nm/min for 15 min.

A ~120 nm thick SiO$_2$ cladding layer was then deposited onto the sample by electron-beam (e-beam) evaporation with substrate temperature set to 50°C. After the deposition, a lift-off process was carried out to remove the excess SiO$_2$ layer attached onto the surface of the annealed i-ZnO:Al(3%) rib waveguides. The SiO$_2$ cladding layer was used as an electrical isolation layer to prevent the lateral diffusion of injection carriers from the rib waveguide. As the refractive index of SiO$_2$ cladding layer ($n \sim 1.45$) is smaller than that of the i-ZnO:Al (3%) rib ($n \sim 2.1$), strong lateral optical confinement can also be achieved. Finally, a layer of n-ZnO:Al (5%) with thickness of ~150 nm was deposited onto the surface of the sample by the FCVA technique to serve as an electron-injection layer. The deposition conditions are the same as that of the ZnO:Al (3%) layer. The carrier concentration of n-ZnO:Al (5%) film was found to be ~10^{21} cm^{-3}. A ~100 nm thick Au (Ni) film was deposited onto the p-GaN:Mg/sapphire substrate [n-ZnO:Al (5%) layer] as the p-type (n-type) metal contact by e-beam evaporation. For the purpose of comparison, another p-GaN/annealed i-ZnO:Al (3%)/n-ZnO:Al (5%) heterojunction laser diode without a rib waveguide structure was also fabricated.

Edge emission spectra of the diode laser with a buried rib waveguide are plotted in Figure 10.20a. For $I \geq 3.2$ mA, sharp peaks at around 387 nm emerge from the edge emission spectrum. For further increase in I, the number and intensity of the lasing peaks also increase. The corresponding surface emission spectra, which exhibit only broad spontaneous emission, are also plotted in the inset of Figure 10.20a. From the corresponding L–I curves shown in Figure 10.20b, it is noted that the intensity of edge emission (threshold current) is enhanced by approximately nine times (reduced from 5 to 3.2 mA) when compared to that without the buried rib waveguide. Furthermore, at a large value of I,

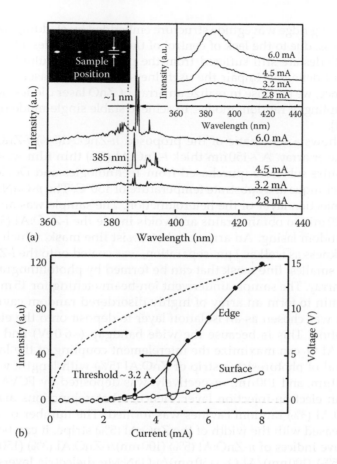

FIGURE 10.20

(a) EL spectra measured from the edge of the laser diode with a buried rib waveguide. The inset shows the corresponding EL spectra measured from the surface of the laser diode. The corresponding edge-emission near-field image is also shown in the inset of the figure. (b) Light-current (open circles-measured from the surface, closed-cirlces-measured from the edge) and light-voltage curves of the laser diode with a buried rib waveguide. (Used with permission from Liang, H.K., Yu, S.F., and Yang, H.Y., Directional and controllable edge-emitting ZnO ultraviolet random laser diodes, *Appl. Phys. Lett.*, 96, 101116. Copyright 2010, American Institute of Physics.)

the lasing peaks were excited electrically at around 387 nm, which is closed to the emission peak of n-ZnO:Al (3%) film. This implies that the electrical confinement of the p-i-n heterojunction inside the annealed i-ZnO:Al (3%) waveguide is improved at a large value of I, and the influence of current crowding is suppressed. Hence, the electrical-to-optical conversion efficiency of the laser diode is strongly enhanced by the presence of the 2 μm rib waveguide embedded inside the SiO$_2$ cladding layer. The near field image of the edge emission is also shown in the inset of Figure 10.20b. Only a single light spot is observed from the edge of the rib waveguide. Hence, these indicate that strong lateral and transverse optical confinement can be achieved simultaneously from the rib waveguide structure, and the edge emission is highly directional.

10.6.3 High-Power and Single-Mode Operation UV ZnO Laser Diodes

The problems of using ZnO polycrystalline thin films for the realization of UV ZnO laser diodes such as high scattering loss and multiple-direction emission of lasing beam have been solved by using ridge waveguide structure embedded by cladding layers; see Section 10.6.2. Nevertheless, due to the lack of control of the random modes, electrically pumped UV ZnO laser diodes are still suffered from the problems of multiple-mode operation. This problem will definitely impair the usefulness of UV ZnO laser diodes in practical applications. Hence, we propose to realize an array of ZnO laser diodes, which has strong interelement coupling of random modes, to achieve stable single-mode operation at high output power [11].

Figure 10.21 shows a schematic of the proposed n-ZnO:Al(5%)/i-ZnO:Al(3%)/p-GaN heterojunction laser array. A ~150 nm thick i-ZnO:Al (3%) thin film was deposited on a p-GaN:Mg/sapphire substrate (purchased from Technologies and Device International Inc.) by FCVA technique at a substrate temperature of 150°C. The p-GaN:Mg layer is also used as a hole-injection layer of the heterojunction. The sample was annealed at 900°C in open air for 20 min to obtain grains and voids inside the i-ZnO:Al (3%) layer to support coherent random lasing. An array of photoresist line mask, which has 1 μm width and 800 nm thickness as well as 1 μm separation, was coated onto the i-ZnO:Al (3%) thin film. This is the smallest line mask that can be formed by photolithography to fabricate the laser diode array. The sample underwent ion-beam etching for 15 min at an etching rate of ~10 nm/min to form an array of highly disordered random cavities. A ~120 nm thick Al_2O_3 film was chosen as an isolation layer to deposit onto the etched surface by e-beam evaporation. This is because the wide bandgap (~6.0 eV) and large refractive index (~1.78) of Al_2O_3 can maximize the interelement coupling of the laser diode array. After the removal of photoresist, a strip of ZnO:Al (5%) with length, width, and thickness of 2 mm, 20 μm, and 100 nm, respectively, was deposited by FCVA technique onto the sample as an electron-injection layer. Hence, a heterojunctions array of 10 highly disordered ZnO:Al (3%) random cavities was formed. The number of random cavities can also be increased with the width of the ZnO:Al (5%) stripe. It can be shown that the effective refractive indices of n-ZnO:Al (5%) (100 nm)/i-ZnO:Al (3%) (150 nm)/p-GaN:Mg and n-ZnO:Al (5%) (100 nm)/Al_2O_3 (150 nm)/p-GaN:Mg dielectric layers are 2.4411 and 2.4374, respectively, at an operating wavelength of ~385 nm. In this case, the penetration length of the evanescent wave into the isolation layer can be estimated to be ~0.8 μm. As the random cavities have a separation of 1 μm, strong interelement coupling should be obtained from this design of laser diode array. Subsequently, a ~100 nm thick Ni film

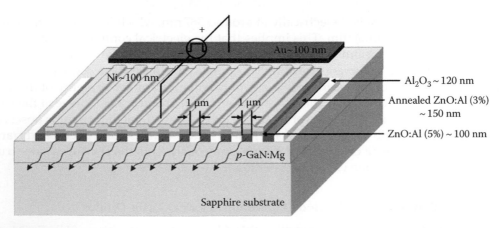

FIGURE 10.21
Schematic of a ZnO random laser diode array. (After Liang, H.K. et al., *Appl. Phys. Lett.*, 97, 241107, 2010.)

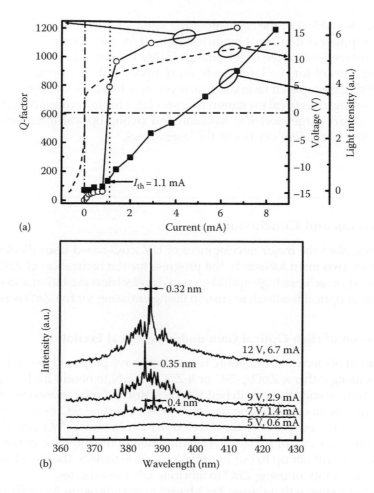

FIGURE 10.22

(a) Voltage-current curve, light-current curve, and Q-factor versus injection current of the ZnO random laser diode array. (b) EL spectra measured from the edge of the laser diode array at different bias conditions. (Used with permission from Liang, H.K., Yu, S.F., and Yang, H.Y., ZnO random laser diode arrays for stable single-mode operation at high power, *Appl. Phys. Lett.*, 97, 241107. Copyright 2010, American Institute of Physics.)

and a ~100 nm thick Au film were coated onto the n-ZnO:Al(5%) strip and p-GaN:Mg substrate, respectively, by e-beam evaporation.

Figure 10.22a plots the V–I and L–I curves as well as Q-factor versus current of the ZnO laser diode array. The diode array was driven by a rectangular pulse voltage source with repetition rate and pulse width of 7.5 Hz and 80 ms, respectively. The diode array exhibits a diode-like rectifying I–V characteristic with a turn-on voltage of ~5 V. It is also observed that a kink appears in the light-current curve for injection current, I, equal to a threshold value of Ith ~1.1 mA. This is consistent with the measured Q-factor of the heterojunction as for $I \sim I$th, the value of Q-factor jumps from ~40 to ~1000 and reaches ~1200 at $I \sim 6.7$ mA. Figure 10.22b also shows the EL spectra of the diode array. There is only one dominant peak emerged from the EL spectra at I equal to or larger than Ith. In addition, FWHM of the lasing peak, $\Delta\lambda$, decreases from 0.4 nm at $I \sim 1.4$ mA to 0.32 nm at $I \sim 6.7$ mA. This implies that the electrical energy is only channeled to the dominant random mode so that $\Delta\lambda$ decreases with the increase of I. On the other hand, although the position of the dominant mode is stable (i.e., the corresponding value of $\Delta\lambda$ approaching the noise limit) at a value of I, hopping of dominant mode is observed for the change of I. This may be due to the carrier-induced refractive index change inside the random cavities as the thermal effect is less significant for the laser diode array under pulsed operation. Therefore, we have verified that the proposed laser diode arrays can achieve RT stable single-mode operation even under high electrical excitation (i.e., $>6 \times I_{th}$). This is because the Al_2O_3 dielectric insulator can achieve large electrical isolation and strong optical interelement coupling between the adjacent random cavities of the laser arrays.

10.7 Discussions and Conclusions

This chapter describes the major development of UV ZnO-based laser diodes for the past decade. There are two main focuses in the progress for the realization of ZnO-based laser diodes: (1) methods to achieve high optical gain under electrical excitation and (2) techniques to realize coherent optical feedback to sustain lasing emission for the ZnO active layer.

10.7.1 Realization of High Optical Gain under Electrical Excitation

For the realization of electrical excitation, researchers have proposed to use heterojunction design such as using either n-ZnO/p-SiC or n-ZnO/p-GaN to obtain high concentration of hole injection from some lattice-matched p-type semiconductor materials. However, the drawback of this method is that the use of expensive p-type SiC or GaN hole-injection layers do not serve the purpose for the development of low-cost ZnO-based laser diodes. Furthermore, the efficiency of radiative recombination at the interface either n-ZnO/p-SiC or n-ZnO/p-GaN is still not up to our expectation. Nevertheless, the use of heterojunction can verify the possibility of using ZnO to fabricate UV laser diodes.

Researchers have shown that p-type ZnO-based materials using As or Sb as the dopants can be the potential candidates for the realization of ZnO-based laser diodes. There is one common configuration for these ZnO-based laser diodes—their active layer is either made of ZnO/ZnBeO or ZnO/ZnMgO QWs structure. The reason is simple; QWs structure can provide much higher optical gain than bulk polycrystalline thin films so that the required concentration of holes to support of stimulated emission under electrical excitation can be lower than that of bulk layer. Hence, if the QW structures of the proposed ZnO-based laser diodes are replaced by ZnO polycrystalline thin films as the active layer, it is doubt whether the ZnO-based laser diodes can still sustain lasing action. Nevertheless, the main problem of using p-type ZnO layer is still their reliability. To the best of the authors' understanding, there is still no stable p-type ZnO layer that can be obtained. Even if ZnO-based laser diodes can be fabricated by using As- or Sb-doped p-type ZnO layer, their lasing characteristics can be deteriorated significantly within a month's time. Therefore, the capability to achieve stable p-type ZnO layer is still a bottom neck for the fabrication of ZnO-based UV laser diodes.

In order to avoid either using p-type ZnO or adopting other lattice-matched p-type materials for the realization of hole-injection layer, we can utilize impact ionization process to generate hole concentrations inside the ZnO active layer of the UV ZnO laser diodes. It has been proposed by using MOS structure for the fabrication of UV ZnO laser diodes on Si or glass substrate. In the MOS structure, "M" is the anode contact metal usually using Au, "O" is the dielectric layer that can be SiO_2 or MgO, and "S" is the *n*-type ZnO. There is no need for the use of *p*-type ZnO or other *p*-type lattice-matched materials to generate holes. Furthermore, the corresponding fabrication cost can be significantly lower than that of the heterojunction design. The working principle is simple: if applied voltage is supplied to the MOS structure, high voltage (high-electric field strength) will be developed across the dielectric layer. As a result, electrons and holes can be generated inside the dielectric layer via impact-ionization process. Then, holes can penetrate through the dielectric layer and drift into the *n*-ZnO layer to recombine radiatively with the electrons. One of the disadvantages of this approach is that the threshold ionization energy is directly proportional to the bandgap of the dielectric layer. This implies that the large bandgap dielectric layer requires high bias voltage to generate holes. In order to reduce the threshold conditions of the ZnO-based laser diodes using MOS structure, one of the alternative methods is to insert another dielectric layer with smaller bandgap into the MOS structure in order to reduce the threshold ionization energy. In this case, the researchers have proposed to fabricate MoOS structure where "o" states for a dielectric layer with smaller bandgap than that of "O." For example, Au/*i*-ZnO/MgO/*n*-ZnO is one of the possible structures with low threshold ionization energy.

10.7.2 Realization of Coherent Optical Feedback to Sustain Lasing Emission

Laser is a device that amplifies light and produces a highly directional, high intensity beam that most often have a very pure frequency or wavelength. Hence, for a device to behave like a laser, it is required that (1) the total emission intensity of the device increases rapidly at and above a small value (i.e., a threshold value) of the excitation intensity and (2) the excitation of sharp peaks (with linewidth of few Å) for the excitation intensity larger than the threshold value. In order to obtain these emission characteristics, the "device" should provide (1) high optical gain (i.e., to amplify spontaneous emission) and (2) coherent optical feedback (i.e., to achieve stimulated emission) simultaneously.

Due to the RT high excitonic binding energy, stimulated emission may also be obtained from ZnO-based materials without the use of optical feedback. Reports have shown that stimulated emission had been observed from ZnO/ZnMgO and ZnO/ZnBeO MQWs thin films grown on lattice-matched substrate at and below RT [5,17–19]. Strong sharp emission peaks can be excited (for the excitation intensities above a threshold value) from the ZnO MQWs thin films even without the presence of optical feedback. The excitation of sharp peaks may be due to the recombination mechanisms of localized excitons, ex–ex scattering, and localized biexcitons. In fact, exciton localization and the related effects on stimulated emission have been frequently observed in II–VI QWs at low temperatures. As the localization effects are only related to some discrete energies, sharp peaks can be emerged from the emission spectrum under excitation. Furthermore, due to the small density of states in the localized states of excitons, low threshold (i.e., low concentration of population inversion is required) can be achieved with QWs structures. As a result, stimulated emission can be obtained from the ZnO QWs even without the presence of optical feedback. Nevertheless, stimulated emission was only observed form ZnO QWs under optical excitation. The difficulty to obtain stimulate emission under electrical excitation may be due to the poor crystal quality of the QWs grown on lattice-mismatched substrate. In this case, the excitonic gain can be significantly deteriorated, and only electron-hole plasma recombination can be survived at RT. Hence, it is not surprising that excitonic stimulated emission is not observed from the current-fabricated ZnO-based laser diodes.

In the recent development of ZnO-based laser diodes, it is necessary to have optical feedback mechanism to obtain coherent amplification of spontaneous emission as electron-hole plasma recombination is still the dominant optical gain mechanism. Furthermore, it must be noted that cleaved facets of ZnO thin-film waveguides are too rough to provide sufficient optical feedback for sustaining UV lasing. In fact, ZnO thin films can form microcavities to support Fabry–Perot resonant modes [20]. If the ZnO film is properly grown on lattice-matched substrate, parallelly arrayed hexagonal ZnO microcrystallites can be formed to achieve laser cavity with {1100} facets. Nevertheless, it is not easy to grow closely packed hexagonal ZnO microcrystallites over a wide area so that Fabry–Perot resonant oscillation is not commonly observed. Alternatively, randomly distribution of hexagonal ZnO microcrystallites (i.e., ZnO grains and voids) can be more easily formed inside the ZnO polycrystalline thin films [14,15]. As the nonuniform distribution of ZnO grains and voids can support random lasing action inside the ZnO polycrystalline thin films, a number of reports have reviewed the successful fabrication ZnO random laser diodes using ZnO polycrystalline thin films as the active layer. However, the disadvantages of multiple-mode operation and wide divergence of the emission beams, which are the nature characteristics of random lasers, limited the use of ZnO polycrystalline thin films for the fabrication of practical laser diodes. However, our recent achievement in fabrication of ZnO random laser diodes has shown that highly directional emission and single-mode operation can be obtained from ZnO polycrystalline thin films. Hence, it is believed that random cavities are the most promising configuration to obtain optical feedback from bulk ZnO thin films.

10.7.3 Conclusions

For the fabrication of UV laser diodes using ZnO instead of p-type GaN, it is necessary to obtain stable and conductive p-type ZnO-based materials. Although ZnO laser diodes can be fabricated using alternative structures as the hole-injection layer, this is not a long-term solution. If reliable p-type ZnO-based materials can be realized, it will be more sensible to use ZnO MQWs as the active layer of the laser diodes. This is because stimulated emission can be obtained from the high excitonic gain of the MQWs without the requirement of optical feedback. Hence, using localized states of excitons can simplify the design of ZnO laser diodes. In addition, the corresponding threshold conditions can be much lower than that using electron-hole plasma recombination as the optical gain. Nevertheless, we still do not know how long we have to wait until reliable and stable p-type ZnO-based materials can be fabricated. MOS structure laser diode using n-ZnO polycrystalline thin film as the semiconductor material is the current most simple and low-cost method available.

Acknowledgment

This work was supported by the Hong Kong Polytechnic University research grant no. 1-ZV6X.

References

1. U. Ozgur, Y.I. Alivov, C. Liu, A. Teke, M.A. Reshchikov, S. Dogan, V. Avrutin, S.J. Cho, and H. Morkoc, A comprehensive review of ZnO materials and devices, *J. Appl. Phys.*, 98, 041301, 2005.
2. E.S.P. Leong and S.F. Yu, UV random lasing action in p-SiC(4H)/i-ZnO–SiO$_2$ nanocomposite/n-ZnO:Al heterojunction diodes, *Adv. Mater.*, 18, 1685–1688, 2006.
3. E.S.P. Leong, S.F. Yu, and S.P. Lau, Directional edge-emitting UV random laser diodes, *Appl. Phys. Lett.*, 89, 221109, 2006.
4. X.Y. Ma, P.L. Chen, D.S. Li, Y.Y. Zhang, and D.R. Yang, Electrically pumped ZnO film ultraviolet random lasers on silicon substrate, *Appl. Phys. Lett.*, 91, 251109, 2007.
5. Y.R. Ryu, J.A. Lubguban, T.S. Lee, H.W. White, T.S. Jeong, C.J. Youn, and B.J. Kim, Excitonic ultraviolet lasing in ZnO-based light emitting devices, *Appl. Phys. Lett.*, 90, 131115, 2007.
6. S. Chu, M. Olmedo, Z. Yang, J.Y. Kong, and J.L. Liu, Electrically pumped ultraviolet ZnO diode lasers on Si, *Appl. Phys. Lett.*, 93, 181106, 2008.
7. H. Zhu, C.X. Shan, B. Yao, B.H. Li, J.Y. Zhang, Z.Z. Zhang et al., Ultralow-threshold laser realized in zinc oxide, *Adv. Mater.*, 21, 1613–1617, 2009.
8. H. Zhu, C.X. Shan, J.Y. Zhang, Z.Z. Zhang, B.H. Li, D.X. Zhao et al., Low-threshold electrically pumped random lasers, *Adv. Mater.*, 22, 1877–1881, 2010.
9. Y. Tian, X.Y. Ma, D.S. Li, and D.R. Yang, Electrically pumped ultraviolet random lasing from heterostructures formed by bilayered MgZnO films on silicon, *Appl. Phys. Lett.*, 97, 061111, 2010.
10. H.K. Liang, S.F. Yu, and H.Y. Yang, Directional and controllable edge-emitting ZnO ultraviolet random laser diodes, *Appl. Phys. Lett.*, 96, 101116, 2010.
11. H.K. Liang, S.F. Yu, and H.Y. Yang, ZnO random laser diode arrays for stable single-mode operation at high power, *Appl. Phys. Lett.*, 97, 241107, 2010.
12. V. Avrutin, D.J. Silversmith, and H. Morkoc, Doping asymmetry problem in ZnO: Current status and outlook, *Proc. IEEE*, 98, 1269–1280, 2010.
13. O. Maksimov, Recent advances and novel approaches of p-type doping of ZnO, *Rev. Adv. Mater. Sci.*, 24, 26–34, 2010.
14. H. Cao, Y.G. Zhao, H.C. Ong, S.T. Ho, J.Y. Dai, J.Y. Wu, and R.P.H. Chang, Ultraviolet lasing in resonators formed by scattering in semiconductor polycrystalline films, *Appl. Phys. Lett.*, 73, 3656–3658, 1998.
15. S.F. Yu, C. Yuen, S.P. Lau, and H.W. Lee, Zinc oxide thin-film random lasers on silicon substrate, *Appl. Phys. Lett.*, 84, 3244–3246, 2004.
16. C. Yuen, S.F. Yu, E.S.P. Leong, H.Y. Yang, S.P. Lau, N.S. Chen, and H.H. Hng, Low-loss and directional output ZnO thin-film ridge waveguide random lasers with MgO capped layer, *Appl. Phys. Lett.*, 86, 031112, 2005.
17. A. Ohtomo, K. Tamura, M. Kawasaki, T. Makino, Y. Segawa, Z.K. Tang, G.K.L. Wong, Y. Matsumoto, and H. Koinuma, Room-temperature stimulated emission of excitons in ZnO/(Mg, Zn)O superlattices, *Appl. Phys. Lett.*, 77, 2204–2206, 2000.
18. H.D. Sun, T. Makino, N.T. Tuan, Y. Segawa, Z.K. Tang, G.K.L. Wong, M. Kawasaki, A. Ohtomo, K. Tamura, and H. Koinuma, Stimulated emission induced by exciton–exciton scattering in ZnO/ZnMgO multiquantum wells up to room temperature, *Appl. Phys. Lett.*, 77, 4250–4252, 2000.
19. J. Cui, S. Sadofev, S. Blumstengel, J. Puls, and F. Henneberger, Optical gain and lasing of ZnO/ZnMgO multiple quantum wells: From low to room temperature, *Appl. Phys. Lett.*, 89, 051108, 2006.
20. Z.K. Tang, G.K.L. Wong, P. Yu, M. Kawasaki, A. Ohtomo, H. Koinuma, and Y. Segawa, Room-temperature ultraviolet laser emission from self-assembled ZnO microcrystallite thin films, *Appl. Phys. Lett.*, 72, 3270–3272, 1998.

Part III

ZnO-Based Electronic Devices and Application

Part III

ZnO-Based Electronic Devices and Application

11

Metal-Semiconductor Field-Effect Transistors and Integrated Circuits Based on ZnO and Related Oxides

Heiko Frenzel, Michael Lorenz, Friedrich-L. Schein, Alexander Lajn,
Fabian J. Klüpfel, Tobias Diez, Holger von Wenckstern, and Marius Grundmann

CONTENTS

11.1 Introduction 370
11.2 ZnO-Based MESFET on Sapphire Substrate 372
    11.2.1 Preparation 372
    11.2.2 Experimental Details 373
    11.2.3 Optimization of the Device Performance Influence of the Gate Material 374
        11.2.3.1 Variation of the Net Doping Concentration and Channel Thickness 377
        11.2.3.2 Geometry Effects 379
    11.2.4 Reliability and Degradation 381
        11.2.4.1 Elevated Temperatures 382
        11.2.4.2 Bias Stress and Light Exposure 383
        11.2.4.3 Long-Term Stability 385
    11.2.5 ZnO-MESFET on Glass Substrates 386
    11.2.6 Comparison of ZnO-MESFET with Competing Technologies 388
    11.2.7 Switching Behavior of ZnO-MESFET 389
11.3 High-Gain ZnO-Based Inverters 392
    11.3.1 Fundamental Inverter 392
    11.3.2 FET-Logic Inverter 395
    11.3.3 Schottky-Diode FET-Logic Inverters 397
    11.3.4 Fundamental Inverter with Integrated Level-Shifting 400
    11.3.5 Comparison of ZnO-MESFET-Based Inverters with Other Technologies 401
11.4 Transparent Rectifying Contacts 402
    11.4.1 Preparation 402
    11.4.2 Transparent Rectifying Contacts for Visible-Blind Ultraviolet Photodiodes Based on ZnO 403
    11.4.3 Transparent MESFET and Inverters 409
    11.4.4 Comparison of Transparent MESFET and Inverter with the Literature 411
    11.4.5 Stability of Fully Transparent ZnO-Based Fundamental Inverter Circuits against Incident Light and Elevated Temperature 413
11.5 MESFET Based on Amorphous Oxides 418
11.6 Summary and Outlook 424
Acknowledgments 426
References 426

11.1 Introduction

Metal-semiconductor field-effect transistors (MESFET) are commonly known from GaAs technology [1] and are widely used for high-speed logic circuits due to their high channel mobility [2,3]. In contrast to GaAs-MESFET, which consist of an n-type implanted channel in a p-type or semi-insulating GaAs substrate, oxide MESFET consist of a thin semiconducting channel layer on an insulating substrate (Figure 11.1a). Their gate structure is a rectifying Schottky contact (SC). However, this technology is a rather new subject in the ZnO community and in transparent electronics. Since the p-type doping of ZnO and consequently the fabrication of bipolar devices still remains a challenge, unipolar devices become more and more interesting. For ZnO thin films, the MESFET technology was demonstrated some years ago by Ryu et al., who used a Ti-gate contact on p-type ZnO [4]. Later, Kandasamy et al. used Pt-gate n-ZnO MESFET for hydrogen gas sensing [5], and Kao et al. used Pt/ Au-gate contacts for their MESFET [6]. However, these reported thin-film ZnO MESFET showed inferior electronic properties with high gate voltage sweeps between 4 and 20 V, very low on/off-ratios below 10, and barely obtained pinch-off and saturation behavior. On the basis of ZnO nanorods, MESFET and logic devices such as OR, AND, NOT, and NOR gates have been realized by Park et al. using Au-gate contacts [7]. Their on/off-ratio was approximately 10^4 and a subthreshold slope between 100 and 200 mV/decade was achieved. However, channel mobilities were neither reported for ZnO thin film nor nanorod MESFET.

ZnO-MESFET performance advanced significantly in 2008 by using Ag_xO-gated MESFET as demonstrated by Frenzel et al. showing an on/off-ratio of 10^8, a gate-voltage sweep of only 3 V, and a channel mobility of 11.3 cm^2/(V s) [8]. Since then, ZnO-based MESFET have been improved, especially with regard to their applicability in low-cost transparent electronics [9]. However, in transparent electronics, still the established technology of metal–insulator–semiconductor field-effect transistors (MISFET) is used. Currently, there are several approaches for transparent MISFET, which are assigned to substitute amorphous-Si transistors in backplanes of active-matrix displays: carbon-nanotube FET (CNTFET), organic FET (OFET), and transparent-oxide MISFET (TMISFET). The different technologies are compared in Table 11.1. Single CNT promise a high channel mobility of 3000 cm^2/(V s) [10]. They are fabricated, however, by means of complex laser-based, plasma-based, or chemical processes, for example, chemical vapor deposition. CNT-networks can then be deposited from solution on large substrates at low temperatures to create CNTFET [11]. The channel mobilities of such transistors are only in the range of 0.5–30 cm^2/(V s) due to interface effects and the highly disordered structure

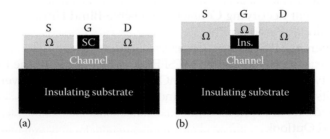

(a)   (b)

FIGURE 11.1
Schematic structure of (a) a MESFET and (b) a MISFET. Both transistors have ohmic source (S) and drain (D) contacts. The gate (G) consists of a Schottky contact (SC) for MESFET (a) and a metal–insulator–semiconductor diode for MISFET (b).

TABLE 11.1

Comparison of Different Transparent-Transistor Technologies with the Current Opaque Si-Transistors

	MISFET					MESFET
Transistor technology	a-Si	Poly-Si	CNTFET	OFET	TMISFET	TMESFET
Transparency	No	No	Yes	Yes	Yes	Yes
Channel mobility (cm²/(V s))	≤1	50–100	0.5–30	≪1	>10	>10
Slope (mV/decade)	~500	200–300	200–2000	600–2200	200–2000	60–200
Off-current (A)	~10^{-12}	~10^{-12}	~10^{-10}	10^{-12}–10^{-9}	<10^{-13}	~10^{-12}
On/off-ratio	10^5–10^7	10^5–10^7	10^2–10^4	~10^5	10^6–10^{10}	10^6–10^9
Gate-voltage sweep (V)	~20	~20	20–100	20–50	10–80	1–3
Fabrication temperature (°C)	~250	~250	100–900	<100	20–350	20–150
Display applications	LCD, etc.	LCD, small org. EL	LCD, OLED, etc.	OLED, e-paper	LCD, OLED, e-paper, etc.	LCD, OLED, e-paper, etc.
Fabrication complexity	Moderate	High	High	Low	Moderate	Low
Bias-stress threshold-voltage shift (V)	~5	~5	>5	1–10	~2	0
Light-induced threshold-voltage shift (V)	2–5	2–5	~2	10–20	2–20	0–0.65
Stability	High	Low	Low	Moderate	High	High

[12–14]. OFET can as well be deposited from solution. But the transistors exhibit very low channel mobilities ≪1 cm²/(V s) and have to be sophistically encapsulated in order to assure stability under ambient conditions [15,16]. TMISFET with amorphous channels can be fabricated at low temperatures (<150°C) by means of sputtering techniques and have higher channel mobilities >10 cm²/(V s) [17–19]. The materials-of-choice are zinc-oxide related: zinc–tin oxide (ZTO), gallium–zinc–tin oxide (GZTO), indium–zinc oxide (IZO), hafnium–indium–zinc oxide (HIZO), and gallium–indium–zinc oxide (GIZO). As can be seen in Table 11.1, amorphous TMISFET show the best performances among MISFET, that is, transparency, high mobilities and on/off-ratios, moderate fabrication complexity, and high stability. They are therefore favored among these transistor techniques.

Compared to the here presented MESFET, the MISFET technology involves some disadvantages. First of all, the insulator has to be deposited by means of highly sophisticated techniques like atomic layer deposition (ALD) [20,21] or plasma-enhanced chemical vapor deposition (PECVD) [22]. Often, even organic insulators are used [23]. They are easy to deposit but cannot be structured by common photolithographic techniques. Instead shadow masks are often used. The thickness of the insulators has to be as large as 50 nm or more to ensure acceptable leakage current densities below 10^{-6} A/cm². For all MISFET, the voltage drop across the insulator leads to higher applied gate-voltage sweeps. Further, high defect and trap concentrations located at the insulator–semiconductor interface influence the electrical properties considerably. The channel mobility is significantly lowered [24], large shifts of the threshold voltage occur under the influence of light and bias stress [25,26] as well as hysteresises.

MESFET do not exhibit these disadvantages due to the absence of any insulator (Figure 11.1) and are therefore a promising approach to low-cost transparent electronics.

11.2 ZnO-Based MESFET on Sapphire Substrate

11.2.1 Preparation

The semiconducting channel layer of the MESFET presented in this section and the MESFET-based inverters (Section 11.3) were deposited by pulsed laser deposition (PLD) on transparent a-plane sapphire substrates. Due to the typical net doping concentration in the range of $10^{18}\,cm^{-3}$, a film thickness of 20–30 nm was chosen in order to ensure, that the space charge region originating from the Schottky-gate contact can be extended over the channel depth by applying an external field and the transistor can be properly switched off. The deposition temperatures range from 630°C to 700°C and the oxygen partial pressure was kept between 0.02 and 0.04 mbar. Note, that the deposition temperature can be drastically reduced using a magnetron sputtering technique to deposit amorphous ZnO-related oxide channels on glass substrates as described in Section 11.2.5. Each sample was pre-characterized using Hall-effect and spectroscopic ellipsometry at room temperature to obtain the free carrier concentration and Hall-effect mobility as well as the channel thickness. After deposition of the channel layer, the processing of the devices was carried out by means of standard photolithography using a *SÜSS MJB3 Mask Aligner*, wet-chemical etching and metallization (Figure 11.2). First, a mesa structure was wet-chemically etched into the ZnO thin film using phosphoric acid to form channels, that ensure the direction of the current from source to drain and crosstalk between different FET is excluded. Second, the ohmic source and drain contacts were formed by dc-sputtering of Au under a pure Ar atmosphere at a pressure of 0.02 mbar. Note, that under these deposition conditions, Au forms an ohmic tunneling contact on ZnO, whereas reactive sputtering or thermal

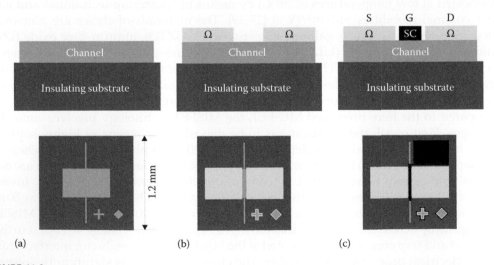

FIGURE 11.2
Processing steps for MESFET (top: side view, bottom: top view). (a) Channel etching, (b) dc-sputtering of ohmic source (S) and drain (D) contacts, and (c) reactive dc-sputtering of gate (G) Schottky contact (SC).

evaporation of Au leads to rectifying Schottky contacts. In a third step, the Schottky-gate contacts were deposited using reactive dc-sputtering of noble metals as Ag, Pt, Pd, or Au under an Ar/O$_2$-flux ratio of 50sccm/50sccm. The oxygen radicals partially oxidize the sputter-target material such that metal as well as metal-oxide particles are deposited on the substrate. For two reasons the reactive sputtering process leads to an improvement of the Schottky contacts with respect to non-reactive sputtered or evaporated Schottky contacts. On the one hand, the negatively charged oxygen ions bombard the ZnO surface during the initial phase of the deposition process which results in an *in situ* plasma surface cleaning. This leads to the removal of contaminations such as hydroxide-induced, highly conductive surface layers that tend to reduce the Schottky barrier height [27]. On the other hand, x-ray photoelectron spectroscopy (XPS) revealed that the contact material (except Au) is oxidized [28]. This coincides with a higher work function and with that leads to higher Schottky barrier heights as investigated on single crystalline ZnO substrates by Allen et al. [29,30].

Due to the partial oxidation of the Schottky-metals, their conductivity is significantly lowered with respect to pure metals. Therefore, it is necessary to deposit a highly conducting capping layer on top of the Schottky contact in order to form an equipotential surface [28].

The so formed Schottky-contact structure is highly rectifying and accurately follows the model of thermionic emission of hot carriers [28,30].

11.2.2 Experimental Details

Figure 11.3 depicts the relationship between the MESFET's output ($I_{SD}(V_{SD})$) and transfer characteristic ($I_{SD}(V_G)$) from which the figures of merit were extracted. The current–voltage measurements were performed using an *Agilent 4156C Precision Semiconductor Parameter*

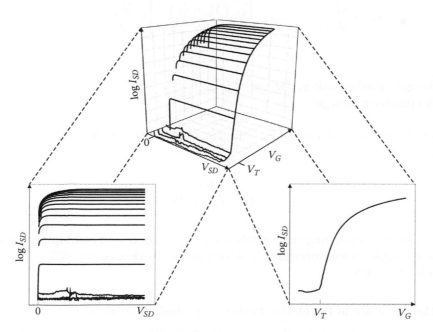

FIGURE 11.3
Relation between output and transfer characteristics of a typical ZnO-MESFET.

Analyzer connected to a *SÜSS semi-automatic wafer-prober* equipped with tungsten probes. All samples were measured (if not differently assigned) under dark conditions in ambient air and at room temperature. From the output characteristic, the pinch-off and saturation behavior of the MESFET can be evaluated. The presented transfer characteristics were recorded for the highest applied source–drain voltage V_{SD} to ensure saturation. From that, the MESFET's on/off-ratio I_{on}/I_{off} and the threshold voltage V_T can be extracted. Note, that the MESFET's V_T differs from the threshold voltage, which is commonly used for MISFET. In the case of MISFET, the threshold voltage is often extracted from the linear extrapolation of the $I_{SD}^{1/2}$ using the equation

$$I_{SD} = \mu C_i \frac{W}{L} \frac{(V_G - V_T)^2}{2},$$

where
  μ is the field-effect mobility
  C_i is the insulator's capacitance
  W/L is the channel's width-to-length ratio

Since for thin-film MISFET and MESFET in particular, this behavior is not ideally linear, we decided to use V_T as shown in Figure 11.3, being an off-voltage. This is the more appropriate model parameter.

The MESFET's channel mobility can be determined either from the output characteristic for $V_{SD} \to 0$ using the drain transconductance g_{D0} or from the transfer characteristic in the saturation regime ($I_{SD} = I_{SD,sat}$) using the saturation forward transconductance $g_{m,sat}$. The relation between both transconductances is

$$g_{D0} = \left. \frac{\partial I_{SD}}{\partial V_{SD}} \right|_{V_{SD} \to 0} = g_{max} \left[1 - \left(\frac{V_{bi} + V_G}{V_P} \right)^{1/2} \right] = \frac{\partial I_{SD,sat}}{\partial V_G} = g_{m,sat},$$

where
  V_{bi} is the gate-diodes built-in voltage
  V_P is the pinch-off voltage

From that, the maximum transconductance g_{max} can be extracted and the channel mobility is then given by

$$\mu_{ch} = \frac{g_{max}}{q(N_D - N_A)d(W / L)}.$$

The values for the net doping concentration $N_D - N_A$ are obtained from quasi-static capacitance-voltage (QSCV) measurements, d is the channel thickness, and W/L is the channel's width-to-length ratio.

11.2.3 Optimization of the Device Performance Influence of the Gate Material

The gate material is the most critical part of the MESFET. In order to obtain MESFET with high on/off-ratios and channel mobilities, but with low threshold voltages, gate materials with a high Schottky barrier height are needed. This can be for example noble metals or in our approach non-insulating metal oxides. The effect of different Schottky-gate materials on the output and transfer characteristics of ZnO-based MESFET has been investigated. Ag, Pt, Pd, and Au were chosen for this purpose [31]. As described in Section 11.2.1, the reactively sputtered metals were partially oxidized, except Au [28]. The so formed metal oxides exhibited Schottky barrier heights as high as 0.95, 0.90, 0.79, and 0.69 V for Ag, Pt, Pd, and Au, respectively, on the thin ZnO channel layer. These values are higher or among the highest reported for Schottky contacts on ZnO [8,32,33]. The Au contact's barrier height is almost as large as the highest reported barrier of 0.71 V [34]. For Pt, a similar range between 0.89 and 0.93 V on hydrogen peroxide treated ZnO was reported [35]. Reactively sputtered Ag was previously reported by Allen et al. [29,30] who observed slightly higher Schottky barrier heights but ideality factors close to unity. The ideality factors for the MESFET samples with Ag, Pt, Pd, and Au were 1.7, 2.04, 1.53, and 2.37, respectively. However, the Schottky contacts in Allen's case were fabricated on ZnO single crystals and exhibited ideality factors close to unity.

Figure 11.4 depicts the influence of the various gate materials on the ZnO-MESFET characteristics. The measured samples were grown under nominally identical growth conditions and exhibited similar channel thicknesses between 20 and 26 nm and net doping concentrations between 1.5×10^{18} and 2.8×10^{18} cm^{-3}. The output characteristics show clear saturation and pinch-off behavior. However, for Pd and Au a gate voltage higher than 0.8 V leads to a shift of the individual output curve toward higher source–drain voltages. In this case, increasing gate leakage currents overcome the source–drain currents. The Schottky gate is in flatband condition. This behavior directly reflects the lower Schottky barriers for Pd and Au. All MESFET are normally-on, having a conductive channel at zero gate voltage. In the on state, they reach source–drain currents of several 10 μA corresponding to 10^{-2} A/cm^2.

The comparison of transfer characteristics (Figure 11.4a) shows a strong field effect for all MESFET. The highest on/off-ratio was achieved for the Ag-gate MESFET exceeding 10^8. Here, the lowest off-current density was in the range of 10^{-9} A/cm^2 and a threshold voltage of $V_T = -1.4$ V was obtained. For Pt, Pd, and Au, gate leakage currents affect the off-current. Nevertheless, the Pt-MESFET still achieved an on/off-ratio of 4.5×10^6, whereas it was 2.6×10^5 for Pd and 1.6×10^3 for Au. The differences in the characteristics are explainable by the slightly different channel thicknesses and first and foremost by the different Schottky diode's barrier heights.

Table 11.2 summarizes the measured MESFET properties for the various gate materials. The channel mobilities for Ag, Pt, and Pd are in good agreement with the Hall-effect mobilities. This is expected from the MESFET theory, because due to the missing insulator, scattering mechanisms at the insulator/semiconductor interface are negligible and the channel mobility equals the Hall-effect mobility of the respective semiconductor [24]. This is, however, not generally valid. The channel mobility can be over- or underestimated under the presence of parasitic effects such as excessive gate currents. This can be seen in the case of Au (Table 11.2), where the channel mobility $\mu_{ch} = 23.9$ cm^2/(V s) is significantly lower than the Hall-effect mobility $\mu_{Hall} = 43.9$ cm^2/(V s). Here, the negative gate leakage current, which was also responsible for the shift of the output characteristic (Figure 11.4e), lowers the channel's forward transconductance and thus μ_{ch} is underestimated. Note, that the obtained Hall-effect mobilities of the MESFET channel layers are generally lower than typical mobilities of (~1 μm thick) ZnO thin films grown by PLD [37]. For the thinner films (~20–30 nm), this is due to the larger influence of a low-mobility layer at the ZnO/substrate interface, higher defect densities and smaller grains [38].

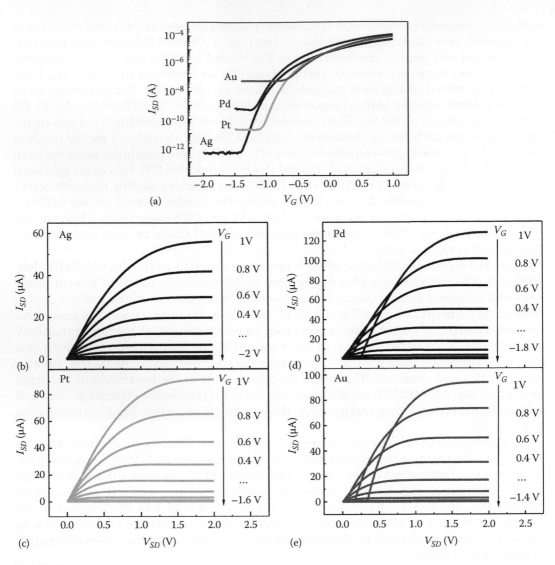

FIGURE 11.4
Comparison of various Schottky-gate materials at room temperature. (a) Transfer characteristics. (b–e) Output characteristics. (From Frenzel, H., ZnO-based metal-semiconductor field-effect transistors, Dissertation, Universität Leipzig, Der Andere Verlag, Tönning, Germany, 2010.)

TABLE 11.2

Overview of the Measured MESFET Properties

Gate Metal	d (nm)	$N_D - N_A$ $(10^{18}\,cm^{-3})$	μ_{Hall} (cm²/ (V s))	μ_{ch} (cm²/ (V s))	On/ Off-Ratio	S (mV/ Decade)
Ag	20	2.8	9.3	11.3	1.2×10^8	81
Pt	26	2.4	12.1	11.4	4.5×10^6	83
Pd	26	2.4	12.1	12.8	2.6×10^5	156
Au	20	1.5	43.9	23.9	1.6×10^3	282

Typically, in the vicinity of the MESFET's threshold voltage, the slope S of the transfer characteristics is exponentially dependent on the gate voltage. For low-power applications, this is an important figure of merit as it affects the switching speed, for example, of inverters. Therefore, S has to be as small as possible. It can be seen from the values in Table 11.2, that for Ag and Pt, S is already close to the theoretical minimum for FET operating at room temperature: $S_{min} = \ln(10)(kT/q) \approx 60\,mV/decade$ with k being the Boltzmann constant, g the elementary charge, and T the absolute temperature. This model is based on electron diffusion through the depletion region [39] and is valid for MESFET as well as depletion-mode MISFET. For the slope S, in both cases, an ideality factor N_S has to be added to S_{min}, which considers the leakage currents for MESFET [40] or insulator capacitances for MISFET [41]. S is minimal for Ag and Pt, because the off-current is low. Close to the threshold voltage, where the channel is almost fully closed and the pinch-off point is moving toward the source contact, the electron injection into the channel is limited by the potential barrier of the depletion region. This leads to the exponential dependence. The injected electrons diffuse through the depletion region to the drain contact. For higher gate voltages, the electrons drift through the conductive part of the channel and S is increasing. For positive V_G, the increasing gate leakage current affects the transfer characteristic. Thus, as for Pd and Au, higher slopes are obtained. Comparing the MESFET's slopes with MISFET from the literature, it can be seen that they are among the best values for oxide FET. A MISFET based on indium oxide and organic pentacene insulator was reported to have as slope of $S = 90\,mV/decade$ [23]. In other reports, where amorphous oxide semiconductors (AOSs) were used, slopes between 108 and 180 mV/decade were reported [42–45].

As the previous results show, Ag-gate MESFET exhibit the highest on/off-ratio and the steepest slopes among oxide-based transistors and are therefore best suited for low-power, mobile applications. The presented MESFET have on-currents in the range of several 10–100 μA. This is sufficient for most applications such as integrated circuits (inverters, voltage amplifiers). For display applications, first and foremost for active-matrix organic-light-emitting-diode (AMOLED) displays, which is a current driven device, a higher on-current in the range of milliamperes is necessary [21]. The on-current can be increased according to the proportionality

$$I_{SD} \propto e\mu_{ch}(N_D - N_A)\frac{dW}{L}$$

either by increasing (a) the channel mobility, (b) the net doping concentration and the channel thickness, or (c) the gate's width-to-length ratio. These possibilities are discussed in the following sections.

11.2.3.1 Variation of the Net Doping Concentration and Channel Thickness

The influence of the net doping concentration was investigated by means of four different sample types: S1 consists of a thick (131 nm) nominally undoped ZnO layer directly grown on the sapphire substrate, S2 is a comparably thick (117 nm) undoped ZnO layer grown on a MgO buffer layer, S3 is a thin (20 nm) undoped ZnO layer on sapphire (identical to the Ag sample in Table 11.3), and S4 is a thin (32 nm) ZnO layer on MgO buffer [8]. The MgO buffer layer is needed to avoid the diffusion of Al from the sapphire substrate into the ZnO during PLD growth. As expected, the net doping concentration of the samples with MgO buffer layer was drastically reduced to $1.5 \times 10^{14}\,cm^{-3}$ for S2; $N_D - N_A$ for S4 was even

TABLE 11.3

Overview of the Measured MESFET Properties

Sample	d (nm)	$N_D - N_A$ (10^{18} cm^{-3})	μ_{Hall} (cm^2/ (V s))	μ_{ch} (cm^2/ (V s))	On/ Off-Ratio	S (mV/ Decade)
S1	131	0.5	15.4	19.1	—	—
S2	117	1.5×10^{-4}	29.4	27.4	4×10^4	432
S3	20	2.8	9.3	11.3	1.2×10^8	81
S4	32	$<10^{-5}$	9.7	—	—	—

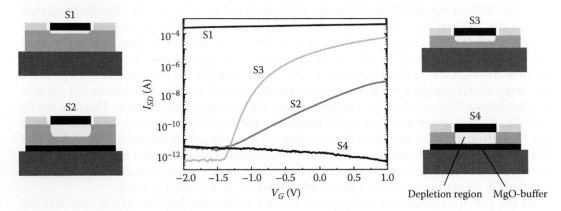

FIGURE 11.5

Transfer characteristics of ZnO-MESFET with various channel doping concentrations and thicknesses. The sketches illustrate the individual sample configuration with (S2, S4) and without (S1, S3) MgO buffer layer and with thick (S1, S2) and thin (S3, S4) channel layer.

not measurable. With reduced $N_D - N_A$, it is necessary to increase the channel thickness in order to ensure that the depletion region can be modulated within sufficient gate voltages. Figure 11.5 compares the transfer characteristic of the four MESFET samples. Only the samples S2 and S3 show a strong field effect. The on/off-ratio of S3 is >10^8 and it can be tuned on and off within a gate-voltage range of only 2.5 V. On the other hand, the current of S2 can only be tuned four decades. Both samples exhibit normally-on behavior having equal threshold voltages of $V_T = -1.5$ V. In contrast to that, S1 can be denoted "static-on" because it did not show any significant field effect within the applied voltage range. The modulation of the depletion region for the obtained carrier concentration of 5×10^{17} cm^{-3} is too small with respect to the channel thickness of S1. The output characteristics of this MESFET (not shown) were linear and did not show any pinch-off or saturation behavior. Compared to S1, S4 can be denoted as "static-off." The channel is insulating and no modulation of the source–drain current could be observed. Nevertheless, for S1, S2, and S3, the channel mobility μ_{ch} was determined. Having linear output characteristics, μ_{ch} for S1 was obtained from the drain transconductance for $V_{SD} \rightarrow 0$. For S2 and S3, μ_{ch} was calculated using the forward transconductance of the transfer characteristic in the saturation regime. The obtained channel mobilities are again in good agreement with the Hall-effect mobilities of the unstructured channel layers. It is noticeable, that both thicker channels have higher mobilities than the two thinner channels. This can be understood, since the dislocation density is larger close to the substrate interface. Therefore, it is favorable for later designs to use a ZnO-MESFET with a thicker channel layer but lower net doping concentration. Thus, higher channel mobilities can be achieved. However, the on/off-ratio of such kind of MESFET has to be improved.

11.2.3.2 Geometry Effects

The previous investigations suggest, that for the used PLD-grown ZnO channels on sapphire, the parameters of the sample S3 ($d \sim 30\,nm$, $N_D - N_A \sim 10^{17} - 10^{18}\,cm^{-3}$, no buffer layer) are most appropriate for MESFET performance. However, for the use as pixel current-driver transistors in AMOLED displays, the on-current has to be increased by 2 orders of magnitude [21]. This cannot be achieved by using higher doping concentrations, since they are already close to the critical Mott concentration of $8.3 \times 10^{18}\,cm^{-3}$ for ZnO [46], where the semiconductor behaves metallic and the expansion of a depletion region is not possible. Therefore, the on-current is increased using MESFET with interdigitated source/drain contacts and meander-shaped gate contacts (Figure 11.6c). The gate length was kept constant at $10\,\mu m$, whereas the width was varied such that the W/L ratio was in the range between $W/L = 10.75$ and $W/L = 700$. Figure 11.6a depicts transfer characteristics of various ZnO MESFET with different W/L ratios. The on-current was increased to a maximum value of 15 mA and an on/off-ratio of 5×10^8 was achieved with $W/L = 700$. It can further be seen from Figure 11.6b that the transconductance, obtained from the forward source–drain current, scales with the W/L ratio as expected from MESFET theory. The channel mobility stays constant at a mean value of $5.8\,cm^2/V\,s$ up to $W/L = 67.5$. For higher W/L ratios, μ_{ch} is lower than the value for standard geometry due to the increasing influence of gate currents.

Another option to increase the device's on-current and switching speed is to increase its channel mobility. This can be done using two ways: improving the crystal quality and using two-dimensional electron gases (2DEG).

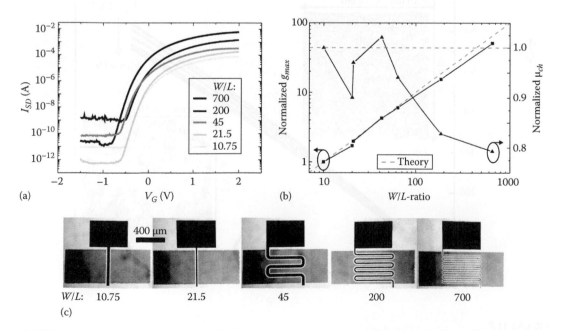

(a)

(b)

(c)

FIGURE 11.6
(a) Transfer curves, (b) W/L-scaling, and (c) optical photographs of ZnO-MESFET for different W/L ratios.

The crystal quality is improved in ZnO thin films that were homoepitaxially grown on ZnO single-crystal substrates [47]. To investigate this, two kinds of samples were fabricated. One sample is a ZnO channel, directly grown on the ZnO substrate. For the other sample, a MgZnO buffer layer was grown between the substrate and the channel in order to reduce impurity diffusion. The MgZnO buffer layer reduces the surface roughness as well as the grain boundaries density of the above grown ZnO channel. The measured properties of these MESFET are summarized in Table 11.4.

It can be seen, that the ZnO channel without buffer layer is more conductive than the one with MgZnO buffer layer, indicating, that the MgZnO is an effective diffusion barrier for impurities from the substrate. Due to the ZnO substrate being not sufficiently insulating, there are higher leakage currents flowing through the substrate. For the same reason, MESFET functionality was only given for the MESFET with buffer layer (Figure 11.7).

A pinch-off or saturation behavior was not distinctive in the output characteristics. The applied gate-voltage between +1 and −3 V did only result in a small modulation of the source–drain current. Although, a variation of the source–drain current was not possible for the sample without buffer layer, the channel mobility μ_{ch} has been determined for both

TABLE 11.4

Overview of Measured Properties for Homoepitaxial ZnO MESFET

Sample	d (nm)	$N_D - N_A$ (10^{18} cm^{-3})	μ_{Hall} (cm^2/(V s))	μ_{ch} (cm^2/(V s))
With buffer layer	~50	0.6	90	18
Without buffer layer	~50	3	150	50

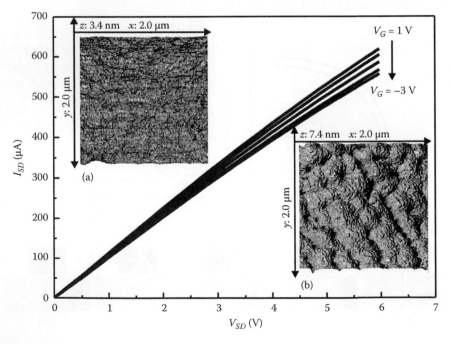

FIGURE 11.7

Output characteristics of MESFET with homoepitaxial channel layer and MgZnO buffer. (a) AFM-graph of the surface of the channel with MgZnO buffer. (b) AFM-graph of the surface of the channel without MgZnO buffer.

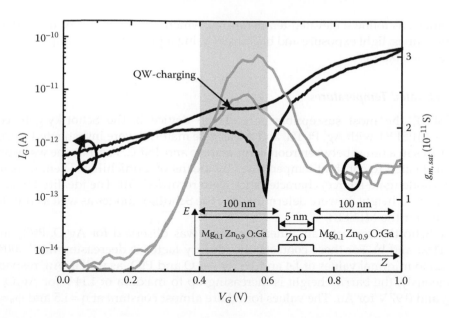

FIGURE 11.8
Current–voltage measurement of the Schottky-gate diode and forward transconductance in the saturation regime of a MESFET with MgZnO/ZnO/MgZnO QW channel. Inset: Scheme of the used QW structure.

samples from the drain transconductance in the linear regime. The sample without buffer layer has the higher crystal quality and thus exhibits a higher channel and Hall-effect mobility. The value of $\mu_{ch} = 50\,cm^2/(V\,s)$ is the highest obtained for ZnO-MESFET so far. However, the Hall-effect mobility was obtained to be three times higher. This is most probably due to ionized impurity scattering, as the fitting of temperature-dependent Hall-effect measurements with a corresponding model (not shown) revealed. The comparably low mobilities for the sample with buffer layer are probably due to higher compensation effects and additional scattering at the ZnO/MgZnO interface. The discrepancy between the mobilities is due to the leakage currents flowing through the ZnO substrate. This results in insufficient saturation and underestimated channel mobilities.

The other approach to increase the channel mobility is the use of a 2DEG. For that, a MgZnO/ZnO/MgZnO quantum well (QW) heterostructure was fabricated and MESFET were processed [9]. The 5-nm thick ZnO-QW was embedded in 100nm thick Ga-doped MgZnO barrier layers. As can be seen in Figure 11.8, charging of the QW has been observed by current–voltage measurements at a voltage of around 0.5 V. The transfer measurements of the corresponding MESFET revealed a sharp increase of the forward transconductance in the same voltage range (Figure 11.8). This implies electron transport through the QW, where the channel mobility is higher than in the surrounding barrier, due to less ionized impurity scattering. However, the source–drain current is rather small due to low carrier injection through the MgZnO:Ga barrier. For that reason, the determination of the channel mobility was not possible. However, the general principle of a ZnO MESFET with QW-channel was proven.

11.2.4 Reliability and Degradation

In later applications the ZnO-based MESFET are exposed to circumstances under which they have to work without a significant loss of performance. Such circumstances are, among others, as follows: elevated temperatures, which can occur, for example, in cars or near to machines; light exposure and bias stress being a problem, for example, in displays; and long-term stability.

11.2.4.1 Elevated Temperatures

For MESFET, the most susceptible part of the device is the Schottky-gate contact. Therefore, MESFET with Ag, Pt, Pd, and Au as gate material were investigated at elevated temperatures in a range between room temperature and 150°C. For that, the waferprober's chuck was heated to constant temperatures by means of a unichiller. Then, transfer and Schottky-diode (gate-source) characteristics were recorded [31]. The ideality factor η and effective barrier height ϕ_B were determined for the Schottky diodes as well as the channel mobility μ_{ch} for the MESFET.

For the Schottky contacts, an annealing effect was observed for Ag_xO, PtO_y, and Au (Figure 11.9a and b). For these contacts, the ideality factor is decreasing until 100°C (for Au: 125°C) to minimal values of 1.4 and 1.6 for Ag_xO and PtO_y as well as Au, respectively. Simultaneously, the barrier height is increasing up to maxima of 1.14 V for Ag_xO, 1.07 V for PtO_y, and 0.97 V for Au. The values for Pd are almost constant at $\eta = 1.5$ and $\phi_B = 0.82$ V

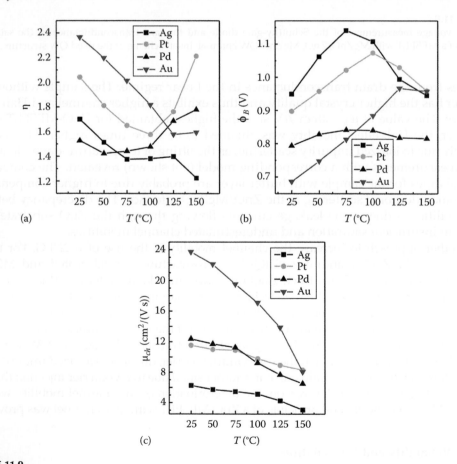

FIGURE 11.9
(a) Ideality factor, (b) Schottky-barrier height, and (c) channel mobility of various metal-gate MESFET at elevated temperatures.

in the temperature range between 25°C and 100°C. An increase of the temperature above 100°C or 125°C leads to increasing ideality factors and decreasing barrier heights. Cooling the devices to room temperature again did not show a recovery of the initial characteristics. The Schottky contacts were irreversibly degraded. This degradation was also observed in the transfer characteristics (not shown), where the off-current of the devices increased with increasing temperatures due to leakage currents flowing over the gate. Figure 11.9c depicts the obtained channel mobilities for elevated temperatures. Here, no annealing effect was observed. Instead, all mobilities were monotonically decreasing with increasing temperatures. This can be explained by an increasing scattering rate at lattice vibrations and by gate-leakage defects. Unfortunately, μ_{ch} decreases more drastically for Au than for all other materials although it showed the best annealing effect in the Schottky characteristics. This is probably due to the *a priori* much higher gate leakage currents leading to a faster degradation of the MESFET. Although the Ag_xO-MESFET showed the smallest channel mobility, its ideality factor and Schottky barrier height and with that the on/off-ratio were the best. Therefore, it was chosen as standard material for further investigations. However, PtO_y should be considered as alternative, because it shows a similar performance as Ag_xO.

11.2.4.2 Bias Stress and Light Exposure

Bias stress measurements were performed on Ag_xO-gated MESFET with an initial on/off-ratio of 1.0×10^8, a channel mobility of $8.3\,cm^2/(V\,s)$, and an off-voltage of $-1.27\,V$ (Figure 11.10b). For MISFET, having an insulating gate, a shift of the threshold voltage can occur during bias stress due to charging of traps at the interface between the gate insulator and semiconductor channel [26]. This is a serious problem of the MISFET technology, because the shifts (that can be as large as several volts) have to be compensated by sophisticated circuits. Without such circuits, for example, a pixel in a display driven by MISFET would get darker the longer the operation time. Therefore, such a shift of V_T has to be minimized. MESFET do not have an insulating gate where traps can be accumulated and charged or discharged. Thus, it is predicted, that the threshold-voltage shift during bias stress is much lower than for MISFET.

To minimize the influence of ambient light in this experiment, the MESFET sample was kept in darkness for 1 week before measurement. The bias stress measurement was performed under dark conditions in ambient air and room temperature. The procedure was as follows: First, a positive gate voltage of $V_G = 1\,V$ and a maximum drain voltage of $V_{SD} = 2\,V$ were applied to the MESFET for 22 h. In this case, the depletion region under the gate contact is minimal and the maximal saturation current of $I_{SD,sat} = 100\,\mu A$ flows through the channel. Transfer characteristics (Figure 11.10a) were recorded in a time interval of 1000 s. Figure 11.10b depicts the off-current and apparent channel mobility. It was observed, that the transfer characteristics of the MESFET did not change within the first 22 h of positive gate stress, that is, the threshold-voltage shift was zero and the off-voltage as well as the off-current stayed constant. The channel mobility decreases slightly from 8.3 to $8.0\,cm^2/(V\,s)$. This behavior indicates a secure long-term performance of the MESFET under high-power conditions. In a second measurement, negative-gate stress ($V_G = -1\,V$, $V_{SD} = 2\,V$) was applied for 22 h. Here, the channel is almost fully closed and a small source–drain current of 1 nA flows. Under these conditions, the off-current decreases immediately by one decade resulting in an on/off-ratio of 10^9, whereas the channel mobility drops to a value in the range of $3.0\,cm^2/(V\,s)$. After that, both values remain constant during the rest of the stress time. Third, a 150 W metal halide lamp was turned on for a few seconds

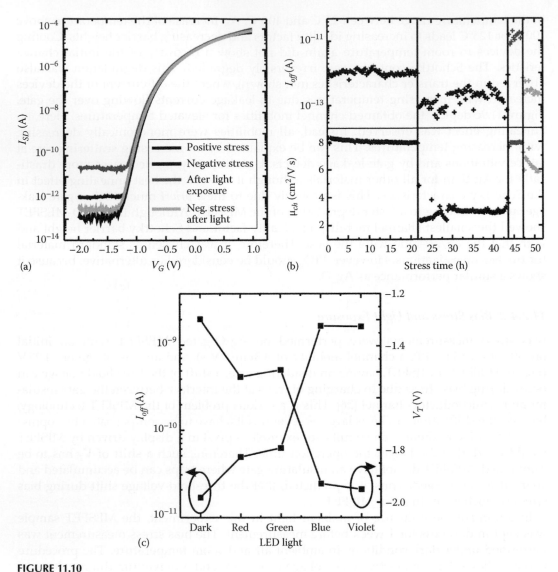

FIGURE 11.10
(a) Transfer characteristics of ZnO-MESFET under bias stress, (b) corresponding time-behavior of the channel mobility and off-current, and (c) light-dependence of ZnO-MESFET.

in order to recharge traps that might be emptied during negative voltage stress. Then, the MESFET was again measured in darkness but without any applied stress voltage for 170 min. During this measurement, the off-current is 10 times higher than the initial value, whereas the channel mobility is again at 8.0 cm²/(V s). This observation indicates a persistent photoconductivity effect. Finally, negative gate-voltage stress ($V_G = -1\,V$, $V_{SD} = 2\,V$) was applied again, which restored the on/off-ratio to a mean value of 3.0×10^8. The channel mobility is monotonically decreasing to a value of 5.8 cm²/(V s) within 250 min.

The effect of persistent photoconductivity during the bias stress measurements can be explained as follows. For the positive gate-voltage, the Schottky-gate contact is in flat-band condition and the depletion-layer depth is zero. No change in the MESFET conductivity is observed because the applied electric source–drain field of ~0.3 kV/cm is not strong enough to charge or discharge a corresponding trap in the channel. Not until a negative voltage is applied at the gate, when the effect of lifting trap states above the Fermi-energy level is strong enough to charge or discharge a trap. The off-current as well as the apparent channel mobility of the MESFET is reduced. After recharging the traps by means of white light, the MESFET characteristics stay constant for the following 170 min in darkness, where no voltage was applied. By applying negative gate-voltage stress, the channel conductivity begins to decrease again.

Apparently, the influence of light plays an important role for the electrical performance of ZnO-based MESFET. Particularly for display applications, the impact of visible light irradiation on the MESFET parameters is interesting. To investigate this behavior in more detail, the MESFET were exposed to light of different-colored LEDs with a wavelength of 625 nm (red), 525 nm (green), 470 nm (blue), and 425 nm (violet), respectively. The colors as well as the intensity of all LEDs (1 mW/cm^2) were chosen according to the standards of the National Television Systems Committee (NTSC). In the first 80 min, the sample was kept in darkness; transfer characteristics were recorded in time intervals of 500 s. During this time, the characteristics remained unchanged and exhibited an off-current of $I_{off}=1.7\times10^{-11}$ A and a threshold voltage of $V_T=-1.3$ V. The channel mobility was $\mu_{ch}=7.8$ cm^2/V s. For the next 130 min, the sample was exposed to red light (Figure 11.10c). The off-current slightly increased to $I_{off}=4.8\times10^{-11}$ A and the threshold voltage shifted to $V_T=-1.52$ V, immediately before it stayed constant for the rest of the stress time. Also under green-light exposure, the values remained in the same range. For the blue and violet light, however, I_{off} and V_T changed significantly to 1.7×10^{-9} A and -1.95 V, respectively. This is due to photo-generated charge carriers, which reduce the expansion of the depletion region. The increasing off-current and decreasing threshold voltage indicate an increasing net doping concentration N_D-N_A. The probability of transitions increases exponentially with increasing photon energy. Note that all light-induced changes in the transfer characteristics of the MESFET were completely reversible and no persistent photoconductivity was observed in this sample after the light was turned off. This is contrary to the observations that were made on ZTO MISFET [25] under the illumination with visible light. There, under irradiation with violet light, threshold-voltage shifts between 2 and 20 V were observed depending on the processing temperature and the zinc/tin composition. The off-voltage shift in MESFET with about 0.65 V is much smaller. For MESFET, a charge trapping in the dielectric plays no role since there is no insulator. However, the small threshold-voltage shift can be attributed to the creation of defects in the channel volume material or at the surface of the channel.

11.2.4.3 Long-Term Stability

The long-term stability was investigated for MESFET comprising a ZnO channel as well as a MgZnO channel with 0.25% Mg content. A comparison of the long-term stability is shown in Figure 11.11. The samples were both equally fabricated without any encapsulation and stored in an exsiccator at room temperature and under ambient air and light. Note, that an encapsulation of the Schottky diodes, for example, with CaHfO$_x$, can significantly improve their performance [48]. Transfer and Schottky characteristics were recorded in intervals of several weeks. The measurements at the ZnO-MESFET showed (Figure 11.11a) that the channel mobility and on/off-ratio only remained at their initial values $\mu_{ch}=27$ cm^2/(V s) and 1.3×10^6 for the first 40 days, before both parameters start to decrease rapidly and saturated after 120 days at values in the range of $\mu_{ch}\sim5$ cm^2/(V s) and $I_{on}/I_{off}\sim10^4$. The channel mobility and on/off-ratio of the MgZnO-MESFET (Figure 11.11c) started at lower initial values $\mu_{ch}=10$ cm^2/(V s) and $I_{on}/I_{off}=2\times10^4$. However, both values stay constant over

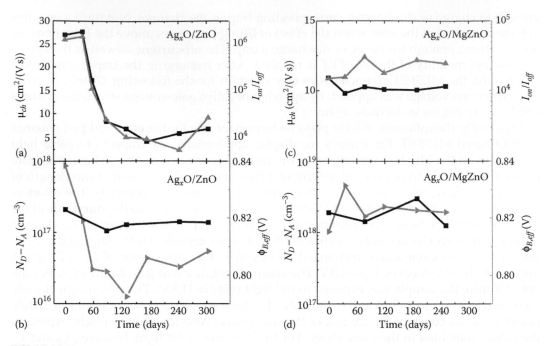

FIGURE 11.11
Long-term stability of (a, c) the channel mobility and on/off-ratio of a ZnO (a) and MgZnO (c) MESFET as well as (b, d) the Schottky barrier height of the gate and net doping concentration of a ZnO (b) and MgZnO (d) MESFET.

the whole period of 250 days. It can be concluded from measurements of the Schottky gates (Figure 11.11b and d) that the net doping concentration is constant for both samples. This implies that a diffusion of Ag from the gate contact does not lead to compensation and is therefore not responsible for this effect [49]. Instead, it can be seen that the Schottky barrier height decreases for the ZnO-MESFET, whereas it stays constant for the MgZnO-MESFET. Here, the decrease of the Schottky barrier height by $\Delta\phi_{B,eff}=0.04\,\mathrm{V}$ is consistent with the obtained decrease of the channel mobility by a factor of 5. This degradation of the Schottky gates is more pronounced for the pure ZnO-channel than for the MgZnO-channel.

11.2.5 ZnO-MESFET on Glass Substrates

Up to now, the discussed ZnO-MESFET were fabricated on sapphire substrate, on which ZnO thin films grow with good crystalline quality with *c*-axis orientation and with a comparatively low density of grain boundaries. However, for that, high-temperature growth methods are needed. On the other hand, for the use of the ZnO-MESFET technology in future low-cost transparent electronics, it is of great interest to fabricate those devices at lower temperatures and on cheaper and more easily scalable glass substrates. In a first step, we compare PLD-grown MESFET with the standard parameters for sapphire substrates with those grown on three different glass substrates: Quartz, Corning 1737, and Corning Eagle XG [32]. In Section 11.5, ZnO-MESFET are investigated, which were grown on glass substrates at room temperature by means of magnetron sputtering.

The used glass substrates consist either of pure silicon oxide in the case of Quartz or alkaline earth boro-aluminosilicates in the case of the two Corning glass substrates. The latter two are commonly used for the fabrication of flat-panel displays. Since glasses are very good insulators and provide no Al-diffusion as sapphire does, the ZnO channel was intentionally doped using a PLD target composed of ZnO with 0.01 wt% Al_2O_3. Using this target for the thin films on glass, the oxygen partial pressure during PLD-growth had to be reduced from 0.02 to 3×10^{-4} mbar in order to obtain adequate crystalline quality and highest channel mobilities. A comparison of atomic force microscopic (AFM) measurements of the ZnO channel surface on sapphire and glass substrates showed differences in the morphology (Figure 11.12). The surface texture of the ZnO on quartz and the borosilicates are similar. On glass, a Volmer–Weber island growth [50] is observed; that is, the interaction between atoms of the film on the surface is larger than the interaction between atoms of the film and the substrate. On the other hand, on sapphire, an epitaxial relation between film and substrate exists. A mixed mechanism between Franck–van der Merwe film growth [50] and island growth occurs. The grain sizes on the glass substrates lie in the range between 30 and 50 nm. The roughnesses of the channel surface on the glass substrates are between 1.28 and 1.46 nm. This is larger compared to sapphire (0.78 nm). Due to the amorphous substrates, the mosaicity of the ZnO thin film on the glass substrates is lower than for the film on sapphire.

Figure 11.12 depicts the transfer characteristics of ZnO-MESFET on glass substrates compared to the reference sample on sapphire. All MESFET on glass show normally-off behavior with threshold voltages of 0 V for quartz, +170 mV for Corning 1737, and +340 mV for Corning Eagle XG. With increasing threshold voltage, the on/off ratio decreases simultaneously. This is due to a reduction of the net doping concentration $N_D - N_A$ from 10^{18} cm^{-3} to the range of 10^{17} cm^{-3} (cf. Table 11.5). Due to the structural quality of the thin ZnO films on glass, Hall-effect measurements did not show a conclusive result. Thus, the channel mobilities could not be compared to the Hall-effect mobilities. As expected, μ_{ch} is highest

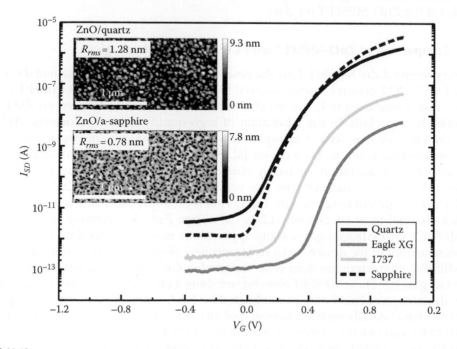

FIGURE 11.12

Transfer characteristics of ZnO-MESFET on various glass substrates compared to sapphire. Inset: AFM-graphs of the channel surface on quartz and sapphire.

TABLE 11.5

Comparison of MESFET Properties on Various Glass Substrates and Sapphire

Substrate	d (nm)	$N_D - N_A$ (10^{18} cm^{-3})	V_T (mV)	μ_{ch} (cm^2/(V s))	On/Off-Ratio (10^5)	S (mV/Decade)
Quartz	30	0.57	0	1.30	4.7	125
C. 1737	25	0.18	170	0.22	2.3	101
C. Eagle XG	24	0.13	340	0.04	0.8	104
Sapphire	15	0.13	0	25.0	26.9	77

for the MESFET on sapphire substrate (25 cm^2/(V s)). It is dramatically reduced for quartz glass substrate (1.3 cm^2/(V s)) and decreases by another one and two orders of magnitude for Corning 1737 (0.2 cm^2/(V s)) and Eagle XG (0.04 cm^2/(V s)), respectively. Apparently, the decreasing on-current for the increasing threshold voltage is responsible for that. Also the on/off-ratios of the MESFET are reduced achieving only the range of 10^5. The lower on-current is caused on the one hand by higher currents over the gate electrode, which lead to a lower forward transconductance and subsequently to a lower μ_{ch}. On the other hand, the electron mobilities are generally reduced on amorphous substrates due to scattering at grain boundaries.

In order to verify the presence of compensating defects, room temperature admittance spectroscopy (AS) on identical ZnO-Schottky diodes were performed [9]. These measurements indicated a large amount of deep defects in the ZnO films on glass. Additionally, secondary ion mass spectroscopy time-of-flight (SIMS-tof) measurements showed an accumulation of the elements Li, Na, K, and B on the surface of the glass substrates, which are the most probable origin of compensating defects leading to the low conductivity (and mobility) of the ZnO-MESFET on glass.

11.2.6 Comparison of ZnO-MESFET with Competing Technologies

The here presented ZnO-MESFET are the best reported so far. A comparison of these ZnO-MESFET with PLD-grown channel material to other MESFET and ZnO-MISFET is given in Table 11.6. There exist only a few other reports about MESFET based on ZnO. Their difficulties lie especially in the fabrication of high-quality Schottky contacts. Although *p*-type ZnO is very rare, Ryu et al. reported MESFET on *p*-type ZnO using Ni as ohmic and Ti as Schottky contacts, respectively [4]. However, their output characteristics show that pinch-off and saturation are barely obtained for large source–drain voltages in the range of 10 V before breakdown. The gate voltage had to be as large as 20 V to switch the MESFET over a current range of only 1 mA. This is most probably due to a high conducting thick channel layer (cf. Section 11.2.3). On *n*-type ZnO, Kandasamy et al. reported a ZnO-MESFET for hydrogen gas-sensing applications using Pt Schottky contacts [5]. The saturation current of the device had a strong positive slope, which does not correspond to ideal MESFET characteristics. Kao et al. reported a ZnO-MESFET using Pt/Au Schottky contacts as gate [6]. Their MESFET also did not show a clear pinch-off and saturation. The source–drain current could only be modulated by 4 mA within a gate voltage range of 4 V. Further electrical details such as channel mobility, on/off-ratio, etc. were not reported for ZnO-MESFET and can therefore not be compared to our MESFET.

MESFET and related logic devices have also been demonstrated on the basis of ZnO nanorods [7]. Metal-organic vapor-phase epitaxy (MOVPE) was used to grow

TABLE 11.6

Comparison of Device Parameters for ZnO-Transistors

Channel	Type	V_T (V)	V_{SD} (V)	On/Off-Ratio (10^6)	μ_{ch} (cm²/(V s))	S (mV/Decade)	References
ZnO	MESFET	−1.4	4	130	11.3	81	This work [8]
ZnO	MESFET	0	2	2.7	25	77	This work [32]
p-ZnO	MESFET	2.5	−10	10^{-5}	—	—	[4]
ZnO	MESFET	—	6	—	—	—	[5]
ZnO	MESFET	−5	2	$<10^{-5}$	—	—	[6]
ZnO	MISFET	-10^3	500	<10	68	—	[51]
ZnO	MISFET	−13	20	<0.1	40	—	[52]
ZnO	MISFET	<−10	10	10^{-5}	5.3	—	[6]
ZnO-QW	MISFET	−7	10	10^{-3}	140	770	[53]
ZnO	MISFET	−7.2	5	$<10^{-3}$	62	—	[54]
ZnO	MISFET	−32	10	100	35	940	[55]
ZnO	MISFET	−15	0.3	$<10^{-2}$	12	—	[56]
ZnO	MISFET	−12	40	10	0.4	—	[57]
ZnO	MISFET	0	5	0.1	0.03	—	[58]
ZnO	MISFET	15	20	0.2	20	1240	[59]
ZnO	MISFET	0	10	0.1	1.7	—	[60]
ZnO	MISFET	−5	10	10	8.4	950	[61]
ZnO	MISFET	15	40	0.9	8	900	[62]
ZnO	MISFET	−10	40	50	11	—	[63]
ZnO	MISFET	−1	7	10^{-3}	0.75	1270	[64]
ZnO	MISFET	<−40	40	10^{-2}	0.36	—	[65]
ZnO	MISFET	−1	20	>100	12.9	—	[66]

single-crystalline ZnO nanorods. These were dispersed on SiO_2 before Au-Schottky contacts were fabricated on them using electron-beam lithography to process MESFET, OR, AND, NOT, and NOR gates, respectively. The nanorod MESFET exhibit a gate-width-normalized transconductance of 20 µS/µm (compared to 4590 µS/µm for the here presented MESFET) and a minimal slope between 100 and 200 mV/decade (compared to 81 mV/decade).

Comparing the ZnO-MESFET presented in this work to other reported ZnO-MISFET shows the advantages of MESFET technology. For MISFET, there is often a tradeoff between applied voltage range, on/off-ratio, and channel mobility. All MISFET exhibit larger, partially much larger voltages than MESFET. However, many of them exhibit larger channel mobilities but their on/off-ratio is only 10^3 or 10^5, respectively.

11.2.7 Switching Behavior of ZnO-MESFET

Usual applications for MESFET, especially based on GaAs, are high frequency or high current tasks. In principle, ZnO has a similar potential for oxide semiconductors due to its comparatively high carrier mobility. However the current growth techniques usually lead to polycrystalline or amorphous films.

A challenge in MESFET technology is the delay of the transistor switching due to traps at the surface, at interface layers or in the bulk material, which limits the working range in the frequency domain [67–70]. Also the charging and discharging of traps in a semi-insulating substrate is a common delay mechanism [71]. When ZnO is grown on insulators like sapphire or glass, this effect can be neglected. However, in a polycrystalline film traps at grain boundaries must be taken into account, as first principle calculations indicate the formation of deep levels at the boundaries [72]. Also impurities are likely present in the layer. These elements can originate from the target material or diffuse in the material from neighboring device parts.

Various measurement techniques are available for the characterization of the switching behavior of transistors [73,74]. One is the transient current analysis, which uses square pulsed voltages to record the current response of the device. It can be distinguished between gate lag and drain lag measurement, depending on whether the gate or the drain voltage is pulsed. The typical current response for both cases is depicted in Figure 11.13. The transients emerge from trapping and detrapping effects of charge carriers. These depend on the electric field at the trap sites, which varies locally in the transistor. Examples for mechanisms leading to such field dependent emission rates are the Poole–Frenkel effect, phonon assisted tunneling, or direct tunneling between the trap states and the conduction band [75]. The measured transient will be a superposition of transients originating from various processes on different locations in the device. The delayed switching of the transistor also causes a continuous change of the electrical potential at the trap sites, despite step-like voltage changes at the gate or drain contacts. This will also lead to a non-exponential transient. The measured currents are usually described by a stretched exponential function instead, which has the form $I(t) = I_0 + \Delta I \cdot \exp(-(t/\tau)^\beta)$, with β characterizing the deviation from the pure exponential function. Surface states can also cause so-called virtual gating, when the charge accumulation at the channel surface induces an additional depletion region. The described mechanisms cause the transient fit parameters like the time constant τ to be dependent on the applied voltages. Thus it not possible to correlate the quantities obtained by the transient current analysis directly to trapping rates and activation energies known from, for example, deep level transient spectroscopy. They can, however, be a valuable tool

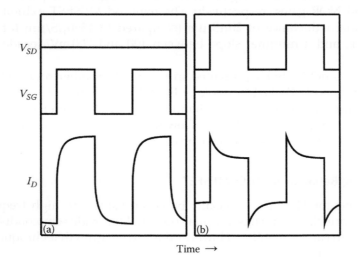

FIGURE 11.13
Schematic depiction of (a) gate lag and (b) drain lag.

when comparing transistors with different geometries, materials or testing the effect of passivation layers.

The frequency dependence of these effects can be measured using sine-shaped voltages while measuring the current amplitude. This method gives fewer details about the trapping processes than transient analysis, but can show whether processes on different timescales are involved.

For ZnO-based MESFET the typical drain and gate lag behavior has been measured [76] (Figure 11.14). It is particularly present in the case of transistors with Ag_xO gates. With the gate materials PtO_y and Au much lower transient amplitudes were observed. In many cases no drain lag at all was observed, which limits either the time constants to <1 ms or the transient amplitudes to <100 nA at an overall drain current around 10–100 µA. Typical average time constants $\langle\tau\rangle = \tau \cdot \Gamma(1 + 1/\beta)$ are between 5 and 50 ms, depending on the measurement conditions as well as on the sample. So far there exists no model to describe these variations.

The significant difference between devices with Ag_xO on the one hand and PtO_y and Au on the other can be attributed to the diffusion of Ag into the channel during the Schottky contact fabrication. This is supported by capacitance measurements and SIMS depth profiles performed on Schottky contacts, which showed diffusion of Ag into the ZnO connected with a decrease in the contact capacitance [49].

To clarify the existence of further trapping mechanisms, frequency dependent measurements can be performed. In Figure 11.15 measurements for two different gate voltage patterns are shown, the first represents the small signal case ($\Delta V_{SG} \sim 0.1$ V) and the second the large signal case ($\Delta V_{SG} \sim 3$ V). Again the different behavior of Ag_xO-gated FET is apparent, as the amplitude drops with rising frequency, while the other two devices show no current decrease. The decline of the currents above 200 kHz is caused by the measurement setup, which has been verified by calibration measurements with ohmic resistors instead of the MESFET. An example for these is shown with the circles in the Figure 11.15.

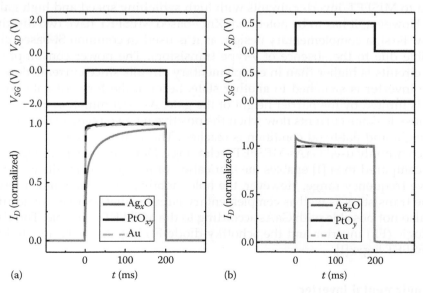

FIGURE 11.14

Gate lag (a) and drain lag (b) measurements on ZnO-MESFET with Ag_xO, PtO_y and Au gates.

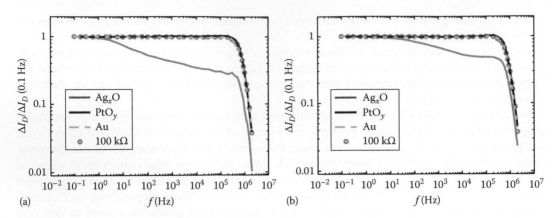

FIGURE 11.15
Small signal ($\Delta V_{SG} \sim 0.1\,\text{V}$) (a) and large signal ($\Delta V_{SG} \sim 3\,\text{V}$) (b) frequency-dependent measurements on ZnO-MESFET.

11.3 High-Gain ZnO-Based Inverters

To demonstrate an application for ZnO-MESFET, inverter circuits were designed and fabricated. The inverter (or NOT gate) is the basic logic function within the Boolean algebra [77]. The logic elements "true" (1) and "false" (0) are realized by a high voltage level and a low voltage level. The output of an inverter is only "true" (1, high voltage), when the input is "false" (0, low voltage) and vice versa. Further logic functions such as AND, OR, NAND, NOR can be linked together to implement more complex functions.

Using ZnO-based MESFET with their higher mobility and lower operation voltages compared to MISFET, inverter circuits with high switching speed and high gain can be realized. However, it has to be noted that ZnO-based MESFET have two disadvantages up to now. First, a complementary design, as it is used in common Si-based circuits, is not possible due to the absence of *p*-type transistors. The power consumption of the unipolar circuits is higher than in complementary circuits since currents flow not only when the inverter is switched to another state. Second, the forward gate voltage of a MESFET is limited by the barrier height of the gate. An accumulation mode is not possible because leakage currents flow when the positive gate voltage exceeds the Schottky barrier height and flat-band condition is reached. These disadvantages are also known from the commonly used GaAs-MESFET technology. Here, the five times higher electron mobility compared to Si [1] enables the fabrication of integrated circuits working in the microwave frequency range. However, the hole mobility in GaAs is much lower, making *p*-type transistors as well as complementary circuits unattractive. An accumulation mode is also not possible with GaAs according to the MESFET principle. The circuits for the FET-logic (FL) inverter and the Schottky-diode FL inverter were adopted from the GaAs-MESFET logic [1].

11.3.1 Fundamental Inverter

Figure 11.16 depicts the circuit and a photograph of a fundamental MESFET inverter consisting of two identical normally-on MESFET connected in series. Both MESFET have

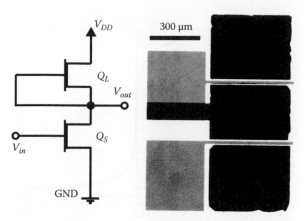

FIGURE 11.16
Circuit and photograph of a fundamental MESFET inverter.

equal width-to-length ratios of $W/L = 21.5$. The source contact of the bottom transistor Q_S (switching transistor) is grounded, whereas the source of the top transistor Q_L (load transistor) is short-circuited with its gate. Thus, the gate voltage of Q_L is constant at $V_{G,L} = 0\,V$. The load transistor is operated either in the linear or the saturation regime, depending on the applied operating voltage V_{DD}. The gate voltage of the switching transistor Q_S denotes the input voltage V_{in} of the inverter. With that, Q_S can be turned on and off. The inverter switches from high-state to low-state when the resistances of Q_L and Q_S are equal. The output voltage V_{out} of the inverter is connected to the common drain respective source/gate contact between the two MESFET.

The characteristic V_{out} over V_{in} is denoted as voltage transfer curve (VTC) of the inverter. It serves as quantification of the implementation of the inverter function as integrated circuit. Its figures of merit are defined as follows.

In the ideal case, the resistances of the transistors are infinite when they are switched off, and equal to zero in the on-state. Consequently, the operating voltage would be equal to the output high level ($V_{OH} = V_{DD}$) and the output low level would be zero ($V_{OL} = 0\,V$). Furthermore, the ideal inverter's switching point is at $V_{SP} = V_{DD}/2$ and the slope of the VTC at this point would be infinite. The inverter's state is only uncertain at this switching point. A comparison between the ideal VTC and a typical real VTC is depicted in Figure 11.17. Due to the finite resistances of the inverter's transistors, the switching of the real inverter takes place in a transition range. The slope of this range equals the gain g of the inverter if it was operated as voltage amplifier:

$$g = \left| \frac{\partial V_{out}}{\partial V_{in}} \right|.$$

The maximum of the gain is denoted as peak gain magnitude (*pgm*). This value is achieved at the switching point of the inverter. The *pgm* is recommended to be as large as possible to achieve maximum amplification and switching speed.

Different from the ideal inverter, the state of the real inverter is not uncertain in only one point (switching point) but in a transition range. The borders of this range are limited by the input low and high voltages V_{IL} and V_{IH}, where $g = 1$. V_{IL} is the maximal input voltage for which the inverter's output level is securely high and V_{IH} is the minimal input voltage

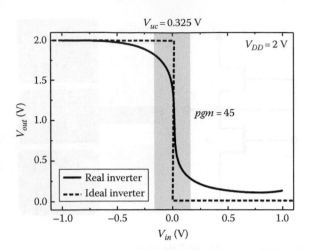

FIGURE 11.17
Voltage–transfer curve of an ideal and a typical real fundamental inverter (FI) based on ZnO-MESFET.

where the output is securely low. The difference and with that the width of the transition range is also called uncertainty voltage:

$$V_{uc} = V_{IH} - V_{IL}.$$

The uncertainty voltage should be as small as possible to ensure that the inverter is in an unambiguous state for a wide voltage range.

The VTC, shown in Figure 11.17, refers to a fundamental inverter (FI) measurement obtained for $V_{DD} = 2\,V$. The inverting function is obvious: For the negative input voltage $V_{in} = -1\,V$, the output is high, whereas it is low for positive V_{in}. The output high level $V_{OH} = 1.999\,V$ is only 1 mV below the value for the ideal case: $V_{OH} = V_{DD} = 2\,V$. The output low level $V_{OL} = 0.111\,V$ is achieved for $V_{in} = 0.78\,V$. For higher input voltages the output voltage starts to increase again, which can be attributed to excessive gate-currents of the MESFET Q_S. As described in Section 11.2.2, these currents lead to a shift of the output characteristics toward higher V_{SD}.

The switching point of $V_{SP} = 0.015\,V$ is close to the theoretical value $V_{SP} = 0\,V$. However, it is positive as it is for all measured MESFET inverters within this work. This systematic deviation can be explained by a higher saturation current at Q_L compared to Q_S, such that for the switching, higher V_{in} have to be applied to level out the transistors conductivities. Probably, contact resistances and/or parasitic capacitances, due to a trapping of charge carriers at the metal-oxide gate, are responsible for this effect. Consequently, the switching point for inverters with Au gates lies at $V_{SP} = 0\,V$.

The uncertainty voltage $V_{uc} = 0.325\,V$ is only a factor of three higher than for the GaAs-MESFET technology ($V_{uc} \sim 0.1\,V$ [1]). A value of $pgm = 45$ was obtained for this inverter at $V_{DD} = 2\,V$.

The measurement of pgm was performed by interpolating the switching region of the VTC with a cubic spline and numerical differentiation. Note, that the differentiation of the raw VTC data (measured with 5 mV steps) would lead to an underestimation of pgm. The value would depend on the measurement steps.

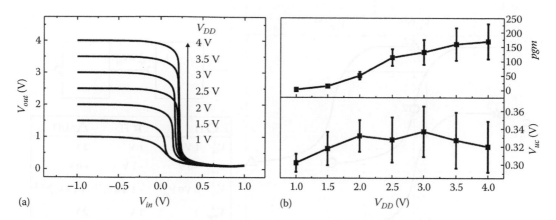

FIGURE 11.18

(a) VTC of a fundamental ZnO-MESFET inverter for various operation voltages V_{DD}. (b) pgm and V_{uc} as function of VDD.

As can be seen in Figure 11.18a, the VTC strongly depends on V_{DD}. For all V_{DD} between 1 and 4 V, the respective V_{OH} is reached, whereas V_{OL} remains constant. A shift of the switching point from $V_{SP}=0$ V to $V_{SP}=0.015$ V is observed between $V_{DD}=1.5$ V and $V_{DD}=2$ V. This is a further hint for activated charges, possibly located at the gate/channel interface.

The VTC is stretched for higher operating voltages. This implies that the steepness and with that the pgm increases, whereas V_{uc} remains constant (Figure 11.18b). The error bars in Figure 11.18b denote the standard derivation of the mean values obtained from 25 inverters on one sample. As can be seen, pgm monotonically increases from $pgm=2.5$ for $V_{DD}=1$ V to a maximum mean value of $pgm=170$ for $V_{DD}=4$ V. Individual inverters even reached a maximum gain of $pgm=250$ at $V_{DD}=4$ V. The uncertainty voltage remains in the range between $V_{uc}=0.3$ V and $V_{uc}=0.34$ V.

Since both resistances of the MESFET in the FI are equal, the switching point is ideally located at $V_{SP}=0$ V. Furthermore, in the presented inverters, normally-on MESFET are used. This leads to a problem if the inverters are intended to be used in integrated circuits. The output voltage of the shown FIs is always positive. In order to obtain a high output voltage (logic 1) from an in series connected second inverter, the first inverter must address a negative voltage to the input of the second inverter. If V_{in} is smaller than the threshold voltage of Q_S, it is turned off, and $V_{out}=V_{DD}$. For positive V_{in} the low output state of the inverter is achieved. Note that if a second inverter is connected in series at the output of the first inverter and if the first inverter's output is low, that is, near zero, the input of the subsequent inverter is within the uncertainty voltage range and its output is therefore insecure. This constellation can for example be problematic for the realization of ring oscillators, where an odd number of inverters are connected in series. Therefore, the VTC needs to be shifted by a level-shifter (LS) either toward higher input voltages or toward negative output voltages. This is schematically shown in Figure 11.19. Three possible ways to implement a level-shifting into the inverter circuit are described in the following sections: the FL inverter, the Schottky-diode FET-logic (SDFL) inverter and the FI with different load and switching transistor.

11.3.2 FET-Logic Inverter

The FL inverter was first developed by van Tuyl and Liechti in 1974 for the GaAs-MESFET technology [3]. They used additional Schottky diodes and a transistor denoted as pull-down

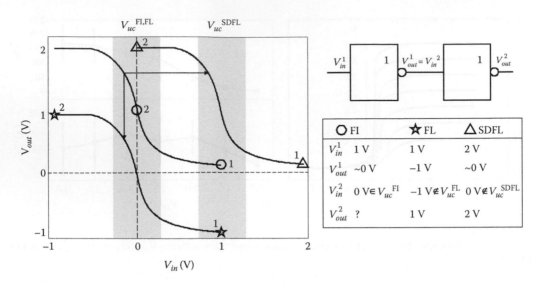

FIGURE 11.19
Schematic shifting of the VTC to ensure unambiguous complementary working points of two inverters in series. The table summarizes such working points for three different inverter types.

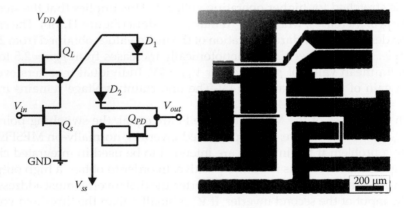

FIGURE 11.20
Circuit design and optical photograph of a ZnO-MESFET-based FL inverter. The width-to-length ratios of Q_S/Q_L and Q_{PD} are $W/L = 10.5$ and $W/L = 5.25$, respectively.

transistor Q_{PD} to adjust the voltage levels. Figure 11.20 depicts a photograph and the circuit design of the ZnO-MESFET-based FL inverter investigated in this work. Two Schottky diodes D_1 and D_2 as well as the pull-down transistor Q_{PD} are connected in series at the output of the FI. The level-shifting is achieved by the voltage drop across the two diodes. Q_{PD} serves as constant current supply for the diodes. In order to ensure that Q_{PD} is in saturation, a negative voltage V_{SS} ($V_{SS} > V_{OL} + |V_T|$) has to be applied. Additionally, the current that flows through Q_{PD} discharges a parasitic load capacity for the low output level of the inverter. The current through Q_L is the sum of the currents through Q_S and Q_{PD} ($I_L = I_S + I_{PD}$). Because, the gate width-to-length ratio of Q_{PD} is half of Q_S and Q_L, the saturation current is also half. Thus, the output characteristic of Q_L is shifted toward negative source–drain currents (therefore the name pull-down transistor) and with that, the VTC of the FL inverter

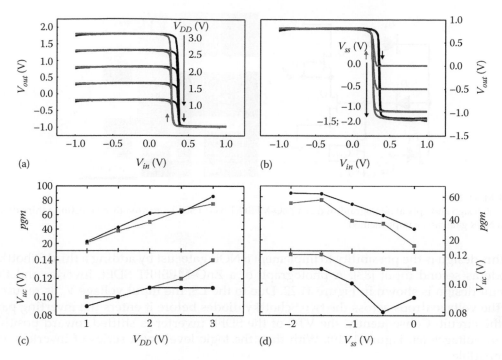

FIGURE 11.21
Voltage-transfer characteristics of an FL inverter (input–voltage cycling). (a, c) VTC, *pgm* and V_{uc} as function of V_{DD} (V_{SS}=−1 V) and (b, d) as a function of V_{SS} (V_{DD}=2 V).

is shifted toward negative output voltages (cf. Figure 11.19). The measurement of a ZnO-FL inverter in dependence of V_{DD} and V_{SS} is shown in Figure 11.21. The VTC is indeed shifted by 1.2 V, which corresponds exactly to the sum of the voltage drops across the two serial ZnO Schottky diodes. Loop measurements, that is, V_{in} swept in a loop from negative voltages to positive voltages and back, were performed. They show a small hysteresis, where the switching point is slightly lower in backward than in forward direction. This implies the influence of interface charges at the load transistor. However, the voltage difference is only $\Delta V \sim 0.09$ V. The level-shifting depends on the applied operation voltage V_{SS} for Q_{PD}. It is however limited to the Schottky diode's voltage drops. The VTC shifts down to V_{out}=−1.2 V down to V_{SS}=−1.5 V and remains constant for lower pull-down voltages. As for increasing V_{DD}, the *pgm* also increases with increasing (negative) V_{SS} (Figure 11.21). It saturates for V_{SS}=−1.5 V and V_{SS}=−2 V, when the level-shifting is limited. The highest gain for the FL inverter is *pgm*=80 at V_{DD}=3 V and V_{SS}=−2 V, which is slightly smaller than for the FI. However, the uncertainty level is significantly reduced with respect to the FI. It lies in a range between V_{uc}=0.08 V and V_{uc}=0.14 V, and shows a small increase for increasing V_{DD} or V_{SS}. Due to the low V_{uc} (which results in less measurement points within the switching region), the determination of the *pgm* as described in Section 11.3.1 probably leads to an underestimation of this value.

11.3.3 Schottky-Diode FET-Logic Inverters

When the LS is connected at the input of the FI, the design is denoted as SDFL inverter. It was first developed by Welch and Eden in the late 1970s [78–82]. A prominent advantage

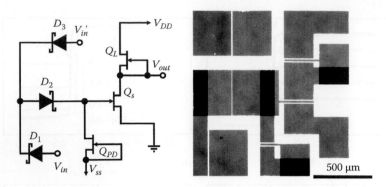

FIGURE 11.22
Circuit design and optical photograph and of a ZnO-MESFET-based SDFL inverter. By connecting a third diode D_3, a NOR gate can be implemented.

of this design is the possibility to implement a NOR gate just by adding a third Schottky diode as second input [83]. A photograph of a ZnO-MESFET SDFL inverter and the circuit design is shown in Figure 11.22. Due to the LS, the input voltage V_{in} is reduced by the voltage drops across the two Schottky diodes before it enters the inverting part of the circuit. Consequently, the VTC of the SDFL inverter is shifted toward positive input voltages (cf. Figure 11.19). With that, the logic levels of a series of inverters are compatible.

Based on the fact that MESFET with MgZnO channel are more stable, reproducible, and reliable than ZnO-MESFET (cf. Section 11.2.4), SDFL inverters with ZnO- and MgZnO-MESFET are compared. Both samples were fabricated with equal process parameters. They exhibit comparable net doping concentrations of $3.4 \times 10^{17} \, cm^{-3}$ (MgZnO) and $2.5 \times 10^{17} \, cm^{-3}$ (ZnO). Note that the yield of working inverters was 90% for MgZnO, whereas it was only 60% for ZnO. Figure 11.23 shows the VTC, corresponding transfer characteristics of Q_S as well as the *pgm* and V_{uc} in dependence of V_{DD} for both samples. Due to the smaller channel thickness of 21 nm for MgZnO compared to 28 nm for ZnO, the main difference between both samples is that the behavior of the switching transistor Q_S is nearly normally-off ($V_T = -0.25 \, V$) for MgZnO and normally-on ($V_T = -1 \, V$) for ZnO. It can be seen, that the performance of the SDFL inverters with V_T closer to 0 V is improved, that is, higher *pgm* and lower V_{uc} are achieved. This can be explained either by higher currents flowing through the normally-on MESFET at $V_G = 0 \, V$ and by the higher slope of Q_S around $V_G = 0 \, V$ for the MgZnO inverter. In this case, a smaller change of the input voltage is adequate to equalize the resistances of Q_S and Q_L.

The VTC for ZnO-SDFL inverter are comparable to that of the FI. Both SDFL inverters achieve the operating voltage ($V_{DD} = 2 \, V$) as high output level. The low output levels are $V_{OL} \sim 0.16 \, V$ and $V_{OL} \sim 0.02 \, V$ for ZnO and MgZnO, respectively. The VTC for MgZnO-SDFL inverter are therefore much closer to the ideal inverter characteristic.

Using the LS with one diode (D_2), the VTC are shifted by $\Delta V_{in} = 0.77 \, V$ for ZnO and $\Delta V_{in} = 0.62 \, V$ for MgZnO. Using two Schottky-diodes ($D_1 + D_2$), the shift is $\Delta V_{in} = 1.50 \, V$ (ZnO) and $\Delta V_{in} = 1.36 \, V$ (MgZnO). Thus, using the two-diode SDFL inverter, adequate logic levels would be for example $(V_{in} | V_{out}) = (2 \, V | 0 \, V)$ and $(0 \, V | 2 \, V)$ (cf. Figure 11.19).

The V_{uc} and *pgm* strongly depends on the applied operating voltage V_{DD} and the use of the level shifter (Figure 11.23). Without level shifter, the ZnO inverter's *pgm* increases by more than 9000% from *pgm* = 1.5 at $V_{DD} = 1 \, V$ to *pgm* = 141 for $V_{DD} = 3 \, V$. The MgZnO

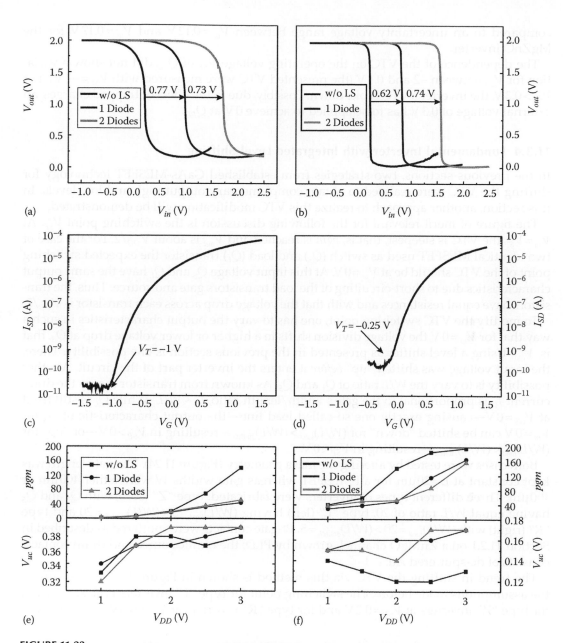

FIGURE 11.23
(a, b) VTC for the SDFL inverters on (a) ZnO and (b) MgZnO. (c, d) Corresponding transfer characteristics of Q_S. (e, f) *pgm* and V_{uc}.

inverter's *pgm* without level shifter increases by almost 500% from *pgm* = 40 at V_{DD} = 1 V to *pgm* = 197 at V_{DD} = 3 V. However, adding the level shifter leads to a significant decrease in the *pgm* of both inverters. The maximum *pgm* values are only *pgm* = 40 and *pgm* = 165 for ZnO and MgZnO, respectively. The decrease can be explained by additional resistances and leakage currents at the Schottky diodes and the pull-down transistor Q_{PD}. The advantage of MgZnO versus ZnO can also be seen in the uncertainty voltages. V_{uc} is drastically higher for the ZnO inverters having values between V_{uc} = 0.32 V and V_{uc} = 0.38 V, compared to an uncertainty voltage range between $V_{uc}=0.12\,V$ and $V_{uc}=0.17\,V$ for the MgZnO inverter.

The dependence of the VTC on the operating voltage V_{SS} of Q_{PD} did not show a variation for V_{SS} between –2 and 0.3 V (the presented VTC were measured with $V_{SS}=-1\,V$). For $V_{SS}>0.3\,V$, the inverter's VTC broke down. Possibly, due to the effect of trapped charges, an external voltage of 0.3 V has to be applied to achieve 0 V at Q_{PD}.

11.3.4 Fundamental Inverter with Integrated Level-Shifting

In the previous sections, two strategies from established GaAs-MESFET technology for shifting the VTC were realized to obtain complementary output high and low levels. In this section, another approach to realize this VTC modification will be demonstrated.

The figure of merit relevant for the following discussion is the switching point V_{SP}. At $V_{in}=V_{SP}$ the VTC is steepest, that is, pgm is reached and V_{out} is about $V_{DD}/2$. For the case of two identical MESFET used as switch (Q_S) and load (Q_L) transistor the expected switching point of the VTC should be at $V_{in}=0\,V$. At this input voltage Q_S and Q_L have the same output characteristics due to short-circuiting of the load transistors gate and source. Thus, the transistors have equal resistances and with that the voltage drop across each transistor is $V_{DD}/2$.

To modify the VTC switching point, one has to vary the output characteristics in such a way that for $V_{in}=0\,V$ the voltage division shifts to a higher or lower voltage drop at Q_S, that is, V_{out}. Using a level shifter as presented in the previous section is one possibility. There, the input voltage was shifted "up" *before* it enters the inverter part of the circuit. Another possibility is to vary the W/L ratio of Q_L and Q_S. As known from transistor theory the drain current I_D is proportional to the gate's width/length ratio, $I_D \sim W/L$. While Q_L is still fixed at $V_{GS}=0\,V$—causing exactly one so-called load line—the output characteristic of Q_S for $V_{in}=0\,V$ can be shifted "down" for $(W/L)_{load}>(W/L)_{switch}$—resulting in $V_{SP}>0\,V$—or "up" for $(W/L)_{load}<(W/L)_{switch}$—resulting in $V_{SP}<0\,V$.

Both cases were tested by altering the mesa geometry (Figure 11.24). The gate length was kept constant at $L=20\,\mu m$ for all MESFET whereas gate widths W were either 400 μm or 100 μm. Three different types of inverters were fabricated: type "Z" (zero) with Q_L and Q_S having equal W/L ratio of 20, type "L" (left) having $(W/L)_{load}=5<(W/L)_{switch}=20$ and type "R" (right) with $(W/L)_{load}=20>(W/L)_{switch}=5$. Pt-gate contacts were sputtered as described in Section 11.2.1 on a ZnMgO channel, grown by PLD; the ohmic source and drain contacts consists of dc-sputtered Au.

The trend in shifting the VTC via this method is shown in Figure 11.25 and confirms the assumptions made before. The switching point for type "L" inverters is at $V_{SP}=-0.2\,V$, for type "Z" inverters at $V_{SP}=0.2\,V$ and for type "R" inverters at $V_{SP}=0.8\,V$. An additional

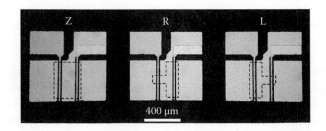

FIGURE 11.24
Photograph of the MESFET inverter with different W/L ratios: (Z) $(W/L)_L=(W/L)_S=20$, (R) $(W/L)_L=20>(W/L)_S=5$ and (L) $(W/L)_L=5<(W/L)_S=20$. The channel mesa is highlighted with a dashed line.

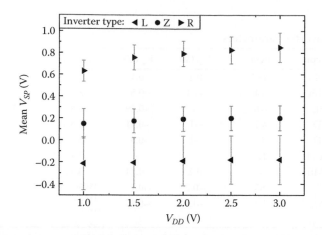

FIGURE 11.25
Mean switching point V_{SP} for various supply voltages V_{DD} and three differently scaled inverters: type L $(W/L)_{load}/(W/L)_{switch}=5/20$, type Z $(W/L)_{load}/(W/L)_{switch}=20/20$, type R $(W/L)_{load}/(W/L)_{switch}=20/5$. Error bars represent the standard deviation of all measured circuits.

result is the different absolute value of the V_{SP} shift relative to type "Z": about 0.4 V for type "L" and 0.6 V for type "R." This can be explained with the relation between the transistor saturation current $I_{D,sat}$ and the gate voltage which is $I_{D,sat} \sim (V_G - V_T)^2$. The larger the $(V_{in} - V_T)$ the larger is the effect on output characteristics and therefore on the VTC. With decreasing V_{DD} the transistors approach the nonlinear regime causing a slight decrease in switching point V_{SP} as Figure 11.25 points out, too. Furthermore, one can see a deviation from expected $V_{SP}=0$ V for type "Z" inverters which is probably due to charge traps located at the gate's metal oxide/semiconductor interface.

Note, that for the scaling concept, an unfavorable increase of the output voltage for high input voltages $(V_{in}>1.2$ V) was observed, which is due to excessive gate currents at the switching transistor for large positive voltages, where the gate diode is strongly in forward direction. Nevertheless it is clearly shown that W/L scaling of MESFET inverter circuits is an alternative possibility to modify the VTC in a predictable way. Advantages in comparison to SDFL are less device area, four instead of five voltage connectors, and obviously the simplicity of the circuitry.

11.3.5 Comparison of ZnO-MESFET-Based Inverters with Other Technologies

Inverters based on MESFET-technology have been demonstrated on a variety of material systems; the most prominent is GaAs. In Table 11.7, various inverter parameters for different materials and types are compared. It is noticeable that the peak gain magnitudes for the respective inverters are not as high as for the MgZnO-MESFET inverter, whereas the uncertainty voltages are in a similar range. For GaAs and AlGaN/GaN high-electron-mobility-transistor (HEMT)-based inverters, the switching speed (which is determined by means of ring oscillators) is of higher importance than gain and uncertainty voltage. The highest gain of 83 at $V_{DD}=5$ V exhibited a fundamental inverter (FI) consisting of CdS nanowires with In/Au [85]. Also a complementary MESFET (CMES) inverter with n-type CdS and p-type Zn_3P_2 nanowires achieved a *pgm* of 20. The complementary technology, which is also reported for Si-MESFET inverters [86] has the advantage of low power dissipation. The static power dissipation of the reported CdS/Zn_3P_2-nanowire inverter, normalized with the switching transistor's gate area, is $P_D=85.7$ pW/μm^2 at $V_{DD}=1$ V. This is,

TABLE 11.7

Comparison of Inverter Parameters

Channel	Type	*pgm*	V_{SP} (V)	V_{uc} (V)	V_{DD} (V)	V_{SS} (V)	References
MgZnO	SDFL	197	0.2	0.13	3	−1	[83]
GaAs	SDFL	~4	~1.1	~0.5	2.5	−1.2	[84]
GaAs	BFL	~6	0	~0.29	2.5	−1.5	[87]
AlGaN/GaN	E-D	4.2	~1	~0.41	2.5	—	[88]
AlGaN/GaN	E-D	4.9	~1	~0.5	3	—	[89]
Diamond	DCFL	1.6	—	—	−2.4	—	[90]
Si	CMES	~5	0.2	~0.14	0.4	—	[86]
CdS-NW	FI	83	~0.2	~0.1	5	—	[85]
nCdS/ p-Zn$_3$P$_2$	CMES	20	−0.5	~0.1	1	—	[91]

SDFL: Schottky-diode FET-logic, BFL: buffered FET-logic, E-D: enhancement-depletion-mode inverter, CMES: complementary MESFET inverter, FI: fundamental inverter.

however, of the same order as that of the MgZnO-based SDFL inverter presented in the present work, where the power dissipation at $V_{DD}=3$ V and $V_{SS}=-1$ V was $P_D=111$ pW/μm^2 for the low-output and $P_D=46$ pW/μm^2 for high-output. The power dissipation is higher for the ZnO-SDFL inverter presented in this work ($P_D=476$ pW/μm^2 for the low-output and $P_D=5000$ pW/μm^2 for the high-output). This is due to one and two orders of magnitude higher currents through the MESFET. However, these values are still low compared to GaAs inverters (34–45 μW/μm^2 [81]) or AlGaN/GaN-HEMT inverters (444 μW/μm^2 [89]).

This comparison is valid for opaque MESFET, only. Note, that the presented ZnO-MESFET-based inverters show the best reported *pgm* and V_{uc} when they are transparent. A comparison of device parameters of transparent inverters (MISFET or MESFET) is separately given in Table 11.10.

11.4 Transparent Rectifying Contacts

With the help of the −afore-presented opaque MESFET and inverters, their distinct advantages with respect to MISFET were shown. However, the broad spectrum of applications in transparent electronics will not be offered until the MESFET are realized fully transparent. This was achieved with the development of transparent rectifying contacts (TRC) that are used as gate contacts [92,93].

11.4.1 Preparation

The samples for the transparent MESFET and inverters were fabricated by PLD as described in Section 11.2.1 using a ZnO target with a small amount of Mg (0.25%). Two pieces of a 50×50 mm^2 sapphire wafer with a nominally 30 nm thick nominally undoped MgZnO film were used. The transparent devices were processed using standard photolithography and lift-off techniques. The ohmic source and drain contacts were deposited either by dc-sputtering of very thin Au under an Ar atmosphere for the inverters or by growing a highly conducting layer of ZnO doped with 3 wt% Al by PLD at room temperature and an oxygen partial pressure of 0.016 mbar for the MESFET. The specific contact resistivity of this layer obtained by a transmission line method [94] was in the range of $10^{-4}\,\Omega\,cm^2$. For the fabrication of the TRC, a modification of the two-step dc-sputtering process for opaque MESFET was used [93]. First, an ultrathin layer of partially oxidized Ag or Pt was reactively sputtered in a mixture of 50% Ar and 50% O_2 at a power of 5W for 30s. With that, the thickness of the metal-oxide layer is reduced to 5nm and the contacts are transparent. The small thickness and oxidation of the first layer leads to an increased sheet resistance. In order to ensure an equipotential surface, a current spreading layer consisting of Au (on Ag_xO) or Pt (on PtO_y) was non-reactively sputtered under a pure Ar atmosphere. This second step is even more important for the ultrathin metal-oxide layers than for the thick opaque layers. The overall thickness of the contact structure is only about 10nm. Nevertheless, scanning force microscopy revealed that the contact layers are closed and smooth with an rms roughness of 1.5nm.

The combination of an ultrathin, rectifying metal-oxide layer with an equally thin conducting capping layer leads to high transparency of the contact under perpetuation of its rectifying function. In later designs, the metallic capping layer can be substituted with a TCO which would increase the transparency significantly. The mean transmission (Figure 11.26) of the complete device structure (substrate, MgZnO channel and TRC) with Ag_xO and PtO_y as TRC in the visible spectral range (400–800nm) is 70% and 60%, respectively. The maximum transmittance for Ag_xO is 74% at 740nm. However, for the green and blue spectral range, Ag_xO shows a decreasing transmission, which is due to its fundamental absorption edge at (2.3 ± 0.2) eV, which was derived from the onset of the absorption coefficient obtained by spectroscopic ellipsometry measurements. Both samples show an absorption edge at 375nm due to the underlying ZnO layer.

11.4.2 Transparent Rectifying Contacts for Visible-Blind Ultraviolet Photodiodes Based on ZnO

With the reduction of the opacity of the Schottky diodes on ZnO, a multitude of new applications are feasible. Aside from using the diode as gate electrode in transistors and circuits constructed out of such devices, the diodes itself can already be applied as rectifiers and detectors for ultra-violet light. Table 11.8 contains a compilation of the structure and the performance parameters of ZnO-based photodetectors, revealing that only very few of them, up to now, are fully transparent ($T > 70\%$).

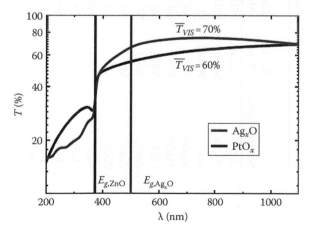

FIGURE 11.26
Transmission spectra of TRC structures comprising Ag_xO or PtO_y.

TABLE 11.8

Overview of ZnO-Based Photodetectors

Semiconductor	Device Type	Electrode Material	External Quantum Efficiency	Responsivity A/W	Bias V	UV–Vis Ratio	Fully Transparent	References
ZnO	Schottky	PEDOT:PSS	~1	0.3	0	1,000	×	[95]
MgZnO	Schottky	PEDOT:PSS	>0.9	0.3	0	>100	×	[96]
ZnO	Schottky	Pt	0.63	0.185	—	—		[97]
MgZnO	MSM	Au	—	0.0001	10	1,000		[98]
MgZnO	MSM	Au	0.0054	0.0012	2	>1,000		[99]
ZnO	MSM	Ag	—	1.5	5	>1,000		[100]
ZnO	MSM	Ni	0.24	0.09	1	>100		[101]
ZnO	MSM	Au	—	30	3	10,000		[102]
ZnO	MSM	Au	—	160	10	—		[103]
MgZnO	MSM	Au/Cr	—	0.37	5	—		[104]
MgZnO	Schottky	Au	—	173	-1.5	>10,0000		[105]
MgZnO	MSM	Au/Cr	—	1200	5	>10,000		[106]
ZnO	MSM	Al	—	400	5	>100		[107]
ITO/ZnO/NiO	Nip	—	—	—	-5	>10	×	[108]
MgZnO	Schottky barrier	Pt	0.17	0.034	0	>1,000		[109]

By pulsed-laser deposition (PLD), an about 200-nm thick Al-doped ZnO film, serving as backside contact [33] and subsequently a 1 µm thick ZnO film were deposited on an as received circular 2 in. a-plane sapphire substrate ($p_{O_2} = 0.02$ mbar, $T_{Growth} = 670°C$). On the as-grown film, TRCs were fabricated by reactive dc-sputtering ($P = 5$ W) of an about 5 nm thick silver oxide (platinum oxide layer) in a 0.02 mbar Ar/O$_2$ atmosphere. Subsequently, the Ag$_x$O (PtO$_y$) contact layers were capped *in situ* with a 5 nm thin gold (platinum) current-spreading layer [28], deposited by dc-sputtering ($P = 5$ W) in a 0.02 mbar pure Ar atmosphere (see Figure 11.27). Molybdenum steel shadow masks defined the circular shape of the sputtered Schottky contacts. The ohmic Al-doped ZnO back contact layer was itself contacted by dc-sputtered gold layers. The transmission spectra were recorded using an UV–vis Spectrometer Perkin Elmer Lambda 40 in the spectral range from 200 to 1100 nm. Current–voltage measurements (*IV*) were performed using an Agilent 4156C Precision Semiconductor Parameter Analyzer under dark conditions in air. The samples were mounted and contacted on TO-18 sockets, including an 80°C—30 min annealing step in air to harden the epoxy resin used for bonding the connecting wires. The capacitance–voltage characteristics were measured with an Agilent 4294A Precision Impedance Analyzer under dark conditions in air. The photocurrent measurements were carried out on the socketed diodes in photovoltaic mode (bias 0 V). A 100 W halogen lamp radiation was monochromatized by a Zeiss SPM 2 monochromator using a LiF-prism. The slit width of the monochromator was 0.5 mm. The photocurrent was amplified by a Stanford Research Systems SR510 lock-in amplifier with a sensitivity of 10^6 A/V. For the determination of the external quantum efficiency and the responsivity, the wavelength-dependent photon flux reaching the diode was determined using a calibrated pyrodetector Oriel 70128.

The transmission spectra of the TRC deposited on as-received fused quartz glass substrates with identical growth conditions as used for the photodetectors are depicted in Figure 11.28. In this way, the zinc oxide absorption is excluded and the absorption of the TRC alone can be deduced also in the ZnO near band edge region. Please note that on zinc oxide the transmission of the TRC differs slightly, due to the different interface reflectivity. In order to exclude this effect, the optical constants of the thin TRC layers need to be determined, which is not part of this work. The transmission of the PtO$_y$ contact depends only weakly on the wavelength of the incident light, whereas the transmission of the silver-based contact decreases remarkably below 550 nm. Probably, the decrease is due to absorption in the silver oxide, as the respective fundamental absorption edge was to be at

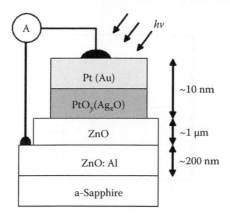

FIGURE 11.27
Schematic representation of the photodetector structure.

2.3±0.2 eV (cf. Figure 11.26). The mean transmission of the Ag_xO and the PtO_y contacts in the visible spectral range are 73% and 75%, respectively. At the ZnO-free exciton energy at room temperature, 51% and 69% of the incident light is still transmitted by the Ag_xO and the PtO_y contacts, respectively. The current–voltage characteristics of the transparent Ag_xO and PtO_y diodes are depicted in Figure 11.29. Thermionic emission is assumed to be the predominant current transport mechanism [29,30]. For $eV \gg k_BT$ the current is given by

$$I = I_S \exp\left(\frac{e(V - IR_S)}{nk_BT}\right) - \frac{V - IR_S}{R_P}. \qquad (11.1)$$

This model considers the series and parallel resistance R_S and R_P of the diode, respectively, and the voltage dependence of the Schottky barrier height by the ideality factor n. T denotes the temperature, k_B is the Boltzmann constant and e the elementary charge. The saturation current $I_S = SA^*T^2 \exp e\Phi_{B,eff}/k_BT$ depends on the effective barrier height $\Phi_{B,eff}$ and the contact area S; the effective Richardson constant A^* of ZnO is 32 A/cm²K²,

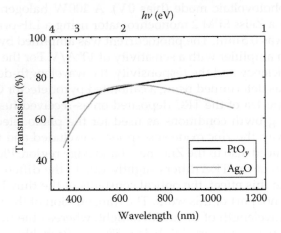

FIGURE 11.28
Transmission spectra of the TRC contact layers on a fused quartz glass substrate; the dotted line indicates the position of the ZnO-free excitons (3.31 eV).

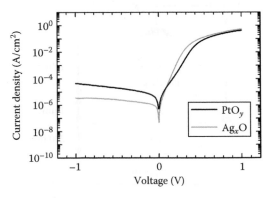

FIGURE 11.29
Current–voltage characteristic of the photodiodes measured under dark conditions.

using $m_e^* = 0.27m_{e,0}$. Using a least-square fit of the forward *IV* characteristics, the diode parameters ideality factor n, series and parallel resistance R_S and R_P and effective barrier height $\Phi_{B,eff}$ were deduced by Equation 11.1. The resulting effective barrier height is 0.73 V for both diodes. The series resistance for the Ag_xO (PtO_y) contact is 260 (450) Ω, respectively. A parallel resistance has been considered for the fit of the data, due to the fact that the reverse current of the diodes exceeds the reverse current predicted by thermionic emission theory (cf. Figure 11.29). The parallel resistance of the PtO_y contact is 1×10^7 Ω compared to 1×10^8 Ω for the Ag_xO contact. A possible explanation is that, for a small part of the contact area, the Pt-capping layer is short-circuited with the semiconductor layer. The ideality factors of the contacts are 1.2 (1.6) for the Ag_xO (PtO_y) contacts. Those values are among the lowest reported values for Schottky contacts on heteroepitaxially grown ZnO. The rectification ratio at ± 1 V is 1.5×10^5 for Ag_xO and 1×10^4 for PtO_y, respectively. The contacts degrade, due to the annealing step included in the socketing process, yielding lower effective barrier heights. For thicker and therefore opaque MESFET gate electrodes, a similar degradation due to annealing of the contacts has been observed [31]. Prior to the annealing step, the effective barrier heights of the Ag_xO (PtO_y) contacts are 0.87 (0.82) V. The mean barrier heights $\Phi_{B,m}$ were determined by capacitance–voltage measurements (*CV*) at an oscillation frequency of 100 kHz, and are larger than the effective barrier heights extracted from the *IV* characteristics. For silver oxide (platinum oxide) $\Phi_{B,m} = 0.92$ V ($\Phi_{B,m} = 0.78$ V) were deduced from the $1/C^2$ plot. Probably, this difference of the barrier heights is due barrier inhomogeneities [28,110]. Current flows preferentially at regions with lower barrier heights; thus, the effective barrier height measured by *IV* is reduced (compare Equation 11.1). In contrast, the mean barrier height probed by *CV* is only affected by the extension of the space charge layer. This is valid under the assumption that the extension of the space charge layer is larger than the length scale of the barrier height fluctuations at the interface. Taking into account a Gaussian distribution of the barrier height, the standard deviation σ of the distribution is given by [110]

$$\sigma = \sqrt{\frac{2kT}{e}(\Phi B, m' - \Phi_{B,eff})}. \tag{11.2}$$

For the silver oxide-based TRC, the barrier height distribution is with $\sigma = 98$ mV significantly broader compared to $\sigma = 50$ mV for the platinum oxide-based TRC. Thus, the effective barrier height of silver can probably be remarkably increased by reducing the width of the barrier height distribution of the silver oxide by optimizing the deposition. On the one hand, such optimization may refer to increase in the interface quality (e.g., by using single crystalline substrates [see Section 11.2.3]) and on the other hand require the optimization of the contact fabrication itself (e.g., by the deposition method, e.g., eclipse PLD [30]) and/ or the deposition conditions (e.g., reduction of droplets or by using a sputter-up system).

The external quantum efficiency

$$\eta_{ext} = \frac{|I_{ph}|}{eA_{opt}} \frac{h\nu}{P} \tag{11.3}$$

and the responsivity $R = e\eta_{ext}/h\nu$ of the photodetectors in photovoltaic mode ($V = 0$) aredepicted in Figures 11.30 and 11.31. A_{opt} is the contact area and P is the incident optical power at the energy $h\nu$. Whereas for photon energies below the band gap of ZnO the photocurrent

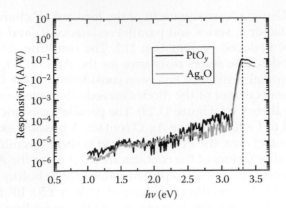

FIGURE 11.30
Spectral responsivity of the photodiodes. The dotted line indicates the position of the ZnO-free excitons (3.31 eV).

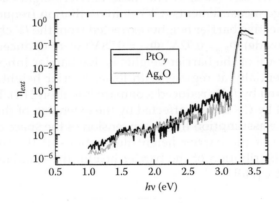

FIGURE 11.31
Spectrum of the external quantum efficiency of the photodiodes. The dotted line indicates the position of the ZnO-free excitons (3.31 eV).

generation is strongly suppressed, for photon energies above the band gap a large fraction of the incident light is converted into an electric current. The UV–vis rejection ratio is for both, Ag_xO and PtO_y, at least 10^3 for the blue, but exceeds 10^4 for the red and green spectral range. The maximum external quantum efficiency and responsivity for the PtO_y (Ag_xO) contact are 32% (22%) and 0.1 (0.07) A/W, respectively. The maxima are reached at 3.31 eV, which is about 60 meV below the fundamental band gap of ZnO. In conclusion, the photocarriers are either generated by ionization of shallow donors [111] or by generation and subsequent thermal dissociation of excitons. The exciton binding energy in zinc oxide is about 60 meV. The normalized detectivity for a Schottky photodiode complying with the thermionic emission model is given by $D^* = R\sqrt{\exp(e\Phi_B / k_BT) / (4eA^*T^2)}$ [95]. Using the latter relation, the maximum values were calculated to be $D^*_{max} = 0.9 \times 10^{11}$ cm $Hz^{1/2}/W$ for the Ag_xO-diode and $D^*_{max} = 1.29 \times 10^{11}$ cm $Hz^{1/2}/W$ for the PtO_y-diode. Recently reported PEDOT:PSS-Schottky contacts on ZnO [95,96] exhibit detectivities, which are about 3 orders of magnitude higher. This is due to a higher barrier height of 1.1 eV and a higher transmittance of PEDOT:PSS on ZnO.

11.4.3 Transparent MESFET and Inverters

Using the previously discussed TRC as gate electrode, transparent MESFET and FI were fabricated and characterized. The source and drain contacts were formed using degenerately doped ZnO:Al grown by PLD at room temperature. Figure 11.32 shows an optical microscopic image of the transparent MESFET and inverter. The mean transmission through the entire device structure reaches 87% through the source/drain contacts and 62% through the PtO_y gate in the visible spectral range.

The electrical characteristics of the transparent MESFET are similar to their opaque analogues. A typical output characteristic of a transparent Ag_xO-gate MESFET is depicted in Figure 11.33a. It shows a clear linear and saturation region within a drain voltage sweep between 0 and 2.5 V and strong dependence on the gate voltage. The curve for $V_G = 1$ V shows a small offset of the source–drain current due to gate currents (cf. Figure 11.4). Figure 11.33b shows transfer characteristics of the same Ag_xO-MESFET and a PtO_y-gate MESFET for a source–drain voltage of $V_{SD} = 2$ V. These normally-on MESFET have on/off-ratios of 1.3×10^5 for PtO_y and 1.4×10^6 for Ag_xO and threshold voltages of $V_T = -1.71$ and -1.26 V, respectively. The lower negative threshold voltage for Ag_xO is probably due to the indiffusion of Ag into the ZnO channel, which serves as compensating defect in the n-type semiconductor [112]. The off-currents are limited by the TRC's leakage current and lie in

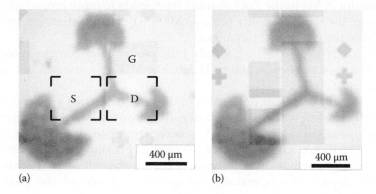

(a)   (b)

FIGURE 11.32
Optical microscopic images of (a) a transparent MESFET and (b) a transparent FI, both with 30 nm MgZnO channel and AgO_x gate.

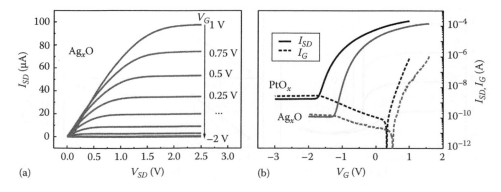

(a)   (b)

FIGURE 11.33
(a) Output characteristic, (b) transfer and gate characteristics of transparent MESFET.

TABLE 11.9

Overview of the Measured Transparent MESFET Properties

Sample	d (nm)	$N_D - N_A$ (10^{18} cm^{-3})	μ_{Hall} (cm^2/ (V s))	μ_{ch} (cm^2/ (V s))	On/ Off-Ratio	S (mV/ Decade)
Ag$_x$O	30	1.4	12.1	11.9	1.4×10^6	120
PtO$_y$	30	3.0	12.1	11.4	1.3×10^5	160

the range of nanoamperes for PtO$_y$ and 100 pA for Ag$_x$O. The on-currents reflect the series resistances of the channels and are limited by the forward gate voltage, for which the TRC is in flatband condition.

The prominent properties of the transparent MESFET are summarized in Table 11.9. As both samples were pieces of one wafer, the thicknesses and Hall-effect mobilities are equal. The lower net doping concentration of the Ag$_x$O sample compared to the PtO$_x$-sample obtained from QSCV measurements can be again explained by the compensation effect of Ag in ZnO. However, the channel mobilities are in good agreement with the Hall-effect mobility. Note, that the reproducibility of the MESFET with PtO$_y$ gate was high [92]. All 38 MESFET on one 10×10 mm^2 chip worked and the standard deviation from the mean channel mobility was only 1.1 cm^2/(V s). This reconfirms the statements in Section 11.2.4 that a small amount of Mg in the ZnO channel layer leads to more reproducible and stable devices. However, the reproducibility of Ag$_x$O MESFET is less distinctive, probably due to Ag-diffusion from the gate into the channel.

Based on the transparent MESFET, a fully transparent FI is demonstrated. An optical microscopic image can be seen in Figure 11.32b. The VTC of a transparent Ag$_x$O inverter is shown in Figure 11.34a.

Also in this case, the excellent properties of the opaque device are successfully transferred to the fully transparent device. For negative input voltages, the output high level approaches the operation voltage $V_{DD} = 4$ V, whereas for positive V_{in}, the output low level is $V_{OL} \approx 0.2$ V. V_{OL} is slightly higher than for the opaque FIs shown in Section 11.3.1. This is due to higher leakage currents over the gate of the forward biased switching transistor.

Figure 11.34b depicts the operating voltage dependence of the peak gain magnitude *pgm* and the uncertainty voltage V_{uc}. The *pgm* value increases drastically with V_{DD} and covers two orders of magnitude within a voltage range of 4 V. The maximum V_{DD} is limited to 4 V due to the increasing probability of breakdown of the

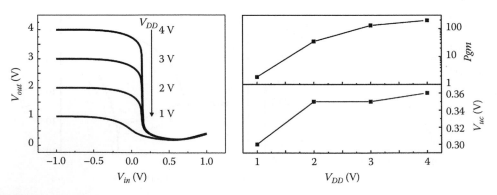

FIGURE 11.34
(a) VTC and (b) peak gain magnitude and uncertainty voltage of a transparent MESFET inverter.

switching transistor Q_S. The uncertainty level V_{uc} shows a step from 0.3 V at $V_{DD} = 1$ V to 0.35 V at $V_{DD} = 2$ V and then remains constant. This is due to a shift of the inverter characteristic by 0.2 V toward positive V_{in} probably caused by a parasitic capacitance at the gate contact of Q_S.

11.4.4 Comparison of Transparent MESFET and Inverter with the Literature

The device parameters of the presented transparent MESFET and inverter are among the best reported for transparent transistors and leading for the transparent inverter. Tables 11.10 and 11.11 provide a literature review of the most important figures of merit for transparent inverters and transistors, respectively.

For the classification of the transparent MESFET within other approaches to transparent transistors, the requirements for transparent electronics formulated by Wager [113] for the on/off-ratio and the channel mobility has to be noted. It is incidentally noted that all referenced transistors and inverters meet the required minimal transmission of 70% in the visible spectral range. The channel mobility has to be larger than 1 cm²/(V s); the on/off-ratio should exceed 10^6. Additional to these requirements and for the sake of low power consumption (e.g., for mobile applications), low electric fields and therefore low crosstalk between individual devices in an integrated circuit, it is suggested to limit the gate voltage sweep (and with that the source–drain voltage sweep) to the range below 5 V. With that, the presented ZnO-based MESFET with Ag_xO gate is the only FET which meets all of these requirements, whereas organic and carbon-nanotube (CNT) FET are up to now not compatible with transparent electronics. CNTs are hard to handle and exhibit low on/off-ratios. Although they can reach high channel-mobilities, their gate voltage sweep is too large. Organic FET suffer from low channel mobility and on/off-ratio, and require high operating voltages. It has to be emphasized that the potentially achievable on/off-ratio of transparent ZnO-based MESFET is not yet fully utilized. It can be improved for example by means of geometry and device design (e.g., higher W/L ratios) as shown in Section 11.2.3.

Three devices in Table 11.11 are nearly fulfilling the requirements: two inorganic oxide MISFET [23,114] and the PtO_y-gated MESFET. Yabuta et al. reported a MISFET based on amorphous $InGaZnO_4$ as channel and Y_2O_3 as insulator [114]. It achieved an on/off-ratio of 10^8 and a channel mobility of 12 cm²/(V s). However, the operating voltage still needs to be reduced. Wang et al. reported a hybrid inorganic/organic bottom-gate MISFET with indium oxide as channel and an organic self-assembled superlattice dielectric having excellent characteristics ($\mu_{ch} = 120$ cm²/(V s), $V_{SD} = 1$ V) [23]. However, the on/off-ratio of 10^5 is a factor of 10 to low. Another disadvantage is that the organic dielectric unfortunately cannot be processed by standard photolithography techniques.

TABLE 11.10

Comparison of Transparent Inverter Parameters

Channel	Type	*pgm*	V_{SP} (V)	V_{uc} (V)	V_{DD} (V)	pgm/V_{DD} (V^{-1})	References
MgZnO	MESFET	196	0.2	0.36	4	49	This work [83]
ZnO	MISFET	~5	~3	~2	10	0.5	[115]
IGO	MISFET	1.5	~8	~12	30	0.05	[116]
IGZO/pentacene	MISFET	56	2.84	~0.9	7	8	[44]
ZTO	MISFET	10.6	~3.3	1.87	10	1.06	[117]

TABLE 11.11

Comparison of Transparent Transistor Parameters

Channel	Type	V_T (V)	V_{SD} (V)	On/Off-Ratio (10^6)	μ_{ch} (cm²/ (V s))	S (mV/ Decade)	References
MgZnO	Ag-MESFET	−1.26	2	1.4	11.9	120	This work [83]
MgZnO	Pt-MESFET	−1.71	2	0.13	11.4	160	This work [83]
ZnO	MISFET	−10	10	10	0.35	~1000	[57]
ZnO	MISFET	~0	20	1.6	1.2	~3000	[118]
ZnO	MISFET	−1	5	0.1	0.97	~800	[58]
ZnO	MISFET	−5	10	1	7	—	[119]
ZnO	MISFET	19	20	0.3	27	1390	[120]
IGOZO	MISFET	3	2	1	80	~1000	[121]
IGO	MISFET	2	20	0.01	7	~2000	[116]
a-IZO	MISFET	−10	20	1	30	300	[19]
a-IGZO	MISFET	1.6	4	0.001	8.3	~1500	[122]
a-IGZO	MISFET	2	20	50	11	200	[123]
a-IGZO	MISFET	1.4	8	100	12	200	[114]
a-IGZO	MISFET	5.9	5	4.9	35.8	590	[124]
a-ZTO	MISFET	−5	40	10	20	~700	[18]
a-ZTO	MISFET	0	4	0.02	28	~300	[125]
SnO₂	MISFET	10	35	0.1	2	4000	[126]
SnO₂-NW	MISFET	−5	1	0.1	120	270	[127]
In₂O₃	MISFET	0.2	1	0.1	120	90	[23]
Pentacene	Org. MISFET	−7	−30	0.5	0.91	2230	[15]
Pentacene	Org. MISFET	1	−20	1	0.38	600	[16]
Pentacene/ CNT	Org. MISFET	0	−40	0.1	0.6	1400	[14]
CNT	MISFET	5	0.5	7	0.5	—	[12]
CNT	MISFET	−20	−5	0.01	0.5	~2000	[13]
CNT	MISFET	0.5	−5	0.01	—	180	[10]
CNT	MISFET	−3	−1	0.0001	26	—	[128]

The advantages of ZnO-based MESFET are also evident regarding the subthreshold slope. Together with the FET reported in [23] with a slope of 90 mV/decade, the here presented MESFET are closest to the theoretical minimum slope of $S = 60$ mV/decade [39,40] for FET operating at room temperature. All other reported transparent FET exhibit much higher slopes which leads to long switching times of circuits based on them. However, the optimization of ZnO-based MESFET is still at its beginning and for opaque ZnO-MESFET, already slopes as low as 77 mV/decade were reported [32].

Among the transparent inverters reported so far, the inverters based on ZnO-MESFET show a superior performance at a considerably low operating voltage of only $V_{DD} = 4$ V exhibiting the highest gain and the lowest uncertainty level (see Table 11.10). With $pgm = 196$, the transparent inverter is ideally suited to be used as transparent voltage amplifier for low signal measurements. The performance of these inverters ($V_{uc} = 0.36$ V) is already comparable with the well-established GaAs-MESFET technology ($V_{uc} \sim 0.1$ V). The advantage of the MESFET-based inverter is obvious compared to the ZnO MISFET-based inverter reported in Ref. [115]. There, the pgm was in the range of 5 and $V_{uc} > 2$ V for

$V_{DD} = 10\,\text{V}$ and only about 60% of V_{DD} was reached as high level. For MISFET most of the gate voltage drops over the insulator, which leads to higher operating voltages and lower gain of the devices.

11.4.5 Stability of Fully Transparent ZnO-Based Fundamental Inverter Circuits against Incident Light and Elevated Temperature

In order to study the stability of our fully transparent MESFET in devices, FIs have been fabricated by combining two transparent, normally-on MgZnO-based MESFET (see Section 11.3.1) [129].

The investigated devices were processed on $10 \times 10\,\text{mm}^2$ pieces cleaved from a $50 \times 50\,\text{cm}^2$ a-plane sapphire substrate. The transparent gate contacts were fabricated as described in Section 11.4.1. The mean device transmission in the visible spectral range exceeds 65% [92]. During current–voltage measurements, the samples were illuminated from the top (facing the electrodes) with red ($\lambda = 628\,\text{nm}$; FWHM $= 20\,\text{nm}$), green ($\lambda = 525\,\text{nm}$; FWHM $= 30\,\text{nm}$), and blue ($\lambda = 435\,\text{nm}$; FWHM $= 25\,\text{nm}$) LED. The irradiance was kept constant at about $10\,\text{W/m}^2$ for all LED, which is comparable to typical indoor irradiation levels. The temperature of the wafer prober's chuck was kept constant at values in the range between 20°C and 150°C.

Figure 11.35 depicts the voltage transfer characteristics (VTC) of a transparent inverter circuit measured under dark conditions and exposed to red, green, and blue light.

Red and green light do not change the VTC, whereas blue light slightly affects the VTC. Generally, the photocarrier generation inducing this effect may originate from interface-states at the gate-electrode or the source–drain electrodes and or absorption of the source–drain or gate electrode material or the channel material. As the blue photon energy is lower than the fundamental band gap of MgZnO of the channel and the ZnO:Al of the source and drain electrodes, carriers are not generated by fundamental absorption in there. In Section 11.4.2, the wavelength-dependent external quantum efficiency of the Ag_xO-photodetectors in photovoltaic mode reveals no specific response for blue light. Those photodetectors were grown on about $1\,\mu\text{m}$ thick ZnO on sapphire, which probably exhibits a better crystalline quality than the about $25\,\text{nm}$ thick MgZnO films grown on sapphire of the inverter circuits investigated here.

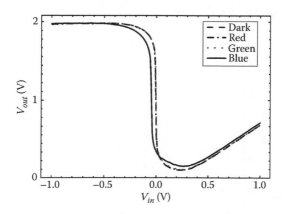

FIGURE 11.35
Influence of visible light on the VTC of transparent MgZnO-based inverter for $V_{DD} = 2\,\text{V}$.

Conclusively, deep trap states in the channel are most probably responsible for the carrier generation. As the absorption is detected only for blue light and not for green or red light, the activation energy of the trap states is between 2.36 and 2.8 eV below the conduction band, thus the photoresponse is induced by deep acceptor states.

As the load transistors' gate is short-circuited with the load transistors' source electrode, no photovoltage is generated there in contrast with the switching transistors' gate. At the latter gate electrode, a photovoltage, which is a reverse bias, is generated. This increases the effective gate voltage due to the fixed input voltage. Thus, the channel conductivity of the switching transistor is increased and the VTC is shifted to lower input voltages (see Figure 11.35). The switching regime shows the most obvious deviations, as the maximum of the derivative of the VTC is located there (see Figures 11.36 through 11.39). The *pgm*-point shifts by about −50 mV due to the illumination by blue light. The *pgm* is reduced by 15% and the uncertainty level is increased by 15% (see Figures 11.38 and 11.39). As expected, for the high V_{out}, no significant dependency on the incident light is observed. The deviation from the dark state is less than half a percent (see Figure 11.36). For input voltages > 0.3 V switching transistors' gate leakage currents yield an increase of the output voltage up to 0.7 V (see Figure 11.35). However, only for blue light, a weak 5% increase of V_{out} for V_{in} = +1 V is observed (see Figure 11.37). The minimum of the VTC is at V_{out} = 0.1 V, which is close to the ideal

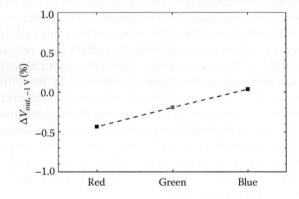

FIGURE 11.36
Light-induced relative change of V_{out} for V_{in} = −1 V.

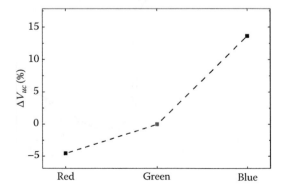

FIGURE 11.37
Light-induced relative change of V_{out} for V_{in} = +1 V.

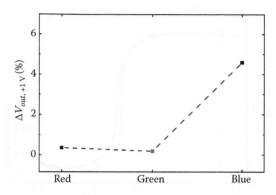

FIGURE 11.38
Light-induced relative change of V_{uc}.

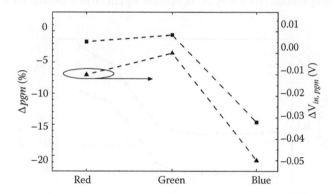

FIGURE 11.39
Light-induced relative change of *pgm* and absolute input voltage shift of *pgm*-point.

value of 0 V. Here, an increase to about 0.15 V is induced by blue light, whereas for red and green light the output voltage remains almost unchanged. Despite of the photo-induced performance reduction, the parameters of the illuminated MESFET inverter circuits are still superior compared to (dark) transparent oxide-based MISFET inverter circuits (see Table 11.10).

In order to determine the thermal stability of the inverter circuits, the VTC was measured between room temperature and 150°C. For clarity the development of the VTC is plotted for only three temperatures in Figure 11.40. The inverter circuits remain operational even for operating temperatures as high as 150°C.

The most pronounced thermally induced change occurs for positive input voltages beyond the switching regime. At high temperatures the output voltage increased significantly as depicted in detail in terms of V_{out} for $V_{in} = +1$ V ($V_{out,+1V}$) in Figure 11.41. Up to about 90°C the low output level remains constantly flat; for higher temperatures it increases considerably. This is due to an increase of the gate current of the switching transistor, as the clear correlation between $V_{out,+1V}$ and $I_{in,+1V}$ indicates (see right axis in Figure 11.41). For opaque Ag_xO-based MESFET the onset of the thermal degradation has been observed in the same temperature range [31] (cf. Section 11.2.4). Conclusively, the current increase is most probably caused by such thermally induced, irreversible degradation of the gate electrode for $T > 90$°C.

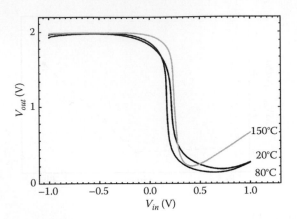

FIGURE 11.40
Influence of elevated temperatures on the VTC of transparent MgZnO-based inverter for $V_{DD}=2\,\mathrm{V}$.

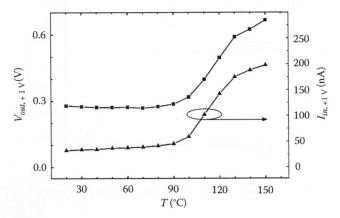

FIGURE 11.41
Temperature-induced absolute change of V_{out} and I_{in} for $V_{in}=+1\,\mathrm{V}$.

A significant effect of the temperature increase is also observed in the switching regime. Below the degradation threshold of about 90°C, the VTC generally idealizes for increasing temperature, namely the *pgm* increases by almost 50% to about 50 and the *pgm*-point ($V_{in,pgm}$) shifts closer to the ideal switching point of $V_{in}=0\,\mathrm{V}$ (see Figures 11.42 and 11.43). Furthermore, the uncertainty level V_{uc} is reduced (see Figure 11.44). The gate degradation also affects the switching regime for temperatures of 90°C and more, as the *pgm* decreases and the switching point shifts toward higher input voltages. At 150°C the *pgm* with about 32 is still above the room temperature value, while the $V_{in,pgm}$ slightly surpasses the room temperature value. In contrast, the gate degradation has no obvious effect on the evolution of the uncertainty level with temperature; it decreases continuously with increasing temperature. The weakest effect of the increased temperature is observed for the high output level regime. The deviation from the supply voltage continuously decreases for increasing temperature, as depicted by V_{out} for $V_{in}=-1\,\mathrm{V}$ ($V_{out,-1\mathrm{V}}$) in Figure 11.45. This is due to an increase of the gate current of Q_S (see right axis of Figure 11.45).

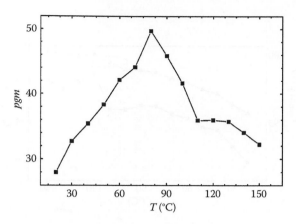

FIGURE 11.42
Temperature dependence of the peak gain magnitude.

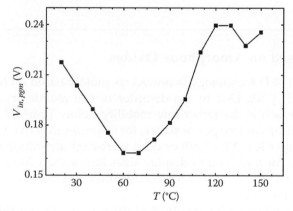

FIGURE 11.43
Temperature dependence of the *pgm*-point input voltage.

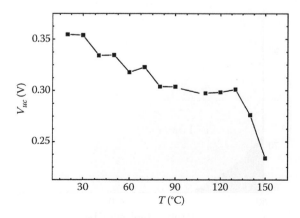

FIGURE 11.44
Temperature dependence of the uncertainty level.

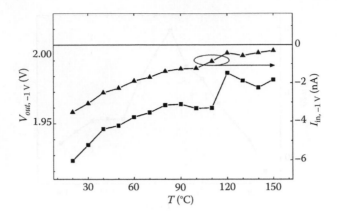

FIGURE 11.45
Absolute change of V_{out} for $V_{in} = -1\,V$ and the respective gate current of Q_s. The line denotes the supply voltage V_{DD}.

11.5 MESFET Based on Amorphous Oxides

Thin-film-transistor (TFT) technology is nowadays mostly based on hydrogenated amorphous silicon (a-Si:H) [130]. Due to the disorder of the atomic arrangement the main obstacle of this approach is the rather low mobility below $1\,cm^2/(V\,s)$ [20,131]. The low mobility drastically limits device performance, for example, in active-matrix liquid-crystal displays (LCD). Future 4K (4096 × 2160) or even ultra-high-definition (7680 × 4320) resolution screens with 3D capabilities and display sizes larger than 70 in. will require carrier mobilities larger than $1\,cm^2/(V\,s)$ in order to have sufficient refresh rate and thus reduced eye strain (Figure 11.46).

In amorphous oxides like GIZO the effect of disorder on the carrier mobility is rather weak. The large metal cation ns-orbitals overlap and lead to well-defined conduction paths

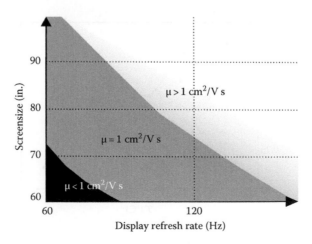

FIGURE 11.46
Requirement on the channel mobility for different screen sizes versus the display refresh rate.

within the semiconductor [132,133]. Also grain boundary scattering is no issue hence electrons in these materials exhibit Hall-effect mobilities μ_{Hall} of up to 60 cm²/(V s) [134].

Another important benefit of AOSs compared to a-Si:H is the possibility to deposit the material at room temperature while maintaining the high mobilities as mentioned previously, whereas silicon needs increased deposition temperatures in order to have comparable mobilities [135]. Deposition temperatures below 200°C are also a necessity for thin-film growth on flexible organic substrates. Furthermore, better uniformity and lower surface roughness of the deposited amorphous oxide thin-film material compared to conventionally used crystallic semiconductor material will lead to an increased homogeneity, for example, of the luminance in large area displays.

The current research on TFTs is based on MISFET that offer a base for almost all driver circuitry in the display industry. When using amorphous oxide channel material the mobility of the TFTs reach several 10 cm²/(V s) [136,137]; however, MISFET require the deposition of an insulator generally resulting in the use of high operating voltages of the devices. Due to an increased trap density on the interface between the semiconductor and the insulator, a more or less pronounced shift in the threshold voltage can be observed in bias stress and long-term measurements [130,138,139].

The MESFET technology offers a simple route for the processing of transistors by omitting the insulating layer and using a Schottky-gate diode as gate. Several advantages have already been discussed.

As has been shown in Section 11.2.5 the MESFET concept is compatible to pulsed-laser deposition grown polycrystalline ZnO on sapphire and on low-cost glass substrates [32]. In this subchapter the effects of the zinc oxide with Ga and In and the subsequent device application in Schottky-gated field-effect transistors is shown.

One hundred and sixty nanometer GIZO thin films were deposited on Corning 1737 glass substrates by radio-frequency magnetron sputtering at room temperature [136]. In the following, two thin-film samples will be compared: one consisting of as-grown GIZO material and one comprising of GIZO being annealed in ambient air for 1 h at 150°C.

2Θ–ω x-ray diffraction patterns (Figure 11.47) as well as *reflection high energy electron diffraction* patterns (not shown here) confirm the assumption that film growth mode is amorphous, since no peaks corresponding to either Ga_2O_3, In_2O_3, or ZnO can be found. This also proves that cluster formation of the respective oxides in the thin film is no issue and a smooth surface with a low roughness (1.1 nm) of the thin films is achieved (see Figure 11.48).

Hall-effect measurements were accomplished to determine the resistivity, the electron mobility μ_{Hall} and the carrier density n (data summarized in Table 11.12). For the specific

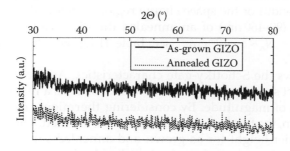

FIGURE 11.47
XRD-measurements on as-grown and annealed GIZO thin films revealing the amorphous structure of the semiconductor.

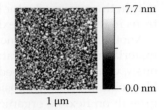

FIGURE 11.48
AFM image of the surface of an as-grown GIZO thin film. Surface morphology is not changed by annealing of the samples and the surface roughness is not affected.

TABLE 11.12

Electrical Properties of the As-Grown and Annealed GIZO Thin-Film Samples Determined from Hall-Effect Measurements

	d (nm)	ρ (Ω cm)	μ_{Hall} (cm^2/(V s))	n (10^{16} cm^{-3})
GIZO RT	120	190	2.6	1.3
GIZO 150°C	120	6.6	20.8	4.5

deposition conditions used herein, annealing has strong effects on the electrical properties of the samples.

The sample annealing causes a reduction of trap state density, a local atomic rearrangement and an improvement of film compactness [136]. Consequently the electrical characteristics of the semiconductor are improved: For as-grown GIZO thin films the resistivity of 190 Ω cm and is reduced to 6.6 Ω cm after the annealing step. The carrier density n and with that the mobility μ_{Hall} of the carriers increases. This is expected for the GIZO material, in contrast to, for example, polycrystalline ZnO, where a reduced carrier concentration n typically corresponds to an increased electron mobility. The effect is attributed to the different carrier transport mechanism. On the one hand in ZnO the mobility is mainly controlled by ionized impurity scattering, thus higher n corresponds to lower μ. On the other hand in GIZO higher n leads to increased μ, since more adjacent metal cation s-orbitals overlap leading to well-defined conduction paths within the semiconductor [133,136].

From the thin film samples, MESFET devices were fabricated employing standard photolithography using lift-off technique (for details refer to Section 11.2.1). For the Schottky-gate contact Ag$_x$O was used.

Quasistatic capacitance–voltage (QSCV) measurements were carried out to derive the net doping density. These measurements were performed with the transistor devices using the source and the gate contact. Figure 11.49 depicts $1/C^2$ versus the applied gate voltage V_G directly obtained from the measurement and the derived net doping density $N_D - N_A$ versus the width of the space charge region w below the Schottky-gate contact. For the bulk region (90–160 nm) of an annealed thin-film sample, the net doping density extracted is $N_D - N_A \sim 5 \times 10^{16}$ cm^{-3} (see Table 11.13). For the as-grown samples it is $N_D - N_A \sim 1 \times 10^{16}$ cm^{-3}.

Figure 11.50a shows the Schottky-gate characteristics. As-grown and annealed thin-film Schottky-contact samples have a similar behavior with a rectification ratio $I_{SG}(2\,V)/I_{SG}(-2\,V)$ of 8 orders of magnitude. By considering thermionic emission to be the only transport mechanism, the forward IV-characteristic was fitted. The extracted parameters of the effective barrier height for as-grown and annealed samples are $\Phi_B \approx 0.95$. This value is well within the range of the highest barrier heights reported (around 1.1 V) for Schottky contacts on ZnO-based materials [28,29]. The ideality factor of the Ag$_x$O Schottky-gate contacts is 2 and 1.9 on as-grown and annealed GIZO material. For an

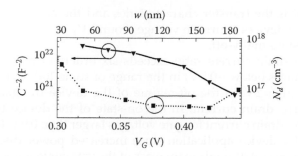

FIGURE 11.49

$1/C^2$ versus the gate voltage directly obtained from QSCV measurement and the calculated net doping density N_D-N_A versus the width of the space charge region.

TABLE 11.13

Comparison of Device Parameters of GIZO-Based MESFET Samples

	d (nm)	W/L ($\mu m/\mu m$)	$N_D - N_A$ (10^{16} cm^{-3})	μ_{ch} (cm^2/ (V s))	$\mu_{ch,med}$ (cm^2/(V s))	V_T (V)	S (mV/ Decade)	On/Off (10^6)
GIZO RT	160	430/20	1	7.3	0.01	−0.5	123	4.4
GIZO 150°C	160	430/10	5	14.1	14.7	−1.9	112	25

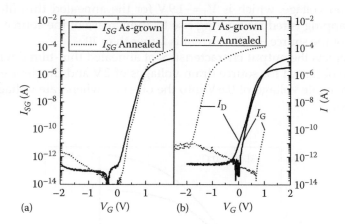

FIGURE 11.50

(a) Schottky gate characteristics and (b) transfer characteristics and gate-leakage characteristics of as-grown and annealed GIZO-based MESFET.

ideal Schottky contact this factor should be close to unity. The discrepancy could stem from inhomogeneities of the Schottky contact's barrier height [28] and the non-ideal front-to-front contact geometry. However, similar values for the effective barrier height and the ideality factor were reported for Ag$_x$O Schottky contacts on polycrystalline ZnO grown on glass and a-sapphire substrates [28,32] (cf. Section 11.2.3), which reflects the stability of the barrier formation of the metal–semiconductor contact toward alloying the zinc oxide with gallium and indium. Overall, reports for Schottky contacts on amorphous oxides are sparse beside the work of Shimura et al. [140] where almost no rectification was reported.

Figure 11.50b depicts the transfer characteristics and the gate leakage currents of the devices. For as-grown devices a turn-on voltage around $V_T = -0.5\,V$ is observed, thus these transistors are almost normally-off.

The on/off source–drain current ratio exceeds six orders of magnitude with the off-current (for gate voltages below $-0.5\,V$) in the range of some 10th of pA. The gate voltage sweep necessary is $\Delta V_G = 2\,V$. In the off-regime of the device the gate current is nearly identical to the source–drain current. In the on-state of the device the leakage current dominates the source–drain current for gate voltages larger than $0.9\,V$. Larger gate voltages should be avoided for device application since increased power dissipation and losses would be the consequence. Although an increase of the gate current is an inherent property of the MESFET working principle, these leakage currents can be drastically reduced by annealing the thin films prior device fabrication. In the on-state of the device (for gate voltages larger than $0.5\,V$) the leakage gate current is less than $10^{-13}\,A$. There is a slight increase of the gate current for decreasing gate voltages, but it still remains similar to the source–drain current in the off-state of the transistor (for gate voltages smaller than $-1.9\,V$). Accordingly, there is very low power consumption [92] of the device when the transistor is in the on-state, making these types of field-effect devices especially favorable in mobile, low-voltage/low-power applications.

As described previously annealing reduces trap states and increases the compactness of the thin film. As indicated by the Hall-effect data (Table 11.12) this leads to an increase of the carrier concentration but also to a much lower resistivity of the annealed thin film being a factor of 30 smaller compared to as-grown samples. Consequently there is a shift to a lower turn-on voltage, which is $V_T = -1.9\,V$ for the annealed thin-film device. Also charge carrier trapping is reduced giving rise to an increase of the source–drain current. The on/off-ratio of the source–drain current is larger than 10^7.

Figure 11.51 depicts the output characteristics of annealed thin-film devices. Very good saturation can be observed for source drain voltages of $2\,V$ and the device switches from the on-state with a gate voltage of $0.6\,V$ into the off-state when gate voltages of less than $-1.8\,V$ are applied.

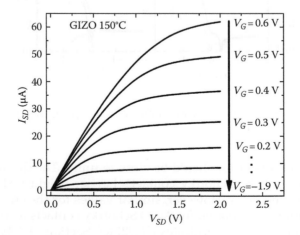

FIGURE 11.51
Output characteristic of an annealed GIZO-based thin-film transistor. Note the very low switching gate voltage and the small source–drain voltage required to obtain saturation.

From the transfer characteristics two important transistor parameters can be determined: The channel mobility $\mu_{ch} = g_{max}/(W/LdeN_d)$ from the maximum of the forward transconductance $g = \partial I_{SD}/\partial V_G$ (see Table 11.2 for W/L and d of the respective sample) and the subthreshold slope S that presents the gate voltage sweep necessary to change the source–drain current by one order of magnitude.

The channel mobility of the particular transistor depicted in Figure 11.50b is $\mu_{ch} = 7.3\,cm^2/(V\,s)$ which is higher than the mobility of the charge carriers determined by Hall-effect measurements. The median value of the channel mobility of all 28 devices on the sample chip is only $\mu_{ch,med} = 0.01\,cm^2/(V\,s)$ and thus over two orders of magnitude lower. This indicates poor reproducibility of devices fabricated on as-grown GIZO material.

By annealing the thin-film sample the channel mobility ($\mu_{ch} = 14.1\,cm^2/(V\,s)$) increases and is almost similar to the Hall-effect mobility ($\mu_{Hall} = 20.8\,cm^2/(V\,s)$) as predicted by MESFET theory. The reproducibility ($\mu_{ch,med} = 14.7\,cm^2/(V\,s)$) of the devices was largely improved, which is indicated in Figure 11.52. Out of 50 devices on the sample chip 32 are operational. Although this is only a yield of 62%, processing issues were limiting a further increase of the reproducibility for this particular sample.

The subthreshold swing is also improved: The value S for annealed thin-film devices is slightly lower ($S = 112\,mV/decade$) compared to as-grown transistor samples ($S = 123\,mV/decade$). Typical values for the subthreshold swing are in the range between 120 and 300 mV/decade for GIZO-based transistors [42,43,136,141] making the values for annealed thin-film devices reported herein the best so far for oxide TFT.

In conclusion the electrical properties of low-temperature processed Schottky-gated TFT with amorphous channel material were investigated. All processing steps of the thin films and the transistors were carried out at room temperature, only limited by an annealing step at 150°C. By means of annealing the thin-film sample an improvement of the electrical characteristics is observed. Due to increased film compactness and reduction of trap states in the annealed thin films the resistivity is reduced by a factor of 30 while the Hall-effect mobility increases by one decade. These properties affect the transistor. Increased channel mobility, reproducibility, and homogeneity of the fabricated devices can be accentuated. Together with the low gate sweep voltages of less than 2.5 V the GIZO-based MESFET could be a reasonable component of mobile gadgets and devices.

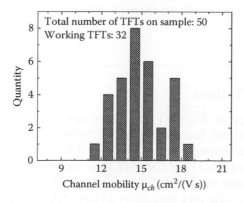

FIGURE 11.52
Histogram revealing the yield and the distribution of devices versus the channel mobility.

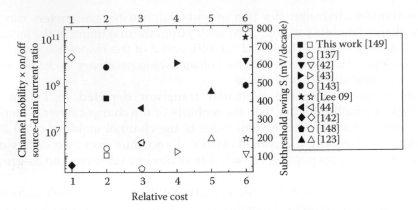

FIGURE 11.53
Mobility multiplied with the respective on/off source–drain current ratio (full symbols) and the subthreshold swing (open symbols), two important device parameters versus the relative fabrication cost. Note that the product of the mobility and the on/off-ratio should be as high as possible, while a lower subthreshold swing S preferable.

Figure 11.53 illustrates the channel mobility multiplied with the on/off current ratio and the subthreshold slope over the relative fabrication cost per device for selected references (see also Table 11.14). Martins et al. [142] reported GIZO-based MISFET on paper substrate. These devices are obviously fabricable at a very low cost, but suffer from low on/off current ratios and high subthreshold swing. Park et al. [143] demonstrated IZO-based transistors on glass substrates, but the highest processing temperature involved was about 450°C making these devices incompatible with the concept of flexible substrates. Na et al. [44] investigated MISFET with GIZO channel on n-Si substrates. Obviously these devices cannot be used in transparent electronics. Most other references have more sophisticated fabrication and growth methods involved increasing the relative cost. But for the application in consumer electronics low fabrication cost combined with reasonable transistor properties are desirable. These requirements are obviously met by the MESFET reported herein so not only further scientific exploration of these devices but also the emergence of these transistors in electronic devices is eligible.

Table 11.14 summarizes selected literature covering low temperature-processed field-effect transistors based on amorphous channel material.

11.6 Summary and Outlook

The well-known concept of MESFET has been introduced to ZnO-based electronics. Furthermore, for the first time, fully transparent Schottky contacts and MESFET have been developed applying an alternative approach toward low-cost transparent oxide electronics. It has been shown that ZnO-based MESFET are superior in important figures of merit compared to the commonly used MISFET technology. They exhibit high on/off-ratios up to the range of 10^8, usual channel mobilities $>10\,cm^2/(V\,s)$, minimal slopes of $\sim 80\,mV/$decade and are tuneable within a gate-voltage sweep of $<3\,V$. The MESFET are stable up to temperatures of $\sim 100°C$ and long-term stable under normal conditions for >300 days. The transfer characteristics showed no voltage shift during bias stress (for 22 h) and only a

TABLE 11.14

Comparison of Amorphous MISFET Device Parameters with MESFET Devices

Growth Method	T_{max} (°C)	Substrate	Channel	d (nm)	V_t (V)	V_{SD} (V)	$I_{on/off}$	μ_{ch} (cm²/Vs)	S (mV/Decade)	References
RFS	150	Glass	GIZO	160	−1.9	2	2×10^7	15	112	This work [149]
RFS/ALD	RT	Glass	IZO	30	6.4	20	10^7	100	800	[137]
RFS/PECVD	150	n-Si	GIZO	10	0.19	20	7×10^7	74	290	[136]
PLD	RT	PET	GIZO	30	1.6	4	10^3	6–9	—	[17]
PLD	400	n-Si	GIZO	40	−0.23	10	$>10^9$	12	<120	[42]
DCS/PECVD	340	Glass	GIZO	30	0.13	10	$>10^9$	9.5	130	[43]
SP	450	Glass	IZO	15	−0.3	20	10^9	6.6	150	[143]
RFS/ALD	300	Glass	GIZO	50	2.5	15	10^{10}	12.2	200	[20]
PECVD	200	Glass	a-Si:H	200	0	10	$>10^7$	<0.59	520	[131]
PLD	RT	Glass	GIZO	50	2	20	5×10^7	11	200	[123]
RFS/ALD	180	Glass	AZTO	25	1	15.5	$>10^9$	10.1	580	[144]
RFS	140	Glass	GIZO	50	1.4	6	10^8	12	200	[114]
RFS/PECVD	150	Glass	GIZO	60	1.2	10	10^8	30	190	[145]
PA-PLD/ALD	350	Glass	ZTO	50	−2	10	10^7	5	—	[21]
PLD	500	n-Si	SGZO	50	6.8	10	10^6	1.8	5000	[146]
RFS	RT	n-Si	GIZO	50	1.2	5	10^7	11.4	181	[44]
RFS	RT	Paper	GIZO	50	3.75	40	10^4	35	2400	[147]
RFS	RT	Paper	GIZO	40	1	15	10^4	40	650	[142]
RFS	RT	p-Si	GIZO	50	0.34	15	10^6	3.6	40	[148]

Growth method of the semiconductor active channel (RFS: radio frequency magnetron sputtering, DCS: direct current magnetron sputtering, ALD: atomic layer deposition, PECVD: plasma-enhanced chemical vapor deposition, PLD: pulsed-laser deposition, SP: solution processing, RIE: reactive ion etching, SP: sputtering, PA-PLD: plasma assisted pulsed-laser deposition).
T_{max} is the maximum temperature used in the fabrication process of the device, the substrate material and active channel material (IZO: indium zinc oxide, AZTO: Al-doped zinc tin oxide, ZTO: zinc tin oxide, SGZO: tin-gallium-zinc-oxide) with the thickness d, V_t the threshold voltage, V_{SD} the source drain voltage, $I_{on/off}$ being the on/off-ratio of the source drain current, μ_{ch} the field-effect mobility and the minimum of the subthreshold slope S.

small shift up to 0.65 V under irradiation with blue and violet light. It has been shown that the MESFET are at least operable at frequencies of 200 kHz. ZnO-MESFET-based inverters revealed a clear inverting behavior with maximal gains of ~200 at an operating voltage of 4 V and low uncertainty levels in the range of 0.1–0.4 V. The inverter's voltage-transfer curves could be shifted by means of FET- and SDFL circuits as well as using FIs with different gate width-to-length ratios. With that, complementary logic levels can be applied to the inverter enabling ring oscillators and more complex logic circuits. Transparent rectifying contacts have been developed and applied to visible-blind photodetectors, transparent MESFET, and inverters. These devices exhibited an overall transmission between 60% and more than 70% in the visible spectral range. In photovoltaic mode, the transparent photodetectors achieve external quantum efficiencies as high as 32% and responsivities as high as 0.1 A/W. The fully transparent MESFET ($I_{on}/I_{off} \sim 10^6$, $\mu_{ch} > 10\,cm^2/(V\ s)$) and inverters ($pgm \sim 200$, $V_{uc} = 0.36$ V) exhibited comparable electrical properties as the presented opaque devices and are therefore among the best or even superior to other reported transparent devices. The transparent inverters keep operational up to temperatures of 150°C. The MESFET technology has also been demonstrated on the basis of amorphous oxides such as gallium–indium–zinc–oxide, which have been fabricated under temperatures below 150°C. These MESFET showed a better performance and higher reliability as their crystalline ZnO counterpart, that is, glass, and are in principle applicable to flexible substrates.

In the next steps, the MESFET's frequency-dependent behavior will be further investigated by means of ring oscillators; the MESFET will be applied as current-driving transistors for active-matrix displays; and the use of passivation layers will be investigated in order to reduce degradation and improve the stress stability. As future challenges, for example, complementary circuits and MESFET on flexible substrates will be considered.

Acknowledgments

The authors thank G. Biehne, M. Hahn, H. Münch, and H. Hochmuth for technical support, J. Lenzner for FEM cross sections as well as C. Sturm, H. Franke, and S. Schöche for spectroscopic ellipsometry measurements. Financial support by Deutsche Forschungsgemeinschaft in the framework of Sonderforschungsbereich 762 "Functionality of Oxide Interfaces" (SFB762) and the Graduate School "Leipzig School of Natural Sciences—Building with molecules and nano-objects" (BuildMoNa) (GS 185/1) is gratefully acknowledged. AL and HvW are grateful for support by the European Social Fund. AL is also grateful for support by Studienstiftung des deutschen Volkes.

References

1. A. S. Sedra and K. C. Smith, *Microelectronic Circuits*, Oxford University Press, Oxford, U.K., 2004.
2. T. Mizutani, N. Kato, S. Ishida, K. Asai, Y. Sakakibara, K. Komatsu, and M. Ohmori, High speed enhancement-mode GaAs MESFET integrated circuits, *Jpn. J. Appl. Phys.* **19**, 329–333 (1980).

3. R. L. van Tuyl and C. A. Liechti, High-speed integrated logic with GaAs MESFETs, *IEEE J. Solid-State Circuits* **9**, 269 (1974).

4. Y. R. Ryu, T. S. Lee, J. A. Lubguban, H. W. White, Y. S. Park, and C. J. Youn, ZnO devices: Photodiodes and p-type field-effect transistors, *Appl. Phys. Lett.* **87**, 153504 (2005).

5. S. Kandasamy, W. Wlodarski, A. Holland, S. Nakagomi, and Y. Kokubun, Electrical characterization and hydrogen gas sensing properties of a n-ZnO/p-SiC Pt-gate metal semiconductor field effect transistor, *Appl. Phys. Lett.* **90**, 064103 (2007).

6. C. J. Kao, Y. W. Kwon, Y. W. Heo, D. P. Norton, S. J. Pearton, F. Ren, and G. C. Chi, Comparison of ZnO metal–oxide–semiconductor field effect transistor and metal–semiconductor field effect transistor structures grown on sapphire by pulsed laser deposition, *J. Vac. Sci. Technol. B* **23**, 1024–1028 (2005).

7. W. Park, J. Kim, G.-C. Yi, and H.-J. Lee, ZnO nanorod logic circuits, *Adv. Mater.* **17**, 1393–1397 (2005).

8. H. Frenzel, A. Lajn, M. Brandt, H. von Wenckstern, G. Biehne, H. Hochmuth, M. Lorenz, and M. Grundmann, ZnO metal-semiconductor field-effect transistors with Ag-Schottky gates, *Appl. Phys. Lett.* **92**, 192108 (2008).

9. H. Frenzel, A. Lajn, H. von Wenckstern, M. Lorenz, F. Schein, Z. Zhang, and M. Grundmann, Recent progress on ZnO-based metal-semiconductor field-effect transistors and their application in transparent integrated circuits, *Adv. Mater.* **22**, 5332–5349 (2010).

10. P.-W. Chiu and C.-H. Chen, High-performance carbon nanotube network transistors for logic applications, *Appl. Phys. Lett.* **92**, 063511 (2008).

11. A. Schindler, J. Brill, N. Fruehauf, J. P. Novak, and Z. Yaniv, Solution-deposited carbon nanotube layers for flexible display applications, *Phys. E Low-Dimens. Syst. Nanostruct.* **37**, 119–123 (2007).

12. E. Artukovic, M. Kaempgen, D. S. Hecht, S. Roth, and G. Grüner, Transparent and flexible carbon nanotube transistors, *Nano Lett.* **5**, 757–760 (2005).

13. T. Takenobu, T. Takahashi, T. Kanbara, K. Tsukagoshi, Y. Aoyagi, and Y. Iwasa, High-performance transparent flexible transistors using carbon nanotube films, *Appl. Phys. Lett.* **88**, 033511 (2006).

14. Q. Cao, Z.-T. Zhu, M. G. Lemaitre, M.-G. Xia, M. Shim, and J. A. Rogers, Transparent flexible organic thin-film transistors that use printed single-walled carbon nanotube electrodes, *Appl. Phys. Lett.* **88**, 113511 (2006).

15. J.-M. Choi, D. K. Hwang, J. H. Kim, and S. Im, Transparent thin-film transistors with pentacene channel, AlO_x gate, and NiO_x electrodes, *Appl. Phys. Lett.* **86**, 123505 (2005).

16. J. B. Koo, J. W. Lim, S. H. Kim, S. J. Yun, C. H. Ku, S. C. Lim, and J. H. Lee, Pentacene thin-film transistors and inverters with plasma-enhanced atomic-layer-deposited Al2O3 gate dielectric, *Thin Solid Films* **515**, 3132–3137 (2007).

17. K. Nomura, H. Ohta, K. Ueda, T. Kamiya, M. Hirano, and H. Hosono, All oxide transparent MISFET using high-k dielectrics gates, *Microelectron. Eng.* **72**, 294–298 (2004).

18. H. Q. Chiang, J. F. Wager, R. L. Hoffman, J. Jeong, and D. A. Keszler, High mobility transparent thin-film transistors with amorphous zinc tin oxide channel layer, *Appl. Phys. Lett.* **86**, 013503 (2005).

19. N. L. Dehuff, E. S. Kettenring, D. Hong, H. Q. Chiang, J. F. Wager, R. L. Hoffman, C.-H. Park, and D. A. Keszler, Transparent thin-film transistors with zinc indium oxide channel layer, *J. Appl. Phys.* **97**, 064505 (2005).

20. H. Lee, G. Yoo, and J. Kanicki, Asymmetric electrical properties of fork *a*-Si:H thin-film transistor and its application to flat panel displays, *J. Appl. Phys.* **105**, 124522 (2009).

21. P. Görrn, F. Ghaffari, T. Riedl, and W. Kowalsky, Zinc tin oxide based driver for highly transparent active matrix OLED displays, *Solid-State Electron.* **53**, 329–331 (2009).

22. M. J. Lee, S. I. Kim, C. B Lee, H. Yin, S. E. Ahn, B. S. Kang et al., Low-temperature-grown transition metal oxide based storage materials and oxide transistors for high-density non-volatile memory, *Adv. Funct. Mater.* **19**, 1587 (2009).

23. L. Wang, M.-H. Yoon, G. Lu, Y. Yang, A. Facchetti, and T. J. Marks, High-performance transparent inorganic-organic hybrid thin-film n-type transistors, *Nat. Mater.* **5**, 893–900 (2006).

24. B. V. Zeghbroeck, Principles of semiconductor devices, http://ece.colorado.edu/~bart/book/welcome.htm, (accessed April, 2012).

25. P. Görrn, M. Lehnhardt, T. Riedl, and W. Kowalsky, The influence of visible light on transparent zinc tin oxide thin film transistors, *Appl. Phys. Lett.* **91**, 193504 (2007).

26. P. Görrn, P. Hoelzer, T. Riedl, W. Kowalsky, J. Wang, T. Weimann, P. Hinze, and S. Kipp, Stability of transparent zinc tin oxide transistors under bias stress, *Appl. Phys. Lett.* **90**, 063502 (2007).

27. B. J. Coppa, R. F. Davis, and R. J. Nemanich, Gold Schottky contacts on oxygen plasma-treated, n-type ZnO(000-1), *Appl. Phys. Lett.* **82**, 400–402 (2003).

28. A. Lajn, H. von Wenckstern, Z. Zhang, C. Czekalla, G. Biehne, J. Lenzner et al., Properties of reactively sputtered Ag, Au, Pd, and Pt Schottky contacts on n-type ZnO, *J. Vac. Sci. Technol. B* **27**, 1769 (2009).

29. M. W. Allen, S. M. Durbin, and J. B. Metson, Silver oxide Schottky contacts on n-type ZnO, *Appl. Phys. Lett.* **91**, 053512 (2007).

30. M. W. Allen, R. J. Mendelsberg, R. J. Reeves, and S. M. Durbin, Oxidized noble metal Schottky contacts to n-type ZnO, *Appl. Phys. Lett.* **94**, 103508 (2009).

31. H. Frenzel, A. Lajn, H. von Wenckstern, G. Biehne, H. Hochmuth, and M. Grundmann, ZnO-based metal-semiconductor field-effect transistors with Ag-, Pt-, Pd-, and Au-Schottky gates, *Thin Solid Films* **518**, 1119–1123 (2009).

32. H. Frenzel, M. Lorenz, A. Lajn, H. von Wenckstern, G. Biehne, H. Hochmuth, and M. Grundmann, ZnO-based metal-semiconductor field-effect transistors on glass substrates, *Appl. Phys. Lett.* **95**, 153503 (2009).

33. H. von Wenckstern, G. Biehne, R. A. Rahman, H. Hochmuth, M. Lorenz, and M. Grundmann, Mean barrier height of Pd Schottky contacts on ZnO thin films, *Appl. Phys. Lett.* **88**, 092102 (2006).

34. C. A. Mead, Surface barriers on ZnSe and ZnO, *Phys. Lett.* **18**, 218–218 (1965).

35. S.-H. Kim, H.-K. Kim, and T.-Y. Seong, Effect of hydrogen peroxide treatment on the characteristics of Pt Schottky contact on n-type ZnO, *Appl. Phys. Lett.* **86**, 112101 (2005).

36. H. Frenzel, ZnO-based metal-semiconductor field-effect transistors, Dissertation, Universität Leipzig, Der Andere Verlag, Tönning, Germany, 2010.

37. H. von Wenckstern, M. Brandt, G. Zimmermann, J. Lenzner, H. Hochmuth, M. Lorenz, and M. Grundmann, Temperature dependent Hall measurements on PLD thin films, *Materials Research Society Symposium Proceedings*, eds. J. Christen, C. Jagadish, D. C. Look, T. Yao, and F. Bertram, Vol. 957, 0957-K03-02 (2007).

38. J. W. Orton, B. J. Goldsmith, M. J. Powell, and J. A. Chapman, Temperature dependence of intergrain barriers in polycrystalline semiconductor films, *Appl. Phys. Lett.* **37**, 557–559 (1980).

39. C. Liang, N. Cheung, R. Sato, M. Sokolich, and N. Doudoumopoulos, A diffusion model of subthreshold current for GaAs MESFETs, *Solid-State Electron.* **34**, 131 (1991).

40. S. Jit, P. K. Pandey, and P. K. Tiwari, Modeling of the subthreshold current and subthreshold swing of fully depleted short-channel Si-SOI-MESFETs, *Solid-State Electron.* **53**, 57–62 (2009).

41. S. M. Sze, *Physics of Semiconductor Devices*, John Wiley & Sons, New York, 1981.

42. K. Nomura, T. Kamiya, H. Ohta, M. Hirano, and H. Hosono, Defect passivation and homogenization of amorphous oxide thin-film transistor by wet O_2 annealing, *Appl. Phys. Lett.* **93**, 192107 (2008).

43. A. Sato, K. Abe, R. Hayashi, H. Kumomi, K. Nomura, T. Kamiya, M. Hirano, and H. Hosono, Amorphous In–Ga–Zn–O coplanar homojunction thin-film transistor, *Appl. Phys. Lett.* **94**, 133502 (2009).

44. J. H. Na, M. Kitamura, and Y. Arakawa, High field-effect mobility amorphous InGaZnO transistors with aluminum electrodes, *Appl. Phys. Lett.* **93**, 063501 (2008).

45. Y.-J. Cho, J.-H. Shin, S. Bobade, Y.-B. Kim, and D.-K. Choi, Evaluation of Y_2O_3 gate insulators for a-IGZO thin film transistors, *Thin Solid Films* **517**, 4115–4118 (2009).

46. M. Brandt, H. von Wenckstern, C. Meinecke, T. Butz, H. Hochmuth, M. Lorenz, and M. Grundmann, Dopant activation in homoepitaxial MgZnO:P thin films, *J. Vac. Sci. Technol.* **27**, 1604–1608 (2009).

47. H. von Wenckstern, H. Schmidt, C. Hanisch, M. Brandt, C. Czekalla, G. Benndorf, G. Biehne, A. Rahm, H. Hochmuth, M. Lorenz, and M. Grundmann, Homoepitaxy of ZnO by pulsed-laser deposition, *Phys. Status Solidi RRL* **1**, 129–131 (2007).

48. H. von Wenckstern, S. Müller, G. Biehne, H. Hochmuth, M. Lorenz, and M. Grundmann, Dielectric passivation of ZnO-based Schottky diodes, *J. Electron. Mater.* **39**, 559–562 (2010).

49. H. von Wenckstern, A. Lajn, A. Laufer, B. K. Meyer, H. Hochmuth, and M. Lorenz, Ag related defect state in ZnO thin films, *Physics of Semiconductors: 29th International Conference on the Physics of Semiconductors Rio de Janeiro*, Brazil, eds. M. Grundmann, M. Caldas, and N. Studart, Vol. 1199, pp.122–123 (2010).

50. D. B. Chrisey and G. K. Hubler (eds.), *Pulsed Laser Deposition of Thin Films*, John Wiley & Sons, New York, 1994.

51. T. I. Suzuki, A. Ohtomo, A. Tsukazaki, F. Sato, J. Nishii, H. Ohno, and M. Kawasaki, Hall and field-effect mobilities of electrons accumulated at a lattice-matched ZnO/ScAlMgO4 heterointerface, *Adv. Mater.* **16**, 1887–1890 (2004).

52. J. Nishii, A. Ohtomo, K. Ohtani, H. Ohno, and M. Kawasaki, High-mobility field-effect transistors based on single-crystalline ZnO channels, *Jpn. J. Appl. Phys.* **44**, L1193–L1195 (2005).

53. K. Koike, I. Nakashima, K. Hashimoto, S. Sasa, M. Inoue, and M. Yano, Characteristics of a $Zn_{0.7}Mg_{0.3}O$/ZnO heterostructure field-effect transistor grown on sapphire substrate by molecular-beam epitaxy, *Appl. Phys. Lett.* **87**, 112106 (2005).

54. S. Sasa, M. Ozaki, K. Koike, M. Yano, and M. Inoue, High-performance ZnO/ZnMgO field-effect transistors using a hetero-metal-insulator-semiconductor structure, *Appl. Phys. Lett.* **89**, 053502 (2006).

55. J. Zhu, H. Chen, G. Saraf, Z. Duan, Y. Lu, and S. Hsu, ZnO TFT devices built on glass substrates, *J. Electron. Mater.* **37**, 1237–1240 (2008).

56. E. Bellingeri, I. Pallecchi, L. Pellegrino, G. Canu, M. Biasotti, M. Vignolo, A. S. Siri, and D. Marré, Crystalline ZnO/$SrTiO_3$ transparent field effect transistor, *Phys. Status Solidi A* **205**, 1934–1937 (2008).

57. R. L. Hoffman, B. J. Norris, and J. F. Wager, ZnO-based transparent thin-film transistors, *Appl. Phys. Lett.* **82**, 733–735 (2003).

58. S. Masuda, K. Kitamura, Y. Okumura, S. Miyatake, H. Tabata, and T. Kawai, Transparent thin film transistors using ZnO as an active channel layer and their electrical properties, *J. Appl. Phys.* **93**, 1624–1630 (2003).

59. E. M. C. Fortunato, P. M. C. Barquinha, A. C. M. B. G. Pimentel, A. M. F. Gonçalves, A. J. S. Marques, L. M. N. Pereira, and R. F. P. Martins, Fully transparent ZnO thin-film transistor produced at room temperature, *Adv. Mater.* **17**, 590–594 (2005).

60. R. Navamathavan, J.-H. Lim, D.-K. Hwang, B.-H. Kim, J.-Y. Oh, J.-H. Yang, H.-S. Kim, and S.-J. Park, Thin-film transistors based on ZnO fabricated by using radio-frequency magnetron sputtering, *J. Korean Phys. Soc.* **48**, 271–274 (2006).

61. H.-H. Hsieh and C.-C. Wu, Scaling behavior of ZnO transparent thin-film transistors, *Appl. Phys. Lett.* **89**, 041109 (2006).

62. P. Barquinha, E. Fortunato, A. Gonçalves, A. Pimentel, A. Marques, L. Pereira, and R. Martins, Influence of time, light and temperature on the electrical properties of zinc oxide TFTs, *Superlattices Microstruct.* **39**, 319–327 (2006).

63. H.-C. Cheng, C.-F. Chen, and C.-Y. Tsay, Transparent ZnO thin film transistor fabricated by sol-gel and chemical bath deposition combination method, *Appl. Phys. Lett.* **90**, 012113 (2007).

64. M. S. Oh, D. K. Hwang, K. Lee, S. Im, and S. Yi, Low voltage complementary thin-film transistor inverters with pentacene-ZnO hybrid channels on AlO_x dielectric, *Appl. Phys. Lett.* **90**, 173511 (2007).

65. S. H. Noh, W. Choi, M. S. Oh, D. K. Hwang, K. Lee, S. Im, S. Jang, and E. Kim, ZnO-based non-volatile memory thin-film transistors with polymer dielectric/ferroelectric double gate insulators, *Appl. Phys. Lett.* **90**, 253504 (2007).

66. D. H. Levy, D. Freeman, S. F. Nelson, P. J. Cowdery-Corvan, and L. M. Irving, Stable ZnO thin film transistors by fast open air atomic layer deposition, *Appl. Phys. Lett.* **92**, 192101 (2008).

67. M. Rocchi, Status of the surface and bulk parasitic effects limiting the performances of GaAs IC's, *Physica B+C* **129**, 119 (1985).

68. P. Harrang, A. Tardella, M. Rosso, P. Alnot, and J. F. Peray, Conductance transient spectroscopy of metal-semiconductor field effect transistors, *J. Appl. Phys.* **61**, 1931 (1987).

69. Y. Fujisaki and N. Matsunaga, The origin of gate hysteresis and gate delay of MESFETs appear under low frequency operation, *Gallium Arsenide Integrated Circuit (GaAs IC) Symposium*, p.235 (1988).

70. H. Sasaki, H. Matsubayashi, O. Ishihara, R. Konishi, and K. Ando, Analysis of gate lag in GaAs metal-semiconductor field-effect transistor using light illumination, *Jpn. J. Appl. Phys.* **34**, 6346 (1995).

71. K. Horio and Y. Fuseya, Two-dimensional simulations of drain-current transients in GaAs MESFETs with semi-insulating substrates compensated by deep levels, *IEEE Trans. Electron Devices* **41**, 1340–1346 (1994).

72. W. Koerner and C. Elsaesser, First-principles density functional study of dopant elements at grain boundaries in ZnO, *Phys. Rev. B* **81**, 085324 (2010).

73. J. Graffeuil, Z. Hadjoub, J. P. Fortea, and M. Pouysegur, Analysis of capacitance and trans-conductance dispersions in MESFETs for surface characterization, *Solid-State Electron.* **29**(10), 1087–1097 (1986).

74. E. Kohn, I. Daumiller, M. Kunze, M. Neuburger, M. Seyboth, T. J. Jenkins, J. S. Sewell, J. Van Norstand, Y. Smorchkova, and U. K. Mishra, Transient characteristics of GaN-based hetero-structure field-effect transistors, *IEEE Trans. Microwave Theory Technol.* **51**, 634 (2003).

75. O. Mitrofanov and M. Manfra, Mechanisms of gate lag in GaN/AlGaN/GaN high electron mobility transistors, *Superlattices Microstruct.* **34**, 33–53 (2003).

76. F. J. Klüpfel, A. Lajn, H. Frenzel, H. von Wenckstern, and M. Grundmann, Gate- and drain-lag effects in (Mg,Zn)O-based metal-semiconductor field-effect transistors, *J. Appl. Phys.* **109**, 074515 (2011).

77. G. Boole, *An Investigation of the Laws of Thought*, Walton and Maberly, London, U.K., 1854.

78. B. M. Welch and R. C. Eden, Planar GaAs integrated circuits fabricated by ion implantation, *1977 Int. Electron Dev. Meeting* **23**, 205 (1977).

79. R. C. Eden and B. M. Welch, Low power depletion mode ion-implanted GaAs FET integrated circuits, *IEEE Trans. Electron Devices* **24**, 1209 (1977).

80. R. C. Eden, GaAs integrated circuits: MSI status and VLSI prospects, *IEEE* **24**, 6 (1978).

81. R. C. Eden, B. M. Welch, and R. Zucca, Low power GaAs digital ICs using Schottky diode-FET logic, *IEEE International Solid-State Circuits Conference Digest of Technical Papers*, February 1978, p.68 (1978).

82. R. C. Eden, Schottky diode FET logic integrated circuit, US Patent 4300064 (1981).

83. H. Frenzel, F. Schein, A. Lajn, H. von Wenckstern, and M. Grundmann, High-gain integrated inverters based on ZnO metal-semiconductor field-effect transistor technology, *Appl. Phys. Lett.* **96**, 113502 (2010).

84. M. J. Helix, S. A. Jamison, C. Chao, and M. S. Shur, Fan out and speed of GaAs SDFL logic, *IEEE J. Solid-State Circuits* **SC-17**, 1226 (1982).

85. R.-M. Ma, L. Dai, H.-B. Huo, W.-J. Xu, and G. G. Qin, High-performance logic circuits constructed on single CdS nanowires, *Nano Lett.* **7**, 3300–3304 (2007).

86. J. Nylander, U. Magnusson, M. Rosling, and P. Tove, Influence of silicon-sapphire interface defects on SOS MESFET behaviour, *Solid-State Electron.* **31**, 1493–1496 (1988).

87. M. Rosenbluth, W. L. Bloss, W. E. Yamada, and B. K. Janousek, Characteristics of GaAs buffered FET logic (BFL) MESFETs and inverters exposed to high-energy neutrons, *IEEE Trans. Nucl. Sci.* **38**, 20 (1991).

88. Y. Cai, Z. Cheng, W. C. W. Tang, K. M. Lau, and K. J. Chen, Monolithically integrated enhancement/depletion-mode AlGaN/GaN HEMT inverters and ring oscillators using CF_4 plasma treatment, *IEEE Trans. Electron Dev.* **53**, 2223 (2006).

89. Y. Cai, Z. Cheng, Z. Yang, C. W. Tang, K. M. Lau, and K. J. Chen, High-temperature operation of AlGaN/GaN HEMTs direct-coupled FET logic (DCFL) integrated circuits, *IEEE Electron Device Lett.* **28**, 328 (2007).

90. M. Itoh and H. Kawarada, Fabrication and characterization of metal-semiconductor field-effect transistor utilizing diamond surface-conductive layer, *Jpn. J. Appl. Phys.* **34**, 4677 (1995).

91. R. M. Ma, L. Dai, C. Liu, W. J. Xu, and G. G. Qin, High-performance nanowire complementary metal-semiconductor inverters, *Appl. Phys. Lett.* **93**, 053105 (2008).

92. H. Frenzel, A. Lajn, H. von Wenckstern, and M. Grundmann, Ultrathin gate-contacts for metal-semiconductor field-effect transistor devices: An alternative approach in transparent electronics, *J. Appl. Phys.* **107**, 114515 (2010).

93. H. Frenzel, A. Lajn, H. von Wenckstern, and M. Grundmann, Transparente gleichrichtende Metall-Metalloxid-Halbleiterkontaktstruktur und Verfahren zu seiner Herstellung und Verwendung, Deutsches Patent Nr. 10 2009 030 045 (2009).

94. T. Jang, J. Kwak, O. Nam, and Y. Park, New contact resistivity characterization method for nonuniform ohmic contacts on GaN, *Solid-State Electron.* **50**, 433–436 (2006).

95. M. Nakano, T. Makino, A. Tsukazaki, K. Ueno, A. Ohtomo, T. Fukumura et al., Transparent polymer Schottky contact for a high performance visible-blind ultraviolet photodiode based on ZnO, *Appl. Phys. Lett.* **93**, 123309 (2008).

96. M. Nakano, T. Makino, A. Tsukazaki, K. Ueno, A. Ohtomo, T. Fukumura et al., $Mg_xZn_{1-x}O$-based Schottky photodiode for highly color-selective ultraviolet light detection, *Appl. Phys. Express* **1**, 121201 (2008).

97. H. Endo, M. Sugibuchi, K. Takahashi, S. Goto, S. Sugimura, K. Hane, and Y. Kashiwaba, Schottky ultraviolet photodiode using a ZnO hydrothermally grown single crystal substrate, *Appl. Phys. Lett.* **90**, 121906 (2007).

98. Z. G. Ju, C. X. Shan, D. Y. Jiang, J. Y. Zhang, B. Yao, D. X. Zhao, D. Z. Shen, and X. W. Fan, $Mg_xZn_{1-x}O$-based photodetectors covering the whole solar-blind spectrum range, *Appl. Phys. Lett.* **93**, 173505 (2008).

99. D. Jiang, C. Shan, J. Zhang, Y. Lu, B. Yao, D. Zhao, Z. Zhang, X. Fan, and D. Shen, Schottky barrier photodetectors based on $Mg_{0.40}Zn_{0.60}O$ thin films, *Appl. Phys. Lett.* **93**, 123309 (2008).

100. S. Liang, H. Sheng, Y. Liu, Z. Huo, Y. Lu, and H. Shen, ZnO Schottky ultraviolet photodetectors, *J. Cryst. Growth* **225**, 110 (2001).

101. S. J. Young, L. W. Ji, S. J. Chang, and Y. K. Su, ZnO metal–semiconductor–metal ultraviolet sensors with various contact electrodes, *J. Cryst. Growth* **293**, 43 (2006).

102. K. W. Liu, J. G. Ma, J. Y. Zhang, Y. M. Lu, D. Y. Jiang, B. H. Li, D. X. Zhao, Z. Z. Zhang, B. Yao, and D. Z. Shen, Ultraviolet photoconductive detector with high visible rejection and fast photoresponse based on ZnO thin film, *Solid-State Electron.* **51**, 757 (2007).

103. B. Angadi, H. C. Park, H. W. Choi, J. W. Choi, and W. K. Choi, Oxygen-plasma treated epitaxial ZnO thin films for Schottky ultraviolet detection, *J. Phys. D Appl. Phys.* **40**, 1422 (2007).

104. T. Takagi, H. Tanaka, S. Fujita, and S. Fujita, Molecular beam epitaxy of high magnesium content single-phase wurzite MgZnO alloys and their application to solar-blind region photodetectors, *Jpn. J. Appl. Phys.* **42**, L401 (2003).

105. G. Tabares, A. Hierro, J. M. Ulloa, A. Guzman, E. Muñoz, A. Nakamura, T. Hayashi, and J. Temmyo, High responsivity and internal gain mechanisms in Au-ZnMgO Schottky photodiodes, *Appl. Phys. Lett.* **96**, 101112 (2010).

106. W. Yang, R. D. Vispute, S. Choopun, R. P. Sharma, T. Venkatesan, and H. Shen, Ultraviolet photoconductive detector based on epitaxial $Mg_{0.34}Zn_{0.66}O$ thin films, *Appl. Phys. Lett.* **78**, 2787 (2001).

107. Y. Liu, C. R. Gorla, S. Liang, N. Emanetoglu, Y. Lu, H. Shen, and M. Wraback, Ultraviolet detectors based on epitaxial ZnO films grown by MOCVD, *J. Electron. Mater.* **29**, 69 (2000).

108. K. Wang, Y. Vygranenko, and A. Nathan, Optically transparent ZnO-based n–i–p ultraviolet photodetectors, *Thin Solid Films* **515**, 6981 (2007).

109. H. Endo, M. Sugibuchi, K. Takahashi, S. Goto, K. Hane, and Y. Kashiwaba, Fabrication and characteristics of a Pt/Mg$_x$Zn$_{1-x}$O Schottky photodiode on a ZnO single crystal, *Phys. Status Solidi C* **5**, 3119 (2008).

110. J. H. Werner and H. H. Güttler, Barrier inhomogeneities at Schottky contacts, *J. Appl. Phys.* **69**, 1522 (1991).

111. H. Oheda, Phase-shift analysis of modulated photocurrent: Its application to the determination of the energetic distribution of gap states, *J. Appl. Phys.* **52**, 6693 (1981).

112. B. D. Ahn, H. S. Kang, J. H. Kim, G. H. Kim, H. W. Chang, and S. Y. Lee, Synthesis and analysis of Ag-doped ZnO, *J. Appl. Phys.* **100**, 093701 (2006).

113. J. F. Wager, Applied physics: Transparent electronics, *Science* **300**, 1245–1246 (2003).

114. H. Yabuta, M. Sano, K. Abe, T. Aiba, T. Den, H. Kumomi, K. Nomura, T. Kamiya, and H. Hosono, High-mobility thin-film transistor with amorphous InGaZnO[sub 4] channel fabricated by room temperature RF-magnetron sputtering, *Appl. Phys. Lett.* **89**, 112123 (2006).

115. J. Sun, D. A. Mourey, D. Zhao, and T. N. Jackson, ZnO thin film, device, and circuit fabrication using low-temperature PECVD processes, *J. Electron. Mater.* **37**, 755–759 (2008).

116. R. Presley, D. Hong, H. Chiang, C. Hung, R. Hoffman, and J. Wager, Transparent ring oscillator based on indium gallium oxide thin-film transistors, *Solid-State Electron.* **50**, 500–503 (2006).

117. D. Heineck, B. McFarlane, and J. Wager, Zinc tin oxide thin-film-transistor enhancement/depletion inverter, *IEEE Electron Device Lett.* **30**, 514–516 (2009).

118. P. F. Carcia, R. S. McLean, M. H. Reilly, and J. G. Nunes, Transparent ZnO thin-film transistor fabricated by RF magnetron sputtering, *Appl. Phys. Lett.* **82**, 1117–1119 (2003).

119. J. Nishii, F. M. Hossain, S. Takagi, T. Aita, K. Saikusa, Y. Ohmaki et al., High mobility thin film transistors with transparent ZnO channels, *Jpn. J. Appl. Phys.* **42**, L347–L349 (2003).

120. E. M. C. Fortunato, P. M. C. Barquinha, A. C. M. B. G. Pimentel, A. M. F. Gonçalves, A. J. S. Marques, R. F. P. Martins, and L. M. Pereira, Wide-bandgap high-mobility ZnO thin-film transistors produced at room temperature, *Appl. Phys. Lett.* **85**, 2541–2543 (2004).

121. K. Nomura, H. Ohta, K. Ueda, T. Kamiya, M. Hirano, and H. Hosono, Thin-film transistor fabricated in single-crystalline transparent oxide semiconductor, *Science* **300**, 1269–1272 (2003).

122. K. Nomura, H. Ohta, A. Takagi, T. Kamiya, M. Hirano, and H. Hosono, Room-temperature fabrication of transparent flexible thin-film transistors using amorphous oxide semiconductors, *Lett. Nat.* **432**, 488–492 (2004).

123. A. Suresh, P. Wellenius, A. Dhawan, and J. Muth, Room temperature pulsed laser deposited indium gallium zinc oxide channel based transparent thin film transistors, *Appl. Phys. Lett.* **90**, 123512 (2007).

124. M. Kim, J. H. Jeong, H. J. Lee, T. K. Ahn, H. S. Shin, J.-S. Park, J. K. Jeong, Y.-G. Mo, and H. D. Kim, High mobility bottom gate InGaZnO thin film transistors with SiO$_x$ etch stopper, *Appl. Phys. Lett.* **90**, 212114 (2007).

125. B. N. Pal, B. M. Dhar, K. C. See, and H. E. Katz, Solution-deposited sodium beta-alumina gate dielectrics for low-voltage and transparent field-effect transistors, *Nat. Mater.* **8**, 898 (2009).

126. R. E. Presley, C. L. Munsee, C.-H. Park, D. Hong, J. F. Wager, and D. A. Keszler, Tin oxide transparent thin-film transistors, *J. Phys. D Appl. Phys.* **37**, 2810–2813 (2004).

127. E. N. Dattoli, Q. Wan, W. Guo, Y. Chen, X. Pan, and W. Lu, Fully transparent thin-film transistor devices based on SnO$_2$ nanowires, *Nano Lett.* **7**, 2463–2469 (2007).

128. Q. Cao, S.-H. Hur, Z.-T. Zhu, Y. G. Sun, C.-J. Wang, M. A. Meitl, M. Shim, and J. A. Rogers, Highly bendable, transparent thin-film transistors that use carbon-nanotube-based conductors and semiconductors with elastomeric dielectrics, *Adv. Mater.* **18**, 304–309 (2006).

129. A. Lajn, T. Diez, F. Schein, H. Frenzel, H. von Wenckstern, and M. Grundmann, Light and temperature stability of fully transparent ZnO-based inverter circuits, *IEEE Electron Device Lett.* **32**(4), 515–517 (2011), DOI: 10.1109/LED.2011.2106193.

130. R. A. Street, Thin-film transistors, *Adv. Mater.* **21**, 2007 (2009).

131. J. J. Huang, C. J. Liu, H. C. Lin, C. J. Tsai, Y. P. Chen, G. R. Hu, and C. C. Lee, Influences of low temperature silicon nitride films on the electrical performances of hydrogenated amorphous silicon thin film transistors, *J. Phys. D Appl. Phys.* **41**, 245502 (2008).

132. H. Hosono, N. Kikuchi, N. Ueda, and H. Kawazoe, Working hypothesis to explore novel wide band gap electrically conducting amorphous oxides and examples, *J. Non-Cryst. Solids* **198–200**, 165–169 (1996).

133. R. Martins, P. Barquinha, I. Ferreira, L. Pereira, G. Gonçalves, and E. Fortunato, Role of order and disorder on the electronic performances of oxide semiconductor thin film transistors, *J. Appl. Phys.* **101**, 044505 (2007).

134. E. Fortunato, A. Pimentel, A. Gonçalves, A. Marques, and R. Martins, Next generation of thin film transistors based on zinc oxide, *Mater. Res. Soc. Symp. Proc.* **811**, E1.9 (2004).

135. E. Fortunato, P. Barquinha, G. Goncalves, L. Pereira, and R. Martins, Oxide semiconductors: From materials to devices, in *Transparent Electronics—From Synthesis to Applications*, eds. A. Facchetti and T. J. Marks, John Wiley & Sons, Chichester, U.K., 2010.

136. P. Barquinha, L. Pereira, G. Gonçalves, R. Martins, and E. Fortunato, Toward high-performance amorphous GIZO TFTs, *J. Electrochem. Soc.* **156**, H161–H168 (2009).

137. E. Fortunato, P. Barquinha, A. Pimentel, L. Pereira, G. Gonçalves, and R. Martins, Amorphous IZO TTFTs with saturation mobilities exceeding $100\,cm^2/Vs$, *Phys. Status Solidi RRL* **1**, R34–R36 (2007).

138. R. A. Street, *Hydrogenated Amorphous Silicon*, Cambridge University Press, Cambridge, U.K., 1990.

139. Y. Kaneko, A. Sasano, and T. Tsukada, Characterization of instability in amorphous silicon thin-film transistors, *J. Appl. Phys.* **69**, 7301–7305 (1991).

140. Y. Shimura, K. Nomura, H. Yanagi, T. Kamiya, M. Hirano, and H. Hosono, Specific contact resistances between amorphous oxide semiconductor In-Ga-Zn-O and metallic electrodes, *Thin Solid Films* **516**, 5899 (2008).

141. M. Grundmann, H. Frenzel, A. Lajn, M. Lorenz, F. Schein, and H. von Wenckstern, Transparent semiconducting oxides: Materials and devices, *Phys. Status Solidi A* **207**, 1437–1449 (2010).

142. R. Martins, P. Barquinha, L. Pereira, N. Correia, G. Gonçalves, I. Ferreira, and E. Fortunato, Write-erase and read paper memory transistor, *Appl. Phys. Lett.* **93**, 203501 (2008).

143. K. B. Park, J. B. Seon, G. H. Kim, M. Yang, B. Koo, H. J. Kim, M. K. Ryu, and S. Y. Lee, High electrical performance of wet-processed indium zinc oxide thin-film transistors, *IEEE Electron Device Lett.* **31**, 311–313 (2010).

144. D. H. Cho, S. Yang, C. Byun, J. Shin, M. K. Ryu, S. H. K. Park et al., Transparent Al-Zn-Sn-O thin film transistors prepared at low temperature, *Appl. Phys. Lett.* **93**, 142111 (2008).

145. J.-M. Lee, W.-S. Cheong, C.-S. Hwang, I.-T. Cho, H.-I. Kwon, and J.-H. Lee. Low-frequency noise in amorphous indium-gallium-zinc-oxide thin-film transistors, *IEEE Electron. Device Lett.* **30**, 505–507 (2009).

146. Y. Ogo, K. Nomura, H. Yanagi, T. Kamiya, H. Masahiro, and H. Hosono, Amorphous Sn-Ga-Zn-O channel thin-film transistors, *Phys. Status Solidi A* **205**, 1920 (2008).

147. W. Lim, E. A. Douglas, S. H. Kim, D. P. Norton, S. J. Pearton, F. Ren, H. Shen, and W. H. Chang, High mobility InGaZnO4 thin-film transistors on paper, *Appl. Phys. Lett.* **94**, 072103 (2009).

148. C. H. Jung, D. J. Kim, Y. K. Kang, and D. H. Yoon, Transparent amorphous In-Ga-Zn-O thin film as function of various gas flows for TFT applications, *Thin Solid Films* **517**, 4078 (2009).

149. M. Lorenz, A. Lajn, H. Frenzel, H. V. Wenckstern, M. Grundmann, P. Barquinha, R. Martins, and E. Fortunato, Low-temperature processed Schottky-gated field-effect transistors based on amorphous gallium-indium-zinc-oxide thin films, *Appl. Phys. Lett.* **97**, 243506 (2010).

12

Growth of ZnO for Neutron Detectors

Eric A. Burgett, Elisa N. Hurwitz, Nolan E. Hertel, Christopher J. Summers,
Jeff Nause, Na Lu, and Ian T. Ferguson

CONTENTS

12.1 Introduction 436
12.2 History of Scintillators 436
    12.2.1 Current Need for Neutron Detection..436
    12.2.2 Current State-of-the-Art Neutron Detectors 437
12.3 Scintillation Detection Theory/Mechanisms 440
    12.3.1 Radiation Interaction Mechanisms 440
    12.3.2 Gamma Ray Interactions and Detection 441
        12.3.2.1 Gamma Ray Interactions 441
        12.3.2.2 Gamma Ray Detection 443
    12.3.3 Neutron Interactions and Detection 443
    12.3.4 Heavy-Charged Particle Interactions and Transport 445
    12.3.5 Radiation Detection 447
    12.3.6 Scintillation Detector Theory 447
12.4 ZnO Scintillators 450
    12.4.1 Previous Work on ZnO Scintillators 450
    12.4.2 Introduction to ZnO 450
12.5 ZnO Growth 452
    12.5.1 Crystal Growth...452
    12.5.2 MOCVD Growth...455
    12.5.3 Hybrid MOCVD ALD Growth..457
    12.5.4 ZnO Bulk Growth 458
12.6 Performance of ZnO Neutron Scintillators 460
    12.6.1 Impact of Growth Conditions on Scintillator Performance.................460
        12.6.1.1 Effect of II–VI Ratio on ZnO Samples 460
        12.6.1.2 Effect of Growth Temperature on ZnO Samples.....................461
        12.6.1.3 Effect of Growth Pressure on ZnO Samples 462
    12.6.2 Impact of Dopants on ZnO Scintillator Performance...........................463
        12.6.2.1 Impact of Nitrogen Doping on Scintillator Performance.................463
        12.6.2.2 Impact of Gallium Doping on Scintillator Performance...................465
        12.6.2.3 Impact of Aluminum Doping on Scintillator Performance..............466
        12.6.2.4 Impact of Boron on Scintillator Performance............................466
        12.6.2.5 Impact of Lithium Doping on ZnO Scintillator Performance 466
        12.6.2.6 Impact of Gadolinium on ZnO Scintillator Performance.................467
    12.6.3 Neutron Detection Performance...470
12.7 Summary and Future Directions 475
References..480

12.1 Introduction

This chapter explores the growth of zinc oxide (ZnO) scintillators for application in neutron detectors. This chapter presents an overview of the theoretical basis for neutron detection and neutron-versus-gamma discrimination. A discussion of the ZnO crystal growth techniques with particular emphasis on bulk melt growth and metalorganic chemical vapor deposition (MOCVD) is addressed. The chapter concludes with a discussion of the effect of dopants on ZnO scintillator growth and performance. ZnO offers a material that is in many ways superior to the current state of the art neutron detectors.

12.2 History of Scintillators

12.2.1 Current Need for Neutron Detection

Current state-of-the-art detectors are still utilizing materials and methods originally developed during the height of the cold war. New detection materials and mechanisms for neutron detection have been limited over the last 50 years [1] with only incremental improvements in designs, new experimental and calculated response matrices, and better pulse-processing electronics. The relatively slow development in neutron detection technology has been, in part, because there has not been a high demand for better radiation detector materials or methods until the last decade.

Homeland security requires the screening for illicit nuclear material and currently employs helium (^{3}He) tubes in portal monitors worldwide [2]. During the cold war, the nuclear threat was always well understood, and only the superpower nations had nuclear weapons. The threat of a terrorist's use of nuclear weapons became the primary concern following the Soviet Union's collapse. Unlike the traditional cold war scenario in which nuclear weapons would be deployed using an intercontinental ballistic missile, a terrorist-detonated nuclear weapon could be smuggled into the country in which it would be deployed. Detecting the special nuclear material (SNM) inside a nuclear weapon has become a very high priority for national security.

Detecting SNM is difficult and includes any fissionable isotope, such as U or Pu, or materials that can be used to increase the yield of a nuclear weapon, such as ^{3}H or ^{6}Li. The difficulty in detecting SNM is that it emits primarily low-energy photons, beta particles, and alpha particles. These kinds of radiation are easily shielded with a small amount of lead or other high Z material. If the U or Pu is assembled into a solid mass, due to its high Z number and high density, it is self-shielding. This prohibits most of the gamma rays from exiting the solid mass. Neutron radiation, however, is not easily shielded and is, therefore, the radiation of choice for active interrogation or detection [3]. Plutonium-242 produces neutrons as a result of its decay, albeit with a low frequency [4]. SNM also has a high cross section for photon-induced neutron production (photonuclear production) or fission induced by gamma rays (photofission) [5]. Last, neutrons can induce fission in the SNM. These neutrons are highly penetrating, can be detected at long distances, and are significantly more difficult to shield.

A number of passive and active neutron detection systems have been proposed to exploit this. A passive neutron detection system is merely an array of gamma and neutron detectors that can have sophisticated software to interpret the data collected by the detectors. An active system utilizes a radiation source, either electron accelerator produced Bremsstrahlung spectra at high energies or neutrons generated from a Deuterium–Tritium (D–T), Deuterium–Deuterium (D–D), or other photoneutron source, to induce fission. Active systems may be either one or two sided. A one-sided system relies on a bank of neutron and gamma ray detectors colocated with the accelerator operating in reflectance mode [6]. A two-sided active interrogation system uses detectors on both sides of the object. This has the added advantage of twice the detection efficiency, but it is not well suited for deployment in a hostile theater such as a battlefield [7]. The two-sided approach is typically utilized in cargo container inspection systems.

Currently, both the active and passive neutron detection systems utilize a moderate and capture detection scheme based on ^{3}He tubes [8]. Congress mandated that portal monitors monitor all ports of entry into the United States and all cargo containers be screened [9]. The increased deployment of portal monitors has contributed to a national and international shortage of ^{3}He, since it comprises 0.000137% of ^{3}He found in nature [10]. ^{3}He is primarily collected during the production of tritium. As tritium decays, it produces a beta particle and transitions to ^{3}He. The main source of ^{3}He production in the United States was a stockpile stewardship program in the maintenance and construction of nuclear weapons. The cessation of tritium reprocessing has reduced the stockpiles of ^{3}He.

The nuclear physics community uses liquid ^{3}He as a cryogenic liquid to cool MRI magnets and superconducting magnets at the world's largest accelerators. These applications further reduce the supply of ^{3}He.

12.2.2 Current State-of-the-Art Neutron Detectors

Neutrons are particularly difficult to detect because, as the name indicates, they are neutral particles. Detecting them requires a material with a large cross section for charged particle production. The lighter atomic (low Z) materials tend to center around a handful of nuclei whose inner spin-orbit coupling is near to either one or a pair of magic numbers. A magic number is a number of nucleons (protons or neutrons) that can be arranged into complete shells within the atomic nucleus [11]. Atomic nuclei with a magic number of nucleons have a higher binding energy and are more stable against nuclear decay. The nuclei traditionally used as neutron targets are H, He, B, Li, Gd, U, In, and Ag. Neutron detection technologies have focused on one of five main categories: gas-filled tubes, solid foil-based detectors, liquid scintillators, inorganic scintillators, and solid-state detectors.

The early gas-filled tubes used ^{10}B-containing compounds and were based on the B(n, α)^{7}Li reaction. The first tube utilized boron trifluoride, a caustic and toxic gas, in a cylindrical or spherical chamber. High voltages, greater than 2000 V, were needed to achieve high gas amplification, which was hindered by high gas fill pressures. Current designs overcome this challenge with ^{10}B-lined cylindrical tubes filled with a safer P-10 gas (90% argon and 10% methane) or pure argon. These tubes still require relatively high voltages.

^{10}B-lined tubes, however, suffer from what is known as the "wall effect." Following the ^{10}B(n, α) reaction at thermal energies, it is likely that one of the two heavy-charged particles (HCPs), either the α particle or the Li nucleus, will strike the wall. The "wall effect" makes a stair-step pulse-height distribution.

When the α particle strikes the wall, the ^{7}Li nucleus deposits all of its energy in the gas and generates the step at 0.84 MeV. When the ^{7}Li nucleus strikes the wall, the α particle deposits all of its energy in the gas and generates the step at 1.47 MeV. Since no ionization occurs in the fill gas of the tube, that energy is lost.

[3]He was also used as a thermal neutron detector, and its low Z number makes it ideal for gamma discrimination. The Q-value for the reaction of $^3He(n, p)$ is quite large (764 keV) [12] with a significant amount of energy absorbed in the gas. Large gas amplifications can be achieved with the dry fill gas. Operating voltages in excess of 3000 V are not uncommon for larger tubes. Gas pressures between 2 and 10 atm are routinely used to raise the detection efficiencies. ^3He tubes are often surrounded by a hydrogenous neutron-thermalizing material (moderator material) such as polyethylene to further increase the detection efficiencies at higher energies. Neutron spectrometers, such as the Bonner sphere spectrometer [13–15], use the ^3He tube. The energy response of the ^3He tube can be varied to produce unique energy responses by varying the moderator thickness. These unique energy responses can be deconvolved through a mathematical unfolding process. This versatility and the ability to discriminate neutrons from gamma radiation are the primary reasons for the popularity of this technology.

Another thermal neutron detector system is based on activation reactions in foils. Some of the first neutron detectors called "Moon detectors" were Geiger–Muller tubes, which had silver foil placed over the windows of the detectors [16]. These tubes were effective, but slow to respond.

Another approach was to incorporate a fissionable material into a gas-filled tube detector, referred to as a fission chamber. Neutrons induce fission, and one of the two fission fragments enters the fill gas of the detector [17]. This technology takes advantage of the large energy produced, as the fission process releases approximately 180 MeV to the two fission fragments. Measurements of the combination of various fissionable and fertile isotopes, such as ^{235}U, ^{238}U, and ^{237}Np, and application of the fission thresholds for these materials yield an estimate of the neutron energy profile [18]. These detectors, however, require the use of fissionable material and must be registered with the state or federal government.

Solid-state neutron detection methods have been investigated over the last 20 years. Solid-state neutron detectors based on silicon diodes were originally coated with a neutron detector material such as ^6Li, ^{10}B, or ^{235}U. More recently, application of microelectronic fabrication methods has enabled the production of pillar, monolithic, and trench silicon diode structures, in which holes, rods, or long trenches are etched in a silicon diode wafer. Using a thermal evaporation process, ^6LiF or ^{10}B is infiltrated into the etched spaces. Infiltrating the material into the diode structure itself allows both charged particles produced in the (n, α) reaction to be captured, which enhances neutron-versus-gamma discrimination. Examples of the trench and pillar Silicon structures are shown below in Figure 12.1 [19,20]. Developers of these detectors have claimed intrinsic detection efficiencies—greater than 50%—for thermal neutrons.

Silicon trench detectors are a promising detection method, but they often suffer from breakdown of the silicon structures, problems infiltrating the ^{10}B into the structure, and difficulties with fabrication. One of the drawbacks of these systems is the rate of charge collection, particularly after infiltration with neutron detector materials. During the infiltration process, it is difficult to ensure that no lithium diffuses into the semiconductor, as ^6Li passivates the p-i-n structure of the semiconductor, rendering it useless.

Another method for detecting neutrons is the scintillation process. Several detector designs around the scintillation process have been conceived including organic and inorganic solutions. Inorganic detectors, such as ^6LiI, are attractive because they readily absorb thermal neutrons. The ^6LiI scintillator has a high Z number, making it also sensitive to gamma radiation. But it is difficult to discriminate gamma radiation from neutron radiation if the crystal is very large. Crystals made of B and Gd have also been proposed, as

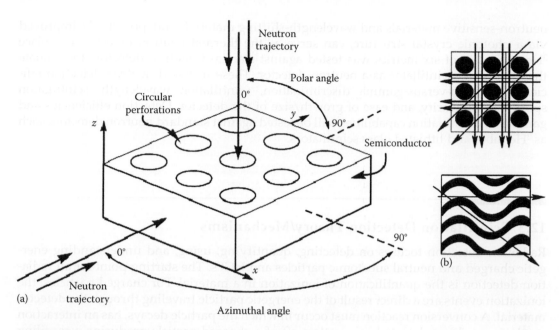

FIGURE 12.1
(a) Pillar structure of [10]B-lined silicon pillars grown at LLNL. (b) Silicon trench detector filled with [10]B designed at Kansas State University. (From Bellinger, S.L., McNeil, W.J., Unruh, T.C., and McGregor, D.S., Angular response of perforated silicon diode high efficiency neutron detectors, *IEEE NSS Conference Record*, Honolulu, HI, 1904, 2007.)

these materials also have high Z number and large gamma ray responses. Highly lithiated glass is an effective neutron detector material. However, scintillators employing this material also suffer from poor neutron-versus-gamma discrimination.

Organic scintillators are made from an aromatic hydrocarbon, most commonly anthracene [22]. These can be either solid single crystals or crystals dissolved in a solvent to form liquid scintillators. More noteworthy organic scintillators with POPOP dyes are being developed, including the BC501a [23], the EJ301 [24], and the original NE213 [25]. These materials utilize the (n, p) scattering cross section of hydrogen as the detection modality. Gamma and neutron radiation can be detected and separated simultaneously using pulse-shape discrimination [26]. Improvements on this system over the years have included capture-gated neutron detection and spectroscopy. Pulse heights are recorded for all of the proton scatters that lead to the capture of a neutron by dissolved [10]B or [6]Li in the scintillator matrix [27]. All organic scintillators are flammable to some extent, and liquid scintillators are particularly flammable, as they are dissolved in toluene, xylene, or naphthalene [28].

There is no currently unified solution to the neutron detection problem to replace [3]He tubes. Gas-filled detectors based on [10]B gases have a low neutron-versus-gamma discrimination, have poor detection efficiencies, and/or are made from a toxic gas. Solid foil-based neutron detection materials have low detection efficiencies and are slow to respond. Fission chambers rely on radioactive material and are relatively small. Inorganic scintillators and glasses suffer from slow speeds and poor neutron-versus-gamma discrimination when grown as large crystalline structures. Organic scintillators, while fast, are made from flammable liquids or, if cast into a solid, cannot discriminate neutrons from gamma rays.

A new neutron detector with properties that meet or exceed the performance of [3]He tubes is needed. It is hypothesized that zinc oxide (ZnO) scintillators, doped with neutron-sensitive materials and wavelength-shifting materials and, potentially, improved via a photonic crystal structure, can serve as the thermal neutron detector to replace ^{3}He. A matrix of six metrics was tested against standard neutron detectors to evaluate the novel ZnO scintillator as a neutron detector. These included neutron detection efficiency, neutron-versus-gamma discrimination, scintillation wavelength, scintillation speed, optical clarity, and ease of growth/size of the detector. Detection efficiencies and gamma-discrimination capabilities will be tested against standard neutron detectors such as ^{3}He tubes and lithiated glass scintillators.

12.3 Scintillation Detection Theory/Mechanisms

Radiation detection focuses on detecting, quantifying, using, and understanding energetic charged and neutral subatomic particles and HCPs. The starting point for all radiation detection is the quantification of ionization in a material. For charged particles, the ionization events are a direct result of the energetic particle traveling through the detector material. A conversion reaction must occur in which the particle decays, has an interaction producing a charged particle, or scatters off of a charged particle producing a recoiling charged particle, to detect neutral particles or indirectly ionizing particles. In the early 1930s, it was found that if HCPs, such as alpha particles, strike special screens, such as zinc sulfide, they produce sparks of light—referred to as scintillation.

Scintillation radiation detection has not changed significantly since that time. Ionization events produce electrons and holes. The electrons are collected either directly or indirectly and digitized, and a semiconductor or gas-filled detector collects the charge directly. A scintillator converts part of the energy to luminescent or phosphorescent light. The light is then collected by a photomultiplier tube (PMT) or avalanche photodiode (APD) and is then digitized.

12.3.1 Radiation Interaction Mechanisms

The measurement of neutrons is always complicated by the presence of background gamma radiation, which may derive from a number of sources, including naturally occurring radioactive material (NORM), terrestrial background radiation, cosmic radiation, and intentional gamma-emitting sources (a nuclear medicine patient). It is necessary to have a detection material and system that is capable of accurately discriminating neutron from gamma radiation for both homeland security and nuclear physics applications.

Screening for SNM with a passive monitoring system requires the detection of 1–10 neutrons in a short amount of time over a background of 10^5 gamma rays [29]. An active interrogation system initially generates a large burst of photons or neutrons. The detector must not be "blinded" by this pulse, as it must be capable of detecting the few neutrons released in response. If the detector has slow pulse rise and decay times, pulses can bleed together. The charge collection and mean migration time of the detector determine its capacity to resolve the pulse heights required to discriminate neutrons from gamma radiation.

The simplest form of neutron-versus-gamma discrimination is in the use of a low level discriminator, as used in ^{3}He tubes. Gamma rays, since they impart little energy into the active volume of the detector, produce small pulse heights. Neutrons, however, produce charged particles, which impart a larger amount of energy. If the sample rate is too slow to resolve the time between pulses, the electronics may add two or more pulses together, the sum of which may look like that of a neutron. If the detector system is too sensitive to gamma rays, it will be "blinded" by the gamma flash and continue to saturate the PMT for a long period of time, limiting the effective time the detectors can measure neutrons.

12.3.2 Gamma Ray Interactions and Detection

12.3.2.1 Gamma Ray Interactions

For a radiation detector to be sensitive to gamma rays, it must be able to detect electrons. Gamma rays are neutral electromagnetic wave/particles, which interact by three main mechanisms, Compton scattering, pair production, and photoelectric absorption. A representative photon cross section is defined as the distribution of events for a particle or radiation being absorbed by a nucleus.

At low energies photoelectric absorption dominates, while at high energies pair production is dominant. In the region between photoelectric absorption and pair production (at a minimum of 1.022 MeV), Compton scattering dominates. In photoelectric absorption, the full energy of the photon is absorbed. An electron in this process is excited from an inner K-shell state and is ejected. This electron carries away the incident photon's energy less the electron's binding energy.

Full energy capture in one single interaction is desirable for gamma ray detection. The reaction cross section reveals large saw tooth step discontinuities, which arise from the electron orbital transitions. K, L, and S shell electron lines can be seen for high Z materials. In low Z materials, such as ZnO, only K shell step discontinuities are visible. Compton scattering is an interaction between the incident gamma ray and the inner shell electrons. After the gamma ray scatters off the electron, and the electron is knocked out of its orbit, the gamma ray exits the reaction with a lower energy. High-energy gamma rays must scatter many times until they reach an energy at which they can be captured via the photoelectric absorption process. The maximum energy that can be lost in the Compton scattering process is shown in Equation 12.1. Here, the incident energy (energy denoted by $h\nu$) of the photon is scattered to the new energy $h\nu'$. m_0c^2 is the rest mass of an electron (0.511 MeV). Equation 12.2 shows the electron's maximum energy after the scattering reaction. The electron that is ejected from the nuclear orbital is known as a "Compton" electron. This multiple scattering process for high-energy gamma rays produces a number of several hundred kiloelectron volt:

$$h\nu'\big|_{\theta=\pi} = \frac{h\nu}{1 + (2h\nu/m_0c^2)} \tag{12.1}$$

$$E_{e-}\big|_{\theta=\pi} = \frac{2h\nu/m_0c^2}{1 + (2h\nu/m_0c^2)} \tag{12.2}$$

Finally, pair production occurs when a gamma ray interacts near the nucleus, and energy is converted to produce an electron–positron pair, which exits the reaction in opposite directions. The reaction conserves energy and momentum. In the photoelectric absorption reaction, the electron leaves the orbitals of the original reaction nucleus, but lacks enough energy to travel very far. The range of this reaction is minimal, but dominant for high Z materials.

The photoelectric absorption reaction dominates most materials from one hundred to a few hundred kiloelectron volt. Above this energy, the photon's wavelength is too small to directly excite the electron. Instead, it interacts ballistically to produce recoil electrons in a scattering reaction.

The primary source of gamma rays, particularly in the accelerator regime, falls in the Compton scattering region. Background radiation from NORM is also dominated by sources of radiation from ~300 keV (radon progeny) to around 1.4 MeV (40 K). The electrons produced in this reaction follow the formula shown in Equation 12.3:

$$E_{e-}\big|_{\theta=\pi} = h\nu - h\nu' = h\nu\left(\frac{(h\nu/m_0c^2)(1-\cos\theta)}{1+(h\nu/m_0c^2)(1-\cos\theta)}\right) \tag{12.3}$$

While the minimum energy required for the pair production reaction is 1.022 MeV, it is slow to increase in cross section until approximately 10 MeV. In this energy range, Compton scattering is the predominant mechanism for most background and accelerator gamma ray generation. Compton electrons have a significant range, from millimeter to centimeter before they fully deposit all of their energy. The range of electrons of various energies in zinc is shown in Figure 12.2.

Energetic electrons (energies greater than 100 eV) travel through material in a unique fashion. There are two primary reactions electrons can undergo to lose energy: (1) a reaction with an atomic electron and (2) deflection by the positive nuclear charge. The direct interaction with atomic electrons is relatively unlikely and occurs only at higher energies where the electrons can overcome the Coulomb repulsive force due to the small size of the two particles and strong electronic repulsion. The larger the number of atomic electrons per unit volume, the more efficient a material is at absorbing the energy of scattered electrons. This is the primary reason high Z and high density materials are such efficient shields and detectors.

As the energetic electron moves, it experiences the repulsive force of all of the atomic electrons in its vicinity. This produces billions of microreactions per millimeter, which

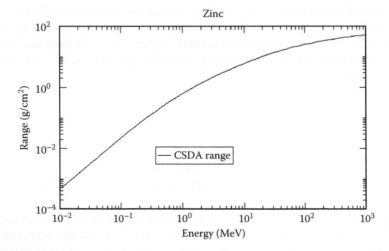

FIGURE 12.2

CSDA to electrons in Zinc. At low energies, the electrons have a range of a few millimeters while at higher energies, 1 MeV and above, the range is in centimeters. (From http://physics.nist.gov/PhysRefData/Star/Text/ESTAR.html.)

causes a very small amount of energy to be lost to the surrounding electrons. Continuous slowing down approximation (CSDA) simulates the transport of electrons through a medium. The particle's energy determines the expected value of the distance it will travel between large energy loss reactions—referred to as the CSDA range.

It is desirable for gamma spectroscopy to determine the energy of the incident gamma ray by using detectors, which can fully absorb the gamma ray and all of its recoil electrons' energies. A combination of high Z materials and large cross-sectional areas/volumes are used to ensure that the total energy is deposited in the detector. The detector sums the energy from Compton scattering to determine the full energy peak (photopeak), which corresponds to the incident gamma energy.

A large detector is required to capture the total energy. In practice, small detectors are used, in which it is probable that some portion of the gamma rays will scatter in the detector volume and is not absorbed. In this case, a unique gamma spectrum feature arises, known as the Compton continuum. Compton scattered electrons are scattered into a continuum of energies based on the scattering angle, which can range from 0 to 2π in the center of mass. The electron is typically stopped in the active detector material, while the gamma ray escapes. If the gamma rays were to be stopped, the energy in the photoelectric absorption would be added to the scattered energy and summed to the photopeak. In very thin scintillators, this effect is even more pronounced. There is almost no photopeak and only a Compton continuum.

A scintillator smaller than the CSDA range of the Compton-scattered electrons gives rise to another anomaly. In these thin scintillators, only a portion of the Compton continuum is deposited. This misshaped continuum is due to the fact that it is not probable for the electron to deposit much energy in the scintillator matrix. This physical trait can be exploited to enhance gamma-versus-neutron discrimination.

12.3.2.2 Gamma Ray Detection

In semiconductor gamma ray detectors, such as those made of high purity germanium (HPGe), cadmium zinc telluride, germanium drifted with lithium (GeLi), or silicon drifted with lithium (SiLi), electrons and corresponding holes or ions are directly collected by a bias applied across the semiconductor's intrinsic region. The reverse-biased junction prevents ambient electrons from crossing the bandgap. These detectors are almost always cooled to some extent, with some of them cooled to liquid nitrogen temperatures or below. When the full energy is collected, the number of electron–hole pairs (EHPs) that are produced directly corresponds to the incident gamma ray energy.

EHPs are not directly collected for scintillation detectors. The electrons are allowed to recombine in the scintillator structure, and this often results in a nonradiative de-excitation. Occasionally, depending on the scintillation efficiency of the scintillator material, the ionized electrons will de-excite radiatively, releasing an optical photon directly corresponding to the energy it lost. There is a reasonably linear relationship between the energy imparted and the number of optical photons produced even though there is no direct collection of ionization events. The nonlinearity of a scintillator becomes a problem at low energies and has been the focus of a significant amount of research.

12.3.3 Neutron Interactions and Detection

Neutrons are similar to gamma ray radiation in that they are both neutral. However, neutrons do not interact with electrons, but with atomic nuclei. They are not affected by the strong electrostatic repulsion of the nucleus as protons would be or by the electron cloud as electrons would be. Neutrons can interact by one of two main channels: scatter events and absorption events. Scatter events can either be elastic (where momentum and energy are conserved) or inelastic (where energy is conserved, but momentum is not). In a special case of absorption for high energy neutrons, multiple neutrons can be ejected as in the $(n, 2n)$, $(n, 3n)$, or $(n,$ multiple $n)$ reactions. These reactions are useful for measuring high-energy neutrons where the cross section for detection is very small. Absorption events can be broken into (1) radiative capture resulting in gamma ray production, (2) a fission reaction where the compound nucleus is so unstable that it splits into two or more fragments, and (3) charged particle production producing alpha particles, protons, or exotic combinations of them. All of the above reactions can be utilized to make an active neutron detector. Preference in creating neutron detectors is given to reactions with large Q values—the energy released in the reaction—which produce highly ionizing particles.

Elastic scattering has been used in neutron detection in organic scintillators, which have high hydrogen content. Neutron elastic scattering off from the hydrogen nucleus ejects the nucleus (a proton), but does not eject gamma radiation. This elastic scattering reaction is the reaction that is measured to enable the neutron-versus-gamma discrimination in liquid and solid scintillators.

Inelastic scattering can be used in conjunction with a gamma ray detector to detect neutrons [31]. Inelastic scattering neutron detectors are not a practical solution for the two most important applications, homeland security and nuclear physics, because they rely solely on gamma ray detection equipment to detect the neutron reaction. For each of these applications, the gamma ray signal resulting from neutron inelastic scattering is small in comparison to the gamma ray backgrounds. This method has been used with prompt neutron gamma activation analysis to determine trace quantities of materials in a sample. The inelastic cross section interactions and radiative capture cross sections result in gamma ray spectra, which are unique to the isotope being investigated.

The radiative capture reaction is commonly used with few specific isotopes, resulting in their immediate de-excitation by electron capture and internal conversion. Internal conversion is a radioactive decay process in which an excited nucleus interacts with an electron from one of the lower energy orbitals, causing the electron to be emitted from the atom. The internal conversion prompts the emission of an Auger electron. This reaction produces numerous characteristic x-rays and Auger electrons, which can be collected for neutron detection. High Z (the atomic number) and high A (the mass number, i.e., total number of protons and neutrons in the nucleus) nuclei such as Gd and Hf are subject to this reaction. In this reaction, Auger electrons with over 7 MeV of kinetic energy can be released [32,33]. Since this reaction pathway results in electron and photon production, large detector volumes are needed to detect these particles. In addition, the signals produced by Auger-electron/internal conversion reactions appear almost identical in pulse-height distribution to those produced by gamma rays. If full energy collection is possible, neutron-versus-gamma discrimination is still possible. Typically, however, only partial energy is collected, which makes the signal almost indistinguishable from gamma radiation.

Novel detection systems such as the capture-gated neutron spectrometer have been built on this system, combining an organic scintillator such as NE213 or BC501a with a Gadolinium solution. This method requires the entire pulse history to be digitized and the user to analyze the pulse history backward in time from the neutron capture reaction.

TABLE 12.1

Reactions, Q Values, and Cross Sections from Selected Neutron
Target Nuclei

Reaction	Q-Value (eV)	Cross Section (in Barns) for Thermal (0.025 eV) Neutrons
$^{10}B + n \rightarrow {}^7Li + \alpha$	2.31	3840
$^6Li + n \rightarrow {}^3H + \alpha$	4.78	940
$^3He + n \rightarrow {}^3H + p$	0.764	5330
$^{235}U + n \rightarrow X + Y$ (fission fragments)	~200	575

The proton recoil events are summed to estimate the original neutron energy. This system, however, is plagued by gamma contamination and must be used in a low count rate field. Otherwise, it is impossible to accurately separate the pulse histories.

The primary mode for detecting neutrons is through the creation of HCPs such as alpha particles, protons, or fission fragments. The (n, p), (n, α), and $(n, \text{fission})$ reactions for the vast majority of isotopes are threshold reactions. Nuclei such as ^3He, ^6Li, and ^{10}B are one neutron away from doubly close-packed low-energy states. Upon capturing a neutron, at least one distinct close-packed state promptly forms inside the nucleus, converting one into two nuclei. These two nuclei, usually consisting of an alpha particle and either a triton or a lithium nucleus, instantly repel each other in an exothermic reaction in the center of mass. The momentum from the incident neutron dictates the direction of the exiting channels. The positive Q-value for the reaction is of paramount importance in neutron detection. This exothermic reaction releases energy to the two charged particles, which can be easily detected. Most thermal neutron detectors utilize one of the three target nuclei above. A table of common neutron target nuclei reactions, their relative cross sections, and their energies can be found in Table 12.1.

12.3.4 Heavy-Charged Particle Interactions and Transport

HCPs generate high-linear energy transfer (LET) radiation. High LET radiation, while experiencing the Coulomb forces of the surrounding atoms, have enough mass and momentum that they tend to travel in a straight line, unlike electrons, and are rarely ever deflected. High LET radiation loses energy in a linear fashion corresponding to the Bethe–Block stopping power formula seen in Equation 12.4:

$$S = -\frac{\partial E}{\partial x}$$

$$-\frac{\partial E}{\partial x} = \frac{4\pi e^4 z^2}{m_0 v^2} NB$$

$$B = Z\left[\ln\frac{2m_0 v^2}{I}\right] - \ln\left(1 - \frac{v^2}{c^2}\right) - \frac{v^2}{c^2}$$

$$R = \int dx = \int \frac{dE}{S} \int_E^0 \frac{dE}{dE/dx} \tag{12.4}$$

where
    S is the stopping power, which is the energy loss per unit length
    e is the charge on an electron
    z is the charge of the particle
    Z is the atomic number
    B is the collection of constants defined in Equation 12.4

The integral of the inverse of the Bethe–Bloch equation from the incident particle energy down to zero determines the range of the high LET radiation described as R earlier. The Bethe–Bloch range equation is shown in Equation 12.4.

The Bethe–Bloch and range equations are directly proportional to the square of the charge on the ionizing particle and its mass. Consequently, heavier and higher-charged particles, such as alpha particles, do not travel as far as their lighter and lower charge counterparts, protons, at equal initial energies. Whereas electrons can travel in the range of millimeters to centimeters, alpha particles and protons of similar energies will only travel a few microns in the same material, that is, similar dimensions to those of a semiconductor device. Shielding of radiation interactions in matter [34] uses the Monte Carlo method to simulate the transport of these charged particles. An example of the range and deflection of a HCP versus an electron in ZnO is shown in Figure 12.3 [35,36].

The energy loss profile of the HCP reveals that it produces more ionization events at the end of the HCP track, while the electron produces more in the beginning (Figure 12.3). The region at the end of the HCP's track, which has a very high ionization density, is called the Bragg peak. If the amount of energy is integrated with respect to thickness, particularly toward the end of the Bragg peak, the energy imparted per micron is significantly lower for electrons than HCPs. This physical difference may be exploited to produce highly efficient neutron detectors with low gamma sensitivities.

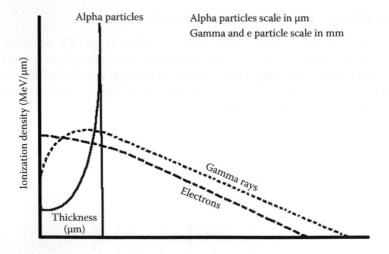

FIGURE 12.3
Range and energy loss profile for electrons, gamma rays, and alpha particles in ZnO. (From Berger, M.J. and Seltzer, S.M., Tables of energy losses and ranges of electrons and positrons, Report No. NASA-SP-3012, National Aeronautics and Space Administration, Washington, DC, 1964; Krane, K.S., *Introductory Nuclear Physics*, Wiley, New York, 1987.)

12.3.5 Radiation Detection

In gas-filled detectors such as ^{3}He, ^{10}BF3, and ^{10}B-lined tubes, the HCPs travel through the fill gas–producing ionization events [37]. These ionization events are directly collected by applying a bias voltage between the tube wall (cathode) and the central anode wire. When increasing this voltage, charge collection moves from a linear collection with no amplification (ionization chamber region) to an amplification region called the proportional region. Gas amplification occurs when the ionized electrons are accelerated with such force that they, too, cause more ionization events. Finally, above the proportional region is the Geiger discharge region. In this region, the voltage and accelerating force on the electron is so strong it produces a Townsend cascade and a Geiger avalanche form. The avalanche completely ionizes the gas along the length of the wire, producing a large pulse.

However, in the linear ion chamber region and the linear proportional voltage region, the detector is considered to be nonparalyzable. This means that if two pulses were to arrive nearly simultaneously, before the first one could be cleared and quenched, the second pulse would add to the first pulse and "ride on top." Sophisticated discrimination electronics can correct for this problem. The true count rate is calculated with Equation 12.5, where n is the true interaction rate, m is the recorded or measured count rate, and τ is the system dead time or time to recover from the pulse:

$$n = \frac{m}{1 - m\tau} \tag{12.5}$$

In the Geiger discharge region, the detector becomes paralyzable. If two pulses appear nearly simultaneously, the detector can no longer resolve both pulses, and the second pulse to arrive is ignored. They can easily become paralyzable due to the long-charge collection times of these detectors. Detectors can also become paralyzable with a slow detector charge collection, a long pulse rise and decay time, and a high rate of radiation interactions. In Equation 12.6 for the paralyzable model, n is the true interaction rate, m is the recorded or measured count rate, and τ is the system dead time or time to recover from the pulse:

$$m = n \cdot e^{-n\tau} \tag{12.6}$$

When designing detector systems, particularly for high count rate applications, it is important to design them in a nonparalyzable fashion. A paralyzable detector will begin to actually decrease in count rate as the count rate increases past its maximum count rate response. It is impossible to determine on which side of the saturation peak the detector is functioning. The correlation between count rates and observed responses for the paralyzable and nonparalyzable detector models can be seen in Figure 12.4.

ZnO's fast pulse rise and decay times facilitate neutron discrimination despite a high rate of radiation interactions, as is discussed in more detail in Section 12.5.4.

12.3.6 Scintillation Detector Theory

When radiation passes through matter, it loses energy through ionization events. The amount of energy required to produce an ion-electron pair is known as the work function. Once ionized, atoms need to de-excite, that is, lose their charge. Gas-filled tube and

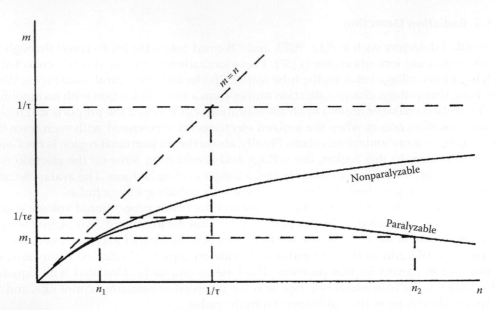

FIGURE 12.4
Paralyzable versus nonparalyzable detector models. (From Knoll, G.F.: *Radiation Detection and Measurements*. 1979. Copyright Wiley-VCH Verlag GmbH & Co. KGaA. Reprinted with permission.)

semiconductor detectors directly collect the charge by means of a high voltage field, which sweeps the charge out. Scintillation detectors, in contrast, allow the atoms or molecules of the scintillator to de-excite in the detector material. Materials that are classified as insulators or semiconductors have electron energy bands where electrons can exist. There are two primary bands—the lower valence band and the upper conduction band. The valence band is formed by electrons bound at lattice sites of the crystalline structure, while the conduction band consists of electrons, which have enough energy to migrate through the crystalline structure. The forbidden zone, where electrons cannot reside, lies between the valence band and the conduction band.

In addition to the valence and conduction bands, doped scintillators possess activator sites that consist of acceptor and donor bands. When a scintillator relies on the energy difference between the conduction and valence bands, the recombination efficiency is typically low and results in a photon far outside the visible range in the deep UV spectrum. A standard band diagram for a semiconductor scintillator is shown in Figure 12.5. Excited nuclei also exhibit a number of energy states in the immediate vicinity of the conduction and valence bands. This is due to the thermal energy in the system. The vibrational states add a small variation in the gap energy. This contributes to the broadening of the scintillation fluorescence peak.

In a traditional inorganic doped scintillator, such as NaI(Tl), LiI(Eu), or CsI(Tl), electrons can drift through the lattice until they encounter an acceptor location. Some of the ionized electrons can de-excite by dropping into acceptors and by releasing their excess energy in quanta of light, the energy of which is less than the bandgap between the valence and conduction bands. This process can be relatively slow due to the long migration times between ionization and deexcitation. However, most electrons deposit their energy into the crystalline lattice nonradiatively as phonons of heat. The different kinds of radiation emitted in nuclear reactions—gamma, scattered electrons, alpha and beta particles, protons, and fission fragments—carry energy from tens of kiloelectron volt to tens of megaelectron volt.

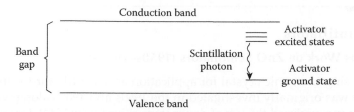

FIGURE 12.5
Band structure of standard semiconductor material scintillator such as CsI(Tl), NaI(Tl), or doped ZnO [37]. Activator excited states provide a lower energy excitation level in the scintillator band structure for electrons. Activator ground states lie above the valence band further narrowing the energy gap. This is particularly important in insulating materials. (From Knoll, G.F.: *Radiation Detection and Measurements*. 1979. Copyright Wiley-VCH Verlag GmbH & Co. KGaA. Reprinted with permission.)

Using the approximation of a 3-eV work function, a 1 MeV particle will produce $\sim 3.3 \times 10^5$ ionization events [38]. If the de-excitation time is a Poisson distribution, very long decay times on the order of tens to hundreds of microseconds are possible [39]. Not all of the excitation events emit optical light pulses. Some electrons fall into traps created by crystalline defects, contaminants in the crystal, and phonon de-excitation. In undoped inorganic scintillators, such as ZnO, BaF, and CsI, the electron preferentially de-excites locally, since there are no activator sites. This produces a very fast scintillation pulse at a large photon energy in the UV range [40]. The bandgap in these structures allows electrons to directly de-excite across the bandgap.

Electrons in organic scintillators may de-excite anywhere along the molecular chain. Organic scintillators are usually faster than inorganic scintillators for this reason [44]. Organic scintillators usually have a benzene ring structure to exploit its singlet and triplet decay modes. The scintillation spectrum of organic scintillators is significantly broader than that of inorganic scintillators due to the number of different singlet and triplet bands, the intersystem crossing of the bands, and the overlapping π-orbital bonds. The number of available de-excitation bands is higher in organic scintillators, which enables higher luminosities than in inorganic scintillators.

The scintillation wavelength of light depends directly on the bandgap scheme of the material. Whereas wide bandgaps generate short UV scintillation pulses, small bandgaps generate long wavelength scintillation pulses closer to green wavelengths. Most scintillator's de-excitation fluorescence wavelengths are in the near UV to blue wavelengths [37]. Longer wavelengths usually result in longer pulse decay times, as exhibited in NaI(Tl) and LaBr$_3$. This is due to the longer mean lifetime of the electron resulting from the elevated temperature of scintillator material, and the electron's ability to reach more shallow acceptor locations. Without donors or acceptors, a scintillator produces light at the band-gap wavelength. Near the band-gap transition, the transmission of light is negligible, and the index of refraction is very high. Consequently, to produce large-volume scintillators like those used in gamma-radiation detection, acceptors and donors are added to produce scintillation photon emissions at longer wavelengths that are not immediately absorbed by the scintillator [41,42].

During evaluation of new scintillation materials, the scintillation spectrum and the efficiency of the light collection system must be taken into consideration. Scintillation light is collected through one or many light conversion tools. Currently, the PMT is the principal technology for scintillation light collection. Other systems, such as APDs, show high promise for PMT replacement.

12.4 ZnO Scintillators

12.4.1 Previous Work on ZnO Scintillators (1950s–1960s)

ZnO is a promising inorganic crystal for application as a scintillator for thermal neutron detection. ZnO was originally investigated in the 1950s and 1960s along with ZnS [43]. In the 1960s, Lehmann explored the use of donor impurities in ZnO [44], for example, preparing Ga-doped ZnO powder. At that time, this was the only form of this material available as large crystal growth had not been achieved. In the early 1990s, interest in ZnO returned as researchers began searching for more efficient ultraviolet LEDs. In this effort, the search for *p*-type ZnO laid the groundwork for developing efficient scintillators.

12.4.2 Introduction to ZnO

Many different impurities, such as Ga, B, and Li, may be used for scintillation applications [45]. Lehmann hypothesized that substituting Zn atoms with Ga atoms would introduce a degenerate donor band overlapping the bottom of the conduction band and result in greater electron production from ionization by high energy radiation [44]. Donor band electrons recombine with ionization holes in the valence band. This results in near band edge emission with a decay time of less than 1 ns. The luminosities were comparatively low, since the scintillation properties were measured in materials in powder form. However, ZnO scintillation properties can be dramatically improved by producing single crystals, due to recent advances in crystal growth technology. An ultrafast scintillator capable of efficiently detecting neutrons can be realized by doping or coating the ZnO crystal with neutron target nuclei. Traditionally, "doping" refers to semiconductor dopant levels up to 0.1% atom fractions [46]. In this discussion, however, dopant quantities can refer to atomic compositions up to 10% atom fraction.

ZnO has a room-temperature bandgap of 3.37 eV (375 nm) and an exciton binding energy of 60 meV [47]. This II–VI semiconductor has been studied for many optoelectric applications, including LEDs, solar cells, and laser diodes. Pure ZnO has optical transparency ranging from 380 to 2500 nm [48]. Interest in ZnO has grown recently due to its large photoresponse and tunable bandgap. Alloying ZnO films with MgO and CdO permits the bandgap to be engineered from 2.3 to 4.0 eV [49]. The exciton binding energy of ZnO (60 meV) is twice that of another wide-band-gap material GaN (28 meV) [50]. The higher exciton binding energy increases the speed of electron-hole recombination during scintillation, which results in bright, room-temperature emission, and decreases electron trapping. This also prevents the ZnO scintillator from having a significant temperature coefficient. The band diagram showing the impact of vacancies on the scintillation properties of ZnO can be found in Figure 12.6 [51].

Measurements with ZnO:In scintillators indicate a pulse full-width half maximum (FWHM) of ~0.65 ns compared to that of a fast plastic scintillator with a pulse FWHM of ~1 ns [52]—faster than all other organic and inorganic scintillators [37]. The ZnO scintillator has a rise less than 1 ns compared to a BC-408 plastic scintillator with a rise time of ~1.5 ns (Figure 12.7). Ultrafast rise and decay times allow scintillators to perform well in high count rate environments.

ZnO can exist in three crystalline forms: a hexagonal wurzite structure, a cubic zincblende, and a rare zinc rocksalt (Figure 12.8). It can also exist in an amorphous form. The rocksalt phase forms when it is alloyed or when the growth pressure is increased. It is

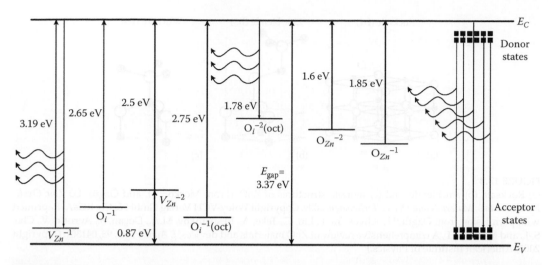

FIGURE 12.6
Band diagram for room temperature undoped ZnO. On the left are donor and acceptor states created by thermal excitation of the material. The lines indicate an energy state created by a vacancy in either oxygen or zinc and substitutional or interstitial defects. Curvy arrows show potential light emission. (Data from Morkoç, H. and Özgür, Ü.: *Zinc Oxide: Fundamentals, Materials and Device Technology.* 2009. Copyright Wiley-VCH Verlag GmbH & Co. KGaA.)

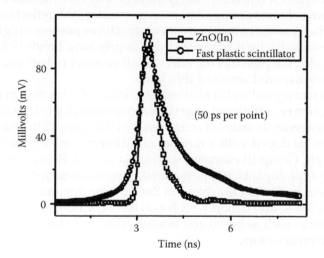

FIGURE 12.7
High speed of response of Cermet's In-doped ZnO scintillator compared to that of a plastic scintillator (BC400 fast plastic). (Reprinted from *Nucl. Instrum. Meth. A*, 505, Simpson, P.J., Tjossem, R., Hunt, A.W., Lynn, K.G., and Munne, V., Superfast timing performance from ZnO scintillators, 82. Copyright 2003, with permission from Elsevier.)

preferable to use wurzite ZnO over zincblende for many reasons, including stability at room temperature and ease of growth [53]. ZnO wurzite phase has lattice constants of $A = 3.25$ and $C = 5.2$ with a C/A ratio of ~1.6—a nearly perfect lattice for a hexagonal structured material (C/A ratio of 1.6333) [48]. ZnO, however, is tolerant to growth on a variety of different materials. Depending on the orientation of the substrate, seed crystal, or lack thereof, different ZnO orientations can be achieved.

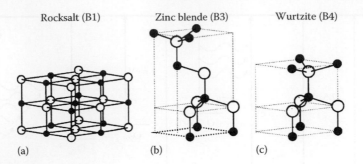

Rocksalt (B1)   Zinc blende (B3)   Wurtzite (B4)

(a)   (b)   (c)

FIGURE 12.8
(a) Rocksalt, (b) zincblende, and (c) wurzite structures of ZnO. (From Morkoç, H. and Özgür, Ü.: *Zinc Oxide: Fundamentals, Materials and Device Technology*. 2009. Copyright Wiley-VCH Verlag GmbH & Co. KGaA; Reprinted with permission from Özgür, Ü., Alivov, Ya. I., Liu, C., Teke, A., Reshchikov, M.A., Doğan, S., Avrutin, V., Cho, S.-J., and Morkoç, H., A comprehensive review of ZnO materials and devices, *J. Appl. Phys.*, 98, 041301. Copyright 2005, American Institute of Physics.)

ZnO, however, is a relatively soft material with a Mohs hardness scale value of only 4.5 [55,56], comparable to window glass, which has a Mohs hardness of 5.5 [57]. It is resistant to most solvents and water, but can be etched by strong bases and acids. It is thermally stable to 1975°C, at which temperature it will begin to decompose. Above 800°C in an atmosphere with oxygen, it will accept oxygen and fill any zinc vacancies. The wurzite and zincblende structures do not possess inversion symmetry (the reflection of the structure becomes the structure), and, consequently, these structures possess a high piezoelectricity. Wurzite ZnO's piezoelectricity is enhanced by its largely ionic bonds [58]. Future work, for example, could exploit the piezoelectric effect for other novel radiation detection methods such as piezoelectric nanorod arrays or thin films.

ZnO is a transparent crystal with a relatively low index of refraction in the visible region. The index of refraction near the band edge rises above three due to its bandgap. In the optical wavelengths, however, its index of refraction is ~1.9 (Figure 12.9) [59,60].

ZnO crystals can be doped with a number of different nuclei. Several *n*-type dopants exist, predominantly, Group III elements, such as Ga and Al. Recent research has focused on the search for *p*-type dopants. *P*-type material, or semiconductor material, which is rich in acceptor locations, is difficult to achieve in ZnO. *P*-type dopants are mostly from Group I nuclei, such as Li, Na, K, and Group V, such as N [61–64]. ZnO can be doped with neutron-detecting target nuclei such as boron and lithium. This provides a unique opportunity to produce new neutron detectors.

12.5 ZnO Growth

12.5.1 Crystal Growth

There are a number of different crystal growth methods for ZnO scintillators. They are broken into two distinct groups: bulk growth and thin-film/nanoparticle growth. Bulk growth can be divided into melt growth, hydrothermal growth, seeded chemical vapor transport (SCVT), chemical vapor transport (CVT), and chemical vapor deposition (CVD) growth. Each growth method was pioneered by a primary commercial company.

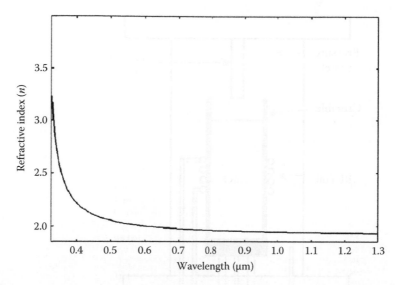

FIGURE 12.9
Index of refraction for ZnO. (From Bass, M. et al., Design fabrication and testing, sources and detectors, radiometry and photometry, *Handbook of Optics*, 3rd edn., Vol. II, McGraw-Hill Professional, New York, 2009.)

SCVT growth is primarily performed by Eagle Pitcher, which has reported growth rates of 40 μm/h in a horizontal tube furnace at 1000°C–1200°C [53,65]. This company uses a zinc source and hydrogen as a carrier gas with an overpressure of water vapor to maintain stoichiometry. In this process, the crystal is seeded by an initial ZnO crystal and grows on a 5.08 cm (2 in.) diameter [66]. The growth rate for this method, however, is constrained and asymptotically approaches a steady-state value lower than the 40 μm/h. Doping, which requires a suitable source capable of sublimation to the gas phase, is difficult with this method.

Chemical vapor transport is similar to SCVT, but there is no seed crystal to nucleate growth. This method usually has a lower crystalline quality due to the random nature of the initial growth. However, it is a low-cost option to produce bulk materials. Using a standard horizontal tube furnace, the growth is conducted by sublimation of zinc under a hydrogen carrier gas. Oxygen or water is used as the other reagent, while monitoring the partial pressure of carrier gas and reactant gases [67].

CERMET, Inc., is the leader in bulk melt-grown materials. CERMET utilizes a patented crystal growth technology, cold wall bulk melt growth at high pressure, to produce doped and undoped ZnO single crystals. An RF heat source induces fields in the starter ZnO material in the water-cooled cold wall chamber. These fields produce Eddy currents, which promote Joulean heating in the material until it melts. The highly refractory melt is contained in a cold wall crucible by a solid thermal barrier that forms between it and the molten material. The cooled material prevents the molten material from directly contacting the crucible cooling surface. This entire melting process is carried out in a controlled gas atmosphere ranging from 1 atm to over 100 atm. This prevents the loss of volatile components as well as the decomposition of compounds into atomic components [68]. CERMET has melted ZnO in 6 in. diameter crucibles, producing kilogram-sized boules from which inch-sized single crystals were cut. A schematic of the ZnO crystal growth furnace is shown in Figure 12.10. High dopant levels can be achieved using this method.

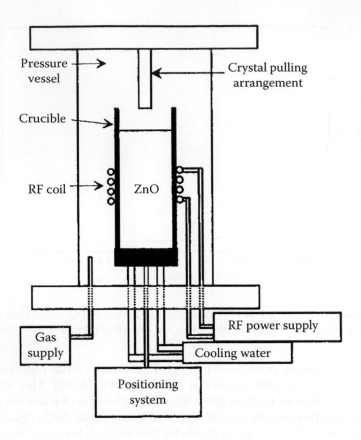

FIGURE 12.10
Schematic of the Cermet high pressure ZnO growth furnace [75]. The ZnO powder is heated up to ~1900°C via pressurized induction melting apparatus, and the melt is contained in a cooled crucible. (Reprinted from *III–Vs Review*, 12, Nause, J., ZnO broadens the spectrum, 28. Copyright 1999, with permission from Elsevier.)

Hydrothermal growth of ZnO consists of a solution using LiOH, NaOH, or KOH as a solvent at very high pressures (150 atm+) and a temperature less than 400°C. ZnO is dissolved under extreme conditions and then allowed to precipitate slowly out of solution as the pressure/temperature is reduced [71]. This growth rate is exceptionally slow (mm per day), and the ZnO contains high concentrations of contaminant nuclei from the solvent [72]. However, there has been a report of exceptional crystalline quality with a FWHM of 8 arc seconds in the XRD plot [73].

Each method described earlier has been investigated for nearly 20 years. While hydrothermal growth has produced the best-reported crystalline quality, it incorporates contaminants undesirable for scintillation applications. Melt growth provides the highest speed and largest boules of usable scintillation material. Melt growth also shows the most promise for doping with neutron target materials and allows for *in situ* doping of almost any practical dopant with relative ease. While it is possible to grow at high growth rates by CVT and SCVT, the cost of fabrication of materials by these methods tends to be prohibitive. Inability to dope the materials makes these methods unsuitable for the fabrication of scintillation detectors.

12.5.2 MOCVD Growth

One promising growth mechanism for ZnO is by metalorganic chemical vapor deposition. MOCVD is an epitaxial growth method of chemical vapor deposition in which the alkyl precursors are attached to an organic molecule. Metalorganics are usually volatile pyrophoric liquid materials, which aid in the growth process. Special growth reactors are required for MOCVD crystal growth and can grow various materials including GaN, GaAs, InP, and ZnO [74]. Subtle changes to the internal components of these reactors enable application of different precursors. The change required to accommodate different growth materials is costly and time consuming, as the tool is disassembled and rebuilt. MOCVD permits very fine control of the structure and dopant levels and can produce a very uniform epitaxial layer. A schematic of the reactor chamber specially designed for ZnO growth is illustrated in Figure 12.11.

The heater is capable of rapid temperature changes and can operate in vacuum as well as extreme environments, such as oxidizing or reducing atmospheres. Different designs of reactors have been proposed over the years. However, the most common one in mass production is the rotating disk model [75]. The rotating disk spins under a shower head, from which the metalorganic precursor is introduced [76]. The growth chamber's pressure may vary between 0.1 and 100 Torr [77,78], and, for ZnO growth, low pressure growth is preferable. The design relies on a stable reactor condition in which laminar flow is maintained over the growth platter [79]. Significant effort is put into designing the vessel head to achieve laminar flow [80]. As the metalorganic precursor is heated, it is cracked, breaking the hydrogen and carbon nuclei and leaving the free metal ion in the gas phase flow. A reactant gas is then flowed simultaneously or in pulsed mode over the growth platter. The reactant gas completes the other half of the compound semiconductor.

In multicomponent semiconductors, several alkyl sources and reactant gases can be pulsed or flowed into the chamber. For example, in ZnO growth, diethylzinc is the

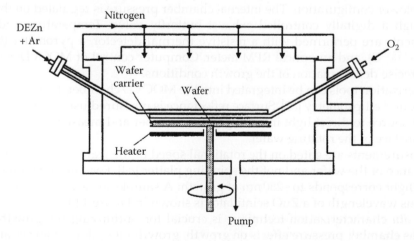

FIGURE 12.11
Schematic of a standard MOCVD growth chamber [69]. The wafer is located on a heated, rotating wafer carrier in an evacuated reactor chamber. Precursors flow into the chamber through their individual inlets. (Reprinted with permission from Gorla, C.R., Emanetoglu, N.W., Liang, S., Mayo, W.E., and Lu, Y., Structural, optical, and surface acoustic wave properties of epitaxial ZnO films grown on (012) sapphire by metalorganic chemical vapor deposition, *J. Appl. Phys.*, 85, 2595. Copyright 1999, American Institute of Physics.)

metalorganic of choice, and either oxygen (O_2), nitrous oxide (N_2O), or nitrogen monoxide (NO) can be used as the reactant gas [81]. Ideally, oxygen or nitrous oxide is typically the best choice. These two gas molecules crack to form free oxygen. Nitrous oxide also produces N_2, which is stable and is not incorporated into the matrix. N_2 is the common carrier gas that is flowed into the growth chamber to maintain laminar flow over the substrate for this very reason. Nitrogen monoxide, however, when cracked leaves free nitrogen. This free nitrogen is absorbed readily into the matrix resulting in nitrogen-doped samples [82]. The reaction for ZnO MOCVD growth using oxygen as the gas can be seen in Equation 12.7. MOCVD growth of ZnO is one of the few completely green reactions producing safe, nontoxic gases as byproducts:

$$(C_2H_5)_2Zn + 7O_2 \rightarrow ZnO + 5H_2O + 4CO_2 \tag{12.7}$$

Dopant gases are ideally metalorganic liquids similar to diethylzinc. Trimethylgallium (TMGa), trimethylindium (TMIn), and diethylboron (DEB) are frequently used for gallium, indium, and boron doping, respectively. These metalorganics are sometimes too efficient at doping in comparison to the ZnO growth. Metalorganic sources are highly volatile, as many have boiling points below 10°C. In ZnO MOCVD, nitrogen (carrier gas) is flowed through the liquid source. The flow rate of a metalorganic through the reactor is determined by partial pressures of the carrier gas and the source through a bubbler. The composition and doping of a material are precisely controlled by varying the source temperatures, the carrier gas flow, and the partial pressures. Bypass lines are often installed to minimize the pressure differential in the bubbler. This allows rapid switching at both constant pressure and constant flow rate between the metalorganic source and carrier gas to minimize spiking in the doping profile.

A series of mass flow controllers, solenoid valves, and pressure controllers/transducers regulate the gas flow of the carrier gas, reactant gas, and the metalorganic sources. Vacuum is maintained by a high volume, high vacuum pump, which usually includes a two-stage vacuum/blower configuration. The internal chamber pressure is regulated on the exhaust side through a digitally controlled exhaust butterfly valve. Temperature and rotation measurements are performed with a resistive thermal detector, a pyrometer (for operation above 700°C), and an optical RPM meter. Computer control of MOCVD components ensures precise determination of the growth conditions.

Characterization tools can be integrated into the MOCVD chamber, for example, surface reflectance and ellipsometry [83]. Surface reflectance is measured using a fiber-optic wideband light source to direct light into the growth chamber and measure the reflected light from the surface of the rotating wafer.

The measurements are gated on the rotational speed of the wafer so as to only measure the reflectance of the wafer and not the spinning platter [84]. For ZnO, one full oscillation at 730 nm light corresponds to ~250 nm of growth. A sample surface reflectance measurement versus wavelength of a ZnO scintillator is shown in Figure 12.12.

This *in situ* characterization technique is crucial for optimizing the growth parameters such as chamber pressure effects on growth, growth rates, dopant levels, and optical quality. The status of the crystal growth process can be easily monitored without the need to stop the reaction, cool the chamber, and bring the chamber back to atmospheric conditions.

Two-dimensional (2D) thin-film growth generates highly uniform films with very few defects at high growth rates and requires a specific range of growth conditions. In contrast, 3D film growth gives rise to nanorod-like structures and polycrystalline materials, which

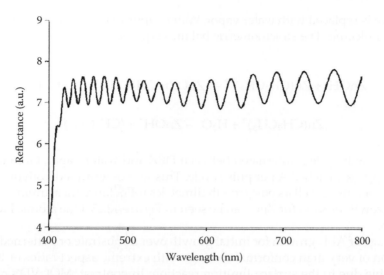

FIGURE 12.12
Reflectance pattern from a thin-film ZnO scintillator.

are impractical for application in scintillators. The crystalline defects and grain boundaries produce trapping locations, longer scintillation decay times, and scintillation outside the bandgap. ZnO growth in the MOCVD process has a very narrow window in which 2D growth can occur and is especially sensitive to the reactor pressure (between 10 and 50 Torr). Outside this region, the metalorganic preferentially adheres to the surface and does not diffuse into a smooth 2D growth. 2D growth requires a reactor temperature between 475°C and 550°C. Above 550°C, the growth rate is reduced below 100 nm/h. Below 475°C, a sufficient quantity of oxygen cannot be cracked. Switching oxygen precursor gases may improve this requirement. Hybrid 2D and 3D ZnO has also been grown [85].

The crystal quality of the MOCVD grown ZnO samples is heavily dependent on the substrate, although ZnO offers significant tolerance to strain the first few layers. Ideally, the lattice mismatch between the substrate and the ZnO structure should be minimized. Sapphire is the preferred growth substrate with a lattice mismatch parameter of approximately 30%. However, despite the large lattice mismatch, ZnO can compensate for it within the first 100 Å, promoting high-quality 2D growth [85]. ZnO has been grown on other substrates, including silicon, silicon oxide, and glass. High quality films can be grown on almost any material including non-wurzite structures and amorphous materials by utilizing techniques described in the following section.

12.5.3 Hybrid MOCVD ALD Growth

Atomic layer deposition (ALD) permits very precise thin-film growth by introducing (or pulsing) one precursor at a time [86]. Following the pulse of the first precursor, a purge pulse of carrier gas clears the chamber. Subsequently, a gas is introduced to "crack" the dangling molecule. Then, a purge gas pulse clears this cracking gas. These four steps are repeated for the second precursor. In contrast to MOCVD, only one gas is in the reactor chamber at a time. To further ensure that uniform thin films are formed, the growth conditions are self-limiting. Leaving a dangling chain exposed on one surface limits the reaction to nucleate atoms one atom thick across the surface. The DEZ source is still used, but the oxygen source is replaced with water vapor. Water vapor reacts with the surface and forms a hydroxide molecule. The stoichiometric balance equations are shown in Equations 12.8 and 12.9:

$$ZnOH^* + Zn(CH_2CH_3)_2^* \rightarrow ZnOZn(CH_2CH_3)^* + (CH_3CH_3) \tag{12.8}$$

$$Zn(CH_2CH_3)^* + H_2O \rightarrow ZnOH^* + (CH_3CH_3) \tag{12.9}$$

The pulsed growth cycle is alternated between DEZ and water vapor. Growth rates were found to be approximately 2 Å per pulse cycle. This was determined both by *in situ* reflectance measurements as well as postgrowth filmetrics reflectance measurements. An image of the ALD growth process for ZnO can be seen in Figure 12.13. One atomic layer is deposited per pulse.

An advantage of ALD growth for initial growth over a substrate or patterned structure is the generation of very high conformal coatings with extreme aspect ratios of 10,000:1. This is only possible due to the surface-limiting reaction. In contrast, MOCVD's rapid growth would cover up the small features of patterned structures and generate a smooth layer on top. ALD by its very nature is a slow growth process and due to the self-limiting reaction usually produces amorphous material, especially, with a large lattice mismatch between the desired material and the substrate.

12.5.4 ZnO Bulk Growth

Bulk and MOCVD grown ZnO samples exist in various shapes, including cubes, cylindrical boules, and thin wafers. Ten-micrometer-thick layered scintillators are also shown.

Figure 12.14 shows various undoped and doped ZnO boules and wafers. All of these wafers were grown at Cermet, Inc.

Figure 12.15 compares the bulk-grown ZnO samples to MOCVD-grown ZnO samples. The samples are of high quality, are optically clear, and will prove to be good, high-efficiency neutron detectors.

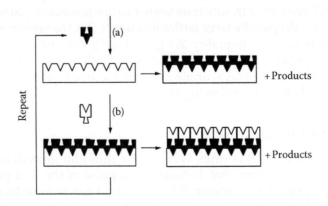

FIGURE 12.13
ALD of ZnO using (a) water molecules and (b) DEZ. (Reprinted with permission from George, S.M., Ott, A.W., and Klaus, J.W., Surface chemistry for atomic layer growth, *J. Phys. Chem.*, 100, 13121. Copyright 1996 American Chemical Society.)

FIGURE 12.14
Large cubes of polished ZnO. The left and right cubes are undoped ZnO while the center cube is lithium-doped ZnO.

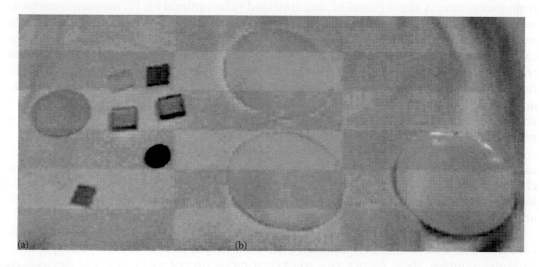

FIGURE 12.15
(a) Small bulk-grown ZnO samples including indium, lithium, cobalt, and undoped ZnO. (b) The 2″ wafers showing MOCVD-grown ZnO.

12.6 Performance of ZnO Neutron Scintillators

In this section, we will review the impact of growth methods/conditions and of various dopants on scintillation detectors created from ZnO. Two particular types of ZnO scintillators were tested: bulk melt-grown ZnO, and MOCVD-grown ZnO. Three classes of dopants were investigated: (1) neutron detector dopant materials, (2) dopants to improve the scintillation wavelength providing acceptor and donor bands, and (3) dopants to improve crystalline clarity and scintillator transparency. These dopants are not mutually exclusive as often the incorporation of dopants impacts two or more of these parameters. Six dopants were studied in particular: lithium, boron, gadolinium, aluminum, gallium, and nitrogen. A preliminary investigation of bulk-grown samples of cobalt-, indium-, and magnesium-doped scintillators eliminated these materials from further consideration. Lithium and gadolinium were studied in both the bulk growth and MOCVD samples. Aluminum, boron, gallium, and nitrogen were only studied in MOCVD samples. Undoped samples were studied to assess the impact of growth conditions on the samples grown with MOCVD.

12.6.1 Impact of Growth Conditions on Scintillator Performance

Three main parameters were investigated to ascertain the optimal growth conditions for MOCVD growth of ZnO scintillators, (1) the II–VI ratio as determined by oxygen and zinc flow rates, (2) temperature, and (3) pressure.

12.6.1.1 Effect of II–VI Ratio on ZnO Samples

The effect of the II–VI ratio on growth is dramatic. Increasing the zinc to oxygen ratio improves the photoluminescence (PL). Moreover, the samples became visibly opaque from excess zinc. A slight excess of zinc appeared to be optimal for PL measurements and resulted in a nearly transparent thin film. When excess zinc is present, there is usually excess carbon trapping due to the inability to form carbon dioxide during growth. The carbon trapping can be seen as defect luminescence in the Blue Band (BL shift). Carbon trapping slows the response by introducing a longer tail in the decay pulse heights in addition to reducing the transparency of the scintillator. Excess oxygen samples became white and did not scintillate nearly as efficiently as those which were slightly oxygen depleted. It should be noted that in this work, we are optimizing the ZnO for scintillation applications rather than for its electronic properties.

Plotted below are the relative PL spectra after varying the ratio between DEZ flow and oxygen flow. Oxygen flow rate was held constant to maintain identical flow patterns in the chamber. Pressure was held at a constant 3 Torr during growth. The impact of the ratio between zinc and oxygen is shown below in Figure 12.16a. A flow rate ratio between 150 and 350–900 was determined to be optimal.

The alpha spectra in Figure 12.16b reveal a larger alpha peak for the 350–900 ratio, indicating higher detection efficiency. However, this sample has a very low transparency. In larger samples, these higher zinc to oxygen ratios do not yield practical devices. At thicknesses greater than 4 µm, these structures would be unusable. A balance between efficiency and optical clarity was reached in the 150–900 ratio.

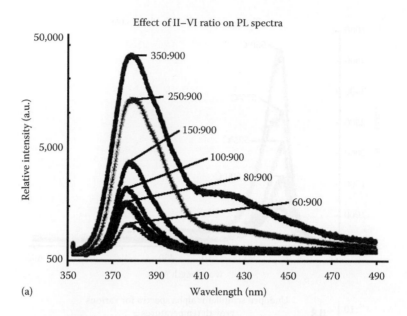

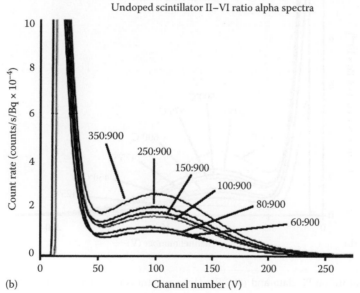

FIGURE 12.16

(a) Effect of growth ratio between zinc flow rates and oxygen flow rates in the MOCVD growth chamber and (b) [230]Th alpha spectra comparison for undoped ZnO.

12.6.1.2 Effect of Growth Temperature on ZnO Samples

Temperatures in the range of 400°C–600°C were studied to determine their effect on growth. Growth rates were found to increase up to 550°C and then noticeably decrease above that temperature. The impact of the temperature on the PL spectra of the scintillators is shown in Figure 12.17. When the temperature is not high enough, not enough oxygen is cracked, and the scintillators come out dark with carbon trapped interstitially.

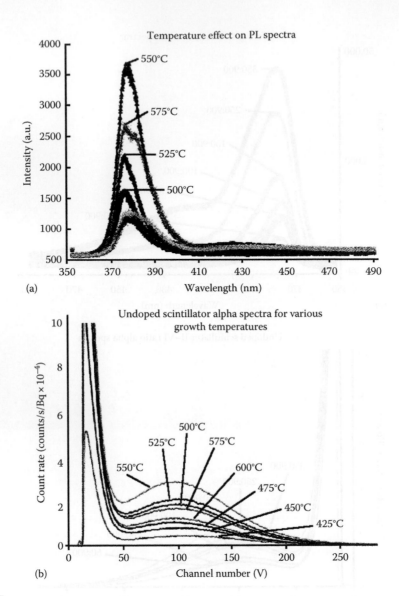

FIGURE 12.17
(a) Effect of temperature on PL data and (b) ^{230}Th alpha spectra comparison for different MOCVD growth temperatures.

Carbon trapping can again be seen in the blue luminescence peak that begins to form around 425 nm. Crystalline quality is also noticeably lower with 3D growth dominating. Good quality samples at reasonable growth rates were maintained at 550°C. Above 575°C, part of the PL band-gap peak begins to shift and broaden at lower wavelengths. At 600°C, the growth rate slows considerably, and the material quality is poor.

12.6.1.3 Effect of Growth Pressure on ZnO Samples

Growth pressure does not have a significant impact on the PL as temperature or the II–VI ratio. It does, however, impact the maximum growth rate and crystalline quality, and this

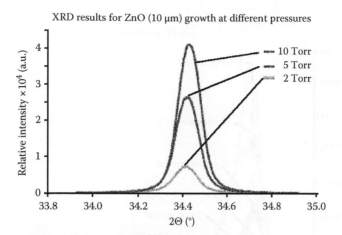

FIGURE 12.18
Effect of pressure in the growth chamber on XRD crystalline quality.

can be seen in high-resolution x-ray measurements. Figure 12.18 shows XRD results for 10 μm thick scintillators in the range of 2–10 Torr.

12.6.2 Impact of Dopants on ZnO Scintillator Performance

12.6.2.1 Impact of Nitrogen Doping on Scintillator Performance

Nitrogen doping was performed during MOCVD of ZnO with ammonia, NH_3. Ammonia, when cracked, provides one nitrogen atom and three hydrogen atoms. When performing nitrogen doping with ammonia, the oxygen flow rate must be increased to compensate for the excess hydrogen present. Nitrogen doping was found to improve the growth of ZnO to >4 μm/h. Crystalline transparency also improved, and samples thicker than 15 μm were still clear. The ammonia was found to act as a surfactant and appears to clean the surface of excess carbon during growth. The structures usually exhibited 3D growth with nanorod-like structures present at a larger degree than undoped scintillators.

There was one large drawback with just nitrogen doping. When using ammonia, the three excess hydrogen atoms had a tendency to become trapped in the crystalline structure. This is the first time free hydrogen atoms were released into the chamber without being attached to carbon nuclei. Hydrogen functions as a deep acceptor in the scintillator and passivates the PL centers by trapping the excited electrons, and no luminescence is observed.

A rapid thermal anneal of the scintillator under an oxygen overpressure was found to restore, to a limited extent, the original PL spectra. Oxygen overpressure and higher temperatures liberate the trapped hydrogen atoms, and oxygen combines with hydrogen to form water vapor. Rapid thermal annealing was completed at 800°C with various time durations and oxygen flow rates of 30–80 sccm to remove the hydrogen. The samples were allowed to cool in an oxygen atmosphere. A two-wafer growth run was conducted under the "optimal" pulsed ammonia growth conditions to prepare samples for the ammonia-doped scintillator tests.

The annealed samples gradually regained the peak in the PL spectra (Figure 12.19a). The wavelength of scintillation was blue-shifted by approximately 10 nm. The peak continued to increase but never reached the undoped level as the sample annealing time was

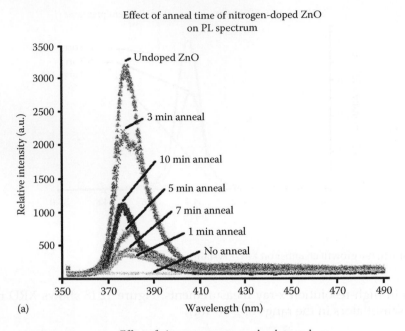

(a)

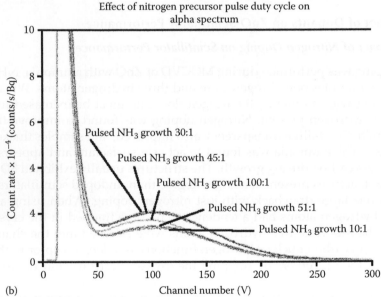

(b)

FIGURE 12.19

(a) Nitrogen-doped samples after annealing under oxygen and (b) ^{230}Th alpha spectra comparison for nitrogen-doped ZnO after 5 min anneal under oxygen at 800°C.

increased. A 5 min anneal at 800°C of the nitrogen doped samples yielded the highest intensity peak in the PL spectrum among the doped samples but never reached the peak of the undoped sample. The samples became oxygen saturated and cloudy with longer anneal times.

The 5 min anneal also resulted in the return of the alpha peak (Figure 12.19b). The peak position shifted to the right indicating a slight increase in light collection.

12.6.2.2 *Impact of Gallium Doping on Scintillator Performance*

Gallium also was investigated in an attempt to shift the scintillator spectrum's peak by introducing additional transition levels. TMG was used as the dopant precursor, and gallium is a far more reactive metalorganic source than zinc. Continuous flow rates often caused the samples to blacken due to excess gallium or the formation of gallium oxide, and so PL was only performed on transparent samples. Ten-micrometer-thick samples were grown, which were not optically transparent, and the PL spectra show an average wavelength shift of 7nm (Figure 12.20a). The peaks with gallium are skewed, indicative

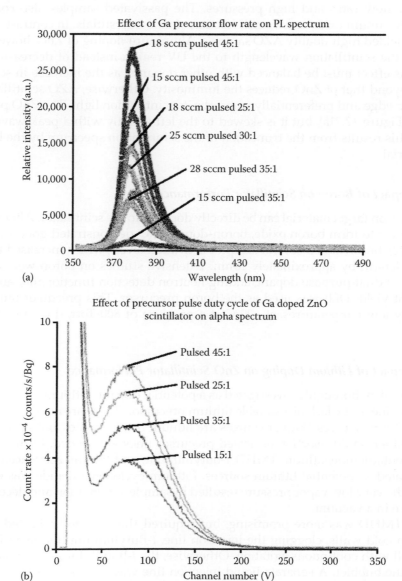

(a)

(b)

FIGURE 12.20
(a) Gallium-doped PL data and (b) ^{230}Th alpha spectra comparison for gallium-doped ZnO.

of the gallium alloying in the ZnO matrix. Only pulsed introduction to the reactor could incorporate gallium into the matrix.

12.6.2.3 Impact of Aluminum Doping on Scintillator Performance

Aluminum-doped ZnO (AZO) is highly transparent, crystalline, and highly conductive. It is commonly used as an ohmic contact for solar-cell work. AZO grows very well on sapphire (Al_2O_3) since both exhibit wurzite structures. AZO can be grown by MOCVD to over 20 μm thick with high crystalline quality with dopant levels of up to 20%. During initial growths, continuous Al precursor flow rates completely passivated the samples, even at low flow rates and high pressures. The passivated samples also consisted of 50%–80% aluminum oxide and a small fraction of zinc interstitials. In contrast, pulsed Al precursor yielded high-quality AZO samples. Aluminum doping of ZnO, however, tends to increase the scintillation wavelength to the UV region instead of decreasing it (blue region). This effect must be balanced with other dopants, as the increase in scintillation bandgap beyond that of ZnO reduces the luminosity. Otherwise, AZO scintillates below its band-gap edge and preferentially reabsorbs its scintillation light. The ZnO peak is still dominant (Figure 12.21a), but it is skewed to the left slightly with a peak wavelength of 376.3 nm. This results from the truncation of the scintillation spectrum by the band edge of the material.

12.6.2.4 Impact of Boron on Scintillator Performance

Boron, a neutron target material can be directly doped into the scintillator. Although boron has a tendency to form boron oxide, boron-doped ZnO demonstrated good PL intensity (Figure 12.22). Boron added several acceptor activation levels, which increased the scintillation wavelength by approximately 10 nm. Extensive studies on boron were conducted because it is a dual-purpose dopant, adding neutron detection functionality and improving the light yield. DEB was used as the boron precursor. This precursor must be kept at extremely low temperatures, at −20°C at a pressure of 800 Torr, due to its low vapor pressure.

12.6.2.5 Impact of Lithium Doping on ZnO Scintillator Performance

Lithium was also thoroughly investigated as a potential dopant. Lithium samples are difficult to grow due to the lack of a suitable lithium precursor. There are no diethyl lithium or dimethyl lithium sources. Similar sources are lithium β-diketonates and lithium TMHD. However, either liquid injection or heated precursor lines are needed. In this work, lithium cyclopentadienide, lithium TMHD, *t*-butyllithium, and lithium acetylacetonate were all investigated as potential lithium sources. Lithium cyclopentadienide was eliminated quickly as the very low vapor pressure resulted in sample sublimation and decomposition of the source in a vacuum.

Lithium TMHD was more promising, but required the lines to be heated, as it condensates on cold walls, clogging the injection line. *t*-Butyllithium dissolved in pentane worked well with continuous agitation. Otherwise, the *t*-butyllithium precipitated at the bottom of the bubbler. A separate liquid injection line was installed for this source but, due to the large flow volume, it was considerably more difficult to control and to produce a transparent crystal. Finally, the lithium acetylacetonate provided a good solution in a heated vapor phase. After a few days, it typically solidified from its original powder form.

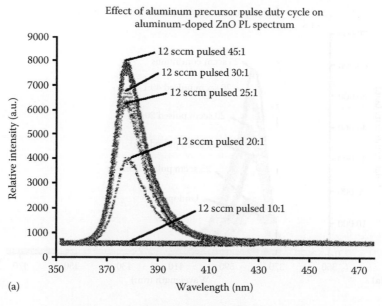

(a)

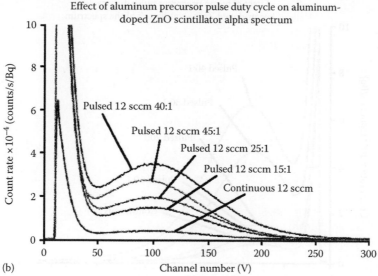

(b)

FIGURE 12.21

(a) Aluminum-doping impact on PL spectra and (b) [230]Th alpha spectra comparison for aluminum-doped ZnO.

Subsequently, this same sample was dissolved in acetone and used as a liquid injection source. The samples grown by MOCVD were comparable to the bulk growth samples prepared by Cermet Inc. But it was difficult to obtain clear lithium-doped samples by MOCVD. The resulting PL spectra are shown in Figure 12.23.

12.6.2.6 Impact of Gadolinium on ZnO Scintillator Performance

Gadolinium was the last major dopant to be investigated in both bulk and MOCVD samples. A Gd(TMHD) source was used as the MOCVD precursor. The bulk growth samples

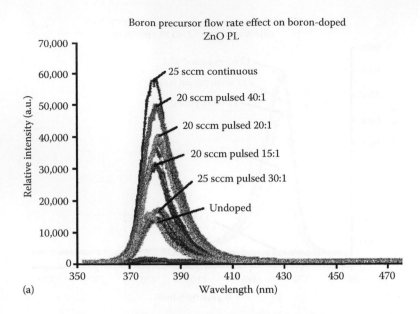

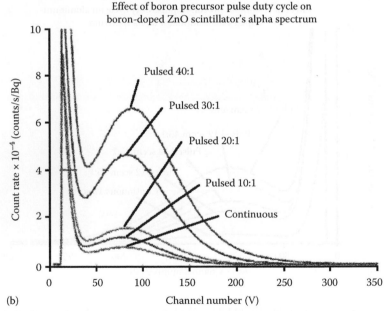

FIGURE 12.22

(a) Impact of boron doping on ZnO PL spectra and (b) ^{230}Th alpha spectra comparison for boron-doped ZnO. Flow across the boron bubbler was 20 sccm at 800 Torr.

were comparable to the MOCVD. Gadolinium, with its extensive electron structure, provides extra acceptor energy states for excited electrons, and it dramatically raises the effective Z number of the material. The 3+ state and Gd bond were studied. The availability of only one MOCVD Gd source limited the study to flow rate variation over the sample. PL demonstrated high intensities from these samples. But the use of gadolinium doping was not conclusive at this time.

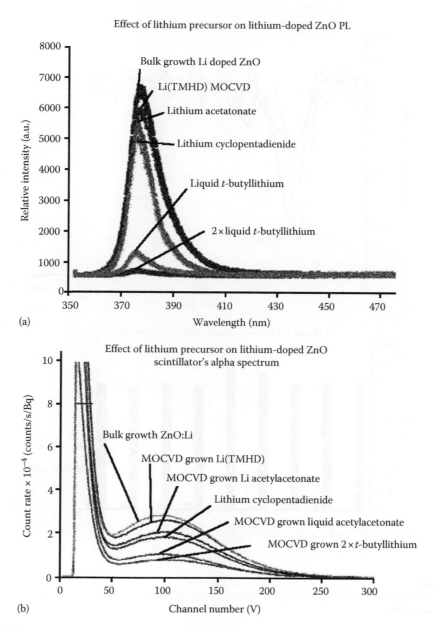

FIGURE 12.23

(a) Lithium-dopant impact on PL light intensity and (b) ^{230}Th alpha spectra comparison for lithium-doped ZnO.

Pulse rise and decay time (Figure 12.24a) distributions were analyzed for all of the scintillators (Figure 12.24b). The pulses were digitized using a 1 GHz sampling oscilloscope. Rise and decay times were measured for all of the samples discussed earlier.

The undoped ZnO scintillators show the fast fall times, but doping ZnO dramatically increases luminescence. Nitrogen-doped ZnO has both the longest fall time and very low-intensity PL peak. Boron-doped and gadolinium-doped ZnO have the highest PL intensities and average rise and fall times.

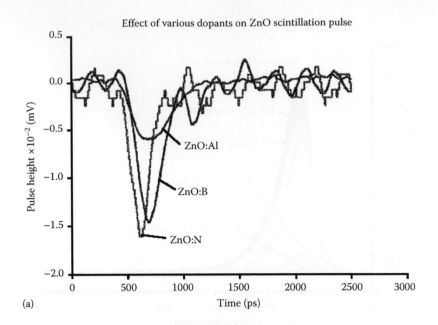

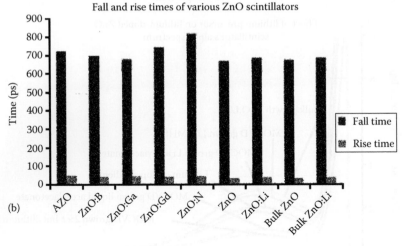

FIGURE 12.24
(a) Pulse height rise and fall times for three scintillators and (b) the rise and fall times of the best production quality scintillators.

12.6.3 Neutron Detection Performance

The ZnO neutron scintillators were tested by (1) measuring their response to high-energy neutrons such as those directly from an (α, n) source (^{239}Pu source of (α, n) neutrons) and (2) developing several MCNP models to test various scintillator configurations. Results from the ZnO scintillator's response in free space recorded by a multichannel MCA demonstrate that undoped ZnO could possibly discriminate between gamma rays and neutrons (Figure 12.25).

MCNPX v2.7d was used to simulate the response of various scintillator configurations to both neutron and gamma ray fields. The energy deposition tally was used to correctly

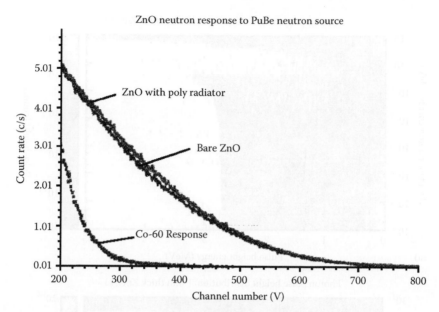

FIGURE 12.25
ZnO neutron detector response to fast neutrons.

calculate the energy imparted to the scintillator, as it takes the effects of secondary particles, such as gamma particles, tritons, and protons, into account. The neutron capture isotope algorithm was used to model the neutron capture reaction. These were then summed and plotted with a 3D contour mapping package in GNUPlot [88].

Configurations of only uniformly doped scintillators, stacks of one-sided, and stacks of two-sided scintillators for a variety of thicknesses were run. Simulation results for the 1 mm samples are shown below in Figure 12.26. Samples thicker than approximately 15 μm demonstrated compromised neutron-versus-gamma discrimination. At 0.5 mm, the signal from the gamma-ray field emitted from a standard fission gamma spectrum is comparable to the signal from the (n, α) reactions in the scintillator. Above 1 mm, the detector is saturated by the gamma ray signal.

To adequately test the neutron detectors, four commercial reference detectors were used (Table 12.2).

The neutron sources were a ^{241}AmBe-driven steady-state graphite pile and a graphite slowing down spectrometer (GSDS). While the steady-state graphite pile provides a simple test platform for thermal neutron detectors, the GSDS provides a means to select energy responses.

The lithiated glass scintillator has a characteristic (n, α) peak centered around channel 215. Integrating the peak and subtracting the integral of the background continuum calculate the neutron count.

The ^{10}B-lined gas-filled tube does not exhibit a discernible (n, α) peak. Unless the incident neutron has very high energy, one of the particles is generally emitted into the detector's wall and, therefore, does not produce a useful ionization event.

The ^{10}BF3 tube has three peaks due to the "wall effect" at approximately channels 250 (recoiling ^{7}Li nucleus), 350 (recoiling), and 600 counts (recoiling α particle and energetic ^{7}Li nucleus). The last peak at channel 900 corresponds to the recoiling ^{7}Li at ground state.

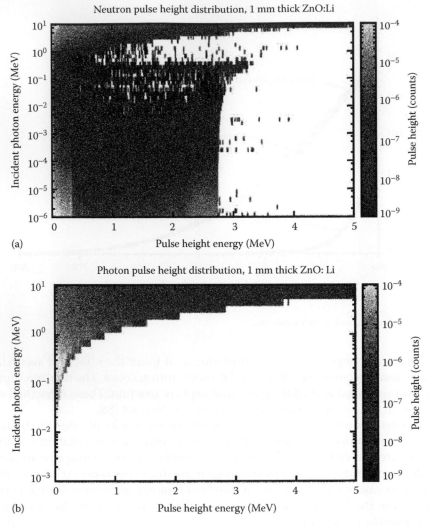

FIGURE 12.26
(a) Pulse-height distribution simulation resulting from neutrons incident on a 1 mm lithium-doped ZnO with a 5 μm ^{6}LiF conformal coating and (b) the pulse-height distribution resulting from various photon energies in a 1 mm thick ZnO-doped scintillator.

TABLE 12.2

Commercial Reference Detectors

Detector Type	Diameter (cm)	Length (cm)
^{3}He tube (LND, Inc.)	0.635	10.16
^{10}B-lined gas filled	2.54	30.48
^{10}BF3	2.54	30.48
Lithiated glass scintillator	5.08	1.27

TABLE 12.3

Dimensions of Novel Neutron Scintillators

Detector Type	Thickness (μm)	Coating
B-doped ZnO	6	10-μm ^{6}LiF-conformal coating
Li-doped ZnO	6.2	10-μm ^{6}LiF-conformal coating
Al-doped ZnO	10	10-μm ^{6}LiF-conformal coating
Ga-doped ZnO	8.1	10-μm ^{6}LiF-conformal coating
N-doped ZnO	9.3	10-μm ^{6}LiF-conformal coating
Gd-doped ZnO	5.9	10-μm ^{6}LiF-conformal coating
Undoped ZnO	10	10-μm ^{6}LiF-conformal coating

The ^{3}He tube pulse-height distribution for thermal neutrons also exhibits the "wall effect." The continuum of charged reaction products lies between channels 300 and 900. It has less of a stair step because the reaction generates particles of similar charge. The total energy peak is centered around channel 1250. The sharp drop in the count rate of the gamma distribution region around channel 100 facilitates neutron-versus-gamma discrimination.

The (*n*, gamma) responses of six novel neutron detectors were measured. Two of the ZnO detectors contained neutron target nuclei (Li and B). The structures of these detectors are summarized in Table 12.3.

The first novel neutron detector to be compared is MOCVD-grown boron-doped ZnO. The (*n*, α) peak results from a combination of B and Li-based reactions and is separated from the gamma radiation region (Figure 12.27b and d). The detector has good neutron-versus-gamma discrimination due to its thin geometry.

The lithium-doped ZnO scintillator exhibits the characteristic (*n*, α) peak also visible in the boron and lithiated glass responses (Figure 12.27a and c). The (*n*, α) peak results from the ^{6}Li-based reaction and can be distinguished from the gamma radiation response. The detector has good neutron-versus-gamma discrimination due to its thin geometry.

Bulk and a ~6 μm thick MOCVD-grown Li-doped ZnO scintillator were tested. The count rate is significantly higher in the bulk sample, since the thicker sample can absorb higher energy and generate a higher number of reactions.

The neutron responses of lithium-doped and boron-doped ZnO detectors were compared (Figure 12.28). Since both Li and B are neutron target nuclei, most of the energy of both particles is collected. The lower intensity of lithium-doped ZnO sample's response corresponds to its lower number of dopants compared to the boron-doped ZnO sample.

The remaining ZnO samples still detect neutrons well, even though they did not contain neutron target nuclei. These samples had a 10 μm ^{6}LiF conformal coating (an evaporative coating process). As a result, they could only collect one of the charged particles produced in the ^{6}Li(*n*, α) reaction due to the isotropic nature of this reaction in the center of mass. Consequently, their responses resemble those of the ^{10}B-lined tube pulse-height distribution.

The AZO exhibits a lower pulse-height distribution because of its lower scintillation light emission (Figure 12.29a). Neutron-versus-gamma discrimination is still possible, but the gamma region of the spectrum is almost a continuum similar to the ^{10}B-lined tube.

Gallium-doped ZnO performed well with acceptable neutron-versus-gamma discrimination. The sample used was 8.1 μm thick with a 10 μm ^{6}LiF conformal coating.

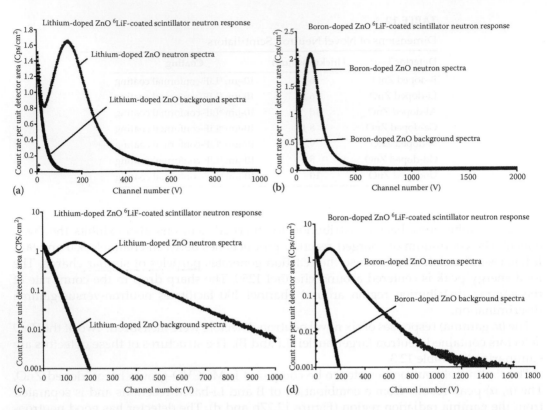

FIGURE 12.27

(n, α) spectra from ^{6}LiF-coated B-doped ZnO (a and c). (n, α) spectra from ^{6}LiF-coated B-doped ZnO (b and d). (a) and (b) have linear vertical axes, whereas (c) and (d) have logarithmic vertical axes. All graphs show the (n, α) peaks clearly.

The background 100 mR/h gamma ray field is easily distinguished from the neutron component of the spectrum (Figure 12.29b).

Nitrogen-doped ZnO performed with acceptable neutron-versus-gamma discrimination (Figure 12.29c). The sample was 9.3 μm thick with a 10 μm conformal coating of ^{6}LiF. Although the sample was annealed for 5 min in oxygen, some of the trapped hydrogen remained and reduced the light emission. The background 100 mR/h gamma ray field is easily distinguished from the neutron region of the spectrum. The total energy peak edge is located at a lower energy due to the significantly lower light emission. While the sample was optically transparent, the lower light yield is not desirable in scintillator performance.

Undoped ZnO performed well with decent neutron-versus-gamma discrimination (Figure 12.29d). The sample was 6.4 μm thick with a 10 μm conformal coating of ^{6}LiF. The background 100 mR/h gamma ray field is easily discerned from the neutron component of the spectra.

Gadolinium-doped ZnO was also tested (Figure 12.29e). The gadolinium samples performed significantly worse than the other lithium-doped samples and exhibited poor neutron-versus-gamma discrimination. The background 100 mR/h gamma ray field is no longer easily distinguished from the neutron region of the spectrum, since the spectra overlap from channel 100 to 300.

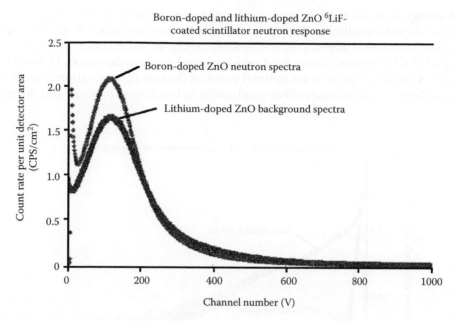

FIGURE 12.28

(n, α) spectra from ^{6}LiF-coated Li-doped ZnO in comparison to B-doped ZnO. The n, α peak is clearly visible.

To quantitatively compare the results of the neutron detection system, the detectors responses to the steady-state thermal neutron flux and the detector count rates as a function of detector area were computed. The neutron detection efficiency per unit area of each of the scintillators is compared in Figure 12.30a. Figure 12.30b shows the comparison of neutron detection efficiency per unit volume.

12.7 Summary and Future Directions

The ZnO scintillator is a good neutron-radiation detector. ZnO can be grown in a number of different ways, including bulk melt growth and MOCVD growth. Thin films can be grown that exhibit equivalent or superior neutron-versus-gamma discrimination to conventional neutron detectors. High thermal neutron cross sections were observed, which produce highly efficient neutron detectors. It is not necessary to observe a (n, α) peak to create a neutron detector with good neutron-versus-gamma discrimination, although this feature should be present in an ideal neutron detector so that the Q-value can be used for neutron-versus-gamma discrimination. The (n, α) peak/region cannot overlap the gamma ray region.

Novel ZnO neutron detectors have been modeled in MCNP5. Using these detectors, growth studies were undertaken successfully to optimize the scintillator structures for both MOCVD and bulk melt growth. The samples that do not possess lithium or boron interstitially mixed into the matrix currently resemble the response of a boron-lined gas-filled tube. This is due to the inability to collect both particles because of the coating layer thickness. Two-sided scintillators can be grown over conformal ^{6}LiF coatings.

The full energy peak for both the triton and alpha particle in the ^{6}Li (n, α) reaction can be collected in doped scintillators. This Q-value could be used for neutron spectroscopy.

It has been shown that doped ZnO scintillators exhibit superior characteristics compared to undoped ZnO in optical clarity, light emission, and neutron detection. However, when doping the scintillator, there is an optimal point of dopant concentration. Past this optimal value, the material becomes either conducting or insulating, opaque, and decreases in

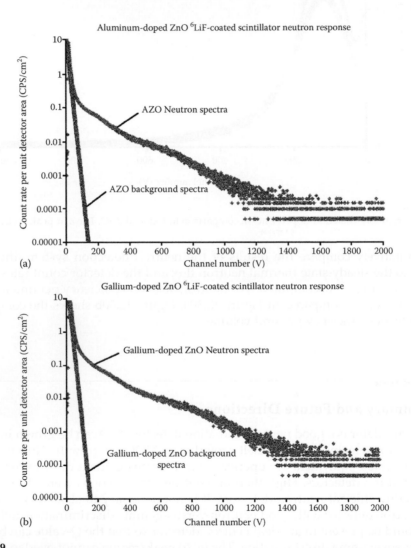

FIGURE 12.29

(n, α) spectra for simple-coated ^{6}LiF structured AZO and the corresponding 100 mR/h gamma ray field (a). (n, α) spectra for simple-coated ^{6}LiF structured gallium-doped ZnO and the corresponding 100 mR/h gamma ray field (b). (n, α) spectra for simple-coated ^{6}LiF-structured nitrogen-doped ZnO corresponding 100 mR/h gamma ray field (c). (n, α) spectra for simple-coated ^{6}LiF-structured undoped ZnO and the corresponding 100 mR/h gamma ray field (d). $n, \gamma >$ conversion electron spectra for uniformly doped gadolinium ZnO and the corresponding 100 mR/h γ ray field (e). Neutron-versus-gamma discrimination is acceptable for all spectra except for the Gd-doped ZnO spectrum. The α spectrum has turned into a continuum, which resembles a ^{10}B-lined tube for all of the plots except for the Gd-doped ZnO spectrum. In the Gd-doped ZnO spectrum, the electron conversion is a continuum resembling a γ-ray response with a slight peak around channel 800.

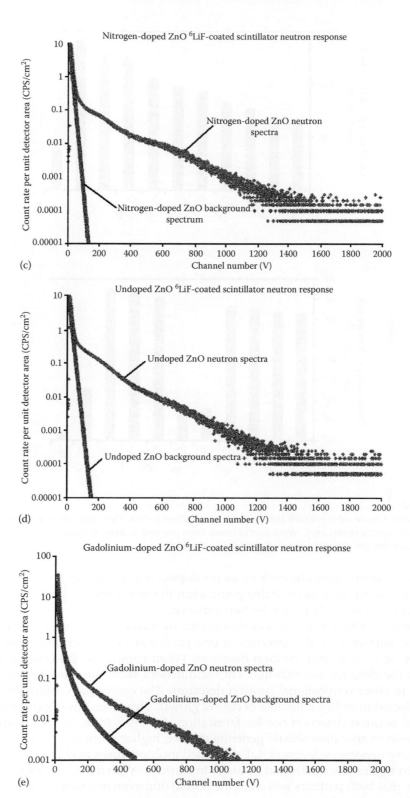

FIGURE 12.29 (continued)

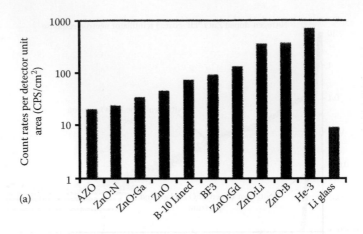

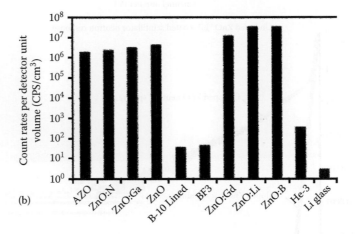

FIGURE 12.30
(a) Comparison of count rates per unit area for neutron detectors tested. The count rate was summed above the zero-value gamma-ray cutoff. (b) Comparison of count rates per unit volume for neutron detectors tested. The count rate was summed above the zero-value gamma-ray cutoff.

performance. Determining the peak value for doping was the key to this work. Increasing the dopant concentration beyond this point, even though in theory, it should have raised detection efficiencies, in fact impedes performance.

The thermal neutron reaction was dominated by surface reactions. Comparing count rates versus surface area, the detectors at one particular layer thickness perform well in comparison to conventional neutron detectors. When comparing the count rate per unit volume of the detector, the thin-film ZnO scintillators show superior neutron detection compared to other conventional neutron detectors. The conformal coating thickness was initially selected to be 10 μm because of ease of growth. However, this thickness was neither optimal for neutron detection nor for fabrication. A switch from thermal evaporation to electron beam evaporation should generate thinner, higher efficiency coatings. One thousand and fifty nanometer layers of ^{6}LiF deposited multiple times over layers of doped ZnO would yield higher efficiencies and be easier to fabricate. With the 10 μm thick coatings, the probability that both particles will enter the scintillator, even in a two-sided device, is low.

By minimizing this thickness to 10–50 nm and increasing the number of layers, it is possible to improve scintillator efficiency.

The MCNPX models of the neutron detectors show that the optimal neutron-versus-gamma discrimination capabilities of the ZnO scintillator are realized when the scintillator thickness is kept as thin as possible. Making stacks of one or two-sided scintillators offset with sapphire substrates (or other nonscintillating/non-Chrenkov radiation emitting materials) is beneficial for increasing the neutron detection efficiency, but effectively separates the responses of each scintillator, minimizing the gamma response. Neutron-versus-gamma discrimination can be achieved at larger thicknesses, but at a noticeable reduction in efficiency if the zero crossing value of the photon response for the low energy cutoff is used.

When examining the models produced in MCNP5 and MCNPX and extrapolating from current data, stacks of three or four of these scintillators can have efficiencies equal to or greater than ³He tubes. The detectors are already equivalent in neutron-versus-gamma discrimination. Comparing the speed of the ³He tube to the ZnO scintillator, the ZnO scintillator is a superior choice. The ZnO scintillator can be grown in large surface area devices producing very efficient neutron detectors, with efficiencies superior to ³He tubes. These novel devices are superior to lithiated glass, ¹⁰BF3 tubes, and ¹⁰B-lined tubes in overall efficiencies and speed. The fabrication cost of these scintillators is relatively low in comparison to that of ³He or other structures. In addition, there exists a direct road to commercialization of these devices through companies such as Cermet, Inc.

This doped ZnO structure shows promise in fulfilling the need for a near term ³He tube replacement for homeland security applications. Either bulk melt-grown materials or MOCVD grown materials are suitable for this endeavor. Bulk melt-grown materials have the promise of large throughput and very thick samples. However, bulk melt growth would not be a good use of limited enriched ⁶Li compounds, but this still needs to be further investigated. Application of conformal coatings of uniform samples is the most economic path to commercialization with bulk melt growth materials. MOCVD grown materials grown with a larger, scaled-up version of the reactor do show promise. Here, rapid layer changes and enriched lithium can be used to their advantage to make thin multilayered stacks of materials. Materials grown by this method could approach the necessary 100% intrinsic efficiency. The single stacks achieved 43% intrinsic efficiencies for thermal neutrons.

Overall, this work has shown that regardless of the growth method, doped ZnO and ⁶LiF-coated ZnO with or without dopants can produce a highly efficient neutron scintillator. Dopant studies were conducted for three types of dopants: (1) those for wavelength shifting, (2) crystal transparency improvement, and (3) neutron detector nuclei. In the crystal transparency regime, aluminum improves the optical clarity of the sample, but shifts the scintillation spectrum toward the UV. Of the wavelength-shifting dopants, gallium and boron exhibit a good blue wavelength shift. However, they produce optically clouded materials because the oxide formation of each of these materials is a white, opaque material. Lithium and boron can be doped as neutron detector nuclei. Lithium produces a more optically clear material. Lithium does blue-shift the scintillation spectrum slightly but is difficult to grow by MOCVD due to the lack of a good precursor. Lithium acetylacetate shows the most promise for the growth of MOCVD lithium-doped ZnO. Dopant materials and scintillation luminosity are inversely proportional. While the dopant materials add beneficial characteristics to the scintillator, high dopant levels degrade the scintillation. A fine balance must be struck between dopant characteristics and scintillation luminescence.

References

1. R. Cooper, I. Anderson, C. Britton, K. Crawford, L. Crow, P. DeLurgio et al. A program for neutron detector research and development. Oak Ridge, TN: Oak Ridge National Laboratory (2003).
2. Committee on the Impact of Selling the Federal Helium Reserve, Commission on Physical Sciences, Mathematics, and Applications, Commission on Engineering and Technical Systems, National Research Council. *The Impact of Selling the Federal Helium Reserve*. Washington, DC: National Academy Press, (2000), ISBN-10: 0-309-07038-4.
3. D. Slaughter, M. Accatino, A. Bernstein, J. Candy, A. Dougan, J. Hall et al. *Detection of Special Nuclear Material in Cargo Containers Using Neutron Interrogation*. Livermore, CA: Lawrence Livermore National Laboratory (2003), UCRL-ID-155315.
4. J.T. Mihalczo, J.K. Mattingly, J.S. Neal, and J.A. Mullens. NMIS plus gamma spectroscopy for attributes of HEU, PU and HE detection. *Nucl. Instrum. Methods Phys. Res. B*, **213**, 378 (2003). http://www.canberra.com/pdf/Products/Neutron-Detection-Counting.pdf
5. M. Nieto, *Resource Note on Photofission of Nuclei, for U-235 and Pu-239 Detection*. Los Alamos, NM Los Alamos National Laboratory, LA-UR-03-2056 (2003).
6. C.E. Moss, M.W. Brener, C.L. Hollas, and W.L. Myers. Portable active interrogation system. *Nucl. Instrum. Methods Phys. Res., Sect. B*, **241**, 793 (2006).
7. J.L. Jones, W.Y. Yoon, K.J. Haskell, D.R. Norman, J.M. Hoggan, C.E. Moss, C.A. Goulding, C.L. Hollas, W.L. Myers, and E. Franco. Photofission-based, nuclear material detection: Technology demonstrations. Idaho Falls, ID: Idaho National Engineering Environmental Laboratory, INEEL/EXT-02-01406, December (2002).
8. R. Norman, J. Jones, J.W. Sterbentz, K.J. Haskell, W.Y. Yoon, Pitas generation III system design report: The developmental prototype, *INL/EXT-08-13798*, January 2008; http://www.canberra.com/pdf/Products/Neutron-Detection-Counting.pdf
9. Congress, United States of America. Security and Accountability for Every Port Act. H.R.4954 1884–1962 (2006).
10. J.K. Tuli. Nuclear wallet cards. National Nuclear Data Center, Upton, New York: Brookhaven National Laboratory (2005).
11. M.G. Mayer, On closed shells in nuclei. II. *Phys. Rev.*, **75**, 1969–1970 (1949).
12. M.K. Kopp, K.H. Valentine, L.G. Christophorou, and J.G. Carter. New gas mixture improves performance of ^{3}He neutron counters. *Nucl. Instrum. Methods Phys. Res.*, **201**, 395 (1982).
13. D.J. Thomas, and A.V. Alevra. Bonner sphere spectrometers—A critical review. *Nucl. Instrum. Methods Phys. Res., Sect. A*, **476**, 12 (2002).
14. E. Burgett. A broad spectrum neutron spectrometer utilizing a high energy bonner sphere extension. Atlanta, GA: Georgia Institute of Technology (2008).
15. B. Wiegel and A.V. Alevra. NEMUS—The PTB neutron multisphere spectrometer: Bonner spheres and more. *Nucl. Instrum. Methods Phys. Res., Sect. A*, **476**, 36 (2002).
16. P.M. Dighe, K.R. Prasad, and S.K. Kataria. Silver-lined proportional counter for detection of pulsed neutrons. *Nucl. Instrum. Methods Phys. Res., Sect. A*, **523**, 158 (2004).
17. R. Rhodes. *The Making of the Atomic Bomb*. New York: Simon & Schuster (1986).
18. S.A. Wender, S. Balestrini, A. Brown, R.C. Haight, C.M. Laymon, T.M. Lee et al. A fission ionization detector for neutron flux measurements at a spallation source. *Nucl. Instrum. Methods Phys. Res., Sect. A*, **336**, 226 (1993).
19. R.J. Nikolic, C.L. Cheung, C.E. Reinhardt, and T.F. Wang. Roadmap for high efficiency solid-state neutron detectors. *Proc. SPIE*, **6013**, 601305 (2005).

20. R.J. Nikolic, Q. Shao, L.F. Voss, A.M. Conway, R. Radev, T.F. Wang, M.Dar, N. Deo, C.L. Cheung, L. Fabris, C.L. Britton, and M.N. Ericson, Si pillar structured thermal neutron detectors: Fabrication challenges and performance expectations, *Proc. SPIE* 8031, 803109 (2011). DOI 10.1117/12.885880.

21. S.L. Bellinger, W.J. McNeil, T.C. Unruh, and D.S. McGregor. Angular response of perforated silicon diode high efficiency neutron detectors. *IEEE NSS Conf. Record*, Honolulu, HI, p. 1904–1907 (2007).

22. J. Birks. *The Theory and Practice of Scintillation Counting*. New York: Pergamon Press (1964).

23. M. Moszynski, G.J. Costa, G. Guillaume, B. Heusch, A. Huck, and S. Mouatassim. Study of n-[gamma] discrimination with NE213 and BC501A liquid scintillators of different size. *Nucl. Instrum. Methods Phys. Res., Sect. A*, **350**, 226 (1994).

24. D. Rochman, R.C. Haight, J.M. O' Donnell, M. Devlin, T. Ethvignot, and T. Granier. Neutron-induced reaction studies at FIGARO using a spallation source. *Nucl. Instrum. Methods Phys. Res., Sect. A*, **523**, 102 (2004).

25. R. Batchelor, W.B. Gilboy, J.B. Parker, and J.H. Towle, The response of organic scintillators to fast neutrons. *Nucl. Instrum. Methods*, **13**, 70 (1961).

26. M.L. Roush, M.A. Wilson, and W.F. Hornyak. Pulse shape discrimination. *Nucl. Instrum. Methods*, **31**, 112 (1964).

27. J.B. Czirr, D.B. Merrill, D. Buehler, T.K. McKnight, J.L. Carroll, T. Abbott, and E. Wilcox. Capture-gated neutron spectrometry. *Nucl. Instrum. Methods Phys. Res., Sect. A*, **476**, 309 (2002).

28. ELJEN Technology, *EJ-301 MSDS Data Sheet*, 1–5 (2010), http://www.eljentechnology.com/index.php/joomla-overview/this-is-newest/71-ej-301.

29. J.T. Mihalczo, J.A. Mullens, J.K. Mattingly, and T.E. Valentine. Physical description of nuclear materials identification system (NMIS) signatures. *Nucl. Instrum. Methods Phys. Res., Sect. A*, **450**, 531 (2000).

30. National Institute of Standards (NISI), Physical Meas. Laboratory, ESTAR, H.H., Anderson, M.J. Berger (Chairman), H. Bichsel, J.A. Dennis, M. Inokuti (vice-chairman), D. Powers, S.M. Seltzer, D. Thwaites, J.E. Turner, and D.E. Watt. http://physics.nist.gov/PhysRefData/Star/Text/ESTAR.html

31. R.P. Gardner, C.W. Mayo, E.S. El-Sayyed, W.A. Metwally, Y. Zheng, and M. Poezart. A feasibility study of a coincidence counting approach for PGNAA applications. *Appl. Radiat. Isot.*, **53**, 515 (2000).

32. R.C. Greenwood, C.W. Reich, H.A. Baader, H.R. Koch, D. Breitig, O.W.B. Schult et al. Collective and two-quasiparticle states in ^{158}Gd observed through study of radiative neutron capture in ^{157}Gd. *Nucl. Phys. A*, **304**, 327 (1978).

33. G. Leinweber, D.P. Barry, M.J. Trbovich, J.A. Burke, N.J. Drindak, H.D. Knox, R.V. Ballad, R.C. Block, Y. Danon, and L.I. Severnyak. Neutron capture and total cross-section measurements and resonance parameters of gadolinium. *Nucl. Sci. Eng.*, **154**, 261 (2006).

34. J.F. Ziegler, M.D. Ziegler, and J.P. Biersack. SRIM—The stopping and range of ions in matter (2010). *Nucl. Instrum. Methods Phys. Res., Sect. B*, **268**, 1818 (2010).

35. M.J. Berger, and S.M. Seltzer. Tables of energy losses and ranges of electrons and positrons. Report No. NASA-SP-3012, Washington, DC: National Aeronautics and Space Administration (1964). (Available from National Technical Information Service, Springfield, VA 22161.)

36. K.S. Krane. *Introductory Nuclear Physics*. New York: Wiley (1987), ISBN: 978-0-471-80553-3.

37. G.F. Knoll. *Radiation Detection and Measurements*. New York: Wiley (1979), ISBN: 047149545X.

38. B.D. Rooney and J.D. Valentine. Calculating nonproportionality of scintillator photon response using measured electron response data. *IEEE Trans. Nucl. Sci.*, **44**, 509 (1997).

39. A. Lundby. Scintillation decay times. *Phys. Rev.*, **80**, 477 (1950).

40. M. Nikl. Wide band gap scintillation materials: Progress in the technology and material understanding. *Phys. Status Solidi A*, **178**, 595 (2000).

41. J.B. Birks and F.W.K. Firk. The theory and practice of scintillation counting. *Phys. Today*, **18**, 60 (1965).

42. P.A. Rodnyi, E.I. Gorohova, S.B. Mikhrin, A.N. Mishin, and A.S. Potapov. Quest and investigation of long wavelength scintillators. *Nucl. Instrum. Methods Phys. Res., Sect. A*, **486**, 244 (2002).

43. S.C. Curran and W.R. Baker. U.S. Atomic Energy Report MDDC, 1296, November 17 (1944); *Rev. Sci. Instrum.*, **19**, 116 (1948).

44. W. Lehmann. Edge emission of n-type conducting ZnO and CdS. *Solid-State Electron.*, **9**, 1107 (1966).

45. M. Pan, J. Nause, V. Rengarajan, R. Rondon, E.H. Park, and I.T. Ferguson. Epitaxial growth and characterization of p-type ZnO. *J. Electron. Mater.*, **36**, 457 (2007).

46. E.L. Hull, R.H. Pehl, N.W. Madden, P.N. Luke, C.P. Cork, D.L. Malone, J.S. Xing, K. Komisarcik, J.D. Vanderwerp, and D.L. Friesel. Temperature sensitivity of surface channel effects on high-purity germanium detectors. *Nucl. Instrum. Methods Phys. Res., Sect. A*, **364**, 488 (1995).

47. C. Klingshirn. The luminescence of ZnO under high one- and two-quantum excitation. *Phys. Status Solidi (b)*, **71**, 547 (1975).

48. H. Morkoç and Ü. Özgür. *Zinc Oxide: Fundamentals, Materials and Device Technology*. Weinheim, Germany: Wiley-VCH Verlag GmbH & Co. KGaA (2009).

49. A. Janotti and C.G. Van de Walle. Fundamentals of zinc oxide as a semiconductor. *Rep. Prog. Phys.*, **72**, 126501 (2009).

50. S. Chichibu, T. Azuhata, T. Sota, and S. Nakamura. Excitonic emissions from hexagonal gan epitaxial layers. *J. Appl. Phys.*, **79**, 2784 (1996).

51. Y. Zhang, J.-G. Lu, Z.-Z. Ye, H.-P. He, L.-L. Chen, and B.-H. Zhao. Identification of acceptor states in Li-N dual-doped p-type ZnO thin films. *Chin. Phys. Lett.*, **26**, 04103 (2009).

52. P.J. Simpson, R. Tjossem, A.W. Hunt, K.G. Lynn, and V. Munne. Superfast timing performance from ZnO scintillators. *Nucl. Instrum. Methods Phys., Sect. A*, **505**, 82 (2003).

53. N.H. Nickel and E. Terukov. *Zinc Oxide—A Material for Micro- and Optoelectronic Applications*. St. Petersburg, Russia: Springer (2004), ISBN-10: 1-4020-3474-1.

54. Ü. Özgür, Ya.I. Alivov, C. Liu, A. Teke, M.A. Reshchikov, S. Doğan, V. Avrutin, S.-J. Cho, and H. Morkoç. A comprehensive review of ZnO materials and devices. *J. Appl. Phys.*, **98**, 041301 (2005).

55. A. Hernandez Battez, R. Gonzalez, J.L. Viesca, J.E. Fernandez, J.M. Diaz Fernandez, A. Machado, R. Chou, and J. Riba. CuO, ZrO_2 and ZnO nanoparticles as antiwear additive in oil lubricants. *Wear*, **265**, 422 (2008).

56. S.O. Kucheyev, J.E. Bradby, J.S. Williams, C. Jagadish, and M.V. Swain. Mechanical deformation of single-crystal ZnO. *Appl. Phys. Lett.*, **80**, 956–958 (2002).

57. Bill Cordua, The hardness of rocks and minerals, *The Lapidary* Digest (1998), http://www.gem-select.com/gem-info/gem-hardness-info.php

58. A. Onodera, N. Tamaki, Y. Kawamura, T. Sawada, and H. Yamashita. Dielectric activity and ferroelectricity in piezoelectric semiconductor Li-doped ZnO. *Jpn. J. Appl. Phys.*, **35**, 5160 (1996).

59. K. Postava, H. Sueki, M. Aoyama, T. Yamaguchi, Ch. Ino, Y. Igasaki, and M. Horie. Spectroscopic ellipsometry of epitaxial ZnO layer on sapphire substrate. *J. Appl. Phys.*, **87**, 7820 (2000).

60. M. Bass, C. DeCusatis, J. Enoch, V. Lakshminarayanan, G. Li, C. MacDonald, V. Mahajan, and E. Van Stryland. *Handbook of Optics*, 3rd edn., Vol. II, Design fabrication and testing, sources and detectors, radiometry and photometry. New York: McGraw-Hill Professional (2009), ISBN-10: 0071498907.

61. Y. Gai, G. Tang, and J. Li. Formation of shallow acceptors in ZnO doped by lithium with the addition of nitrogen. *J. Phys. Chem. Solids*, **72**, 725 (2011).

62. S.S. Lin, Z.Z. Ye, J.G. Lu, H.P. He, L.X. Chen, X.Q. Gu, J.Y. Huang, L.P. Zhu, and B.H. Zhao. Na doping concentration tuned conductivity of ZnO films via pulsed laser deposition and electroluminescence from ZnO homojunction on silicon substrate. *J. Phys. D Appl. Phys.*, **41**, 155114 (2008).

63. L. Xu, F. Gu, J. Su, Y. Chen, X. Li, and X. Wang. The evolution behavior of structures and photoluminescence of K-doped ZnO thin films under different annealing temperatures. *J. Alloys Compd.*, **509**, 2942 (2011).

64. P. Nunes, E. Fortunato, P. Tonello, F. Braz Fernandes, P. Vilarinho, and R. Martins. Effect of different dopant elements on the properties of ZnO thin films. *Vacuum*, **64**, 281 (2002).
65. K.F. Nielsen. Growth of ZnO single crystals by the vapor phase reaction method. *J. Cryst. Growth*, **3/4**, 141 (1968).
66. M. Haupt, A. Ladenburger, R. Sauer, K. Thonke, R. Glass, W. Roos, J.P. Spatz, H. Rauscher, S. Riethmüller, and M. Möller. Ultraviolet-emitting ZnO nanowhiskers prepared by a vapor transport process on prestructured surfaces with self-assembled polymers. *J. Appl. Phys.*, **93**, 6252 (2003).
67. V. Avrutin, G. Cantwell, J. Zhang, J.J. Song, D.J. Silversmith, and H. Morkoç. Bulk ZnO: Current status, challenges, and prospects. *Proc. IEEE*, **98**, 1339 (2010).
68. S. Wang. Method and apparatus for zinc oxide single boule growth. U.S. Patent 7279040B1 United States of America, issued October 9, 2007.
69. C.R. Gorla, N.W. Emanetoglu, S. Liang, W.E. Mayo, and Y. Lu. Structural, optical, and surface acoustic wave properties of epitaxial ZnO films grown on (012) sapphire by metalorganic chemical vapor deposition. *J. Appl. Phys.*, **85**, 2595 (1999).
70. J. Nause. ZnO broadens the spectrum. *III-Vs Rev.*, **12**, 28 (1999).
71. E. Ohshima, H. Ogino, I. Niikura, K. Maeda, M. Sato, M. Ito, and T. Fukuda. Growth of the 2-in-size bulk ZnO single crystals by the hydrothermal method. *J. Cryst. Growth*, **260**, 166 (2004).
72. D.F. Croxall, R.C.C. Ward, C.A. Wallace, and R.C. Kell. Hydrothermal growth and investigation of Li-doped zinc oxide crystals of high purity and perfection. *J. Cryst. Growth*, **22**, 117 (1974).
73. M.J. Suscavage, M. Harris, D. Bliss, P. Yip, S.-Q. Wang, D. Schwall et al. High quality hydrothermal ZnO crystals. *MRS Internet J. Nitride Semicond. Res.*, **4**, G3.40 (2004).
74. M. Razeghi. *The MOCVD Challenge: Volume 2: A Survey of GaInAsP-GaAs for Photonic and Electronic Device Applications.* Boca Raton, FL: CRC Press (1995).
75. A.N. Jansen, M.E. Orazem, B.A. Fox, and W.A. Jesser. Numerical study of the influence of reactor design on MOCVD with a comparison to experimental data. *J. Cryst. Growth*, **112**, 316 (1991).
76. A. Gurary, P.T. Fabiano, D.R. Voorhees, and S. Beherrell. Induction heated chemical vapor deposition reactor. U.S. Patent 6368404 B1, issued April 9, 2002.
77. B.P. Zhang, N.T. Binh, K. Wakatsuki, Y. Segawa, Y. Yamada, N. Usami, M. Kawasaki, and H. Koinuma. Pressure-dependent ZnO nanocrystal growth in a chemical vapor deposition process. *J. Phys. Chem. B*, **108**, 10899 (2004).
78. R. Tena-Zaera, J. Zuniga-Perez, C. Marti'nez-Tomas, and V. Munoz-Sanjose. Numerical study of the ZnO growth by MOCVD. *J. Cryst. Growth*, **264**, 237 (2004).
79. K.F. Jensen, D.I. Fotiadis, and T.J. Mountziaris. Detailed models of the MOVPE process. *J. Cryst. Growth*, **107**, 1 (1991).
80. H. Moffat and K.F. Jensen. Complex flow phenomena in MOCVD reactors: I. Horizontal reactors. *J. Cryst. Growth*, **77**, 108 (1986).
81. Y. Liu, C.R. Gorla, S. Liang, N. Emanetoglu, Y. Lu, H. Shen et al. Ultraviolet detectors based on epitaxial ZnO films grown by MOCVD. *J. Electron. Mater.*, **29**, 69 (2000).
82. W. Xu, Z. Ye, T. Zhou, B. Zhao, L. Zhu, and J. Huang. Low-pressure MOCVD growth of p-type ZnO thin films by using NO as the dopant source. *J. Cryst. Growth*, **265**, 133 (2004).
83. S. Nakamura. Analysis of real-time monitoring using interference effects. *Jpn. J. Appl. Phys.*, **30**, 1348 (1991).
84. J.M. Olson and A. Kibbler. In situ characterization of MOCVD growth processes by light scattering techniques. *J. Cryst. Growth*, **77**, 182 (1986).
85. A. Ohtomo, R. Shiroki, I. Ohkubo, H. Koinuma, and M. Kawasaki. Thermal stability of supersaturated $Mg_xZn_{1-x}O$ alloy films and $Mg_xZn_{1-x}O/ZnO$ heterointerfaces. *Appl. Phys. Lett.*, **75**, 4088 (1999).
86. S.J. Lim, S. Kwon, and H. Kim. ZnO thin films prepared by atomic layer deposition and RF sputtering as an active layer for thin film transistor. *Thin Solid Films*, **516**, 1523 (2008).
87. S.M. George, A.W. Ott, and J.W. Klaus. Surface chemistry for atomic layer growth. *J. Phys. Chem.*, **100**, 13121 (1996).
88. T. Williams and C. Kelly. GNUplot, an interactive plotting program (2011) http://www.gnuplot.info/docs_4.4/gnuplot.pdf

13

Amorphous In-Ga-Zn-O Thin Film Transistors: Fabrication and Properties

Toshio Kamiya and Hideo Hosono

CONTENTS

13.1 Introduction 486
13.2 Oxide Semiconductors and Oxide TFTs: History 487
13.3 Electronic Structure and Carrier Transport Specific to Oxide Semiconductors 488
13.4 Material Design Concept of AOS with High Mobility 491
13.5 Roles of Constituent Elements and Other AOS Materials 494
13.6 Film Growth 498
13.7 Atomic Structure of AOSs 498
13.8 Electronic Structure of AOSs 500
13.9 Optical Properties of AOSs 504
13.10 Electron Transport in AOSs 505
13.11 Subgap Defects 506
13.12 Doping, Oxygen Deficiency, and Hydrogen 509
13.13 Advantages of AOS 512
13.14 Fabrication of AOS TFTs 513
    13.14.1 TFT Structures and Fabrication Process 513
    13.14.2 Operation Characteristics of a-IGOZ TFT 514
    13.14.3 Unique Annealing Effect 514
    13.14.4 Uniformity 515
    13.14.5 Stability 516
13.15 Progress in AOS TFT as the FPD Backplane 518
    13.15.1 History of Prototype Displays 518
    13.15.2 New Display Structure 519
    13.15.3 Flexible Displays 520
    13.15.4 Transparent Displays 520
    13.15.5 System Integration 521
    13.15.6 Mass Production 521
13.16 Perspective as a Chapter Summary 522
    13.16.1 TFT Fabrication Using Solution Processes 522
    13.16.2 *P*-Type Oxide TFT and CMOS Circuits 525
References 525

13.1 Introduction

Amorphous oxide semiconductor (AOS) is an alternative of crystalline oxide semiconductor and now expected to be most promising for channel materials of thin-film transistors (TFTs) in next-generation flat-panel displays (FPDs) such as active-matrix (AM) fast/jumbo-size/ultrahigh-resolution liquid-crystal displays (LCDs), organic light-emitting diode displays (OLEDs), electronic papers (e-papers), and flexible large-area electronic devices. Since the first demonstration of an AOS TFT using an amorphous In-Ga-Zn-O (a-IGZO) channel [1], many chemical compositions have been proposed to date; however, a-IGZO has been studied most intensively and its mass production started in March, 2012 [2,3]. As will be explained later on, AOSs have superior properties to crystalline semiconductors such as ZnO for large-area microelectronics devices (giant microelectronics) because they exhibit large carrier mobilities >10 cm^2 (V s)$^{-1}$ even if these are fabricated at room temperature (RT); therefore, AOS TFTs and their devices are applicable to flexible devices fabricated on organic plastic substrates and these devices have much better performances than those fabricated using conventional semiconductors such as hydrogenated amorphous silicon (a-Si:H) and organic semiconductors (Figure 13.1). Further, AOS TFTs are oxides of metal cations with closed-shell electronic configurations and have large bandgaps >3.0 eV, which enable to develop completely transparent electronic circuits. In this chapter, we review brief history, fundamental properties, TFT technology, and device

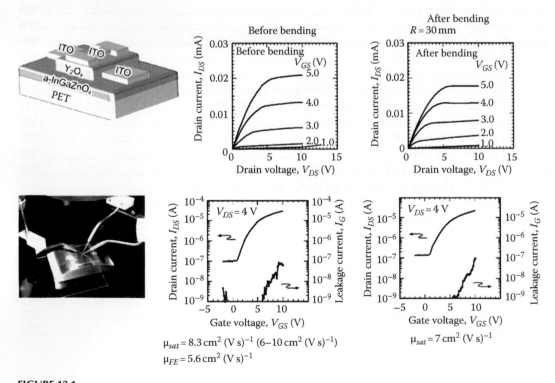

$\mu_{sat} = 8.3\,\text{cm}^2\,(\text{V s})^{-1}\ (6–10\,\text{cm}^2\,(\text{V s})^{-1})$
$\mu_{FE} = 5.6\,\text{cm}^2\,(\text{V s})^{-1}$

$\mu_{sat} = 7\,\text{cm}^2\,(\text{V s})^{-1}$

FIGURE 13.1

First AOS TFT using a-IGZO channel. It is fabricated on flexible PET substrate using transparent ITO electrodes and so is transparent and flexible. Large TFT mobility >7 cm^{-2} (V s)$^{-1}$ is maintained even after the bending test at the curvature radius of 30 mm.

applications of AOSs in particular for a-IGZO. Here, we recommend some reference papers and books for helping know more details [4–13].

13.2 Oxide Semiconductors and Oxide TFTs: History

Figure 13.2 summarizes the history of oxide semiconductor and their TFTs. The research of oxide TFTs started in mid-1960s but had almost disappeared in open-accessible literatures after that until 1990s. The next oxide TFT reappeared in 1996 as an epitaxial SnO_2 TFT combined with a ferroelectric gate and became active in 2000s due to expectation that polycrystalline ZnO (poly-ZnO) can produce semiconductor active layers even at low temperatures including RT by keeping reasonable Hall mobilities >10 cm^2 (V s)$^{-1}$. This feature is considered to be very promising for low-temperature, large-area devices such as solar cells (SCs) and FPDs. Oxides are already used as transparent window electrodes and such oxides are called transparent conducting oxides (TCOs). However, active layers in semiconductor devices require low carrier density, low defect density, but satisfactory high mobilities; these kinds of materials are now called transparent oxide semiconductors (TOSs) instead of TCOs. As mentioned previously, good TCOs are required to improve devices such as SCs and FPDs because better transparency/electrical conductivity improves energy conversion efficiency in SCs and reduces power consumption and improves picture quality in FPDs; on the other hand, good TOS active layers, for example, in TFTs, improve device size/resolution, electrical current drivability, operation speed, and so on and enable

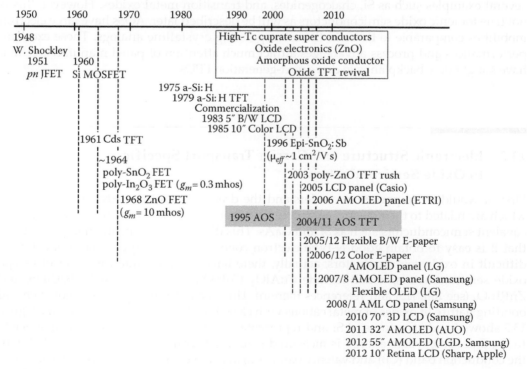

FIGURE 13.2
History of oxide semiconductors and TFTs.

applications to next-generation FPDs such as jumbo-size/ultrafine/fast LCDs and OLEDs. These applications are expanding to higher frame rates, 3D displays, flexible displays, and transparent displays.

Research on oxide TFTs had been restricted to polycrystalline materials until late 2004. However, polycrystalline N-type oxide semiconductors represented by ZnO and SnO_2 are sensitive to environmental atmospheres because adsorption of oxygen-related species on the surface/grain boundaries (GBs) traps carrier electrons and desorption induces a reverse process. As known from the fact that these features have been utilized for gas sensors, these properties are very common and likely intrinsic to N-type oxide semiconductors. Long-term stability, which is the major obstacle for commercialization, also comes from intrinsic nature of surfaces and GBs in polycrystalline oxide semiconductors. Due to this reason, notwithstanding that poly-ZnO was expected for channel materials in TFTs, it has been revealed that these still have many issues such as degradation in mobility and stability probably due to the GB issues.

This obstacle was practically resolved by using AOSs in place of polycrystalline ones in 2004. AOS was first proposed as a new class of TCO and reported to have very different properties from conventional amorphous semiconductors, for example, large electron mobilities >10 cm^2 (V s)$^{-1}$, in 1995 and published in 1996 [14]. Then, RT fabrication and operation of AOS TFT were demonstrated in 2004 [1]. Amorphous semiconductors have outstanding advantages in device fabrication processes over crystalline semiconductors. Also, they are free from the GB issues unlikely poly-Si and poly-ZnO. These features make them very favorable for large-area devices such as TFTs.

It was widely believed that the electron transport properties in amorphous semiconductors are largely degraded from those in corresponding crystalline ones on the basis of several examples such as Si, chalcogenides, and transition metal oxides. However, this is not true for ionic oxide semiconductors as will be described later. AOSs have large electron mobilities comparable to those in the corresponding crystalline analogs. These excellent performances and process advantages attract much attention of panel manufactures that have sought new backplanes to drive next-generation FPDs.

13.3 Electronic Structure and Carrier Transport Specific to Oxide Semiconductors

First, it would be important to understand the drawback and the advantage of oxides, which are related to their electronic structures largely different from those of conventional covalent semiconductors such as Si and GaAs. This difference results in an important fact that it is easy to obtain good N-type electron conduction but P-type hole conduction is difficult in oxide semiconductors; actually, there have been a limited number of P-type oxide semiconductors, which include $CuAlO_2$, $CuInO_2$, $SrCu_2O_2$, $LaCuOS$, $BaCuFS$, and $ZnRh_2O_4$ (see Ref. [15] and references therein). This is a natural consequence of chemical bonding nature in oxides of metal cations with closed-shell electron configurations. Figure 13.3 shows band structures of Si and representative oxide semiconductors such as SnO_2, In_2O_3, and ZnO, where energy is measured from the Fermi level (E_F). It is seen that both the conduction band (CB) and valence band (VB) in Si have large band dispersion and thus small carrier effective masses m^* (see, e.g., Ref. [16]). The small m^* is favorable to attain large carrier mobilities μ because of the relation $\mu = e\tau/m^*$. On the other hand, although the CB

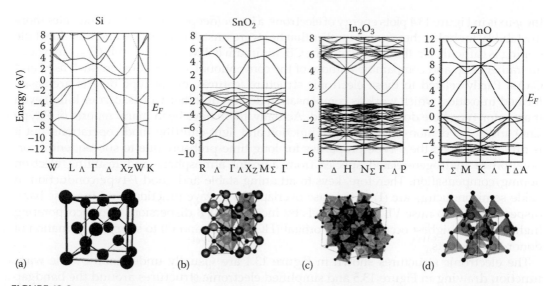

FIGURE 13.3
Band structures and crystal structures of (a) Si and (b–d) representative crystalline oxide semiconductors (b: SnO_2, c: In_2O_3, d: ZnO). The energy is measured from the Fermi level (E_F).

dispersions are large in the oxide semiconductors, the VB dispersions are rather small and the VB bands are almost flat, indicating that the electrons in oxides can have large mobilities but holes are rather localized with small carrier mobilities. Note that ZnO has exceptionally large VB band dispersions due to a O 2p – Zn 3d interaction [17] and is considered as a candidate for a good *P*-type oxide semiconductor if high-density and stable hole doping can be possible by keeping low defect density and reasonable high hole mobility.

Another important reason why *P*-type conduction is difficult in oxides is that the VB maximum (VBM) energy levels (E_{VBM}) are very deep (7–8 eV from the vacuum level (E_{vac}). *c.f.* E_{VBM} of Si is ~6 eV) due to the strong electron affinity of O ions as seen in Figure 13.4 [18]. Because

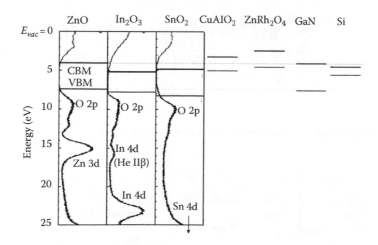

FIGURE 13.4
Band alignment diagram of representative crystalline oxide semiconductors compared with conventional semiconductors. The black solid lines indicate the CB minimum and the gray ones the VBM energy levels. The dashed lines are guide for eyes to compare with those of ZnO.

the *y*-axis in Figure 13.4 plots energy of electrons, a lower (deeper) energy level indicates more unstability for holes; therefore, it implies that holes are unstable in oxides, and therefore hole doping is difficult. On the other hand, the CB minimum (CBM) energy levels (E_{CBM}) in oxides are rather deep and comparable to those of the conventional semiconductors such as Si and GaAs. Partly related to these electronic structure features, donor-type defects are easily formed in oxides, which work as compensation defects (in other words, carrier killer) even if acceptor-type hole doping is achieved. Actually, there have been many reports of *P*-type ZnO and some demonstrate good p/n junction and light-emitting diode operations, but it is difficult to keep the *P*-type properties for long time probably due to subsequent incorporation of hydrogen-related species from ambient atmosphere and resulting electron doping/compensation. Therefore, keys to attaining stable and good *P*-type conduction in oxide semiconductors are (1) to increase overlap of VB wave functions and increase band dispersions, (2) to raise VB energy levels by increasing VB dispersions and incorporating higher-energy highest occupied ionic orbital (HOIO) ions, and (3) to suppress formation of donor-type defects.

The electronic structures shown in Figure 13.4 are visually understood by the wave function drawing in Figure 13.5 and simplified electronic structures around the bandgap in Figure 13.6. For example in Si, CBM and VBM are made of anti-bonding (sp³ σ*) and bonding states (sp³ σ) of Si sp³ hybridized orbitals and its bandgap is formed of the energy splitting of the σ* − σ levels (Figure 13.6a). By contrast, CBM and VBM in oxides are usually formed of different ionic species as visually seen in Figure 13.5. Although the CBM of GaAs still has large contributions of both the cation and the anion, the VBMs of GaAs and ZnO are formed almost solely by the anions (i.e., the VBM wave functions are concentrated

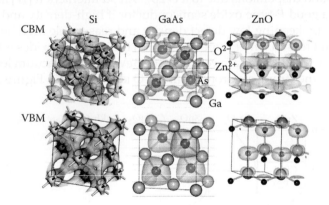

FIGURE 13.5
Wave functions of representative semiconductors such as Si, GaAs, and ZnO.

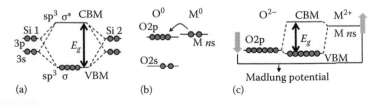

FIGURE 13.6
Simplified electronic structures of (a) Si and (b and c) TCOs/TOSs.

	IIb	IIIb	IVb
3		Al (3+) 113	Si (4+) 92
4	Zn (2+) 154	Ga (3+) 127	Ge (4+) 108
5	Cd (2+) 180	In (3+) 149	Sn (4+) 126

FIGURE 13.7
Radii of the lowest unoccupied ion orbitals of metal cations. The values are in picometers and those in the parentheses indicate corresponding ionic charges.

on the As and O ions). In the case of ZnO, which has the highest ionicity among these semiconductors, even the CBM is made mainly from the cation's orbitals although the anions still show some contribution.

This feature originates from the Madelung potential [15]. When a metal atom (*M*) and an oxygen atom are apart in vacuum, the energy levels of HOIO are not different largely as illustrated in Figure 13.6b, and they are stable in the neutral atom states. When they come close, charge transfer occurs due to the different electron affinities. As explained previously, the neutral states are more stable in vacuum without an extra interaction, but the ionized ions form negative electrostatic potential at the cation site and positive potential at the anion sites (Madelung potential), which consequently stabilizes the ionized states in ionic materials (Figure 13.6c). The Madelung potential lowers the energy levels in the oxygen ions and raises those in the metal cations (the gray arrows in Figure 13.6c). Therefore, CBM is mainly made of the empty s orbitals of the metal cation and VBM is of fully occupied O 2p orbitals. Thus, the large bandgap and the transparency of oxide semiconductors come from the large Madelung potential. Due to this electronic structure and the large Madelung potential, the location of VBM is so deep and the overlap of the wave functions between neighboring O ions are small. However, the situation is opposite for the CBM, which is composed of metal cation's orbitals and is located at a rather low (i.e., deep) energy level. In addition, the wave functions of heavy cations are spatially spread (Figure 13.7 [19]) and make large overlaps with neighboring cations, which results in the large CB dispersions and small electron effective masses 0.25–0.35 m^* and reasonably large electron mobilities ~200 cm^2 (V s)$^{-1}$ (Table 13.1).

13.4 Material Design Concept of AOS with High Mobility

What happens if crystalline TOS materials become amorphous? In an amorphous state, structural randomness concentrates on an energetically weak structural unit. In most amorphous materials, structural randomness appears prominently in bond angle distribution. *When the bond angle has a large distribution, how the effective mass (in other words, the transfer rate between neighboring cation's orbitals) is altered for carrier electrons?* Here, we discuss

TABLE 13.1

Carrier Transport and Band Structure Parameters of Oxide Semiconductors in Comparison with Conventional Semiconductors

	Mobility (cm^{-2} (V s)$^{-1}$)	Effective Mass (m_e)	Band Gap (eV)	Optical Transition
C [16]	$\mu_e = 2400$	$m_e = 0.2$	5.47	Indirect
	$\mu_h = 2100$	$m_h = 0.25$		
Si [16]	$\mu_e = 1500$	$m_{et} = 0.98$, $m_{el} = 0.19$	1.12	Indirect
	$\mu_h = 500$	$m_{ht} = 0.49$, $m_{hl} = 0.16$		
Ge [16]	$\mu_e = 3900$	$m_{et} = 0.82$, $m_{el} = 1.64$	0.66	Indirect
	$\mu_h = 1900$	$m_{ht} = 0.28$, $m_{hl} = 0.04$		
GaAs [16]	$\mu_e = 8500$	$m_e = 0.067$	1.42	Direct
	$\mu_h = 400$	$m_h = 0.082$		
ZnS [20]	$\mu_e = 230$	$m_e = 0.28$	3.76	Direct
	$\mu_h = 40$	$m_{h//c} = 1.4$, $m_{h\perp c} = 0.49$		
ZnSe [20]	$\mu_e = 400$	$m_e = 0.160$	2.70	Direct
	$\mu_h = 110$	$m_h = 0.75$		
GaN [20]	$\mu_e = 380$	$m_E = 0.19$	3.36	Direct
		$m_h = 0.60$		
6H-SiC [20]	$\mu_e = 480$	$m_{e\perp c} = 3-69$, $m_{e//c} = 0.48$	3.02	Indirect
	$\mu_h = 50$	$m_{h\perp c} = 1.85$, $m_{h//c} = 0.66$		
In$_2$O$_3$ [21–24]	$\mu_e = 160$ [25]	$m_e = 0.30 \sim 0.44$ (Sn doped) [26]	2.89 [27]	
			3.75 [28,29]	Indirect [30]
SnO$_2$ [31,32]	$\mu_e = 260$ [33]	$m_e = 0.39$ [33]	$E_{g\perp c} = 3.57$,	Direct forbidden [36]
		$m_{e\perp c} = 0.299$, $m_{e//c} = 0.234$ [34]	$E_{g//c} = 3.93$ [35]	
ZnO [37–40]	$\mu_e = 180$ [41]	$m_e = 0.27$ [41]	3.37	Direct [42]
	$\mu_{e\perp c} = 150$ [20]	$m_h = 0.59$ [26]		
	$\mu_{e//c} = 167$ [20]	$m_e = 0.28 \sim 0.33$ (Ga doped) [26]		
a-In$_2$O$_3$ [43]	$\mu_e = 36$		3.3	
a-2CdO·GeO$_2$ [44]	$\mu_e = 12$ [44]	$m_e = 0.33$ [45]	3.4 [45]	
a-2CdO·PbO$_2$ [46,47]	$\mu_e = 10$ [47]	$m_e = 0.30$ [45]	1.8 [45]	
a-InGaO$_3$(ZnO)$_m$ ($m = 1 \sim 4$) [48]	$\mu_e = 13 \sim 21$ [48]	$m_e = 0.34$ ($m = 1$) [49]	3.0 – 2.85 [48]	
a-Zn-In-O [50,51]	$\mu_e = 30$ [52]			

two cases for this question: (1) covalent semiconductors and (2) ionic semiconductors. In case (1), the magnitude of the overlap between the empty orbitals of the neighboring atoms is very sensitive to the variation in the bond angles and forms different energy electronic states at different positions. As a consequence, rather deep localized states would be created at somewhat high concentrations and thereby the drift mobility would be largely degraded due to carrier scattering/trapping by these defects.

On the other hand, the magnitude of the overlap in case (2) is different largely depending on the choice of metal cations; if the spatial spread of the s orbital (Figure 13.7) is larger than the inter-cation distance, the magnitude of the wave function overlaps is insensitive to the bond angle distribution because the s orbitals are isotropic in shape. As a consequence, we may anticipate that these ionic amorphous materials consisting of heavy post-transition cations have large band dispersion, small carrier effective mass, and large mobility comparable to those in the corresponding crystals. If the spatial spread of the metal s orbital is small, such a favorable situation is not expected. The spatial spread of the s orbital of a metal cation is primarily determined by the principal quantum number (n) and is modified by the charge state of the cation as shown in Figure 13.7. It shows that the radius of the outermost wave function (i.e., that of the lowest unoccupied ion orbital, LUIO) increases with increasing n, and large radii are realized, for example, in Zn^{2+}, Cd^{2+}, and In^{3+}, which form good TCOs such as ZnO, CdO, and In_2O_3. Hereafter, we take the value of n as a measure of the spatial spread of the metal cation s orbital.

Thus candidates for good TOS having large electron mobilities comparable to those of the corresponding crystals are transparent oxides constituting of *post transition metal cations with an electronic configuration* $(n-1)d^{10}ns^0$, *where* $n \geq 5$. Note that a transition metal cation with an open shell electronic configuration is ruled out as a candidate because it is not transparent due to optical absorptions arising from d–d transitions. In the case of crystal-line oxide semi-conducting oxides, this requirement is relaxed to be $n \geq 4$ as exemplified by ZnO (Zn^{2+} has the $[Ar](3d)^{10}(4s)^0$ configuration), because crystalline materials have much regular and compact structures than amorphous oxides.

Figure 13.8 illustrates the comparison of orbitals in (Figure 13.8a and b) Si and (Figure 13.8c and d) a post transition metal (PTM) oxide between (Figure 13.8a and c) crystalline (c-) and (Figure 13.8b and d) amorphous (a-) states. The significant reduction of the electron mobility in the a-Si from c-Si may be understood intuitively from the figure, whereas moderate mobility in c-PTM oxides is maintained even in the amorphous state. Conventional amorphous semiconductors such as a-Si:H exhibit much deteriorated carrier transport

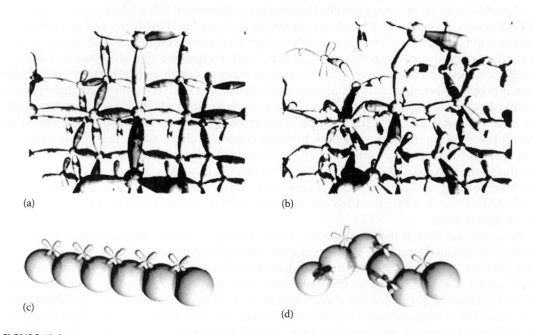

(a)

(b)

(c)

(d)

FIGURE 13.8
Schematic orbital drawings of carrier pathways in (a) crystalline Si, (b) amorphous Si, (c) crystalline oxides, and (d) amorphous oxides.

properties than associated crystalline materials. This is because the chemical bonds in the covalent semiconductors are made of sp^3 or p orbitals that have strong spatial directivity. Therefore, the strained chemical bonds in amorphous structures form rather deep and high-density localized states below CBM and above VBM, causing carrier trapping (Figure 13.8a and b). By contrast, as CBMs of oxides are made of spherically spread s orbitals of metal cations, and their overlaps with neighboring metal s orbitals are not altered largely by disordered amorphous structures; therefore, electronic levels of CBM are insensitive to local strained bonds, and electron transport is not affected significantly (Figure 13.8c and d). This is a reason why AOSs exhibit large electron mobilities even in amorphous structures.

As for stability of doped electron, the CBM levels of the heavy PTM oxides are deep. As a consequence, transparent amorphous oxides following this hypothesis is being capable of electron-doping and have a large electron mobility comparable to the corresponding crystalline phases.

13.5 Roles of Constituent Elements and Other AOS Materials

Among AOSs, a-IGZO has extensively been studied as a semiconducting channel layer of transparent TFTs. Figure 13.9 summarizes electrical properties (Hall mobility μ_{Hall} and carrier density N_e) for the films in the In_2O_3–Ga_2O_3–ZnO system [53]. All the films were deposited at the same condition, that is, on SiO_2 glass substrates at RT under an oxygen partial pressure of 1.0 Pa. Although pure In_2O_3 and ZnO films exhibited large Hall mobilities of ~34 and ~17 cm^2 (V s)$^{-1}$, respectively, they are crystalline even if the films are deposited at RT. Moreover, it is not easy to control N_e down to <10^{17} cm^{-3} in these films without compensation doping. Pure Ga_2O_3 forms amorphous films but carrier doping, that is, formation of a shallow donor by an oxygen vacancy, is very hard irrespective that we examined a wide range of deposition conditions. Thus, the end-member materials in this ternary system are not appropriate because of local nonuniformity due to the GB issues, no stable amorphous phase and/or the difficulty in carrier doping. As known in glass science, incorporation of aliovalent and different size cations is effective to enhance amorphization, and it is much favorable to introduce network forming cations. Indeed, stable amorphous phases are formed in the binary systems of In_2O_3–Ga_2O_3 (a-IGO) and ZnO–Ga_2O_3 (a-GZO), and in the ternary system of In_2O_3–Ga_2O_3–ZnO (a-IGZO). Both μ_{Hall} and N_e rapidly decrease with increasing the Ga^{3+} ion content, for example, μ_{Hall} in the a-IGO films decreases from ~25 cm^2 (V s)$^{-1}$ at N_e ~10^{20} cm^{-3} to ~1 cm^2 (V s)$^{-1}$ at ~10^{18} cm^{-3} as the Ga^{3+} ion content increases from 30% to 50%.

However, we should note that the μ_{Hall} values in Figure 13.9 are not maximum potential of these materials because μ_{Hall} largely depends on N_e in AOS due to the presence of potential barriers arising from structural randomness as seen in Figure 13.10. Carrier mobility strongly depends on carrier concentration, and large mobilities are obtained at carrier concentrations larger than a threshold value (e.g., ~10^{18} cm^{-3} for a-IGZO) [1,54,55]. However, introduction of high-density carriers (e.g., >10^{20} cm^{-3}) is very difficult in the larger Ga content films. This result indicates that large mobility is not easily obtained in the a-IGZO films with large Ga contents. This feature is unfavorable for TCO applications, but it would not be a disadvantage for semiconductor device applications because the difficulty in carrier doping by oxygen deficiency in turn suggests better controllability and stability of

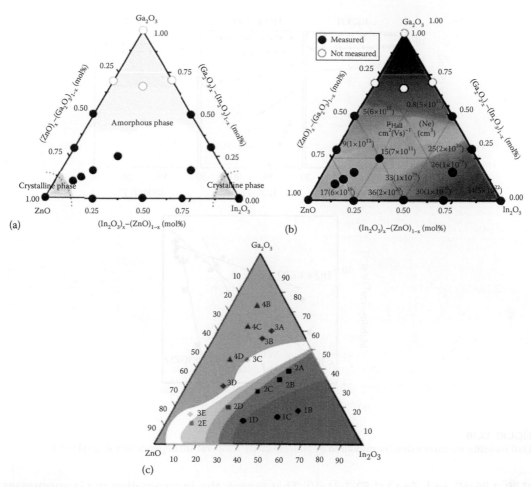

FIGURE 13.9
Structures and carrier transport properties of AOSs in In_2O_3-Ga_2O_3-ZnO ternary system. (a) Phase stability, (b) Hall mobility (in cm^2 (V s)$^{-1}$) and carrier density (in the parentheses, in cm^{-3}), and (c) resistivity.

carrier concentration, in particular at low concentrations, and favorable for semiconductor device applications. Even if high-density doping is difficult by choosing deposition condition, it is still possible to induce high-density carriers by external electric field if TFT structures are employed, which may make it possible to utilize the potential large mobilities that may be available at large carrier concentrations.

Here, let us discuss the roles of the constituent ions in the AOSs. As would be known from the aforementioned discussion on electronic structure, In, Ga, Zn, and Sn are the major constituents of good TCOs, and their unoccupied s orbitals form the electron transport paths also in AOSs [1,14]; therefore, major portion of these ions are the fundamental requirement to obtain AOSs with large electron mobilities (*mobility enhancer*). To know the role of the other ions, we should compare, for example, a-IZO and a-IGZO (Figure 13.11). As reported in Ref. [53], a-IZO has larger electron mobilities than a-IGZO, but it is much difficult to control at low electron concentrations required for TFTs (e.g., far below $10^{17}\,cm^{-3}$). As clarified from bond energies (E_b) calculated by first-principles density-functional theory (DFT) [56], Ga-O bonds are much stronger ($E_b = 2.04$–$2.48\,eV$) than In-O

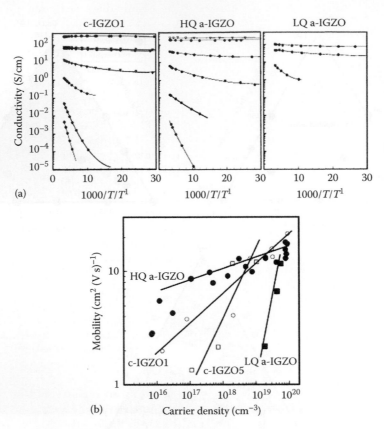

FIGURE 13.10
Hall mobility vs. carrier density measured by Hall effect. (a) Temperature dependences and (b) at RT.

(1.70–1.86 eV) and Zn-O (1.52–1.84 eV). That means, the incorporation of Ga suppresses the formation of oxygen deficiency and consequent generation of mobile electrons. On the other hand, the incorporation of much Ga content deteriorates the electron mobility as seen in Figure 13.9b. Therefore, addition of appropriate amount of a *stabilizer ion* that forms a strong chemical bond with oxygen ions is important to obtain stable AOS materials and TFTs. Canon surveyed stabilizer ions (Mg, Al, Si, Ti, Ge, Mo) in the In-Zn-O system by the combinatorial approach and reported a-In-Si-Zn-O would be an alternative candidate of a-IGZO [7]. It should be noted that one would obtain very large TFT mobility if he fabricates a TFT using poor/unstable channel/gate insulator materials, for example, as seen in the a-IZO TFT case in Figure 13.11c [57]. However, the obtained TFT mobility exceeds 100 cm^2 (V s)$^{-1}$ and is extraordinarily larger than Hall mobility (<40 cm^2 (V s)$^{-1}$ at Ne = 10^{19} cm^{-3} [58]). As known from the fundamental semiconductor physics, TFT mobility should not exceed drift mobility (approximately equals Hall mobility) [16]. In this case, we can see large hysteresis in TFT transfer characteristics, indicating that the channels, gate insulator and/or their interface have high-density charge traps, which alter the actual gate capacitance. So the obtained mobility is not intrinsic physical quantity and apparent one. In such a case, one should be very careful to employ these values by considering whether these values are appropriate for performance evaluation as practical semiconductor devices, for example, in terms of high frequency operation, reproducibility, and stability.

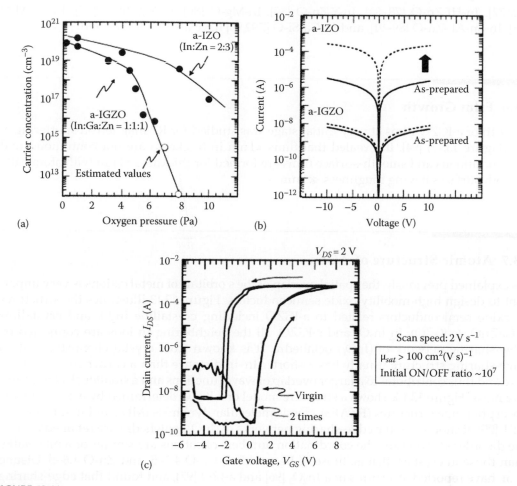

FIGURE 13.11
Transport property characteristics of AOS without Ga (i.e., a-IZO) in comparison with that with Ga (a-IGZO). (a) Electron density as a function of oxygen partial pressure during deposition, (b) current–voltage characteristics as a function of exposure time to air, and (c) large hysteresis in a-IZO TFT with extraordinary large apparent mobility >100 cm²(V s)⁻¹.

Other than a-IGZO, many AOS materials have been reported to date. As will be seen in the following, In, Zn, and Sn have been employed for mobility enhancers, Ga, Al, Si, Y, Mg, and Hf for stabilizers, and Zn for a *network former*. Oregon State University and Hewlett-Packard Company reported high-temperature fabrication of a-Zn-Sn-O (a-ZTO) TFTs [59] and In-Zn-O (a-IZO) TFTs [60] (hereafter, the chemical formulas are represented in order of mobility enhancers, stabilizers, and network formers). These TFTs exhibited very high mobilities up to 55 cm² (V s)⁻¹ although they were subjected to high temperature annealing at 300°C–600°C, giving us expectation that AOS TFT technology will range to high-speed circuit applications. For RT-fabricated AOS TFTs, Canon Inc. first succeeded in fabricating a-IGZO TFTs at RT by RF magnetron sputtering [61]. Combinatorial approaches have been employed to screen multi-component AOS systems including the In-Ga-Zn-O system [62] and the ZTO system [63]. Other materials including a-IZO [64,65], a-In-Ga-O (a-IGO) [66], a-Sn-Ga-Zn-O (a-SGZO) [67,68], Sn-Al-Zn-O [69,70], In-Zr-Zn-O [71], In-Sn-Zn-O

[72–77], In-Hf-Zn-O [78–84], In-Y-Zn-O [85], In-Mg-O [86], In-Mg-Zn-O [87], In-Sn-Al-O [88], In-Sn-Al -Zn-O [89–91], and In-Ga-Si-O [7,92,93] have also been reported.

13.6 Film Growth

Growth of a-IGZO films at very initial stage was studied for RF magnetron sputtering at RT (Figure 13.12) [94]. It revealed that films <1 nm in thickness are not continuous, and that continuous and smooth-surface films are formed for thickness >1 nm with atomically smooth surfaces having roughness ≪1 nm.

13.7 Atomic Structure of AOSs

As explained previously, the spatial overlap of the s orbitals of metal cations is very important to design high-mobility oxide semiconductors. Figure 13.13 illustrates the structures of oxide semiconductors related to a-IGZO including crystalline In_2O_3 and crystalline $InGaZnO_4$ (c-IGZO). In In_2O_3 and c-IGZO, all the neighboring In ions are connected by edge-sharing network of (InO_6) octahedra. It is known from crystal chemistry that an edge-sharing structure usually has a shorter In–In distance than a corner-sharing structure and thus contributes to a larger overlap of wave functions and a smaller electron effective mass. Figure 13.13c shows a structure model of a-IGZO determined by extended x-ray absorption fine structures (EXAFS), classic molecular dynamics (MD), and first-principles DFT [95]. It shows that the edge-sharing networks of (InO_n) polyhedra are retained even in the disordered structure (the coordination numbers of cations are similar or a bit smaller than those in crystals, that is, those of In are 4.5–6, Ga-O 4.3–5, and Zn-O 4.6–5. Utsuno et al. have reported structures of a-In_2O_3 [96] and a-IZO [97], and found that edge-sharing network structures increase and In-O-Zn corner-sharing structures decrease with increasing the In content. In c-$InGaZnO_4$, x-ray diffraction (XRD) structure analyses reveal that

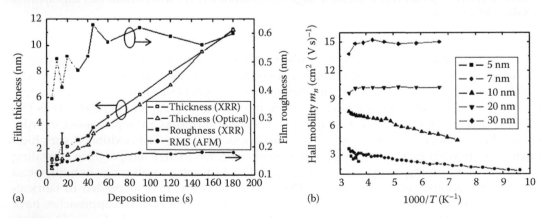

FIGURE 13.12
Growth of a-IGZO films by RF magnetron sputtering at RT on SiO_2 glass and SiO_2/c-Si wafer substrates. (a) Film thickness and surface roughness as a function of deposition time.

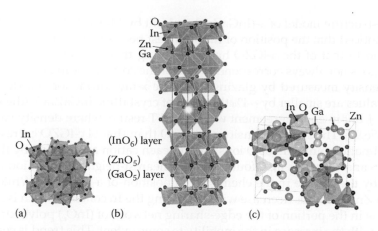

FIGURE 13.13
Structures of In-based oxide semiconductors. (a) Crystalline In_2O_3(bixbyite), (b) crystalline $InGaZnO_4$, and (c) amorphous $InGaZnO_4$.

Zn and Ga ions share the same crystallographic site of a fivefold coordination trigonal-bipyramid $(Zn,Ga)-O_5$ but the XRD results are just an averaged structure. Omura et al. reported that the Zn ions take off-center tetrahedral sites in the trigonal-bipyramidal site, while Ga ions take the center position in $c-InGaZnO_4$ by DFT [98], which is in reasonable agreement with the common agreement of crystal chemistry (i.e., Zn^{2+} ions usually take fourfold coordinated sites). Cho et al. confirmed a similar result by EXAFS for a-IGZO that the coordination structure of a Zn ion is distorted tetrahedral and suggest that this is the reason why holes are localized on the Zn atoms [98].

Figure 13.14 shows structure characterizations of a-IGZO. In XRD patterns, a-IGZO films do not exhibit sharp diffraction peaks assignable to a crystalline phase but have two halos around 32° and 56° ("a-IGZO(meas)"). These are different largely from those of a-Si:H (22° and 52°, "a-Si:H") and SiO_2 glass (22°, "glass"), and can be used for determining the amorphous phase. The halo positions of a-IGZO are reproduced well by the XRD pattern simulated

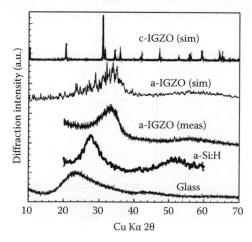

FIGURE 13.14
XRD patterns of related materials.

based on the structure model of a-$(InGaZnO_4)_{270}$ built by MD in Ref. [6,95] ("a-IGZO(sim)"). It should be noticed that the position of the strongest peak of c-IGZO ("c-IGZO(sim)") does not correspond to that of the a-IGZO halo, indicating that the halo position in an amorphous phase does not always correspond to that of the corresponding crystal.

The film density measured by glazing-incidence x-ray diffraction (GIXRD) was ~6.1 g cm^{-3}. These values are smaller by ~4% than that of crystalline $InGaZnO_4$ (the x-ray density is 6.379 g cm^{-3}), but in good agreement with the DFT results where density varies from 5.8 to 6.1 g cm^{-3} [56,95]. The smaller density of a-IGZO than that of c-IGZO corresponds to the smaller coordination numbers of O ions around each cation as revealed by the aforementioned EXAFS analyses. It is also found that the nearest-neighbor (NN) ion distances are not changed by the change in the chemical composition of a-IGZO [6]. While, the second NN In-(In,Ga,Zn) distances decrease with increasing the In content, which is accompanied by the increase in the portion of the edge-sharing network of (InO_n) polyhedra and would be associated with the increase in the mobility to some extent. This trend is consistent with the earlier discussion that CBM in a-IGZO is mainly made of In s orbitals, and the In–In distances affect the CBM dispersion and consequently the electron mobility.

Here, it should be noted that these EXAFS analyses give unrealistic small coordination numbers (CNs) for the second NN ions, for example, the In–In CNs are 0.7–0.9. This would be due to the limitation of EXAFS analyses for amorphous materials and these values are not real, because the CNs obtained by the DFT calculations give more reasonable values, for example, the In–In CNs >3 at the In–In distances larger than 0.37 nm for a-IGZO with the In:Ga:Zn chemical composition ratio of 1:1:1. This discrepancy comes from the decoherence of the EXAFS signal caused by the disordered coordination structures, and the aforementioned a-IGZO model built by DFT consistently explains these results [6].

13.8 Electronic Structure of AOSs

There have been several reports on theoretical calculations of electronic structures and defects for IGZO. Those of c-IGZO have been reported in Refs. [98,100–104], and a-IGZO in Refs. [56,95,103,105–107]. The pseudo-band structure of a-$InGaZnO_4$ is shown in Figure 13.15b compared with that of c-$InGaZnO_4$ in Figure 13.15a [6,106]. Figure 13.16 shows those of c-Si

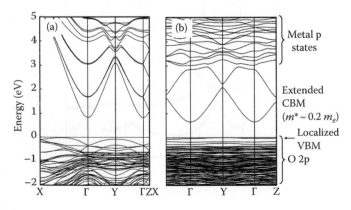

FIGURE 13.15
(Pseudo-)band structures of (a) c-$InGaZnO_4$ and (b) a-$InGaZnO_4$.

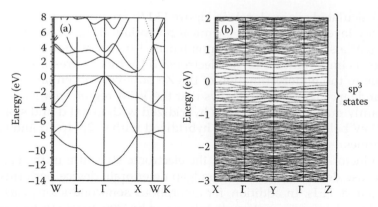

FIGURE 13.16
(Pseudo-)band structures of (a) c-Si and (b) a-Si.

and a-Si. Note that we use the term "pseudo-band structure" because an amorphous material does not have a periodic structure and the band theory is not applied in an exact sense; however, even for amorphous materials, pseudo-periodic calculations are still useful to know the magnitude of transfer integral, band dispersion and effective mass (discussion of local effective mass in amorphous should be referred to [108,109]). These show that the CB band width and dispersion of a-IGZO are very similar to those of c-IGZO. Indeed, the effective masses calculated from the curvatures at the CBMs provide 0.28 m_e for c-IGZO and 0.30 m_e for a-IGZO. On the other hand, for Si, all the bands become almost flat in a-Si, implying that electrons and holes are strongly localized. A similar trend is observed also in a-IGZO in the VB bands and the high energy deep CB bands at >3.0 eV. These localized bands have similar characteristics. That is, the bands in Si are composed of sp^3 hybridized orbitals, the VB bands in IGZO of O 2p, and the deep CB bands in IGZO of metal p orbitals. On the other hand, the CBM bands in IGZO are made mainly of spherical s orbitals of the metal cations. Thus, these results support the aforementioned idea that the CBM bands in metal oxides made of spherical s orbitals are insensitive to disordered amorphous structures, but sp^3 and p orbitals, which have strong spatial directivity, are sensitive to a disordered structure and form localized states.

Total and projected density of states (PDOSs) are shown in Figure 13.17b in comparison with a hard x-ray photoemission spectrum (HX-PES) of an a-IGZO film [110].

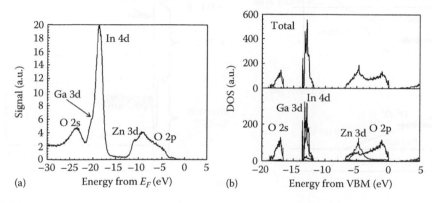

FIGURE 13.17
Density of states in a-IGZO (a) measured by HX-PES and (b) calculated by DFT.

Note that the calculated energy levels in Figure 13.17b are underestimated because DFT provides energy levels as a kind of chemical potentials [111], which are smaller than ionization potentials. In addition, it is pointed that incorporation of Coulomb repulsion interaction is important for Zn 3d electrons [17], which overestimates Zn 3d – O 2p interaction due to the underestimation of the Zn 3d level and enhances underestimation of the bandgap. Figure 13.17b shows that the VBM region in a-IGZO (and also in c-IGZO) is mainly made of O 2p orbitals hybridized with metal d and s orbitals. Zn 3d appears at ~11 eV below E_F and strongly hybridized with O 2p. In 4d and Ga 3d appear at similar energies, −18 to −20 eV.

For semiconductor device applications, the electronic structure in the bandgap is also important because defect states in the bandgap deteriorate device performances in particular for SCs and TFTs. In addition, for amorphous semiconductors, carrier transport properties and mechanisms are strongly influenced by structural randomness and these are related to the electronic structures in and near CBM. Figure 13.18 compares these electronic structures of a-Si:H and a-IGZO that have been clarified to date [4,5]. The detailed features are explained in the following.

Figure 13.19 shows a HX-PES spectrum of an a-IGZO film around the bandgap. It exhibits extra DOSs above the VBM (near-VBM states) and beneath the CBM (near-CBM states). It has been revealed that the near-VBM states have rather high densities, for example, $>5 \times 10^{20}$ cm^{-3} with an energy width of ~1.5 eV [110,112]. Much smaller, but similar structures are reported also in crystalline (Zn,Mg)O [113]. For a-IGZO, DFT calculations of oxygen-deficient a-IGZO models suggest that a possible origin of these levels is an oxygen vacancy structure with a free space comparable to the size of an oxygen ion [56,105,106]. These deep states explain why a-IGZO TFTs do not show an inversion operation because the high-density occupied states in the bandgap pin the Fermi level when negative gate bias is applied to TFTs and inhibit to induce mobile holes in the VB. The near-VBM states are important also for optical properties and photoconduction because it causes photoexcitation by subgap energy photons and results in subgap optical absorption as will be explained later on. Angle-dependent HX-PES measurements revealed that the near-VBM

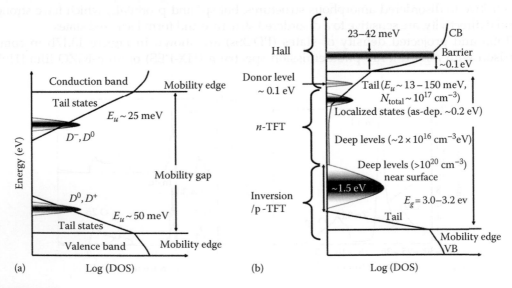

FIGURE 13.18
Schematic models of subgap density of states in (a) a-Si:H and (b) a-IGZO.

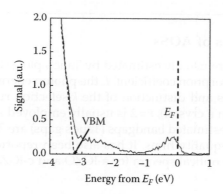

FIGURE 13.19
HX-PES spectrum of a-IGZO around the bandgap.

states are concentrated in the top-most surface region ~1 nm in thickness for high-quality films, which would be formed by exposure to vacuum.

It should also be noticed that rather high-density subgap states exist below the CBM. It is also observed in ZnO [114] and a-IZO [115], but it would be reasonable because a-IZO has high-density free electrons. However, those in a-IGZO are ~5×10^{19} cm^{-3} and do not show correlation with free electron densities measured by the Hall effect [112]. In addition, if such high-density defects exist below the CBM, they pin the Fermi level under application of positive gate bias and deteriorate the TFT operation, which is confirmed by device simulations. Because the densities of the near-CBM states decrease in the top-most surface region and also exhibit correlation with photoconductivity [112,116], it is tentatively considered that these signals come from electrons excited by x-ray to CB.

HX-PES also provides electron affinity and ionization potential of a-IGZO films, which are important for device simulations and designing electronic devices such as SCs and TFTs. Figure 13.20 shows that these values obtained from the capacitance–voltage (C–V) method and from HX-PES give good agreement, and the electron affinity (i.e., E_{CBM}) is similar to that in ZnO. It also tells that built-in potential with P-type c-Si is as large as 0.44 eV [117].

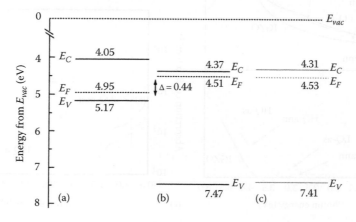

FIGURE 13.20
Electron affinity and ionization potential of a-IGZO and band alignment with (a) c-Si and (b,c) a-IGZO. (b) Estimated from C–V method on a-IGZO/c-Si junction and (c) from HX-PES.

13.9 Optical Properties of AOSs

Bandgap values of AOSs are usually estimated by Tauc's plot [118], which has the form of $\alpha E = [B(E - E_g)]^r$ (α is an absorption coefficient, E the photon energy, and B and r constants). Assuming parabolic bands and distinction of the k-selection rule for optical transitions (this must be considered in a crystal), $r = 2$ is usually employed as plotted in Figure 13.21. It has been found that the estimated bandgaps (Tauc's gaps) are 3.0–3.2 eV for- a-IGZO and tend to be larger for high-quality films. It has also been reported that the Tauc–Lorentz model [119] fits well to the optical spectra for a-IGZO and c-IGZO [106,120] (Figure 13.22).

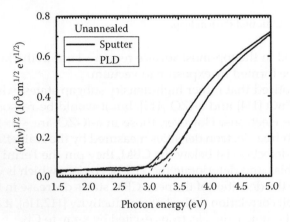

FIGURE 13.21
Optical absorption spectra of a-IGZO films deposited by RF magnetron sputtering and PLD at RT plotted in terms of Tauc's plot.

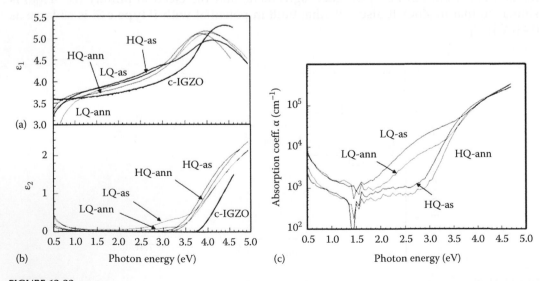

FIGURE 13.22
Optical spectra of various a-IGZO films. For notation of HQ, LQ, -as, and -ann, see Ref. [112]). (a) Real part and (b) imaginary part of optical dielectric functions. (c) Optical absorption spectrum.

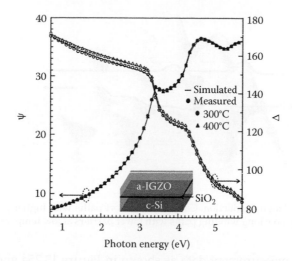

FIGURE 13.23

Spectroscopic ellipsometry analyses of a-IGZO/c-Si solar cells. It provides thicknesses of reaction SiO_2 layers to be negligible (0.02 nm) for 300°C-annealed device but 4.8 nm for 400°C-annealed one.

Such optical models can be used for detecting thin reaction layers in devices, for example, for a-IGZO/c-Si SCs as shown in Figure 13.23 [117].

Optical spectra also provide information about subgap DOS similar to the HX-PES spectra. The a-IGZO films in Figure 13.22 have bandgaps 3.0–3.2 eV, but observable optical absorptions are found in the photon energy range from 2.0 to the bandgap (Figure 13.22b and c). It has been found that absorption spectra follow the Urbach law with Urbach energies of ~150 meV in optical absorption spectra [109]. Also, effective masses are obtained. Effective mass of a-IGZO (In:Ga:Zn = 1:1:1) is estimated to be ~0.34 m_e from free carrier absorption analyses on highly doped a-IGZO films [49]. This value is similar to that of c-IGZO (0.32 m_e). Similar results are also available from the increase of the bandgap in highly doped films (Burstein–Moss effect) as described in Ref. [4].

13.10 Electron Transport in AOSs

a-IGZO exhibits peculiar carrier transport properties as already seen in Figure 13.10, for example, (1) electron mobility increases with increasing the free electron density, and (2) the maximum Hall mobilities are similar to those of corresponding crystals [1,54,55,121]. The reason for case (2) has been explained previously. The peculiar behavior (1) is opposite to single-crystalline semiconductor cases; in general, carrier mobility decreases with increasing the carrier density because higher doping causes larger ionized impurity scattering and consequently smaller carrier mobility.

This peculiar behavior is explained by a percolation conduction model [54,55]. As illustrated in Figure 13.24, a distribution of potential barriers is formed above CBM due to the disordered amorphous structure. Electrons take shorter transport paths at high temperatures even if these paths have high potential barriers (the path in Figure 13.24a), while these take longer winding paths with lower barriers because the electrons do not have enough thermal energies (the path in Figure 13.24b). This model reproduces the temperature (T)

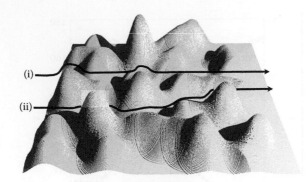

FIGURE 13.24
Illustrative model of percolation conduction. Carriers flow short paths with higher potential barriers at high temperature (i), while they flow long winding paths with lower barriers at low temperature (ii).

dependences of Hall measurement data as shown in Figure 13.25a and b. The $\sigma - T$ data are better fitted to $\log \sigma - T^{-n}$ ($n =$ e.g., 1/4) behaviors. Such behaviors are often explained by Mott's variable range hopping (VRH), but the percolation model is more likely in this case because VRH is not applicable to a case in which Hall voltage signals follow the standard Hall effect theory. It is also found that Hall voltage anomaly would appear in the low-doped a-IGZO films in the low temperature regions, for example, <120 K as seen in the deviation of the calculated electron densities from the measurement values (Figure 13.25a).

The analyses of the $n_{Hall} - T$ data provide that the donor levels are 100–150 meV for c-IGZO and 110 meV for a-IGZO [54] below CBM, and the percolation conduction model indicates that the potential barriers have average heights 40–120 meV and distribution widths 20–30 meV as plotted in Figure 13.25c.

The percolation model is also consistent with μ–N_e relationships obtained by I–V and C–V combined analyses (Figure 13.26) [122]. It revealed that μ–N_e relation does not depend on annealing condition of a-IGZO films and is universal for a-IGZO films with the same chemical composition. The universal curve follows the relation of the analytical percolation conduction model $\mu(N_e) = \mu_0 \exp(-e\phi_{eff}(N_e)/k_B T)$ [54] with a small correction in the effective potential barrier height.

13.11 Subgap Defects

Ionic AOSs have several common properties, which are not seen in conventional amorphous semiconductors. The first feature is their large electron mobilities such as 10–40 cm² (V s)$^{-1}$, which are higher by >2 orders of magnitude than that in a-Si:H (Figure 13.27 [123,124]). The second feature is that a degenerate state is easily attained, which is totally different from the other amorphous semiconductors. For instance, c-Si is easily changed to the degenerate state by carrier doping (~10^{18} cm^{-3}), but no such state has been attained in a-Si:H [125] to date. That is, carrier conduction takes place by hopping or percolation between localized tail states in conventional amorphous semiconductors. This is the reason why drift mobility in the amorphous state is so small compared with that in the crystalline state. On the other hand, in post transition metal oxides, E_F can exceed the mobility

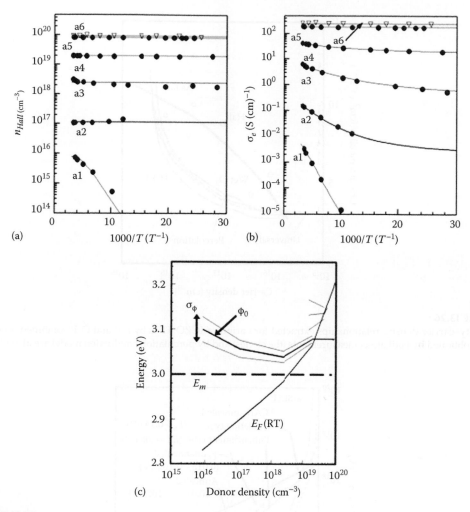

FIGURE 13.25
Analyses of temperature dependences of the Hall effect data based on percolation conduction model. (The symbols) measured and (the solid lines) simulated results of (a) electron density and (b) conductivity. (c) Electronic structure in the CB.

gap easily by carrier doping, leading to band conduction. It is considered that this striking difference originates from that in chemical bonding nature of these materials, that is, strong ionic bonding with spherical wave functions is much favorable to form shallow tail states having small density of states.

As explained previously, a clear advantage of oxide semiconductors is that subgap DOSs are very small compared to conventional covalent semiconductors. This feature is also explained by the electronic structure of oxides. Figure 13.28a shows a simplified electronic structure of a dangling bond in silicon. The bandgap of Si is made of energy splitting of bonding–anti-bonding states; therefore, the energy level of the non-bonding state is located near the middle of the bandgap. Because it is occupied by one electron, such a mid-gap defect traps both an electron and a hole and deteriorates TFT operation for both N- and p-channel TFTs. This is the reason why hydrogen passivation is necessary to remove the

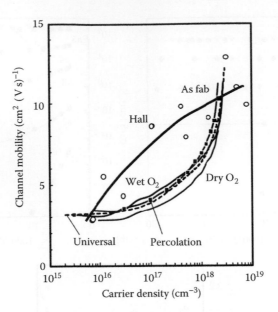

FIGURE 13.26
Mobility–carrier density relationships extracted for various a-IGZO TFTs by *I–V* and *C–V* combined analyses. Those obtained by Hall effect measurements (the circles) and by percolation conduction model are also plotted.

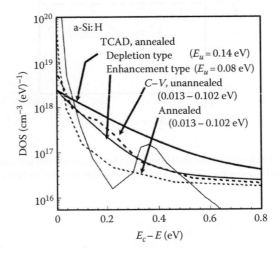

FIGURE 13.27
Subgap DOSs of a-IGZO obtained by device simulations (TCAD) and a *C–V* method (*C–V*). Unannealed and annealed TFTs were examined by the *C–V* method, and depletion-type and enhancement-type TFTs annealed at 400°C by the device simulations. That of a-Si:H is also shown for comparison.

mid-gap defect states out into CB and VB (Figure 13.28b). In the case of oxides, CB and VB are made mainly of different ionic species; therefore, even an oxygen deficiency (V_O) is formed, the non-bonding state of a neighboring metal cation is located in CBM or near CBM (Figure 13.28c). In this case, V_O generates two free electrons in CB and works as a shallow donor, which is a well-known mechanism for doping in *N*-type TCOs. However, as revealed by theoretical calculations, such a mechanism is not valid for all TCOs and

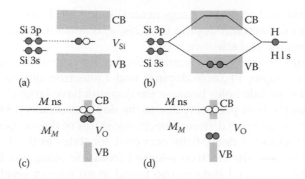

FIGURE 13.28

Simplified electronic structures of defects in Si and oxides. (a) Si vacancy in c-Si forms a singly occupied dangling state near the mid gap, (b) hydrogen passivation by Si-H removes the mid-gap state (hydrogen passivation), (c) an oxygen vacancy in an oxide forms a non-bonding state in or near CB, and (d) in general, an oxygen vacancy state is relaxed to deep bandgap but is fully occupied by two electrons.

oxide semiconductors because the energy levels of such V_O states are often deepened even closer to VBM (Figure 13.28d) [56,105,126–128].

13.12 Doping, Oxygen Deficiency, and Hydrogen

In crystalline semiconductors and a-Si, carrier doping is explained by substitutional doping, where an atom with a different number of valence electrons substitutes a host atom and thus dopes an electron or a hole. On the other hand, the substitution doping is not applied to conventional amorphous materials because an amorphous structure cannot define a "lattice site" and thus "substitution" does not have valid meaning; a-Si:H is an exception in this sense because the Si atoms take well-defined fourfold coordinated sites even in the amorphous network. However, in AOS, as seen in the a-IGZO model built by DFT, such well-defined ion sites are not found at least up to now.

This situation would be understood by a simple case of Al-doped ZnO. In ZnO, Al^{3+} takes a Zn^{2+} site and generates an electron e^- to maintain the charge neutrality. Note that it is also possible to maintain the charge neutrality by incorporating a half of an oxygen ion $1/2O^{2-}$ into ZnO; however, ZnO is a crystal and such an excess O^{2-} site can be only at interstitial sites. DFT calculations reveled that such an interstitial oxygen is not stable [127] and not formed even when Al^{3+} or another donor dopant is doped; therefore, substitution of Zn^{2+} with Al^{3+} generates e^-. However, in AOS, it is very difficult to keep stoichiometry of chemical composition, and the oxygen composition is also changed by aliovalent ion doping as well as deposition and annealing atmospheres/conditions. That is, substitution of Zn^{2+} with Al^{3+} in a-IGZO may simultaneously cause oxygen incorporation by a reaction a-$(InGaZnO_4)_N + Al => a-In_N Ga_N (Zn_{N-1}Al)O_{4N+1/2}$, which satisfies the charge neutrality without generating electrons.

This fact in turn indicates that electron doping is possible just by forming oxygen deficiencies, which is the same model of residual free electrons in TCOs. This is done by reducing oxygen partial pressure during film deposition (P_{O2}) or post-deposition thermal annealing in a reducing atmosphere. Alternatively, ion implantation such as $H^+ / H_2^+ / Li^+$

is also an effective way to change the charge neutrality [6,129]. Thus substitution doping is not a valid concept to understand doping in AOSs, and counting formal changes of constituent ions [6] is a simple and effective way.

This situation would be similar to chemical doping employed in organic semiconductors; doping of, for example, PF_6(hole-doping) and I_3(electron-doping) to polyacetylene (amorphous) generates mobile holes because the dopants have large electron affinity, and electron transfer from the host polyacetylene to the dopants occurs [130].

DFT calculations provided two different electronic states of the oxygen-deficient a-IGZO [56,105–107]. One group forms deep, fully occupied V_O^0 states at 0.4–1 eV above the VBM (Figure 13.29a). In this case, the electron released from the oxygen deficiency is trapped at the V_O site. Another localized state is also found at an energy level close to the CBM ($\sim$0.6 eV below E_{CBM}), suggesting that the formation of a deep level V_O state breaks the coherency of the CBM band and forms a localized state near CBM. We speculate that this would be a candidate for electron traps near E_C observed by the device simulations and the C–V measurements [123,124]. The formation energies of these deep oxygen deficiencies $E_f(V_O)$ are also calculated by the DFT results [56]. It gave the formation energies of 2.9–4.7 eV. It also shows that the $E_f(V_O)$ value depends on coordination structure around V_O, that is, if more In ions are coordinated, $E_f(V_O)$ is smaller. It is explained by the bond energies of In-O (1.70–1.86 eV), Ga-O (2.04–2.48 eV), and Zn-O (1.52–1.84 eV) [56]. This result supports the idea of the "stabilizer ion" because incorporation of Ga ions increases the bond breaking energy for creating oxygen deficiency.

It should be noted that not all oxygen deficiency form such a deep trap levels mentioned previously. The other oxygen deficiency group forms a shallow donor state as seen in Figure 13.29b, and most of their $E_f(V_O)$ are smaller than those of the aforementioned localized V_O cases. In this case, the oxygen deficiency does not form a large free space, and therefore a "V_O site" cannot trap an electron, which results in the doping of free electrons to CB. This difference implies that some large free space like a vacancy site is necessary to form an electron trap level deep in the bandgap.

It is also known that incorporation of hydrogen generates free electrons in a-IGZO as seen in Figure 13.30. Thermal annealing in a 3% H_2 gas increases the free electron density (N_e) even at 150°C, but >200°C is required to obtain the maximum N_e [6]. Introduction of high N_e at RT is possible by H_2^+ implantation [129] and hydrogen plasma [94], which enables

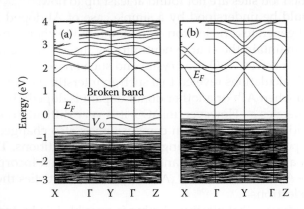

FIGURE 13.29
Pseudo-band structures of two types of oxygen-deficient a-IGZO. (a) Definitive V_O site exists with a large free space and form deep states in the bandgap, and (b) V_O structure is averaged out, a large free space is not found, and E_F is in the CB due to electron doping.

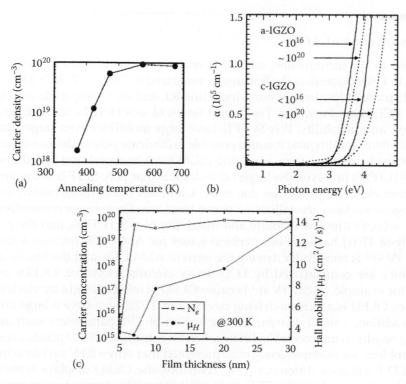

FIGURE 13.30
Examples of hydrogen doping to a-IGZO. (a) Annealing in a 3%H$_2$/97% N$_2$ gas at varied temperature, (b) H$_2^+$ implantation at RT, and (c) hydrogen plasma treatment at RT.

altering N_e without affecting the atomic structures of the host AOSs. DFT results indicate that incorporation of an H atom into a-IGZO always forms an −OH bond (Figure 13.31) and generates a free electron [56]. That is to say, neutral hydrogen atoms are unstable and are stabilized in the form of −OH, that is, H$^+$, by releasing a free electron. This situation totally differs from that in a-Si:H where H^0 makes a Si–H bond without generating a free electron.

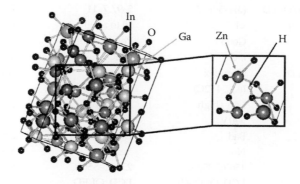

FIGURE 13.31
Local structure in hydrogen-doped a-IGZO built by DFT. Stable −OH bonds are formed to an O ion coordinated with three cations.

13.13 Advantages of AOS

As explained in the introduction, a-IGZO is now expected for channel materials of TFT backplanes in FPDs because of the features summarized in Table 13.2. AOS TFTs are formed on glass substrates at low temperatures including RT, and thus compatible with the current a-Si:H TFT LCD manufacturing. The largest issues of a-Si:H TFTs are their low mobility <1 cm^2 $(V s)^{-1}$ and instability. Poly-Si TFTs have large mobilities even larger than 100 cm^2 $V s^{-1}$ and excellent stability, and it is also possible to fabricate poly-Si devices on inexpensive glass substrates by excimer laser annealing (ELA) (such materials are called low-temperature poly-Si [LTPS]); however, the largest drawback is that poly-Si TFTs show unacceptable distribution in electrical properties due to the GB issues (short-range nonuniformity) and the inhomogeneous laser crystallization over a large area (long-range nonuniformity).

These drawbacks (the low mobility and instability of a-Si:H TFTs, and the poor uniformity of poly-Si TFTs) have not been critical issues for AM-LCDs because a low mobility of ~0.5 cm^2 $(V s)^{-1}$ is enough for driving the current AM-LCDs, and the instability and the nonuniformity are compensated by LCD driver circuits. However, OLEDs need larger mobilities, for example, >4 cm^2 $(V s)^{-1}$ because OLED pixels emit light by electrical current injection (i.e., OLED is a current-driving device) and TFTs must have a large current drivability. In addition, a small distribution/fluctuation of TFT parameters such as threshold voltage (V_{th}) results in unacceptable difference in brightness of OLED pixels, causing a serious *mura* problem, for example, Jeong et al. presented that only ±0.1 V variation in V_{th} causes 16% of OLED luminance difference [131]. Therefore, the OLED displays reported to date using a-Si:H, poly-Si, and organic TFTs must employ complex compensation circuits having four or more TFTs (expressed as, e.g., 4T2C and 5T2C circuits as shown in Table 13.2, which indicate that one circuit involves four or five transistors and two capacitors), even though it is still considered that these TFTs will not be used in future mass production of OLED displays due to the aforesaid problems.

TABLE 13.2

Comparison of a-Si:H, Poly-Si, and Amorphous Oxide TFTs

	a-Si:H	Poly-Si (LTPS/HTPS)	Amorphous Oxide
Generation	>10G	4G/8G?	8G
Channel	a-Si:H	ELA/SPC	a-InGaZnO$_4$
TFT Masks for LCD/OLED	(3)4–5/6–7	5–9/ 7–11	4–5/6–7
Mobility (cm^{-2} $(V s)^{-1}$)	<1	30 to >100	1–20(100?)
TFT uniformity	Good	Poor/better	Good
TFT polarity	*n*-ch	CMOS	*n*-ch
Pixel circuit for OLED	Complex (e.g., 4T2C)	Complex (e.g., 5T2C)	Simple ($2T+1C$)
Cost/yield	Low/high	High/low	Low/high
Vth shift	>10 V	<0.5 V	<1 V
Light stability	Bad	Good	Better than a-Si
Circuit integration	No	Yes	Yes
Process *T*	150°C–350°C	250°C–550°C	RT – 400(600)°C
Display mode	LCD, OLED(?)	LCD, OLED	LCD, OLED, E-paper
Substrate	Glass, metal, plastic	Glass, metal, plastic	Glass, metal, plastic
Solution process printing	No	Laser annealed	270°C–400°C

In addition, even for AM-LCD, the situation is changing rapidly; it is reported that the low mobility of a-Si:H TFTs cannot drive jumbo-size LCDs operating at high frame rates such as >120 Hz (e.g., ≫55″) [4,132,133]; the required mobility will be much higher as the display resolution becomes higher, the frame rate faster and the panel size larger.

AOS TFTs have much larger mobilities >10 cm^2 (V s)$^{-1}$ and applicable to these next-generation FPDs. Further, it has been revealed that AOS TFTs are much stabler than a-Si:H and organic TFTs.

13.14 Fabrication of AOS TFTs

13.14.1 TFT Structures and Fabrication Process

TFTs are fundamental building blocks for large-area microelectronics (giant microelectronics) such as FPDs and system-on-glass/panel. Furthermore, fabricating low-temperature TFTs will allow flexible large-area electronic devices to be developed. These devices are flexible, lightweight, shock resistant, and potentially affordable, which are inevitable for large, economic, high-resolution displays, wearable computers, and paper displays. Further, when combined with "transparent circuit technology," TFTs can integrate display functions even on winding windshields of automobiles.

Figure 13.32 illustrates typical device structures used for AOS TFTs. For research purposes, the bottom gate, top contact structure fabricated on a SiO$_2$/c-Si wafer is very convenient due to its easy fabrication process. However, the gate electrode is common for all the TFTs and it is not used for practical displays and circuits. For FPDs, the inverted staggered structure, which is the same as the current a-Si:H TFTs, has been widely employed due to its compatibility with the current FPD manufacturing. The inverted staggered structure is further classified into (1) etch-stopper type and (2) channel etch type.

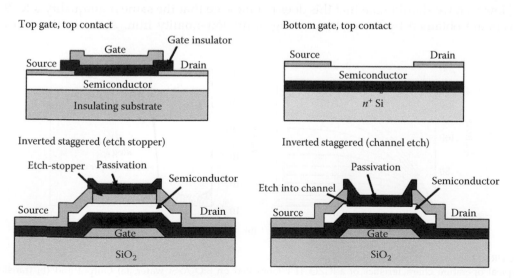

FIGURE 13.32
Typical device structures used for AOS TFTs.

13.14.2 Operation Characteristics of a-IGOZ TFT

Figure 13.33 shows a typical output characteristic (drain–source current (I_{DS}) – voltage (V_{DS})) measured at various gate biases (V_{GS}) and transfer characteristics (log I_{DS} – V_{GS}) at various V_{DS} measured at RT in air. The saturation regime field-effect mobility (μ_{sat}) was estimated to be ~10 cm^2 (V s)$^{-1}$. The transfer characteristic showed a low off-current less than 10^{-13} A, giving a large current on-to-off ratio $I_{on/off} > 10^{10}$. Another important feature is that a-IGZO TFT exhibits a small subthreshold voltage swing (dV/$d$log I_{DS}, S value) of ~100 mV per decade, which is much smaller than that of a-Si:H TFT (~400 mV per decade) and comparable to that of LTPS TFTs. The S value is important for TFT operation because it determines a minimum V_{GS} ($\Delta V_{GS,on}$) to switch a TFT from an off state to an on state by $\Delta V_{GS,on} > S \cdot \log_{10} I_{on/off}$, which corresponds to 3.2 V for a-Si:H TFTs and 0.8 V for a-IGZO TFTs if $I_{on/off} = 10^8$ is employed. S values also provide a convenient way to estimate the subgap trap density at the Fermi level D_{sg} by $S = \ln 10 \cdot (k_B T/e)(1 + eD_{sg}/C_{OX})$, and the trap densities at the interface/surface and in the channel bulk can be separated by examining the channel thickness dependence of S value [94,134].

13.14.3 Unique Annealing Effect

AOS TFTs have a large process allowance due to their unique annealing behaviors. The performance of unannealed a-IGZO TFTs fabricated at RT depends largely on deposition condition of the a-IGZO layer. If the channel is deposited at an unoptimized condition (low-quality, LQ a-IGZO, see Ref. [112] for more detail about notation), the TFT exhibits poor performance (see "LQ, unannealed" in Figure 13.34); however, the TFT performance is improved to that prepared under the optimized condition ("HQ, unannealed") just by annealing ("LQ, annealed") at an appropriate temperature (e.g., 300°C–400°C) far below the crystallization temperature of AOS (>500°C [53]). Although it is difficult to detect structural change by annealing for the HQ a-IGZO layer, some densification is found for the LQ a-IGZO layers [6]. Anyway, this result suggests that the deposition condition is not critical to obtain good TFT characteristics as long as an appropriate thermal annealing is performed.

However, we should note that this does not indicate that the same high-quality a-IGZO layers are obtained by thermal annealing of the low-quality film. Actually, the optical

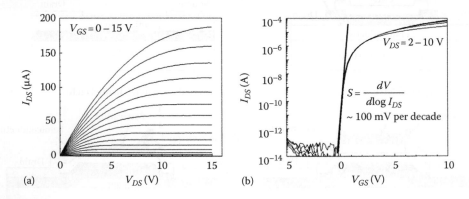

FIGURE 13.33
Typical operation characteristics of a-IGZO TFT fabricated on SiO$_2$/c-Si wafer. (a) Output and (b) transfer characteristics.

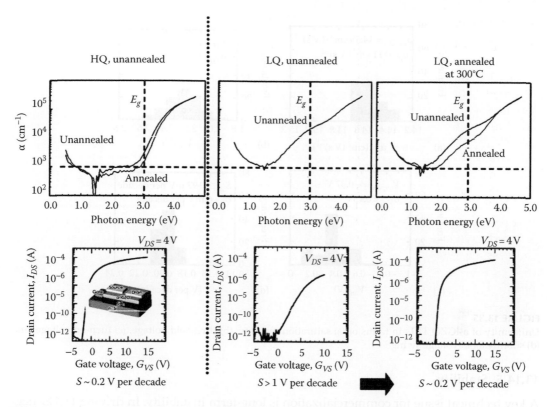

FIGURE 13.34
Unique annealing effect of a-IGZO TFTs. HQ and LQ indicate "high-quality" and "low-quality" a-IGZO films as used in Ref. [112]. The upper panel shows transfer characteristics of a-IGZO TFTs, and the lower panel optical absorption spectra in logarithmic plot before and after thermal annealing at 400°C.

absorption spectra (also those in Figure 13.22) showed that the subgap optical absorption is much higher for the annealed LQ films than for the unannealed HQ film, but the TFT characteristics are similar as seen in Figure 13.34. This is because oxide TFTs are n-channel ones and insensitive to the deep fully occupied defects like the near-VBM defects observed by HX-PES in Figure 13.19 that are the origin of the subgap absorption; however, the deep defects, which are inert for the TFT characteristics, are important to improve photoresponse of TFTs as will be seen below.

13.14.4 Uniformity

It is reported that a-IGZO TFTs have some distribution in operation characteristics if those fabricated at RT are not subjected to appropriate thermal annealing, but exhibit excellent uniformity after these treatments [136]. Figure 13.35 shows the performance histograms of a-IGZO TFTs which were fabricated on a glass substrate by a conventional sputtering and subsequently annealed [135]. About 100 TFTs were fabricated in a $1 \, \text{cm} \times 1 \, \text{cm}$ area of an a-IGZO thin film. The TFTs exhibit excellent uniformity and high average performance. The value of μ_{sat} resides within the standard deviation σ_μ of $0.11 \, \text{cm}^2 \, (\text{V s})^{-1}$ (0.76% of the average value). It strongly suggests that the a-IGZO TFTs essentially have a good short-range uniformity and are advantageous in integrated circuits.

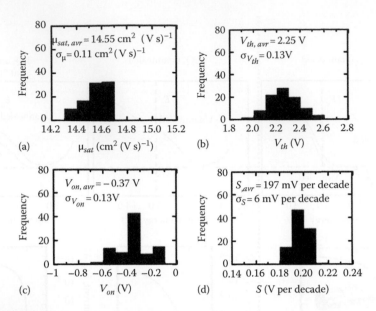

FIGURE 13.35
Uniformity of a-IGZO TFTs in terms of (a) saturation mobility, (b) threshold voltage, (c) turn-on voltage, and (d) subthreshold voltage swing.

13.14.5 Stability

A key technical issue for commercialization is long-term instability. In driving LCDs, negative bias is applied under light illumination in most of the time, while light is illuminated only under positive bias for OLEDs. Thus, negative bias—light illumination stability (NBIS)—is important for LCDs and constant current positive bias stability (CCS) for OLEDs.

First, we should note that uniformity and stability of a-IGZO TFTs are poor for unannealed TFTs [136,137]. The degradation of unannealed a-IGZO TFTs occurs both as threshold voltage shift (ΔV_{th}) and increase in S value, which suggests increase in subgap defects near CBM [137]. In addition, these degradations are recovered by aging at RT in air with long time constants over tens of hours [138]. The degradation in S values disappears and the recovery behaviors become much gentler after thermal annealing.

Therefore, as explained earlier, most of a-IGZO TFTs in the prototype displays have been subjected to thermal annealing at ≥300°C. Note that works to reduce the annealing temperature are undergoing, and lower annealing temperature would be satisfactory for devices that do not require severe stability and uniformity. Instability, however, remains even for annealed a-IGZO TFTs. It has been reported that ΔV_{th} follows a stretched exponential law [139]. Most of ΔV_{th} under positive bias stress tests show positive shifts and are explained by the following origins, for example, trapping of positive charges in (1) a gate insulator, (2) a channel—gate insulator interface, (3) a bulk channel region, and (4) a back channel surface as well as (5) creation of acceptor-type deep traps (i.e., unoccupied defects that can accept extra electrons).

A part of the instability is explained by adsorption and desorption of oxygen-related species from atmosphere and improved by employing a dense passivation layer on top of the back channel [140,141]. We found that incorporation of H_2O vapor in an annealing O_2 gas (wet annealing) is effective to improve the TFT characteristic, stability, and uniformity [136,137], and combination with a passivation layer almost resolved the CCS instability problem [142].

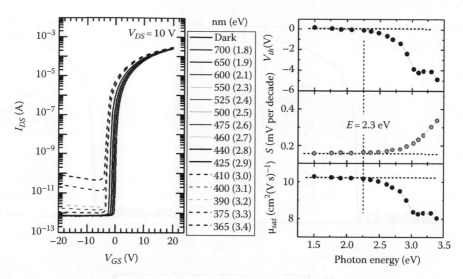

FIGURE 13.36
Photoresponses of a-IGZO TFT as a function of excitation photon energy.

As for light illumination instability, photoresponses of AOS TFTs have been reported for a-Zn-Sn-O [143], and a-IGZO [144,145]. a-IGZO TFTs respond to photon energies >2.3 eV (Figure 13.36 [134]), which are lower than the bandgap (3.0–3.2 eV). Light illumination causes increase in the off-current and negative parallel shifts of V_{th}. The threshold photon energy corresponds well to the energy levels of the subgap DOS above VBM observed by HX-PES in Figure 13.19 and the subgap optical absorption in Figures 13.22 and 13.34; therefore, it is attributed to excitation of electrons from the deep subgap states to the conduction band. NBIS instability is examined by separating the effect of negative bias and light illumination in Figure 13.37 [142]. It reveals that application of negative bias only or light illumination only does not cause a large negative ΔV_{th}, but their combination results in the serious negative ΔV_{th}. In addition, elimination of the VBM defects at ~2.3 eV below CBM by forming a passivation layer removed the NBIS instability by subgap photons up to 2.9 eV [142]. These results indicate that the NBIS caused by subgap photon excitation is due to the following process: (1) generation of electron–hole pair at the back channel surface via the near-VBM states, (2) diffusion of holes or other positive charges like H^+ to the depletion layer, and (3) trap of the holes/positive charges by the defects at the gate–insulator–channel interface (Figure 13.38a). An analytical model considering this process explains well the degradation behavior as a function of excitation photon energy and corresponding penetration depth as seen in Figure 13.38b, which also provides a possible hole mobility of ~0.01 cm² (V s)⁻¹ [146].

Similar photoresponses are also observed in steady-state photocurrent measurements on a-IGZO films (Figure 13.39 [147]). Photoconductivity spectra as a function of excitation photon energy decay with Urbach energies of ~0.24 eV for unannealed a-IGZO and ~0.16 eV for annealed a-IGZO. Although fast recombination (<0.1 μs for unannealed a-IGZO and 0.2 μs for annealed a-IGZO) dominates in a-IGZO [148], reversible and very slow photoresponses with time constants more than 1000 s also exist, which are attributed to relaxation of metastable donor states with activation energies 0.9–1.1 eV [147]; a similar model has also been proposed by Takechi et al. [92].

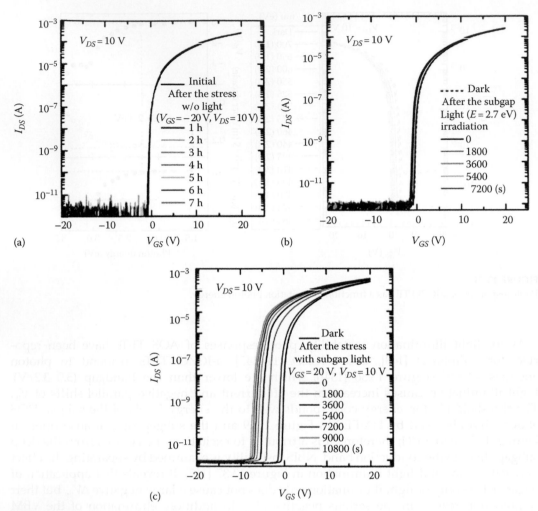

FIGURE 13.37
Stability of a-IGZO TFT under different combination of negative bias and light illumination. (a) Negative bias without light illumination (NBS), (b) light illumination without bias at the photon flux of ~10^{14} cm^{-2} s^{-1}, and (c) NBIS at the photon flux of ~10^{14} cm^{-2} s^{-1}.

13.15 Progress in AOS TFT as the FPD Backplane

13.15.1 History of Prototype Displays

Development history of a-IGZO FPDs until 2008 is summarized in Refs [4,5] and some are seen in Figure 13.40. Toppan Printing Co. Ltd. first reported an AM display using AOS TFTs as a flexible black-and-white (BW) e-paper in 2005 [152]. LG Electronics reported the first AM-OLED display in 2006 [152]. Following them, Samsung SDI and Samsung Advanced Institute of Technology (SAIT) reported AM-OLED displays in 2007 [153,154]. The largest displays in mid 2011 is 70″ AM-LCD operating at 240 Hz frame rate facilitated with 3D vision presented by Samsung Electronic (SEC) at FPD International 2010.

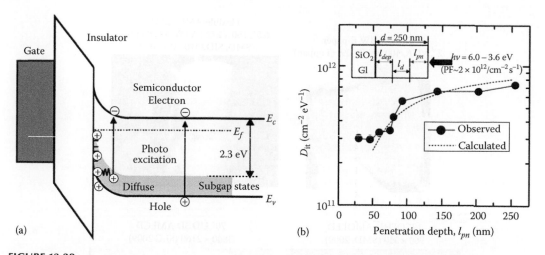

FIGURE 13.38
Model of NBIS instability. (a) Schematic electronic structure and (b) estimated defect density created by NBIS as a function of penetration depth of excitation photons along with the result of diffusion model analysis.

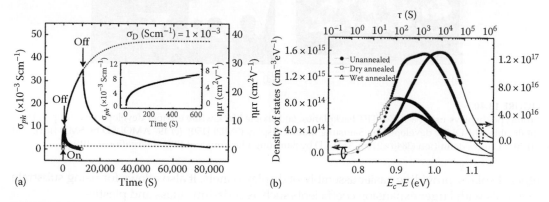

FIGURE 13.39
(a) Slow and reversible photoresponses in a-IGZO films and (b) corresponding electronic states.

13.15.2 New Display Structure

An innovative e-paper display structure called "front drive" was recently proposed by Ito et al. of Toppan Printing Co. Ltd. [156]. In conventional color AM e-papers and displays, a color filter array is formed on a front plane and a TFT array on a back plane. Therefore, fine alignment of these planes through liquid crystal (LC) or E-Ink microcapsules is necessary to avoid color misfit; however, the thickness of LC is only 4–6 μm, while that of the E-Ink microcapsules are much larger 40–50 μm, making the horizontal alignment much difficult. This problem will be much critical for flexible displays because bending the display inevitably causes misalignment of the components on the front and the back planes. The TFT array should be on a backplane for conventional a-Si:H TFTs because a-Si:H is nontransparent. However, AOS TFTs can be highly transparent and provides high transparency even if they are formed in a front plane; thus, the aforementioned difficulty in alignment is resolved by directly forming AOS TFT arrays integrated on the front color filter arrays. This is a first demonstration of a device structure benefited from optical transparency of AOS. This structure is applicable to

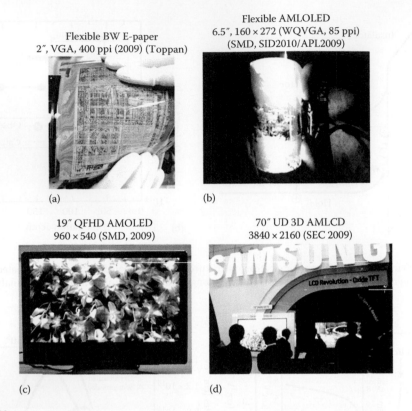

FIGURE 13.40
Prototype displays using a-IGZO TFT backplanes. (a) Flexible black-white e-paper developed by Toppan [149], (b) flexible AM-OLED developed by Samsung mobile display (SMD) [150], (c) 19″ AMOLED by SMD [151], and (d) 70″ 3D ultra definition (3840×2160) AMLCD by Samsung Electronics at FPD International 2010.

other displays, providing easier assembly of display panels. It also allows for using substrate materials with larger expansion coefficients such as soda-lime glass and plastics.

13.15.3 Flexible Displays

Flexible displays have also been reported by several companies. The first one is the BW e-paper reported by Toppan Printing [152]. They have developed more sophisticated ones such as a larger size of 5.35″ for a 150 ppi (pixel per inch) resolution BW e-paper and the world-finest resolution of 400 ppi for a 2″ BW e-paper [149]. The first flexible FPD was presented by LGE as a flexible AM-OLED (3.5″, QCIF+) fabricated on stainless steel foil [156]. Samsung Mobile Display (SMD) reported a very flexible 6.5″ WQVGA AM-OLED fabricated on a polyimide substrate, which is bendable at a curvature radius of 2 cm [150].

13.15.4 Transparent Displays

Another interesting application is a transparent electronic device. Toppan proposed a novel idea of front-drive structure by utilizing the transparency of AOS TFTs as explained earlier [156]. Real transparent applications would be transparent displays. DENSO demonstrated a transparent display as monocolor passive-matrix (PM) inorganic EL displays. Transparent OLED displays have been developed by replacing metal electrodes to (semi-)

transparent ones, and now TDK Microdevices started mass production of semi-transparent OLED display for mobile phones [158]. At SID2010 Exhibition, a 46″ transparent LCD driven by a-Si:H TFTs with an integrated touch panel was presented by Samsung LCD, a 47″ window TV by LG Display, and a 14.1″ transparent OLED driven by LTPS TFTs with transparency up to 38% by SMD. In 2011, Samsung Electronics announced mass production of 22″ transparent LCD [159].

These transparent displays do not employ AOS TFTs and their aperture ratios are low, for example, ~35% for AM-OLED. It is expected that employing transparent AOS TFTs will improve the aperture ratio. To date, Samsung SDI reported 4.1″ transparent dual-emission QCIF AM-OLED using a-IGZO TFTs [160], but transparency was ~20%. LGE and ETRI have developed 2.5″ QCIF+ transparent OLED displays using ZnO TFT backplanes with the panel transmittance of 60% [161,162] and displayed 1.5″ QQVGA transparent AM-OLED driven by AOS TFTs with transparency of 45% at IMID2009. AUO displayed a 2.4″ transparent OLED driven by a-IGZO TFTs with an integrated touch panel [163].

13.15.5 System Integration

AOS TFTs are also expected for developing the peripheral circuits because the largest Hall mobility of AOS exceeds $50\,cm^2\,(V\,s)^{-1}$, and some groups have reported TFT mobilities larger than $100\,cm^2\,(V\,s)^{-1}$, for example, by employing a dual-gate TFT [164], which would cover some peripheral circuits. The dynamic operation characteristics of a-IGZO TFTs have been examined using benchmark circuits such as ring oscillators (RO) and display pixel circuits. Canon Inc. developed a faster five-stage RO using a-IGZO TFTs and demonstrated 410 kHz oscillation, by which they proved by circuit simulations that the dynamic characteristics of the a-IGZO TFTs are consistent with the static ones [165]. The fastest AOS RO was reported by SAIT with the delay time of 0.94 ns per stage [164]. Flexible integrated circuits have also been studied, which was first reported as flexible five-stage RO using a-IGZO TFTs by a National Taiwan University group [166].

Small size TFTs have also been examined for faster operation aiming at application to integrated high-density memory devices. SAIT reported that a-IGZO TFTs are down-scalable to the channel length of 50nm without suffering from the short-channel effect [167]. High-frequency operation up to 180MHz is reported for 1 μm long a-In-Zn-O (a-IZO) TFTs [168].

These features provide more sophisticated glass-based devices called "System-On-Glass (SOG)" or "System-On-Panel (SOP)," in which electronic circuits such as pixel drivers and other peripheral ones are integrated with TFT arrays on the same glass substrates. The first integrated FPD was presented by SEC as a gate driver-integrated 15″ AM-LCD [169]. Semiconductor Energy Laboratory (SEL) integrated the source and drain drivers in 4″ QVGA AM-LCD [170] and AM-OLED [171].

Hitachi, Ltd. reported very low-voltage (1.5 V) operation of AOS TFTs [172] and develops 13.56MHz RFID tag operating by 1nA of driving current [173]. Further, applications of AOS have also been extended to memories; SAIT reported nonvolatile memories using an a-IGZO floating-gate [174–176]. Resistivity switching memory (ReRAM) using a NiO memory node layer [177] and a IGZO memory node layer [178] controlled by a-IGZO TFTs. Ferroelectric memory (FeRAM) using an organic ferroelectric memory node is also demonstrated [179].

13.15.6 Mass Production

For mass production of FPDs, large-area fabrication techniques are necessary to be developed. Although the first AOS TFT was fabricated by PLD, most of AOS TFTs are fabricated

TABLE 13.3

Mask Steps for a-Si LCD and a-IGZO OLED

	5-Mask a-Si LCD [184]	4-Mask a-Si LCD [184]	a-IGZO OLED [186]
1.	Gate metal (GM)	GM	GM
2.	Channel	S/D/Channel	Channel
	Gate insulator (G.I.)		
	(SiN_x/a-Si/n^+ a-Si)		
3.	Source/drain (S/D)	Passivation/G.I.	C/H, ESL, Via hole
4.	Passivation (SiN_x)	Pixel	S/D
5.	Pixel (ITO)		Passivation
6.			Anode
7.			Pixel

by RF/DC sputtering. Some companies have made actions to develop large-size sputtering apparatuses, for example, SEL reported a 3.4″ QHD 326 ppi AM-OLED display fabricated by a 3.5G process ($600 \times 720\,mm^2$) for a-IGZO TFTs [180], and AUO uses a 6G sputtering target, which produced the 37″ [181]. More recently, ULVAC Inc. reported good uniformity of a-IGZO TFTs over a G4 size ($730 \times 920\,mm^2$ area) using a multi-cathode AC sputtering system [182] and also developed a G8-size a-IGZO sputtering system [183].

For fabrication processes, reduction of the number of photolithography masks is important to minimize fabrication cost. The present a-Si:H LCDs employ four- to five-mask processes (Table 13.3), where one mask step can be reduced by using a gray mask (or called a half-tone mask). LGD presented a 6.4″ VGA a-IGZO AM-LCD fabricated by a four-mask process at FPDI2009, and AUO also develops a five-mask process [185]. SMD presented a seven-mask process [186], but it is for AM-OLED with an etch-stopper structure and essentially comparable to those of the a-Si:H processes.

13.16 Perspective as a Chapter Summary

13.16.1 TFT Fabrication Using Solution Processes

Next technical challenge in AOS TFTs will be the development of a nonvacuum process [187]; therefore, oxide TFT fabrication using solution processes is attracting much attention. There are several reasons for this trend. First is cost reduction because vacuum facilities are expensive but all solution and atmospheric processes can remove them. It is also expected that usage efficiency of reagents would be better in solution processes than in vacuum. Second is strong demand for fabrication of TFTs on plastic substrates. The last is the material characteristics of AOS, that is, an oxide semiconductor is a unique material that can be formed by heating in an ambient atmosphere among various semiconductors, and high TFT performance comparable to that based on crystalline semiconductors can be expected even using AOSs as described previously. Although research on this subject is rapidly increasing as of 2011, examination of device reliability has not performed yet.

Table 13.4 summarizes the representative AOS TFTs fabricated by solution processing. A research group of HP and OSU reported a pioneering work in 2007. Recently, University

TABLE 13.4

Developments of Solutions Process for AOS TFTS

Material	Mobility (cm²(V s)⁻¹)	Method	Solution	Temp. (°C)	References
In-Zn-O					
a-IZO	16.1	SC	$ZnCl_2$, $InCl_3$	600	D.-H. Lee et al., *Adv. Mater.* 19, 843 (2007)
a-IZO	7.4	IJ	$ZnCl_2$, $INCl_3$	600	D.-H. Lee et al., *Adv. Mater.* 19, 843 (2007)
a-IZO	7.3	IJ	$Zn(OAc)_2$, $In(OAc)_3$	500	C.G. Choi et al., *Electrochem. Solid-State Lett.* 11, H7 (2008.)
a-IZO	2.13	SC	$ZnCl_2$, $InCl_3$	500	C.-K. Chen et al., *J. Display Technol.* 5, 509 (2009)
a-IZO	6.6	SC	$Zn(OAc)_2$, $In(OAc)_3$	450	K.-B. Park et al., *Electron Dev. Lett.* 31, 311 (2010)
a-IZO	1.54	SC	$Zn(OAc)_2$, $In(NO_3)_3$	400	S. Jeong et al., *Adv. Mater.* 22, 1346 (2010)
IZO	3.3	SC	Commercially available coating solution	400	S.-M. Yoon et al. *Electrochem. Solid-State Lett.* 13, H141 (2010)
a-IZO	5.0	SC		450	M.K. Ryu et al., *J. Soc. Inf. Display* 18, 734 (2010)
IZO	0.54	SC	$Zn(OAc)_2$, $In(NO_3)_3$	300	C.Y. Koo et al., *J. Electrochem. Soc.* 157, J111 (2010)
Zn-Sn-O					
ZTO	1.1	SC	$Zn(OAc)_2$, $Sn(OAc)_2$	500	S. Jeong et al., *J. Phys. C* 112, 11082 (2008)
ZTO	16	SC	$ZnCl_2$, $SnCl_2$	600	Y.-J. Chang et al., *Electrochem. Solid-State Lett.* 10. H135 (2007)
a-ZTO	5	SC	$Sn(OAc)_2$, $SnCl_2$	500	S.K. Park et al., *Electrochem. Solid-State Lett.* 12, H256 (2009)
a-ZTO	14.1	SC	$Zn(OAc)_2$, $SnCl_2$	500	S.-J. Seo et al., *J. Phys. D* 42. 035106 (2009)
ZTO	0.7	SC	$Zn(OAc)_2$, $Sn(OAc)_2$	550	Y. Jeong et al., *J. Electrochem. Soc.* 156, H808 (2009)
ZTO	28	SC	$ZnCl_2$, $SnCl_2$	600	B.N. Pal et al., *Nat. Mater.* 8, 898 (2009)
ZTO	1.06	SC	$Zn(OAc)_2$, $Sn(OAc)_2$	500	K. Song et al., *J. Mater. Chem.* 19, 8881 (2009)
ZTO	5.5	SC	$Zn(OAc)_2$, $SnCl_2$	300	S.-J. Seo et al., *Electrochem. Solid-State Lett.* 13, H357(2010)
ZTO	4.98	IJ	$ZnCl_2$, $SnCl_2$	500	Y.-H. Kim et al., *Proc. IMID'2010* (2010)
ZTO	27.3	SC	$ZnCl_2$, $SnCl_2$	500	C.-G. Lee et al., *Appl. Phys. Lett.* 96, 243501 (2009)
In-Ga-O					
IGO	9	IJ	$Ga_7In_6[(\mu3\text{-}OH)_6(\mu\text{-}OH)_{18}(H_2O)_{24}(NO_3)_{15}]$ made from $Ga(NO_3)_3$ and $In(NO_3)_3$	600	Z.L. Mensinger et al., *Angew. Chem.* 47, 9484 (2008)

(continued)

TABLE 13.4 (continued)

Developments of Solutions Process for AOS TFTS

Material	Mobility (cm²(V s)⁻¹)	Method	Solution	Temp. (°C)	Reference
In-Ga-Zn-O					
IGZO	0.96	SC	$Zn(OAc)_2$, $Ga(NO_3)_3$, $In(NO_3)_3$	450	G.H. Kim et al., *J. Electrochem. Soc.* 156, H7 (2009)
IGZO	1.25	SC	$Zn(OAc)_2$, $Ga(NO_3)_3$, $In(NO_3)_3$	400	G.H. Kim et al., *Appl. Phys. Lett.* 94, 23501 (2009)
a-IGZO	3-16	SC	$Zn(OAc)_2$, $Ga(NO_3)_3$, $In(NO_3)_3$	400	D. Kim et al., *Appl. Phys. Lett.* 95, 103501 (2009)
a-IGZO	0.85	SC	$Zn(OAc)_2$, $Ga(NO_3)_3$, $In(NO_3)_3$	400	S. Jeong et al., *Adv. Mater.* 22, 1346 (2010)
a-IGZO	0.52	SC	$Zn(OAc)_2$, $Ga(NO_3)_3$, $In(NO_3)_3$	500	Y.-H. Kim et al., *IEEE Trans. Electron. Dev.* 57, 1009 (2010)
a-IGZO	0.05	IG	$Zn(OAc)_2$, $Ga(NO_3)_3$, $In(NO_3)_3$	400	D. Kim et al., *Jpn. J. Appl. Phys.* 49, 05EB06 (2010)
a-IGZO	2.07		—	450	Y.-C. Lai, et al., *Proc. IDW'09*, 1677 (2009)
a-IGZO	0.86	SC	$Zn(OAc)_2$, $Ga(NO_3)_3$, $In(NO_3)_3$	450	G. Kim et al., *Phys Status Solidi A* 207, 1677 (2010)
IGZO	0.46	SC	$Zn(OAc)_2$, $Ga(NO_3)_3$, $In(NO_3)_3$	550	[85]
IGZO	1.94	SC	$Zn(OAc)_2$, $Ga(NO_3)_2$, $In(NO_3)_3$	500	[79]
IGZO	2.3	SC	$Zn(OAc)_2$, $Ga(NO_3)_3$, $In(NO_3)_3$	95	Y.-H. Yang et al., *IEEE Electron. Dev. Lett.* 31, 329 (2010)
a-IGZO	1.3	SC	$Zn(OAc)_2$, $INCl_3$, $GaCl_2$	600	Y. Wang et al., *J. Sol-Gel Sci. Technol.* 55, 322 (2010)
a-IGZO	5.4	SC	—	270	M. Ito et al., *Proc. IMID'2010* (2010)
Others					
a-IZTO	30	IJ	$InCl_3$, $ZnCl_2$, $SnCl_2$	600	D.-H. Lee et al., *J. Mater. Chem.* 19, 3135 (2009)
AIO	19.6	SC	$Al(acac)_3$, $In(OAc)_3$	350	Y.H. Hwang et al., *Electrochem. Solid-State Lett.* 12, H336 (2009)
a-ITO	10–20	SC	$InCl_3$, $SnCl_4$	250	H.S. Kim et al., *J. Am. Chem. Soc.* 131, 10826 (2009)
a-GZO	0.009	SC	$Zn(OAc)_2$ $Ga(NO_3)_3$	300	S. Jeong et al., *Adv. Mater.* 22, 1346 (2010)
YIZO	1.92	SC	$Zn(OAc)_2$, $In(NO_3)_3$ $Y(NO_3)_3$	550	[85]
AITO	13.3	SC	$In(OAc)_3$ $SnCl_2$ $Al(acac)_3$	500	[88]
a-ZITO	10–100	SC	$Zn(OAc)_2$, $InCl_3$, $SnCl_2$	400	M.G. Kim et al., *J. Am. Chem. Soc.* 132, 10352 (2010)
HIZO	1.94	SC	$Zn(OAc)_2$, $HfCl_4$ $In(NO_3)_3$	500	[79]
a-IZO	0.05	SC	$Zn(OAc)_2$, $In(NO_3)_3$, $Zr[OC(CH_3)_3]_4$	500	J.H. Kim et al., *Proc. IMID'2010* (2010)
ITaZO	0.37	SC	$Zn(OAc)_2$, $In(NO_3)_3$, $Ta[OC(CH_2)_3CH_3]_5$	500	J.H. Kim et al., *Proc. IMID'2010* (2010)

of Cambridge group reported a-IZO/IGZO TFTs with mobilities ~10 cm^2 (V s)$^{-1}$ fabricated at 230°C [188] and Toppan reported fabrication of AOS TFTs (the solution is supplied by EVONIK, but the composition is not disclosed) with mobilities 5.4 cm^2 (V s)$^{-1}$ by annealing at 270°C for a few minutes [189].

The ultimate goal of this research is to fabricate high performance TFTs on plastics or polymer-coated papers leading to flexible or paper displays. The keys to the success would be in developing innovative precursors and processing methods beyond the traditional sol–gel or solution processes. This issue is the almost the same as how dense thin films comparable to those obtained by vacuum processes are fabricated at low temperatures.

13.16.2 *P*-Type Oxide TFT and CMOS Circuits

Another technical challenge in oxide TFTs is to realize complementary (CMOS) circuits only by oxide TFTs because CMOS circuits can produce very low-power consumption devices. As already explained based on electronic structures of oxides, most of good TOSs are *N*-type, and *P*-type oxide semiconductors are limited to, for example, $CuAlO_2$, $SrCu_2O_2$, LaCuOS, ZnO, SnO, and Cu_2O. Although more than 10 *P*-type TOSs have been reported to date, no p-channel TFT had been realized up until 2008 irrespective of few trials since the challenge of Schockley in 1948 [190]. This fact comes from a general trend that *P*-type oxides have high-density subgap traps.

Among the *P*-type TOSs, only SnO and Cu_2O have produced operating TFTs. Matsuzaki et al. fabricated epitaxial thin films and obtained Hall mobility of ~100 cm^2 (V s)$^{-1}$ at the hole concentration of ~10^{13}cm^{-3}, which is comparable to that in single crystalline Cu_2O. However, Cu_2O-based TFT did not operate enough and the estimated field-effect mobility remained ~0.1 cm^2 (V s)$^{-1}$ [191,192]. After that, poly-Cu_2O TFTs have been reported [193]. Higher mobility of 4.3 cm^2 (V s)$^{-1}$ is reported for poly-Cu_2O TFTs by employing a HfON gate insulator [194].

Ogo et al. [195,196] reported a p-channel TFT with the mobility of 1.4 cm^2 (V s)$^{-1}$ employing SnO (not SnO_2) as the active layer. This is a first demonstration of p-channel oxide-TFT with a mobility >1 cm^2 (V s)$^{-1}$. Poly-SnO TFTs were reported in Refs. [197–199]. Bipolar doping in SnO was reported in Ref. [200] in early 2011, and, subsequently, ambipolar operation of SnO TFT and its invertor circuit has been realized [201]. Improvement of TFT performance and extension to amorphous material systems are next challenges.

References

1. K. Nomura, H. Ohta, A. Takagi, T. Kamiya, M. Hirano, and H. Hosono, Room-temperature fabrication of transparent flexible thin film transistors using amorphous oxide semiconductors, *Nature* **432**, 488 (2004).
2. Sharp says to make smartphone panels at TV panel plant, msnbc.com, http://www.msnbc.msn.com/id/38319942/ (April 21, 2011).
3. Asahi Shimbun Company, Schedule for iPad3 delayed; Order of high-resolution oxide-TFT LCD has been cancelled [in Japanese], Asahi Shimbun Company, http://techon.nikkeibp.co.jp/article/NEWS/20110822/196371/# (*Tech-On!*, web newspaper, August 23, 2011).
4. T. Kamiya, K. Nomura, and H. Hosono, Present status of amorphous InGaZnO thin-film transistors, *Sci. Technol. Adv. Mater.* **11**, 044305 (2010).

5. T. Kamiya and H. Hosono, Material characteristics and applications of transparent amorphous oxide semiconductors, *NPG Asia Mater.* **2**, 1522 (2010).

6. T. Kamiya, K. Nomura, and H. Hosono, Origins of high mobility and low operation voltage of amorphous oxide TFTs: Electronic structure, electron transport, defects and doping, *J. Disp. Technol.* **5**, 273 (2009).

7. H. Kumomi, S. Yaginuma, H. Omura, A. Goyal, A. Sato, M. Watanabe et al., Materials, devices, and circuits of transparent amorphous-oxide semiconductor, *J. Disp. Technol.* **5**, 531 (2009).

8. H. Hosono, K. Nomura, Y. Ogo, T. Uruga, and T. Kamiya, Factors controlling electron transport properties in transparent amorphous oxide semiconductors, *J. Non-Cryst. Solids* **354**, 2796 (2008).

9. H. Hosono, T. Kamiya, and M.O Hirano, Function cultivation of transparent oxides utilizing built-in nanostructure, *Bull. Chem. Soc. Jpn.* **79**, 1 (2006).

10. J.-Y. Kwon, D.-J. Lee, and K.-B. Kim, Review paper: Transparent amorphous oxide semiconductor thin film transistor, *Electron. Mater. Lett.* **7**, 1 (2011).

11. J.F. Wager, D.A. Keszler, and R.E. Presley, *Transparent Electronics*, Springer, Heidelberg, Germany (2008).

12. A. Facchetti and T.J. Marks, *Transparent Electronics*, Wiley, Chichester, West Sussex, United Kingdom (2010).

13. D.S. Ginley, D.C. Paine, and H. Hosono, *Handbook of Transparent Conductors*, Springer, NY (2010).

14. H. Hosono, N. Kikuchi, N. Ueda, and H. Kawazoe, Working hypothesis to explore novel wide band gap electrically conducting amorphous oxides and examples, *J. Non-Cryst. Solids* **198–200**, 165 (1996).

15. T. Kamiya and H. Hosono, Electronic structures and device applications of transparent oxide semiconductors: What is the real merit of oxide semiconductors? *Int. J. Appl. Ceram. Technol.* **2**, 285 (2005).

16. S.M. Sze, *Physics of Semiconductor Devices*, 2nd edn., John Wiley & Sons, Inc., NY (1981).

17. S.-H. Wei and A. Zunger, Role of metal d states in II-VI semiconductors, *Phys. Rev. B* **37**, 8958 (1988).

18. T. Kamiya and M. Kawasaki, ZnO-based semiconductors as building blocks for active devices, *MRS Bullet.* **33**, 1061 (2008).

19. H. Mizoguchi, T. Kamiya, S. Matsuishi, and H. Hosono, A frontier of transparent conductive oxides: Germanate system, *Nat. Comm.* **2**, 470 (2011).

20. O. Madelung, *Semiconductors: Data Handbook*, 3rd edn., Springer-Verlag, Berlin, Germany (1991).

21. G. Rupprecht, Untersuchungen der eletrischen und lichtelektrischen leitfähigkeit dünner indiumoxydschichten, *Z. Phys.* **139**, 504 (1954).

22. H.J.J. Van Boort and R. Groth, Low-pressure sodium lamps with indium oxide filter, *Philips Tech. Rev.* **29**, 17 (1968).

23. M. Mizuhashi, Electrical properties of vacuum-deposited indium oxide and indium tin oxide films, *Thin Solid Films* **70**, 91 (1980).

24. C.A. Pan and T.P. Ma, High quality transparent conductive indium oxide films prepared by thermal evaporation, *Appl. Phys. Lett.* **37**, 163 (1980).

25. R.L. Weiher, Electrical properties of single crystals of indium oxide, *J. Appl. Phys.* **33**, 2834 (1962).

26. H. Fujiwara and M. Kondo, Effects of carrier concentration on the dielectric function of ZnO:Ga and In_2O_3:Sn studied by spectroscopic ellipsometry: Analysis of free-carrier and band-edge absorption, *Phys. Rev. B* **71**, 075109 (2005).

27. A. Walsh, J.L.D.F. Da Silva, S.-H. Wei, C. Korber, A. Klein, L.F.J. Piper et al., Nature of the band gap of In_2O_3 revealed by first-principles calculations and x-ray spectroscopy, *Phys. Rev. Lett.* **100**, 167402 (2008).

28. R.L. Weiher and R.P. Ley, Optical properties of indium oxide, *J. Appl. Phys.* **37**, 299 (1966).

29. I. Hamberg and C.G. Granqvist, Evaporated Sn-doped In_2O_3 films: Basic optical properties and applications to energy efficient windows, *J. Appl. Phys.* **60**, R123 (1986).

30. H. Odaka, S. Iwata, N. Taga, S. Ohnishi, Y. Kaneta, and Y. Shigesato, Study on electronic structure and optoelectronic properties of indium oxide by first-principles calculations, *Jpn. J. Appl. Phys.* **36**, 5551 (1997).

31. N.S. Murty, G.K. Bhagavat, and S.R. Jawalekar, Physical properties of tin oxide films deposited by oxidation of $SnCl_2$, *Thin Solid Films* **92**, 347 (1982).

32. R. Banerjee and D. Das, Properties of tin oxide films prepared by reactive electron beam evaporation, *Thin Solid Films* **291**, 149 (1987).

33. C.G. Fonstad and R.H. Rediker, Electrical properties of high-quality stannic oxide crystals, *J. Appl. Phys.* **42**, 2911 (1971).

34. K.J. Button, D.G. Fonstad, and W. Dreybradt, Determination of the electron masses in stannic oxide by submillimeter cyclotron resonance, *Phys. Rev. B* **4**, 4539 (1971).

35. R. Summitt, J.A. Marley, and N.F. Borreilli, The ultraviolet absorption edge of stannic oxide, *J. Phys. Chem. Solids* **25**, 1465 (1964).

36. V.T. Agekyan, Spectroscopic properties of semiconductor crystals with direct forbidden energy gap, *Phys. Status Solidi A* **43**, 11 (1977).

37. E.E. Hahn, Some electrical properties of zinc oxide semiconductor, *J. Appl. Phys.* **22**, 855 (1951).

38. T. Minami, H. Nanto, and S. Tanaka, Highly conductive and transparent zinc oxide films prepared by RF magnetron sputtering under an applied external magnetic field, *Appl. Phys. Lett.* **41**, 958 (1982).

39. T. Minami, H. Sato, H. Nanto, and S. Tanaka, Group III impurity doped zinc oxide thin films prepared by RF magnetron sputtering, *Jpn. J. Appl. Phys.* **24**, L781 (1985).

40. T. Minami, ZnO-based transparent conducting oxide, *Oyo Butsuri* **61**, 1255 (1992) [in Japanese].

41. A.R. Hutson, Hall effect studies of doped zinc oxide single crystals, *Phys. Rev.* **108**, 222 (1957).

42. C.K. Yang and K.S. Dy, Band structure of ZnO using the LMTO method, *Solid State Commun.* **88**, 491 (1993).

43. J.R. Bellingham, W.A. Phillips, and C.J. Adkins, Electrical and optical properties of amorphous indium oxide, *J. Phys. Condens. Matter* **2**, 6207 (1990).

44. N. Kikuchi, H. Hosono, H. Kawazoe, K. Oyoshi, and S. Hishita, Transparent, conducting, amorphous oxides: Effect of chemical composition on electrical and optical properties of cadmium germinates, *J. Am. Ceram. Soc.* **80**, 22 (1997).

45. K. Shimakawa, S. Narushima, H. Hosono, and H. Kawazoe, Electronic transport in degenerate amorphous oxide semiconductors, *Philos. Mag. Lett.* **79**, 755 (1999).

46. H. Hosono, Y. Yamashita, N. Ueda, H. Kawazoe, and K. Shimidzu, New amorphous semiconductor: $2CdO\cdot PbO_x$, *Appl. Phys. Lett.* **68**, 661 (1996).

47. M. Yasukawa, H. Hosono, N. Ueda, and H. Kawazoe, Novel transparent and electroconductive amorphous semiconductor: Amorphous $AgSbO_3$ film, *Jpn. J. Appl. Phys.* **34**, L281 (1995).

48. M. Orita, H. Ohta, M. Hirano, S. Narushima, and H. Hosono, Amorphous transparent conductive oxide $InGaO_3(ZnO)_m$ (m ≤ 4): A Zn 4s conductor, *Philos. Mag. B* **81**, 501 (2001).

49. A. Takagi, K. Nomura, H. Ohta, H. Yanagi, T. Kamiya, M. Hirano, and H. Hosono, Carrier transport and electronic structure in amorphous oxide semiconductor a-$InGaZnO_4$, *Thin Solid Films* **486**, 38 (2005).

50. T. Minami, T. Kakumu, and S. Takata, Preparation of transparent and conductive In_2O_3-ZnO films by radio frequency magnetron sputtering, *J. Vac. Sci. Technol. A* **14**, 1704 (1996).

51. T. Minami, S. Takata, and T. Kakumu, New multicomponent transparent conducting oxide films for transparent electrodes of flat panel displays, *J. Vac. Sci. Technol. A* **14**, 1689 (1996).

52. E. Fortunato, A. Pimentel, A. Gonçalves, A. Marques, and R. Martins, High mobility amorphous/nanocrystalline indium zinc oxide deposited at room temperature, *Thin Solid Films* **502**, 104 (2006).

53. K. Nomura, A. Takagi, T. Kamiya, H. Ohta, M. Hirano, and H. Hosono, Amorphous oxide semiconductors towards high-performance flexible thin-film transistors, *Jpn. J. Appl. Phys.* **45**, 4303 (2006).

54. T. Kamiya, K. Nomura, and H. Hosono, Electronic structures above mobility edges in crystalline and amorphous In-Ga-Zn-O: Percolation conduction examined by analytical model, *J. Disp. Technol.* **5**, 462 (2009).

55. T. Kamiya, K. Nomura, and H. Hosono, Origin of denite Hall voltage and positive slope in mobility-donor density relation in disordered oxide semiconductors, *Appl. Phys. Lett.* **96**, 122103 (2010).

56. T. Kamiya, K. Nomura, and H. Hosono, Subgap states, doping and defect formation energies in amorphous oxide semiconductor a-InGaZnO$_4$ studied by density functional theory, *Phys. Status Solidi A* **207**, 1698 (2010).

57. T. Kamiya, H. Hiramatsu, K. Nomura, and H. Hosono, Device applications of transparent oxide semiconductors: Excitonic blue LED and transparent flexible TFT, *J. Electroceram.* **17**, 267 (2006).

58. A.J. Leenheer, J.D. Perkins, M.F.A.M. van Hest, J.J. Berry, R.P. O'Hayre, and D.S. Ginley, General mobility and carrier concentration relationship in transparent amorphous indium zinc oxide films, *Phys. Rev. B* **77**, 115215 (2008).

59. H.Q. Chiang, J.F. Wager, R.L. Hoffman, J. Jeong, and D.A. Keszler, High mobility transparent thin-film transistors with amorphous zinc tin oxide channel layer, *Appl. Phys. Lett.* **86**, 013503 (2005).

60. N.L. Dehuff, E.S. Kettenring, D. Hong, H.Q. Chiang, and J.F. Wager, R.L. Hoffman, C.-H. Park, and D.A. Keszler, Transparent thin-film transistors with zinc indium oxide channel layer, *J. Appl. Phys.* **97**, 064505 (2005).

61. H. Yabuta, M. Sano, K. Abe, T. Aiba, T. Den, H. Kumomi, K. Nomura, T. Kamiya, and H. Hosono, High-mobility thin-film transistor with amorphous InGaZnO$_4$ channel fabricated by room temperature RF-magnetron sputtering, *Appl. Phys. Lett.* **89**, 112123 (2006).

62. T. Iwasaki, N. Itagaki, T. Den, H. Kumomi, K. Nomura, T. Kamiya, and H. Hosono, Combinatorial approach to thin-film transistors using multicomponent semiconductor channels: An application to amorphous oxide semiconductors in In-Ga-Zn-O system, *Appl. Phys. Lett.* **90**, 242114 (2007).

63. M.G. McDowell, R.J. Sanderson, and I.G. Hill, Combinatorial study of zinc tin oxide thin-film transistors, *Appl. Phys. Lett.* **92**, 013502 (2008).

64. N. Itagaki, T. Iwasaki, H. Kumomi, T. Den, K. Nomura, T. Kamiya, and H. Hosono, Zn-In-O based thin-film transistors: Compositional dependence, *Phys. Status. Solidi. A* **205**, 1915 (2008).

65. B. Yaglioglu, H.Y. Yeom, R. Beresford, and D.C. Paine, High-mobility amorphous In$_2$O$_3$–10wt%ZnO thin film transistors, *Appl. Phys. Lett.* **89**, 062103 (2006).

66. E. Presley, D. Hong, H.Q. Chiang, C.M. Hung, R.L. Hoffman, and J.F. Wager, Transparent ring oscillator based on indium gallium oxide thin-film transistors, *Solid State Electron.* **50**, 500 (2006).

67. E.M.C. Fortunato, L.M.N. Pereira, P.M.C. Barquinha, A.M.B.D. Rego, G. Gonçalves, A. Vilà, J.R. Morante, and R.F.P. Martins, High mobility indium free amorphous oxide thin film transistors, *Appl. Phys. Lett.* **92**, 222103 (2008).

68. Y. Ogo, K. Nomura, H. Yanagi, T. Kamiya, M. Hirano, and H. Hosono, Amorphous Sn-Ga-Zn-O channel thin-film transistors, *Phys. Status. Solidi. A* **205**, 1920 (2008).

69. D.H. Cho, S. Yang, C. Byun, J. Shin, M.K. Ryu, S.-H.K. Park et al., Transparent Al–Zn–Sn–O thin film transistors prepared at low temperature, *Appl. Phys. Lett.* **93**, 142111 (2008).

70. J.K. Jeong, S. Yang, D.-H. Cho, S.-H.K. Park, C.-S. Hwang, and K.I. Cho, Impact of device configuration on the temperature instability of Al-Zn-Sn-O thin film transistors, *Appl. Phys. Lett.* **95**, 123505 (2009).

71. J.-S. Park, K. Kim, Y.-G. Park, Y.-G. Mo, H.D. Kim, and J.K. Jeong, Novel ZrInZnO thin-film transistor with excellent stability, *Adv. Mater.* **21**, 329 (2008).

72. M.K. Ryu, S. Yang, S.-H.K. Park, C.-S. Hwang, and J.K. Jeong, High performance thin film transistor with cosputtered amorphous Zn–In–Sn–O channel: Combinatorial approach, *Appl. Phys. Lett.* **95**, 072104 (2009).

73. M.K. Ryu, S. Yang, S.-H.K. Park, C.-S. Hwang, and J.K. Jeong, Impact of Sn/Zn ratio on the gate bias and temperature-induced instability of Zn-In-Sn-O thin film transistors, *Appl. Phys. Lett.* **95**, 173508 (2009).

74. G.-S. Heo, Y. Matsumoto, I.-G. Gim, J.-W. Park, G.-Y. Kim, and T.-W. Kim, Deposition of amorphous zinc indium tin oxide and indium tin oxide films on flexible poly(ether sulfone) substrate using RF magnetron co-sputtering system, *Jpn. J. Appl. Phys.* **49**, 035801 (2010).

75. S.-Y. Han, D.-H. Lee, G.S. Herman, and C.-H. Chang, Inkjet-printed high mobility transparent—Oxide semiconductors, *IEEE J. Disp. Technol.* **5**, 520 (2009).

76. W.-S. Cheong, J.-Y. Bak, and H.S. Kim, Transparent flexible zinc–indium–tin oxide thin-film transistors fabricated on polyarylate films, *Jpn. J. Appl. Phys.* **49**, 05EB10 (2010).

77. J. Liu, D.B. Buchholz, R.P.H. Chang, A. Facchetti, and T.J. Marks, High-performance flexible transparent thin-film transistors using a hybrid gate dielectric and an amorphous zinc indium tin oxide channel, *Adv. Mater.* **22**, 2333 (2010).

78. C.-J. Kim, S. Kim, J.-H. Lee, J.-S. Park, S. Kim, J. Park et al., Amorphous hafnium-indium-zinc oxide semiconductor thin film transistors, *Appl. Phys. Lett.* **95**, 252103 (2009).

79. W.H. Jeong, G.H. Kim, H.S. Shin, B.D. Ahn, H.J. Kim, M.-K. Ryu, K.-B. Park, J.-B. Seon, and S.Y. Lee, Investigating addition effect of hafnium in InZnO thin film transistors using a solution process, *Appl. Phys. Lett.* **96**, 093503 (2010).

80. E. Chong, K.C. Jo, and S.Y. Lee, High stability of amorphous hafnium-indium-zinc-oxide thin film transistor, *Appl. Phys. Lett.* **96**, 152102 (2010).

81. J.S. Park, T.S. Kim, K.S. Son, K.-H. Lee, J.S. Jung, W.-J. Maeng et al., High-performance and stable transparent Hf–In–Zn–O thin-film transistors with a double-etch-stopper layer, *IEEE Electron. Device Lett.* **31**, 1248 (2010).

82. D.-H. Son, D.-H. Kim, J.-H. Kim, S.-J. Sung, E.-A. Jung, and J.-K. Kang, Low voltage, high performance thin film transistor with HfInZnO channel and HfO2 gate dielectric, *Electrochem. Solid-State Lett.* **13**, H274 (2010).

83. J.S. Park, T.S. Kim, K.S. Son, K.-H. Lee, W.-J. Maeng, H.-S. Kim et al., The influence of SiO$_x$ and SiN$_x$ passivation on the negative bias stability of Hf–In–Zn–O thin film transistors under illumination, *Appl. Phys. Lett.* **96**, 262109 (2010).

84. H.-S. Kim, K.-B. Park, K.S. Son, J.S. Park, W.-J. Maeng, T.S. Kim et al., The influence of sputtering power and O$_2$/Ar flow ratio on the performance and stability of Hf–In–Zn–O thin film transistors under illumination, *Appl. Phys. Lett.* **97**, 102103 (2010).

85. H.S. Shin, G.H. Kim, W.H. Jeong, B.D. Ahn, and H.J. Kim, Electrical properties of yttrium–indium–zinc-oxide thin film transistors fabricated using the sol–gel process and various yttrium compositions, *Jpn. J. Appl. Phys.* **49**, 03CB01 (2010).

86. H. Koide, Y. Nagao, K. Koumoto, Y. Takasaki, T. Umemura, T. Kato, Y. Ikuhara, and H. Ohta, Electric field modulation of thermopower for transparent amorphous oxide thin film transistors, *Appl. Phys. Lett.* **97**, 182105 (2010).

87. G.H. Kim, W.H. Jeong, B.D. Ahn, H.S. Shin, H.J. Kim, H.J. Kim, M.-K. Ryu, K.-B. Park, J.-B. Seon, and S.-Y. Lee, Investigation of the effects of Mg incorporation into InZnO for high-performance and high-stability solution-processed thin film transistors, *Appl. Phys. Lett.* **96**, 163506 (2010).

88. J.H. Jeon, Y.H. Hwang, B.S. Bae, H.L. Kwon, and H.J. Kang, Addition of aluminum to solution processed conductive indium tin oxide thin film for an oxide thin film transistor, *Appl. Phys. Lett.* **96**, 212109 (2010).

89. D.-H. Cho, S. Yang, C. Byun, M.K. Ryu, S.-H.K. Park, C.-S. Hwang, S.M. Yoon, and H.-Y. Chu, Transparent oxide thin-film transistors composed of Al and Sn-doped zinc indium oxide, *IEEE Electron Device Lett.* **30**, 48 (2009).

90. S. Yang, D.-H. Cho, M.K. Ryu, S.-H.K. Park, C.-S. Hwang, J. Jang, and J.K. Jeong, High-performance Al–Sn–Zn–In–O thin-film transistors: Impact of passivation layer on device stability, *IEEE Electron Dev. Lett.* **31**, 144 (2010).

91. S. Yang, D.-H. Cho, M.K. Ryu, S.-H.K. Park, C.-S. Hwang, J. Jang, and J.K. Jeong, Improvement in the photon-induced bias stability of Al–Sn–Zn–In–O thin film transistors by adopting AlOx passivation layer, *Appl. Phys. Lett.* **96**, 213511 (2010).

92. K. Takechi, M. Nakata, S. Yamaguchi, H. Tanabe, and S. Kaneko, Amorphous In–Sn–Si–O thin-film transistors having various Si compositional ratios, *Jpn. J. Appl. Phys.* **49**, 028002 (2010).

93. E. Chong, S.H. Kim, and S.Y. Lee, Role of silicon in silicon-indium-zinc-oxide thin-film transistor, *Appl. Phys. Lett.* **97**, 252112 (2010).

94. L. Shao, K. Nomura, T. Kamiya, and H. Hosono, Operation characteristics of thin-film transistors using very thin amorphous In-Ga-Zn-O channels, *Electrochem. Solid State Lett.* **14**, H197–H200 (2011).

95. K. Nomura, T. Kamiya, H. Ohta, T. Uruga, M. Hirano, and H. Hosono, Local coordination structure and electronic structure of the large electron mobility amorphous oxide semiconductor In-Ga-Zn-O: Experiment and ab initio calculations, *Phys. Rev. B* **75**, 035212-1–035212-5 (2007).

96. F. Utsuno, H. Inoue, I. Yasui, Y. Shimane, S. Tomai, S. Matsuzaki, K. Inoue, I. Hirosawa, M. Sato, and T. Honma, Structural study of amorphous In2O3 film by grazing incidence X-ray scattering (GIXS) with synchrotron radiation, *Thin Solid Films* **496**, 95–98 (2006).

97. F. Utsuno, H. Inoue, Y. Shimane, T. Shibuya, K. Yano, K. Inoue, I. Hirosawa, M. Sato, and T. Honma, "A structural study of amorphous In2O3–ZnO films by grazing incidence X-ray scattering (GIXS) with synchrotron radiation", *Thin Solid Films* **516**, 5818–5821 (2008).

98. H. Omura, H. Kumomi, K. Nomura, T. Kamiya, M. Hirano, and H. Hosono, First-principles study of native point defects in crystalline indium gallium zinc oxide, *J. Appl. Phys.* **105**, 093712-1–093712-8 (2009).

99. D.-Y. Cho, J. Song, K.D. Na, C.S. Hwang, J.H. Jeong, J.K. Jeong, and Y.-G. Mo, Local structure and conduction mechanism in amorphous In-Ga-Zn-O films, *Appl. Phys. Lett.* **94**, 112112 (2009).

100. M. Orita, H. Tanji, M. Mizuno, H. Adachi, and I. Tanaka, Mechanism of electrical conductivity of transparent InGaZnO4, *Phys. Rev. B* **61**, 1811 (2000).

101. W.-J. Lee, E.-A. Choi, J. Bang, B. Ryu, and K.J. Chang, Structural and electronic properties of crystalline InGaO3(ZnO)m, *Appl. Phys. Lett.* **93**, 111901 (2008).

102. M. Takahashi, H. Kishida, A. Miyanaga, and S. Yamazaki, Theoretical analysis of IGZO transparent amorphous oxide semiconductor, *Proceedings of the 16th International Display Workshop*, AMD8-2 (2008).

103. A. Walsh, J.L.F. Da Silva, and S.-H. Wei, Interplay between order and disorder in the high performance of amorphous transparent conducting oxides, *Chem. Mater.* **21**, 5119 (2009).

104. J.E. Medvedeva and C.L. Hettiarachchi, Tuning the properties of complex transparent conducting oxides: Role of crystal symmetry, chemical composition, and carrier generation, *Phys. Rev. B* **81**, 125116 (2010).

105. T. Kamiya, K. Nomura, M. Hirano, and H. Hosono, Electronic structure of oxygen deficient amorphous oxide semiconductor a-InGaZnO4-x: Optical analysis and first-principle calculations, *Phys. Stat. Solidi (c)* **5**, 3098 (2008).

106. T. Kamiya, K. Nomura, and H. Hosono, Electronic structure of the amorphous oxide semiconductor a-InGaZnO4–x: Tauc–Lorentz optical model and origins of subgap states, *Phys. Stat. Solidi A* **206**, 860–867 (2009).

107. H. Omura, T. Iwasaki, H. Kumomi, K. Nomura, T. Kamiya, M. Hirano, and H. Hosono, First-principles calculation for effect of impurities on electronic states of amorphous In-Ga-Zn-O, *Mater. Res. Soc. Symp. Proc.* **1109**, 120–125, 1109-B04-02 (2009).

108. J. Singh and K. Shimakawa, *Advances in Amorphous Semiconductors*, Taylor & Francis, London, U.K., 2003.

109. S. Kivelson and C.D. Gelatt, Jr., Effective-mass theory in noncrystalline solids, *Phys. Rev. B* **19**, 5160 (1979).

110. K. Nomura, T. Kamiya, E. Ikenaga, H. Yanagi, K. Kobayashi, and H. Hosono, Depth analysis of subgap electronic states in amorphous oxide semiconductor, a-In-Ga-Zn-O, studied by hard x-ray photoelectron spectroscopy, *J. Appl. Phys.* **109**, 073726 (2011).

111. J.F. Janak, Proof that dE/dni = ei in density-functional theory, *Phys. Rev. B* **18**, 7165 (1978).

112. K. Nomura, T. Kamiya, H. Yanagi, E. Ikenaga, K. Yang, K. Kobayashi, M. Hirano, and H. Hosono, Subgap states in transparent amorphous oxide semiconductor, In–Ga–Zn–O, observed by bulk sensitive x-ray photoelectron spectroscopy, *Appl. Phys. Lett.* **92**, 202117 (2008).

113. T. Ohsawa, I. Sakaguchi, N. Ohashi, H. Haneda, H. Ryoken, K. Matsumoto, S. Hishita, Y. Adachi, S. Ueda, H. Yoshikawa, and K. Kobayashi, Formation of compensated defects in zinc magnesium oxides assignable from diffusion coefficients and hard x-ray photoemission, *Appl. Phys. Lett.* **94**, 042104 (2009).

114. B. Li, Y. Adachi, J. Li, H. Okushi, I. Sakaguchi, S. Ueda, H. Yoshikawa, Y. Yamashita, S. Senju, K. Kobayashi, M. Sumiya, H. Haneda, and N. Ohashi, Defects in ZnO transparent conductors studied by capacitance transients at ZnO/Si interface, *Appl. Phys. Lett.* **98**, 082101 (2011).

115. T. Shibuya, M. Yoshinaka, Y. Shimane, F. Utsuno, K. Yano, K. Inoue, E. Ikenaga, J.J. Kim, S. Ueda, M. Obata, and K. Kobayashi, Electronic structural analysis of transparent In2O3–ZnO films by hard X-ray photoelectron spectroscopy, *Thin Solid Films* **518**, 3008–3011 (2010).

116. Kyeongmi Lee, Kenji Nomura, Hiroshi Yanagi, Toshio Kamiya, and Hideo Hosono, "Photovoltaic properties of n-type amorphous In-Ga-Zn-O and p-type single crystal Si heterojunction solar cells: Effects of Ga content, submitted to *Thin Solid Films* **520**, 3808 (2011).

117. K. Lee, K. Nomura, H. Yanagi, T. Kamiya, and Hideo Hosono, Electronic structure and photovoltaic properties of n-type amorphous In-Ga-Zn-O and p-type single crystal Si heterojunctions, *Electrochem. Solid State Lett.* **14**, H346–H349 (2011).

118. J. Tauc, Amorphous and liquid semiconductors, Plenum, New York (1979).

119. G.E. Jellison, Jr. and F.A. Modine, Parameterization of the optical functions of amorphous materials in the interband region, *Appl. Phys. Lett.* **69**, 371–374 (1996); Erratum, ibid, p. 2137.

120. D. Kang, I. Song, C. Kim, Y. Park, T.D. Kang, H.S. Lee, J.-W. Park, S.H. Baek, S.-H. Choi, and H. Lee, Effect of Ga/ In ratio on the optical and electrical properties of GaInZnO thin lms grown on SiO₂/Si substrates, *Appl. Phys. Lett.* **91**, 091910 (2007).

121. K. Nomura, H. Ohta, K. Ueda, T. Kamiya, M. Hirano, H. Hosono, Carrier transport in transparent oxide semiconductor with intrinsic structural randomness probed using single-crystalline InGaO3(ZnO)5 films, *Appl. Phys. Lett.* **85**, 1993–1995 (2004).

122. M. Kimura, T. Kamiya, T. Nakanishi, K. Nomura, and H. Hosono, Intrinsic carrier mobility in amorphous InGaZnO thin-film transistors determined by combined field-effect technique, *Appl. Phys. Lett.* **96**, 262105 (2010).

123. H.-H. Hsieh, T. Kamiya, K. Nomura, H. Hosono, and C.-C Wu, Modeling of amorphous InGaZnO4 thin film transistors and their subgap density of states, *Appl. Phys. Lett.* **92**, 133503-1–1335.3-3 (2008).

124. M. Kimura, T. Nakanishi, K. Nomura, T. Kamiya, and H. Hosono, Trap densities in amorphousInGaZnO4 thin-film transistors, *Appl. Phys. Lett.* **92**, 133512-1–133512-3 (2008).

125. W.E. Spear and P.G. LeComber, Doping of hydrogenated amorphous silicon, *Solid State Commun.* **17**, 1193 (1975).

126. F. Oba, S.R. Nishitani, S. Isotani, H. Adachi, and I. Tanaka, Energetics of native defects in ZnO, *J. Appl. Phys.* **90**, 824–828 (2001).

127. S.J. Clark, J. Robertson, S. Lany, and A. Zunger, Intrinsic defects in ZnO calculated by screened exchange and hybrid density functionals, *Phys. Rev. B* **81**, 115311 (2010).

128. S.B. Zhang, S.-H. Wei, and A. Zunger, Intrinsic n-type versus p-type doping asymmetry and the defect physics of ZnO, *Phys. Rev. B* **63**, 075205 (2001).

129. H. Hosono, N. Kikuchi, N. Ueda, H. Kawazoe, K.-I Shimidzu, Amorphous transparent electroconductor 2CdO•GeO2: Conversion of amorphous insulating cadmium germanate by ion implantation, *Appl. Phys. Lett.* **67**, 2663–2665 (1995).

130. H. Shirakawa, E.J. Louis, A.G. MacDiamid, C.K. Chang, and A.J. Heeger, Synthesis of electrically conducting organic polymers: Halogen derivatives of polyacetylene, (CH), *J. Chem. Soc. Chem. Commun.* 578–580 (1977).

131. J.K. Jeong, Y.-G. Mo, M.K. Ryu, and S. Yang, Study on the stability degradation mechanism of oxide thin film transistors under the bias-thermal stress condition, *International Workshop on Transparent Amorphous Oxide Semicond 2010*, Tokyo, Japan, January 25–26, 2010.

132. J.Y. Kwon, J.S. Jung, K.S. Son, T.S. Kim, M.K. Ryu, K.B. Park, Y.S. Park, S.Y. Lee, and J.M. Kim, GaInZnO TFT for active matrix display, *Digest of the 15th International Workshop on Active-Matrix Flatpanel Displays and Devices*, p. 287, 2008.

133. Y. Matsueda, Required characteristics of TFTs for next generation flat panel display backplanes, *Digest of Int. Transistor Conf.'* **10**, 314 (2010).

134. K. Nomura, T. Kamiya, and H. Hosono, Interface and bulk effects for bias-light-illumination instability in amorphous-In-Ga-Zn-O thin-film transistors, *J. Soc. Inf. Display* **18**, 789–795 (2010).

135. R. Hayashi, M. Ofuji, N. Kaji, K. Takahashi, K. Abe, H. Yabuta, M. Sano, H. Kumomi, K. Nomura, T. Kamiya, M. Hirano, and H. Hosono, Circuits using uniform TFTs based on amorphous In–Ga–Zn–O, *J. SID* **15/11**, 915–921 (2007).

136. K. Nomura, T. Kamiya, H. Ohta, M. Hirano, and H. Hosono, Defect passivation and homogenization of amorphous oxide thin-film transistor by wet O2 annealing, *Appl. Phys. Lett.* **93**, 192107-1–192107-3 (2008).

137. K. Nomura, T. Kamiya, M. Hirano, and H. Hosono, Origins of threshold voltage shifts in room-temperature deposited and annealed a-In–Ga–Zn–O thin-film transistors, *Appl. Phys. Lett.* **95**, 013502-1–013502-3 (2009).

138. K. Nomura, T. Kamiya, Y. Kikuchi, M. Hirano, and H. Hosono, Comprehensive studies on the stabilities of a-In-Ga-Zn-O based thin film transistor by constant current stress, *Thin Solid Films* **518**, 3012–3016 (2010).

139. E.D. Tober, J. Kanicki, and M.S. Crowder, *Appl. Phys. Lett.* **59** 1723, 1991; J.-M. Lee, I.-T. Cho, Lee J.-H. and Kwon H.-I., *Appl. Phys. Lett.* **93** 093504, 121 (2008).

140. J.K. Jeong, H.W. Yang, J.H. Jeong, Y.-G. Mo, and H.D. Kim, Origin of threshold voltage instability in indium-gallium-zinc oxide thin film transistors, *Appl. Phys. Lett.* **93**, 123508 (2008).

141. S. Yang, D.-H. Cho, M.K. Ryu, S.-H. K. Park, C.-S. Hwang, J. Jang, and J.K. Jeong, Improvement in the photon-induced bias stability of AlSnZnInO thin lm transistors by adopting AlOx passivation layer, *Appl. Phys. Lett.* **96**, 213511 (2010).

142. K. Nomura, T. Kamiya, and H. Hosono, Highly stable amorphous In-Ga-Zn-O thin-film transistor realized by active passivation, *Appl. Phys. Lett.* **99**, 053505 (2011).

143. P. Görrn, M. Lehnhardt, T. Riedl, and W. Kowalsky, The influence of visible light on transparent zinc tin oxide thin film transistors, *Appl. Phys. Lett.* **91**, 193504 (2007).

144. T.-C. Fung, C.-S Chuang, K. Nomura, H.-P.D. Shieh, H. Hosono, and J. Kanicki, Photofield-effect in amorphous In-Ga-Zn-O (a-IGZO) thin-film transistors, *J. Inf. Display* **9**, 21–29 (2009).

145. K. Takechi, M. Nakata, T. Eguchi, H. Yamaguchi, and S. Kaneko, Comparison of ultraviolet photo-field effects between hydrogenated amorphous silicon and amorphous InGaZnO4 thin-film transistors, *Jpn. J. Appl. Phys.* **48**, 010203 (2010).

146. K. Nomura, T. Kamiya, and H. Hosono, Stability and high-frequency operation of amorphous In-Ga-Zn-O thin-film transistors with various passivation layers, *Thin Solid Films* **520**, 3778 (2011).

147. D. H. Lee, K.-I. Kawamura, K. Nomura, T. Kamiya, and H. Hosono, Large photoresponse in amorphous InGaZnO and origin of reversible and slow decay, *Electrochem. Solid State Lett.* **13**, H324–H327 (2010).

148. S. Yasuno, T. Kugimiya, S. Morita, A. Miki, F. Ojima, and S. Sumie, Correlation of photoconductivity response of amorphous In–Ga–Zn–O films with transistor performance using microwave photoconductivity decay method, *Appl. Phys. Lett.* **98**, 102107 (2011).

149. M. Ito, C. Miyazaki, N. Ikeda, and Y. Kokubo, Transparent amorphous oxide TFT and its application to electronic paper, *The Sixteenth International Workshop Active-Matrix Flatpanel Displays and Devices S-2*, Nara, Japan (1-3 July, 2009).

150. J.-S. Park, T.-W. Kim, D. Stryakhilev, J.-S. Lee, S.-G. An, Y.-S. Pyo, D.-B. Lee, Y.G. Mo, D.-U. Jin, and H. K. Chung, Flexible full color organic light-emitting diode display on polyimide plastic substrate driven by amorphous indium gallium zinc oxide thin-film transistors, *Appl. Phys. Lett.* **95**, 013503 (2009); J.W. Seo, J.-W. Park, K. S. Lim, S.J. Kang, Y.H. Hong, J.H. Yang, L. Fang, G.Y. Sung, and H.-K. Kim, Transparent flexible resistive random access memory fabricated at room temperature, *Appl. Phys. Lett.* **95**, 133508 (2009).

151. H.D. Kim, J.-S. Park, Y.G. Mo, and S.S. Kim, Oxide thin film transistors for use as next generation active matrix backplanes, *Digest of International Meeting on Information Display 2009*, Seoul, Korea, p. 35, October 12–16, 2009.

152. M. Ito, M. Kon, T. Okubo, M. Ishizaki, and N. Sekine, A flexible active-matrix TFT array with amorphous oxide semiconductors for electronic paper, *Proc. Int. Display Workshop'05*, p. 845, 2005.

153. H.N. Lee, J.W. Kyung, S.K. Kang, D.Y. Kim, M.C. Sung, S.J. Kim, C.N. Kim, H.G. Kim and S.T. Kim, Current status of, challenges to, and perspective view of AM-OLED, *Proc. Int. Display Workshop 2006*, p. 663, 2006.

154. J. K. Jeong, M. Kim, J. H. Jeong, H. J. Lee, T. K. Ahn, H. S. Shin, K. Y. Kang, H. Seo, J. S. Park, H. Yang, H. J. Chung, Y. G. Mo, and H. D. Kim, "4.1" Transparent QCIF AMOLED display driven by high mobility bottom gate a-IGZO thin-film transistors, *7th Int. Meet. Inf. Displ.*, p. 9-4, 27–31, 2007, Daegue, Korea.

155. J.Y. Kwon, K.S. Son, J.S. Jung, T.S. Kim, M.K. Ryu, K.B. Park, J.W. Kim, Y.G. Lee, C.J. Kim, S.I. Kim, Y.S. Park, S.Y. Lee, and J.M. Kim, 4 inch QVGA AMOLED display driven by GaInZnO TFT, *7th International Meeting of Information Display*, P. 9-3, Daegue, Korea, 27–31, 2007.

156. M. Ito, M. Kon, C. Miyazaki, N. Ikeda, M. Ishizaki, Y. Ugajin, and N. Sekine, "Front Drive" display structure for color electronic paper using fully transparent amorphous oxide TFT array, *IEICE Trans. Electron* **E90-C**, 2105–2111 (2007).

157. M-C. Sung, H-N. Lee, C.N. Kim, S.K. Kang, D.Y. Kim, S-J. Kim, S.K. Kim, S-K. Kim, H-G. Kim, and S-t. Kim, *7th International Meeting of Information Display*, Daegue, Korea, p. 9-1, August 27–31, 2007.

158. Semi-transparent OLED panel, the first mass production by TDK, Nihon Keizai Shimbun, May 31, 2011.

159. J. Donelan, Samsung electronics begins mass production of transparent LCD panel, *Inf. Display* **27**, 3 (April 2011).

160. K. Jeong, M. Kim, J.H. Jeong, H.J. Lee, T.K. Ahn, H.S. Shin, K.Y. Kang, H. Seo, J.S. Park, H. Yang, H.J. Chung, Y.G. Mo, and H.D. Kim, "4.1" transparent QCIF AMOLED display driven by high mobility bottom gate a-IGZO thin-film transistors, *7th International Meeting of Information Display*, p. 9-4, Daegue, Korea, 27–31, 2007.

161. S.-H.K. Park, C.-S. Hwang, J.-I. Lee, S.M. Chung, Y.S. Yang, L.-M. Do, and H.Y. Chu, Transparent ZnO thin film transistor array for the application of transparent AM-OLED display, *Digest of SID2006*, San Francisco, CA, p. 25, June 4–9, 2006.

162. S.-H.K. Park, C.-S. Hwang, M. Ryu, S. Yang, C. Byun, J. Shin, J.-I. Lee, K. Lee, M.S. Oh, and S. Im, Transparent and photo-stable ZnO thin-film transistors to drive an active matrix organic-light-emitting-diode display panel, *Adv. Mater.* **21**, 678 (2009).

163. H.-H. Hsieh, T.-T. Tsai, C.-Y. Chang, H.-H. Wang, J.-Y. Huang, S.-F. Hsu, Y.-C. Wu, T.-C. Tsai, C.-S. Chuang, L.-H. Chang, and Y.-H. Lin, 2.4-in. AMOLED with IGZO TFTs and inverted OLED devices, *Digest of SID2010*, Seattle, WA, p. 140, May 23–28, 2010.

164. S. I. Kim, C.J. Kim, J. C. Park, I. Song, S.W. Kim, H. Yin, E. Lee, J.C. Lee, and Y. Park, High performance oxide thin film transistors with double active layers, *International Electron Devices Meeting*, San Francisco, CA, December15–17, 2008. doi: 10.1109/IEDM.2008.4796617.

165. M. Ofuji, K. Abe, H. Shimizu, N. Kaji, R. Hayashi, M. Sano, H. Kumomi, K. Nomura, T. Kamiya, and H. Hosono, Fast thin-film transistor circuits based on amorphous oxide semiconductor, *IEEE Electron. Dev. Lett.* **28**, 273–275 (2007).

166. H.-H. Hsieh, C.-H. Wu, C.-C. Wu, Y.-H. Yeh, H.-L. Tyan, and C.-M. Leu, Amorphous In_2O_3-Ga_2O_3-ZnO thin film transistors and integrated circuits on flexible and colorless polyimide substrates, *Proceedings of SID'08*, Los Angeles, CA, p. 11, 2008.

167. I. Song, S. Kim, H. Yin, C.J. Kim, J. Park, S. Kim, H.S. Choi, E. Lee, Y. Park, Short channel characteristics of gallium–indium–zinc–oxide thin film transistors for three-dimensional stacking memory, *IEEE Electron. Dev. Lett.* **29**, 549–552 (2008).

168. Y.-L. Wang, L.N. Covert, T.J. Anderson, W. Lim, J. Lin, S.J. Pearton, D.P. Norton, J.M. Zavada, and F. Renc, RF characteristics of room-temperature-deposited, small gate dimension indium zinc oxide TFTs, *Electrochem. Solid State Lett.* **11**, H60–H62 (2008).

169. J.-H. Lee, D.-H. Kim, D.-J. Yang, S.-Y. Hong, K.-S. Yoon, P.-S. Hong, C.-O. Jeong, H-S. Park, S.Y. Kim, S.K. Lim, S.S. Kim, World's largest (15-inch) XGA AMLCD panel using IGZO oxide TFT, *Digest of SID2008*, Los Angeles, CA, p. 625, May 18–23, 2010.

170. T. Osada, K. Akimoto, T. Sato, M. Ikeda, M. Tsubuku, J. Sakata, J. Koyama, T. Serikawa, and S. Yamazaki, Development of driver-integrated panel using amorphous In-Ga-Zn-Oxide TFT, *Digest of 16th International Workshop on Active-Matrix Flatpanel Displays and Devices,* Nara, Japan, p. 3-3, July 1–3, 2009.

171. T. Osada, K. Akimoto, T. Sato, M. Ikeda, M. Tsubuku, J. Sakata, J. Koyama, T. Serikawa, and S. Yamazaki, Development of driver-integrated panel using amorphous In-Ga-Zn-Oxide TFT, *Digest of SID2009*, San Antonio, TX, p. 184, May 31–June 5, 2009.

172. T. Kawamura, H. Uchiyama, S. Saito, H. Wakana, T. Mine, M. Hatano, K. Torii, and T. Onai, 1.5-V Operating fully-depleted amorphous oxide thin film transistors achieved by 63-mV/dec subthreshold slope, *Digest of International Electron Devices Meeting*, San Francisco, CA, pp. 1–4, December 15–17, 2008.

173. H. Ozaki, T. Kawamura, H. Wakana, T. Yamazoe, and H. Uchiyama, Wireless operations for 13.56-MHz band RFID tag using amorphous oxide TFTs, *IEICE Electron Exp.* **8**, 225–231 (2011).

174. H. Yin, S. Kim, C.J. Kim, I. Song, J. Park, S. Kim, and Y. Park, Fully transparent nonvolatile memory employing amorphous oxides as charge trap and transistor's channel layer, *Appl. Phys. Lett.* **93**, 172109 (2008).

175. S.-H. Rha, J.S. Junga, J.H. Kima, U.K. Kima, Y.J. Chunga, H.-S Junga, S.Y. Leeb, and C.S. Hwanga, Amorphous oxide semiconductor memory using high-k charge trap layer, *ECS Trans.* **33**, 375 (2010).

176. H. Yin, S. Kim, H. Lim, Y. Min, C.J. Kim, I. Song, J. Park, S.-W. Kim, A. Tikhonovsky, J. Hyun, and Y. Park, Program/erase characteristics of amorphous gallium indium zinc oxide nonvola- tile memory, *IEEE Trans. Electron Dev.* **55**, 2071 (2008).

177. M.-J. Lee, S.I. Kim, C.B. Lee, H. Yin, S.-E. Ahn, B.S. Kang, K.H. Kim, J.C. Park, C.J. Kim, I. Song, S.W. Kim, G. Stefanovich, J.H. Lee, S.J. Chung, Y.H. Kim, and Y. Park, Low-temperature-grown transition metal oxide based storage materials and oxide transistors for high-density non-vola- tile memory, *Adv. Mater.* **19**, 1587–1593 (2009).

178. M.-C. Chen, T.-C. Chang, S.-Y. Huang, S.-C. Chen, C.-W. Hu, C.-T. Tsai, and S.M. Sze, Bipolar resistive switching characteristics of transparent indium gallium zinc oxide resistive random access memory, *Electrochem. Solid State Lett.* **13**, H191–H193 (2010).

179. S.-M. Yoon, S.-H. Yang, S.-W. Jung, C.-W. Byun, S.-H.K. Park, C.-S. Hwang, G.-G. Lee, E. Tokumitsu, and H. Ishiwara, Impact of interface controlling layer of Al2O3 for improving the retention behaviors of In–Ga–Zn oxide-based ferroelectric memory transistor, *Appl. Phys. Lett.* **96**, 232903 (2010).

180. T. Osada, K. Akimoto, T. Sato, M. Ikeda, M. Tsubuku, J. Sakata, J. Koyama, T. Serikawa, and S. Yamazaki, Development of driver-integrated panel using amorphous In-Ga-Zn-Oxide TFT, *Digest of 16th International Workshop Active-Matrix Flatpanel Displays and Devices 2009*, Nara, Japan, July 1–3, 2009.

181. M.-C. Hung, W.-T. Lin, J.J. Chang, P.L. Chen, C.-Y. Wu, C.-J. Lin, H.-L. Chiu, C.-Y. Huang, and Y.-C. Kao, Process development and reliability study of a-IGZO thin film transistor, *International Workshop on Transparent Amorphous Oxide Semiconductors 2010*, Tokyo, Japan, January 25–26, 2010.

182. T. Kurata, Y. Yanagi, T. Isobe, Y. Akamatsu, M. Arai, J. Kiyota, S. Ishibashi, and K. Saito, Characteristics of IGZO TFT fabricated on large-area substrates, *57th Spring Meeting of Japan Society of Applied Physics 17a-TL5*, Kanagawa, Japan, March 17–20, 2010.

183. D. Kobayashi, Y. Takei, T. Isobe, Y. Kurata, M. Arai, J. Kiyota, A. Ishibashi, T. Ohno, Y. Sato, S. Sato, J. Nittaand and K. Takahashi., Growth technique of InGaZnO TFT on large-size sub- strates, *ULVAC Tech. J.* **73**, 8 (2010) [in Japanese].

184. Y. Ukai, Thin-film transistor (in Japanese, Hakumaku Transistor Gijutsu no Subete) (2007) [in Japanese].

185. M.-C. Hung, W.-T. Lin, C.-H. Tu, Y.-C. Kao, C.-Y. Wu, J.-J. Chang, and P.-L. Chen, Employ present 5 masks a-Si TFT design and process flow to realize 5 inch VGA oxide TFT-LCD panel, *Digest of AM-FPD2010*, 3-4 (2010).

186. Y.G. Mo, M. Kim, C.K. Kang, J.H. Jeong, Y.S. Park, C.G. Choi, H.D. Kim, and S.S. Kim, Amorphous oxide TFT backplane for large size AMOLED TVs, Digest of SID2010, 1037 (2010).

187. D.-H. Lee, Y.-J. Chang, G.S. Herman, and C.-H. Chang, A general route to printable high-mobility transparent amorphous oxide semiconductors, *Adv. Mater.* **19**, 843–847 (2007).

188. K.K. Banger, Y. Yamashita, K. Mori, R.L. Peterson, T. Leedham, J. Rickard, and H. Sirringhaus, Low-temperature, high-performance solution-processed metal oxide thin-film transistors formed by a 'sol–gel on chip' process, *Nat. Mater.* **10**, 45 (2010).

189. Transparent oxide semiconductor TFT for large-size TV by solution process, Nikkan Kogyo Shimbun, March 25, 2011.

190. W. Shockley and G.I. Pearson, Modulation of conductance of thin films of semi-conductors by surface charges, *Phys. Rev.***74**, 232 (1948).

191. K. Matsuzaki, K. Nomura, H. Yanagi, T. Kamiya, M. Hirano, and H. Hosono, Epitaxial growth of high mobility Cu2O thin films and application to p-channel thin film transistor, *Appl. Phys. Lett.* **93**, 202107-1–202107-3 (2008).

192. K. Matsuzaki, K. Nomura, H. Yanagi, T. Kamiya, M. Hirano, and H. Hosono, Effects of post-annealing on (110) Cu2O epitaxial lms and origin of low mobility in Cu2O thin-film transistor, *Phys. Stat. Sol. A* **206**, 2192–2197 (2009).

193. E. Fortunato, V. Figueiredo, P. Barquinha, E. Elamurugu, R. Barros, G. Gonçalves, S.-H.K. Park, C.-S. Hwang, and R. Martins, Thin-film transistors based on p-type Cu2O thin films produced at room temperature, *Appl. Phys. Lett.* **96**, 192102 (2010); Erratum, ibid, p. 239902.

194. X. Zou, G. Fang, L. Yuan, M. Li, W. Guan, and X. Zhao, Top-gate low-threshold voltage p-Cu$_2$O thin-film transistor grown on SiO$_2$/Si substrate using a high-κ HfON gate dielectric, *IEEE Electron Dev. Lett.*, **31**, 827 (2010).

195. Y. Ogo, H. Hiramatsu, K. Nomura, H. Yanagi, T. Kamiya, M. Hirano, and H. Hosono, p-channel thin-film transistor using p-type oxide semiconductor, SnO, *Appl. Phys. Lett.* **93**,032113 (2008).

196. Y. Ogo, K. Nomura, H. Yanagi, T. Kamiya, M. Hirano, H. Hosono, Amorphous Sn-Ga-Zn-O channel thin-film transistors, *Phys. Stat. Sol. A* **205**, 1920–1924 (2008).

197. H. Yabuta, N. Kaji, R. Hayashi, H. Kumomi, K. Nomura, T. Kamiya, M. Hirano, and H. Hosono, Sputtering formation of p-type SnO thin-film transistors on glass toward oxide complimentary circuits, *Appl. Phys. Lett.* **97**, 072111 (2010).

198. L.Y. Liang, Z.M. Liu, H.T. Cao, Z. Yu, Y.Y. Shi, A.H. Chen, H.Z. Zhang, Y.Q. Fang, and X.L. Sun, Phase and optical characterizations of annealed SnO thin films and their p-Type TFT application, *J. Electrochem. Soc.* **157**, H598 (2010).

199. E. Fortunato, R. Barros, P. Barquinha, V. Figueiredo, S.-H.K. Park, C.-S. Hwang, and R. Martins, Transparent p-type SnOx thin film transistors produced by reactive rf magnetron sputtering followed by low temperature annealing, *Appl. Phys. Lett.* **97**, 052105 (2010).

200. H. Hosono, Y. Ogo, H. Yanagi, and T. Kamiya, Bipolar conduction in SnO thin films, *Electrochem. Solid State Lett.* **14**, H13–H16 (2011).

201. K. Nomura, T. Kamiya, and H. Hosono, Ambipolar oxide thin-film transistor, *Adv. Mater.* **23**, 3431, in print (2011).

Index

A

Active neutron detection system, 437
ALD, *see* Atomic layer deposition (ALD)
Al-doped ZnO transparent conducting
    layers, 51–52
$Al_xZn_{1-x}O$ thin films, 43–44
Amorphous oxide semiconductor (AOS)
    advantages of, 512–513
    atomic structure, 498–500
    carrier transport
        band alignment diagram, 489
        bandgap, 490
        band structures and crystal
           structures, 488–489
        oxide *vs.* conventional semiconductors,
           491, 492
        wave functions, 490
    doping, oxygen deficiency, and hydrogen,
        509–511
    electronic structures
        bandgap, 502
        electron affinity and ionization
           potential, 503
        hard X-ray photoemission spectrum,
           502–503
        pseudo-band structure, 500–501
        subgap density of states, 502
        total and projected density of states,
           501–502
    electron transport, 505–508
    fabrication of
        annealing effect, 514–515
        operation characteristics, 514
        stability, 516–519
        TFT structures and process, 513
        uniformity, 515–516
    flat-panel displays backplane
        e-paper display structure, 519–520
        flexible displays, 520
        history of, 518, 520
        mass production, 521–522
        system integration, 521
        transparent displays, 520–521
    growth, 498
    history of, 487–488
    material design with high mobility

        deep localized states, 492
        orbital drawings, carrier pathways, 493
        strained chemical bonds, 494
    optical properties, 504–505
    roles of constituent elements
        hall mobility *vs.* carrier density, 494, 496
        mobility enhancer, 495
        multi-component systems, 497–498
        structures and carrier transport
           properties, 494–495
        transport property characteristics, with
           vs. without Ga, 495–497
    subgap defects, 506–509
Atomic layer deposition (ALD)
    advantages, 239
    feature, 238
    quantum dots in SiO_2 nanoparticle layer
        cross-sectional TEM image, 251, 252
        fabrication, 250–251
        HRTEM image, 251, 252
        photoluminescence spectrum, 252, 253
        quantum confinement effects, 250, 252
    UV electroluminescence
        n-ZnO/*i*-ZnO/*p*-SiC heterojunction LEDs,
           261–266
        n-ZnO/*p*-Gan heterojunction LEDs,
           254–261
    ZnO epilayers on *c*-sapphire substrates
        cross-sectional TEM image, 239, 240
        crystalline structure, 239
        electron-hole plasma, 244
        HRTEM image, 240–242
        light emission characteristics, 242
        spontaneous emission
           photoluminescence spectrum,
           242, 243
        stimulated emission photoluminescence
           spectrum, 242, 243
        XRD pattern, 239, 240
    ZnO thin films
        amorphous glass substrates, 244–250
        stable *p*-type, 266–270

B

Bethe–Block stopping power formula, 446–447
[10]B-lined gas filled tubes, 437

C

Chemical sensor
    characteristics, 154–155
    dynamic responses, 158–159
    electron trapping and detrapping
        process, 155
    oxidizing gases, 155
    reductive gases, 156
    spill-over effect, 158
    transconductance, 156–157
Chemical vapor transport (CVT) crystal
        growth, 453
Coherent optical feedback, laser diodes,
        363–364
Conductive-glass/ZnO-nanowire/organic-
        semiconductor/metal structure,
        288–291
Crystal growth, ZnO Scintillators
    Cermet high pressure furnace, 453, 454
    CVT, 453
    groups, 452
    hydrothermal growth, 454
    melt growth, 454
    SCVT, 453

D

Dye-sensitized solar cells (DSSCs)
    performance, 160–161
    protecting and electron-blocking layer, 161
    schematic representation, 160
    surface recombination, 162

F

FET-logic MESFET inverter
    circuit design, 396
    origin, 395–396
    SDFL
        advantage, 397–398
        circuit design, 398
        origin, 397
        voltage transfer characteristics, 398–400
        ZnO *vs.* MgZnO, 398–400
    voltage-transfer characteristics, 396–397
Fission chamber neutron detectors, 438
Flexible light emitting diodes (LEDs), 295–297
Fundamental MESFET inverter
    circuit diagram, 392, 393
    with integrated level-shifting, 400–401
    peak gain magnitude, 393–395
    stability, 413–418

transparent
    figures of merit, 411
    light-induced relative change, 414–415
    temperature effects, 415–418
    thermal stability, 415
    voltage transfer characteristics, 413–416
voltage transfer curve, 393–395

G

GaN-ZnO heterojunction laser diodes, 343–345
Gas sensor, *see* Chemical sensor
Glass/PEDOT:PSS/polymer-semiconductor/
        ZnO-nanowire/metal structure,
        291–293

H

Heteroepitaxial nanohexagons, p-type SiC
        substrates
    cathodoluminescence spectra, 120–121
    growth of
        cross-sectional TEM micrographs,
            108–109
        energy disperse X-ray analysis, 109–110
        mosaic structure, 109–110
        orientation relationship, 108
        pole figure, 111
        reciprocal lattice mapping, 112
        self-ordering phenomena, 107
    low temperature photoluminescence, 121–122
    micro-photoluminescence, 122–123
    mosaic structures, 125
    nonradiative recombinations, 124
    panchromatic image, 121
    preparation of
        structural quality improvement, 106–107
        sublimation epitaxy, 106
    recombination mechanism, 123
    structural characterization
        Bragg-filtered image, 117–118
        closed domain and emanating threading
            dislocation, 116–117
        crystallites, 113–114
        Fourier-filtered image, 117–118
        microstructure and surface
            morphology, 112
        misfit dislocations, 120
        Moire fringes, 114–115
        selected area diffraction pattern, 114
        strain component, 118–119
    temperature-dependent
        photoluminescence, 122

Heterojunction light-emitting diodes, *see* Light-emitting diodes (LEDs)
Heterostructures
   atomic arrangement, 93
   biaxial strain and stress kinetics, 93
   cathodoluminescence spectra, 104–105
   donor states-deep acceptor states
      transitions, 103–104
   energy band diagram, 91–92
   energy cost, 96
   imbalanced interface chemistry, 99
   in-plane and out-of-plane strains, 94–95
   interface arrangements, 99–100
   interface quality, 105
   lattice constants, wurtzite and zinc
      blende, 86–87
   lattice deformation mechanism, 94
   misfit dislocations formation energy, 97
   morphologies, 101
   nonradiative processes, 98
   pulsed laser deposition, 89, 91
   rectifying behavior, 90
   schematic diagram, 88–89
   SEM images, 102–103
   spin-coating pyrolysis, 88
   staggered-type alignment, 100–101
   stoichiometry, 106
   strain relaxation, schematic model, 95–96
   substrate temperature effect, 88, 90
   time-resolved photoluminescence, 98
   typical I–V characteristic, 103–104
   valence and conduction band offsets,
      100–101
Hybrid light-emitting diodes (LEDs)
   n-ZnO-nanowire/*p*-Si
      electroluminescence spectra, 285, 299–300
      energy-band diagram, 284, 285
      fabrication process, 302–303
      photoluminescence spectra, 300–301
      p+-Si/*p*-CuAlO$_2$/*n*-ZnO nanorod LED,
         285, 286
      schematic representation, 297, 298
   n-ZnO nanowires and *p*-organic
      semiconductors
      bendable LEDs, 293–297
      conductive-glass/ZnO-nanowire/
         organic-semiconductor/metal, 288–291
      glass/PEDOT:PSS/polymer-
         semiconductor/ ZnO-nanowire/metal,
         291–293
      structure of, 303–305
   n-ZnO/*p*-GaN heterostructure
      energy band diagram, 280, 281

      energy-dispersive x-ray spectrum, 283, 284
      flip-chip package process, 283
      photoluminescence spectra, 283, 284
      room-temperature electroluminescence
         spectra, 281–282
      schematic representation, 280, 281
      white-light LEDs, 283
   n-ZnO/*p*-SiC heterojunction, 287
Hybrid MOCVD ALD growth, ZnO
      Scintillators, 457–458
Hydrothermal crystal growth, 454
Hydrothermal method, *see* Solution-based
      chemical synthesis

I

In-Ga-Zn-O thin film transistors (TFT)
   advantages of, 512–513
   atomic structure, 498–500
   carrier transport
      band alignment diagram, 489
      bandgap, 490
      band structures and crystal
         structures, 488–489
      oxide *vs.* conventional semiconductors,
         491, 492
      wave functions, 490
   doping, oxygen deficiency, and hydrogen,
      509–511
   electronic structures
      bandgap, 502
      electron affinity and ionization
         potential, 503
      hard X-ray photoemission
         spectrum, 502–503
      pseudo-band structure, 500–501
      subgap density of states, 502
      total and projected density
         of states, 501–502
   electron transport, 505–508
   fabrication of
      annealing effect, 514–515
      operation characteristics, 514
      stability, 516–519
      TFT structures and process, 513
      uniformity, 515–516
   flat-panel displays backplane
      e-paper display structure, 519–520
      flexible displays, 520
      history of, 518, 520
      mass production, 521–522
      system integration, 521
      transparent displays, 520–521

growth, 498
history of, 487–488
material design with high mobility
  deep localized states, 492
  orbital drawings, carrier pathways, 493
  strained chemical bonds, 494
optical properties, 504–505
roles of constituent elements
  hall mobility *vs.* carrier density, 494, 496
  mobility enhancer, 495
  multi-component systems, 497–498
  structures and carrier transport
    properties, 494–495
  transport property characteristics, with
    vs. without Ga, 495–497
subgap defects, 506–509
Inverters
  FET-logic
    circuit design, 396
    origin, 395–396
    voltage-transfer characteristics, 396–397
  fundamental (*see* Fundamental MESFET
    inverter)
  parameter comparison, 401–402
  SDFL
    advantage, 397–398
    circuit design, 398
    origin, 397
    voltage transfer characteristics, 398–400
    ZnO *vs.* MgZnO, 398–400

L

Laser diodes; *see also* Random laser diodes
  As-doped *p*-type ZnO-based materials,
    349, 350
  Au/*i*-ZnO/MgO/*n*-ZnO structure
    band alignment, 353, 354
    electroluminescence spectra, 355
    transportation mechanism, 354–355
  Au/MgO/n-ZnO structure
    band alignment, 353, 354
    carrier transportation mechanism,
      353, 354
    electroluminescence spectra, 355
    lasing characteristics, 355
  coherent optical feedback, 363–364
  n-ZnO/MgO/*p*-GaN heterojunction
    electroluminescence spectra, 352, 353
    energy band diagram, 352
  optical gain realization, 362–363
  Sb doped *p*-type ZnO-based materials,
    350–351

Light-emitting diodes (LEDs)
  crystal structure parameters, 178
  epitaxial growth, *n*-ZnO/MgO/TiN/*n*-Si(111)
    heterostructure
    cross-sectional HRTEM, 225
    crystal quality, 225
    current–voltage characteristics, 226–227
    double Schottky barrier, 226
    light output *vs.* injection current, 227, 229
    photoluminescence spectrum, 226, 228
    pulsed laser deposition, 224–225
    X-ray diffraction spectrum, 225–226
  n-ZnO/*i*-ZnO/*p*-SiC heterojunction
    Burstein-Moss effect, 263
    current *vs.* voltage characteristics, 262
    photoluminescence spectra, 262, 263
    room-temperature electroluminescence
      spectra, 266
    schematic structure, 261
  n-ZnO/*n*-GaAs heterostructure
    cross-sectional TEM, 230
    current–voltage characteristics, 231–232
    electroluminescence and
      photoluminescence spectra, 232, 233
    electron energy-loss spectroscopy,
      230–231
    energy band diagrams, 232, 234
    indium doped ZnO thin film, 229–230
    light output intensity *vs.* injection
      current, 232–234
  n-ZnO/*p*-GaN heterojunction
    band diagram, 260, 261
    cross-sectional TEM image, 257–259
    current *vs.* voltage characteristics, 254, 255
    Hall effect measurement, 254–255
    high-angle annular dark-field
      scanning transmission electron
      microscopy, 258–260
    HRTEM images, 258, 259
    photoluminesence spectrum, 255
    room-temperature electroluminesence
      spectrum, 255–256
    schematic structure, 254
    threading dislocations, 257
    X-ray diffraction patterns, 256, 257
  n-ZnO/*p*-SiC heterojunction
    band diagram, 264, 265
    current *vs.* voltage characteristics, 262
    room-temperature electroluminescence
      spectra, 263–264
    schematic structure, 261
  n-ZnO/SiOx/*n*-Si and *n*-ZnO/SiOx/*p*-Si
    heterostructure

cross-sectional TEM, 220
current–voltage characteristics, 220–221
energy band diagram, 221–224
schematic diagram, 220
p-GaN film/n-ZnO film
electroluminescence spectra, 184
magnetron sputtering, 186
nonradiative center, 181
parameters, 182
pulse laser deposition, 185
vapor cooling condensation
technique, 183
p-GaN film/n-ZnO nanoscale array
angle-distribution, 201–203
annealing, 207–208
applied voltages, 201–202
electrodeposition, 207–208
forward bias voltages, 198–199
hydrogen treatment, 204–205
injection currents, 203–204
optical microcavity, 204
parameters, 197–198
reverse-bias voltage, 200–201
schematic illustration, 198–199
single mode lasing emission, 204–205
time response, intensity, 209
Zn(NO₃)₂ and HMT aqueous solution, 206
p-GaN film/single n-ZnO nanowire, 196–197
p-GaN film/ZnO nanowire array/n-ZnO
film, 210–211
p-GaN/i-ZnO/n-ZnO
electron beam evaporation deposition,
188, 191
energy band diagram, 180, 181
growth method, doping, 186–187
i-layer thickness, 194–195
injection currents, 195–196
magnetron sputtering, 187, 189
nanorod heterostructures, 190, 192
plasma-assisted molecular beam epitaxy
technique, 186, 188
radiative excitonic recombination, 193
rib waveguide, 192–194
schematic configurations, 189, 192
structure, 186
ultraviolet electroluminescence, 188, 190
vapor cooling condensation
technique, 189, 191
p-GaN/n-ZnO
energy band diagrams, 179
nanoradiative recombination center, 180
ZnO nanowire array/GaN film/ZnO
nanowire array, 211–212

M

Melt crystal growth, 454
MESFET, *see* Metal-semiconductor field-effect
transistors (MESFET)
Metal–insulator–semiconductorfield-effect
transistors (MISFET)
active-matrix displays, 370
amorphous device parameters, 425
bias stress, 383
disadvantages, 371
schematic structure, 370
thin film transistors, 419
threshold voltage, 374
Metalorganic chemical vapor deposition
(MOCVD)
applications
sacrificial layers, 50–51
transparent conducting layers, 51–52
growth on sapphire substrates
nanorod arrays, 8, 11–17
nanostructures, 23–34
nanotube arrays, 17–23
nanowall networks, 5–8
schematic diagram, 4–5
thin films
Al-doped ZnO (AZO), 43–44
annealing, 34–43
MgₓZn₁₋ₓO, 44–49
ZnO scintillators, 455–457
Metal-semiconductor field-effect transistors
(MESFET)
amorphous gallium–indium–zinc
oxide(GIZO)
channel mobility, 423–424
electrical properties, 419, 420
Hall-effect measurements, 419–420
on/off source–drain current ratio, 422
output characteristics, TFT, 422
quasistatic capacitance–voltage
measurements, 420, 421
Schottky-gate characteristics, 420–421
x-ray diffraction patterns, 419
device parameters comparison, 388–389
disadvantages, 392
inverters
advantages, 412–413
FET-logic, 395–397
fundamental, 392–395, 400–401
parameter comparison, 401–402
SDFL, 397–400
schematic structure, 370
switching behavior, 389–392

transparent
classification, 411
figures of merit, 411, 412
transparent-transistor technologies *vs.*
opaque Si-transistors, 371
TRC (*see* Transparent rectifying
contacts (TRC))
ZnO on glass substrate, 386–388
ZnO on sapphire substrate
bias stress measurements, 383, 384
channel mobility, 374
channel thickness variation, 377–379
crystal quality, 379–381
display applications, 377
elevated temperatures, 382–383
geometry effects, 379–381
light exposure, 385
long-term stability, 385–386
low-power applications, 377
measured properties, 375, 376
net doping concentration, 377–379
output and transfer characteristics,
373–374
photoconductivity effect, 384–385
preparation, 372–373
Schottky-gate materials, 375, 376
transconductances, 374
two-dimensional electron gases,
379, 381
$Mg_xZn_{1-x}O$ thin films, 44–49
MISFET, *see* Metal–insulator–
semiconductorfield-effect transistors
(MISFET)
Mohs hardness, 452
Moon detectors, 438

N

Nanohexagons, SiC substrates, *see*
Heteroepitaxial nanohexagons, p-type
SiC substrates
Nanorod arrays
MOCVD growth on sapphire substrates
cross-sectional SEM images, 13–14
density and surface roughness, 12–13
3D island morphology, 12
intensity ratio, 15
optical properties, 13–14
photoluminiscence spectra, 15–16
SAED pattern, 17
SEM images, 11
TEM, 15–16
XRD, 13, 15

Nanostructures
dye-sensitized solar cells
performance, 160–161
protecting and electron-blocking
layer, 161
schematic representation, 160
surface recombination, 162
gas/chemical sensor
characteristics, 154–155
dynamic responses, 158–159
electron trapping and detrapping
process, 155
oxidizing gases, 155
reductive gases, 156
spill-over effect, 158
transconductance, 156–157
growth of
chemical vapor deposition, 137–138
solution-based chemical
synthesis, 135–137
vapor-liquid-solid process, 134–135
LEDs (*see* Light-emitting diodes (LEDs))
MOCVD growth on sapphire substrates
columnar grain microstructure and grain
size, 27–28
crystal quality, growth rate, and surface
roughness, 27, 29
EDS analysis, 29, 32
FE-SEM images, 32, 34–35
formation, 29, 31
growth mechanism, 30–31, 33
optical properties, 25, 27
photoluminiscence spectra, 28, 30
SEM images, 23–25
X-ray photoelectron spectroscopy
(XPS), 32, 34
XRD pattern, 24, 26
p-GaN/*n*-ZnO heterojunction
principle, 180–181
photoconductivity properties
schematic illustration, 150–151
Schottky contact, 154
UV photodetection, 153, 155
UV sensor, 151–152
photoluminescence properties
bound exciton lines, 146–147
energy levels, native defects, 147–148
exciton emission, 149
surface excitons, 150
UV luminescence and visible emission,
146, 148
p-type SiC substrates (*see* Heteroepitaxial
nanohexagons, p-type SiC substrates)

solution-growth (*see* Solution-grown n-type nanostructures)

transport properties
  device characteristics, 142
  electronic conduction, 138
  electron tunneling, 144
  etching, 146
  passivation, 145
  surface dynamics, 144
  thermionic emission theory, 141–142
  transient drain current characteristics, 145
  types, 139–140
  variable-range-hopping mechanism, 138–139

Nanotube arrays
  MOCVD growth on sapphire substrates
    photoluminiscence spectra, 22, 24
    SAED and HRTEM, 22–23
    schematic diagram, 17–18
    SEM image, 17, 19–20
    three-step process, 17–18
    XRD, 20–22

Nanowall networks
  MOCVD growth on sapphire substrates
    energy-dispersive x-ray spectroscopy, 7
    flow rates dependence, 8, 10
    growth rate *vs.* flow rate, 7–8
    room temperature photoluminscence spectra, 8, 10
    SEM images, 6
    surface and structural morphologies, 8, 9

Nanowires
  charge transport, 73–74
  doping, 67
  electronic characterization, 71–72
  optical properties, 74–76
  organic photovoltaics applications, 76–77
  structural characterization
    anomalous X-ray diffraction, 70
    NMR spectra, 67–68
    optical density, 69
  synthesis
    aspect ratio, undoped wires, 63
    doping species and surfactant, 66
    hexagonal base plate formation, 64
    morphological control, 61
    ripening-based mechanism, 64–65
    scanning electron micrographs, 62
    solvent power, 62
    surface instabilities, 66
    surfactants, 63
    trioctylamine, 62

Neutron scintillators
  crystal transparency regime, 479
  detection performance
    commercial reference detectors, 471–473
    count rates per unit area/volume comparison, 475, 478
    MCNPX v2.7d simulation, 470–471
    pulse-height distribution simulation, 471, 472
  dopants impact
    aluminium doping, 466, 467
    boron doping, 466, 468
    gadolinium doping, 467–470
    gallium doping, 465–466
    lithium doping, 466–467, 469
    nitrogen doping, 463–464
  impact of growth conditions
    II–VI ratio effect, 460, 461
    pressure effect, 462–463
    temperature effect, 461–462
  MCNP5 models, 475, 479
  MCNPX models, 479
  neutron detector nuclei, 479
  novel
    dimensions of, 473
    neutron *vs.* gamma discrimination, 473–474
    (n, gamma) responses, 473
  wavelength-shifting dopants, 479

NOT gate, *see* Inverters

O

Optical gain, laser diodes, 362–363

P

Passive neutron detection system, 436–437

p-GaN/*i*-ZnO/*n*-ZnO heterojunctions
  electron beam evaporation deposition, 188, 191
  energy band diagram, 180, 181
  growth method, doping, 186–187
  i-layer thickness, 194–195
  injection currents, 195–196
  magnetron sputtering, 187, 189
  nanorod heterostructures, 190, 192
  plasma-assisted MBE technique, 186, 188
  radiative excitonic recombination, 193
  rib waveguide, 192–194
  schematic configurations, 189, 192
  structure, 186
  ultraviolet electroluminescence, 188, 190
  vapor cooling condensation technique, 189, 191

Q

Quantum dots in SiO₂ nanoparticle layer
  cross-sectional TEM image, 251, 252
  fabrication, 250–251
  HRTEM image, 251, 252
  photoluminescence spectrum, 252, 253
  quantum confinement effects, 250, 252

R

Random laser diodes
  GaN-ZnO heterojunction, 343–345
  polycrystalline ZnO thin films
    advantage, 356
    coherent backscattering, 346–347
    metal-oxide-semiconductor structure,
      347–348
    postgrowth annealing, 345, 346
    SEM images, 346
    X-ray diffraction patterns, 346
  SiC-ZnO heterojunction, 340–342
  ultraviolet ZnO
    directional and controllable edge-
      emitting, 357–359
    high-power and single-mode
      operation, 359–362

S

Sacrificial layers, 50–51
Schottky-diode FET-logic (SDFL) inverter
  advantage, 397–398
  circuit design, 398
  origin, 397
  voltage transfer characteristics, 398–400
  ZnO *vs.* MgZnO, 398–400
Scintillators
  dopant studies, 479
  gamma ray detection, 443
  gamma ray interactions, 441–443
  heavy-charged particle interactions and
    transport, 445–446
  ⁶LiI scintillators, 438–439
  neutron interactions/detection, 443–445
  organic, 438–439
  paralyzable *vs.* nonparalyzable detector
    models, 447, 448
  radiation detection, 440, 447
  radiation interaction mechanisms, 440–441
  scintillation detector theory, 447–449
  special nuclear material detection, 436
  ZnO

band diagram, 450, 451
  bulk growth, 458, 459
  crystal growth, 452–454
  crystalline forms, 450, 452
  hybrid MOCVD ALD growth, 457–458
  MOCVD growth, 455–457
  Mohs hardness, 452
  neutron scintillators (*see* Neutron
    scintillators)
  plastic scintillator *vs.* indium doped,
    450, 451
  previous work, 450
  refraction index, 452, 453
SDFL inverter, *see* Schottky-diode FET-logic
  (SDFL) inverter
Seebeck test, 267
Seeded chemical vapor transport (SCVT)
  crystal growth, 453
SiC-ZnO heterojunction laser diodes, 340–342
Silicon trench detectors, 438
Solid-state neutron detectors, 438
Solution-based chemical synthesis, 135–137
Solution-grown n-type nanostructures
  charge transport, nanowires and nanowire
    ensembles, 73–74
  doping, 67
  electronic characterization, 71–72
  optical properties, nanowire meshes, 74–76
  organic photovoltaics applications, 76–77
  structural characterization
    anomalous X-ray diffraction, 70
    NMR spectra, 67–68
    optical density, 69
  synthesis
    aspect ratio, undoped wires, 63
    doping species and surfactant, 66
    hexagonal base plate formation, 64
    morphological control, 61
    ripening-based mechanism, 64–65
    scanning electron micrographs, 62
    solvent power, 62
    surface instabilities, 66
    surfactants, 63
    trioctylamine, 62
Special nuclear material (SNM) detection, 436
Switching behavior, ZnO-MESFET, 389–392

T

Thermal neutron detector system, 438
Thin films
  Al$_x$Zn$_{1-x}$O, 43–44
  amorphous BeZnO (a-BeZnO)

optical bandgap engineering, 318, 319
Rutherford backscattering spectrometry, 318
on amorphous glass substrates
atomic force microscopy, 245, 246, 249
photoluminescence spectra, 247–250
schematic representation, 245
XRD patterns, 245, 246
BeZnO
dislocations, 314
energy bandgap, 311, 312
hybrid beam deposition method, 311
lattice parameters, 313
photoluminescence spectra, 316, 317
transmittance spectra, 311, 312
X-ray diffraction patterns, 311–313
$Mg_xZn_{1-x}O$, 44–49
random laser diodes, polycrystalline ZnO
advantage, 356
coherent backscattering, 346–347
metal-oxide-semiconductor structure, 347–348
postgrowth annealing, 345, 346
SEM images, 346
X-ray diffraction patterns, 346
stable *p*-type
ALD technique, 267
Hall-effect measurements, 267–269
hole concentration, 270
mobility, 270
photoluminescence spectra, 269, 270
resistivity, 270
seebeck test, 267
X-ray photoelectron spectroscopy, 268, 269
ZnBeMgO
vs. BeZnO films, 321
structural and optical properties, 322
UV-transmittance spectra, 323
X-ray diffraction patterns, 320
Thin film transistors (TFT)
In-Ga-Zn-O
advantages of, 512–513
amorphous (*see* Amorphous oxide semiconductor (AOS))
atomic structure, 498–500
carrier transport, 488492
doping, oxygen deficiency, and hydrogen, 509–511
electronic structures, 500–503
electron transport, 505–508
fabrication of, 513–519
flat-panel displays backplane, 519–522

growth, 498
history of, 487–488
material design with high mobility, 492–494
optical properties, 504–505
roles of constituent elements, 494–498
subgap defects, 506–509
MISFET, 419
Transparent rectifying contacts (TRC)
pulsed laser deposition, 402–403
visible-blind ultraviolet photodiodes
current–voltage characteristics, 405, 406
external quantum efficiency, 407, 408
spectral responsivity, 407, 408
transmission spectra, 405, 406

U

Ultraviolet (UV) electroluminescence
n-ZnO/*i*-ZnO/*p*-SiC heterojunction LEDs
Burstein-Moss effect, 263
current *vs.* voltage characteristics, 262
photoluminescence spectra, 262, 263
room-temperature electroluminescence spectra, 266
schematic structure, 261
n-ZnO/*p*-GaN heterojunction LEDs
band diagram, 260, 261
cross-sectional TEM image, 257–259
current *vs.* voltage characteristics, 254, 255
Hall effect measurement, 254–255
high-angle annular dark-field scanning transmission electron microscopy, 258–260
HRTEM images, 258, 259
photoluminesence spectrum, 255
room-temperature electroluminesence spectrum, 255–256
schematic structure, 254
threading dislocations, 257
XRD patterns, 256, 257
n-ZnO/*p*-SiC heterojunction LEDs
band diagram, 264, 265
current *vs.* voltage characteristics, 262
room-temperature electroluminescence spectra, 263–264
schematic structure, 261
Ultraviolet ZnO random laser diodes; *see also* Random laser diodes
directional and controllable edge-emitting, 357–359
high-power and single-mode operation, 359–362

V

Visible-blind ultraviolet photodiodes
    current–voltage characteristics, 405, 406
    external quantum efficiency, 407, 408
    spectral responsivity, 407, 408
    transmission spectra, 405, 406

W

Wall effect, 437
White light-emitting diode
    cross-sectional TEM, 230
    current–voltage characteristics, 231–232
    electroluminescence and photoluminescence
        spectra, 232, 233
    electron energy-loss spectroscopy (EELS),
        230–231
    energy band diagrams, 232, 234
    indium doped ZnO thin film, 229–230
    light output intensity *vs.* injection current,
        232–234
Work function, 447

Z

ZnBeMgO alloys
    bandgap, 325, 327–328
    vs. BeZnO films, 321
    bowing parameters, 324–325
    bulk modulus, 324
    CASTEP package, 326
    formation energy, 327–328

lattice constant, 324
    natural band alignment, 325–326
    optoelectronic applications, 333–334
    pulsed laser deposition method, 319
    room temperature Raman spectra, 322
    special quasirandom structure (SQS)
        approach, 324
    transmittance spectra, 319–320
    UV-transmittance spectra, 323
    valence band maximum alignment, 325, 326
    X-ray diffraction patterns, 320–321
ZnBeO alloys
    band gap, 319, 324–325
    bulk modulus, 324
    energy bandgap, 311, 312
    hybrid beam deposition method, 311
    lattice constant, 324
    lattice parameter, 313, 314
    natural band alignment, 325–326
    optical bandgap engineering, 318, 319
    optoelectronic applications
        multiple quantum well devices, 331, 332
        solar-blind UV photodetectors, 332–333
        UV laser diodes, 330–331
        UV LED devices, 329–330
    photoluminescence spectra, 315–318
    Rutherford backscattering spectrometry, 318
    transmittance spectra, 311, 312
    Vegard's law, 313
    X-ray diffraction
        measurements, 311, 313
        rocking curve, 313, 315

Printed and bound by CPI Group (UK) Ltd, Croydon, CR0 4YY

24/10/2024

01778286-0017